Human Adaptation

Human Adaptation

The Biosocial Background

Yehudi A. Cohen
editor

Routledge
Taylor & Francis Group
LONDON AND NEW YORK

First published 1974 by Transaction Publishers

Published 2017 by Routledge
2 Park Square, Milton Park, Abingdon, Oxon OX14 4RN
711 Third Avenue, New York, NY 10017, USA

Routledge is an imprint of the Taylor & Francis Group, an informa business

Library of Congress Catalog Number: 2010031069

Library of Congress Cataloging-in-Publication Data

Cohen, Yehudi A.
Human adaptation : the biosocial background / Yehudi A. Cohen.
p. cm.
Originally published : Chicago : Aldine Publishing Company, 1968.
Includes bibliographical references and index.
ISBN 978-0-202-36384-4 (alk. paper)
1. Anthropology. 2. Human evolution. I. Title.

GN29.C568 2010
301--dc22

2010031069

ISBN 13: 978-0-202-36384-4 (pbk)

To Those Who Didn't Make It

As Well As

Those Who Did

CONTENTS

PREFACE TO SECOND EDITION

THIS BOOK CONTAINS 44 selections that are intended to serve as an introduction to physical anthropology, archeology, and linguistics from the point of view of the processes of adaptation. The organization of these selections attempts to strike a balance between biological and prehistoric cultural adaptations to provide coherence for the study of man's evolution. Several selections—notably those in connection with linguistic adaptations—deal with living people in order to shed light on earlier evolutionary processes. Slightly more than half of the selections deal with biological evolution, although, as will be seen, it is sometimes difficult to determine where biological adaptation leaves off and cultural adaptation begins. The two are inseparable. In two other books, *Man in Adaptation: The Cultural Present* and *Man in Adaptation: The Institutional Framework,* I bring together similar collections of readings that represent an attempt to provide coherence for the study of cultural evolution and the institutional, psychological and ideological dimensions of the adaptive strategies that have characterized human societies from the earliest known forms of social life to the present.

My purpose in this book is to focus on the concept of adaptation instead of presenting an encyclopedic and exhaustive compilation or readings that reflect my personal tastes or the standards of current popularity. In making these selections I was guided by at least five criteria.

1. As I combed the anthropological literature I was primarily concerned with the relevance of a piece to the theme of adaptation. Many illuminating and elegant articles, which could easily be included in any collection of anthropological readings, were excluded because they dealt with other problems. In several instances it was necessary to go beyond the traditional boundaries of anthropology in search of selections that would throw needed light on different facets of adaptation.

My bias in selecting and organizing these readings is clearly evolutionary. Although it is difficult to study the processes of human adaptation without an evolutionary perspective, some anthropologists feel that such a separation can be made, and my intent is to draw attention to a point of view around which anthropological materials can be organized rather than to present or to defend an argument. Hence wherever it was feasible and whenever a choice existed, I chose a selection that dealt with the problem of adaptation instead of one that was devoted simply to an evolutionary argument, so that attention can be focused on the processes of adaptation without necessarily adopting the evolutionary point of view.

2. I selected papers that are empirically grounded; that is, papers that can be considered to be within the "case-study approach." Clearly, as in the emergence of human speech, this is not always possible, and in such instances I have relied on speculative and imaginative writings that most clearly seem to fit the available data. The study of evolution lends itself to controversy, and some selections included here clearly contradict each other.

3. In choosing among alternatives that were relevant to the theme of adaptation and within the case-study approach, I allowed some of my personal tastes to intrude. Where choices had to be made, I elected to include selections that displayed wit and imagination. Scientists cannot always display a sense of humor, especially where social behavior is concerned, but when possible, I operated in the belief that a humorless social science has lost its viability.

4. Without necessarily confining myself to the traditional categories of anthropology (a healthy disrespect for custom is sometimes an adaptive necessity if a science, like a society, is to grow), I tried to view the major subject matters of anthropology through the lenses of adaptation. A few areas that are parts of the anthropological tradition (such as art and music) have not yet been viewed in this way, and therefore they have not been touched on in this volume.

5. Finally, I have attempted to highlight the major methodological problems in the analysis of human evolution and behavior. If these problems were not made explicit in the selection, I have tried to point them out in my editorial introductions.

■ These editorial introductions appear in sans serif type (as here) and are set off from the text of the selections themselves by squares like those preceding and ending this paragraph. To assist the reader who wishes to explore more thoroughly the topics that are covered in this book, I have included suggestions for further reading at the end of each of the introductions to parts and selections. The reader may note that this book is one of three volumes. *Man in Adaptation: The Cultural Present* deals with the process of adaptation in living cultures, beginning with nomadic foragers and concluding with some aspects of contemporary civilization; *Man in Adaptation: The Institutional Framework* deals with behavior within an evolutionary and adaptational framework. But the volumes have been organized so that each can stand separately. ■

The final form of this book reflects in certain ways the economic realities with which I have been confronted by my publisher. Many more selections were eligible for inclusion according to my criteria, but I had to limit its scope to keep its price within the reach of many readers (and there were a few instances in which permission to reprint could not be secured). For the same reason, I have omitted all footnotes and bibliographical references; the reader who wishes to consult the sources that were used by the authors of these selections can easily refer to the original publications. Throughout the book, the accents and diacritical marks that normally appear in certain foreign names and phrases have been deleted at the request of the publisher, for reasons of economy and simplicity. Similarly, a number of tables, charts, and illustrations that appeared in the original articles have been deleted, where they were not essential to the flow of the argument; all such omissions have been marked with ellipses in the usual way. With these few exceptions, the selected articles and passages are reprinted without abridgement. In a few cases where authors used highly technical terminology, I have supplied definitions in footnote form.

Although it is customary in prefaces and forewords to express the most important kinship obligations last, I want to reverse this form and note my obligation for the atmosphere created and sustained by my wife, Rhoda Cohen, which has made this book possible. The work on this book and on its companion volume often continued late into the night and into weekends and holidays. My family has repeatedly assured me there never was need to apologize for this, and were it otherwise, I doubt that the work would have been completed.

This work began about 1962, when a friend gave me a copy of Sir P. B. Medawar's *The Future of Man* (New York: Basic Books, 1960). I found the ideas in the book (see Selection 4) very exciting as I explored their relevance for the study of social organization. As time passed, I noticed that I was relying more and more on the ideas and implications of Medawar's book in my teaching and in my research. When some of my students expressed their appreciation for this point of view, especially in introductory courses, and asked for further reading about adaptation, I began to think of collecting a small group of selections.

From the time I received Medawar's book to the time I completed this preface, I have amassed many special intellectual debts. One aspect of adaptation in modern culture is the steady attenuation of culturally provided mechanisms for repaying or reciprocating such obligations, and unfortunately there is no way in which I can properly express my sense of debt to the more than 40 scientists whose work is represented here. I not only want to thank them for their courtesy in allowing me to reprint their work and for their patience as we corresponded about this project but also to express my admiration for the quality of thought and the excitement of imagination that became more and more evident every time I read and re-read each paper. I have learned much from each of them.

My deepest thanks go to the anonymous readers of Aldine Publishing Company who went out of their way to make helpful suggestions and to offer their encouragement. Many times I went back to read their assurances that this was a worthwhile project. To Alexander J. Morin, publisher of Aldine, I offer a non-ritual salute for his work, assistance, and emboldening words, which were above and beyond the call of duty.

I also take this opportunity to express my appreciation to J. Ralph Audy and Warren G. Kinzey for their helpful suggestions, which, in addition to being a source of many ideas, saved me from making some embarrassing errors. I thank Martin A. Baumhoff and Phyllis Jay for directing me to several important sources that might otherwise have escaped my attention. I also express my appreciation to Mrs. Patricia Rademaker for her conscientiousness and diligence in handling the mass of administrative details that were involved in the preparation of this book and for relieving me of all concern in this matter.

In preparing the second edition of this book, I benefited greatly from the help of Gloria Levitas and Bernard G. Campbell. I especially want to thank H. Dean Kedenberg for his valuable assistance.

SURVIVAL THROUGH ADAPTATION

E. B. WHITE

Scientists recently reported we Yankees possess the most powerful index fingers in the world. And to what do they attribute this digital superiority? You guessed it. The telephone dial!

—*Stuffer with a bill from the New England Telephone Company.*

How did the fiddler crab (the male),
Uca pugnax, acquire his outsize claw,
Picking his preposterous fights in muddy Lilliput
Among the stilts of mangrove in the vast
Intertidal zone, hurling his challenges,
Waving the enormous tiny disproportionate
Claw (which slowly grew by being waved), flashing
His semaphore across the stinking flats?
How? A matter of survival. The necessities of life.

And the frog, *Rana* (the male), with his notorious
Phallic thumb. Whence came this strange appendage?
Where else but from sheer necessity, to hold
His slippery mate for the exact, the correct,
Performance of his offices.

And now I, Man (the male), most splendid
Of the Mammalia, *Homo sapiens,* longtime
Darling of the New England Telephone Company
With my enormous index finger, perfectly adapted
To my environment, I, endlessly phoning
Hurling my challenges, quarrelling, conducting
My courtships, picking my fights long distance
With this great incredible index finger, the
Most powerful in the world. Now I, a New Englander,
Join the company of the immortals: the frog,
The crab, who slowly changed in subtle ways
In order to survive. I, too. Hey!
Cock-a-doodle-doo! Watch me dial!

INTRODUCTION

THE PURPOSE OF this book is to provide a guided introductory tour in physical anthropology, linguistics, and archeology. It is selective and focuses on the role of adaptation in man's attempts to transcend the restrictions of his biological makeup and natural habitats. As every experienced tourist knows, different guides stress their own interests and themes, and this holds true in anthropology and in many other fields. The basic plan or structure of the discipline is constant—for example, it is divided into physical anthropology, archeology, and cultural anthropology—but the perimeters of the various parts sometimes shade off into each other. Within each subdiscipline (or intellectual neighborhood) different themes or points of interest will attract the attention and excite the imagination of different guides and audiences.[1]

A person who confronts for the first time the vast amount of material that makes up the discipline of anthropology can easily be overwhelmed. He can soon become bogged down in apparently unrelated details, methods, problems, findings, and interpretations, and he may conclude that the science is but a thing of shreds and patches to which every investigator has made his own whimsical and impetuous contributions.

I have focused on the concept of adaptation in bringing these selections together because this concept provides a unifying theme that makes it possible to bridge the seemingly disparate interests of anthropologists—from the emergence of *Homo sapiens* out of his nonhuman primate ancestry to the development of huge metropolitan areas in modern states. I do not contend that adaptation is the only unifying theme in anthropology, but only that it is one way of making sense out of the many different things that anthropologists do.

Anthropology, in addition to being a scientific discipline, is a point of view about man's history and his culture. Perhaps one way of conceptualizing this point of view is to assert that *Australopithecus,* who lived about 1.5 million years ago, is as important in man's development as Napoleon, that the aboriginal populations of the world are as important for an understanding of man as the civilizations of Greece and Rome.

1. Parts of this Introduction parallel, and to some degree repeat, the Introduction to *Man in Adaptation: The Cultural Present* and *Man in Adaptation: The Institutional Framework.* I also draw on several editorial and interpretative essays in those volumes. Nevertheless, I suggest that the reader of the other volumes also read those Introductions because they differ in emphasis and sometimes in content.

Underlying the anthropological study of man is the principle that there is a reality, or several realities, to which man must adapt if he is to survive, reproduce, and perpetuate himself. Populations must adapt to the realities of the physical world; they must do this not only in terms of acquiring a livelihood, constructing shelter, and designing clothing but also they must maintain a proper "fit" between their biological makeup and the pressures of the various niches of the world in which they seek to live. Social groups—where culture is found—must develop adaptive mechanisms in the organization of their social relations if order, regularity, and predictability is to exist in patterns of cooperation and competition and if they are to survive as viable units. By *culture* I mean here the energy systems, the objective and specific artifacts, the organizations of social relations, the modes of thought, the ideologies, and the total range of customary behavior that is transmitted from one generation to another by a social group and that enable it to maintain life in a particular habitat. Thus the institutions of society also must be seen as an integral part of a society's adaptation. When an individual is born he enters a world he did not make but with which he must come to terms. He must learn to live with the established means of coping with the natural habitat, with institutions and foreordained value systems, and the like, if he is to succeed socially and biologically.

When we speak about adaptation in a particular manlike or human form we are concerned with the adaptations that have been achieved by the species as a whole at a particular stage of evolution or by populations within the species. The adaptiveness of any individual or series of individuals is merely a reflection or a special case of the adaptation that has been achieved by the group of which he is a member. The study of individual adaptation involves quite different concepts, as is illustrated in Hans Selye's *The Stress of Life* (New York: McGraw-Hill, 1956).

Every culture has its standards of social success, just as organic nature has its standards of success in terms of physical survival and reproduction. Neither set of standards can be considered an achievement of the moment; both are products of history. This book is a collection of studies that, from the point of view of adaptation, deal with the relationship between biology and culture in man's emergence and evolution.

An organism's adaptation is its *relationship* to its habitat. Thus the concept of adaptation is historical: when we say that a population is adapting we mean that it is altering its relationship to its milieu to make that milieu a more fit place in which to live or to make itself more fit to live in that milieu. The growth of a coat of fur by a population of mammals in a cold climate is adaptive because it alters the animals' relationship to the milieu; the historical process out of which the Eskimos' unique clothing developed, enabling each man to live in a private quasi-tropical climate, also is adaptive because it alters the relationship of the group to the milieu. A change that does not affect a population's relationship to the milieu, such as embroidery on a parka or moccasin, is not an adaptation. (Some anthropologists maintain that embroidery is adaptive because it provides visual variety and thus alters the milieu. If it is used as insignia to distinguish specialized groups in the organization of labor or some other phase of social relations, this also might be important in the adaptive process. Otherwise, I think that a blanket extension of the concept loses sight of the fact that artistic or decorative activity may not facilitate the reproductive and survival capacity of the group, which is the essence of adaptation.)

Similarly, when we say that a population is adapted to its habitat we mean that it has achieved, and maintains, a viable relationship with that habitat. This adaptation assures the population's survival and efficient functioning in the sense of doing-its-job-in-nature. The achievement of this type of viable relationship always results from modifications in the habitat or in the population of organisms over a long period of time; it is never achieved in one generation or in a single genetic stroke. The historical aspect of the process of adaptation is underscored by the phenomenon of biological evolution—the process of sequential change that we see in the forms and behavior of living things. The fact of evolution demonstrates that adaptation also has taken place; if adaptation had not characterized survival of living forms, evolution could not have occurred.

I have avoided the use of the word *environment* because of my conviction that much more terminological rigor is necessary in the study of adaptation than has characterized it heretofore. An environment is made up of several parts and it is necessary to distinguish among them. By *environment* I mean the total system of components that interact with each other and characterize a group or population. (I am using *environment* in much the same way that many persons use *ecosystem,* but I prefer the former because it is more inclusive.) The physical or natural habitat is only one component of an environment, although it is of primary importance. In the study of biological adaptation, the species that occupies a habitat must be regarded as a distinct component of the environment because it alters the habitat by the use that it makes of it; each species places its unique stamp on the habitat that it seeks to exploit by altering the "balance of nature" in it. The same considerations apply to man, but in a still more complex way, since human organizations of social relations—especially political institutions—constitute still another component in the environment.

Among nonhuman animals, adaptation takes place principally by means of genetic mutation. In man, however, adaptation is accomplished by cultural means, principally through the harnessing of new sources of energy for productive ends and through the organizations of social relations that make it possible to use these energy systems effectively. Examples of such sources of energy are the bow and arrow, the spear, the digging stick or hoe, the plow and draft animals, steam, and electricity. Whenever groups of people introduce a new source of energy into a habitat they create a new environment. Among the most important adaptive accompaniments of a new source of energy as a component of the environment are changes in the organizations of social relations that make it possible to use the energy system.

Thus "adaptation" and "environment" overlap to a considerable extent. But an environment contains more than a habitat and a group and its adaptation; it also contains the adjustments among the elements of the adaptation. Adaptation and adjustment must be regarded as distinctive processes. Adaptation refers to the processes by which a population or group alters its relation to its habitat. Especially in human societies, however, some changes in a group's customary behavior do not appreciably affect its relations with its habitat, and these changes must be considered separately from adaptive changes. A very good example of this is the development of linguistic variations within a group that speaks the same language (e.g., "boid" instead of "bird"). Another example of adjustive change is found in the sphere of etiquette (e.g., when men abandon the ritual of doffing their hats to women).

It is necessary to distinguish "adaptation" from "adjustment" not only for purposes of clarity but also because even the adjustments within an adaptation alter the group's environment—however subtly—when it is conceived as a total system of components. Thus the use of *boid* denotes the continued existence of social-class differences in a society, and a system of social classes (or castes) appears to be an important aspect of adaptation in societies that rely on advanced agriculture or industrialization; but such organizations of social relations can function effectively without linguistic differences. Changes in male-female etiquette denote important changes in the roles of women in the division of labor in society and in their political status, but these roles are not directly affected by relinquishments of ritual postures. The abandonment of such rituals may reinforce the new roles that are enjoyed by women, but they do not produce them.

When we study adaptation in anthropology we are concerned (as in other aspects of anthropological inquiry) with populations, not with individual organisms or persons. More specifically—and to place this in realistic perspective—we speak about populations but we study individuals. There are two principal reasons for this, and they are closely related to each other. The first is a practical consideration; the second reason is theoretical.

On the practical side, which is especially important in connection with man's prehistoric biological adaptations, we possess only one skeleton—and sometimes only part of a skeleton—as the representative of an entire species. This is not a comfortable scientific position, but science, like the rest of life, involves risks. If we are limited to only one skeleton or fossil, we have no other choice than to assume that it is representative of a population. In other words, anthropologists often have to adapt to the realities of their world and abstract or generalize from a single person to an assumed population.

This procedure is not confined to anthropologists who study fossil remains, it is also at the core of the methods of scientists who have entire living populations potentially at their disposal. But it is difficult, if not impossible, to study *all* Eskimos and European Canadians to learn whether there is a difference in blood flow in the hands of each group, or to study *all* residents of the Andes at every altitude to learn how Peruvians have adapted biologically to the relative lack of oxygen at 17,000 feet. Nor can an anthropologist who studies a living society interview all members of the group about every aspect of life. Instead, the investigator chooses a sample he considers representative of the entire group; then, after studying it, he attempts to present a picture of the population as a whole with respect to the problem he is trying to understand.

The theoretical reason is itself closely tied to practical expediency. Adaptation is a process, a relationship between a phenomenon and its source, and a forward movement over time and under different circumstances. No single organism—not even the sum of a group of organisms—can give direct evidence of an evolutionary process. The molars of *Australopithecus* (one of our forebears) cannot tell us about the climatic and dietary changes in East Africa that apparently were the undoing of his cousin, *Paranthropus;* no spearman can tell us why he does not use the bow and arrow even though he is familiar with its use; the expanded chest or elevated hemoglobin of an Andean Indian cannot explain how these residents of high altitudes compensate for the reduced supply of oxygen; an excavated irrigation network cannot by itself tell us very much about the relationships between villagers and the rulers of a state.

Usually we focus on populations rather than upon individuals in the study of adaptation because it is the population that survives and perpetuates itself. In the study of biological adaptation the relevant population is the breeding population because it is the vehicle for the gene pool, and we analyze the gene pool because it is the mechanism of evolutionary change. When we study human cultural adaptation we focus on the social group because it carries the culture, and culture is the mechanism of evolutionary change for humans.

What the anthropologist seeks to do with the information he gathers from the representatives of the group under study is to abstract from the data and portray a system or set of principles that help explain the phenomena he has found—the relationships among factors that led to the emergence and survival of a particular species, the relationships between stature and the use of spears or bows and arrows, the total physiological structure that enables men to survive at great heights, the stimulus of a state organization to the construction and maintenance of large-scale irrigation networks. The study of individual phenomena is only a means to understanding a process.

The adaptations man has achieved are the most advanced of all forms of life because man has a set of specialized tools of adaptation that are unparalleled by those of other forms: his culture. Man's culture is an adjunct to the human architecture, superimposed on his genetically determined organ systems. Thus culture has made man's mechanisms of adaptation even more specialized than his physiological tools. Viewed in the context of the total evolution of life, man's culture is a revolutionary addition to his architecture and his most powerful instrument for adaptation. Man is now able to adapt himself—through his cultures—to different habitats long before genetic mechanisms do this for him; he does not have to await genetic modifications in his constitution.

I.
ISSUES AND CONCEPTS IN THE STUDY OF ADAPTATION

WE BEGIN WITH the gene, the basic biological mechanism in the process of adaptation. If, however, we look at the total course of human evolution—and it does not matter from which end we look—one theme emerges with remarkable clarity: Man's history seems to show that he has striven unceasingly to free himself from the confining limitations of his genetic endowment. He also gives the impression of having tried to free himself from the restrictive limitations of his natural habitats. It is necessary to say "seems to show" and "gives the impression" to avoid an attribution of deliberateness, consciousness, or forethought to the species. These are themes that emerge from a historical perspective.

Man has been able to transcend the limitations of his genetic endowments and natural milieus because of his cultures. The study of adaptation deals not only with man's relationships to his habitats but also, and more important, with his ability to come to terms with and exploit changing habitats. His capacity for culture—unmatched anywhere else in nature—has made this possible. But while man's capacity for culture contains his greatest potential for adapting to changing habitats and although he is able to adapt himself to different habitats without having to await genetic modifications in his constitution, to assume that man's cultural adaptations have supplanted his biological adaptations would be a serious misconception. The adaptive process in man is as much biological as it is cultural. Man's biological evolution has not ceased, although students of evolution agree that it has probably slowed down drastically.

One of the principal reasons for the deceleration of man's biological evolution —and consequently for the greater importance of cultural evolution in adaptive advances—has been the steadily increasing number of variations within the species, that is, its growing heterogeneity. This genetic diversity provides the species as a whole with greater variety and flexibility in equipment needed for coping with varied and changing physical and social habitats. Man's genetic mixture is increasing worldwide not only because of medical advances that make it possible for people with potentially lethal genes to remain alive in increasingly greater numbers but also because of increased contact between members of different societies and genetic groups ("races"). Intercultural relationships normally are accompanied by sexual relationships, and the resultant genetic mixture provides the species as a whole with adaptive strength inasmuch as a frequent consequence of genetic mixture is heterosis, or hybrid vigor. (Politicians who see a sinister plot behind this would have to admit that the hybrid corn, peas, and beef on their dinner tables also are products of this plot to dilute their delusionary "racial purity.")

Thus the genetic heterogeneity of the human species is not only a product of the general nature of human culture but also one of the principal sources of strength of the species. These are among the principal themes stressed by Dobzhansky in Selection 2.

The purpose of life is to maintain life; adaptation is life. Like all other living forms, man constantly seeks to maintain and improve his adaptations to the habitats in which he lives. But the nature of man's adaptations and the instruments he applies in achieving them are more complex than those of all others in organic nature because they are a web of biological and cultural responses. Neither can be understood without the other, because, in fact, neither exists without the other. Man is simultaneously a biological and a cultured being.

No adaptation is permanent or static, because biological change never stops and because no environment remains unchanged. Similarly, as we see when we turn to adaptation in living cultures in the accompanying volumes, social systems constantly seek new adaptive strategies because social change never stops and no sociocultural environment remains unaltered. When the natural or social milieu changes, a population must seek a new adaptation, that is, a new relationship to new pressures and demands in its altered environment. Nor is adaptation a single phenomenon, act, or trait, nor even a series of discrete phenomena, acts, or traits. Adaptation refers to the relationship of the population of organisms as a total functioning system to a particular habitat.

The study of individual adaptation requires concepts that are very different from the concepts that are involved in the adaptations of populations, and it is necessary to distinguish between the two. Basic to this is the difference between genotype and phenotype: Genotype refers to genetic endowment, and phenotype refers to the actual and observable characteristics of individuals. The phenotype is determined by the genotype, although there is no assurance that a particular phenotype will necessarily emerge from a given genetic endowment. For example, if a person is genetically endowed with a potential of six feet in height, this predisposition will not necessarily become manifest if the individual's diet is inadequate for full growth.

More important, the interaction between a genotype and the population group as a whole—which is often referred to as the deme—will frequently determine the survival value of a genotype. For example, some kinds of dwarfism are the phenotypical manifestations of genetic mutations, and individuals who manifest this genetic trait may be capable of doing almost everything that nondwarfs can do. Despite this ability, however, the deme might recurrently remove this genetic strain from the population by, let us say, developing strong prejudices that make it almost impossible for dwarfs to marry, which in turn reduces the probability of the reproduction and perpetuation of this genetic strain in the population. Every deme, like every habitat, contains selective pressures that favor and disfavor the survival of different genotypes. One of the principal differences between nonhuman and human demes is that, in the latter, the pressures of the deme usually are cultural; those of the habitat usually are physical.

It is often difficult, however, to draw a sharp line between cultural and physical pressures (a major theme in Selection 4), and one of the most notable examples of this difficulty is provided by diabetes mellitus. Before the discovery of insulin it was almost impossible for many diabetics to survive long enough to reproduce and thus transmit the genetic predisposition for the disease in the population. (In some cases it was possible to survive with this illness by drinking an ounce of whiskey hourly, an adaptation that surely must have horrified temperance groups.) The discovery of insulin was a cultural intervention upon "nature"; a reversal of preexisting forces in natural selection, it introduced a new physical element (insulin) into the environment. Was this new potential for diabetics cultural or physical? It was both, which suggests that—at least in connection with *Homo sapiens*—the distinction between cultural and physical pressures in adaptation is spurious. The group or deme as a whole, in relationship to its habitat, by and large sets the standards of adaptation—of who shall live and who shall die, of who shall reproduce and who shall not transmit a genotype.

Adaptation is often equated with success, especially in popular thought. It is of course true that adaptations are successful insofar as they produce a viable "fit" between a population of organisms and their habitat; however, to focus exclusively on the successful aspects of any adaptation is to miss some of the most important characteristics of the adaptive process.

One of the themes that will be stressed in this section (especially in Selection 4) is that all adaptations are limited. There are several sources of this characteristic of adaptation. First, any adaptation enables a particular organism to function efficiently in a particular habitat, not in just any habitat. This can be seen at a relatively elementary level. An organism that has achieved a viable relationship to a particular habitat would be maladapted if this habitat changed radically, and the organism would face extinction. Thus, as almost every empirical study demonstrates, adaptation never occurs in the abstract but only in relation to specific habitational pressures and challenges. There is no such thing as an adapted organism; there are only organisms that have adapted to particular habitats.

The limitation on any adaptation stems from its historical aspect, that is, from the fact that adaptation is part of the evolution of life. Although various architectural changes are adaptive at a particular moment in history and in a particular milieu, they may prove fatal in the long run. This principle must be qualified slightly and placed in a sharper perspective of time.

It must always be kept in mind that habitats change. Sometimes the natural milieu undergoes significant alterations, such as those produced by recessions of glaciers, dessication, or increase in rainfall. In the case of man the sociocultural environment frequently changes, including modifications in the natural world. An adaptation that provides the necessary precondition for future evolutionary changes is not in itself fatal in the long run. It will only prove fatal if—despite its original effectiveness—the habitat changes to such an extent that the initial adaptation proves to be completely ineffective under the new conditions and leads to the extinction of the lineage before evolutionary modifications can occur.

Before the emergence of culture, genetic mechanisms governed the adaptation of organisms to their respective habitats. Populations that were genetically fit for their immediate habitats survived; those that were less fit did not survive and did not leave offspring. However, those that survived and reproduced did not necessarily represent terminal points in evolution; instead, they often continued to evolve into more highly adapted forms.

One of the significant advantages in the adaptation of any viable form is the possibility of further adaptations. In other words, every viable form in nature contains the potential for its own evolutionary replacement. Successful adaptations are expendable; they are sacrificed in the service of further evolution and more advanced adaptations. The adaptation that is achieved by a viable form can prove fatal to it in the long evolutionary run if it loses out to its more highly adapted descendants in the competition for the resources of their immediate habitats.

Another limitation on every adaptation is that while it facilitates the survival and perpetuation of the group, it creates disadvantages for at least some members of the group. The popular expression in daily social life, that one "pays a price for everything," is also relevant to the process of biological adaptation. This requires some attention, because disadvantages seem to be an inherent aspect of adaptation.

The inauspicious accompaniments of every adaptation can be seen in specific modifications in the course of evolution. When our earliest ancestors achieved erect posture and bipedal locomotion, this adaptation probably was won at the expense of earlier capacities. They sacrificed speed, among other things, and thus were at a defensive and offensive disadvantage. That man survived these costly disadvantages was due to the ability, which he won at almost the same time, to fashion and improve tools, and later to make (or at least preserve) fire and to construct shelter, as well as many other cultural innovations. The frequency with which members of our species are afflicted with backaches of all sorts is a corollary of our adaptive erect bipedalism.

P. B. Medawar (Selection 4), in his analysis of the gene for the sickle-cell trait, provides a superb model for understanding and further exploring the undesirable accompaniments of adaptations. He demonstrates that, in any population in a malarial area in which this gene is found, individuals who inherit it from only one parent (those who are heterozygous for the gene) are endowed with an inborn resistance to one of the most virulent forms of malaria; those who inherit the gene from both parents (those who are homozygous for the gene) are likely to die from sickle-cell anemia; and those who do not inherit the gene from either parent are highly susceptible to the virulent form of malaria. This is the only gene action to which this model has been applied in the context of large populations within the species. Its more extensive application will certainly expand the horizons of the study of adaptation.

The concept of disadvantages in adaptation leads to another problem that goes to the heart of all biological functioning but has, unfortunately, been neglected. To understand this more fully it is necessary to lay bare a fundamental, implicit premise that is shared by many Westerners—the attitude that all "diseases" or other onerous conditions are unqualifiedly "bad" and undesirable. Such a notion probably has its origin in the archaic belief that clinical symptoms are supernatural retributions for sinful behavior; it is not unusual for a sick person to ask: "What have I done to deserve this?" and chapels often are considered as integral a part of hospitals as operating rooms. Western scientists have only recently begun to transcend the limitations of popular notions of the significance of clinical symptomatology and to free themselves from the quasi-religious assumptions that lead most Westerners to look at all illness with abhorrence or at least with apprehension.

An important question that arises from such considerations is: What adaptive "illnesses" must populations of *Homo sapiens* develop in order to function efficiently on other planets or on the oceans' floors? Needless to say, if we continue to apply quasi-religious standards and the norms of sea-level environments and continue to believe that all illnesses are unequivocally undesirable, we will not be able to plan for life in extraterrestial environments; or if we plan on the basis of primordial concepts of adaptation, the results are liable to be disastrous.

Some writers have suggested the hypothesis that carriers of some genetically produced diseases, such as diabetes and hypertension, may possess unique advantages over their fellow men. I do not suggest that carriers of a genetically based disease suffer no disadvantages, but I believe we should shift in our thinking about such phenomena away from an exclusive focus on their injurious aspects toward a consideration of their possibly adaptive elements. It is only by exploring and uncovering the adaptive side of illnesses that such advantages will prove exploitable in planning for the species' colonization of extraterrestial habitats as well as for individual lives in contemporary settings. It is largely in these terms that the study of adaptation is one of the major frontiers of science.

As we progress in keeping alive tens of thousands of individuals with potentially lethal genes—diabetics, children with a potentially fatal predisposition to measles, individuals predisposed to poliomyelitis, hypertensives, individuals with Rh disease, and the like—we must explore all of the possibly adaptive advantages in these genetic predispositions so that they can be exploited for the benefit of the affected individuals as well as for the species as a whole.

The study of biosocial adaptation often carries significant implications for social, economic, and political problems confronting modern societies; I draw attention to these where appropriate. While biological evolution in *Homo sapiens* has been greatly slowed, the rate of cultural evolution has steadily accelerated. The increasing rapidity of transformations in human culture and social organization is observable not only in economic and political institutions but also in technological and scientific advances that promise to alter the entire genetic background of man and these changes will entail profound legal and religious problems.

To illustrate, biological engineers have been very successful in applying biological principles to many plants and animals. They can develop varieties that are beautiful, strong, nutritious, sociable, swift, tasty, large, docile, or a combination of several agreed-upon virtues. Would similarly desirable results ensue if genetic knowledge were applied to the human species? Recent dramatic discoveries in molecular biology have opened up new possibilities for the manipulation of hereditary material. An important question is, of course, "Who will be the engineer?" More fundamentally, a question that has been lost sight of in the vortex of emotions that this issue arouses is, "For what kinds of physical and social habitats will this engineering be undertaken?" Since no adaptation emerges in a vacuum, but only in relation to a particular habitat, and assuming that it will be possible to develop different varieties or new sub-species of *Homo sapiens,* for which particular physical and social milieus will these varieties be developed?

For instance, some geneticists believe there may be a correlation between the human XYY genotype and antisocial behavior. Assuming this to be true—and the evidence is far from conclusive—and assuming that it will someday be possible to detect such a genotype at birth and alter it, is it valid to assume that the XYY genotype will be maladaptive under all social conditions? Even assuming that gene therapy may someday alleviate some human diseases, what will be the short-range and long-term side effects of gene therapy in different habitational settings, especially those that have not yet been even imagined? If human life may be created artificially, what will the consequences be for say inheritance laws and incest taboos? But there is a deeper question: Assuming that genetic engineering will be possible, why should it be done?

The reader who wants a broader base for understanding adaptation in *Homo sapiens* can consult several excellent works, one of which is Theodosius Dobzhansky's *Mankind Evolving: The Evolution of the Human Species* (New Haven: Yale University Press, 1962). This book has already assumed the proportions of a classic in the thinking of many anthropologists. In this book the inseparable relationship between cultural and biological adaptations has received one of its clearest expositions, and many of the concepts in the foregoing introduction rely directly on Dobzhansky's work. Dobzhansky's article "Evolution at Work" (*Science,* 127 [1958]: 1091-98) explores the notion that architectural changes in a species which are adaptive at a particular moment in history may prove fatal in the long run.

Another eloquent statement that seeks to express the relationship between biological predispositions and human cultural patterns is a paper by Lionel Tiger and Robin Fox, "The Zoological Perspective in Social Science" (*Man,* 1 [1966, N.S.]: 75-81), and later we will return to some of the concepts that are outlined in this paper. Tiger has further spelled out some of these concepts in a short paper "Patterns of Male Association" (*Current Anthropology,* 8 [1967]: 268-69).

Also of direct relevance to some of our issues and discussions are *Process and Pattern in Evolution,* by Terrell H. Hamilton (New York: Macmillan, 1967); *Three Papers on Human Ecology* (Mills College Assembly Series, 1965-66); *The Biology of Human Adaptability,* edited by Paul T. Baker and J. S. Weiner (Oxford: Clarendon Press, 1966); and *Human Biology: An Introduction to Human Evolution, Variation, and Growth,* by G. A. Harrison, J. S. Weiner, J. M. Tanner, and N. A. Barnicot (Oxford: Clarendon Press, 1964). Among the works that stress processes of adaptation in the individual are Hans Selye's *The Stress of Life* (New York: McGraw-Hill, 1956) and Ludwig von Bertalanffy's *Problems of Life: An Evaluation of Modern Biological Thought* (London: Watts, 1952). Both of these authors are giants in the fields connoted in the titles of their books. The most recently published work in this area, and perhaps the most exciting intellectually, is Bernard Campbell's *Human Evolution: An Introduction to Man's Adaptations* (Chicago: Aldine, 1966, second edition, 1974).

Several sources may be consulted in regard to questions of biological engineering. Genetic manipulation is discussed in *Come, Let us Play God,* by Leroy G. Augenstein (New York: Harper & Row, 1968). In *The Future of Man* (New York: Mentor, 1959), P. B. Medawar (Selection 4), speculates on the genetic qualities of future generations of our species. "Gene Therapy for Human Genetic Disease?" by Theodore Friedmann and Richard Roblin (*Science,* 175 [3 March 1972]: 949-55) discusses the difficult scientific and ethical problems raised by proposals for genetic manipulation in humans. The difficulties associated with the hypothesized relationships between genetic predisposition and behavior are explored in "Behavioral Implications of the Human XYY Genotype," by Ernest B. Hook (*Science,* 179 [12 January 1973]: 139-50).

1. GENES, CULTURE, AND MAN

GEORGE W. BEADLE

Reprinted from Columbia University Forum *(Fall 1965), 3(3): 12-16. George W. Beadle is a Nobel Laureate (1948) and served as President of the University of Chicago. Before he assumed the latter post, he was Professor of Biology and Chairman of the Division of Biology at the California Institute of Technology. Since 1969, he has been William E. Wrather Distinguished Service Professor at the University of Chicago.*

■ The basic unit of observation and analysis—the starting point—in the study of adaptation is the deme, or breeding population. However, not only are there many instances in which we are forced to make inferences about a deme from one individual—as in the examination of many fossil remains, where we are confined to only one skeleton—we also must remember that the basic biological mechanism of adaptation is to be found in the individual. This mechanism is the gene. Thus, before we go on to explore the fates of different genetic predispositions in whole populations in relation to the habitats that they seek to inhabit, it is necessary to have some picture of what a gene is and how it works. It must be added that this picture will probably change very soon, because biologists' knowledge of the gene is increasing at a remarkable pace.

Beadle tells us with remarkable clarity, in this first selection, what is currently known about the gene. Equally important, he sets the tone for a concern that will reappear in several subsequent selections. Even so "pure" a science as the study of genes cannot be divorced from considerations of social policy. This is not to say that all biologists are agreed on what the policies of society should be in any sphere, but, as students of the processes by which life is sustained, they are committed to the tenet that life *should* be sustained. Reproduction is necessary to the sustenance of life, but it is not sufficient. A favorable environment—social as well as physical—also is indispensable. If the environment cannot be improved, the population itself must undergo change. If neither occurs, life in that environment will become extinct. One does not necessarily have to agree with all of Beadle's conclusions to appreciate the pressing importance of the questions he raises.

Because of the rapid rate at which biological scientists are making advances in "breaking the genetic code"—facilitated in large measure by Beadle's work—caution must be exercised in making suggestions for further reading about genetic structure. Especially in connection with the gene, today's truth can easily become tomorrow's outdated insight. However, two works can be consulted. Isaac Asimov's *The Genetic Code* (New York: Orion Press, 1963) is a short introduction in layman's terms—to the extent that this is possible—that is extremely well written. Also very readable and worthwhile is *The Language of Life: An Introduction to the Science of Genetics* by George W. Beadle and Muriel Beadle (Garden City, N.Y.: Doubleday, 1966). One of the most recent reports on advances in deciphering the code for the synthesis of proteins is "The Genetic Code: III" by F. H. C. Crick *(Scientific American* [October, 1966], 215 [4]: 55-62). ■

MAN IS THE product of two kinds of interdependent evolution that for convenience and simplicity we call biological and cultural. The latter is of far greater significance in man than in any of his fellow creatures on earth. In biological inheritance and evolution, we are not basically different from other species, whether these be submicroscopic viruses, bacteria, protozoa or more complex plants and animals; in cultural characteristics there is an enormous gap between us.

Let us first consider biological inheritance. In all living beings, this depends in large degree on the generation-to-generation transmission of giant nucleic acid molecules of one or the other of two kinds: deoxyribonucleic acid, DNA, or ribonucleic acid, RNA. In addition, as John Moore and Ruth Sager of Columbia University and others have shown, cytoplasm is also essential, directly or indirectly; even the simplest viruses cannot multiply outside of cellular hosts, all of which have specific cytoplasms that have continuity from one generation to the next.

Let me illustrate by describing what happens in man. Each of us develops from a tiny, almost microscopic, spherical cell, the fertilized egg. It is surrounded by membranes. Inside, there is a layer of jelly-like, viscous cytoplasm and in the center is a nucleus. In the nucleus are 46 chromosomes. In the chromosomes, there are genes. And the genes are made of DNA. Today we know an enormous amount about DNA. Our knowledge began to accumulate almost a century ago with the work of a biochemist named Miescher. Piece by piece we began to understand the structure of giant DNA molecules, which can be likened to a submicroscopic three-dimensional jigsaw puzzle. Finally, in 1953, the pieces fell into place in the skillful hands of James Dewey Watson and Francis H. C. Crick, working together at Cambridge University. Several of the key clues to the final solution were provided by Professor Erwin Chargaff at Columbia. The working out of the detailed structure of DNA is one of the great achievements of biology of the twentieth century, comparable in importance to those of Darwin and Mendel of the previous century. The Watson-Crick structure immediately suggested how DNA carries information, how it replicates or copies itself with each cell generation, how it is used in development and function, and how it undergoes the mutational changes that are the basis of organic evolution.

DNA is composed of units called nucleotides, molecules—subunits of larger molecules—that are made of five kinds of atoms: carbon, hydrogen, oxygen, nitrogen and phosphorus. There are thirty-some atoms in each of these nucleotides and there are four kinds of nucleotides in DNA, arranged in long chain-like molecules, many hundreds or thousands strung together and paired in what is called an antiparallel manner, that is, one chain runs up and paired with it is a chain that runs down. This pairing is accomplished through hydrogen bonding.

The sequence of the nucleotide units in the molecule is a kind of molecular code which spells out information. In the fertilized egg, there are approximately five billion nucleotides strung together in the nucleic acid of the genes that determine the final human organism. How much information does this represent? Encoded in letters of the alphabet, it would be sufficient to fill about a thousand volumes, 600 pages per volume, 500 words per page. If one were to take all the nucleic acid from all the cells that gave rise to all the people on earth today (three billion) it would make a cube about 1/8 of an inch on a side, and represent some three trillion volumes of information—about 60,000 times as much information as has been printed since Gutenberg invented the press. This molecular language is quite amazing. There are only three letters in the alphabet, all of the words turn out to be three-letter words, and there are only 64 words in the dictionary.

The DNA molecule replicates with each cell division as illustrated in Figure 1.1. Each half of the double molecule makes a new complement, resulting in two double molecules, each an exact copy of the original. This is the basis of all biological reproduction at the molecular level. Is it true? The fact is, it is true. And it has been demonstrated in many ways. Kornberg has been able to get DNA molecules to replicate in a test tube. Meselson and Stahl have shown by isotope labeling that the molecule does in fact do exactly what Watson and Crick predicted it should do. And more recently, Spiegleman has shown that a related nucleic acid, RNA, does exactly the same thing in a cell-free test tube system. So the Watson-Crick hypothesis is well substantiated.

How is this information used in making a human being out of a single egg cell? We do not know everything about it by any means, but it is quite clear now that the DNA of the nucleus transfers information to a related nucleic acid, RNA. RNA molecules which are called messenger RNA move out into the cytoplasm of the cell, outside the nucleus, where, in association with submicroscopic particles called ribosomes, they collect the amino acids that are to make protein molecules. The amino acids are collected in the right sequence to make a protein that corresponds in its amino acid sequence to the coding of the DNA of the nucleus and the messenger RNA. For a long time many physicists, chemists, biologists, and mathematicians worked on this coding problem. A few years ago, a young biochemist at the National Institutes of Health, Marshall Nirenberg, found the clue to breaking this code by using synthetic RNA. Practically all of the 64 code words have now been identified as to their meaning in terms of protein synthesis. There are words that specify the 20 kinds of amino acids. There are words that indicate the beginning of the message and others that indicate the end of the message.

The raw materials of evolution are mutations, and the Watson-Crick structure suggests how mutations occur in DNA. In this copying mechanism, occasionally there is a mistake—the equivalent of a typographical error. Indeed, we know now that all of the kinds of errors that can be made in typing a message in our alphabet can be made in DNA. There can be additions, deletions, transpositions, and substitutions of letters. And all of these are known genetically. These mutations may occur spontaneously, be induced by high energy radiation, or be produced by chemical means. Mutations in general tend to be unfavorable, and, in man, the unfavorable mutations lead to what we call genetic disease; there are hundreds of these known. However, I should like to point out that mutations, like typographical errors, are favorable or unfavorable not in an absolute sense, but in terms of the environmental context and also in terms of the genetic context.

For example, a bacterium which acquires a mutation that makes it resistant to penicillin is obviously favored in the presence of penicillin; but in the absence of penicillin, it has a lower reproductive rate and the mutation is unfavorable. In the same sense, if a unit of DNA is omitted at one point, then adding one nearby is a favorable mutation because it restores a reading frame and most of the message is read correctly. Either mutation is unfavorable if it occurs alone, but if it occurs as a second one, it may be favorable.

Mutations are the basis of evolution in the sense that those that are favorable are multiplied by Darwinian natural selection and those that are unfavorable are eliminated. To give an analogy, suppose you had a typist copying and recopying the Gettysburg Address without ever correcting or proofreading. If she averages one mistake in a thousand letters—there are about 1,100 letters in the Gettysburg Address—she will make somewhat more than one error per typing of the Gettysburg Address. In a thousand typings, she will have accumulated something over 1,000 typographical errors, and the Gettysburg Address will be a mess. Now change the analogy and put an inspector behind the typist. If every time she makes a mistake that makes the Gettysburg Address less good, he throws it away and says, "Retype the original message," of course errors will never accumulate.

But, you say, what about positive evolution? Let us change the analogy a little more and give the inspector judgment. Every time he sees a typographical error that is unfavorable, he throws it away, but if it is neutral or favorable, he passes it; the typist will now accumulate favorable mutations. But you will say, "How can you improve the Gettysburg Address?" That is very difficult indeed, which leads to a conclusion. The better an organism is adapted to its environment, the less probable it is that a random mutation will improve it. And conversely, the less fitted it is to the environment, the more likely is a random mutation to improve it.

If in biological inheritance we are not unlike other creatures, how, then, are we different? Most significantly in the degree of elaboration of our nervous systems, especially the brain. The gap between our brain and that of the nearest nonhuman relatives is a very wide one indeed. Ours and his are both constructed in large part according to DNA instruction, in a manner that we are far from under-

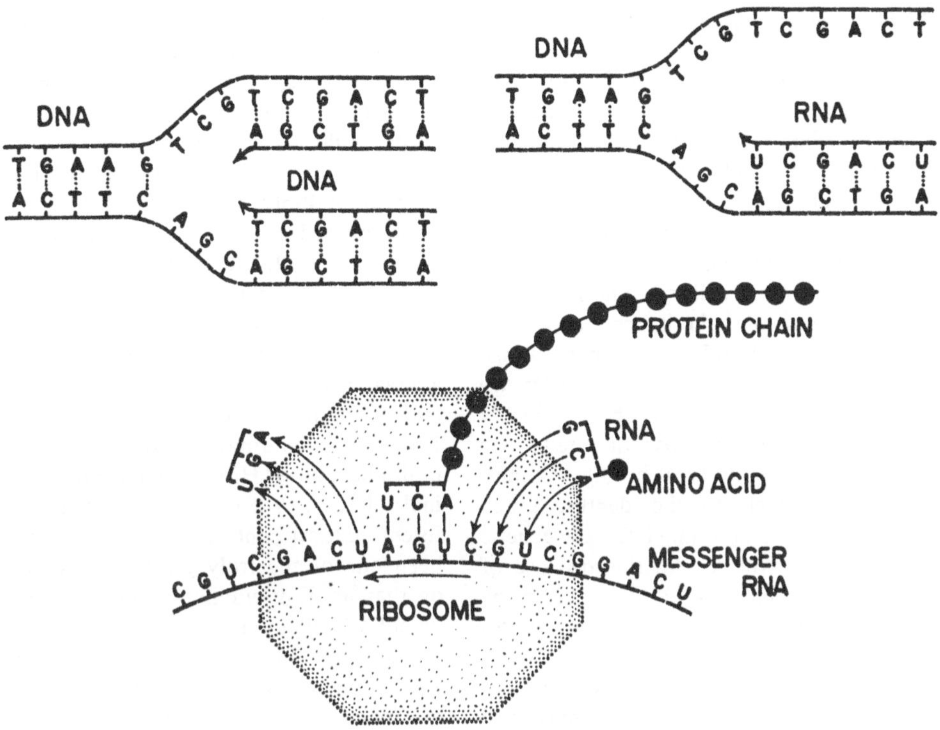

FIGURE 1.1: TRANSLATING THE GENETIC CODE

The Watson-Crick model of the DNA molecule consists of two complementary helical chains, the links of which are the four different nucleotides that contain the genetic information. A nucleotide is made up of one sugar molecule plus one phosphoric acid molecule plus one of the four bases that are the gene's code "letters": adenine (A), thymine (T), guanine (G) and cytosine (C); these also serve to bind the two chains together, A always bonding with T and G with C. The drawing at upper left shows the DNA molecule duplicating itself. In a way not yet entirely understood, the two chains part at one end of the molecule, and as the molecule unwinds and the chains continue to separate (from right to left in this schematic drawing), each chain serves as a template on which a new complementary chain is formed. In the same fashion, in the drawing at right, one chain of DNA serves as a template to form a molecule of RNA. Again each base determines its complement and hence the order of the code "letters" in the forming RNA molecule, with the exception that in RNA uracil (U) takes the place of T as the bond of A. The bottom drawing shows how messenger RNA determines the sequence of amino acids that form protein molecules. Each one of the 20 amino acids is specified by a three-letter RNA sequence, and thus the four-letter "language" of DNA is translated into the 20-letter language of proteins. Messenger RNA, having received its information from a DNA template in the nucleus of the cell, passes into the cytoplasm, where in association with a ribosome it can be matched with a different type of RNA that specifies a particular amino acid. The ribosome apparently acts as a sort of vise for positioning amino acids in the growing protein chain. In the drawing, the RNA molecule UGA, having matched with the ACU "word" of the messenger RNA molecule and added its amino acid to the protein chain, is just leaving the ribosome; UCA matches with AGU to add a different amino acid; GCA will match with CGU to add the next link to the protein chain. When another three-letter combination signifies that the "message" is over, the protein molecule will be complete.

standing. But ours, unlike his, has evolved in a highly special way through feedback of cultural inheritance.

Some two million or more years ago, Louis Leakey's *Homo habilis,* or whatever you prefer to call him, began to make and use simple tools, of which only crudely chipped stone choppers have survived, along with his bones. Tools certainly gave *Homo habilis* great selective advantage over his less cultured relatives. This cultural inheritance accelerated biological inheritance, and man was off on a cultural evolutionary binge never experienced by any other species we know. At first it was slow—but probably continuous. Civilizations rose and fell—one after another—rising to great heights at times, and dropping back to the depths, much faster than can be accounted for by genetic change, I am convinced. All this because of the evolution of that DNA-directed brain, by which men could make bigger and better artifacts.

Our brains, like those of our ancestors and related species, contain what we may call built-in information that regulates many of our bodily functions, like breathing, circulating blood, instinctive behavior and so forth. But unlike them, it contains a vast amount of "put-in" information that is our cultural inheritance.

Our ability to put information into the brain, record it, remember it, rearrange it, and retrieve it, is what makes possible our awareness, our insight, our reason, our communication—all of which has led to language, art, religion, literature, technology, and science. This ability has brought us now to a knowledge of how we have evolved. It has given us the power to direct the evolution of plants and animals to our ends and the knowledge to control our own evolutionary future—both the biological and the cultural components of it. Are we prepared to use that knowledge intelligently and wisely? I have grave doubts.

First, consider biological evolution. We can reduce genetic disease by negative eugenic methods and we do—to some extent. Except for relatively rare dominant genetic diseases, the process is so discouragingly slow as to be almost ineffective. Each of us probably has recessive genes for half a dozen or more genetic diseases. Carried to the extreme, negative eugenic methods would mean that few of us would have children. On the other hand, we know that positive eugenics, that is, preferential multiplication of the fittest, can be accomplished in man. We do it regularly with plants we cultivate and with animals we domesticate. The unanswered questions are: What do we want in man? Who is to decide? And there is another question: What parts of a great many desirable traits are genetically and what parts are culturally determined? Few of us would have advocated preferential multiplication of Hitler's genes through germinal selection. Yet who can say that in a different cultural context Hitler might not have been one of the truly great leaders of men, or that an Einstein might not have been a diabolical villain. I prefer to believe that in the absence of much more understanding than we now have, we will do best to preserve maximum genetic diversity. That is, aside from those who are clearly and significantly genetically defective, let all segments of the species multiply equally rapidly—but at an overall rate that our resources can sensibly support.

The circumvention of genetic disease by nongenetic methods is called euphenics. Phenylketonuria or PKU, galactosemia, cretinism, retina blastoma, and many other diseases are diseases that through medical intervention or diet control can be circumvented in the sense that we make the individual normal but do not change the underlying genetic constitution. If such individuals are made completely normal, including reproductively, and if indeed they do reproduce normally, these diseases will increase in frequency. The increase will be slow because it is a function of the mutation ra e, which i something like one in 100,000 per generation or one in a million per generation. This slow rate of increase is fortunate in one sense, for it means we have much time to think about the problem, and we need it. But it is unfortunate in another sense, for in no one generation will the increase be dramatic enough so that we will, in fact, think about it. There is one simple solution, to persuade people affected with genetic disease to adopt children rather than have them in the normal manner. According to popular press reports seen recently, another series of euphenic advances promises that worn-out organs will be replaced by artificially cultured counterparts before too long. Already we have artificial kidneys, artificial veins and arteries, partly artificial hearts. We welcome such advances, but there is one difficulty, and I think it is a serious one. The rapidly increasing social cost of these endeavors is great indeed. How much further can it increase?

Let me turn to the possibility of what we might call cultural euphenics. Unlike biological inheritance, our cultural inheritance begins anew each generation. In the absence of cultural information put into the brain, none of us would speak or write or sing in any intelligent way, or build even the simplest tool. We would revert half a million years or more in the cultural sense. Still, in another generation it could all be restored. There is almost no inherent limit to the speed with which culture can be lost, improved, or otherwise modified. It can spread over the face of the earth, nowadays, in days or in months. One of its characteristics is that in each generation it is acquired beginning at birth, or even before.

We have only recently come to recognize how much culture goes in during the earliest years. Professor Benjamin Bloom of the University of Chicago has shown that about one-half of general intelligence is acquired by the age of four, and about one-half of normal school learning is acquired by about the age of nine. The implications for the culturally deprived, the culturally impoverished and changing nations is great indeed. The trick is to break or supplement early in life, in an acceptable way, the normal parent-to-child chain of cultural transmission.

In education, we are inclined to believe quite firmly, lies the road to a better life. But, as has been pointed out, there are cultures in which education is considered a hurdle to be overcome in order to get a job. Cultural euphenics has unlimited potential in changing such attitudes. It is safe, for it is easily reversed, unlike eugenic change.

What about genetic differences in ability to acquire specific cultural patterns? Obviously, differences in ability to acquire culture do occur among individuals. But do such differences occur in significant degree among large populations? That is quite a different question. If one were to take 1,000 healthy babies of Australian aborigines and put them in our cultural context at birth, with our tradition of parental affection and all the rest, would they be as successful in acquiring our cultural pattern as a similar sample of our children? Or, would the converse lead to an equal degree of change in a single generation of our children in their cultural context? We simply do not know the answer. No culture-free test of ability to acquire a given cultural pattern has ever been devised and the experimental approach that I mentioned has never been made.

It seems quite clear to me that we can and must do a far better job of up-grading our cultural environment. I see no need, while doing so, to be obsessed with a desire for cultural uniformity. Cultural diversity may be as important for man's future evolution as is genetic diversity.

2. GENETIC DRIFT AND SELECTION OF GENE SYSTEMS

THEODOSIUS DOBZHANSKY

Reprinted from Mankind Evolving *(New Haven: Yale University Press, 1962). Theodosius Dobzhansky is Adjunct Professor of Genetics at the University of California at Davis. Before he came to the United States, he was Lecturer in Genetics at the University of Leningrad from 1924 to 1927. Before going to the University of California, he was Professor at the Rockefeller University. A past president of several scientific societies, he holds many honorary degrees and has received awards for his research, including the National Medal of Science, awarded by the President of the United States in 1965.*

■ It is a gross oversimplification—and therefore misleading—to say that the gene is the principal mechanism of human adaptation and let the matter go at that. Not only do human genes carry codes and messages for behavioral capacities from parents to children, they also carry messages from the prehuman primate past. The evidence from the study of evolution points conclusively to the principle that a great many of our capacities or potentials for social life were established in primates prior to the emergence of *Homo sapiens* on the evolutionary scene. These inheritances, some of which will be discussed in later selections, are in turn retransmitted and reinherited in every human generation.

Despite these commonalities between *Homo* and his ancestors, there are profound differences. These differences are of many kinds, including revolutionary alterations in locomotion, in the hand, and in the capacity for culture and communication.

In studying the most rudimentary processes of human adaptation the basic concept has to be the genetic makeup of a population, its gene pool. But the human gene pool is always changing. There are two principal reasons for this. First, human populations are in constant flux; they are constantly dividing—partly as an adaptation to habitational and social pressures—and recombining. Recent archeological and paleontological explorations make clear the fact that even the earliest men had contacts with alien groups, sometimes thousands of miles from their home bases.

This leads to the second reason why a gene pool is always changing. Whether combinations and recombinations of genes from different populations will flourish or become extinct depends on how they fit into different habitats. It is clear from the following discussion of "genetic drift," by Dobzhansky, that there is no such thing as a hospitable or inhospitable environment as such; nor are there adaptive or maladaptive genetic predispositions in the abstract. Instead, and just as populations are in dynamic relationship with each other, genetic and habitational predispositions constantly act upon each other. Sometimes these interactions are beneficial and lead to the survival of the group; at other times the perpetuation of life is impossible. One of the conclusions that can be drawn from the study of this aspect of human adaptation is that the "dichotomy" of heredity and "environment" is false; neither affects the perpetuation of life independently of the other. Furthermore, as Dobzhansky observes in this selection, no gene operates by itself; instead, it is an indivisible part of a total gene system. In addition, as he states, human culture exerts a profound influence on man's genetic constitution. With the emergence of *Homo sapiens*, then, the genetic and cultural systems must be seen as indivisible.

The concepts Dobzhansky explores in this selection are examined in greater detail in his *Mankind Evolving* (New Haven: Yale University Press, 1962). In *Human Evolution* (Chicago: Aldine, 1966), Bernard Campbell suggests that the notion of genetic drift may have less utility than many biologists and anthropologists have supposed. (Campbell's work also includes a complete and up-to-date bibliography.) ■

IT WOULD BE dogmatic to insist that all race differences must have been produced by some form of direct or indirect selection. But what are the other possibilities? Many, in fact a majority, of racial traits in man seem to have no influence on the fitness of their carriers. Would you be better or worse off if your nose were a little longer or shorter or your lips a little thinner or thicker? One human polymorphism which has different incidence in different races is the ability (dominant) or inability (recessive) to taste phenylthiocarbamide (PTC), a chemical substance not known to occur in nature. What, then, was the gene for tasting or not tasting PTC doing in human populations before chemists synthesized this substance about forty years ago?

At this point we must proceed with the greatest caution. There is no gene "for" tasting this or any other chemical substance or "for" anything but causing the organism to develop in cooperation with all the other genes it has. The precise roles which different genes and their different alleles play in development are usually unknown, and our inability to detect easily the influence of certain genes on fitness does not mean that such influence does not exist. The story of the blood groups is instructive in this respect. For many years it seemed that people derived no advantage and suffered no disadvantage from the blood type they happened to possess. But it has been discovered that the incidence of O blood is slightly but significantly higher in persons with duodenal ulcers than in the controls (members of the same population who do not suffer from these ulcers); the incidence of A-type blood is higher in the victims of stomach cancer than in persons free of stomach cancer; and people who develop certain forms of goiters are nontasters of PTC more frequently than those without goiters.

All this does not mean that we have now explained why different races differ in the frequencies of the blood groups and PTC tasting (as some people seem to have concluded a bit rashly). Not only are many O blood carriers free of ulcers, people with A blood without cancers, and nontasters of PTC without goiters, but these diseases also occur in the carriers of other genotypes. There is also no reason to believe that duodenal ulcers are less dangerous or less frequent in countries that have high frequencies of O blood than in those with low frequencies, or that it is any less dangerous to have a cancer of the stomach in countries in which A bloods are frequent than where they are rare. (Note added in proof: Vogel, Pettenkofer, and Helmbold have made an interesting attempt to correlate the frequencies of A bloods in human populations with prevalence of epidemics of plague and of syphilis, and of O bloods with that of smallpox. This matter evidently requires further study.)

A possible mechanism bringing about differences in gene frequencies between populations is so-called random genetic drift. If not acted upon by mutation or selection, the frequency of any gene in the gene pool of a population will tend to remain constant from generation to generation. It should now be added that an absolute constancy would be expected only in ideal, infinitely large populations; the smaller a population, the fewer breeding individuals it contains, the greater will be the "drift," the chance deviations, up and down, from a constant frequency in different generations. Genetic drift, worked out mathematically chiefly by Sewall Wright, is a rather abstruse matter in detail, but its principle can be explained very simply. Suppose that a person with a name as rare and outlandish as mine raises a family with ten sons. If this family continues to live in New York City the name will still be rare, but if the family happens to move to a country hamlet the name may become a relatively frequent one there. Conversely, one family called Smith moving out of the hamlet may make that name not represented there at all, while one Smith family leaving New York makes no appreciable difference in the frequency of Smiths in the city.

Early mankind was probably an array of more or less small endogamous bands or tribes, living as food gatherers or food collectors. In some parts of the world, as in aboriginal Australia, such tribes still exist or recently existed. Some tribes may have been from time to time reduced by environmental hazards, starvation, or disease to small numbers of persons or may even have been annihilated entirely. Conversely, other tribes that happened to stumble upon favorable circumstances, such as an unoccupied territory, increased in numbers, spread, and gave rise to large populations and new tribes. Birdsell made most interesting estimates of the probable rates of such expansion. The continent of Australia was most likely populated by small bands of people who entered from the north, the first invaders coming perhaps some thirty-two thousand years ago. Assuming that the original band numbered twenty-five persons, Birdsell reckons about twenty-two centuries as the time needed for their descendants to spread throughout the entire continent. Using a slightly different set of assumptions, he arrives at seventy-seven centuries for the expansion of the Australopithecines from their assumed ancestral home in South Africa to the rest of Africa, southern Asia, and Java.

A small band of migrants will evidently bring not the entire gene pool of the population whence they originated but only a small sample or slice of it. This sample may easily be atypical, like the rare surname in the above example: Some genes of the original population may not be included in the sample, while others may be overrepresented. The genes brought by the migrants will be impressed upon the new colony, tribe, and eventually, the race descended from the immigrant foundation stock. Just how great the differences between the original and the descended races caused by random genetic drift are likely to be has been disputed ground in evolutionary theory for nearly thirty years.

The dispute arose in large part through a misunderstanding. No one has seriously contended that genetic drift alone, without natural selection, can bring about major evotionary changes. The model of the evolutionary process proposed some three decades ago by Sewall Wright has been that of a species subdivided into numerous isolated or semi-isolated endogamous colonies or bands, some of them with continuously small populations, say a few dozen families, and others reduced only from time to time to such small numbers. Random genetic drift together with natural selection among the colonies may lead to "evolutionary inventions," i.e., to the formation of new and adaptively valuable gene constellations. The evolutionary process in a species consisting of such colonies is likely to be more rapid and effective than in a large unitary species.

The form of genetic drift that may have been important in human evolution can be visualized as follows. A small band of people, coming upon an unoccupied territory, expands and forms a tribe, a group of tribes, and eventually a new race. A useful new technique or an invention may conceivably cause such expansion even without migration to new territories. As pointed out above, the sample of the gene pool contained in the expanding band or tribe may and, in fact, almost always will include only a fraction of the genetic variability present in the original race whence the migrants sprang. Now, we must remember that in organic development the genes do not act independently of each other; instead the genes of an individual or a population form an interdependent system.

For example, the adaptive value of a gene for muscularity or corpulence or A or B blood groups depends on what other genes an individual or a population contains and, of course, on the environment. A gene may be useful in combination with some, neutral with other, and harmful with still other genes. Natural selection acting on a population tends to bring about the most favorable composition of the gene pool. The gene variants present in a population must, therefore, be adapted not only to the environment but also to each other; they must form a *coadapted gene system*. When a small group gives rise to a new population or race, the process of coadaptation by natural selection leads to reconstruction and emergence of a new gene pool. It is possible that the genes for the O, A, and B blood groups have different frequencies in different human populations because they are more favorable in combination with the other genes of some of these populations than with genes in other populations.

The process of coadaptation of gene systems by natural selection has been observed in the experiments of Dobzhansky and Pavlovsky on the fly *Drosophila pseudoobscura.* Very briefly the story is as follows. Two geographic races of the fly (coming from California and from Mexico or from California and Texas) were intercrossed and second generation hybrids were obtained. Ten groups of twenty individuals each were taken at random and used as founders of new populations or "colonies." The populations were maintained for about a year and a half (about twenty fly generations) in specially constructed population cages under uniform laboratory environments. Because of the high fertility of the flies, the experimental populations, though descended from a small number, were not small—they fluctuated between about 1,000 and 4,000 adult individuals.

Natural selection acted on these populations, making them adapted to live in their, of course highly artificial, environment. The remarkable result was that the ten experimental populations were found to diverge genetically and at the conclusion of the experiment had different genetic constitutions. This seems at first sight a paradox. Why should a genetic divergence occur among populations derived originally from the same source and kept in a uniform environment? The answer is that each of the ten groups of twenty founder individuals contained a somewhat different assortment of genes. Natural selection did, so to speak, its best with the genetic materials at its disposal in each population. Since these materials were diverse, so were the results of the evolutionary changes induced by natural selection. Such a divergence may have easily been important in human evolution. Populations derived from small groups of founders or colonizers diverged genetically, because they received somewhat different gene complements.

LEVELS OF TECHNOLOGY AND GENE FREQUENCIES

Hulse has pointed out that in 1600 there were some three million persons of British stock, while at present there are at least 150 million, a fiftyfold increase. World population has increased only about sixfold during the same period. The parts of Europe containing high proportions of blond people held about 3 per cent of the world population in 1600, while at present people originating from this area comprise about 12 per cent of mankind. Hulse correctly points out that, racists to the contrary notwithstanding, these figures do not necessarily prove either a biological superiority of the Blond Beast, or a genetic excellence of the British stock. Technological progress or stagnation may enhance or reduce the frequencies of the genes of culturally active or sluggish groups of people no less effectively than more strictly biological advantages or defects.

Hulse calculates further that between 1600 and 1950 the populations of Eastern Europe (Russia and Poland) increased from 15 to 180 million, i.e., 1,100 per cent, while those of Central and Western Europe increased from 85 to 375 million, i.e., 341 per cent. Most of the former populations contain above 15 per cent of the gene for the B blood group and the latter usually less than 15 per cent. B blood has thus become increased in frequency. Hulse interprets the differential population growth in this case as a consequence of improvement of agricultural techniques and particularly the introduction of a new staple food—potatoes. Conversely, oppressed castes and races may fail to perpetuate their genes. Between 1792 and 1861 about 571,000 Negro slaves were imported in Cuba, and yet in 1861 there were only 603,000 persons of African descent on that island. The situation in Chile was even more extreme; Negro slaves were imported during the eighteenth and early nineteenth centuries, yet at present African genes are virtually nonexistent in that country. The chief reason is apparently that, because of the remoteness of the country from slave markets, almost exclusively male slaves, considered more valuable, were imported, and they were unable to find mates either among the white or the native Indian population.

Cultural factors have thus changed and continue changing the genetic composition of mankind. Caution is called for in interpreting these changes as due to natural selection of the usual biological sort. Such an interpretation would be correct only if it were shown that cultural and technological changes are in turn set on foot by genetic differences among populations. This is, of course, a widespread assumption, pleasing to some people but certainly not scientifically validated or even particularly probable on theoretical grounds. We must face the fact that the causes of genesis, advancement, and deterioration of civilizations remain unsolved, despite the efforts of some of the best intellects that mankind has produced. The possibility that genetic factors are involved in these phenomena should not be dismissed dogmatically, but it seems certain that they were not the only and not even the most potent determining factors involved.

3. HOW TO BE HUMAN

WILLIAM W. HOWELLS

Reprinted from Mankind in the Making *(Garden City, N.Y.: Doubleday, 1967). William W. Howells is Professor of Anthropology at Harvard University and Curator of Somatology at the Peabody Museum. In addition to human evolution, his particular interest lies in detailed studies of physical variation in the human population, both living and as represented by crania. He has recently been using multivariate analysis in these studies to find the more important aspects of such variation in both recent and prehistoric populations.*

■ Cutting across man's commonalities with his nonhuman ancestors, he has many traits that are unique in organic nature. It appears, for example, that he alone is capable of thinking of himself in the third person. More important, it appears probable that he alone is able to think of himself as unique. Many religions try to validate these feelings that man has about himself (although few of them can agree on the nature of this uniqueness). One expression of this sense of uniqueness, especially in Western society, is the belief that if *Homo* became extinct he would eventually reappear (or be re-created), which is almost like saying that we are an inextricable part of the design of the universe.

A popular story that was supposed to have made the rounds among members of the United Nations told that, after mankind had destroyed itself in a moment of what it considered an expression of progress, one monkey turned to another, saying "Well, what do you say, do we start the whole mess all over again?" Howells makes the point in this selection that the results probably would be vastly different. (To interject a personal conviction—on which, I should hasten to add, I can never be proved wrong—I believe that the records of man's adaptive histories show that he is incapable of destroying himself completely.) Howells' brief stroll in the woods of scientific imagination yields an important lesson: The condition and nature of man, presently and in the future, cannot be understood apart from his history. This history teaches us that man is in some respects indeed unique, but so is every species in nature.

Howells raises many more questions in this selection than can be answered, but a voluminous literature is available to the reader, beginning with Charles Darwin's *On the Origin of Species* (Cambridge, Mass.: Harvard University Press, 1964). A very lucid and thoughtful work is Loren Eiseley's *The Firmament of Time* (New York: Atheneum, 1966), which contains a short but pertinent bibliography. The advanced reader can consult *The Meaning of Evolution,* by George Gaylord Simpson (New Haven: Yale University Press, 1949) and *Behavior and Evolution,* a collection of essays edited by Anne Roe and George Gaylord Simpson (New Haven: Yale University Press, 1958). Both of these books cover almost all of the areas that are included in the present collection of readings, in addition to some of the questions raised by Howells. ■

"WHEN I CONSIDER the heavens, the work of Thy hands," says the Eighth Psalm, "What is man, that Thou art mindful of him? . . . Thou hast put all things under his feet: All sheep and oxen . . . the fowl of the air, and the fish of the sea . . . O Lord our Lord, how excellent is Thy name in all the earth!"

How excellent! David, extolling the uniqueness of man, might have extended more appreciation to the rest of nature. We know now, better than the Psalmist knew, the meaning of the animals under man's feet. We know how slow was our coming, how long our preparation. We know, through millions of years, what powerful forces brought us out. We see what boons we had in skeletons, jaws, limbs, warm blood. Homely things, but with them any mammal is a testament of evolutionary creation. Add fingers and acute eyes, and you have something like a monkey. Add true hands and legs and you have the beginnings of a man. Then natural selection made our brains, and everything we do with them.

All this came gradually, in due course. Each patient step made possible the next. Each step gave rise to hundreds of kinds of successful animals, perfect in their own ways. Man himself could only appear when a very high organization had been attained. For hands and a big brain would not have made a fish human; they would only have made a fish impossible. Man's own trail, among the many trails in evolution, was well defined: he had to be a mammal and he had to be a primate.

Was he inevitable? Mankind long thought so, and Genesis entertains no doubt that the world was made for us. Being human, we find it hard to see things any other way. Not long ago it was shocking to ask whether man had evolved from other animals. It is doubtless equally shocking to ask whether man arrived simply as a matter of luck.

Actually it would be a futile question. One cannot pass judgment on such a thing without going straight on to the origins of the universe. It does not matter here what your religious views are, for man must be looked on as an extraordinary achievement of design and organization. The universe itself is built on laws of matter. These are the foundation for laws of life; and on such a basis, we know, man at last came on the scene. He was not there to begin with, and he took a long time coming. But he came, and he expresses all the fullness of the possibilities of the organization of life as we understand it. How much of this is "chance"? There is no present answer.

I am not really trying to be profound, and I have only a few pages left in any case. Talking about chance, I am merely wondering if man was forced to take the shape he did. If we had it to do over again, would we choose all the same forks in the road, or would we turn out differently? Perhaps science fiction has the answer.

Doubtless there are other "men" in space. Led by the astronomers, we now face possibilities of life elsewhere at which the mind boggles. We know there are many billions of stars in our own particular galaxy, the Milky Way. Galaxies like ours are grouped in clusters, thousands to a cluster, and such clusters go on and on, out of reach of the telescopes, by the hundreds of millions. Planets to live on? Dr. Harlow Shapley has figured it out roughly. Not many stars have planets around them, he thinks; perhaps one in a million. As in our own system, few planets are of the right kind, with water, days and nights, favorable temperature and chemicals, and so on. Perhaps one good place in about a trillion stars, all told. But these are really splendid odds, considering the number of stars; Dr. Shapley ends by guessing that there are about a hundred million worlds where higher life has actually been forged by evolution. "We are not alone." Intelligent beings abound in the universe, most of them far older than we ourselves.

We can try to imagine what such people are like. Here we get little help from the comic books, which only show us flying saucers, manned by flabby little web-footed goblins with knobs on their heads. We must be strictly scientific, starting from scratch and assuming nothing about the beings we are studying, except that they are "intelligent." But this at once means that they are "human," in the sense that they have culture, like ourselves: they communicate ideas to one another, and create things jointly. Otherwise intelligence means nothing. And we could never communicate with them, if they could not already communicate with each other.

Furthermore, we might just as well imagine them on a favorable world something like ours, in matters of temperature, gravity, some atmosphere and land surface, and so on, since it seems to be the kind most suitable for life. Very well. We have intelligent, communicating creatures, on a faraway Earth. Are they anything like us? I think they are.

They might be considerably rearranged. They might "see" things we only "feel," like heat wave-lengths; or "feel" things we "hear," and so on. Their "bones," or whatever props them up, might be differently placed in relation to blood vessels and nerves. But they would have these things. Communicating, creating creatures must have motion—they could not be like trees, with little power to act and exert force. So they would have to be self-contained, moving about and getting their fuel like the animals of this world. They would need structure, and a nervous organization probably using electrical nerve impulses. They would need a liquid transportation system; we can hardly suppose that nourishment flows through their veins in the form of breakfast cereal. And so they must have begun their evolution, as we did ours, in a liquid medium, say water.

So much for raw material. What about design? Are the strangers round, or flat, or pointed? Do they smell through tailfins and see through a radiator grill? The chances are excellent that they have a head end (which implies a tail end as well), because almost all the animals of this world which can move alertly and exert any real force are built on this same plan. It puts the senses where they will do the most good. And the main center of the nervous system is close by, so that sense impulses will not have to make a long journey aft to a brain in the rump. Therefore our men will have heads. And, like the more efficient of our own animals, they will doubtless do their eating at this end as well. So, if they are going to be intelligent, our distant cousins had better be plotted something like a vertebrate or an insect of our own vicinity.

But they will do well not to imitate insects, for many reasons. All the cooperation of insect life—warrior ants, worker bees—is printed on the nervous system of each single insect. They act without benefit of ideas, only from instinct. Their social "ideas" come from natural selection alone, not from thought. So intelligent creatures will have made a choice, early in evolution, of a nervous system which is more open to fresh impressions: a brain which can learn. Eventually such a brain will become large. Come to think, it will make quite a lump somewhere, probably in the head.

While we are roughing out shape, we might ask whether such a species will come in more than one form. Bees and ants do. And caterpillars turn into moths. Above all, the great majority of living things come in two sexes. It is one of nature's most popular ideas, and not for the first reason that will occur to you. Sex ensures a great plenitude of new gene combinations, different in each of the offspring, through the coming together of genes from two parents. This is one of the keystones of evolution, as you have seen. We on earth have long ago made all our arrangements on the basis of two sexes. Perhaps our extraterrestrial equivalents have three sexes, like German nouns. Or more; who is to say? Certainly two is most likely. It is an almost universal rule here below, with no complaints.

We now have a Thing with one head and two sexes. (Two heads are *not* better than one; making up a single mind is more than most of us can do, as it is.) We had better assume that It goes on land. The water is far less promising as a medium for creation and communication, though perhaps not impossible. As for air, the birds have found that it does not work out well, if the goal is intelligence. Birds are beautiful but stupid, having been obliged to put their brain development largely in the service of coordination of movement, for flying.

Being on land, these men will have limbs as well as a head. They must have limbs, because they must have hands. If we can learn anything from our evolution, it is that we had to be able to *do* things to become human. And our whole struggle was the getting and freeing of hands to do them with. Surely, we would not have had large brains without them.

The hands will be better with fingers on them. There need not be five; perhaps a few more or less will serve. But you may have been struck by the fact that we have decidedly kept all five of the bunch handed down to us by *Seymouria* and his forefathers. So five seems like a good number, perhaps a minimum. Therefore look for plenty of fingers, on the ends of two arms. Two arms, not three, because the creatures should be symmetrical like us; and not four, because coordination would probably be too difficult for efficiency. Centipedes have to run their arms in teams.

Now for a big question. Will they be

standing up and walking around like us? They would not look very human otherwise. But might they not have two hands and *four* legs, that is, three pairs of limbs? Insects have. Do we do anything well that a centaur could not do better? We are reasonably content with our own seating arrangements, but perhaps we are making the best of a bad job. You know already why we go around on only two legs: if we wanted hands to use, we had no choice.

The choice existed once, in the early fishes. But the lobe fins and amphibians chose to keep only four limbs out of an original larger stock of fins. Unimaginative beasts, what was good enough for a bedstead was good enough for a labyrinthodont. And so these feckless ancestors nearly slammed the door on hands entirely, since later hand-users (or wing-users) among the vertebrates had to manage by balancing on the two remaining legs. Supposing that ancient vertebrates had found some simple use for an extra pair of fore-limbs, like the insects, while they still had the chance. Then these forelimbs might have continued being sufficiently adaptive for evolution to hang onto, as their possessors came out on land. Had this happened, we might all have avoided the problems which turn up in the blueprints of bipeds. There might, in fact, have appeared on earth many intelligent, hand-using, four-footed animals. So I will lay a small bet that the first men from Outer Space will be neither bipeds nor quadrupeds, but bimanous quadrupedal hexapods. (I have just invented that last word, in the hope that it means "six limbs.")

Finally, will they be big or small? Who will make a nice pet for whom, when we finally meet? Here, of course, gravity would keep weight down. But, on this earth at least, an intelligent animal must not be too small. Its brain—or whatever it operates on—must be absolutely large. And there is no reason to suppose that the basic cells and fibers of the brain could be substantially smaller than the same things here, so there should be no saving of mass. Furthermore, small animals—the active, highly organized kind, like mammals—do not live as long as large ones, not long enough to afford them the luxury of behavior which is largely uninherited and must be learned very slowly. So other men should not be much less bulky than we are.

Perhaps they should be more so. We have not become bigger, ourselves, as we got brainier, and J. B. S. Haldane pointed out long ago the hazards of being too big. The giants in *Pilgrim's Progress*, or any story book giants, were amateurishly engineered. Being the same shape as ourselves, they would have collapsed in a heap, with broken thighs; their legs were not big enough to support their bulk. For, when an animal gets really large, its legs must become disproportionately thick and strong to keep up with its mass, its cubic content. This happens in elephants. Man, as a biped, is probably as big as is safe for him. Bipedal dinosaurs, it is true, got much larger, like Tyrannosaurus, but at the cost of being decidedly bottom-heavy, with great haunches and a fat tail. So the men of elsewhere, if bipeds, are probably no giants. But if they have four feet to hold them up, then they might well be as big as a horse, or larger, and still be both intelligent and maneuverable.

So, on the whole, perhaps mankind as represented by ourselves is a pretty good model for intelligent creatures. However, this does not tell us whether we had to happen at all. I have said what a "human" being might be like in other worlds than ours. But what about the chances of men coming into existence again, not elsewhere, but on this very planet? Supposing, in a moment of idiot progress, we really killed ourselves off. Would *Homo* rise again?

Man came from an Australopithecine, which in turn came from some simpler hominid, perhaps like *Ramapithecus*. Those animals are gone; man has competed them into the grave. There are still apes. They might do for a fresh start, but I strongly suspect they are too specialized, and too busy looking for fruit in the forest, to turn to freer use of hands. Monkeys? Just possibly, if something made it worthwhile for a species to stand up. The new men might then have tails. But, in fact, the monkeys have made no move to mimic the hominoids, or human ancestors, during about thirty-five million years.

No other higher mammals of this earth will serve. Horses, dogs, elephants, all are deeply committed to being what they are. The next try would have to come from a tree shrew, laboriously repeating all of primate history. And before little *Tupaia* could put forth progressive descendants now, the world would have to be swept clean of the kind of competition which might overwhelm them on the way up. This means: get rid of most higher mammals, above all, rats, cats, and monkeys.

If he fails us, we (or rather our carbon copies) are done for. The remaining links of progress are now missing links; the good chances are gone. The mammal-like reptiles gave out long ago; and getting something human from the specialized creatures in the next ranks is hopeless: birds, snakes, frogs. The fishes? Lobe fins, with the makings of lungs and limbs, were put out of business eons ago by the ray fins, who can never leave the sea. The main army of fishes has gone well past the fork that once led to the land. Only the lungfish remain, waiting in mud for the rain to come again, and the coelacanth, so deep in the ocean that he dies in shallow water.

We might need brand-new "vertebrates." Well, then, eradicate the fish, who rule the seas as we rule the land and who are not likely to stand aside while nature experiments with ridiculously crude forerunners of ostracoderms once more. Conceivably life would have to start afresh. In that case, wipe out everything that moves, to keep the necessary simple molecules from being eaten as they form. So all in all our hopes for repetition are not good, and we had better stay the hand that drops the bomb.

All this musing is not really science. I am taking it as a path to walk around our object, mankind, while we view his place in nature. We are inclined to swing back and forth between extremes in what we say about him. One moment he is a specklet in space and time, a nothing among millions of planets harboring life. The next, we are awed by his complexity of structure and his primacy on earth. The Psalmist, not too certain, seems to favor the second view, and I heartily agree, having no disposition to write about specklets. I hope this book does something toward putting man in perspective, for its moral is that our roots are long and deep.

4. THE MEANING OF FITNESS AND THE FUTURE OF MAN

P. B. MEDAWAR

Reprinted from The Future of Man, © *1959 by P. B. Medawar, Basic Books, Inc., Publishers, New York, 1960, pp. 27-14, 95-103. Sir Peter Brian Medawar is a Nobel Laureate (1960). He is Director of the Clinical Research Centre, Harrow, England. His books include* The Art of the Soluble, Induction and Intuition in Scientific Thought, *and* The Hope of Progress.

■ Man continues to evolve biologically and culturally. Like the dichotomy between heredity and environment, the dichotomy between biological and cultural evolution is false. Man is so biologically generalized that it is doubtful the species could have survived without the development of culture. In purely biological terms, there is hardly a habitat in which human life could be perpetuated if *Homo sapiens* had not developed culture. This is why the very definition of *Homo sapiens* must include culture.

But human culture could not have developed without the emergence of man's unique biological capacities, those that made his mental, speech, motor and other functions possible. Dobzhansky (in Selection 2) stressed the adaptive function of genetic variability and diversity. Medawar, in addition to discussing the concepts of fitness and continuing human evolution, in the following selection stresses another side of man's relationship to his habitat: A genetic predisposition has disadvantages for members of the population alongside its adaptive potentials. Thus Medawar and Dobzhansky complement each other. The latter focuses on the total gene pool of a population in relation to its habitat while the former concentrates on the relationships between a particular genetic predisposition and its habitat.

In the first part of this selection, Medawar adroitly presents two different environmental backdrops for genetic predispositions; one is "social" (England's National Health Service) and the other "natural" (lethal malaria). The social component of the environment can be regarded as one aspect of a society's continuing adaptation, but it can also be viewed as a "given" in the total environment of that population; it is as potent a selective factor in its habitat as the prevalence of lethal malaria is in another. It is an example of a social system in which man's attempts to transcend the restricting limitations of his natural milieu—for example, in answering the question of who shall live and who shall die—have attained their greatest fruition thus far.

In the second part of this selection, Medawar carries his analysis to a very important conclusion: One of the most important premises of all human culture, especially at its highest and most successful levels of adaptation, is that "nature does not always know best." He also, rightly, warns against facile—and socially dangerous—attempts to transpose the principles of organic evolution into laws of social relations, as in so-called social Darwinism, which almost invariably are used to stifle civil liberties and to impede continuing social reorganizations. His statement is eloquent testimony to the tenet that science need not be dissociated from human values.

Medawar's thinking about the processes of adaptation is elaborated in his article "Problems of Adaptation" *(New Biology,* 11 [1951]: 10-26), and his book, *The Uniqueness of the Individual* (New York: Basic Books, 1957). Also of direct relevance is Clifford Geertz's article "The Transition to Humanity" (in *Horizons of Anthropology,* edited by Sol Tax [Chicago: Aldine, 1964]). One of the many provocative ideas in the latter paper is Geertz's hypothesis that man is "not just the producer of culture but, in a specifically biological sense, its product." ■

THE MEANING OF FITNESS

AT THE BEGINNING of my first lecture I mentioned some of the questions I was hoping to answer, and among them were these two: Is there any real reason to suppose that advances in medicine and hygiene are undermining the fitness of the human race?

And is man potentially capable of further evolution, or must we suppose that his evolution has now come to an end? In the course of trying to answer these questions I shall be obliged to use the words *fitness, inheritance,* and *evolution,* but to use them in narrow or unfamiliar ways. Scientists do sometimes tend to *brandish* these special usages at the layman, as if they had a sort of inner rightness; but it would be more gracious, and would reveal a better sense of language, if they apologized for them or explained them away.

In everyday speech *fitness* means suitability or adaptedness or being in good condition; *evolution* means gradual change, with the connotation of unfolding; and as for *inheritance,* we may hope to inherit money, rights, or property; we might inherit, too, a mother's eyes or a grandfather's gift for fiddling. These are the meanings (there is no need to say the "proper" meanings) of fitness, evolution, and inheritance—the meanings for which scientists chose them when they were struggling to put their conceptions into words. In the course of time those conceptions have become clearer—more choosy, if you like—but the words which embody them have remained the same. The change that has gone on is sometimes described by saying that scientists give the meanings of words a new precision and refinement: a fair statement, were it not for the implication that they extract the true or pure meaning from crude ore. The innocent belief that words have an essential or inward meaning can lead to an appalling confusion and waste of time. Let us take it that our business is to attach words to ideas and definitions, not to attach definitions to words.

The idea scientists now have in mind when they speak of "fitness" can be explained like this. All the people alive 100 years from now will be our descendants, but not all of us will be their ancestors. In retrospect, therefore, it will be possible to give us scores or marks according to the share we took in being the ancestors of those future people; and those who took a larger share will be described as fitter than those who took a lesser share. The word *fitness,* then, has come to mean *net reproductive advantage,* and students of heredity, geneticists, do not deliberately use it in any other sense. One hears bitter complaints about this newer use of *fitness,* because it neglects so much of what is deeply important in human life: for example, the influence of good or evil people who happen to have no children but who are so obviously fit or unfit members of society in everything except this narrow genetic sense. But the contempt we may feel for the word must on no account be transferred to the idea that it embodies, an idea which has a central place in modern evolutionary thought.

The argument that advances in medicine and hygiene are undermining the overall fitness of mankind is based on the belief that there is a hereditary or genetic element in all human ills and disabilities, even if it amounts to no more than a predisposition. This is known to be true of some diseases and not known to be false of any, so there can be no disagreement here. In its simplest form, the argument then runs as follows: because of the discovery of insulin, antibiotics, and so on, we are preserving for life and reproduction people who even ten years ago might have died. We are therefore preserving the genetically ill-favored, the hereditary weaklings, who can intermarry with and therefore undermine the constitution of normal people; and as as result of all this, mankind is going downhill.

If by "going downhill" is meant "declining in biological fitness," with the implication that mankind will probably die out, this argument is simply a museum of self-contradictions. It is true that we preserve for life people who even ten years ago might have died; but then we do not live ten years ago: We live today. It is also true that if some disaster were to destroy the great pharmaceutical industries to which diabetics and the victims of Addison's disease literally owe their lives, then a great many of them might die; but what could be deduced from this, except the lunatic inference that people who might conceivably die tomorrow might just as well be dead today? So let me put the argument in a form in which it might be put by a humane and intelligent person. He might say something like this:

> I live in a country with a National Health Service, and the effect of this is that, in a sense, I myself suffer from diabetes and rheumatoid arthritis, and so on—from mental deficiency too. Of course my sufferings are only economic, in the sense that it is my taxes that help to pay the bill; but, as a result of them, I can afford to have only two children, though I very much wanted three. Now I am a sound and healthy person, and though I'm all for helping other less lucky people, it is clear that what you call their "biological fitness" is being bought at the expense of mine.

There are two arguments here, and they cannot be considered apart. The first is that people of a genetically sound constitution are being crowded out by the inferior. My spokesman was too humane to resent the idea that the inferior should survive and have children, but he saw some danger in the fact that the population of the future would contain fewer of his descendants because it would contain more of theirs. The second point he makes is that inborn resistance to a disease can be taken as evidence of a *general* soundness of body, of fitness in some rounded and comprehensive sense; so that even if the people he described as unlucky could all be cured of their particular disabilities, there would still be a deep-seated, though hidden, deterioration of mankind.

The arguments I have just outlined are serious and respectable, but they are not generally valid; they may sometimes represent the very opposite of the truth. Consider one of the forms of inborn resistance to a very severe form of malaria, subtertian malaria. It is now known that people can enjoy a definite inborn resistance to subtertian malaria if their red blood corpuscles contain something between 30 and 40 per cent of an unusual form of hemoglobin, hemoglobin S as opposed to hemoglobin A. One of the consequences of possessing hemoglobin S is that the red blood cells tend to collapse if deprived of oxygen; they become sickle shaped instead of remaining rounded, and people whose blood behaves in this way are said to show the "sickle cell trait." Sickle cell trait can be found in parts of Africa, in some Mediterranean countries, and in parts of India—always in places where malaria has been or still is rife. It is not a disabling condition, so its victims should not be said to "suffer" from it; and, in any event, it confers a high degree of resistance to the multiplication of the malaria organism in the blood.

This sounds like a splendid example of Nature's ingenuity in coping with a particularly murderous disease, malaria, without the help of these new-fangled drugs; but until one knows the rest of the story one cannot appreciate how devilishly ingenious it is.

The formation of hemoglobin S instead of A is due to an inborn difference of a particularly uncompromising kind, in the sense that if a person is genetically qualified to produce hemoglobin S, by possessing the appropriate "gene" or genetic factor, then he surely will. People who show sickle cell trait do so because they have inherited the gene that changes hemoglobin A to S from one, and only one, of their parents. But when two such people marry and have children, one quarter of their children, on the average, will inherit that gene from both their parents; all their hemoglobin, instead of only part of it, will be of type S; and as a result of this they usually die early in life of a destructive disease of the blood known as "sickle cell anemia." This highly successful form of inborn resistance to malaria therefore makes it certain that a number of children will die.

The situation as a whole can be set out in the form of a balance sheet or equation. In some parts of the world where malaria is rife, people with sickle cell trait are the fittest people. Alongside them are, on the one hand, normal people, whose hemoglobin is wholly of type A; and, on the other hand, the victims of sickle cell anemia, whose hemoglobin is wholly of type S. The proportion in which these three classes occur adjusts itself automatically to a pattern in which the loss of life due to malaria and to sickle cell anemia nicely counterbalances the gain that is due to the greater fitness of those with sickle cell trait. Nevertheless, in malarial regions, populations which possess this genetic structure are fitter than populations which do not.

Essentially the same explanation will account for the widespread occurrence in certain parts of Italy of the disease known as Cooley's anemia and, not impossibly, for the otherwise paradoxically high incidence of a certain fatal inborn disease of the pancreas in Great Britain and elsewhere. In all such cases it may turn out that there is, or recently has been, some special advantage in having inherited from one parent, and one parent only, the genetical factor which produces such disastrous effects when it is inherited from both.

The moral of this story—though morality seems to have little to do with it—is that mankind will improve if we stamp out inborn resistance to malaria by stamping out malaria itself. Sickle cell anemia is in fact disappearing from the Negro population of America at about the rate we should expect if malaria had ceased to be a scourge to it 200 or 300 years ago. So the only good thing about inborn resistance to malaria is inborn resistance to malaria. It does *not* reveal any general soundness of constitution; and this is just the opposite to what my imaginary spokesman supposed. It is simply not true to say that advances in medicine and hygiene must cause a genetical deterioration of mankind. There is more to be feared from a slow decline of human intelligence, but that is a different matter: *if* it is happening, it is because the rather stupid are biologically fitter than those who are innately more intelligent, not because medicine is striving to raise the biological fitness of those who might otherwise be hopelessly unfit.

This question of a possible decline of intelligence is very important, and I shall devote my fifth lecture to it; but, having referred to mind, and defects of mind, it is essential to make this point. Some forms of idiocy and imbecility are congenital. *Congenital* is a vague word, but I use it here to refer to an idiocy which follows from an inborn defect of the genetic make-up as it was laid down at the moment of conception. This defect represents an inborn difference from other people, but it is no more a property of the genetic make-up *as a whole* than the inborn difference between people of blood-groups A and B. One particular form of imbecility, now known as phenylketonuria, is the effect of a single, particular, and accurately identified inborn error of metabolism. In point of inheritance it is essentially similar to another disturbance of metabolism, alkaptonuria, the most serious effect of which is usually no worse than a darkening of the urine after it is exposed to air. To suppose, then, that congenital imbecility pointed to some general inborn inadequacy or degeneracy is nonsense—ignorant and cruel nonsense, too. Our ambition should be to *cure* phenylketonuria, for it is an illusion to suppose that congenital afflictions are necessarily incurable; and if eventually we do cure phenylketonuria, we shall in no sense be conniving at a genetical degradation of mankind.

In a later lecture I shall mention one form of gross mental defect, mongolism, which is by no means so simple in origin as phenylketonuria. It is the result of a damaging genetic accident involving a whole chromosome, and it is not at all easy to see how it might be cured. But it *is* an accident, a particular accident, one which happens more often to children of older mothers; it is not to be thought of as an outward fulfilment of some inner degeneracy of a family stock. When you come to think of it, all defects of the genetic constitution must have an accidental or unpremeditated or casually intrusive quality—*epiphenomenal* is the word; for it is impossible, indeed self-contradictory, that any animal should have evolved into the possession of some complex and nicely balanced genetic make-up which rendered it unfit. It is this fact that justifies our always hoping to find a cure.

Inheritance was the second of the three words of which I said that biologists use them in special or unfamiliar ways. Just what is inherited when geneticists speak of inheritance? It is becoming increasingly popular to say that a child inherits certain genetical *instructions* about how his growth and development are to proceed. This sounds like an ordinary metaphor—very apt, no doubt, but perhaps misleading; but I do not think *metaphor* is quite the word. The idea of genetical instruction has come into use because, under the influence of telephone engineers and higher mathematicians, we now recognize a general, abstract similarity between all kinds of different ways of transmitting information. The passage of genetical instructions from parents to children is a particular concrete example of the more general idea of an act of communication, and just as valid an example of that idea as the information which we transmit in writing, by telephone, or by direct speech.

A gramophone record is a solid object which contains instructions about what particular sounds a reproducing apparatus is to utter. Genetical instructions are also conveyed by solid objects, in this case chromosomes; and the specificity of the instructions—their property of being this instruction and not that—is a specificity of chemical structure: Different molecular patterns convey different information, just as different sinuosities of the grooves of a record embody instructions about making different sounds. The discovery that nucleic acids are the substances that embody genetical information is to my mind the most important discovery in modern science, but I shall not argue the point because nothing turns on whether you agree with me or not.

Genetical instructions are sometimes strict and uncompromising, in the sense that they can be carried out in one way only, if at all. The nature of one's blood-group or hemoglobin is strictly governed by one's genetic constitution, with little or no opportunity for compromise. But much more often the instructions allow a certain latitude in their execution—we differ from one another partly because we received different genetical instructions from our parents, but partly because, from conception onwards, our surroundings have acted upon us differently, and have therefore affected the way in which those instructions are carried out.

The theory of evolution is a theory which declares that genetical instructions change in character in the course of time. In my first lecture I said that a population had a certain age structure, revealed by classifying its members into groups by age. In the same way a population has a genetic structure, a particular pattern of genetic make-up; just as its members are of many different ages, so also are they of many different genetic kinds. *Evolution* is a change in the genetic structure of a population—but a systematic change, a change with a definite direction or consistent trend.

We have seen such changes happening in our own lifetimes. In many hospitals bacteria have come to resist the action of penicillin, because bacteria genetically qualified to develop that resistance, at one time a tiny minority, have become the prevailing type. Likewise the genetic structure of some populations of moths has altered: dark forms have now become the prevailing forms among the sooty vegetation of an industrial countryside. For all we know to the contrary, changes of this kind are the rudiments of greater evolutionary changes; and it was Darwin's theory, you remember, that they have come about because the different members of a species are endowed with different degrees of fitness: They leave more, or fewer, descendants, as the case may be, and if this happens the genetic structure of the population as a whole must clearly change.

This is all very well as far as it goes, but if we are ever to get a complete understanding of evolution we must obviously try to arrive at a complete theory of inborn variation: what forms does it take, what makes it possible, how does it happen, how is it maintained?

An analytical theory of inborn variation is one which will explain it in terms of the properties of chromosomes and nucleic acids, the substances which convey and embody genetical instructions. Let me give an example of what I mean. We can be sure that, identical twins apart, each human being alive today differs genetically from any other human being; moreover, he is probably different from any other human being who has ever lived or is likely to live in thousands of years to come. The potential variation of human beings is enormously greater than their actual variation; to put it another way, the ratio of possible men to actual men is overwhelmingly large. What mechanisms provide for the stirring about and shuffling and recombining of genetical information that makes this virtually endless diversity possible?

The most ancient and perhaps the most fundamental mechanism or stratagem that serves this function is that which is known to geneticists, in one of its forms anyway, as "crossing-over." Crossing-over is the swapping of parts between two chromosomes—a process which can occur when they have a certain general correspondence of structure; the effect of it is to combine, or recombine, genetical information in novel ways. One day biochemists and biophysicists will tell us what properties of the chromosomes make it possible for this swapping to occur. Then again, mutation, the birth of a newly variant gene, is an important process in evolution. We must ask what property or properties of the materials of heredity make mutation possible. This is the *kind* of question we must ask if we are ever to understand the pattern and progress of evolution. Let me ask another such question. If we look back upon the course of evolution we can see that within many of its lines, within many grand pedigrees of descent, there has occurred a process of becoming more complex, or, as zoologists say, more "advanced." Mammals are more "advanced" than fishes; insects are more "advanced" than worms. In the long view there has been an increase in the complexity of the genetical instructions which, so to speak, authorize an animal to be whatever it is.

If we merely confine ourselves to talking about degrees of fitness, the process seems gratuitous; what properties of the hereditary material make it understandable that it should have occurred? An explanation can only be groped after, but one kind of explanation might run something like this. The molecules of nucleic acid are of the sort that chemists describe as "polymeric": They repeat the general pattern of their structure lengthwise, and can therefore build upon themselves to increase in length. They have also the crucially important property of lending themselves to duplication, because after various chemical maneuvers two similar molecules can be formed where there was only one before. There are many other more subtle properties of this kind; for example, the ability to break up and rejoin, to increase in length by letting in new stretches between the ends. Taken all together, these properties amount to what might be called a *repetitiousness* of nucleic acids and chromosomes, a readiness to become manifold or luxuriant or to elaborate upon genetic information—it is difficult to know which word to use. It is of the physical nature of nucleic acids that they can offer up for selection ever more complex sets of genetical instructions, can propose ever more complex solutions of the problem of remaining alive and reproducing. Every now and again one of these more complex solutions will be accepted, and so there is always a certain pre-existing inducement or authority for evolution to have what, in retrospect, we call an "upward trend."

All this is extremely lame and halting; but, as I said in my first lecture, it is a more useful way of trying to explain the phenomenon than by talking about a "vital force" of some kind which inspires organisms to advance in evolutionary history. All I am asking is: What material properties of chromosomes and nucleic acids qualify them for the functions which they do in fact discharge?

However that may be, human beings are the outcome of a process which can perfectly well be described as an advancement; and the second of the two questions I put at the beginning of this lecture was, in effect, where could we go from here?

First let me say that, even in the last fifty years, profound changes have occurred in human populations which are certainly not evolutionary changes. In some countries, for example, the average rate of growth and development has been and still is steadily going up. In the Scandinavian countries the average age of onset of first menstruation has declined between four and six months per decade for the last seventy years. In this country the height of adolescent boys has gone up by about three-quarters of an inch per decade, and their greatest height is reached by about eighteen or nineteen, instead of by twenty-five or twenty-six. These changes have been brought about by better nurture and nourishment, particularly in the first five years of life. The well-fed class may be nearing the end of this process, but the average will continue to change until the less well-fed can catch them up.

Other changes have happened in human history that might conceivably be evolutionary changes. There was once a music-hall joke of uncertain import which turned on what Mr. Gladstone may or may not have said in (I hope I am right in saying) 1858. One of the things he *did* say or imply in 1858 was that color vision may have developed in mankind since the days of Homer; for he told us that Homer's world, as Homer described it, was almost colorless; and he might have added that color blindness does not seem to have been referred to in writing before 1684. But I understand that the poverty of color words in Greek and other ancient languages is to be construed as lack of sensibility, not lack of sensitivity; as lack of perceptiveness, not of ability to perceive; and though full color vision *might* have evolved within recorded history, there is no good evidence that it has done so.

But evolutionary changes, as I defined them, have occurred repeatedly in human history. The rise and fall of the genetic factor responsible for sickle cell anemia is one, and in later lectures I shall mention others. These are comparatively trivial changes. Could man evolve *radically*, or be made to evolve radically, in future? I have left this question to the end because it is utterly pointless and distracting. The answer, to be delivered with every inflection of impatience, is, yes indeed. The necessary conditions are satisfied: A luxuriance of inborn diversity, a system of mating that maintains it, and an unspecialized structure as the zoologist uses that word, a structure which does not in itself commit human beings to any one way of life. From the point of view of genetical evolution, human beings have retained an amateur status.

But in fussing over the nature of some great metamorphosis which might conceivably happen, but which could only happen in real life if we were to be the victims of a sustained and consistent tyranny tens of centuries long, we may forget to ask a really important question: what changes *are* happening in the genetic structure of human populations as a result of forces acting upon us now? I stand by my original decision not to attempt to predict these changes or to discuss their consequences. My question is, what kind of knowledge and understanding must we acquire about mankind, and about genetics generally, if we are to identify and predict such changes; and this is essentially what my next three lectures will be about. In my last lecture I shall give a still more cogent reason for saying that the question I put—*can* man evolve as animals may yet evolve?—is pointless, because he has in fact adopted a new kind of biological evolution (I emphasize, a biological evolution) to which a great deal of what I have said in this lecture does not apply.

THE FUTURE OF MAN

Just as a hereditary system is a special kind of system of communication—one in which the instructions provide for the issue of further instructions—so there is a specially important kind of hereditary system: one in which the instructions passed on from one individual to another change in some systematic way in the course of time. A hereditary system with this property may be said to be conducting or undergoing an *evolution*. Genetic systems of heredity often transact evolutionary changes; so also does the hereditary system that is mediated through the brain. I think it is most important to distinguish between four stages in the evolution of a brain. The nervous system began, perhaps, as an organ which responded only to elective stimuli from the environment; the animal that possessed it reacted instinctively or by rote, if at all. There then arose a brain which could begin to accept instructive stimuli from the outside world; the brain in this sense has dim and hesitant beginnings going far back in geological time. The third stage, entirely distinguishable, was the evolution of a nongenetical system of heredity, founded upon the fact that the most complicated brains can do more than merely receive instructions; in one way or another they make it possible for the instructions to be handed on. The existence of this system of heredity—or tradition, in its most general sense—is a defining characteristic of human beings, and it has been important for, perhaps, 500,000 years. In the fourth stage, not clearly distinguishable from the third, there came about a systematic change in the nature of the instructions passed on from generation to generation—an evolution, therefore, and one which has been going at a great pace in the past 200 years. I shall borrow two words used for a slightly different purpose by the great demographer Alfred Lotka to distinguish between the two systems of heredity enjoyed by man: *endosomatic* or internal heredity for the ordinary or genetical heredity we have in common with other animals; and *exosomatic* or external heredity for the nongenetic heredity that is particularly our own—the heredity that is mediated through tradition, by which I mean the transfer of information through nongenetic channels from one generation to the next.

I am, of course, saying something utterly obvious: Society changes; we pass on knowledge and skills and understanding from one person to another and from one generation to the next; a man can indeed influence posterity by other than genetic means. But I wanted to put the matter in a way which shows that we must not distinguish a strictly biological evolution from a social, cultural, or technological evolution. *Both* are biological evolutions; the distinction between them is that the one is genetical and the other is not.

What, then, is to be inferred from all this? What lessons are to be learned from the similarities and correspondences between the two systems of biological heredity possessed by human beings? The answer is important, and I shall now try to justify it. The answer, I believe, is almost none.

It is true that a number of amusing (but in one respect highly dangerous) parallels can be drawn between our two forms of heredity and evolution. Just as biologists speak in a kind of shorthand about the "evolution" of hearts or ears or legs—it is too clumsy and long-winded to say every time that these organs participate in evolution, or are outward expressions of the course of evolution—so we can speak of the evolution of bicycles or wireless sets or aircraft with the same qualification in mind: They do not really evolve, but they are appendages, exosomatic organs if you like, that evolve with us. And there are many correspondences between the two kinds of evolution. Both are gradual if we take the long view; but on closer inspection we shall find that novelties arise, not everywhere simultaneously—pneumatic tires did not suddenly appear in the whole population of bicycles—but in a few members of the population; and if these novelties confer economic fitness, or fitness in some more ordinary and obvious sense, then the objects that possess them will spread through the population as a whole and become the prevailing types. In both styles of evolution we can witness an adaptive radiation, a deployment into different environments: there are wireless sets not only for the home, but for use in motorcars or for carrying about. Some great dynasties die out: airships, for example, in common with the dinosaurs they were so often likened to; others become fixed and stable: toothbrushes retained the same design and constitution for more than a hundred years. And, no matter what the cause of it, we can see in our exosomatic appendages something equivalent to vestigial organs; how else should we describe those functionless buttons on the cuffs of men's coats?

All this sounds harmless enough. Why should I have called it dangerous? The danger is that by calling attention to the similarities, which are not profound, we may forget the *differences* between our two styles of heredity and evolution; and the differences between them are indeed profound. In their hunger for synthesis and systematization, the evolutionary philosophers of the nineteenth century and some of their modern counterparts have missed the point. They thought that great lessons were to be learned from similarities between Darwinian and social evolution; but it is from the differences that all the great lessons are to be learned. For one thing, our newer style of evolution is Lamarckian in nature. The environment cannot imprint genetical information upon us, but it can and does imprint nongenetical information which we can and do pass on. Acquired characters are indeed inherited. The blacksmith was under an illusion if he supposed that his habits of life could impress themselves upon the genetic make-up of his children; but there is no doubting his ability to teach his children his trade, so that they can grow up to be as stalwart and skillful as himself. It is because this newer evolution is so obviously Lamarckian in character that we are under psychological pressure to believe that genetical evolution must be so too. But although one or two biologists are still feebly trying to graft a Lamarckian or instructive interpretation upon ordinary genetical evolution, they are not nearly so foolish or dangerous as those who have attempted to graft a Darwinian or purely elective interpretation upon the newer, nongenetical, evolution of mankind.

The conception I have just outlined is, I think, a liberating conception. It means that we can jettison all reasoning based upon the idea that changes in society happen in the style and under the pressures of ordinary genetic evolution; abandon any idea that the direction of social change is governed by laws other than laws which have at some time been the subject of human decisions or acts of mind. That competition between one man and another is a necessary part of the texture of society; that societies are organisms which grow and must inevitably die; that division of labor within a society is akin to what we can see in colonies of insects; that the laws of genetics have an overriding authority; that social evolution has a direction forcibly imposed upon it by agencies beyond man's control—all these are biological judgments; but, I do assure you, bad judgments based upon a bad biology. In these lectures you will have noticed that I advocate a "humane" solution of the problems of eugenics, particularly of the problems of those who have been handicapped by one or another manifestation of the ineptitude of Nature. I have not claimed, and do not now claim, that humaneness is an attitude of mind enforced or authorized by some deep inner law of exosomatic heredity: There are technical reasons for supposing that no such laws can exist. I am not warning you against quack biology in order to set myself up as a rival peddler of patent medicines. What I do say is that our policies and intentions are not to be based upon the supposition that Nature knows best; that we are at the mercy of natural laws, and flout them at our peril.

It is a profound truth—realized in the nineteeth century by only a handful of astute biologists and by philosophers hardly at all (indeed, most of those who held any views on the matter held a contrary opinion)—that nature does *not* know best; that genetical evolution, if we choose to look at it liverishly instead of with fatuous good humor, is a story of waste, makeshift, compromise, and blunder.

I could give a dozen illustrations of this judgment, but shall content myself with one. You will remember my referring to the immunological defenses of the body, the reactions that are set in train by the invasion of the tissues by foreign substances. Reactions of this kind are more than important: they are essential. We can be sure of this because some unfortunate children almost completely lack the biochemical aptitude for making antibodies, the defensive substances upon which so much of resistance to infectious disease depends. Until a few years ago these children died, because only antibiotics like penicillin can keep them alive; for that reason, and because the chemical methods of identifying it have only recently been discovered, the disease I am referring to was only recognized in 1952. The existence of this disease confirms us in our belief that the immunological defenses are vitally important; but this does not mean that they are wonders of adaptation, as they are so often supposed to be. Our immunological defenses are also an important source of injury, even of mortal injury.

For example, vertebrate animals evolved into the possession of immunological defenses long before the coming of mammals. Mammals are viviparous: The young are nourished for some time within the body of the mother, and this (in some ways) admirable device raised for the first time in evolution the possibility that a mother might react immunologically upon her unborn children—might treat them as foreign bodies or as foreign grafts. The hemolytic disease that occurs in about one new-born child in 150 is an error of judgment of just this kind: it is, in effect, an immunological repudiation by the mother of her unborn child. Thus the existence of immunological reactions has not been fully reconciled with viviparity; and this is a blunder—the kind of blunder which, in human affairs, calls forth a question in the House, or even a strongly worded letter to *The Times.*

But this is only a fraction of the tale of woe. Anaphylactic shock, allergy, and hypersensitivity are all aberrations or miscarriages of the immunological process. Some infectious diseases are dangerous to us not because the body fails to defend itself against them but—paradoxically—because it does defend itself: In a sense, the remedy *is* the disease. And within the past few years a new class of diseases has been identified, diseases which have it in common that the body can sometimes react upon its own constituents as if they were foreign to itself. Some diseases of the thyroid gland and some inflammatory diseases of nervous tissue belong to this category; rheumatoid arthritis, lupus erythematosus, and scleroderma may conceivably do so too. I say nothing about the accidents that used to occur in blood transfusions, immunological accidents; nor about the barriers, immunological barriers, that prevent our grafting skin from one person to another, useful though it would so often be; for transfusion and grafting are artificial processes, and, as I said in an earlier lecture, natural evolution cannot be reproached for failing to foresee what human beings might get up to. All I am concerned to show is that natural devices and dispositions are highly fallible. The immunological defenses are dedicated to the proposition that anything foreign must be harmful; and this formula is ground out in a totally undiscriminating fashion with results that are sometimes irritating, sometimes harmful, and sometimes mortally harmful. It is far better to have immunological defenses than not to have them; but this does not mean that we are to marvel at them as evidences of a high and wise design.

We can, then, improve upon nature; but the possibility of our doing so depends, very obviously, upon our continuing to explore into nature and to enlarge our knowledge and understanding of what is going on. If I were to argue the scientists' case, the case that exploration is a wise and sensible thing to do, I should try to convince you of it by particular reasoning and particular examples, each one of which could be discussed and weighed up; some, perhaps, to be found faulty. I should not say man is driven onwards by an exploratory instinct and can only fulfill himself and his destiny by the ceaseless quest for Truth. As a matter of fact, animals do have what might be loosely called an inquisitiveness, an exploratory instinct, but even if it were highly developed and extremely powerful, it would still not be binding upon us. We should not be *driven* to explore.

Contrariwise, if someone were to plead the virtues of an intellectually pastoral existence, not merely quiet but acquiescent, and with no more than a pensive regret for not understanding what could have been understood; then I believe I could listen to his arguments and, if they were good ones, might even be convinced. But if he were to say that this course of action or inaction was the life that was authorized by Nature; that this was the life Nature provided for and intended us to lead; then I should tell him that he had no proper conception of Nature. People who brandish naturalistic principles at us are usually up to mischief. Think only of what we have suffered from a belief in the existence and overriding authority of a fighting instinct; from the doctrines of racial superiority and the metaphysics of blood and soil; from the belief that warfare between men or classes of men or nations represents a fulfillment of historical laws. These are all excuses of one kind or another, and pretty thin excuses. The inference we can draw from an analytical study of the differences between ourselves and other animals is surely this: that the bells which toll for mankind are—most of them, anyway—like the bells on Alpine cattle; they are attached to our own necks, and it must be *our* fault if they do not make a cheerful and harmonious sound.

5. NATURAL SELECTION IN MAN: SOME BASIC PROBLEMS

L. S. PENROSE

Reprinted from Natural Selection in Human Populations *edited by D. F. Roberts and G. A. Harrison (Pergamon Press, 1959). Until his death in 1972, L. S. Penrose served as Professor at the Kennedy-Galton Centre of Harperbury Hospital in Hertfordshire, England. He was the author of various books, including* Outline of Human Genetics, On Objective Study of Crowd Behaviour, *and* Influence of Heredity on Disease.

■ The historical perspective of genetic evolution provides us with a valuable framework within which to assess what is known about human evolution and adaptation. It also provides clues to guide us into a vast domain that is not yet fully charted. Anthropological thought, like genetics, is deeply rooted in Darwin's original concepts, which require that both make use of lessons that can be drawn from the past. Continuing a theme developed in other selections, Penrose provides a frame of reference within which we may observe the compatibility of the two sciences. More important, he illustrates how a century of exploration into genetic processes has utilized Darwin's concepts for understanding human evolution and adaptation and shows the promise this exploration holds for future inquiry.

Penrose begins by noting how the scientific advances in genetics during the past century help us to pinpoint some of the central evolutionary problems in a more definitive way. His principal goal here is to specify these problems and to assess their implications for *Homo sapiens* in modern society. Using two basic genetic concepts—human polymorphism and natural selection—Penrose clarifies these problems by posing five questions.

The first three questions deal with the recognition and frequency of genes, their effect on the individual, and how they got into circulation. The next question is whether gene frequencies are stable, and on the basis of his discussion of these four questions, Penrose poses his fifth question, dealing with future trends. Using the case of human intelligence as an example of human polymorphism, he suggests there are genetic dynamics that maintain the equilibrium of this human trait. In subsequent selections we return to the idea that human polymorphism has a vital long-term significance for adaptability and success in human evolution.

Penrose observes that although human culture has freed *Homo sapiens* from many habitational constraints, this by no means implies that the species has been released from the influences of natural selection. People have modified their habitats and have altered their position in respect to nature. The result is not freedom from selection; rather, it suggests a transfer of the push for selection from certain genotypes to others. For instance, physical prowess and strength may have been necessary for survival at previous stages of development; now we rely on tools and weapons, and this suggests that selection has favored those who are most capable of inventing and making tools and weapons. As we see in future selections, the rudiments of culture existed long before the emergence of the genus *Homo;* nevertheless, *sapiens* evolved. There is no reason to assume that natural selection has ceased to operate or that it will ever cease. As Fox reminds us in Selection 6, man is an animal, albeit a "cultural animal." Just as different expressions of polymorphism tend to balance each other through the process of adaptation, our concepts of the species *Homo sapiens* as a biological entity and as a cultural innovator must be balanced if we are to appreciate the process of our own evolution.

A vast literature deals with the processes of natural selection. Some of the readings that were suggested in connection with previous selections also are pertinent here: Simpson's *The Meaning of Evolution* (New Haven: Yale University Press, 1949), Theodosius Dobzhansky's *Mankind Evolving* (New Haven: Yale University Press, 1962), Campbell's *Human Evolution* (Chicago: Aldine, 1966, second edition, 1974), and Darwin's works. A popular and well-written introduction to the subject is Julian Huxley's *Evolution in Action* (New York: Harper, 1953). A very succinct article, in which Dobzhansky explores the ideas in this paper, is his "The Present Evolution of Man" *(Scientific American,* 204 [September, 1960]: 206-17), and also "Man and Natural Selection," *(American Scientist,* 49 [1961]: 285-99). In an age in which claims to national maturity frequently are cast in terms of the ability to play a kind of hopscotch with nuclear and thermo-nuclear weapons, which leads to the introduction of new manmade selective pressures (such as radiation), a very worthwhile article is H. J. Muller's "Radiation and Human Mutation" *(Scientific American,* 199 [November, 1955]: 58-68). ■

THOUGH THE NOTION that man had evolved from an ape-like animal by natural selection was accepted with Darwin's ideas nearly a century ago, the possibility of observing natural selection at work in our every day life was not at first generally appreciated. Although he was greatly interested in human polymorphism, as shown by normal and abnormal hereditary variants, Darwin paid more attention to competition between species or tribes than between individual types. He assumed, in 1871, with Wallace and Galton, that man, by his intelligence, had to a great extent altered the force of natural selection and, at times, reversed its direction. In primitive communities, the more intelligent members would, in the long run, succeed better than the inferior ones and therefore leave more numerous progeny; but in civilized societies, members of weak bodily and mental constitution are better able to propagate their kind.

The subject of natural selection in man was taken up in a comprehensive way by Karl Pearson in a series of papers beginning in 1894. He tried to estimate the ranges of fertility in different communities and subgroups and to discover how much of this variation was heritable. He discussed the question in its political aspects and concluded that socialism did not necessarily produce degeneration. Later on, in 1912, he became more cautious and expressed the view that only a very thorough eugenic policy could possibly save our race from the evils which must follow from the antagonism between natural selection and medical progress.

In recent years we have become more preoccupied with the effects of selection on character determined by specific genes. The way was opened by the theoretical work of Fisher, Haldane and Wright on the gene frequency changes in populations under selection. At the present time attention has shifted to more practical problems because of the great advances in serology and chemistry which have enabled so many human genes to be precisely identified.

Our problems are now more definite than formerly; we can ask questions to which we may expect to get factual answers. The basic questions we ask seem to group themselves as follows:

1. What genes can be recognized in the human population, and who are their frequencies?
2. What are their precise effects?
3. How did the genes get there, that is, when did the mutations take place which brought them into circulation?
4. Are these gene frequencies stable? Are the genes maintaining themselves, increasing in frequency or are they diminishing?
5. What are the trends of these processes in quantitative traits?

Only in the last question do we trespass on the ground which was formerly so attractive for human biological speculation.

1. *What genes can be recognized in the human population, and what are their frequencies?*

Emphasis was originally directed towards human characters which could be measured with a ruler. Such traits did not segregate and were not suitable for genetical analysis except by quantitative statistical methods. The physical characters, which actually segregated, were rare and, because they were regarded as abnormal, they were excluded from anthropological surveys. Polydactyly and brachydactyly were traits of this kind and their incidences could easily have been ascertained, in populations all over the world, had the effort been made. As it is, we have extensive data on the frequency distributions of body measurements, cephalic indices, hair colors and eye colors, which are common variations but nonsegregating.

Investigators tend to use the tools which come to hand conveniently and the characters studied continue to change accordingly from time to time. The transition from physical measurements to blood groups, initiated by the Hirszfelds, is now being succeeded by a shift from serology to chemistry of serum proteins. The development began with the introduction of techniques of chromatography and electrophoresis. Pauling's successful attack on the chemistry of hemoglobins has been followed by further advances by Ingram. The field was again expanded by Smithies with the introduction of starch gel methods and the detection of new serum protein polymorphisms. These movements have been advantageous to anthropology because, by technical advances, we get nearer to the gene and, thus, each new survey gives more accurate information than the last about the genetical compositions of the different human populations studied.

2. *What are their effects?*

It is insufficient merely to count genes, we must also ascertain what effects these genes have on the individuals who carry them. Color changes are dramatic but not always easy to connect with single gene chemical differences or enzyme deficiencies, except perhaps in albinism. With stature it is even more difficult, except in the extreme cases of dwarfs. Looked at from the opposite angle, hidden chemical differences, as in the blood groups or serum proteins, often do not appear to have much selective significance. The antigenic incompatibilities produced between mother and fetus took a long time to discover but they are among the best established selective effects of polymorphic traits. Other effects of blood group peculiarities on health will take a long time to pin down with equal precision. On the whole we should not expect common genes to show marked associations with diseases.

3. *How did the genes get there, that is, when did the mutations take place which brought them into circulation?*

Tied together with the problems of gene frequency and gene action is the problem of gene origin. We presume that each allele, good or bad, must have originated some time in the past. There is a tendency to think of most genetic variation as having arisen quite recently but this may not be so. A common view is that mutation is going on at all loci fairly regularly at a rate of $1/10^5$ or $1/10^6$ per generation and that this rate is maintained equally in all populations. Against this view, however, there are many points to consider.

Mutation has not been directly observed in any of the typical biochemical traits or any blood group system. Thus it may be much rarer than supposed. The known cases of mutation, such as achondroplasia and retinoblastoma, may be atypical. Then there are curious distributions of genes, very rare in the world population but common in certain isolated populations. Pentosuria and Tay-Sachs disease in Jewish groups are classical examples; the D blood-group peculiarity; acatalasaemia in Japan, and so on. The suggestion here is that the causal mutations are examples of excessively rare or unique occurrences. With the enormous increase in population numbers through the centuries it is also possible that some more widely spread known mutants, such as that for thalassaemia, may have arisen only once in the distant past.

Moreover, we cannot, in our present state of ignorance, assume that mutants occur regularly at the same locus, even though they are apparently repeated; the work of Benzer on bacteriophage makes us hesitate to assume that, in such cases, it is necessarily exactly the same point on the DNA chain which is altered on each occasion. So that there is no guarantee that a gene identified in one part of the world for a blood group, an anemia or a malformation is exactly the same as another. How this consideration will affect anthropology remains yet to be seen.

4. *Are these gene frequencies stable? Are the genes maintaining themselves, increasing in frequency or are they diminishing?*

The question of evolution, which depends upon instability of gene frequencies, ultimately is reduced, in formal genetics, to the action of selection upon genotypes. Most of the work of Fisher, Wright and Haldane was concerned with deducing formal laws about rate of entry and extinction of genes which had good or bad effects in theoretical populations. Changes were expected to be very slow, on the whole, and "fixation" of genotypes in the form of favorable homozygosis was the ultimate destination of selections at each locus. Although the possibility of stable equilibrium, which would keep common allelic genes in circulation permanently, was discussed, it was not considered likely to be very significant, perhaps because it had no obvious bearing on evolution, for changes are slowed down by stable polymorphism. It would be well here just to mention the different kinds of *genic equilibrium* which may be found.

Neutral equilibrium. This occurs when there is no selective advantage for any genotype. With random mating this gives the Hardy-Weinberg rule.

Stable equilibrium. There are two distinct types. In the first type, selection is balanced by mutation. This we find, presumably, in such a condition as achondroplasia where infertility is very marked and nearly all cases arise by fresh mutation. Increase of fitness or of mutation would raise the level of incidence though this must remain very low. The disease will not ordinarily be considered as a case of polymorphism because it is so rare. Secondly, there is the important type of stable equilibrium, produced if homozygotes are relatively less favored than the heterozygotes in an allelic system. The main significance of such a situation is that, by this means, apparently unfavorable genes can be maintained in the population. In the homozygote, the gene can be bad or lethal but, in the heterozygote, it can be good and a balance is struck at the appropriate gene frequency. The effects of mutation, on such a system, only make a slight adjustment to the point of equilibrium.

Some principles are worth emphasizing here:

Only a very slight heterozygous advantage is required to balance even a lethal homozygote, in fact it needs to have a proportional advantage equal to the gene frequency. One per cent increase in fertility of carriers of phenylketonuria will more than balance the gene's selective disadvantage in homozygotes. In the main, it will be very difficult to detect the advantage of a heterozygote which keeps a given polymorphism in equilibrium in the human population. However, this should be possible in the exceptional case of a common lethal situation, such as that produced by the sickle cell hemoglobin, where the gene is supposed to be lethal in the homozygote.

It is useless to expect other types of selective advantage to act as stabilizers unless we are prepared to specify rather exceptional circumstances. For instance, family replacement is a favorite idea. Suppose that, for every recessive idiot, the parents have another, extra, child, this is not enough to preserve the balance: Instead they must have four more children. Again blood group A may be a part cause of gastric cancer; if so it must be only the AA homozygote which is damaging, balanced, presumably, by the OO homozygote, who is subject to ulcers. This would help to keep antigen A stable in the population for then the AO heterozygote would have an advantage. The effect, even if proved, would be extremely slight because of the late ages of onset associated with these diseases. Another example is color blindness. For stable equilibrium, a hypothetical advantage for color blind men could be balanced by a disadvantage for color blind women.

Unstable equilibrium. Some situations are known which produce unstable states and one which even produces unstable equilibrium. Disadvantage of the heterozygote produces instability and this is known to occur in the case of maternal-fetal incompatibility. The fetus which is heterozygous is at risk. In 1943, Haldane pointed out that, at the gene frequency of 50 percent, when the homozygotes were equally good and the heterozygote was bad, there was equilibrium but of an unstable type. In this region, even with the frequency of 40 percent, like the Rhesus gene *d*, changes would be very slow. A rare immunizing antigen, unbalanced by mutation or heterozygous advantage of some kind, would be very rapidly selected against, like an ordinary bad dominant trait, and would disappear in relatively few generations. Correspondingly, a very common antigen would tend to become universal. Perhaps there has been selection of this kind against antigen B, in favor of O, while A, nearer to 50 percent, has held its own throughout the world.

The effect of natural selection, besides being influenced by the relative fitnesses and frequencies of the genotypes concerned, is also altered by mating systems. On the whole, inbreeding and assortative mating tend to speed up selection because they boost the numbers of homozygotes. They also tend to prevent stable equilibrium and, in their presence, higher degrees of heterozygous advantage are required to produce stability. Thus, in human populations it is unwise to assume stability until all the factors have been looked into and, if possible, measured.

Some situations, which are difficult to understand, arise in connection with common or semi-lethal traits believed to represent single gene homozygotes; for example, cystic fibrosis of the pancreas, with an incidence 1/1000, is typical and perhaps also some kinds of anencephaly. How could the genes for such diseases have reached a high level like 1/30 unless there was a fabulous mutation rate? Personally, I think the answer is that the heterozygotes, at one time or in one climate, in famine or in pestilence, were at a huge advantage but that the genes are now on their way out, moving too slowly to be noticeable.

5. *What are the trends of these processes in quantitative traits?*

With quantitative traits, investigation is not unpromising because here, although the genic background eludes us still, trends can be easily measured and observed. There is a high degree of polymorphism noticeable in the metrically simple but genetically complex traits, like stature, weight at birth and intelligence, even allowing plenty of play for variations due to environmental factors. What preserves this variation, insofar as it is genetical? We naturally look for evidence of stable equilibrium. One possibility is mutation of genes causing both tall stature and short stature, for example, and the balancing of opposing trends. Another and much more powerful influence could be the practical disadvantages of too great height or too small stature. Since, in a metrical trait, the extremes of the scale are mainly produced by homozygotes, selection against extreme measurements is equivalent to heterozygote advantage.

There is little direct evidence about stature and fitness, in the sense of individual fertility, but the position is clearer in relation to intelligence measurement. Here we know for certain that the very low levels are associated with lethality and that, in practice, the high levels are associated with an easily measurable degree of infertility as compared with medium or slightly lower than normal levels. The result of this is to maintain a fairly stable polymorphism of intelligence variation in human populations. It seems odd that this very striking situation was not appreciated by authorities who have confidently but, so far, incorrectly predicted rapid decline in the intelligence levels of populations in consequence of fertility differentials; any such adverse effect would be most strongly buffered by the tendency to genetical stability. In general, the multiplicity of polymorphic systems which exists in man probably has an evolutionary value in the long run. Fisher once aptly called the variance in genetical traits the "energy" of the species and polymorphism, in this sense, stores energy to combat future environmental changes.

The belief that natural selection in man has been abolished by civilization, socialism, hygiene or whatever it may be, depends upon superficial reasoning. What has happened is that the force has been altered and transferred at certain points from one genotype to another. Going back to Darwin (1871), the lower animals must have their bodily structures modified in order to survive under changed conditions. "They must be rendered stronger or acquire effective teeth or claws in order to defend themselves from new enemies; or they must be reduced in size so as to escape detection and danger," whereas man invents weapons, tools and various stratagems, by means of which he procures food and defends himself. In doing this he has reduced the selective values of one set of genes and increased the advantages of others. This change is clearly seen with genes whose reason for existence was to fight infectious diseases now controlled by other means. The genes no longer give an advantage. For example, the thalassaemia trait will probably slowly diminish in frequency now that there is little malaria. Cure of homozygous Cooley's anemia would, however, block this decline.

The cure of specific known dominant hereditary diseases will not have much effect on the population because the genes concerned are rare usually and slight increases in the prevalence of curable hereditary diseases will not be a biological catastrophe.

The main force of natural selection now seems to be directed towards defects present before birth and leading to failure of development or of function. Even in highly civilized countries, like the United Kingdom or the United States, nearly half of all zygotes formed are unfit in the crude sense of failure to reproduce and it may be assumed that this failure is, to a significant degree, attributable to the genes carried by them. By piecing together evidence from many sources, I have estimated that early prenatal loss accounts for at least 15 percent; then 3 percent of the remainder are stillborn, 2 percent are counted as neonatal deaths and 3 percent more die before reaching maturity. Of the survivors, 20 percent do not marry, and, of those who do, 10 percent remain childless. In view of the large extent, and the persistence, of this loss and the rarity of observed mutation, it seems probable that selection is, for the most part, acting on homozygotes at both ends of the scale, keeping the population in genetical equilibrium. There may be some multiple allelic or pseudoallelic systems present with all the homozygotes lethal. For practical purposes such systems are ineradicable and, if the lethal effect is shown at early embryonic stages only, they will not be harmful.

Consideration of the mechanism of balanced polymorphism goes far towards understanding the problem of selection in man under civilized conditions. The processes involved are complex but we can confidently infer that changes in gene frequency are likely to be slow. Physical anthropologists may be reassured that, considered within historic times, the degrees of population mixtures will probably continue to be accurately measured by gene frequencies.

6. THE CULTURAL ANIMAL

ROBIN FOX

Reprinted from Man and Beast: Comparative Social Behavior *edited by John F. Eisenberg and Wilton S. Dillon (The Smithsonian Institution Press, 1971). Robin Fox is professor of Anthropology at Rutgers University. He has done field work among the Keresan Indians and the Tory Islanders. His principal interests are in structuralism, kinship, and the relationship between biology and human behavior. These are reflected in his publications, which include* From the Keresan Bridge, Kinship and Marriage, *and (with Lionel Tiger)* The Imperial Animal.

■ When anthropologists study human biology, physical evolution, language, or the organization of prehistoric and contemporary societies, the common thread running through all their investigations is human adaptation. All anthropologists, regardless of the subdisciplines they represent, recognize that culture is the adaptation that singularly distinguishes the species *Homo sapiens* from all others. Thus when anthropologists investigate aspects of human biology, they wish to learn which elements in the species' biological make-up constrain and provide potentials for human culture. This entails two cross-cutting modes of inquiry: the study of the biological makeup of the species as a whole that affects culture generally, and the biological components or adaptations that enable particular groups (micropopulations) to maintain effective relationships with their particular habitats. (The latter will be our focus in Part IV.)

In this selection, Fox—who is a social anthropologist—recognizes that we cannot take the easy way out and think of humans exclusively either in biological or social terms; we must, instead, regard them as *biosocial.* The individual starts cultural life by trying to achieve biological survival as best he can within the limits of his technology and natural habitat. Thus institutions are created to try to assure a steady supply of food and to satisfy needs for sexuality, social relations, control, and a sense of being. Regular means for bringing children to adulthood must be established, for amusement, healing, and so forth.

Fox describes these processes with the metaphor of biological programming, suggesting that this programming was established during the prehuman past, in the capacities for social adaptation that we have inherited from our ancestors. Fox's use of the terminology of computer technology is well taken. A program does not determine the answers a computer will provide; information has been stored and arranged on the computer's tapes, and the answers it may give are determined by the kinds of questions that are fed into the program. A program, then, is a set of limits and potentials in the form of information, and if a question is not put into the computer, the programmed information will not be forthcoming. If the information itself is not stored in the computer, no amount of questioning will elicit answers from it. One of the examples Fox cites is that of initiation ceremonies. He stresses that these rituals are not "instinctive," but are the products of a biologically determined predisposition that is elicited by particular conditions.

Not all anthropologists accept this view, as we will see. But whether or not it is adopted, the quest for biological constraints and potentials for human culture involves some of the most important questions facing contemporary anthropology. For instance, culture—like all other phenomena—must be explained in terms of sources outside itself; otherwise, we end up with a series of tautologies in which something is used to explain itself. The point of view set forth in this selection represents a radical departure from conventional anthropological ways of thinking about human culture and social change, but even this approach is not without its own dangers of tautology; for example, Fox asserts that "man . . . is political; and he is political because he is that kind of primate—terrestrial and gregarious." Much more serious work needs to be done, especially in the collection of hard data, before we can be fully satisfied with the results of this approach.

Fox also observes that the differences among societies usually stressed by cultural anthropologists are not as great and not as important as the similarities among people, in courtship behavior, male bonding, political arrangements, and so forth. He is correct in suggesting that many cultural anthropologists have neglected important uniformities in human behavior and social arrangements, but it should also be remembered that the study of differences and of uniformities is equally legitimate; either position to the total exclusion of the other is incompatible with the scientific spirit.

Strenuous objection to the point of view represented here has been raised in some quarters on political grounds of one sort or another. For example, it is sometimes maintained that this approach to human culture legitimates a politically conservative philosophy—that if we are programmed for male dominance or political inequality, there is no point in trying to change these patterns. There are two rebuttals to this. First, the data about male dominance and political inequality are unclear and inconclusive; we must know what the facts really are before any conclusions are drawn one way or another. Thus, we need to know more about social behavior before we may say anything conclusive about the nature of human biology, and Washburn dwells on this in Selection 7. Second—as usually happens when political questions arise—the fact is often overlooked that any point of view in the social sciences may be used for politically conservative purposes, if that is one's goal. For instance, Konrad Lorenz, who is justly famous for his work on imprinting in geese, used the results of his scientific work to justify Nazi legal restrictions against intermarriage with "non-Aryans," but it cannot be said that we should therefore censor research on imprinting. Likewise, the data of cultural anthropology may also be used for either conservative or radical purposes.

For a sociological critique of the concept of territoriality, see Chapter 5 of *The Social Construction of Communities,* by Gerald Suttles (Chicago: University of Chicago Press, 1972). A concise defense of the biological approach to the study of human culture is "On the Importance of Cultural and Biological Determinants in Human Behavior," by J. Merritt Emlen (*American Anthropologist,* 69 [1969]: 513-14).

In "Cultural Factors Affecting the Study of Human Biology" (*Human Biology,* 26 [1954]: 77-79), Stanley M. Garn discusses the problems presented by the fact that it is difficult to gather "culture-free" data about human biology. The paper by Tiger and Fox ("The Zoological Perspective in Social Science") that was mentioned in the introduction to this section is highly pertinent to the problems raised here.

A discussion of Konrad Lorenz's political and scientific philosophy is in "The *Human* Nature of Human Nature," by Leon Eisenberg, *Science,* 176 (14 April 1972): 23-28. A general review of the political significance of anthropological data is in *The Rise of Anthropological Theory,* by Marvin Harris (New York: Cromwell, 1968). ■

PRIMITIVE MYTHOLOGIES testify to the enduring fascination of man with the problem of his own relationship to the natural world. For *Homo* is burdened with being *sapiens,* and one thing this *sapientia* drives him to is a ceaseless and almost passionate inquiry about his status—what T. H. Huxley aptly called *An Enquiry into Man's Place in Nature.* And like Darwin and Huxley, the primitive seeks an answer to the eternal paradox: We are obviously part of nature, and in particular we are part of the animal world, and yet we are set apart from nature by the very fact of knowing that we are part of it. Not only does no other animal know it is going to die, but no other animal knows it is alive—in any sense in which we would normally use the word "know." And no other animal concerns itself with the problem of its own uniqueness. But man is obsessed with it. He is forever seeking to define himself—a task as yet uncompleted—and to do so he has to establish the boundaries between himself and the animal world.

In their mythologies the primitives solve the problem in various ways, most usually by having man descend from an animal ancestor, or rather, various groups of men descend from various animal ancestors. We might like to think that this represents an anticipation of Darwin, but unfortunately most primitives believe in acts of special creation, so we have to disqualify them. One of these acts of creation, however, is usually the clue to the essential difference: be it language, fire, the art of cooking, rules about incest, and so on, that are the diacritics of humanity. We do not communicate, convert energy, eat, or breed quite like the animals, and hence we make that crucial breakthrough from nature to culture, and become the cultural animal.

Not only do we become cultural, we become divine. In many of our ego-boosting mythologies we do not differ simply in degree from the animals, we differ in kind. It is not some simple attribute—like the ability to make fire—but the possession of a divine spark that renders us *in essence* different, that carves out a gulf between us and "brute creation." Here again we cannot seem to settle the matter, and much argument, as we know, ensues about where brute creation stops and the divine human starts. Any human group is ever ready to consign another recognizably different human group to the other side of the boundary. It is not enough to possess culture to be fully human, one must possess *our* culture. Even universalistic religions, which were happy to define man as an animal with a soul, were often not too sure by what criterion one recognized the possession of the *anima,* and categories of *Homo sylvestris* and *Homo feralis* were invented to take care of marginal cases. But at least in the western world this definition sufficed and for many still suffices) until the eighteenth-century savants began to look down on such arguments as perhaps too emotional, and substituted reason as the defining characteristic of man. Linnaeus, to whom we owe our pretentious zoological title, was very much a child of the eighteenth century. Souls were not to be trusted, it seems, since one never knew quite what they were up to, and animals may very well have them. But brute creation did not have reason and that was obvious enough. Soulful our furry friends may be, but rational they are not. They could probably adore God, but they could not understand Pythagoras.

Darwin undermined this stance as much as the position of the religiously orthodox. He noted what in fact many predecessors, including Linnaeus himself, and even Immanuel Kant, had noted—the striking anatomical similarity between ourselves and the rest of the order *Primates* and ultimately between ourselves and the rest of the vertebrates. What Darwin added was theory that could explain how this striking relationship came about, other than by some whim of the Almighty or by Lamarckian effort of will. Now this caused many people other than Bishop Wilberforce to feel that human dignity and uniqueness were in danger. That great anticleric Samuel Butler castigated Darwin for "banishing mind from the universe." He had blurred the distinctions that we had assumed were inviolable. We had emerged gradually from the animal world by a natural process, not suddenly by a supernatural one. The moral was plain to the soul merchants and the reason merchants alike: We in fact differed only in degree and not in kind from our cousins. The reaction was interesting. The anatomical argument was quickly adopted and became its own kind of orthodoxy. Despite a few skirmishes the battle was over before it was fought.

The anatomist W. E. LeGros Clark said recently that it is astonishing to think people ever doubted the anatomical continuity between ourselves and the other primates. And indeed it does seem absurd today, to the extent that when I am faced with an unrepentant fundamentalist I confess I am unable to cope with him. I have no ready-made arguments for defending the self-evident, and so fare badly, thus confirming his worst fears about the conspiracy of the ungodly.

In the hundred years between the appearance of *The Origin of Species* and today, a large (although still too limited) amount of fossil evidence has come to light documenting the gradual transition that what Darwin saw *must* have happened, even in the absence of direct evidence. As far as anatomy was concerned then, the case rested. But human behavior was somehow exempted from the same rubric.

Darwin published in 1873 his most remarkable work, *The Expression of the Emotions in Man and Animals.* Whatever we may think of his specific conclusions, his message was clear enough: In many areas of behavior we show great similarities to our cousins; their behavior, like their anatomy, has evolved through the process of natural selection, ergo, so has ours. Anatomy and behavior, structure and function, were of course intimately linked, and what was true for one was true for the other.

Even in the biological sciences, the impact of this line of thinking was not immediate, and it is only comparatively recently that biologists have been investigating in a serious way the evolution of animal-behavior systems. The reasons for this are not our concern here, although the historians of science should be working on them. But one reason we should note: Investigations of animal behavior really got going under the aegis of Pavlovian-style behaviorism, which is not evolutionary in orientation and has scant respect for anything that is claimed to be innate. A similar reaction (or was it even a reaction?) happened in the social sciences. Darwin had blurred the distinctions all right, and even reason did not appear to be so firmly enthroned now, so anthropology took on the role of "Defender of the Faith" in human uniqueness and weighed in with cultrue as the defining characteristic. Of course, as with their predecessors, they were never able to define very clearly what this was.

So, as in the older myths of those other primitives, the nature of our uniqueness remains something of a mystery. Very roughly, "culture," in anthropological parlance, refers to traditional modes of behaving and thinking that are passed on from one generation to another by social learning of one kind or another. We get a little uneasy when told that animal communities also have "traditions" that get passed on, so we retreat into symbols. Culture is couched in symbols and it is by means of these that it is passed on. Preeminent among the symbol systems is language, and when all else fails we can cling to language. "By their speech ye shall know them"—and to this we will return later.

The social and behavioral sciences thus side-stepped Darwin's challenge. This maneuver was aided by a number of developments. Behaviorism dominated psychology, and "instinct theory" fell into disrepute. Behaviorism was rigorous and "scientific" while instinct theory—primarily under McDougal—seemed nothing more than a kind of thesaurus of human attributes. The eugenics movement which put such store in biological aspects of behavior became more and more entangled with racism, and any attempt to show that there were important biological components in behavior was regarded as incipient racism, and still is in many quarters. In sociology, the "social Darwinists" also fell into disrepute. They were not really Darwinians in the sense in which I am using the term; they simply used analogies from Darwinian biological theory and applied them, usually wrongly, to social processes. Their wrongheaded use of evolutionary doctrines to support the excesses of laissez-faire capitalism eventually sent them into oblivion. With them, the proverbial baby went out with the proverbial bath water. Henceforth, any explanation of a social phenomenon that was "Darwinian" or "biological" was ipso facto erroneous in the social sciences, and that was that. Marx and Durkheim dominated sociology, and while the latter had problems with the autonomy of the subject, his doctrine that the social must be explained in terms of the social and not reduced to any lower level (like the biological, of course) held almost complete sway.

Anthropologists continued to pronounce. Sir Arthur Keith even set a limit below which culture was impossible. The brain, he said, had to reach a size of 750 cubic centimeters before any fossil primate could be considered a "man." This gave substance to the anthropological belief that culture was, in the words of Malinowski, "all of a piece." One never found people with religion but no language, or law but no religion, and so on. If they had one they had all, and it must have happened at that point when the brain reached the size necessary for culture to "occur."

One can immediately see the similarities between this and the Catholic doctrine as ennunciated in the encyclical *Humani generis* where it is allowed that man may have evolved in body a la Darwin, but insisted that at some point an immortal soul was injected into the painfully evolving body. God would have had to wait, it seems, until his chosen primate had crossed Keith's "cerebral rubicon" before doing anything so presumptuous. Anthropologists, also, were almost maniacally preoccupied with explaining cultural differences. They were really not very interested in what made men man, but in what made one lot of men different from another lot.

As a student I had the litany chanted at me: Biological universals cannot explain cultural differentials. And of course at one level they cannot. Muslims, I was told, take off their shoes to go into church while Christians take off their hats. Now find me a biological explanation for that! I was never sure I wanted to find any kind of explanation for it. It seemed to me a pretty arbitrary thing. And anyway, what explanation was I offered? I will not bore you with the answer since I do not want to shake your faith in anthropology too much. But I will confess that even in those salad days I was plaguing my teachers with the question: If we do not really know what biological universals there are, then how can we study the cultural differentials in the first place? How to study the variables without the constants? In response, I was told that biological universals were simply primitive drives like hunger and sex. The fact that sex was universal did not explain why some cultures were polygynous and others monogamous. Maybe not, I thought, but it might explain why they were all adulterous. After all, sex is a very complex business, and might it not be that behavior resulting from it was more than just these rather gray and amorphous urges that I was presented with. Look at the courtship of birds and animals, for example. Ah, came back the answer, but that is *genetic,* whereas human courtship is *cultural.*

When all this was going on—some fifteen years ago in London—I had no ready answer. Anyway, I wanted to pass my exams. It all depends, I thought to myself and in secret, upon what you want to explain. All human cultures have some kind of courtship ceremonies, and when you look at them they look very much alike despite the different cultural trappings. If all you want to explain is why in America girls wear their date's fraternity pins while in Fiji they put hibiscus flowers behind their ears, that is fine. But (a) it does not seem worth explaining, and (b) there probably is no explanation in any scientific sense—it is just what they do. These are simply ways of getting the same courtship job done, and the interesting thing to me is the universality of various similar symbolic devices. Has each culture independently invented the idea that the girl should declare her allegiance in this kind of way, or is there perhaps something more subtle about courtship than we imagined, something uncultural, something unlearned?

To be fair to anthropology, it was fighting on several fronts and often shot at the wrong targets. The "no links between biology and culture" argument was partly an attack on the racists who wanted to explain seeming inequalities between cultures as a result of biological differences. Again the baby went out with the soapsuds when anthropology strenuously set its face against *any* connection between culture and biology, even at the universal level. At best—as for example in the work of Malinowski—culture could be seen as a response to a rather drab set of "biological imperatives," but then this kind of Malinowskian functionalism soon fell into disrepute as well, since it did nothing to explain cultural differences. While I am on with my catalog of complaints, let me add that as far as I could see, for all its obsession with cultural differences, anthropology in fact did nothing to explain them. What it did was to take cultural differences as given and use them to explain other things—largely other cultural differences. All the things that might have explained cultural differences, such as racial variation, environment, history, and diffusion, were at one time or another ruled out of court.

All in all, for a variety of ideological reasons, the anthropological profession, along with psychology and sociology, kept the world safe for humanity by refusing to allow that anything about culture could be "reduced" to biology, and hence kept the gap between us and the brutes nicely wide. We were the "cultural animal" all right, but stress was entirely on the cultural while the animal was relegated to a few odd things like blinking, sucking, feeling hungry, and copulating. Ninety-nine percent of our behavior, it was held, was "learned" and hence cultural. And what was more there was no limit to what could be learned. The human infant was a *tabula rasa* on which culture imprinted itself, and the subsequent behavior of the infant was therefore wholly a matter of which particular culture had been imprinted on it. The differences between cultures in their beliefs, behaviors, and institutions was so great that any considerations of common biological traits were totally irrelevant.

We get to the crux of what I want to say by raising the question: Were not anthropologists suffering from ethnographic dazzle? I borrow the term from linguistics where "orthographic dazzle" refers to the difficulty some people have of sorting out pronunciation from spelling. In some respects and at some levels—the levels of beliefs, of formal institutions—cultures are dazzlingly different. Why are the Japanese the way they are, as opposed to the Americans, the Russians, the Hottentots, and so on? This is a fascinating question. But as I have said, the anthropological answer is rather lame "they are different because they do things differently." Mostly anthropology tells us about the *consequences* of doing things differently, and tells it very well indeed. But are societies and cultures really very different at the level of forms and processes? Or are they not in some ways depressingly the same? Do we not time after time in society after society come up with the same processes carried out under a variety of symbolic disguises? I think we do, and if we can get past the cultural or ethnographic dazzle we can see that this is so. Thus, if you look at the behavior of what my colleague Lionel Tiger has called *Men in Groups*, you find that whatever the overt cultural differences in male-group behavior at the level of symbolism, actual practices and beliefs, and even emotional and other expressive features, in society after society one thing stands out: Men form themselves into associations from which they exclude women. These associations vary in their expressed purposes but in many of their processes they are remarkably uniform. A seemingly bewildering variety of male behavior can be reduced in effect to a few principles once this is grasped.

Similarly I have tried to show that the seemingly endless variety of kinship and marriage arrangements known to man are in fact variations on a few simple themes. The same can be said of political arrangements, which, despite their cultural variety, are reducible to a few structural forms. Once one gets behind the surface manifestations, the uniformity of human behavior and of human social arrangements is remarkable. None of this should surprise a behavioral zoologist; we are, after all, dealing with a uniform species divided into a number of populations. This being a species of rather highly developed mammals one would expect a lot of local differences in traditions between the various populations, but one would expect these differences to reflect species-specific units of behavior. Thus, every species has a complex of social behavior made up of recognizable units—a complex which distinguishes it from other species—but these units may well be put together in different ways by different populations adapting to different environments. But one does not find a baboon troop, for all its ingenuity, adapting like a herd of horses and vice versa. The baboons can only adapt with the material at hand in their stock of behavior units, and the same is true of man.

The degree of flexibility in human populations is obviously greater, but a great deal of it is at the symbolic level. We can tell the story in many different ways, but it is the same old story we are telling. And if we depart too far from the plot—which we have the capacity to do—the result may well be a truly dramatic chaos. For this is man's hang-up. Unlike the baboon or the horse, we can imagine things that are different from the plot laid down for us, and we can put our dreams into practice. The question then is, will the dream work? If you accept that all behavior is culturally learned and that man can learn anything, then the answer is yes. The only limit is human ingenuity. We can invent any kind of society and culture for ourselves. If you believe, as by now it should be obvious I do, that we have a species-specific repertoire of behavior that can be combined successfully only in certain ways, then the answer is no. There are definite limits to what this animal can do, to the kinds of societies it can operate, to the kinds of culture it can live with. But there is no end to its dreams and its fantasies. While its social behavior may have strict limits, its imagination has none.

I have jumped here to my conclusion without detailing the whole route. Let me postpone the latter for a little longer to press home this point. We mentioned earlier that language was the chief characteristic of our species, the crucial distinguishing feature. This is true and can be used to illustrate the point. It is now well established that the capacity for language acquisition and use lies in the brain on the one hand and in the speech organs on the other, and overall in the complex relations between the two. Linguists like Chomsky and psychologists like Lenneberg argue that the capacity for grammatical speech is somehow "in" the brain, and matures as the child matures. Thus every human child has the capacity for grammatical speech, and is ready, as it were, to be programed with whatever actual grammar its culture provides. Now we know that the latter are many and astonishing in their variety, and that their variation is arbitrary.

There is no "explanation" why the English say "horse," the French "cheval," and the Germans "pferd." There is no explanation why any particular pattern of sounds signifies an object or action (with the possible exception of onomatopoeia). This is quite arbitrary. Nevertheless, the speech patterns of all languages are known to operate on a few basic principles which linguists have worked out, and the semantic patterns may well also be reducible in this way, once what Chomsky calls the "deep structures" of all languages are known. Once these are discovered, we can write the "universal grammar" which will tell us the few principles upon which all actual grammars rest. We can do this because despite the enormous variety of "surface grammars" they all are doing the same job and are constrained to do it in a limited number of ways. Thus no language exists that a linguist cannot record with the universal phonetic alphabet in the first place, and analyze with universally applicable techniques of semantic analysis in the second. We can invent artificial languages based on binary signals, or other codes, which require different "grammars," but "natural languages" all can be broken down, first into phones, then phonemes, then morphemes, then lexemes, and so on up to the higher levels of grammaticality.

The rest of culture is probably like this. The potential for it lies in the biology of the species. We have the kinds of cultures and societies we have because we are the kind of species we are. They are built out of our behavioral repertoire and are analyzable into its elements and their combinations. Like language, the capacity for specific kinds of behavior is in us, but exactly how this will be manifested will depend on the information fed into the system. The system here is the behavior potential of the individual; the information is the culture he is socialized in. But in the same way as he can only learn a language that follows the normal rules of grammaticality for human languages, so he can only learn a grammar of behavior that follows the parallel rules in the behavioral sphere. Of course in either case he can try departing from normal grammaticality, but in either case he will then get gibberish, linguistic or behavioral.

We generally do not try to manipulate language because the matter is out of our hands, but with behavior we are continually producing gibbering illiterates, and until we understand the deep structure of the behavioral grammar within which we weave our cultural variations, we will continue to do so. No one wants to produce linguistic gibberish since verbal communication breaks down; but we constantly produce behavioral gibberish and then wonder why social communication breaks down. The answer of those who believe that anything is possible since everything is cultural is: Try to invent yet more and more different langauges with any kind of grammaticality you can think of.

My answer is: Find out how the universal grammar works and then bring changes within that framework; invent new behavioral languages that do not violate the principles of basic grammaticality.

At least two monarchs in history are said to have tried the experiment of isolating children at birth and keeping them isolated through childhood, to see if they would spontaneously produce a language when they matured. The Egyptian Psammetichos, in the seventh century B.C. and later James IV of Scotland, in the fifteenth century A.D. Both, it seemed did not doubt that untutored children would speak, although King James' hope that they would speak Hebrew was perhaps a little optimistic.

I do not doubt that they *could* speak and that, theoretically, given time, they or their offspring would invent and develop a language despite their never having been taught one. Furthermore, this language, although totally different from any known to us, would be analyzable by linguists on the same basis as other languages and translatable into all known languages. But I would push this further. If our new Adam and Eve could survive and breed—still in total isolation from any cultural influences—then eventually they would produce a society which would have laws about property, rules about incest and marriage, customs of taboo and avoidance, methods of settling disputes with a minimum of bloodshed, beliefs about the supernatural and practices relating to it, a system of social status and methods of indicating it, initiation ceremonies for young men, courtship practices including the adornment of females, systems of symbolic body adornment generally, certain activities and associations set aside for men from which women were excluded, gambling of some kind, a tool- and weapon-making industry, myths and legends, dancing, adultery, and various doses of homicide, suicide, homosexuality, schizophrenia, psychosis and neuroses, and various practitioners to take advantage of or cure these, depending on how they are viewed. I could extend the list but this will suffice.

In short, the new Adam and Eve would not only produce, as our monarchs suspected, a recognizable human language, but a recognizable human culture and society. It might not be in content quite like any we have come across: Its religious beliefs might be different, but it would have some; its marriage rules might be unique (I doubt it), but it would have them and their type would be recognized; its status structure might be based on an odd criterion, but there would be one; its initiation ceremonies might be unbelievably grotesque but they would exist; its use or treatment of schizophrenia might be bizarre, but there it would be. All these things would be there because we are the kind of animal that does these kinds of things.

In the same way, in a zoo one can rear infant baboons who know nothing of the state in which their ancestors and wild cousins lived, and yet when they reach maturity they produce a social structure with all the elements found in the wilds and of which they have no experience. Their capacity to produce a unique "language" is of course much more limited than that of our hypothetical naive group of humans, but in both cases the basic grammaticality of behavior will be operative. In the same way that a linguist could take our Garden of Eden tribe and analyze its totally unique language, so an anthropologist would be able to analyze its totally unique kinship system or mythology or whatever, because the basic rules of the universal grammar would be operating.

(Actually in the interests of accuracy I should add a rider here to the effect that the experiment might be impossible to perform. It is one of the ground rules of the universal behavioral grammar of all primates—not just humans—that if you take young infants away from maternal care at a critical period they will grow up to be very disturbed indeed and may well perpetuate this error by mal treating their own children in turn. Thus our experiment may well produce a group of very maladjusted adults and the whole thing founder rather quickly. But at least this gives us one element of the universal system: Some method has to be found of associating mother and child closely and safely during certain critical periods. If isolated during critical periods, not only can the animal not learn *anything* at all, it loses the potential to learn at any other time. It has to learn certain things at certain times—true of language and of many other areas of behavior.)

To return to our human tribe developed *de novo* in our experimental Eden: What I am saying may not seem very remarkable but it goes against the grain of the anthropological orthodoxy. Without any exposure to cultural traditions our tribe would develop *very specific* and highly complex patterns of behavior, and probably very quickly—within a matter of a few generations, once they had developed a language. They would do so for the same reason that the baboons produce a baboon social system in captivity—because it is in the beast. And it is not just a very general capacity that is in the beast—not just the capacity to learn, and to learn easily—which is all the culturalists need to assume, it is the capacity to learn some things rather than others, and to learn some things easily rather than others, and to learn some rather specific things into the bargain.

This is a very important point. I am not positing that initiation ceremonies or male rituals are instinctive, in any old sense of that term. I am positing that they are an outcome of the biology of the animal because it is programed to behave in certain ways that will produce these phenomena, given a certain input of information. If this input does not occur then the behavior will not occur or will occur only in a modified or distorted form. (This is, in fact, more like the modern theory of instinct, but to go into the ramifications of this would take too long.) The human organism is like a computer which is set up or "wired" in a particular way. It is thus in a state of readiness—at various points in the life cycle —to process certain kinds of information. The information has to be of a certain type, but the actual "message" can vary considerably. If the information is received, then the computer stores it and uses it to go on to the next task. If you confuse the system, the machine very easily breaks down and might even blow the fuses if you really mix up the program. Of course to push this analogy to its logical conclusion we would have to have computers feeding each other information to simulate the human situation. Only when they were synchronized would the total system run properly.

This is—although very crude—a different model from that of the old "instinctivists" or of the "behaviorists." To the instinctivists, behavior resulted simply from the manifestation of innate tendencies which in interaction produced such things as territorialism, maternal behavior, or acquisitiveness. To the behaviorists the infant was, as we have seen, a *tabula rasa* and behavior ultimately was the result of learning via conditioning. (Psychoanalysis leans to the instinctivist end, but is a special case in some ways.) Culturalists in anthropology and the social sciences share the *tabula rasa* view and see all behavior above the primitive-drive level as a result of the learning of a particular culture. My view sees the human organism as wired in a certain way so that it can process information about certain things like language and rules about sex, and not other things, and that it can only process this information at certain times and in certain ways. This wiring is geared to the life cycle so that at any one moment in a population of *Homo sapiens* there will be individuals with a certain "store" of behavior at one stage of the cycle giving out information to others at another stage, the latter being wired to treat the information in a certain way. As an outcome of the interaction of these individuals at various stages, certain "typical" relationships will emerge. This may seem either tortuous or obvious, but I can assure you it *is* a different way of viewing human behavior and social structure than the orthodox one. The orthodox view says: When in trouble, change the program because we can write any program we want to. What we should say is: When in trouble, find out what is in the wiring, because only then will we know what programs we can safely write.

The culturalists only acknowledge a very general "capacity for culture," if they acknowledge anything at all about the general characteristics of the species. To them, all culture is pure human invention and is passed on from generation to generation by symbolic learning. Thus, logically, it follows that if ever this store of culture should be lost, it is improbable that it would be invented again in the same forms. Thus something as specific as totemism and exogamy—that old anthropological chestnut—has to be seen in this view as a pure intellectual invention, and it would be unlikely that it would be invented again. I do not think so. I think my tribe with no experience of any other human culture and no knowledge of totemism and exogamy would produce both very quickly, and what is more these phenomena would be immediately recognizable and analyzable by anthropologists. In fact those anthropologists like Tylor who argued for the "psychic unity of mankind" were acknowledging a similar position. They argued that such customs had not been invented in one area and diffused throughout the world, but were stock responses of the human psyche to external pressures. They were somehow reflections of "human nature," a phrase that we have been discouraged from using. The argument for psychic unity also had to face the "constants can explain variables" charge, but we have dealt with that one already.

The psychic unity argument, however, was never pushed as far as I am pushing it. The universal psyche for these anthropologists had no specific content; it was a capacity to do human things, but most of its proponents would have maintained it was a general learning capacity and that culture was invented. As I have said, at one level this is true—at the level of specific content—but we must not be dazzled by this into ignoring those basic processes and forms that crop up with regular monotony. (Here we must not slip into the error of thinking that *universal processes* necessarily produce *uniform results*. Far from it. This is not true even in the plant kingdom and is even less true in the animal.)

I can now return to the problem of the route by which I reached this conclusion. The question that had been plaguing me throughout my undergraduate career was really, "How do we know what's in the wiring and how did it get there?" For we can find out really what it is all about only if we know how it was constructed and to what end it was produced. It is no good trying to use an analog computer as if it were a digital computer since they were designed for different uses. The answer to this should have been obvious, and soon became so. What is in the wiring of the human animal got there by the same route as it got into any other animal—by mutation and natural selection. These "great constructors" as Konrad Lorenz calls them, had produced remarkable end products in the social behavior of all kinds of animals, reptiles, birds, fishes, and insects. And it is here that the message of Lorenz and his associates becomes important: Behavior evolves just as structure evolves and the evolution of the two is intimately linked.

Now we are back to Darwin's principle from which this paper started and from which anthropology so disastrously departed at the turn of the century. What the behavioral zoologists (in Europe usually known as ethologists) showed us was that units of behavior evolve on the same principle as units of anatomy and physiology—that a head movement that was part of a bird's innate repertoire of actions has an adaptive significance as great as the evolution of the wing itself. The head movement may be precisely the thing that inhibits the attack of another bird, for example, and over the millennia has become "fixed" as a signal recognized by the species as an inhibitor. Even if one does not accept that humans have "instincts" of this kind (I think they have a few but not many), the point is well taken that one should look at behavior as the end product of evolution and analyze it in terms of the selection pressures that produced it. If we have this marvelous flexibility in our learning patterns then this is a feature of the biology of our species and we should ask *why* we have this flexibility. What selection pressures operated to bring about this particular biological feature? Our enormous dependence on culture as a mode of adaptation itself stands in need of explanation, for this too is a species-specific characteristic, and it gets us into trouble as much as it raises us to glory. It is a two-edged weapon in the fight for survival, and the simple brutes with their instinctive head-wagging may well live to have the last laugh.

But the brutes, it has transpired, are not so simple. When one looks at our cousins, the other primates, the complexity of their social behavior is amazing. One thing the ethologists taught us to do was to compare the behavior of closely related species in order to get at the "proto-behavior" of the group of animals concerned. The continuing flow of excellent material on nonhuman primates in the wild shows us how many and subtle are the resemblances between ourselves and our simian relatives. Wider afield, the growing science of animal behavior shows that many mammals, and vertebrates generally, have social systems which duplicate features of our own society, and in which similar processes occur, and even similar social pathologies. Lorenz showed how aggression was the basis of social bonding; Wynne-Edwards postulated that the "conventionalized competition" which controlled aggression was itself rooted in the control of numbers; Chance demonstrated that among primates the elementary social bond was that between males rather than that between males and females, and so on. The politics of macaque monkeys suggests that Aristotle was right: Insofar as he is a primate, man is by nature a political animal. (It is significant that quoting this, the phrase *by nature* is often omitted.) Ants can have societies, but ants cannot have politics. Politics only occurs when members can change places in a hierarchy as a result of competition. So man is more than social; he is political, and he is political because he is that kind of primate —terrestrial and gregarious.

As a consequence it becomes more and more obvious that we have a considerable animal heritage and hence a great store of comparative data to draw on in making generalizations about our own species. It forces upon us this observation: If we find our own species displaying certain patterns of social behavior that duplicate those of other similar species—depending, of course, on the level of similarity—we will often need to say only that these patterns are what we would expect from a terrestrial primate, a land-dwelling mammal, a gregarious vertebrate, or whatever. Of course some aspects of these patterns and a great deal of their content will be unique, but this is only to say that they will be species-specific. Every species is unique since it is the end product of a particular path of evolution.

The real question—What is the nature of the uniqueness?—brings us back to where we started. And we cannot answer that question until we know what we have in common with all other species, and with some other species, and with only closely related species. Thus the argument that we differ from all other species as a result of the triumph of culture over biology I find false, because culture is an aspect of our biological difference from other species. It is the name for a kind of behavior found in our species which ultimately depends on an organ, the brain, in which we happen to have specialized. Thus differences between ourselves and other primates for example do not stem from the fact that we have in some way *overcome* our primate natures, but stem from the fact that we are a different kind of primate with a different kind of nature. At the level of forms and processes we behave culturally because it is in our nature to behave culturally, because mutation and natural selection have produced this animal which must behave culturally, must invent rules, make myths, speak languages, and form men's clubs in the same way as the hamadryas baboon has to form harems, adopt infants, and bite his wives on the neck.

But why culture? Why did our simian ancestors not content themselves with a much less flexible, and perhaps at the time less vulnerable, way of coping with nature's exigencies? This is where another strand of evidence comes in—the material on human evolution. In Darwin's day this was practically nonexistent and even now it is relatively meager. But we now can trace with some confidence the general picture of man's evolution over at least four million years, and we have evidence that the hominoid line may well go back over thirty million. This is not the place for a detailed exposition of what we know, and all I can do is to point out some of the implications of our new knowledge.

You will remember that Sir Arthur Keith set the limit of brain size below which was mere animal at 750 cc. The modern human brain averages about 1400 cc, roughly twice that of Keith's minimum. The brain of the chimpanzee is roughly 400 cc, that of the gorilla 500 cc. Now to cut a long story short, the modern discoveries have shown that hominids have existed for at least two million years and probably longer, and that at that early date in their evolution they were indulging in activities that imply the existence of cultural traditions, even if of a rudimentary form. The most striking evidence of this is the existence of tool-making industries first in bone and horn and then in stone (wood does not survive but was undoubtedly used) in small-brained hominids in East and South Africa. Two millions years ago, our ancestors with brain sizes ranging from 435 cc to 680 cc—little better, you see, than the gorilla—were doing very human things, cultural things, before having reached the Rubicon. They were hunting, building shelters, making tools, treating skins, living in base camps and possibly many other things (speaking languages perhaps?) that we cannot know directly, while their morphology was still predominantly ape-like and their brains in some cases smaller than the modern apes. What was not ape-like about them was their dentition and their bipedal stance. In these features they were well launched on the road to humanity since both reflect the adaptation to a hunting way of life which differentiates these animals from their primate cousins.

You will note that I say these "animals"—I might just as easily have said these "men"—and this is the moral of the story. What the record of evolution shows is no sharp break between man and animal that can be pinpointed at a certain brain size or anything else. What it shows is a very gradual transition in which changes in locomotion led the way and in which the brain was something of a sluggard. The pelvis of the Australopithecinae—those man-apes of South and East Africa—is strikingly human and totally unlike anything in an ape, because these were bipedal creatures; but the brain was, if anything, smaller than that of a gorilla—an animal not noted for its cultural achievements.

The moral goes deeper. Once launched on the way to humanity through bipedalism, hunting, and the use of tools, our ancestors became more dependent on their brains than their predecessors had been. If they were going to survive largely by skill and cunning and rapid adaptation to the changing circumstances of the Pleistocene epoch, then a premium was put on the capacity for cultural behavior about which we have been speaking. Man took the cultural way before he was clearly distinguishable from the animals, and in consequence found himself stuck with this mode of adaptation. It turned out to be very successful, although for a while it must have been touch and go. But because he became dependent on culture, mutation and natural selection operated to improve on the organ most necessary to cultural behavior, namely the brain and in particular the neocortex with its important functions of association and control. Those animals, therefore, that were best able to be cultural were favored in the struggle for existence. Man's anatomy, physiology, and behavior therefore are in large part the *result* of culture. His large and efficient brain is a consequence of culture as much as its cause. He does not have a culture because he has a large brain; he has a large brain because several million years ago his little-brained ancestors tried the cultural way to survival. Of course, the correct way to view this is as a "feedback" process. As cultural pressures grew, so did selection pressures for better brains, and as better brains emerged, culture could take new leaps forward thus in turn exerting more pressures, and so on.

Again this is an over-simplified account, and the actual picture of the evolution of the brain is much more complex. But in essence it is true, and for our immediate purposes enough to make the point that our uniqueness is a biological uniquenesses and that culture does not in some mysterious sense represent a break with biology. Our present biological makeup is a consequence among other things of cultural selection pressures. We are, therefore, biologically constituted to produce culture, not simply because by some accident we got a brain that could do cultural things, but because the cultural things themselves propelled us into getting a larger brain. We are not simply the *producers* of institutions like the family, science, language, religion, warfare, kinship systems, and exogamy; we are the *product* of them. Hence, it is scarcely surprising that we continually reproduce that which produced us. We were selected to do precisely this, and in the absence of tuition our mythical tribe would do it all over again in the same way. It is not only the *capacity* for culture then that lies in the brain, it is the *forms* of culture, the universal grammar of language and behavior.

This then is how it all got into the wiring of the human computer. Once we know these facts about human evolution there is no great mystery in principle about the production of culture by human beings and the relative uniformity of its processes. There are many mysteries of fact which will never be solved since we can only infer the behavior of fossil man and never observe it. But in principle, once we accept that culture is the major selection pressure operating on the evolution of human form and behavior, and that it has produced an animal wired for the processing of various cultural programs, then the problem of the uniqueness of man becomes a problem on the same level as the problem of the uniqueness of any other animal species.

Putting together the insights of the ethologists and the students of human evolution, we can scan the behavior of related species for aspects of behavior that are common to all primates, and beyond that we can look to mammals and vertebrates for clues. For in the process of evolution we did not cease to be primates or mammals. In fact, as Weston LaBarre has said, part of our success lies in exaggerating certain mammalian tendencies rather than in losing them—length of suckling, for example. Much of our behavior and in particular our social arrangements can be seen as a variation on common primate and gregarious mammalian themes. Certain "unique" aspects—such as the use of true language—can be investigated for what they are, biological specializations produced by the unique evolutionary history of the species.

This perspective enables us to look at human society and behavior comparatively without any necessity to propound theories of the total and essential difference between ourselves and other animals. It puts the obvious uniqueness into perspective and does not allow us to lose sight of our commonality with the animal kingdom. We are the cultural animal all right, but both terms should be given equal weight, and one does not contradict the other. For the last time then, let me say that culture does not represent a triumph over nature, for such a thing is impossible; it represents an end product of a natural process. It is both the producer and the product of our human nature, and in behaving culturally we are behaving naturally.

To bury this issue once and for all, at least to my own satisfaction, let me add a word about the nature of culture as opposed to the nature of instinct. It is often said that man has "lost" all his instincts. I think this is a bit too extreme. If we might paraphrase Oscar Wilde: To lose some of one's instincts is unfortunate, to lose all of them smacks of carelessness. No species could afford to be that careless. But it is true that in terms of innate mechanisms which produce items of behavior complete at their first performance and relatively unmodifiable by experience, man has very few. Instead, it is often claimed, he has intelligence, foresight, wisdom and the like, and the enormous capacity to learn. Now in ditching instinct and opting for intelligence man took something of a risk, since instinct does provide a surety of response that has been evolved from trial and error over millions of years. Ant societies are much better organized and more efficient than any human societies and are driven wholly on instinctive mechanisms. But again instinct has its costs. It is too rigid. Changed circumstances cannot be met by a rapid adjustment in behavior, and insects and animals heavily dependent on instinct have to wait for processes of genetic change to effect changes in the instincts themselves before they can adjust. The higher we go up the phyletic scale the less true this is of course and with man least true of all. Thus there is a cost-benefit analysis involved in the shedding of innate instincts in favor of more complex modes of behaving.

The crux of the matter is this: Even if a species sheds its dependence on instincts, it still has to do the same things that instincts were designed to do. To put it into our earlier language, culture has to do the same job that instinct had been doing. This is another paradox, I suppose, but an intriguing one, because to get culture to do the same jobs as instinct had been doing, one had to make cultural behavior in many ways like instinctive behavior. It had to be unconscious so that it did not require thought for its operation, it had to be "automatic" so that certain stimuli would automatically produce it, and it had to be common to all members of the population.

If we look at our cultural behavior, how much of it is in fact intelligent and conscious, and how much is at that unthinking automatic response level? The answer, of course, is that the vast majority of our behavior is the latter, absorbed during our socialization and built into our patterns of habitual thought, belief, and response. Habit indeed is, as William James said, the great flywheel of society. Anthropologists speak of covert or unconscious culture to refer to this iceberg of assumptions, values, and habitual responses. And sitting over all of them, of course, is the great evolutionary invention of conscience, superego, moral sense, or whatever you want to call it. The sense of guilt, of having broken the taboos, the rules, the laws of the tribe, keeps most of us in line most of the time. Conscience is an empty cannister that culture fills, but once filled, it becomes a dynamic controller of behavior. Most of our behavior, however, never even rises to the point where conscience and the sense of guilt need to step in. We do what we do from habit, even down to tiny little details of gestures and twitches of the facial muscles. Most of this we never think about, but we rapidly recognize when other people are not behaving "normally" and we lock them up in asylums as lunatics. Think only of the example of the man walking down the road in the rain without a raincoat, smiling, shoulders back, head facing the sky. Clearly a madman. He should be hunched, hurrying, and looking miserable, with his jacket collar up at least.

The genius of nature here stands revealed and the paradox is resolved. Of course most of our learned cultural behavior in fact operates almost exactly like instinct and, as we have seen, this has to be the case. The customs and usages of the tribe, although not instinctive themselves, had to do the same jobs as instincts and hence had to be built into the automatic habit patterns of the tribal members, with guilt as a safeguard. (This was not foolproof, but neither is instinct itself.) So the same effect is achieved, and those habits which have proved useful in survival become part of the behavioral repertoire of the people. But—and here is the genius bit—these habits can be changed within a generation. One does not have to wait for the long process of natural selection to operate before these quasi-instinctual behaviors can be modified. They can be modified very rapidly to meet changing circumstances. Thus one has all the benefits of instinctive behavior without waiting for instincts to evolve. At any one time the rigidity of cultural habits will be just as invulnerable to change as any instinct—as we well know if we reflect on the persistence of traditions—and habits are very conservative. Since most of them are passed on by means other than direct tuition, they tend to persist over generations despite changes in overt education. But they can be changed relatively rapidly compared with the time span needed for changes in genetic material.

Thus man can make rapid adjustments without anarchy (which does not mean that he always does so).

Here again we see learned, cultural behavior as yet another kind of biological adaptation. At this level, other species also display behavior of the same kind, and the higher in the scale they are, the more dependent they become on habits transferred over the generations by learning rather than instincts transferred in the genetic code. But always we must keep in mind that this is not a sharp distinction. The code is not silent about learning and habits. If the position taken here is anywhere near corect, then instructions about habitual behavior are as much in the code as instructions about instinctive behavior.

The model of behavior sees the human actor as a bundle of potentialities rather than a *tabula rasa:* potentialities for action, for instinct, for learning, for the development of unconscious habits. These potentials or predispositions or biases are the end product of a process of natural selection peculiar to our species. One consequence of this view is that much of the quasi-instinctive cultural behavior of man can be studied in much the same way and by much the same methods as ethologists study the truly instinctive behavior of other animals. Many strands of investigation seem to be leading in this direction at the moment.

The kind of overall investigation that emerges from this theoretical position would utilize primarily three kinds of data: Data from human behavior both contemporary and known in history; data on animal behavior, particularly that on wild primates; data on hominid evolution with special attention to the evolution of the brain. Eventually data from genetics—molecular genetics, behavior genetics, and population genetics—will have to be included. But this is perhaps jumping ahead too far. At the moment the best we can say is that we should be prepared to use genetic data when we become sophisticated enough to incorporate it.

We began with the theme of human uniqueness and should end with my point that our uniqueness has to be interpreted in the same way as the uniqueness of any other species. We have to ask "How come?" How did culture get into the wiring? How did the great constructors operate to produce this feature, which, like everything else about us is not anti-nature, or superorganic, or extrabiological, or any of the other demagogic fantasy-states that science and religion imagine for us? Darwin did not banish mind from the universe as Butler feared; indeed, he gave us a basis for explaining how mind got into the universe in the first place. And it got there—as did every other natural and biological feature—by natural selection. The tool-making animal needed mind to survive; that is, he needed language and culture and the reorganization of experience that goes with these. And having got the rudiments and become dependent on them, there was no turning back. There was no retreat to the perilous certainty of instinct. It was mind or nothing. It was classification and verbalization, rules and laws, mnemonics and knowledge, ritual and art, that piled up their pressures on the precarious species, demanding better and better brains to cope with this new organ—culture—now essential to survival. Two related processes, thought and self control, evolved hand in hand, and their end product is the cultural animal, which speaks and rules itself because that is the kind of animal it is; because speaking and self-discipline have made it what it is; because it is what it produces and was produced by what it is.

7. BEHAVIOR AND THE ORIGIN OF MAN

S. L. WASHBURN

Reprinted from The Rockefeller University Review *(January-February, 1968) pages 10-18. Sherwood L. Washburn is Professor of Anthropology at the University of California, Berkeley. His principal interest is the teaching of human evolution. In research he has developed experimental methods for the analysis of problems of human evolution. He has done field work on monkey behavior, and has attempted to relate structure and function through the study of evolution of behavioral systems.*

■ In introducing the previous selection, I observed that Robin Fox—who stresses the need to focus on the biological background of social behavior—is a social anthropologist. Sherwood Washburn is a physical anthropologist; many regard him as the dean of primatological studies in anthropology. In contrast to Fox, Washburn emphasizes the importance of focusing on social organization to understand the biological background of behavior. Whether this contrast reflects the worn adage that the grass always seems greener in other yards or whether it represents the ironies that help to brighten a discipline, it may serve as an example of the principle that every phenomenon must be understood in terms of sources outside itself. Fox, whose primary commitment is to the study of social organization, looks to biology for its origins; Washburn, primarily concerned with the study of physical evolution, calls for greater attention to its social background.

Washburn begins his article with the observation that if we are to understand the origins of *Homo sapiens*, we must consider the relationship between biology and behavior because the two are inseparable. He maintains that all primates—including man—are social animals, and their behavior is expressed in their social systems; the behavior of any species cannot be appreciated in terms of its biological make-up alone. Washburn suggests that the success of a species' social system is what determines the course of evolution.

Against this background, Washburn turns to consider our ancestors' descent from tree- to ground-living. Using recently gathered information about "knuckle-walkers," like the chimpanzee, Washburn suggests that knuckle-walking, regarded as an anatomical and behavioral adaptation, makes it easier to understand the transition from ape to *Homo sapiens*. It is to be noted, incidentally, that the "knuckle-walking" hypothesis is not accepted by all students of evolution and there is considerable controversy surrounding it. But the descent from the trees, as Washburn notes, was bought at the price of the safety and refuge of the forest, and the transition must have required considerable time. In the course of evolving this new adaptation, the loss of the safety afforded by the trees represented a selective pressure for bipedalism, for biological capacities for more developed tool use, and for the evolution of the brain, which permitted the development of skills for living in a social system that maximized the expression of biological capacities.

In "The Transition to Humanity," Clifford Geertz develops the idea that man is not only the producer of culture but, in a biological sense, its product (in *Horizons of Anthropology*, edited by Sol Tax, pp. 37-48 [Chicago: Aldine, 1964]). An intriguing book that casts into doubt many accepted notions about the role of sexuality in human evolution is *The Nature and Evolution of Female Sexuality*, by Mary Jane Sherfey (New York: Vintage, 1973). A source of many interesting hypotheses is Jules Henry's "Culture, Personality, and Evolution" *American Anthropologist*, 61 [1959]: 221-26), in which he maintains that the distinctions between biology and personality (in the cultural sense) are more apparent than real. Although they hardly exhaust the additional sources that can be consulted, two very different approaches have been advanced by cultural anthropologists: Margaret Mead's views can be gleaned from some of her papers in *Anthropology: A Human Science—Selected Papers, 1935-1960* (New York: Van Nostrand, 1964) and from her books, *Coming of Age in Samoa* (New York:

Modern Library, 1953), *Growing Up in New Guinea* (New York: William Morrow, 1962), and *Male and Female* (New York: New American Library, 1964); Weston La Barre's *The Human Animal* (Chicago: University of Chicago Press, 1954) presents a different point of view. ■

THE UNDERSTANDING of human evolution comes from three different sources: from general evolutionary theory, from the fossils, and from the behavior and biology of the living primates. There has been great progress in each of these since the time of Charles Darwin (1871) and Thomas Huxley (1863). The nature of the modern evolutionary synthesis was clearly stated by Julian Huxley in 1963. Now, knowledge of the genetic code gives an understanding of the basic chemical nature of the process of heredity and evolution. In contrast to the single skullcap of a fossil human, all that was available to Huxley in 1863, today there are hundreds of fossils connecting present populations with ancestral ones many thousands of years old. Although many of these are "bones of contention," the fossil record is unmistakable evidence that evolution has taken place. In the third source, the study of behavior, major field studies have replaced myths and travelers' tales, which in the nineteenth century were the sole source of information on the behavior of monkeys and apes.

SOCIAL SYSTEMS AND SURVIVAL

As noted in many of the events celebrating the hundredth anniversary of the *Origin of Species*, the synthetic theory of evolution is remarkably close to Darwin's, and many competing theories (orthogenesis, acquired characters, mutationism, etc.) have been eliminated as the nature of the evolutionary process has become understood. In 1863 Huxley stated, "Whatever part of the animal fabric—whatever series of muscles, whatever viscera might be selected for comparison—the result would be the same—the lower Apes and the Gorilla would differ more than the Gorilla and Man," and this statement is supported by the latest cytological and biochemical studies. Evolutionary theory states that the genetic variability of populations is ordered by selection. Certain phenotypes are more successful, leave more offspring, and this changes gene frequencies, the code of the DNA. It is behavior that determines success and among primates behavior is nurtured, controlled, and ordered by the social system. It is the social system that brings the various physiological functions into relation with the essential adaptive problems of life. In this sense, the effort to understand human evolution is the attempt to reconstruct the ways of life of our ancestors, and language, cooperation, and emotion are problems just as important to this understanding as a behavior, such as locomotion, that leaves direct evidence in the bones.

RUNNING AND SOCIAL LIFE

The importance of considering adaptation may be illustrated with three quite different examples. Patas monkeys are the greyhounds of the primate world. These elongated, fast-running cousins of the arboreal genus *Cercopithecus* live in the savanna, frequently a considerable distance from trees. Rapid running was believed to be an adaptation to this environment, but, when danger threatens, only the single adult male runs away, while the females and young freeze in the grass. The freezing to avoid danger is characteristic of some arboreal monkeys, but on the ground the arboreal adaptation would be suicidal, if it were not for the new decoy behavior of the adult male. The ground life of the patas monkeys is made possible not only by the locomotor adaptation, but also by a social life in which the behaviors are sharply distinguished by age and sex. The locomotor adaptation is successful only because it is a part of a particular social system.

FIGHTING AND HEAD STRUCTURE

To take a quite different kind of example, the head of *Homo sapiens* is well balanced on the vertebral column, and lacks a projecting face. In fossils with larger faces the head may be much less well balanced, and this has been extensively studied. There has been extended controversy over the question of the balance of the head in the genus *Australopithecus,* and what light, if any, this feature throws on bipedal locomotion. But it takes very little muscle to balance the head, and a minimal change in position would allow man to grow a much larger face. The problem takes an entirely different form if we consider what the animals are doing with their faces and how these actions are related to their way of life. In addition to eating with their faces, Old World monkeys and apes fight with them. In the males, large canine teeth and jaw muscles are correlated with big neck muscles, and the sex differences are maximal in ground-living forms in which the males defend the group against predators. The small nuchal area in *Australopithecus* probably has nothing to do with the balance of the head, but is additional evidence, supporting that of the teeth, that these forms did not fight with their faces, that the functions of dominance in the group and protection of the group had been transferred from teeth to tools. Consideration of the social functions of the face leads to a different interpretation not only of the face but also of the neck, and the study of structures as they function in the social group suggests many correlations that have not been made apparent by other approaches.

THE THREAT

A third example is communication of threats. In some monkeys the eyebrow is pulled back, exposing a light upper eyelid. This light color greatly increases the visibility of the gesture, making it obvious for a long distance. The brow motion may be coupled with motions of the ears and erection of the hair on the shoulders. Length of mane, ear muscles, and eyelid color are linked in the gesture of threat. Most of the primate characters that have been considered secondary sexual features are probably parts of the anatomy of agonistic display and fighting, rather than being concerned primarily with sexual attraction. These three examples show the importance of understanding social structure in evaluating running, of fighting in interpreting the structure of the head, and of threat in correlating apparently unrelated features. They demonstrate the need for considering behavior and adaptation when considering the evolution of anatomical features.

Tiger and Fox have stressed the importance of a zoological perspective in the social sciences. The reverse relationship should also be emphasized. Understanding the social system is the key to the adaptive nature of many anatomical natures. For example, it has long been known that man lacks the premaxillary-maxillary suture. The function of this suture is to allow for the growth that makes space for the large canine tooth. Reduction of the tooth is a prerequisite for the loss of the suture, and the reduction of these teeth is a consequence of the evolution of tools. Understanding the suture is important both biologically and socially. Examples could easily be multiplied, but the essential point is that the evolutionary meaning of apparently simple structures depends on the way they function and especially their function in a social group under natural conditions. If the synthetic evolutionary theory is accepted, it is impossible for physical anthropology to pursue its objectives without considering adaptation and behavior.

SEX AND SOCIAL BEHAVIOR

The central role of social behavior in adaptation and socialization—and in giving meaning to biology—may be seen in the nature of the social group. It was once thought that primate society was held together by sexual activity, and that the loss of a breeding season in primates was the key to their lasting social group. But it is now known that many monkeys do have sharply defined breeding seasons. In both the Japanese macaques and the Gibraltar macaques mating is in the fall and births are in the spring, giving the infants the maximum time to mature before their first winter. However, in spite of a limited breeding season, the social group continues throughout the year. (According to Japanese workers, there is much more troop organization in chimpanzees than is described by Reynolds and Reynolds or Goodall. A final judgement must await field work in other localities and the full publication of the reports from the Japan Monkey Center.) Where ecological adaptation favors a small foraging group, the number of adult males may be reduced to one, but, because of the male's nonsexual functions in defense of females and young, feeding groups always include at least one adult male.

Castrated males behave normally in most social situations, and this shows the importance of combining experiment with field work. Natural behavior is much too complex to be analyzable by observation alone. The idea that sex is the force that holds primate society together is refuted by the behavior of male rhesus monkeys, which may shift from one troop to another in the breeding season. Lindburg describes such shifts and notes that the males that moved were those that copulated most frequently. Even one alpha male, which appeared to be in control and receiving all the rewards a monkey can, shifted to an adjacent troop. Just as estrus disrupts the female's usual social relationships, so the male sex drive may be socially disruptive, rather than binding.

MOTHER-INFANT BONDS

The continuity of the social group depends on many factors, and one of the principal ones appears to be the lasting bonds between females and their young. In the few studies in which individual animals are known and where they have been studied for several years, siblings are observed to continue to associate with their mothers; this relation shows in patterns of grooming, resting, feeding, and probably sleeping. The importance of these persisting relations is that they give order to the society in addition to that which depends on the dominance of adult males. MacRoberts has shown that juvenile monkeys that have no protector (no mother, adequately older sibling, or interested adult male) have a difficult time gaining access to food and grooming, and are exposed at the periphery of the group. It becomes clear that even without any food sharing, a protected position in the group structure is of great practical importance to the juvenile primate. The importance of continuing relations between chimpanzee mothers and their young is described by van Lawick-Goodall.

BIOLOGY-BEHAVIOR

In my opinion, the importance of predation has been greatly underestimated with respect to its effect on group structure. Schaller has shown that more than 20 per cent of leopard scats and over 6 per cent of tiger scats contained langur hair. These figures imply a high rate of predation, and the sleeping habits of the monkeys and apes give clear evidence of the importance of this behavior. In short, it is so important for monkeys and apes to be social if they are to survive that they easily learn the fundamental behaviors. As Hamburg has put it, the behaviors that are essential for survival must be easily learned and pleasurable to the individuals concerned. It is not a question of biology against learning; the evolutionary process has produced (through selection) a fit between the biology of a species and the behaviors essential for its survival. Whatever particular form their social life may take, the monkeys and apes are profoundly social because social life has been a fundamental adaptive mechanism for millions of years. The tremendous importance of early learning, demonstrated by the experiments of Harlow and Harlow and others, shows the interrelations between environment and biology that have evolved in a feedback relation with a social environment. It is not a matter of either biology or learning but of the evolution of a biology that makes the learning of social behavior inevitable under normal circumstances.

It might be thought that the study of the behavior of the contemporary primate has little to offer to the interpretation of the bones of our ancestors, to the study of evolution. But, however important those bones may be to the paleontologist, from the point of view of evolution, they were important only when they were parts of living animals. The understanding of evolution comes from the appreciation of the life of past populations. To show the importance of behavior, and particularly of the field studies, I want to reconsider the problem of our ancestors' descent to the ground. The traditional explanation of human bipedal locomotion is that forests were getting smaller and certain arboreal apes were forced to become ground livers and bipeds; descent to the ground and the origin of bipedalism have been considered as parts of the same evolutionary event. But man's close relationship with the African apes suggests that a quite different evolutionary sequence is possible. Both chimpanzees and gorillas are knuckle-walkers. The gorilla spends its time on the ground, and only juveniles climb to any extent. The chimpanzee's diet consists largely of fruit obtained in trees, but the animals move from feeding area to feeding area on the ground. Our closest primate relatives are adapted to life on the ground, and it is possible that our ancestors were similarly adapted to knuckle walking. The Asiatic apes (gibbon and orangutan) rarely come to the ground and have no structural adaptations for ground locomotion; the anatomical adaptations that make knuckle walking possible are confined to the chimpanzee and the gorilla.

Traditionally, the apes have been viewed as primarily arboreal creatures, and their anatomy interpreted almost entirely as adaptation to arboreal life. Field studies show that this is not the case and that a sharp distinction must be made between the still-arboreal apes of Asia and the African knuckle-walkers. It appears that our ancestors were arboreal apes for many millions of years, that they then shared a common knuckle-walking stage with the ancestors of the chimpanzee and gorilla, and that only later did they become bipeds. The possibility of this order is suggested by field studies and by the close relationship of man with the genus *Pan*—shown by chromosomes, serum proteins, and albumins. Obviously, it cannot be proved at the present time that our ancestors went through a behavioral stage comparable to that of the living chimpanzee or gorilla, but there is a good chance that they did, and, if so, speculations about our ancestors' coming to the ground, the origin of their bipedalism and tool using, and their living in the savanna *all* need revision.

CAUSE FOR DESCENT

As noted above, the traditional explanation for our ancestors' coming to the ground was that because the forests were becoming smaller they had to come to the ground. However, this explanation does not account for the ground-living knuckle-walkers, which are limited to the forests. The larger the forests, the greater the extent of the forest floor and forest edge, and the more habitat for knuckle-walkers; the fossils of *Ramapithecus* (including *Kenyapithecus*) suggest a form with a range extending from Africa well into Asia, when the habitat was suitable for a tree-living form. Further, at least four different kinds of Old World monkeys have become primarily ground-living, and this appears to be an adaptation to new areas to avoid competition with closely related forms. If the evolution of the apes is viewed with an understanding of the behavior of contemporary monkeys, it appears probable that the many small apes known to have lived in the African Miocene became extinct through competition with African arboreal monkeys.

The advantage of the large size of the chimpanzee and gorilla is that they can compete with monkeys, but the price they pay is the necessity of spending much of the time on the ground. In Asia, where the numerous and highly successful monkeys of the subfamily Cercopithecinae were absent, the small apes—gibbons—continued, and the Colobinae became highly diversified, filling the ecological niches occupied by the Cercopithecinae in Africa. (This assumes that the baboons and macaques are closely related and are of African origin. The very limited variability of *Colobus* in Africa may be compared to the great variability of leaf monkeys in Southeast Asia.) The emphasis on human evolution and on the apes as our nearest relatives has obscured the fact that, compared to monkeys the apes are highly unsuccessful. The comparison holds whether the criterion of success is the number of genera, species, or individuals, or the number of substantially different ways of life.

In the evolution of locomotion, the interpolation of a knuckle-walking stage makes the transition from ape to man far easier to understand. Man moves slowly, compared with competing mammals. The predators that might prey on man can run faster than he can, and the ungulates, the principal prey of our ancestors, are far fleeter. It has always been difficult for me to see how a creature less fleet than man could have survived at all. But knuckle walking provides a possible intermediate condition that we know is moderately successful. The essential problem in bipedal locomotion of the human kind is that the hind legs must combine the functions of both the forelegs and the hind legs of the quadrupedal gait. Knuckle walking provides a kind of intermediate condition in which, if selection were for more bipedalism, the long arms might be used less and less in locomotion. The problem of being either a quadruped or a very inefficient biped (as are monkeys) is eliminated. It should be remembered that human arms are very long and that some individuals fall well within the range of variation of the contemporary knuckle-walkers; the earliest known human hand (found by Leakey at Olduvai) shows many of the features of a knuckle-walker.

THE MANIPULATIVE PRIMATE

It has long been recognized that the significance of bipedal locomotion was in freeing the hands for the use of tools—that it was the adaptive success of what the hands were doing that dominated locomotor evolution. Here again field studies offer new ways of understanding. It has been noted that monkeys of the genus *Cebus* appear to learn object manipulation easily, but Thorington has shown that extensive manipulation of twigs and branches is a normal part of the feeding behavior of these monkeys. The reason they learn manipulative behavior easily in the laboratory is that it is part of their normal behavior repertoire. This raises the point discussed by Hall—that some object manipulation is not necessarily a mark of great intelligence. As Oakley has stressed, it is skill in object use and the use of a wide variety of improved objects for different purposes that is unique to man. And, as the case of Cebus shows, the ease and conditions of learning the manipulative skills must evolve.

EVOLUTION OF MANIPULATION

A perspective on the problem of learning to use objects may be given in the following way. Primate hands evolved from claw-bearing paws early in the Eocene. For fifty million years in many families of primates, dozens of genera, and hundreds of species, hands were used in feeding, grooming, and locomotion, but, in all this time and in all these manipulative creatures, substantial use of objects as an adaptive mechanism evolved only once. Negative evidence suggests that a very special set of circumstances is needed to account for the beginnings of tool using, and that digital dexterity, although necessary, is a very small part of the explanation. The special situation is knuckle-walking.

Now, next to man, it is the knuckle-walking chimpanzee, among the primates, which performs the greatest amount of object manipulation. Chimpanzees build nests; use sticks to get termites, ants, and honey; use stones to break nuts; use both sticks and stones in agonistic displays; and throw both overhand

and underhand. (In spite of all the speculation on the origin of tools, the importance of the use of objects in agonistic displays as one type of origin only became apparent after recent field studies. The use of objects in agonistic display provides situations in which the effectiveness of the stone or stick might be discovered. Field studies suggest that the traditional notion of discovery, an event in a relatively limited time and place, is biologically wrong. Although agonistic displays may provide one kind of origin of object using, as shown by the chimpanzee and the gorilla, millions of displays over millions of years may have been required to provide the basis for the discovery.)

It would be a remarkable coincidence indeed if the genus that is closest to us in chromosomes, body chemistry, and anatomy should, through chance alone, also be the most similar in object manipulation (particularly when one remembers how unusual this kind of behavior is). Chimpanzees walk bipedally when carrying objects, and field observations fit the experiments and theories of Hewes. Knuckle walking permits a kind of carrying that is impossible for quadrupedal monkeys. In the knuckle-walking position the flexor side of the fingers is up, and the animal can transport objects without this interfering with the normal gait. An ancestral form that was more bipedal than the contemporary chimpanzee could have used hands to assist in locomotion and also to carry. Knuckle-walking gives an intermediate stage between not carrying and full bipedal carrying with the hands free. Gorilla mothers must help their infants cling for approximately six weeks, and chimpanzee mothers often help the infant in the first months. Here is clear evidence that using one hand to help the infant is much less of a handicap to the knuckle-walkers than to a monkey.

THE TOOL-USERS

Observations of the contemporary primates show that coming to the ground has taken place several times, and that many species of primates that are primarily arboreal spend some time on the ground, with no particular anatomical adaptations to this way of life. Field studies show that it is living away from the safety and refuge provided by trees that necessitates substantial anatomical and social adaptations. In the baboons and geladas protection is provided by the fighting ability of one or more males. The adaptation is anatomical (large canine teeth, jaw muscles, neck muscles, etc.), social (large males defending small females and young), and temperamental (aggressive males). The whole social-anatomical complex is practiced in play, to which the males have devoted hours each day over a period of years before the canine teeth erupt and serious fighting takes place. This investment in the playful preparation for fighting may be contrasted to the minimal play with objects. Even chimpanzees do not practice hitting with sticks or throwing stones and, without long practice, skillful performance is impossible. Juvenile humans wrestle and play in very ape-like ways but, in addition, manipulative skills are part of the play, and frequent repetition of manipulative acts is a normal part of play. If a species is to be away from trees and has to rely on stones or sticks for defense, the object must be in the hands ready for use, and it must be used skillfully. The change from fighting with arms and teeth to fighting with objects requires changes in learning and in play, and the acquisition of new skills. The knuckle-walking theory suggests that this long-continued transitional stage between ape and man may have taken place on the forest floor and on the edge of the forests, where flight to the trees would still be possible. According to this theory, living out in the savanna away from trees would have come long after ground living, knuckle walking, and tool making and using.

Just as the arboreal monkeys that evolved into baboons had to make major adaptations to live in the savanna, so our ancestors had to adjust to a very different life from that on the forest floor. Here important fossils give direct evidence that some of our ancestors had made the transition by two million years ago. The foot found by Leakey at Olduvai is remarkably human. The great toe is in line with the other toes, showing that the climbing adaptation for escape into the trees had been lost. In this same deposit there are stone tools, hands that show many ape-like features, and animal bones that suggest these creatures were supplementing a vegetarian diet with hunting. If protection of the group had been by the kind of fighting and social organization seen in savanna-living monkeys, very large canine teeth would be expected, but precisely the opposite is found. The canine teeth of *Australopithecus* are small—no larger than those of many later members of the genus *Homo*. The location of many *Australopithecus*-bearing sites in the savanna, stone tools, animal bones, small canine teeth, the foot, and other fragmentary limb bones, all suggest a form that had been a ground-liver and tool-user for a very long period of time, and this deduction from the anatomical and behavioral evidence is supported by the latest dating of the fossil remains.

Potassium-argon dates show that the *Australopithecus* stage of human evolution lasted for more than a million years (confirmed by fission track, and the stratigraphy of the Olduvai deposits), and my best guess is that our ancestors were small-brained bipeds, making stone tools and living in the savanna for well over two million years before *Homo erectus*. Of course, the fact that *Australopithecus* lived in the savanna does not prove that species of this genus might not have lived in the forest as well. If *Meganthropus* of Java is to be considered in the genus and, perhaps, of the same large species that has been found at Swartkrans and Olduvai, the species must have adapted to a wide variety of habitats. I doubt that a bipedal, tool-using hunter was limited to a narrow ecological niche.

In any event, there was very little evolution of tools during those two million years. Traditionally, object improvement was thought of as a stage that followed simple object use. However, termiting by chimpanzees involves both selection of material and improvement by removing any side growth that would hinder putting the stick in the hole. Again, field studies suggest that the postulated order does not correspond to reality, and that even the amount of object use seen in chimpanzees necessitates careful choice of material and some measure of improvement to fit the material to the task. It has been argued that animals with a brain the size of *Australopithecus could not* do things which chimpanzees are, in fact, doing! If the transition from an ape-like ancestor to *Homo erectus* took a minimum of five million years, we can obtain some notion of the time it took to evolve a biology that could learn to be human. To put the matter differently, the biology and the successful way of life evolved in a feedback relationship. The success of initial tool using, perhaps only slightly more advanced than that seen in the contemporary chimpanzee, led to selection for the biology that made tool-using possible, and study of the brain of *Homo sapiens* shows that large areas are associated with hand skills. The reason tool making evolved so slowly was that the brain had to evolve before the skills of *Homo erectus* were anatomically possible.

INTELLIGENCE AND SOCIAL COMPLEXITY

According to this view, the explanation of human evolution is to be sought in the feedback relation between successful behavior and the biology that makes the behavior possible. The most important changes are in the brain, which makes it possible to learn how to be human. We learn skills so easily that it is hard for us to appreciate the immense gap that separates us from the nonhuman primates. This gap is due not only to differences in the brain, but also to the development of human skills in a social system that encourages, trains, and permits the expression of the biological abilities in ways that are quite impossible for the nonhuman primates.

Ecology has been stressed as an important factor in the origin of tool using. I would stress the social environment, and the evolution of a social system in which skillful object use was rewarded. For example, in cultures where spear throwing is important, children practice this art in play. The children see the importance of the act in their parents' lives, and throughout the extended childhood of man throwing is practiced with adult encouragement. Laughlin has described how Aleut children must be taught and must practice from early childhood in order to be able to spear from a kayak. In nonhuman primates the absence of training and social reward means that the tool-using potential is never achieved. As Le Gros Clark has pointed out, chimpanzees can be trained to perform tasks that are beyond their unaided capacities; in other words, a human brain can guide chimpanzee practice and reward it to develop a new skill, but without human perception, training, and reward, the chimpanzee's performance is limited by both its biology and its social system.

The point is not only that the brain ultimately makes language and complex social life possible, but also, and at a much simpler level, that the perception of what is possible must evolve along with the biology that makes the actions possible. For example, in all the nonhuman primates the range of a social group is, by human standards, extremely limited. Groups of baboons and gorillas may range over parts of as much as fifteen square miles, but the ranges of most primates are far more restricted. These animals have excellent vision and can certainly see food and water beyond a normal range. They are not limited by locomotor ability because in foraging and in the process of daily walking they travel a total distance that is much longer than the length of their ranges. They are limited only by the extent of area they can know and occupy usefully. The human use of hundreds of square miles is the result of human intelligence and a human way of life, not of the greater efficiency with which man moves compared to a baboon.

It is easy to see the relation of motor skills to adaptation and to see how the adaptive success of skills could stand in a feedback relation with the brain. It is more difficult to appreciate that the same is true for other mental abilities, because we are limited by our biological nature. Let me try to clarify the matter by suggesting how an intelligence test might be constructed if the professors who designed it were rats. The maze would be designed to be solved by tactile hairs on the sides of the snout and many clues would be largely for the sense of smell. Tests would be given in the dark, when everyone knows one is most active and intelligent. We would all score zero, but our primitive primate ancestors could have solved that test. Evolution has not created intelligence in some abstract way; rather, our intelligence is based on particular abilities that evolved along with a succession of ways of life. What we think is normal is what succeeded in the ways of life of our ancestors, and the ability to control rage and sex and to cooperate is part of the biological basis for human social life.

In summary, human biological abilities are the result of the success of past ways of human life. Through the feedback relation between behavior and biology, the human gene pool is the result of the behaviors of times past. From the short-term point of view, human biology makes cultures possible. It poses problems and sets limits. But from the long-term evolutionary point of view it was the success of social systems that determined the course of evolution. Whether we study a bone or a stone tool, it had importance only as it mediated in the success, or failure, of behavior.

Field studies offer new ways to appreciate anatomy. They enrich understanding of the social group. They suggest that each stage in the behavioral evolution of man must be reconsidered in the light of our knowledge of our closest living relatives, the knuckle-walkers.

II. INCIPIENT ADAPTATIONS: MONKEYS AND APES

WHAT IS THE basis of the fascination that apes, gorillas, baboons, and other nonhuman primates hold for us? What draws large crowds to the cages of these primates at zoos? Is it that they are so like us and yet so different that the uncanniness of it draws us like steel filings to a magnet? Is it that most of us sense they represent an important aspect of our ancestry and that we hope to learn something about ourselves by watching them? It is that we are intrigued by their freedom and abandon within their resemblance to us—or our resemblance to them—and that we secretly (or not so secretly) envy them?

Perhaps there is some truth in all of these. But appearances—especially in the unnatural setting of a zoo—can be misleading, and especially to untrained eyes. In their original, free-ranging states nonhuman primates reveal much more formal social relationships and consistent patterns of behavior than can be seen in their captive state. However, careful study of their behavior in cages, as well as in their natural habitats, reveals a basic fact that is often overlooked: There are vast differences among monkeys, apes, gorillas, baboons, and other nonhuman primates. It is impossible to generalize about them as an overall group.

This fact becomes more strikingly apparent when they are studied carefully and systematically in their free-ranging and natural state because then it is possible to observe the ways in which they adapt to pressures in different habitats. Such observations add a more striking dimension to the differences among nonhuman primate genera: Even within each genus there are important variations that correspond to variations in habitational pressures. Thus, there are differences not only between baboons and gorillas but also among baboons in their social adaptations to different habitations.

One of the premises of any study of history is that the past is part of the present. This premise has a very specific meaning in studying the evolution of human adaptation: Human capacities and potentials for development were established in the past. The emphasis here is on potentials, which, depending on a variety of conditions, might or might not lead to further evolutionary developments. We have inherited capacities from our primate ancestors, not specific patterns of behavior and adaptation.

Biology and culture are so thoroughly interwoven in man that it is often difficult to determine where one leaves off and the other begins. Primatology—the study of our simian forebears—enables us to evaluate some of our hypotheses about the relationship of culture to biology in the adaptive process. In the broadest terms and on the basis of a comparative study of man and his ancestors, we can begin to formulate some hypotheses about the minimal biological conditions that culture must meet and about man's biological capability. Primatology will not provide us with conclusive answers; it can only assist in pointing us in the right directions. The answers to our questions about possibilities and impossibilities can only come from the study of man himself.

What are some of the cultural capacities or predispositions we might have inherited from our nonhuman precursors? Most people would assume, for example, that the rule of "women and children first" is a human cultural invention, but observations of the social organization of baboons suggest otherwise. Among the baboons described by DeVore and Washburn, when a troop is threatened by predators its members almost automatically assume a formation—not too dissimilar from that of an American infantry squad—that assures the safety of the females and the young. A comparison of many human groups with these baboons and other nonhuman species makes it possible to speculate that the rule of "women and children first" is not entirely a human cultural invention but is part of our inheritance from an evolutionary past.

How, exactly, are we to phrase this hypothesis? Shall we put it exclusively in terms of masculine strength and dominance—in terms of the fact that females often are "weaker" and preoccupied with the care of the young? These considerations certainly are part of the picture, but they seem to be insufficient. To rely exclusively on such hypotheses suggests not only that we are attributing human rationalizations to nonhuman animals but, perhaps more important, implies that we are thinking of man as exclusively social and without strong biological influences in his cultural activities.

There are several ways of looking at the fact that man is a biosocial being. In one view he starts his cultural life by trying to adapt to his inner biological milieu as best he can, that is, within the limits of his technology and habitat. He thus creates institutions in an attempt to assure a steady supply of food—the first task of man is to live—and to satisfy his needs for sexuality, social relationships, control, sense of being, and the like. He establishes regular means for shaping his children's minds in ways that will perpetuate his culture (man's desire to perpetuate his way of life is often taken for granted but the universality of this urge needs to be investigated and understood) and for amusing himself. But such an approach overlooks the possibility that there is a continuity from *Homo sapiens'* forebears to the present, and, in doing this, it assumes that man is a unique creation of nature with no ties to other living forms. Although we might enjoy thinking of ourselves in this way and be comforted by the thought, the evidence contravenes such an ideology.

There is another way of looking at man's essential biosocial nature. In the words of Tiger and Fox, whose article on "The Zoological Perspective in Social Science" has already been mentioned, man is "genetically programmed." Tiger and Fox (who disclaim any connection between their names and their scientific interests) suggest that the combination of evidence from the past and the present leads to the conclusion that much of man's social behavior is genetically programmed, that many of his patterns of living were established as limits and potentials in his nonhuman past, and that these patterns were transmitted as stored information to be elicited under different conditions. Although we have long known that the varieties of human behavior and social organization have limits, anthropologists have tended to focus almost exclusively on the differences, without—as Tiger and Fox note—studying the limits themselves. Is it not possible, they ask, that the limitations on the varieties of man's behavior and organization are genetically preordained? Their hypothesis leads them to speculate that if we join what we know about living people with what we have recently learned about nonhuman animals, we can open new avenues in the study of the evolution of human culture and social organization. They suggest that this might lead to some fundamental changes in our assumptions about the nature of man.

Similarly, as this conceptualization suggests, man's genetic programming consists of a set of potentials and limits. Its responses will be determined by the nature of the demands made by the cultural and physical habitats. Thus we cannot know what adaptations man is capable of—and of what he is incapable—until his genetic programming has been challenged by specific habitational stimuli.

As noted above, social scientists have tended to focus almost exclusively on the varieties of human behavior and social organization. But are these variations understandable in their own terms only, without reference to man's habitats and biological constitution—that is, his genetic programming? Are not the variables in human functioning basically variations on genetic predispositions? Might we not understand more about language and communication generally, family and household organizations, kinship and other aspects of social relationships, modes of perception and cognition, if we placed these variations in the context of man's inherited predispositions? Would we not thereby understand much better the limits of variability—what man is incapable of—and thus the essential nature of man? Tiger and Fox are explicit: these questions should not become the exclusive concern of social science. No science can exist very long if it asks only one kind of question and if it limits the imagination. What is being proposed, instead, is that questions such as these be assigned their rightful place in order to further our understanding of man. These questions require that we reexamine some of our most cherished scientific beliefs. But this is just one example of adaptation; scientists must adapt, as must foragers and farmers.

Now let us return to the rule of "women and children first" and try to uncover some of the implications of this conceptual framework for human social relationships. The pattern of adult male protection of females and children among baboon groups is inseparable from the dominance of adult males within the troop. Of course, there is no evidence of a "suffragette" movement among nonhuman primates. Does this mean that demands for social equality by women under modern sociocultural conditions are in contravention of our genetic programming? Simple-minded "social Darwinists" would of course argue affirmatively, but an evolutionary perspective suggests otherwise.

It will be recalled that one of the conclusions of the study of adaptation in evolutionary terms is that man gives the impression of trying to free himself from the confining limitations of his genetic endowments and habitats. Although we may now believe that we are genetically programmed for the rule of "women and children first," it must be remembered that this is a *biological* predisposition, which has often been used as a rationalization of the *social* principle of "women and children last." The latter tenet certainly had considerable adaptive value—for preserving and furthering life—in premodern societies, but the conditions of modern life make it possible to transcend, or at least modify to a considerable degree, the limitations that inhere in this genetic programming.

Similarly, Tiger and Fox suggest aspects of masculinity that are genetically programmed. Cultural anthropologists commonly maintain that "masculinity" and "femininity" are determined almost exclusively by the ways in which cultures shape the minds of their growing members. Tiger's and Fox's suggestion leads to the further hypothesis that basic features of "masculine personality" are to be found in all societies, each group placing its own stamp on this common male-psychological predisposition. Can the same be hypothesized for women? In other words, is there also a basic feminine personality structure that is an inheritance from our nonhuman primate past? Are there features of personality that all women share, which are further cross-cut by personality traits that are imposed (or elicited) by experiences in particular social groups? Where does biology leave off and cultural influence begin?

I have stressed these questions in order to provide an example of one approach in exploring the relevance of behavior among nonhuman primates to *Homo sapiens*. There are many other related questions, such as population size and density and family organization.

A major concern in modern social planning, especially in complex industrial societies, is the problem of population size, density, and crowding. How much crowding can human beings tolerate? What is the relationship between population density—and tolerance of it—and habitat? In other words, what are the limits on variability in population patterns in the species? By and large, anthropologists have focused on the relationships of population size and density to rainfall, v il le food, levels of technology, and the like. Here, too, we observe the implic assumption that *Homo sapiens* is unique in nature and operates according to principles that are not found among other animals. But is it not possible that human social behavior in relation to population dispersion and density can be seen on a continuum from the nonhuman primate past to the cultural present, and that if we are to carry out any intelligent planning for the future—which is a realization of one of our evolutionary potentials—we must know what limits have been imposed on us by our genetic programming and to what extent we might be able to transcend these limitations?

Among nonhuman animals the size and organization of groups and the nature of social relationships are determined almost exclusively by habitational conditions. For example, forest animals (including primates) who subsist primarily on insects and who are primarily active by night are generally characterized by solitary habits and dispersed populations (Lepilemurs of east and west Malagasy [Madagascar] are among the best-known examples). By contrast, lemurs who live in the forest but subsist on fruit or leaves and are active by day or at dusk and at twilight live in small family parties that are based on a single male. These animals occupy territories in groups and display defensive behavior toward neighboring groups. One of the reasons for the differences between *Lepilemurs* and *Lemurs* is that an insectivorous diet requires individual hunting but fruit or leaf-eating habits make it possible for small groups to congregate and exploit a territory in common.

Once primates begin to edge out of the forest to live in the open they display commensurate changes in behavior and population size. For example, *Lemur macaca* and gorillas, which live in the forest or on its fringes, are active by day in multi-male groups that usually are small, and they subsist on leaves, fruit, and stems. Other primates (such as vervet monkeys and some baboons) that live at the forest fringe or on plains with scattered trees usually are vegetarian but occasionally are carnivorous. They live in multi-male, medium-to-large groups. Primates that live on grasslands or dry plains (such as *Rhesus* monkeys and *Papio* baboons) also are found in medium-to-large groups—in which each reproductive unit usually is dominated by one male—that alternate between vegetarian and omnivorous diets. Life on the plains requires large units because these animals face large numbers of predators, and there is safety in numbers.

Although we cannot here consider the ways in which man—through his capacity for culture—has transcended the limitations imposed by his habitats on the size and organization of social groups (this will be one of our major concerns in later sections), we will note that there are correspondences between man's changes in his relationships to natural habitats and his social organizations. This suggests another important continuity that has been handed down from our nonhuman ancestors. If this continuity is as real as it is apparent, more extensive comparisons of nonhuman and human groups with respect to the relationships between social organization and habitational conditions might increase our understanding of man's potentials in this regard. Of course, more factors affect the nature of human social relationships than those of the habitat, but it is nevertheless possible to examine the latter as one set of variables.

Important in this connection are the implications of an experiment that is taking place on a massive scale—though it is not deliberate experimentation—and that should be evaluated in the light of man's genetically programmed limits and potentials. In the United States, because urban areas are permitted to grow and expand in density and size without control, gigantic metropolitan areas extend for hundreds of miles and contain millions upon millions of people. By contrast, Mexico and the Soviet Union have adopted codes that forbid the introduction of new industries into their capital cities, and they also have discouraged the migration of workers into these cities. Because the study of primate species and human social systems demonstrates that there is a relationship between population density and social organization, this experiment-in-nature—uncontrolled expansion in some societies and controlled growth in others—is a demonstration of man's abilities to regulate his relationships to the habitat on a scale and to an extent previously unknown. Can we not apply some of our scientifically derived concepts to a comparison of such societies in order to learn the consequences of controlled and uncontrolled growth for patterns of social relationships, including family organization, disease and epidemiology, and individual productivity and creativity? Which social policy is more adaptive—in the sense of perpetuating life—to modern conditions? Many other examples could be cited, but the point I stress is that there is much to be learned in this connection; it is another social science frontier that should be explored.

My last example of the importance of primatological studies for the study of human adaptation concerns family organization. As we see in Selection 12, "families"—as we conceptualize them—seem to develop as an aspect of the organization of social relations among nonhuman primates in response to food shortage. When food is scarce, single-male groups seem to become the basic social units, each male providing for his dependent females and offspring. Anthropologists have long maintained that the human family is primarily an economically oriented institution, whatever its other functions. Hence it appears that man's potential for family organization is genetically programmed and is not unique in nature. If this is indeed the case, might we not begin to reconsider the role of the family in habitats of abundance and affluence? Much of our contemporary ethic concerning the family arose in medieval times, when economic scarcity was the rule of life for most people. Is it possible—in light of the evidence of the evolutionary record—that the family is no longer adaptive under modern conditions?

The reader who wishes to broaden his knowledge of primate behavior can consult several sources. A very good paper is Irven DeVore's "The Evolution of Social Life" (in *Horizons of Anthropology,* edited by Sol Tax [Chicago: Aldine, 1964]); DeVore is one of the leading investigators in this field. A highly technical review of current knowledge about the social organization of nonhuman primates in relation to environmental conditions is a paper by J. H. Crook and J. S. Gartlan, "Evolution of Primate Societies" (*Nature,* 210 [June 18, 1966]: 1200-03). DeVore has edited an excellent collection of original studies by several students of primate patterns, *Primate Behavior: Field Studies of Monkeys and Apes* (New York: Holt, Rinehart and Winston, 1965). The reader who is interested in the history of primatological studies should consult Sir Solly Zuckerman's pioneering work, *The Social Life of Monkeys and Apes* (London: Kegan Paul, Trench, and Trubner, 1932). But it is not only nonhuman primates that are of relevance; the reader who wishes an even broader introduction to this field should be familiar with W. C. Allee's *The Social Life of Animals* (New York: Abelard-Schuman, 1951) and W. H. Dowdeswell's *Practical Animal Ecology* (London: Methuen, 1959).

8. ADAPTIVE FUNCTIONS OF PRIMATE SOCIETIES

HANS KUMMER

Reprinted from Primate Societies: Group Techniques of Ecological Adaptation, *pp. 39-68 (Chicago: Aldine Publishing Company, 1971). Hans Kummer is presently with the Zoological Institute at the University of Zurich. His principal research interest is in the social organization of terrestrial primates in a comparative and evolutionary framework. At present he is investigating the spatial arrangement of individuals in primate groups.*

■ Self-perpetuation is the first concern of every individual of every species. In this selection Kummer concentrates on two central problems—feeding and defense against predation—that affect this requirement of evolutionary success. As with all matters of adaptation, these must be examined with the context of a species' habitat. The milieu in which any primate resides contains both natural food resources and carnivorous predators for whom the primates are themselves a food resource.

Against this habitational backdrop, Kummer leads us through a description of group behavior that is highly adaptive. Kummer examines the group size and structure of several species showing how and why they are distinctive. The hamadryas baboon, for example, adopts three different group sizes in coping with its savannah habitat, the one-male group and the band for feeding purposes in different situations, and the troop for defense against nighttime predation. As Kummer observes, the one-male foraging group among the hamadryas baboons suggests that pressures from predators seem to be somewhat less important for this species than the problem of feeding.

The patas monkeys—for which the pressures of predation seem to constitute the principal problem—represent another adaptation to savannah life. In contrast to the continuity of hamadryas groups, the patas monkey unit relies primarily on a capacity for dispersal. While the one-male group of the patas constitutes a foraging unit similar to that of the hamadryas, the baboons fuse into a single unit for protection against predation whereas the patas disperse, with the male trying to lead the predator away from the rest of the group. Significantly, the male patas monkey has the greatest running speed of all primates; this enables him to distract predators from the females and young and then himself evade his hunter.

A lucid introduction to the study of primate adaptation is provided by Sherwood L. Washburn and David A. Hamburg in "The Study of Primate Behavior" (in *Primate Behavior: Field Studies of Monkeys and Apes,* edited by Irven DeVore [New York: Holt, Rinehart and Winston, 1965]), in which the authors propose a taxonomy of primates based on patterns of behavior in contrast to other taxonomies that are based on anatomical structure. The reader will find the rest of Kummer's *Primate Societies* (from which this selection is taken) very rewarding. *Social Groups of Monkeys, Apes and Men,* by Michael Chance and Clifford Jolly (New York: Dutton, 1970) discusses these primate groups from a somewhat different perspective. The detailed studies of communication, social organization, and mother-infant relations in *Primate Ethology: Essays on the Socio-Sexual Behavior of Apes and Monkeys,* edited by Desmond Morris (Garden City, New York: Doubleday Anchor, 1969) also deserve attention. Two excellent collections of papers, both edited by Phyllis Jay Dolhinow (New York: Holt, Rinehart and Winston, 1968 and 1972) are *Primates: Studies in Adaptation and Variability* and *Primate Patterns.* ■

THE TERM "GROUP" IN PRIMATOLOGY

IN DEFINING THE primate group we must accept the limitation of criteria which transform every anthropological term when it is applied to animals. All subjective phenomena of group life—like "identification" or "identity of aims"—are lost for the study of animal groups, because their members lack the symbols for their expression. Definitions of animal groups are restricted to so-called objective criteria because they are the only criteria available. Only the student of human societies is in the enviable position of using and comparing both kinds of characters, what people do and what they feel.

Most students of primate behavior would agree that there are two proven parameters of describing the grouping tendencies of a population: the distribution of its individuals in space, and the frequencies and types of directed communicative acts among them. The use of these parameters may be illustrated with the one-male groups within a troop of hamadryas baboons. Repeated estimates of distance show that the mean distance between a female and her male in the resting troop is about 2 feet, whereas the average distance between two troop members chosen at random is about 10 yards. The members of a one-male group obviously follow each other closely and thus remain together within the crowd of a hundred or more troop members. Even so, a photograph of a troop rarely reveals the one-male groups as spatially separate units, since usually no empty space is left between them. The parameter of communication, however, shows that the one-male groups are distinct units even with regard to their nearest spatial neighbors: The adults of a resting one-male group socially interact with others in about 50 of 100 observation minutes. Of these 50 minutes only 3 contain interactions with troop members outside the actor's own one-male group; the other 47 minutes are spent in interactions *within* the group (Figure 8.1). Thus, a primate group can be defined as a *number of animals which remain together in or separate from a larger unit and mostly interact with each other.* Usually, one is also interested in a group's stability over time, that is, in the persistence of the two parameters.

This definition merely permits us to identify and describe primate groups. Description must be followed by investigations on causation and function, two aspects of living things which biologists fervently try not to confuse. This chapter deals exclusively with the function of groups: It asks what group life does for the species, not why and how it comes about.

GROUP SIZE AND THE RESOURCE UNIT

Among the few well-quantified facts on primate social life is the size of their groups, and we may begin by asking how the number of individuals in the group may affect its success in daily life.

To survive, one must exploit resources and avoid injury. For a monkey or ape, this includes drinking and eating, on one hand, and avoiding predators and using safe roosting lairs, on the other. Among primates, every group member has to feed, drink, flee, and climb a tree for himself. There is no sharing or passing of gathered food or prepared shelter. Mutual assistance is negligible or absent. Only infants are carried and nursed. In such an economy, the group seems materially useless.

However, before the primates can react to resources and dangers, these must be located, and this is group life's first important function. Primates do not exchange resources, but they exchange information about them. For example, the discovery of a small waterhole by one baboon is immediately communicated to his neighbors by the unique posture of the drinking animal: head low, rump and tail in the air. The agitated hand movements of an animal digging up preferred food also attract his fellows. But discoveries of one group member benefit the others not only there and then. During a time of drought, an old male may lead the way to a distant pond which he remembers from a long-past visit in a similar situation. It may not be accidental that primate males, who are generally more active than females in leading the group, are also more likely to travel far and alone, especially when they are young.

How does group size affect the success of sharing information? It seems at first that

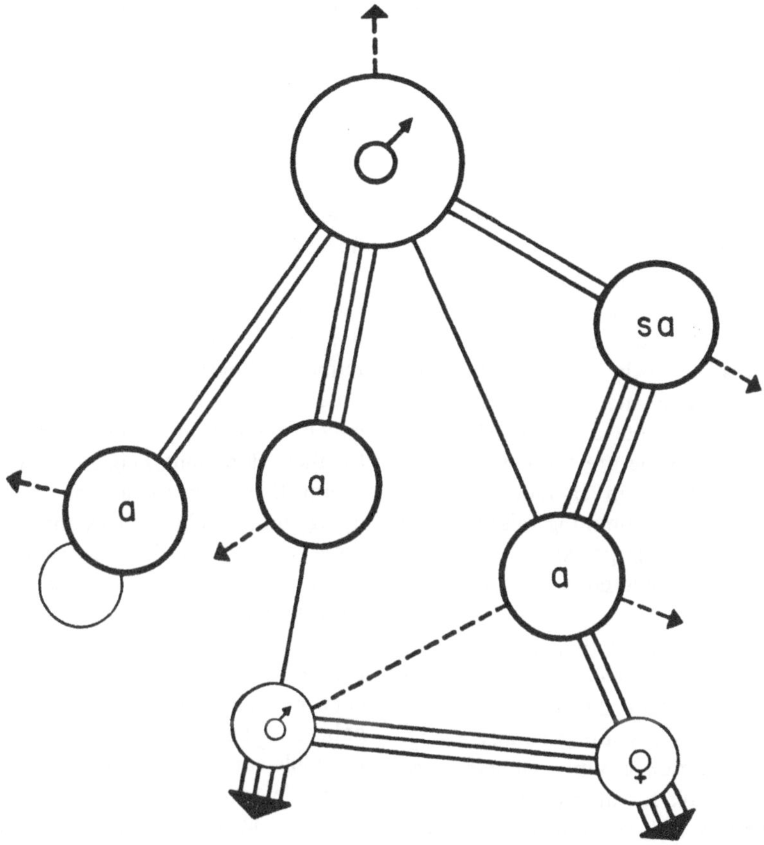

FIGURE 8.1

Sociogram of a hamadryas one-male group. The number of lines connecting the individuals represents the frequency of social interactions among them. The broken arrows indicate that the male (♂) and his adult (a) and subadult (sa) females interacted with outsiders in less than 3% of the observation minutes. In contrast, the two juveniles (bottom of graph) scored 40% in interactions with other groups.

the larger the group the better, since the amount of shared information would be roughly proportional to the number of members. But actual group sizes indicate that there must be limiting factors. The most important among them is probably the distribution of resources in the habitat. In discussing this, we shall for the moment disregard all factors except food.

When a monkey finds a small immobile food item, such as a nut, he will feed only himself; there will be no use broadcasting the discovery. If such nuts were regularly distributed over the entire range, no individual would ever learn anything about the location of nuts that the others would not know already. There would be no information to share. This may be one reason for the relatively small group size of primates in arboreal forest species, where food seems more evenly scattered than in savannas. If, however, the nuts occur only in certain spots and in quantities that could feed more than one monkey, then the group will benefit from his sharing such information.

My first thesis is, therefore, that foraging in groups is adaptive only if food appears in clusters or units that can feed more than one animal. We approach the question of group size by asking *how many* individuals can simultaneously feed on such a food unit. There is no point in attracting twenty group members to a little bush that can feed only three.

There are three ways in which primates appear to solve this problem. First, most species use no calls when they find food. Thus, primates who find food attract the attention of only those group members who can see them at that moment. The message spreads slowly and reaches the entire group only if the food is abundant enough to last for quite a while. The only exception known to me are the chimpanzees of the Budongo forest in Uganda. The tremendous calling and drumming raised in this population when a large stand of fruit-bearing trees is discovered seem to attract other parties.

The second mechanism that limits clustering is that feeding primates tolerate others only beyond a certain distance, usually a few feet. The distances are generally recognized by the inferior group members, who will, therefore, look for another food source when the first is fully occupied. This technique is satisfactory in a relatively rich habitat where food units are not too far apart from each other. It is a poor solution in a barren area where food sources are widely separated. The inferior animals would be left outside and hungry at each food source the group would pass.

In such conditions, a third solution is appropriate. It consists in adapting the size of the foraging group to the size of the food units by means of social organization. My thesis is that under harsh feeding conditions—and this is where we must expect the most precise adaptive fit—group size will closely approach the maximum number of animals that can simultaneously feed on the most crucial type of food unit.

The actual group sizes, however, vary widely even within the same species and in the same type of habitat, and each group maintains its size through all its activities for many months. These data are not very helpful. There are, however, a few species that *change* the size of their group according to their type activity, such as the chimpanzees, the geladas and the hamadryas. Here, we may find some clues.

In hamadryas baboons, the correlation between group size and resource unit is relatively clear. In their Danakil habitat, the flowers and beans of small acacia trees are the main food source. One regularly observes that an isolated tree is picked by a single one-male group, about five animals. This number allows the baboons to keep their usual feeding distance of several yards, at which low-ranking animals are not impeded by their dominant neighbors. The one-male group thus appears as the "single-tree-foraging unit." The groves of ten or more large acacias on the larger river beds are usually occupied by one band at a time. In the dry season, waterholes become the critical resource unit. These river ponds are then situated miles apart, but most of them cary enough water for a hundred or more baboons. Though hamadryas troops rarely assemble on their daily route, and never assemble to drink during the rains, they do assemble at waterholes in dry months.

The second important observed function of troops is their optimal use of sleeping-cliffs. Hamadryas cliffs, a large resource unit, are scarce in some areas, where, unless the population makes full use of them, roosting-space becomes a population-limiting factor. In such areas, hamadryas form troops of up to 700 animals, aggregations that are much larger than suitable for any foraging unit in the area. Comparisons of troop size in areas with many and few cliffs bear this out: In the Awash Valley, where food is scarce but canyon cliffs are almost continuous, troops are one-fifth to one-tenth as large as the troops in the nearly cliffless country around Dire Dawa, where the food situation is improved by agriculture. Although these hypotheses await quantitative testing, we may speculate that there would be no troops among the hamadryas if their roosting-lairs were small and widely separated, each accommodating but a few animals. Anubis baboons, who have sufficient roosting-trees in their typical habitats, do not form troops but make up groups the size of a hamadryas band. When drought forces anubis baboons to use the same waterhole, the groups approach each other with caution. The society of hamadryas baboons thus includes social units of three sizes, each seemingly fitted for a particular type of resource unit.

Patas monkeys offer another example of the relationship between group size and the size of the roosting facility. A population studied in Uganda lives in a savanna with no cliffs or high trees. The patas monkeys must roost on trees which apparently are too small even for their one-male groups. Instead of merging for the night like hamadryas, the patas one-male group splits up to sleep singly or in pairs over a wide area of the savanna. If patas one-male groups meet at waterholes, they threaten and chase each other. Their social behavior is programmed for dispersal, while that of hamadryas baboons is designed for both dispersal and fusion. The gelada baboons of the treeless Ethiopian highlands are of flexible group size. In this species, the society is organized on two levels, the large troop and its component one-male groups and all-male groups. In areas and seasons of good food supply, the geladas tend to form troops. When food is scarce, the troops split up into the smaller units which live and forage independently.

Chimpanzees are the third species that rapidly alters group size by fusion and fission. According to Reynolds, this system "allows dispersion and aggregation to exploit seasonal variations in availability and distribution of fruit," which would be compatible with our thesis. Chimps are, nevertheless, a special case in that they are the only known forest primates that form a fusion-fission society. It is difficult to see why forest-living monkeys are not equally flexible but stick to one-level societies. The answer may lie in the causal instead of the functional dimension. As I shall show for the hamadryas, multi-level societies are difficult for monkeys to realize. The presence of smaller units in a larger group is a constant source of conflict, since units and subunits are stable, rigidly closed, and potentially hostile toward each other. Fissions and fusions are possible only along strictly determined seams. Among the apes, with their disposition for tolerance and truly open groups, foraging units change composition freely. This difference between monkeys and apes may perhaps explain why fusion-fission societies occur only in those monkey species where a harsh environment enforces them by unusually strong selective pressures.

In summary, deduction and examples indicate that group size is favorably adjusted to the size and dispersal of resource units. If the same habitat offers needed resources in units of varying size, it seems that its inhabitants would fare best with a flexible organization. There are several possible explanations of the fact that most primate species live in one-level groups of constant size. For example, resources may be evenly dispersed; or only one type of resource unit may be scarce enough to require an adaptation of group size; or the phylogenetic heritage of the species may not include the behavioral mechanisms that permit a multilevel society, in which case group size would approach a compromise between the requirements of several resources.

THE GROUP AND ITS PREDATORS

Predators are much more important in the ecology of nonhuman primates than in the life of man. While man belongs to the hunters, the smaller, unarmed monkeys are among the hunted. Terrestrial primates like the baboons cannot afford a division of labor that would separate the sexes for different foraging tasks. Without the presence of their large, canine-toothed males, the females probably could not forage without losses to predators.

The problem of predation is most critical for terrestrial, open-country primates. We know of two major techniques they have evolved for dealing with predators: the baboon way and the patas way. Among the usually rather noisy baboons, the emphasis is on discovering the predator, rather than avoiding discovery by it. Once contact with the predator is made, baboons may attack and mob it rather than fleeing indiscriminately and under every circumstance.

Concerted, defensive actions by a baboon group against jackals, dogs, or leopards have been observed repeatedly. Typically the large adult males interpose themselves between the predator and the rest of the group, barking at and displaying their impressive teeth by opening their jaws widely. In the Amboseli National Park, the Altmanns observed a leopard as it jumped from the undergrowth into a group of yellow baboons at the edge of a waterhole: "The baboons sprang away, then turned on the leopard, barking loudly as several members of the group ran at the leopard. At one moment, the dominant male was closest to the leopard. Faced with this mass attack, the leopard turned and ran." An adult male, a subadult male, and a juvenile male were wounded in this encounter, but all recovered. In another instance a chacma baboon group was threatened by dogs. The large adult males immediately interposed themselves between the troop and the attackers: "It was not uncommon for a single dominant male to maim or kill three or four large dogs before retreating in the direction taken by the troop."

What, theoretically, is the optimal group size for these defense tasks? Unlike food, water, or roosting sites, the predator is mobile and can become a danger for many group members within a few minutes. The discovered predator merely shifts its place, whereas a resource unit, once discovered, is exploited and thus becomes irrelevant to other group members. This explains why the approach of a predator, in contrast to the presence of food, is signalled at the maximum distance the baboon's voice can carry. For the task of discovery, the ideal "anti-predator unit" is very large, since every pair of eyes helps and the more baboons reached by the discoverer's warning bark, the better. Often a discovered predator immediately leaves the informed group and tries its luck somewhere else. Its chances will be better if its prey is found in many small groups that are out of communication range rather than in one large group.

The task of mobbing and discouraging the predator also seems to favor the large groups of 50 or more generally formed in most baboon species. However, the reader will without difficulty find a score of variables that a detailed model would also have to include, such as the number and technique of cooperating predators, their visibility, the advantage of temporary large groups that can feed the existing predators with diseased or weak members alone, the negative effects of needless alarm over great distances, and so forth. Within the size range of baboon groups under savanna conditions, however, we may assume that the larger group is a safer group. Predator pressure thus will tend to increase the group sizes elected for resource conditions. The small foraging units of hamadryas baboons support our impression from the field that predator pressure in the semi-desert is less important than the food problem. But we nevertheless observe that the one-male groups forage together in larger, safer units wherever food conditions permit.

This pattern differs from that of the patas monkeys of Uganda studied by Hall. The small, isolated one-male groups of these tawny, long-limbed monkeys are silent and furtive. They do not face an approaching danger as a group. Instead, the alerted male climbs into the upper branches of a tree from which he scans the area. From this exposed vantage point, his size and the white color of his thighs make him rather conspicuous. His further behavior in the presence of a human observer suggests that his role is not only to watch for danger, but also to divert attention from the group. He noisily and conspicuously bounces on the branches, making them shake vigorously. He may also descend from the tree, produce a "whoo-wherr" growl, gallop very close past the observer, or run across the savanna far away from the intruder and the group. As the male engages in such distracting displays, the females and juveniles silently remain in their places, often lying flat on their bellies in the grass. Their last resort in avoiding predators, however, is their tremendous running speed. In this respect, the patas monkeys clearly beat the baboons and possibly all other primates.

The patas technique of avoiding predators is to see but not be seen. While the first is best accomplished by a large group, the second function requires silence, a high degree of dispersal, and small groups. That the patas follow the second course is probably because their resource conditions favor small groups. The loss in the number of eyes is compensated by the degree of watchfulness of the male, who seems to be more alert than any single baboon male. Both in captivity and in the wild, the patas male usually stays at the periphery of the group, sometimes at a considerable distance, and often facing away from it.

Among the ground-living primates, the patas monkeys' well-defined roles of attention-diverting males and hiding females is unique. A somewhat similar role distribution is found only in some arboreal species of guenons *(Cercopithecus).* For example, when a one-male group of *Cercopithecus nictitans* is disturbed, its male emits loud calls and approaches the disturbance, while his group withdraws from him and the danger. Gabonese hunters provoke the approach of the group male by shaking branches and thus improve their chances of shooting him. Like the patas male, the nictitans male diverts the danger rather than leading the group.

The baboon way and the patas way of dealing with predators are relative, not rigid specializations. One day in the Awash Park in Ethiopia, we observed a large group of anubis baboons running in a dense pack to cross an open plain. The baboons are not fast or persistent runners, so after only about 400 yards, an adult female dropped out in the middle of the plain, apparently neither wounded nor sick. Unable to flee at our approach, she chose the alternative technique of crouching quietly in the concealing grass like a patas. On the other hand, adult patas males in West Africa have been observed actively defending their groups. In one case, a patas male chased a jackal which carried an infant patas in its fangs. The jackal soon dropped its prey, and the infant, apparently unharmed, was retrieved by a patas female.

These examples of foraging, roosting and predator-avoidance suggest that large or small groups can be ecologically advantageous in one functional context but maladaptive in another. When a population flexibly shifts between different group sizes, we can expect these group sizes to relate to several important factors in the environment. If only one type of group is maintained, its size may fit a single most important condition, or it may be a compromise.

Primatologists used to formulate hypotheses relating an average group size of species and the overall character of their habitat, such as "forest" or "savanna." Unless the various and perhaps conflicting requirements of these habitats are identified, such hypotheses are questionable. Most of them have not been supported by recent data. For example, ground-living species generally form larger groups than species living mainly in trees, but there are exceptions, such as the extremely terrestrial patas monkeys with their small groups, or the forest-living *Cercopithecus l'hoesti,* which is more terrestrial but lives in smaller groups than other *Cercopithecus* species in the forest. Several arboreal monkeys tend to form mixed groups, consisting of two or three groups of different species. Apparently, some ecological factors favor large groups even in the forest. The tiny arboreal talapoin monkey and the large terrestrial drills are forest primates that have both been observed in groups numbering more than a hundred.

COORDINATION OF INDIVIDUAL ACTIVITIES IN THE GROUP

A major ecological value of the primate group is communication about resources—regardless whether the information concerns the location of a strand of grass ears that will be gone a minute later or the position of a cliff that will remain valid information for a long time. Though the information is carried and passed on, the resources themselves are not. With the exception of defense against predators, each animal has to take care of his own needs, but it must do so within the group, and this is not quite so easy as it first seems.

Imagine a group living in an ideal forest habitat where food, water, and safe places to sleep occur in small portions that are densely and regularly distributed over the area. Under such conditions, each animal finds what it requires within a few steps. It does not have to leave the group to satisfy any need. It can eat, drink, and sleep at any time, quite independently of what others are doing.

Take now a similar group in a different habitat, where resources are clustered and the clusters are far apart—so far, in fact, that communication between neighboring resource sites is impossible. When a group is feeding at one site, a thirsty member cannot go to the nearest water without losing the group. In this situation, group life demands that all its members do the same thing at the same time, that they adhere to the same schedule. There is a possible danger here: If an animal does not drink when its group seeks out the waterhole because it is not thirsty yet, it will suffer from thirst before the group reaches water again, unless it separates from the group.

Group life under such conditions requires a secondary adaptation that makes the primary adaptation of group life possible: It depends on a behavioral mechanism that induces the individual to do as his fellows do. Such a mechanism, called *social facilitation* by ethologists, is widespread among all species. It makes young chicks eat more when they peck in company, birds fly off in flocks, humans get angry or yawn when they see others do so. Social facilitation synchronizes activities which, as such, could very well be carried out individually and at different times. It can be observed or expected whenever it is advantageous that every group member take up the activity of the majority.

Social facilitation should not be confounded with the much more difficult behavior of *imitation*. When a human yawns in response to another's yawn, he does not need to concentrate on the performance of his partner in order to know how to yawn. He has the genetic program for yawning, and he could yawn even if nobody ever showed him how it is done. That in social facilitation the releasing stimulus coincides with the released response is, as it were, accidental and does not improve the quality of the response. (The reader may by now begin to yawn himself with nothing to act on but the written word.) In social facilitation nothing is learned, and the response can be unconscious. In contrast, imitation is the copying of a novel or otherwise improbable act for which the imitator has at best small fragments of a genetic program. Learning a difficult dance is an example. Imitation requires close and most probably conscious attention. Most birds and mammals show social facilitation, but imitation is found only among humans, the great apes, and, to a very limited degree, in monkeys.

One would expect the highest degree of synchronization in vital activities such as fleeing; and indeed, a startled response of a baboon is taken up by his neighbors within a fraction of a second. Feeding, drinking, and beginning to move spread at a more leisurely rate. Warning calls and mobbing behavior are also "contagious," but juveniles and females do not always participate. A monkey scanning the more distant surroundings from a tree will usually provoke no similar behavior unless he signals a discovery. Thus the degree of coordination in the group seems related to two factors: how critical it is that every individual perform the activity, and the time available for the response to take place.

In an extreme case, a single group member can take over a "task" for the entire group. Leading the group and scanning the distant surroundings are social functions that make group life most advantageous. In terms of the group's time budget, it would be inefficient if most members constantly sat on barren termite hills and looked for leopards. One or two scanners plus the limited alertness of their foraging fellows must suffice. In these cases, group members differentiate their action patterns; they assume "roles."

It is interesting to ask whether there is a counterpart to social facilitation in such one-for-all tasks, a *social inhibition* that would reduce an animal's tendency to perform an activity when it sees another group member already attending to it. With respect to temporary roles like sentry activities, the question has not yet received specific attention by field students, but it is known that permanent roles in the group are subject to such mechanisms. When a group has a leader, another male may not lead the group or scan the surroundings until the day that the leader dies, but when this happens, he may take over the leader's role within hours. One must assume that the previous leader's activities inhibited leading behavior in other members of the group. Such discouraging effects are quite obvious in some social roles. Macaque mothers or hamadryas harem males go so far as aggressively preventing others from sharing their roles toward their infants and mates.

Thus, social order provides a complete scale of coordinating mechanisms, ranging from strong encouragement to violent discouragement of "doing likewise," depending, it seems, on the number of animals that a task requires.

DOMINANCE

The extreme form of social inhibition is known as *dominance*. Its ecological function is to clarify the situation when the same action cannot be carried out by more than one group member. When a resource unit—a fruit-bearing twig or a sleeping-ledge—is so small that only one animal can use it, the more dominant animal will take it.

The term "dominance" is widely used to describe a particular type of order in organized groups. Its most general criterion is the fact that an animal consistently and without resistance abandons his place when approached by a more dominant group member, a sequence called "supplanting." In primates, older and stronger individuals supplant the weaker ones, and males are generally dominant over females of the same age. Each group member has to learn his rank of dominance, at least within his own sex and age class. The ultimate means of clarifying positions is fighting.

The consistent ability of some animals to supplant others has several effects on the individual's ecological prospects within the group. The dominant primate can displace his inferiors from the best feeding sites and the safest sleeping places. Thus, baboons sometimes supplant low-ranking group members from grass plants that have already been dug up, by a subtle approach-avoidance sequence without any threatening gestures. This advantage is usually of little importance, since primate vegetable food occurs mostly in small bits scattered over an area that can accommodate all members of the foraging party. With large items of food, however, dominance becomes decisive. The young antelopes that baboons sometimes kill are almost exclusively eaten by the adult males, and fighting over such prey is frequent. The inability of baboons to share food is a behavioral characteristic that probably prevents them from shifting to hunting as a way of life. In contrast, chimpanzees, who also kill and eat small mammals on occasion, do beg each other for parts of the prey and are sometimes successful in obtaining them.

Even with a vegetable diet, the effects of dominance become critical when the total available amount of food falls short of the group's needs. There is a wealth of data, mainly from studies of animals in captivity, demonstrating that under such conditions the dominant animals will take the food while the inferior ones will suffer from the shortage. Dominance will then probably favor the adults and sacrifice juveniles and infants. This seems adaptive, since the experienced and reproductively active adult is more valuable to the group than an easily replaceable youngster. But a food shortage is also critical for the females, placed at a disadvantage by size and dominance factors, and far less replaceable than the youngsters. At present, we can see two possibilities for counter-designs that would protect the females. One speculation, put forward by Crook, suggests that the one-male group of the open-country species is a way of freeing the females as much as possible from food competition with males. Among patas and geladas, the females live with a single male, who is essential for breeding and protection. The supernumerary males of these species form purely male groups that forage outside the home range of the one-male group.

Another speculation can be based on the fact that primate males are heavier than females in most species living in meager habitats. Among hamadryas baboons, females and juveniles often save themselves from a male attack by rushing onto a limb so thin that the heavy male will not dare to follow. Smaller animals are obviously carried by lighter limbs than heavy ones. Even after a 40-pound hamadryas male has exploited an acacia tree for flowers or seeds, a 20-pound female will still find food on it. A future field study may investigate the hypothesis that, in a species that obtains much of its food on the ground, arboreal diet becomes the critical food source of females and juveniles because it reverses the advantage size has on the ground.

The great sexual dimorphism of terrestrial primates is usually interpreted as an adaptation to defense against predators. This is probably the major selective factor in a species like the geladas, which have never been seen to feed on trees. The argument is less convincing for the terrestrial and equally dimorphic drills and mandrills; in their dense forest habitats, they are never forced to face a predator on the ground. In these species and in the savanna baboons, the weight-in-trees factor may have contributed to dimorphism.

The dominance order has certain ecological advantages, as in substituting a smoothly functioning order of priority for endless quarrels about resources. Another function of dominance is the dispersal of the group members in space. The less dominant members avoid entering the area around a dominant animal. This tends to scatter the group, so that foraging individuals will look for food at distances that keep them from interfering with each other. At larger distances, however, dominance seems to reverse its spacing effects. The British ethologist M. R. A. Chance has pointed out that the non-dominant group members tend to anticipate the movements of those higher in rank. In order to do so, they frequently look up and check on the whereabouts of the most dominant members, even following them at a distance to keep them in sight. This focused attention, together with the protectiveness of the dominant animal, make him an attractive figure at distances of five or more yards. Because of the great attention paid him by the rest of the group, the highest in rank becomes a potential leader whose actions are likely to influence others. This hypothesis, again, awaits experimental testing.

The most intricate effects of dominance are not ecological but purely social. They are causal rather than functional aspects of primate societies, and thus fall outside the frame of this chapter. Ecologically, the position of the dominant animal is of little significance unless it is associated with leading or protective functions. Consequently, thinking in terms of dominance is currently being replaced by thinking in terms of roles. A role, however, should not be described merely as what an animal does, but rather as an individual, group-oriented function. Thus, Hall likens the patas male to a "watchdog" of his group. This extension of the laboratory-bred term of dominance opens the way for fresh research on the mechanisms of role distribution and role differentiation, on mechanisms more subtle than open competition, and on the possibility that group members may even compete for roles that benefit the group, not just contribute to their own survival.

LEADING THE GROUP

The ecological function of leadership is an obvious example of a social role in primates, although it has not yet attracted topical research projects. Ecologically, the importance of leadership increases with the distance between resource units. In the tiny range of a titi monkey group (*Callicebus moloch*), where food trees are evenly scattered, leadership is hardly important because one layout of the daily foraging trip is about as profitable as another. The group traverses its whole range, which is less than 110 yards in diameter, several times a day anyway.

The situation differs greatly for some populations of chimpanzees and hamadryas baboons. The latter cover an average of eight miles a day, a distance greater than that of any other primate species investigated. The roosting-cliffs of hamadryas are miles apart, as is the case for their watering places in the dry season. The home range of a typical troop has dense stands of acacia bush, the staple food source, though these are separated by long stretches of meager, stony grassland that is completely dry for several months of the year. Under these conditions, the effort of foraging may exhaust the time and energy of the weaker troop members. The six to 12 miles of daily travel covered by our troops in the Danakil plain may be nearly at the upper limit of a small juvenile's capacity. Tolerably long routes that begin and end at a cliff, touch on water at least once, and lead through one or two satisfactory feeding areas are not too many.

Thus daily foraging of this kind calls for a function analogous to planning. This in turn requires that at least some troop members are informed about the whereabouts and present conditions of the resources. To obtain the ideal design of the route, the slightly differing information of all members would have to be pooled and evaluated. This, however, is impossible, since primates cannot communicate about distant food groves and waterholes and their recent condition. As already mentioned, all a primate can do to inform the departing troop of a profitable location is to indicate, by the direction of his glances and shifts, the direction which he intends to take, and possibly the strength of his motivation to go there.

An alternate procedure would be to follow one well-informed leader. Surprisingly, this appears to be a rare solution among primates. Among anubis and yellow baboons, for example, travel is initiated by many individuals, males and females in turn. Leadership by one adult male is more common in species with one-male groups. Among geladas and hamadryas baboons, the one-male group is commonly led by its male, but even here, females may temporarily take the lead. In the one-male groups of the strongly dimorphic patas monkeys, the females direct and initiate group travel more often than not while the male takes up his peripheral lookout positions.

Mountain gorillas provide the clearest example so far known of leadership concentrated in one individual. The single silverback male is usually in the lead position when the group travels rapidly. Before departure, the leader sometimes "employs a characteristic posture which apparently serves as a signal to the other members of the group, indicating his imminent departure. He faces in a certain direction and stands motionless for as long as ten seconds with front and hind legs spread farther from each other than usual." On occasion, he emits a few short, forceful grunts, to which the group responds by moving in his direction.

As a rule, however, primate groups are led jointly by several adults. This means that the actual route is the result of compromise. Surprisingly, it is nearly impossible to discover signs of conflict between group members with different directional intentions. In the face of ecological necessity, primates have as yet never been observed to dispute the course of their action by even the mildest threat. Traveling group members can nevertheless tend in different directions and follow their designs quite inflexibly. I once observed such a struggle in a small hamadryas band that consisted of only two one-male groups, led by two males, whom I had named Circum and Pater, respectively. The band departed from its cliff shortly after 7 a.m. The younger of the two males, Circum, immediately "proposed" a northerly direction. My field notes contain the following account:

07.30 Circum utters a contact grunt and goes *northward* along the river bed. Again the entire party follows for some 20 yards and then stops.

07.31 Circum again rises; he briefly looks back at Pater and then goes another 30 yards northward. No one follows. He stops, comes back until he is only 20 yards away from his closest female and sits down. All the while Pater has been watching him.

07.32 Circum rises and begins to move *west*, straight across the river bed. Only his youngest female follows him. After a few seconds both come halfway back and sit down.

07.33 Circum sets out again, this time in a *southwest* direction. Now, Pater rises, and the whole party follows Circum in the same marching order as above.

07.40 On the left bank, Circum again begins to point northward. The others follow him. After 110 yards they all sit down, then they climb an acacia and begin to eat the blossoms.

07.58 Pater climbs down and sits next to the trunk facing northward. Circum immediately climbs down to him; Circum stops a second near Pater while the two of them turn their faces toward each other. Then Circum continues further north. Pater allows females and young to pass and then follows at the rear. Twenty yards ahead, everyone stops.

08.07 Circum rises, looks back and proceeds. All follow. After a few yards he swerves to the right, leads the troop back to the right bank, and continues northward. At this, Pater slowly moves to the front of the column. Ten yards ahead, he overtakes Circum and as he reaches the front, turns westward back to the left bank. The females and young swing around him to the west like a rope. Circum, on the rope's end, continues northward a few paces but then follows the rest across the river. For the first time the marching order is reversed: Pater, Pater's only female, Circum's females, Circum. The young are scrambling about at the side of the column. Pater now leads the party westward and then to the southwest. (Note that this was the direction in which he ultimately followed Circum at 07.33.)

During most of the day, Circum tried to lead the band to the north, where the rest of the troop had gone a few days before. He was adamantly opposed by the older and more influential Pater, who insisted on a trip to the southwest. As a result, the trip assumed a peculiar zig zag pattern, but the two males never showed any signs of impatience or aggression, nor did the two groups separate. At 2 p.m., Circum finally abandoned his northward trend and preceded the band in the direction that seemed to correspond with the intentions of the older male.

Quite another question is how information on the current state of food sources is obtained. When howler monkeys arrive at a point of choice in an unfamiliar area, all the adult males simultaneously explore alternate parts of the arboreal paths. "When one of them finds a suitable route, he will give the deep clucking vocalizations; then the females and young begin to follow him slowly, and associated males cease their exploratory behavior and fall in line with the moving column of animals." Thus the less familiar parts of the howler monkey's route at least are not decided upon before departure; the alternatives are explored on the spot. In a howler territory, which is only about 650 yards in diameter, this is an acceptable solution. For hamadryas baboons, the distances to be covered by exploring males would obviously exceed the tolerable; the explorers, moving far beyond contact with the waiting troop, would have to return to make any clucking noises. Aside from the time factor, it is uncertain whether the dominance-oriented baboons would be capable of a howler-type decision process, which, in effect, seems to choose the most attractive pathway regardless of the dominance status of its discoverer.

Chimpanzees, however, seem to use the design of the exploring party efficiently. I have already quoted Vernon Reynolds, who reports that feeding chimps in the Budongo Forest of Uganda can attract other parts of the community over distances up to 2 miles. Parties of chimpanzees meeting at a tree with abundant food engage in a chorus of loud calls and resounding beats against the plank buttresses of trees. Reynolds suggests that their vigorous drumming communicates the food location to other parties in the area.

In the semi-desert, the extent of a daily trip exceeds the carrying distance of even the loudest bark of an exploring party. In choosing the direction of their departure, hamadryas baboons have to rely on the information gathered on preceding trips. We do not know how and by whom the sites are explored and remembered, but we know that on most mornings different males of the troop strive in different directions. The decision is made by a long process in which most adult males of the troop participate while the rest seem unconcerned. I have already described how a hamadryas troop prepares for departure during its morning rest. The troop performs slow on-the-spot movements, changing its shape like an undecided ameba. Here and there, males move a few yards away from the troop and sit down, facing in a particular direction away from the center. Pseudopods are generally formed by the younger adult males and their groups. For some time, pseudopods protrude and withdraw again, until one of the older males in the center of the troop rises and struts toward one of the pseudopods. At this, the entire troop is alerted and begins to depart in the indicated direction. Detailed observation reveals that two male roles are involved in leading the hamadryas troop: the younger initiators who "propose" certain directions and the deciders who choose among the proposed directions. Accordingly, we can talk about the "I-D system"; whereas the troop as a whole pays little attention to the initiators, it immediately follows a decider. The I-D system is also found among the forest-living anubis baboons of Uganda. In this population, however, the deciders are adult females, the males being restricted to the role of initiators.

Interestingly, the I-D system is not a rigid trait of hamadryas organizations. In 1968, when we restudied a troop of hamadryas baboons which had shown the I-D process very clearly in 1961 and in 1964, we found almost no trace of the former routine. Instead of protruding and retracting pseudopods, the troop now departed without much ado from its resting place above the cliffs. Males seemed to disagree less on the direction of the troop's departure. We have no explanation for this change. There is certainly no need to postulate a new tradition. The food situation may have changed or, possibly, an influential leader could now lead the troop without much resistance from other males. Field studies have amply demonstrated that primate species show geographical variations in their organization. Our experience indicates that even the same troop can change its social behavior according to the set of qualities of its actual members. A group process can take a different form in a new generation without any underlying changes of gene pool or tradition, simply because each generation of group members is a new combination of old traits.

In the most complete ecological report on a primate species now available, Stuart and Jeanne Altmann indicate that the yellow baboons of Amboseli National Park may change the utilization of their home according to what they have experienced in certain locations: "One [sleeping] grove . . . was deserted after two members of the group were killed therein by a leopard. Although this grove had been used as a sleeping grove on 13 of the 57 preceding evenings . . . , it was not used once in the subsequent 68 nights for which we have sleeping grove data."

Apart from selecting a nourishing and safe travel route, the group faces the task of maintaining contact even when visibility is poor. When a hamadryas band travels intermittently in a long scattered column, each family male tends to sit between shifts on an elevated spot from which he can see both the preceding and the following males. When a male moves on, the following unit leader usually comes up and sits in exactly the spot just vacated by his companion. This relay behavior seems to support group cohesion in dense bush. Soft contact grunts communicate mutual location among baboons foraging a few yards from each other in dense vegetation. When scanning the surroundings in high grass, monkeys and apes routinely stand on their legs in an upright posture, which they can do for many seconds. The adult males of baboons and of most forest monkeys utter loud calls in response to other parties and groups of their species. Since the calls are highly specific for the species, they carry information on who is where and thus maintain intergroup contact.

9. OBSERVATIONS ON THE ECOLOGY AND SOCIAL BEHAVIOR OF THE MOUNTAIN GORILLA

GEORGE B. SCHALLER and JOHN T. EMLEN, Jr.

Reprinted from F. Clark Howell and François Bourlière (Eds.), African Ecology and Human Evolution *(Chicago: Aldine Publishing Company, 1963). George B. Schaller is Research Associate at the Institute for Research in Animal Behavior of the New York Zoological Society and the Rockefeller University. His main research interests are the study of the ecology and behavior of birds and mammals, and particularly predator-prey relations. He is the author of* The Year of the Gorilla, The Mountain Gorilla: Ecology and Behavior, *and* The Deer and the Tiger: A Study of Wildlife in India. *John T. Emlen, Jr., is Professor of Zoology at the University of Wisconsin. He and his students have worked particularly with problems of population dynamics and behavioral adaptations in birds and mammals, combining field studies with laboratory experiments and analyses.*

■ The many studies that have been conducted by observers of nonhuman primates indicate conclusively that it is impossible to generalize about these primates as a group. It seems there are at least as many adaptations as there are genera in the subhuman primate world. As in the rest of organic nature, most primate groups are so completely adapted to their habitats—that is, so highly specialized—that there are no naturally selective forces that demand or favor further adaptations or evolutionary changes. Such groups represent terminal points in evolution.

In this selection Schaller and Emlen describe such an evolutionary dead end: The lush and damp forest, with abundant food, inhabited by the mountain gorilla. They characterize this situation as an evolutionary dead end because the habitat contains no selective pressures for the gorillas to develop the use of tools or manipulative skills. This habitat also is largely responsible for the relative stability of gorilla groups: Because there are no food shortages there is no need for the group to subdivide recurrently to form single-male reproductive (family) units, as do the hamadryas baboons who live in a habitat in which there is relative food scarcity (see Selection 12).

This selection is especially valuable for those who are reading this kind of material for the first time because the authors present two excerpts from their field journals. These give some idea of the kinds of observations primatologists make and the ways in which they record them.

Schaller has three excellent chapters in Irven DeVore's *Primate Behavior* (New York: Holt, Rinehart and Winston, 1965): "The Behavior of the Mountain Gorilla," "Behavioral Comparisons of the Apes," and a brief chapter on "Field Procedures." Schaller's chapters are an excellent introduction to his book on *The Mountain Gorilla: Ecology and Behavior* (Chicago: University of Chicago Press, 1963) and to his more popular volume, *The Year of the Gorilla* (Chicago: University of Chicago Press, 1964). ■

THE PURPOSE OF this paper is to present a brief summary of personal observations on the ecology and social life of the mountain gorilla *(Gorilla gorilla berengei),* and to a lesser extent on the chimpanzee *(Pan troglodytes schweinfurthii)* and orangutan *(Pongo pygmaeus),* as a basis for some comments on hominid origins and social behavior.

The gorilla data are based on a six-month ecological and distributional survey in the eastern Congo and Uganda from February to July, 1959, followed by fifteen months of intensive field work by Schaller in selected localities. The detailed observations on social behavior were made on a population of ten groups in an undisturbed section of the Virunga Volcanoes of Albert National Park, Congo, unless otherwise stated. Gorillas were encountered 306 times and observed directly for 457½ hours. Most animals were recognized individually and several groups were habituated to the presence of the observer.

Two weeks were spent exclusively on chimpanzees, primarily in the Maramagambo and Budongo Forests of Uganda. Contact with orangutans was limited to 4 encounters with 8 animals in Sarawak over a two-month period. Full descriptions of gorilla and other ape behavior have been published elsewhere; this paper presents only those aspects which have the greatest interest from a comparative primate standpoint.

The study was financed by the National Science Foundation and the New York Zoological Society, with the latter institution acting as sponsor. Local sponsors of the expedition were the Institute of the Parks of the Congo, Makerere College in Uganda, and the Sarawak Museum. To these institutions, and the numerous persons who helped us, we are extremely grateful.

ECOLOGY

Mountain gorillas are forest animals. Over the major portion of their range they inhabit lowland rain forest and at somewhat higher altitudes they occupy physiognomically similar mountain rain forest. Bamboo, which occurs at altitudes of about 8,000 to 10,000 feet along the top of the western edge of the rift valley and in the Virunga Volcanoes, is a habitat of secondary importance. Other vegetation types are very local. For example, the rather open mountain forest stands of *Hagenia abyssinica*, which provided the best conditions for observations on gorillas, and the zone of giant senecios and lobelias above 11,500 feet, where gorillas occasionally penetrate upward to 13,500 feet, are found primarily in the Virunga Volcanoes.

Even though the climate and vegetation vary from distinctly tropical in the Congo basin to temperate in the mountains, the habitats utilized by gorillas are similar in being lush and damp with an abundance of food in the form of vines, leaves, bark, pith, and some fruits. (No evidence of meat-eating was found.) The most conspicuous physical variation of the animal over this extensive altitudinal range is in the length of hair. The main behavioral variations are in the species of plant eaten and the location of nests. In the lowland rain forest, for example, the pith from the stems of *Aframomum* and banana are two of the most important food items; in the Kayonza forest of Uganda, which is a mountain rain forest, such vines as *Momordica foetida* and *Basella alba* furnish the bulk of the diet; and in the *Hagenia* forest of the Virunga Volcanoes, at an altitude of 10,000 feet, *Galium simense* and *Peucedanum linderi* are the two chief forage plants among 29 species eaten. Although gorillas show considerable ecological adaptability, including extensive utilization of secondary forests and field borders, they have remained entirely within the humid forests; plains and dry woodlands constitute effective barriers to dispersal.

Orangutans and chimpanzees differ considerably from the gorilla in their habitat relations and preferences. Orangutans appear to be largely dependent on primary or old secondary rain forest; chimpanzees, on the other hand, are highly adaptable, occurring in equatorial and mountain rain forests, in narrow gallery forests, in open dry woodlands, and where forest adjoins grassland they may even penetrate the latter for short distances. Thus the chimpanzee appears to be the most adaptable of the great apes with regard to habitat.

LOCOMOTION

Gorillas are primarily quadrupedal and terrestrial. Extensive bipedal locomotion is rare in nature: we saw it only three times over distances ranging from 15 to 60 feet. Although the animals climb readily into trees to feed, nest, and for other activities, all of them, but especially the adults, do so with relative caution. They rarely run or jump around. When danger threatens they descend and flee on the ground, which to a lesser degree is also true of the chimpanzee but not the orangutan.

Chimpanzees are likewise quadrupedal, but only semiterrestrial. In agility and climbing ability they are only exceeded by the gibbon and siamang among the apes. They run and jump through the trees with great rapidity. One excited animal was seen to jump 30 feet from a nest to the ground, and leaps from branch to branch sometimes exceeded 8 feet. They may spend considerable time on the ground, especially in dry woodland habitats.

Orangutans rarely descend trees. They are fairly slow and careful quadrupedal climbers whose top speed usually does not exceed two to three miles per hour.

While it may be appropriate to class the apes as brachiators on anatomical ground, it must be emphasized that in the wild only the gibbons and siamangs brachiate extensively; all of the great apes are essentially quadrupedal. We never saw gorillas brachiate in the wild, although they were occasionally seen to hang by their arms and reach for another branch while their legs hung free; chimpanzees have been reported to brachiate for short distances along a horizontal branch; and orangutans were observed to brachiate briefly at times but their main mode of progression was by quadrupedal climbing. Such commonly repeated statements in the literature as: "Only the anthropoid apes among the Primates are full brachiators, that is, their movements in the trees are exclusively arm-swinging movements and never quadrupedal," should thus be modified.

GROUP SIZE, COMPOSITION, AND CHANGES IN COMPOSITION

The ten gorilla groups under close observation in the Virunga Volcanoes varied from 5 to 27 animals each, with some subsequent changes that raised the number of one group temporarily to 30. Average group size was 16.9 animals. It was noted, however, that groups in other areas tended to be on the average somewhat smaller, often consisting of only 3, 4, 5, and 6 animals each, a fact which seems to be at least in part correlated with habitat and predation by man.

Each group contained at least one silverbacked or adult male (about 10 or more years old), but some groups held 2, 3, or even 4 for varying periods of time. Adult and subadult females varied from only one in the smallest groups to as many as 12 in the largest ones. In addition most larger groups contained one or more blackbacked or subadult (?) males (about 6 to 10 years old) and a variable number of juveniles (about 3 to 6 years) and infants (0 to 3 years). Lone blackbacked and silverbacked males were commonly found, sometimes over 20 miles from the nearest group.

Gorilla groups are quite cohesive in that the members rarely drift far from each other. The diameter of feeding or resting groups is usually 200 feet or less, and only infrequently are the animals spread over 300 to 400 feet of forest.

Changes in size and composition occur quite commonly in some groups but only rarely in others. Two illustrations of the type of changes occurring in gorilla groups are given below.

1. Group IV was first observed on March 12, 1959, and detailed records of the composition of this group were collected periodically from August, 1959 to August, 1960.

Composition (August, 1959): 4 silverbacked males, 1 blackbacked male, 10 females, 3 juveniles, 6 infants (1 infant was born on March 12, 1959) = 24.

August 28 to August 30: A peripheral silverbacked male apparently left, and then rejoined the group.

September 9: A new silverback joined the group. The peripheral male of August left again but rejoined by September 22.

September 18-20: An infant was born.

Between October 2, 1959 and January 11, 1960: The peripheral male and the No. 2 male in the hierarchy left.

A new silverbacked male and two females, both with infants, were added. They prob-

ably represent a small group which joined. An infant was born in late December.
April 24, 1960: A silverbacked male, who has been with the group at least since August, 1959, left.
April 25, 1960: An infant was born but died two days later.
Between May 1 and 15: A new silverbacked male joined the group and remained at least to May 24.
Between May 24 and August 12: The male who joined in early May left.

Thus in the course of 12 months at least 7 different silverbacked males associated with the group, but of the 4 present in August, 1959 only the dominant male remained one year later. Three infants were born and two females with infants were added. At the last encounter the composition was: 3 silverbacked males, 1 blackbacked male, 12 females, 3 juveniles, 10 infants = 29.

2. Group VII, which was studied in greater detail than any other group from October, 1959 to September, 1960, showed relative stability.

Composition (October, 1959): 1 silverbacked male, 2 blackbacked males, 6 females, 4 juveniles, 5 infants = 18.
Early February, 1960: An infant was born.
Between February 14 and March 16, 1960: An unknown female with infant joined the group, raising the group total to 21. (This was the only instance in which a single female was noted to join another group.) No lone males associated with this group.

These examples indicate that established groups are fairly stable social units but that they may be augmented from time to time by the addition of individuals or other groups. Lone males may freely come and go in some groups, but apparently not in others.

No long-term observations on chimpanzees or orangutans were made, but the information obtained points to some differences. Whereas gorillas tend to remain in rather cohesive groups, chimpanzees are less closely knit. In one group of seemingly only two animals, the male and female were 300 feet apart. In another case a female with infant was apparently alone in a gallery forest. On the other hand, one temporary aggregation comprising 2 and perhaps 3 groups was seen, and probably at least 50 animals were spread over about a half mile of forest.

The orangutan forms small and seemingly unstable social groups. A perusal of the scanty literature and personal observation revealed that groups of more than 4 are rare, and that lone animals, both male and female, and groups of 2 and 3 are the most common, and even these appear unstable. For example, a female with small infant, which Schaller observed in Sarawak, fed alone in a tongue of forest. An examination of her nest site of the previous night revealed that a subadult had accompanied her, but it could not be found. Natives maintained they had also seen a male with the group two days previously, a statement which was corroborated by an examination of other nest sites. The most prevalent group combinations appear to be single adult females with one or 2 subadults; pairs with one male and female; and groups of 2 or 3 subadults.

HOME RANGE AND TERRITORY

Observations on peaceful interactions between distinct groups and the great areas of overlap of group range indicates that gorillas have no territory in the sense of an area exclusively held or defended against others of the same species. However, gorilla groups do restrict their activities to definite home ranges. The same range may be frequented by several groups in part; six different groups were seen in one small section of forest during the period of the study. Daily tracking of groups, sometimes for as long as 25 consecutive days, revealed average daily movements of a little less than half a mile, with variation from about 300 feet to over 3 miles. The route taken and distance traveled varied from day to day and did not follow a set pattern. Home ranges of groups were on the order of about 10 to 15 square miles each.

SUBGROUPS AND JOINING OF GROUPS

Subgrouping or the temporary splitting of one group into two distinct units is a rare phenomenon in gorillas. Only one instance was noted, and in this the group split in two and united again later in the same day. Circumstantial evidence from trails and incomplete encounters pointed to several other instances of temporary splitting, but in any case it is an infrequent event.

Carpenter proposes that the agonistic territorial behavior which he observed between neighboring groups of several primate species is typical of primate societies, and on this basis suggests that merging of distinct groups does not occur. Overlapping of home ranges brings gorilla groups into close contact on occasion. In 12 definite meetings noted during the study, the groups usually passed each other slowly at distances of 300 feet or less, or sat near each other for varying lengths of time without actual interchange of individuals. However, partial or complete mixing of the two groups occurred at least 4 times. Once 2 groups remained near each other for 3 days before they were seen to mingle briefly. Most joining of groups was of brief duration, one day or less, but we have evidence that in one case two groups united for several months. Seemingly reliable data from an observer in Uganda indicates that the survivors of a small group joined another one after the dominant male died. Agonistic displays in the form of threatening stares or bluff charges were seen between the dominant males of the respective groups on two occasions, but actual fighting was never observed.

INTRAGROUP RELATIONS

Although every gorilla has the possibility of interacting with every other member of its group, numerous factors tend to reduce the number of direct contacts between the animals. Forage is plentiful, competition for mates is apparently lacking, and numerous dominance interactions are simply avoided by circumventing situations which might produce them. In addition, adult gorillas possess temperaments which can best be described as self-contained or aloof, and physical contact with others is usually not sought. Thus, for example, mutual grooming was noted only .28 times per hour of observation, and dominance .23 times.

DOMINANCE

Certain individuals commonly claim prerogatives in right of way along a trail and to a choice of sitting place and their right is recognized by others. Dominance was seen only once in competition for food, and it apparently did not feature in sex, although observations on the latter were limited to two copulations. In most dominance situations the subordinate animal merely received a look or a light touch with the back of the hand from the dominant one. Such signals are not disputed and no dominance fighting was ever observed.

To a large extent dominance is correlated with body size. All silverbacked males are dominant over all other animals; all females and blackbacked males are dominant over all juveniles and infants not in contact with their mothers; and among and between infants and juveniles size likewise seems to be the main correlate. If more than one silverbacked male is present in the group, there is a hierarchy among them. Females, however, seem to lack a definite hierarchy.

LEADERSHIP

The dominant male is also the leader, and, therefore, the focal point of the whole group. His leadership is almost absolute, for all other animals in the group, except sometimes the peripheral males, watch and respond to his every action—when he rests or nests, they do; when he travels, they follow. Differences in temperament and behavior of the dominate male thus affect the action of the whole group.

SEXUAL BEHAVIOR

Only two copulations and one invitation to copulate were observed. In one instance the female invited the male by presenting her rump, and once a female mounted a male briefly before copulation. All the males involved were subordinate individuals, and the dominant males never interfered aggressively. No instances of playmounting or any other type of sexual behavior were noted. The paucity of observations suggests that the sexual drive in gorillas is low. Chimpanzees, on the other hand, are more easily aroused to sexual activity if data from captive and semi-wild individuals are valid for comparison.

MUTUAL GROOMING

Mutual grooming is rare between adult gorillas. No male was ever seen to groom another adult and only once was a male groomed by a female. Grooming between females is also infrequent, and they invite grooming by others only of areas of the body which they cannot themselves reach with ease. The invitation to grooming is indicated simply by presenting a certain part of the body to another animal. Thus mutual grooming among adults seems to have little or no social significance, merely a utilitarian one.

Most grooming activity involves a female and a younger animal. Females groom juveniles and infants readily; and occasionally a juvenile grooms a female, where the activity apparently functions in establishing social contact.

MOTHER-INFANT RELATIONSHIP

The development of one infant in the wild was traced from the day of birth to the age of 1½ years and that of several others for nearly one year. The ages of older infants and juveniles were estimated and are based primarily on comparative size. Infants are entirely helpless at birth and they must be supported by the arms of the female until about 3 months old, an age when they usually begin to ride on their mother's back for short periods. The first tentative crawling movements away from the mother occur between the age of 3 to 4 months and by 5 and 6 months other infants are actively sought in play. By one year of age, the infant wanders freely among the resting members of a group and it forages for its nourishment, with milk being now of secondary importance. By 1½ years it readily climbs around in trees high above the ground. By 2 years it may spend considerable time with other group members and even travel under its own power if the general movement is slow. Although the infant is more or less intimately attached to its mother to the age of about 3 years, the bond is not necessarily broken completely even then. Some juveniles frequently sit by a female, are groomed by her and sometimes sleep with her in the same nest until they are about 4 years old. This relationship persists in spite of the fact that lactation ceases by 1½ or perhaps as late as 2 years and the female may have a new infant. In other words, a loose mother-child relationship continues well after the female has ceased to provide food and protection. Some male gorillas become independent of the group at about 6 years of age when they leave to lead a lone life in the forest, a habit never noted in females.

COMMUNICATION

Gorillas emit about 20 distinct sounds several of which apparently carry direct meaning or serve to attract attention to the performer and permit the communication of further information by means of gestures. Thus, a short, sharp *uh-uh* given by a male when females quarrel may cause them to subside. If, however, the male emits the same sound when the group is quietly resting or feeding all members first look at him and then face the direction which occupies his attention. Vocalizations in their daily routine are relatively infrequent and gestures usually serve to co-ordinate their actions and movements. The mere act of a male walking a distinctive way in a certain direction suffices to indicate that he is leaving. There was no evidence of displacement or productivity, both considered by Hockett as distinctive of human communication systems; in fact, in their means of communication gorillas show no greater elaboration than many mammals. Early man could easily have co-ordinated his whole social existence by such simple means of communication as those employed by the gorilla. Even communal endeavors like game drives require only simple gestures as illustrated by the hunting techniques of wolves, hyenas and other carnivores.

NESTS

All great apes build nests, the orangutan exclusively in trees, the chimpanzee usually in trees, and the gorilla often on the ground. A detailed analysis of the nests indicates that the basic method of construction is the same in all three genera. Nests were not built in any special pattern, except that the branches were broken toward the animal to create a platform. They never included a constructed canopy, but gorillas were observed occasionally to nest beneath the leaning bole of a tree thus avoiding the rain. Gorillas never used the same nest two nights in succession. The breaking in of a few branches was an easy matter and usually required only one-half to 5 minutes.

Although gorillas nest both on the ground and in the trees, the percentage of ground to tree nests varies from area to area and is, at least in part, correlated with the availability of suitable trees in which to construct secure platforms.

Location	*Total number of nests checked*	*Percent of nests on the ground*
Lowland rain forest (near Utu)	110	21.8
Mountain rain forest (Kayonza Forest)	179	53.6
Hagenia forest (Virunga Volcanoes)	2,488	97.0

As with other monkeys and apes, gorillas defecate wherever they are throughout the day and night, and this includes the cup of their nest on which they lie. The amount of dung deposited by a group per day is such that repeated use of a bedding area would soon make it unlivable. The first dwellers of caves or other creatures having a home base, whether they be ape or man-ape, must have learned to refrain from defecating until they had moved away from or to a certain part of the main living quarters, in the manner of various present-day rodents and carnivores.

CURIOSITY AND TOOL-USING

Gorillas are extremely curious to ascertain the nature of strange animate objects, and they approach a lone, quiet person sometimes very closely. But they seem to lack a comparable curiosity to investigate strange inanimate objects in their environment. A tin can, intestines of an antelope, a rucksack, paper—all were objects which lay directly in the path of gorilla groups on occasion without being handled. The animals showed little interest in fondling, pulling, or in tearing objects, a trait so prominent in chimpanzees in zoos.

Although Beatty and Merfield observed the use of tools by free-living chimpanzees, we saw no such behavior in gorillas. Young gorillas play with leaves and sticks and sometimes carry them around, and nesting material and food is occasionally transported by the animals for 15 to 20 feet, but adults showed no inclination to handle anything for the sake of manipulation alone. It should not be supposed, however, that quadrupedal gorillas have difficulty in handling objects, or in carrying anything such as a tool for long distances. Infants are supported by the female with one arm for at least 3 months and the female does quite well on her 3 limbs. The position of the fingers in quadrupedal locomotion is such that a stick or stone could be transported by the animals as easily as they sometimes carry a stalk of celery. Gorillas spend many hours daily on their haunches feeding or just sitting with their hands entirely free to manipulate objects. They can throw well with speed and accuracy in captivity. Gorillas, it would seem, are in many ways admirably preadapted to a life in which tool-using and hunting for meat could play a prominent role.

It may be conjectured that the gorilla's failure to develop tool-using is related to the ease with which it can satisfy its needs in the lush forest habitat. The forest is for these powerful vegetarians an evolutionary dead-end road in that there is no selective advantage for improvement of manipulative skills or mental activity along the lines which characterized human evolution. There is no reason to carry a tool if vegetable food is abundant everywhere, and no preparation of the food is required beyond stripping or shredding with the teeth and fingers. There is, it would seem, little selective pressure to try anything new or to improve on the old. Need for special manipulation involving tools might more likely arise in a harsh and marginal environment where selective premium is placed on mental activity and new modes of fulfilling bodily requirements. Man must have evolved in such a habitat, and of the apes today perhaps only the chimpanzee possesses the combination of need, potential, and habitat where the genesis of tool-using and hunting for meat might occur.

DISCUSSION

The descriptive and comparative data presented in this paper raise three points which bear on current views of social evolution in primates and man. All of these points emphasize the folly of premature generalization and the importance of a broad outlook in selecting the guidelines for theorizing.

1. The great apes present several contrasts in their social behavior and organization with what has been observed in other primates and what has come to be regarded by several scientists as the basic primate pattern from which man's social traits evolved. Territorialism is, for example, present in the gibbon but certainly not in the gorilla and probably not in the chimpanzee. The temporary joining of distinct groups occurs in gorillas and both Nissen and this study suggests that it happens also in chimpanzees. The relations between gorilla groups are quite peaceful as contrasted with those described for several other primate societies. Grooming between adult gorillas is primarily utilitarian, rather than intensely social or even of secondary sexual importance as in some of the monkey species which have been studied. Sex behavior, which is sometimes thought to be the main cohesive force in certain primate groups, has apparently no such encompassing function in gorillas. Thus, territorial aggression, group segregation, and sexual competition—three cornerstones of this hypothetical primate pattern—do not wholly apply to the gorilla. In fact, the question should be raised as to whether any present-day prosimian, catarrhine, or pongid pattern can, at this stage, be regarded as the one from which the human pattern developed. Caution should, therefore, be exercised in using data on the behavior of one or a few species as the basis for constructing theories of social evolution in primates, including man, as was attempted by Sahlins.

2. Marked differences in social behavior exist among the apes. Social groupings vary from the almost groupless orangutans, where aggregations of 1, 2, or 3 animals are most common, to the "family" of 2 to 6 gibbons, to the cohesive gorilla groups of 2 to 30 animals, and finally to the more loosely integrated chimpanzee groups which may aggregate up to 50 animals. Ecologically they vary from dependency on tall rain forest trees to adaptability to open, dry woodland. Temperamentally they vary from the exuberant chimpanzee to the reserved gorilla and the seemingly phlegmatic orangutan. The gorilla appears to have a low sexual drive, the chimpanzee a higher one. Information now available indicates that generalizations for the apes can be made only in very broad terms. A proper evaluation of similarities and differences must await further studies on all species.

3. The great apes, though structurally and mentally the closest living counterparts of man, are less like man in their adaptive responses and social behavior than many less closely related animals. Localization of behavior around a specific home or den is, for instance, prominent in many carnivores and rodents, as well as in birds and fishes. Social units based on a male-female pairing bond of considerable stability, and co-operative division of labor in the care of offspring are found in some carnivores and birds, but none of these attributes is markedly developed in apes.

All three of these points emphasize that social behavior and social systems in animals are more malleable in evolution than morphology, and that phylogenetic relationship as determined by structure is a relatively poor basis for inferring similarities of behavior. Distinctiveness of response to the environment has, in fact, been shown to possess marked survival value in the interaction of closely related and potentially competitive species. Accordingly, the search for evidence on the course of behavioral and social evolution followed by our prehuman ancestors might profitably be focused on ecological equivalence rather than phylogenetic relationship. The findings of archaeologists and paleontologists on the physical, biotic, and social environment of prehistoric hominids should be examined against a background of knowledge of behavioral adaptations in a wide variety of living animals. Such information, coupled with derived data on the sensory, motor, and integrative equipment of fossil forms, should provide a reasonably sound basis for constructing hypotheses on the evolution of human social behavior and social systems.

One theory of human social evolution suggests that the stable male-female relationship in human society is attributable to monogamy, the disappearance of estrus, and the persistence of sexual drive as opposed to the violently hierarchical dominance behavior and intense competition for mates which is supposedly characteristic of nonhuman primate groupings. An equally legitimate and logical theory can be advanced on what we know of gorilla behavior. Here the competition for mates is apparently held in abeyance not by physical dominance and suppression but by a simple reduction in the sex drive; the disruption of breeding groups is prevented not by fighting but by a general lowering of agonistic responses to subordinate and intruding lone males. It is entirely possible that *Australopithecus* and other hominids lived in groups similar to those of gorillas, held together by social bonds in which sex and aggression played secondary roles.

It would be a mistake, however, to assume that all species of prehistoric man were alike in temperament, behavior, and social life. If the closely related chimpanzee and gorilla show differences, why not *Australopithecus* and *Paranthropus?* The basis social structure of the early hominids may have varied tremendously as a result not only of their mode of life, about which we know a little, but also of temperament, sexual drive, or aggressiveness, about which we know nothing.

Hypotheses on the paths of social evolution which our prehuman ancestors followed are beyond the reach of experimental proof and will always remain speculations. If, however, they are to be satisfying to our sense of logic and effective guides for the interpretation of new discoveries, it is highly important that they be kept flexible and constantly in tune with pertinent information from all sources.

10. THE BEHAVIOR OF CHIMPANZEES IN THEIR NATURAL HABITAT

JANE VAN LAWICK-GOODALL

Reprinted from the American Journal of Psychiatry, *Vol. 130, 1973, pp. 1-12. Copyright by the American Psychiatric Association, 1973.* *Jane van Lawick-Goodall received her doctorate in ethology from Cambridge University. She has achieved world renown for her field studies of free-living chimpanzees in the Gombe Stream National Park in Tanzania; since 1967, she has been Scientific Director of the Gombe Stream Research Centre. She holds Visiting Professorships in the Department of Psychiatry and the Program in Human Biology at Stanford University and in Zoology at the University of Dar es Salaam. She is an Honorary Foreign Member of the American Academy of Arts and Sciences.*

■ The chimpanzee's place on the evolutionary ladder as man's closest relative makes this species the best model for insight into the origins of *Homo sapiens* and for experimental research. The evidence for the chimpanzee's proximity to our species comes from chromosomal structure, blood protein, immunity responses, DNA, and neuroanatomy. This selection is a sketch of Van Lawick-Goodall's intensive study of the chimpanzee for more than a decade, an exploration of this species in its natural setting in which she pioneered.

The author is here principally concerned with the study of the evolution of individual behavior, in aggression, sex, status, cooperation, and so forth. This concern must be seen as complementary to, but distinct from, the study of social organization. The activities of a group as such are not the simple sum of the behaviors of its individual members. Among humans, for example, engagement in warfare or in a particular economic organization cannot be explained exclusively in terms of the aggressiveness of individuals or of their proclivities to cooperation or competitiveness. Likewise, while it may be true that there is an inherited evolutionary tendency to establish hierarchical arrangements of dominance and submission in human societies, this cannot explain why some societies are egalitarian and others are not and why the general history of our species reflects a progressive elaboration in social organization. Both individual and group are equally legitimate objects of investigation, but they must be kept separate so that we may more carefully explore the degree in which they affect each other and the extent to which they may operate independently of each other.

Van Lawick-Goodall's findings reveal a chimpanzee infant and juvenile dependency on the mother that strikingly parallels that of humans. This lengthy dependency facilitates the learning of manipulative skills and social relations that are necessary for an animal to become a full-fledged adult member of the society. Family relations constitute strong and life-long bonds between a female and her offspring and among siblings (this is elaborated on by Reynolds in Selection 13). Another striking similarity between humans and chimpanzees is their stressful adolescence.

Chimpanzees are omnivorous, and they sometimes hunt cooperatively. They are the only known nonhuman primates who share food. Also, as far as we know chimpanzees are the only nonhuman animals who make and use tools. In many respects their vocal and nonvocal modes of communication give the impression of highly simplified versions of human social interaction. Though they exhibit little competition for either food or sex, chimpanzee groups are nevertheless characterized by hierarchies of dominance and submission. This status system is often interpreted by observers as providing order and stability in the animals' social organization. This may be true, but does it necessarily mean that there may not be order and social stability without dominance and submission?

Similarities, however, should not close our eyes to differences. As Van Lawick-Goodall concludes, a decade's study of chimpanzees has not only pointed to their kinship with humans, it has served to accentuate our own evolutionary uniqueness. Chimpanzees are vocal and communicative, but they have not come close to developing human language.

They are very affectionate, but without the depth and intensity of which humans are capable. They exhibit the beginnings of a sense of identity as we know it, but only humans have evolved the perception of what is possible and only they have woven this future orientation into the intricate fabrics of their cultures.

A highly readable account of Van Lawick-Goodall's research, together with more extensive observations of the chimpanzees that she has studied, is her *In the Shadow of Man* (New York: Dell, 1971). A more technical account is in her monograph, "The Behavior of Free-Living Chimpanzees in the Gombe Stream Area," *Animal Behavior Monographs,* 1 (Part 3) (1968): 161-311. Though superseded in part by her findings, another important study of Chimpanzees is *Budongo: An African Forest and its Chimpanzees,* by Vernon Reynolds (Garden City, New York: Natural History Press, 1965). Paralleling mother-child relations among chimpanzees is similar behavior among vervet monkeys, reported in "Play-mothering: The Relations between Juvenile Females and Young Infants among Free-ranging Vervet Monkeys," by Jane B. Lancaster (*Folia Primatologica,* 15 [1971]: 161-82). ■

IN 1960 AT THE INSTIGATION of the late Dr. L. S. B. Leakey, I began a longitudinal study of free-living chimpanzees (*Pan troglodytes schweinfurthi*) in the Gombe National Park, Tanzania, East Africa. This park comprises a narrow stretch of rugged, mountainous country running for some ten miles along the eastern shores of Lake Tanganyika and inland three miles or less to the tops of the peaks of the rift escarpment. The rift is intersected by many steep-sided valleys, which support permanent streams. In the valleys, riverine gallery forest is found; between the valleys the slopes are often more open, supporting deciduous woodland. Many of the higher ridges and peaks are covered only with grass. The area supports a population of between 100 and 150 chimpanzees.

This chimpanzee population is divided into communities of individuals who recognize and may interact with each other. Within such a community, which may include up to 50 or so individuals, chimpanzees mostly move about in small temporary associations, the membership of which is constantly changing as individuals or groups of individuals split off to move about alone or to join other associations. These groups may be all males, they may be females and youngsters, or they may be combinations of different age-sex classes. Chimpanzees, especially males, often move about quite on their own.

Some individuals in the community meet only when attractions, such as a local abundance of food or a female in oestrus, happen to draw them together; others meet more often; and some show strong bonds of mutual attraction and associate very frequently—traveling, grooming, feeding, and resting together. A mother and her dependent offspring is the one association that may remain stable over a period of years, though such a family unit frequently moves about with other associations.

In the wild, chimpanzees probably always live in male-dominated societies. Individuals of a community who frequently associate show a fairly well-defined dominance hierarchy, while among chimpanzees who meet only seldom the relative social status may be less clear-cut. As yet there is little information on relationships between individuals of different communities, but we do know that at least some chimpanzees may penetrate the home range of a neighboring community and peacefully travel, feed, or mate with its members.

Chimpanzees are promiscuous in that a female, during oestrus, may be mated by many males and no stable pair-bonds are formed. However, young females in particular may move with the same adult male (sometimes possibly copulating exclusively with him) during successive periods of oestrus.

Chimpanzees are omnivorous, feeding mainly on a variety of plant materials, especially fruits, but also consuming many insects, occasional bird's eggs or fledglings, and sometimes actively hunting and killing medium-sized mammals.

These apes follow no set route, day after day, in their search for food. Within a fairly large home range (which may be 20 square miles or more) they are nomadic, sleeping close to where dusk finds them. They typically move on the ground when traveling, although they do spend a good deal of time in the trees both during feeding and at night. They construct quite elaborate nests for sleeping: each individual typically makes a new nest every night, except for youngsters of up to five or six years (sometimes older), who share one with their mothers.

In the wild, a female chimpanzee does not give birth until she is at least 12 or 13 years old, and she has only one infant every four or five years. Life expectancy in the wild is not yet known but is probably between 40 and 50 years. The longevity record for a captive chimpanzee is about 47 years.

From 1963 onward, observations of social interactions between the different individuals of the nomadic community were considerably facilitated by the establishment of an artificial feeding area where bananas were offered to chimpanzees passing by. This attraction of the chimpanzees to a specific area has enabled detailed longitudinal records on the behavior of approximately 70 different individuals of both sexes and all ages. Since 1965, a growing team of investigators has been contributing to the understanding of the behavior of the Gombe Stream chimpanzees. Hugo van Lawick has built up a unique documentary film and still-photograph record of their behavior.

The need for a study extending over a great many years is due not only to the fact that the chimpanzee has a long life expectancy, or that there is a fairly high mortality rate (due mainly to disease and injury), but also to the fact that these apes are highly individualistic, differing markedly from one another in behavior as well as appearance. Thus a great many individuals must be studied in depth, through as much of their life cycles as possible, before one can make meaningful generalized statements about chimpanzee behavior. The understanding of individual differences is, in fact, one of the principal aspects of our research.

SIGNIFICANCE OF CHIMPANZEE RESEARCH FOR UNDERSTANDING HUMAN BEHAVIOR

The chimpanzee is man's closest living relative. Recent biochemical research has already revealed striking similarities in, for example, the number and form of the chromosomes, the blood proteins, immune responses, and DNA. Neuroanatomical research has shown that the structure and circuitry of the chimpanzee brain is closer to that of the human brain than is that of any other living primate. Behavioral research, in the field and in the laboratory, has highlighted remarkable similarities in this sphere also. Taken together, these findings suggest that, at some point in the distant past, man and chimpanzee shared a common ancestor. If this is true, we may assume that characteristics shared by modern man and modern chimpanzee were present in our stone-age ancestors.

Field studies provide information on the ways in which the structure and behavior of the species in question are adapted to its environment. Thus, since the chimpanzee is man's closest relative, and since he lives in an environment similar to that in which early man is thought to have emerged, an understanding of his behavior may well shed new light on the behavior of early man. This is of utmost significance to those concerned with human evolution—those who are trying to understand how and why man has become what he is today. As Hamburg has pointed out, some aspects of our behavior today—for example, our tendency toward violent action—seem unsuitable for the world in which we now live. Such tendencies, however, were undoubtedly shaped, over millions of years, to ensure the survival of our ancestors in a very different kind of world—a world that was in all probability far more like the world of the chimpanzee. We desperately need a greater understanding of aggression in man, and an appreciation of its evolutionary history and significance would be quite helpful.

Laboratory studies of the chimpanzee, under controlled conditions, can be very meaningful in understanding some aspects of human behavior. This ape is the best experimental model for those investigating human mental disorders. It is possible, for instance, to create in a chimpanzee a condition resembling human psychosis or one closely similar to some human depressions. But in order to work in a meaningful way toward a cure for the depressed or psychotic chimpanzee subject, it is essential to have access to a wealth of background knowledge concerning the normal behavior of the species and the conditions under which abnormal behavior is likely to develop.

Finally, increasing scrutiny of chimpanzee behavior is likely to pinpoint crucial areas in need of careful study in human subjects. Compared with human society, chimpanzee society involves extremely simple cultural traditions. The individual chimpanzee expresses his underlying motivations in a relatively straightforward way, with little masking of his responses. Thus it will be possible to tease out the biological roots of certain aspects of chimpanzee behavior despite its complexity; it may then be rewarding to reexamine the behavior in question in man to see whether similar factors may be involved.

SOME SIGNIFICANT LONG-TERM FINDINGS AT GOMBE

I should first like to outline some of the interesting findings that our longitudinal study is revealing, particularly those relating to the different stages of the life cycle in wild chimpanzees.

PERIOD OF INFANT AND JUVENILE DEPENDENCE ON THE MOTHER

One of the striking findings to emerge from the study at Gombe is the length of the period of infant and juvenile dependence on the mother. The infant relies completely on his mother for food, transport, and protection until he is at least six months old. He may then take his first unsteady steps and begin to ingest minute amounts of solid food. However, riding on the mother continues to be his normal method of getting from place to place until, sometime in the fourth year, he begins to make increasingly longer journeys under his own steam. Milk continues to be his major source of nourishment for at least two years and possibly longer.

Youngsters are not finally weaned until their fifth or sixth year in most instances. The youngest to be weaned, to date, was four and a half years old: one, about six and a half years old at the time of this writing, is still not finally weaned. Juveniles may continue to sleep in a nest with their mothers after being weaned. Usually they start sleeping in their own nests during their sixth or seventh year; this often coincides with the birth of a younger sibling.

During the final stages of weaning, a young chimpanzee may go through a period of apparent depression during which he frequently reverts to earlier forms of infantile behavior such as clinging to his mother during travel and maintaining much closer contact with her than previously. His frequency of play is likely to decrease at such a time, and he may appear listless and apathetic.

These symptoms were especially pronounced in one youngster, Flint, the son of a very aged female, Flo. Flint was weaned early, at four and a half years, toward the end of his mother's pregnancy. He went through a period of depression, as outlined above, during which he constantly solicited social grooming from Flo and, when she rejected any of his other demands (e.g., wanting a share of her food), was likely to fly into wild tantrums, screaming and hitting the ground—even, on occasion, attacking his mother. When the new infant was born Flint recovered somewhat: he stopped riding Flo's back, pestered her less, and in general seemed more lively (though he did continue to push into his mother's nest each night with his infant sister). A few months later, however, he once more became lethargic, once more demanded constant grooming, and reverted to riding his mother's back even though the new infant was clinging beneath. These symptoms showed no signs of abating until, when she was about six months old, Flint's sibling died. After this Flint became much more active, but he has remained abnormally dependent on his old mother and is still, at the age of eight years and three months, sleeping in her nest. He has traveled about without Flo for more than an hour or so only on very rare occasions.

Similar symptoms of depression have appeared in other five- and six-year-olds for a few months immediately following the birth of a sibling, when they therefore no longer have first claim on the caretaking responses of the mother. In some cases this appears to be offset by the intrinsic fascination of the new baby for the juvenile sibling.

The significance, for the youngster, of the affectionate bond with his mother was illustrated dramatically when three individuals, all between three and four years old, lost their mothers. Two of these were "adopted" by elder siblings with whom they traveled and slept at night. But in spite of this, and in spite of the fact that they seemed well integrated into the chimpanzee society, both became increasingly listless and both showed declining frequency of play during the first few months. Subsequently the behavior of one of them, Merlin, became increasingly abnormal and he developed a number of unusual patterns and stereotypes of the kind associated with early social deprivation in the laboratory chimpanzee. These included rocking back and forth, hanging upside down and motionless for minutes on end, and pulling out his hairs individually during self-grooming. In addition he showed more submission and aggression and spent more time in social grooming than is normal for youngsters of his age. In some social responses and in some tool-using techniques, Merlin's behavior appeared to show some regression. He finally died of a paralytic disease but was so emaciated by then that it was almost certain that he would have died anyway.

The second orphan, Beattle, was adopted by an older and more experienced female than Merlin's sister. Beattle not only traveled and slept with her sister but was also permitted to ride about on her guardian's back, a luxury denied to Merlin, who was probably too heavy for his sister. Beattle's condition gradually improved and, about a year after her mother's death, her behavior was comparable with that of other youngsters her age. We might speculate that the added social security she derived from close physical contact with an experienced, almost-adult female who knew how to behave in moments of social excitement helped to minimize the psychological shock caused by her mother's death. We cannot, however, draw any firm conclusions from so inadequate a sample.

The third three-year-old had no elder sibling and after her mother's death wandered about for the most part quite alone. She quickly became lethargic, stopped playing almost entirely, and two months after her mother's death stopped visiting the feeding area. She was not seen again and was finally presumed dead.

During its sixth or seventh year a juvenile is increasingly likely to become accidentally separated from its mother. Initially this usually results in obvious distress on the part of the child (and sometimes on the part of the mother, too). The lost youngster starts to whimper and then scream as it scans the countryside in all directions, often from the top of a tall tree. Perhaps it is only after a series of such accidental separations that the juvenile itself finally initiates a brief bout of independent travel. The young male may start to move about in groups without his mother during his seventh or eighth year, but he does not normally spend more than a few days at a time away from her until his ninth or tenth year. The female may remain almost constantly with her mother for even longer.

The long period of dependency on the mother may be considered adaptive in the chimpanzee, as in man, in relation to social learning. In simpler forms of life, much behavior is almost entirely genetically coded, although at all levels individual experience undoubtedly plays some role in the development of behavior. But as the mammalian brain becomes increasingly complex, culminating in the brain of higher primates and of man himself, social learning plays a vastly more crucial role. The adult chimpanzee lives in a complex society: he must learn to recognize 30 or more individuals in order to react appropriately when he meets them. In addition to appreciating the status of each of these individuals in relation to his own, he must also know how the presence or absence of a high-ranking associate may affect his own or his companions' ranking in the hierarchy in a given group. Chimpanzees have an elaborate system of communication with each other, and their behavioral repertoire also includes such complex patterns as cooperative hunting and tool using. Learning undoubtedly plays a major role in the development of a youngster, and the years when he associates closely with his mother are undoubtedly to his advantage: he can rely on her to react appropriately to individuals of high or low rank, to assist him in times of stress, and to lead him to appropriate food sources. Initially, when the mother cares for his every need, the infant can direct most of his energy into exploring his physical and social environment, and even when he becomes a juvenile and must to some extent fend for himself, he can still spend much time, under the benign leadership of his mother, in gradually acquiring the skills and competences that will fit him for adult life.

It has been shown experimentally that nonhuman primates are able to learn through direct observation of the behavior of others, and the wild chimpanzee infant certainly has much opportunity for learning of this sort. An infant often watches intently while his mother, or another older individual, is engaged in tool using, nest making, feeding, and a whole variety of social behaviors. Subsequently he may be seen to imitate the actions he has observed—either immediately, as when he picks up a grass tool just discarded by his model and endeavors to use it in a similar manner, or a short while later, as when a youngster watches an adult male performing a charging display and, when things are calm again, repeats some of the display movements himself, often in a seemingly playful context. Behaviors of this sort may be practiced time and time again. Thus, while a gradual maturation of locomotor and manipulative patterns undoubtedly plays a vital role in the development of many of the complex activities of the adult chimpanzee, it is almost certain that learning, through both trial and error and direct observation of models, is also a very significant factor.

AFFECTIONATE BONDS IN THE CHIMPANZEE FAMILY

When we refer to a family in chimpanzee society, we mean a mother and her offspring of different ages, together with her daughters' offspring. There is no "father" as such. The male, after playing his role in the conception of an infant, has no further part in the raising of an offspring since, as I already mentioned, no permanent pair bonds develop between male and female chimpanzees. Our study has revealed, however, that the affectionate bonds between mothers and their offspring and between siblings are sometimes strong and long-lasting.

As we have seen, the young male associates very closely with his mother until he is nine or ten years of age. Moreover, all five of the males we observed whose mothers were alive during their adolescence continued to associate with them frequently during that period. We have now been able to make detailed observations on the relationships between three old females and their socially mature sons—that is, males more than 15 or 16 years of age. One of these females had two such sons, the others one each. All these young males associated quite frequently with their mothers, and during such times social grooming between mother and son was frequent. Moreover, on a few occasions mothers were observed to hurry to the assistance of adult as well as adolescent sons; similarly, sons occasionally assisted their mothers. Sade has reported similar lasting bonds between mothers and sons in the rhesus monkey population on Cayo Santiago.

Of interest is the fact that we have not yet observed a sexually mature male try to mate with his mother. Our sample size is too small for conclusions to be drawn about this, since it involves only one mother with two adult sons and another with one. However, in one instance the mother, during four days of oestrus, was mated by every other mature male in her group with the exception of her two adult sons, who were also in the group. Some inhibition of mother-son mating has been observed in two other longitudinal studies of primate societies, Japanese monkeys and rhesus macaques.

Females tend to remain closely associated with their mothers for even longer than males but to date we have been able to follow the development of a relationship between a mother and her daughter into the latter's adulthood in only one case; in other cases either the mother or the daughter died before the daughter became adult.

Of the five mother-daughter pairs we have been able to follow through the juvenile and at least the early adolescent period, two mothers showed affectionate and protective behavior toward their daughters similar to that described for the mother-son relationship. However, the other three mothers were far less tolerant of their female offspring than of their sons, and all three daughters showed fear of their mothers in some contexts, especially in feeding situations. Nishida and Kawanaka describe two adolescent females who initially associated frequently with their mothers but who, when they became sexually receptive, began to travel about independently of their mothers for much of the time.

In the case where we have been able to document the relationship with a mother and her daughter from the latter's infancy to social maturity (the old female Flo and her daughter Fifi), the bond between the two has always been of a very relaxed and tolerant nature. In 1971 Fifi had her first infant (our major hallmark for social maturity in a female), so that for the first time we now have the opportunity to study the development of bonds between grandmother and grandson and between nephew and uncle in wild chimpanzees. Fifi still associates frequently with her mother and youngest brother, Flint, and the relationship between these individuals is presently being carefully studied by M. Hankey.

The extension of the affectionate bond between a mother and her offspring beyond weaning means that when the mother has a new infant this youngest member of the family is likely to have a great deal of contact with his older sibling, who will usually still be semidependent on the mother. He will also have (though to a lesser extent) contact with older independent siblings during those times when they associate with the mother. Thus, as Sade has pointed out, the parent-offspring relationship may ramify into other relationships of potential importance.

We have already seen that an orphaned infant may be adopted by an older sibling; this was the case even when the caretaker (of an infant of 14 months) was a juvenile *male*. This orphan was too young to survive without her mother's milk, but if we examine the three cases in which the orphans were between three and four years old we find that one of those adopted by a sister survived to become a normal youngster; the second, although he eventually died, nevertheless lived for 18 months after the death of his mother. The third, with no elder sibling, disappeared and almost certainly died within three months.

Observations on two pairs of brothers suggest that long-term bonds, similar to those observed between a mother and her son, may typically be formed between brothers. Similar close bonds are found in rhesus monkey brothers.

To date we have been able to study the relationship between only one pair of chimpanzee sisters, the eldest of whom is close to social maturity and the other approaching adolescence. These two associate with their mother almost all the tim and the bond between all three is very close.

Other than juvenile-infant pairs, we have been able to document in detail the relationships between brothers and sisters in only three families, and these data suggest that the bonds between siblings of different sex tend to become weaker as the individuals mature.

Mating does occasionally occur between brothers and sisters, but it is extremely rare. One young female (Fifi) repeatedly ran off screaming when she was first approached by her two brothers in a sexual context, although she was quick to respond to the courtship displays of other males. Eventually both brothers did achieve sexual intercourse with their sister, but thereafter they were observed to copulate with her only a few times during the two years prior to her pregnancy. One other young female was observed to be mated by her elder brother only a few times during 12 or more periods of oestrus; another adolescent female, who has been sexually receptive for over a year, has not yet been observed mating with her elder brother.

ADOLESCENCE

Adolescence is another area about which our longitudinal observations are yielding interesting information. This period of the life cycle is considered by some to be solely culturally determined and therefore unique to man. However, chimpanzees (as well as most other nonhuman primates) show both physiological and behavioral changes around puberty and during the following few years; this makes it appropriate to distinguish a period of adolescence. It commences just prior to puberty (approximately eight years of age) and ends when the individual reaches social maturity (about 12 or 13 years of age in the female and some two years later in the male).

The female, as we have seen, tends to associate with her mother for an even longer period of time than the male. Since a mother is quite likely to be nursing an infant during her daughter's adolescence (and even if she is not, she will certainly associate with other females and infants from time to time), it is unnecessary for the adolescent female to leave her family group in order to gain experience concerning her future role as a mother. Some adolescent females are fascinated by small infants and spend much time playing with, grooming, or carrying them. A female with this "maternal" approach is likely to become very preoccupied with her own infant sibling. Other females seem less interested, particularly for a while after the onset of regular sexual cycling.

The female sexual cycle is characterized by menstrual bleeding and periodic swelling and deturgescence of the anogenital region. The increase in genital swelling coincides with a very marked increase of attraction and receptivity to males. The average length of the cycle is about 34 days. Maximum genital swelling averages six to seven days; menstruation occurs between six and 12 days from the start of detumescence. These swellings first appear as a very slight turgidity of the clitoris in a seven- or eight-year-old female and, with each successive month, gradually get larger. Some five to ten months prior to menarche (between eight and a half and ten years of age for captive chimpanzees, according to Asdell and Riopelle) the adolescent female develops a much larger swelling and suddenly becomes sexually attractive to adult males.

Females show a great deal of individual variation in their initial responses to the sexual advances of mature males. Since courtship typically comprises many gestures that occur also in aggressive contexts, some females are extremely fearful initially and may try to escape, screaming, when a male approaches for copulation. Other females seem to take sexual approaches as a matter of course.

For several months prior to her first period of receptivity, one adolescent female (Fifi) frequently remained very close to older females who showed genital swellings, apparently in order to be on hand when they were mated. She would then either jump on the back of the other female and press her own genital area as close to the male's penis as she could or else go round behind him, turn her back on him, and press her genital area against his rump during copulation. As might be expected, she was quick to respond to the slightest courtship gesture when she herself became receptive for the first time.

In many groups of nonhuman primates, exchange of genes between groups occurs when males from neighboring ranges change groups. This is well documented for rhesus monkeys, Japanese monkeys, and baboons, and it almost certainly occurs in gorillas. Some male chimpanzees at Gombe have sometimes been absent from the feeding area for up to two months; during such times they may have visited neighboring communities and possibly mated with their females. If that were so, however, one would expect males from neighboring communities to occasionally mingle with individuals of our habituated group, but there is no evidence that this has occurred.

Transfer of females from one group to another is known to occur occasionally among baboons and among gorillas. Among chimpanzees, however, there is much evidence that females, principally adolescent females, play a major role in widening the gene pool. At Gombe there are a number of records of adolescent females temporarily leaving their home communities during periods of oestrus and mixing and mating with males of neighboring communities. In at least two cases, young "stranger" females have gradually become integrated into our community of habituated chimpanzees, although to date no habituated adolescent female has permanently left our area. Nishida and Kawanaka have now recorded 39 cases of females transferring from one community to another during oestrus; the majority of these females were adolescent.

It is especially interesting to note that it is the adolescent female herself who appears to initiate the change in range; she does not normally seem to be forced to leave her home area by threatening behavior on the part of males. In our own species, it is very often the women who leave their natal groups and move to live with their new husbands in villages—or even countries—that may be quite strange to the women concerned.

Adolescence often seems to be a stressful period for the male chimpanzee. He has a growth spurt after which he tends to become more aggressive, particularly toward females. By the time he is about ten years old he is able to dominate many females who a few years earlier were themselves able to subdue him with ease. At the same time he must learn to behave with increasing caution to avoid arousing the aggression of the mature males, some of whom are quick to threaten him for behavior they tolerated when he was a mere juvenile. Nevertheless, despite the fact that he may become increasingly fearful of these older males, he often seems to deliberately choose to associate with them. Many of his first journeyings without his mother are made with adult males.

When he is with older males the adolescent tends to occupy a peripheral position in the group. He desists from sexual activity with a popular female—or at least waits for a quiet moment when the big males are resting peacefully. He sits a few yards away from a group of males who are grooming each other but usually does not dare to join them. He often feeds at a slight distance from his superiors. Yet he can often be observed watching the older males very intently.

After associating for a while with his superiors, and particularly if he has been the victim of their aggression, the adolescent male moves away from them and either travels about for a while on his own or returns to his mother if she is still alive. Indeed, his relationship with his mother may well be one of the most stabilizing factors at this time of social change for, despite his larger size and more aggressive behavior to females in general, his behavior toward her remains remarkably constant.

At some period during his adolescence, the young male may engage in a number of status conflicts if there is another male of similar age and social rank as himself with whom he associates quite frequently. These conflicts mainly involve bluff, but may sometimes lead to physical attacks. The charging display appears to have special significance in this context. This is a typically male performance that occurs in a variety of situations, principally when a male arrives at a new food source, when he meets up with other chimps after a separation, or when he is frustrated in the attainment of some goal (when, for instance, he is inhibited from feeding or mating due to the proximity of a superior male). During the display he runs slowly or very fast, on all fours or upright, with hair bristling; he may drag or throw branches, hurl rocks, sway branches, leap through the trees, or (in the case of two individuals) beat his chest like a gorilla. This display enables the performer to look larger, more powerful, and more dangerous than he may really be; it appears to be a useful technique in the acquisition of a higher social status. The more frequent and the more impressive the display, the more rapidly the adolescent may rise in the hierarchy. Since such a display only seldom involves a physical attack on a rival it is adaptive in that it enables a chimpanzee to acquire status without any great risk of injury to himself.

During the final years of adolescence the male gradually begins to threaten and occasionally attack the lower-ranking males of the adult male hierarchy. When he is able to subdue even one of them consistently he can be considered part of their hierarchy and thus socially mature.

DOMINANCE

The past ten years of research at Gombe have brought to light some interesting facts concerning dominance, particularly with regard to those qualities which enable a chimpanzee to acquire the alpha position in his community. In 1963 Goliath, a powerful and aggressive male in his prime (perhaps about 25 years of age) was the alpha male. He had a spectacular charging display during which he covered the ground very fast indeed, dragging and occasionally hurling branches. Early in 1964, however, Goliath was displaced from his top-ranking position in the community by an older and much less robust male, Mike. Mike apparently accomplished this by means of bluff and, without doubt, superior intelligence.

In 1963 Mike was among the lowest-ranking of all the adult males, frequently threatened or even attacked by most of the others. Then he began to incorporate empty four-gallon kerosene cans into his charging displays. Other males had occasionally seized hold of such a can (from my camp area) during a display, but it seemed that only Mike was able to profit from the experience and deliberately use cans to enhance his performance. He soon learned to keep up to three cans ahead of him as he ran, very fast, hitting or kicking them along as he went. He made a great deal of noise; other males as well as females and youngsters rushed out of his way.

After charging past a group of males a few times and scattering them, Mike then sat with his hair erect; the other males, previously his superiors, approached him with gestures of submission or appeasement, crouching before him, touching, kissing, or grooming him. Mike rose to become alpha male of his community in about four months without engaging in any actual attacks on other males that we observed. He maintained that position even when we took the cans away (for we too disliked the noise) for the following six years.

Mike, in his turn, was relieved of his top-ranking position by a younger, very large, and extremely aggressive male, Humphrey. For about a year previously Mike, who was getting old, had appeared uneasy in his alpha position and was being ignored during displays by some of the younger males. Only one fight was observed between Mike and his successor, although other incidents may have occurred prior to this. Unlike Goliath, who had maintained a very high-ranking position for several years after losing his alpha rank, Mike dropped rapidly to a low position in the hierarchy. This was possibly due to the fact that at the time of the takeover, he was very much older than Goliath had been.

In chimpanzee society dominance is something of a conundrum. The usual interpretation of the phenomenon is that it enables a high-ranking individual to have prior access to desirable foods, females, or resting places. Competition over these resources is rare in chimpanzee society, at least at Gombe. Possibly a high-ranking position is desirable because, once an individual has attained it, he need no longer fear threat or attack from too many other chimpanzees of higher rank. Whatever the underlying motivation or ultimate reward, however, many chimpanzees do seem preoccupied with raising their social status, while others are less concerned and tend to keep out of the way during times of social excitement.

The story of Mike's rise to alpha male provides an excellent example of the way in which chimpanzees typically rely on threat or bluff rather than actual physical violence in their interactions with each other. Fights do occur, but they are generally brief; even when they appear vicious, they seldom result in observable physical injury. Moreover, after a fight the victim ordinarily approaches the aggressor showing postures and gestures that have been labeled submissive or appeasing and in response to which the aggressor usually reaches out to touch or pat or kiss the subordinate. This reassurance behavior serves to calm the agitated victim and helps to ensure generally relaxed and peaceful relationships among those individuals who frequently associate in the community.

SIMILARITIES IN CHIMPANZEE AND HUMAN BEHAVIOR

I have outlined various aspects of the life cycle of wild chimpanzees that it has been possible to document because of the long-term nature of our study. As knowledge is gradually accumulated about a variety of different monkey and ape species, it is possible to trace certain evolutionary trends of increasing complexity, culminating in man. Some of the data I have discussed so far clearly illustrate the high position of the chimpanzee on this evolutionary ladder. If, for instance, we take a primitive New World monkey, the marmoset, we find that infants are weaned at six months and attain sexual maturity at 14 months. The gestation period is 140 days, and estimated longevity in captivity about ten years. If we next take the rhesus monkey we find that weaning takes place any time between six months and a year, or sometimes a little later. The monkey is reproductively mature at about four years of age, though it does not acquire full size until about ten years. The average gestation period is 164 days and the life expectancy in captivity about 30 years. Now we come to the chimpanzee, where the youngster is not weaned until he is about five years old and is dependent on his mother for another three or four years, where reproductive maturity is not reached until seven to eight years in the male and (in the wild) about 11 or 12 years in the female. Moreover, the male does not become socially mature for some five or six years after puberty and is unlikely to be reproductively effective in his society until he is at least 12 years old and probably older. The longevity record is almost 50 years in captivity. In all these characteristics the chimpanzee can be seen to be very close indeed to man.

I should now like to discuss some of the behavior patterns of chimpanzees that strikingly resemble some patterns in man.

COOPERATIVE HUNTING AND FOOD SHARING

As mentioned earlier, chimpanzees are omnivorous. They quite frequently hunt fairly large prey animals such as the young of bushpig, bushbuck, and baboons, and young or adult monkeys of various species. Sometimes the capture of prey is a very individual and opportunistic event: one chimpanzee happens upon a suitable victim, seizes and kills it, and begins to feed. On other occasions, however, male chimpanzees may hunt in a group and show behavior that can clearly be labeled as cooperation. For instance, one chimpanzee may cautiously creep up a tree toward a potential victim, such as a young baboon that has become slightly separated from its troop, while other male chimpanzees stand around the base of this tree and also near trees that might act as escape routes for the quarry. Not until the victim becomes aware of danger and tries to make a break will the chimpanzees on the ground leave their positions and converge on the fleeing baboon.

Meat is a delicacy for chimpanzees, and normally a great deal of excited calling accompanies the successful capture of prey. This attracts the attention of any other chimpanzees in the vicinity and they typically converge upon the hunters and cluster around, either watching intently or actually begging from the chimpanzees who have acquired shares of the carcass. When begging, a chimpanzee may hold his hand to the mouth of an eating individual in the hope that he will be given the remains of the mouthful, or he may hold out his hand toward the other, palm up, in the typical begging gesture of man.

Often begging behavior is rewarded. This is significant since food sharing among adults has not been recorded in any other nonhuman primate in the wild. After chewing for a while on a mouthful of meat and leaves (leaves are almost always eaten with meat) a chimpanzee will nearly always allow a begging individual to take it. Sometimes he will permit others to chew on the carcass at the same time as himself, or let them detach pieces for themselves. Occasionally he will actually tear off a piece himself and hand it to the one who is begging.

TOOL USING AND TOOL MAKING

With respect to his tool-using behavior, the chimpanzee comes considerably closer to man than does any other living primate. Tool using in animals has always fascinated people because, for many years, man was typically referred to as *the* tool-using animal. A variety of animals (including invertebrates) do use objects as tools if we define tool using as "the use of an external object as a functional extension of mouth or beak, hand or claw, in the attainment of an immediate goal." It should be emphasized, however, that tool-using ability in itself does not necessarily indicate any special kind of intelligence. The Galapagos woodpecker finch uses a cactus spine to probe insects from crevices in the bark. This is fascinating behavior, but it does not make this bird more intelligent than the ordinary woodpecker, which uses its long beak and tongue for the same purpose. The Galapagos finch employs a behavioral adaptation, the woodpecker a morphological one.

The point at which tool using in a nonhuman animal acquires significance, when viewed in relation to the evolution of tool use in man, is when a species or an individual within that species can adapt its ability to manipulate objects to a wide variety of purposes and, in particular, when it can use an object spontaneously to solve a completely novel problem. The chimpanzee has shown itself capable of using a wide variety of objects for a wide variety of purposes, both in the wild and in captivity. Moreover, if an object is not suitable for the specific purpose for which it is to be used, the chimpanzee will modify it accordingly, so that he may be said to show at least the beginnings of toolmaking behavior.

Our research at Gombe has shown that one community of chimpanzees uses four different kinds of objects—grasses, sticks, leaves, and rocks—in a wide variety of different contexts. Sticks are used to plunge into ant or bee nests during feeding and occasionally as levers to enlarge the opening of such a nest. Sticks were used by chimpanzees to try and force open banana boxes at our feeding area. They may also be used during aggressive displays to intimidate other chimpanzees, baboons, or humans. One infant "clubbed" an insect on the ground. Grass stems are used when the chimpanzees are feeding on termites; they are also used as "investigation probes" to touch objects that the chimpanzees cannot reach with their hands or of which they are afraid. The end of the grass is then sniffed. Sticks and grasses may be trimmed to size and made more suitable in any of the above contexts. Leaves are used to wipe dirt off the body, and also as a "sponge" to sop up water in a hollow in a tree trunk when the chimp cannot reach the fluid with his lips. Here again the object is modified: before dipping the leaves into the water the chimpanzee crumples them by briefly chewing them. This makes his sponge considerably more absorbent and effective. Rocks are used as missiles during aggressive incidents and may be thrown forcefully and with good aim. One infant pounded an insect on the ground with a rock that he held in his hand. Once a chimpanzee used a twig to pick his nose; another individual used one to pick at something in her teeth.

There is not space here to detail the tool-using performances of laboratory- and home-raised chimpanzees, although some of them are very impressive. However, man is still the only primate to "make tools to a regular and set pattern," and he is the only creature who relies on tools for his very survival.

NONVERBAL COMMUNICATION PATTERNS

Perhaps it is in the repertoire of postures and gestures which form the nonverbal communication system that some of the most striking behavioral similarities between man and chimpanzee are found. Chimpanzees may bow, kiss, hold hands, touch and pat each other, embrace, tickle one another with their fingers, bite, punch, kick, scratch each other, and pull out one another's hair. Not only do many of these movements look remarkably like many of the expressive postures and gestures of our own species, but the contexts likely to elicit the behaviors may be strikingly similar in the chimpanzee and man.

When an adult chimpanzee is frightened he may reach out to touch or even embrace another who happens to be nearby; this physical contact often has a calming effect on the gesturer. If both individuals were frightened, then both may be reassured. Similar contact-seeking behavior occurs commonly in man in stressful situations.

When two or more chimpanzees become suddenly excited—if, for instance, they suddenly come across a large amount of a favored food—they often exhibit intense contact-seeking behavior, embracing, kissing, holding hands, and patting each other, all while screaming loudly. Eventually they calm down sufficiently to start feeding. This behavior may be similar to that shown by a human child who, when told of a special treat, may fling his arms around the bearer of good news. When peace was declared in London after World War II, the streets were full of people embracing, clasping hands, kissing, laughing, and crying.

After a young chimpanzee has been threatened or attacked, his need for reassuring physical contact with the aggressor is sometimes well illustrated. As he approaches the adult, still screaming, his behavior is tense; his limbs are flexed so that he is close to the ground and his movements are often jerky. As he gets close to the aggressor he may actually turn and move away for a few steps, but then he turns back again as his desire for contact apparently overcomes his desire to flee. When he finally reaches his superior he adopts a submissive posture, crouched close to the ground, still screaming and with his teeth and gums exposed in a wide grin of fear. In response the other chimpanzee will usually reach out and touch or gently pat the youngster, sometimes continuing for almost a minute until the subordinate quiets and gradually relaxes. Here again we find a similar pattern in man. In most societies people who have been quarreling will "make up" by means of physical contact, whether this be the cuddling and petting of a child or marital partner, or a more formal clasping of hands or brief pat on the back.

At this point I should like to emphasize that while the form of reassurance behavior looks very similar in man and chimpanzee—not only with respect to the gestures themselves and the context in which they may occur, but also in the relaxing effect of the physical contact on the submissive individual—the motivations underlying the behavior are probably quite different in man and chimpanzee. This is a fascinating and complex subject that I hope to discuss in detail elsewhere.

When two chimpanzees meet after a separation they may show gestures and postures that strikingly resemble some forms of human greeting. In such a context chimpanzees may bow, kiss, touch, or pat one another, hold hands, or embrace. A male may chuck a female or an infant under the chin. In chimpanzee society, reunion after separation often involves behaviors that serve to reestablish or emphasize the relative social status of the individuals concerned; originally, greeting behavior in man undoubtedly served a similar function. Indeed, this is still the case on some formal occasions, though human greeting behavior has for the most part become ritualized in our different cultures.

There are also similarities in the aggressive behaviors of man and chimpanzee. A quick upward jerk or a waving of the arm serves as a threat in both species, as may a level stare directed unwaveringly at a subordinate. A chimpanzee may adopt an upright posture when threatening and wave his arms above his head as he advances on the object that has elicited his aggression. He may throw objects overarm or underhand toward his victim. He may brandish a stick or make deliberate clubbing movements. Attacking chimpanzees may bite, hit, and kick their opponents. Female chimpanzees in particular sometimes scratch at and occasionally pull out handfuls of hair from their victims' heads.

THE UNIQUENESS OF THE HUMAN PRIMATE

It will have become apparent that even while I have been stressing the similarities between man and his closest living relative, many points of difference have been touched on. In fact, the longer I study chimpanzees and the more amazed I become at the extent of the behavioral similarities in the chimpanzee and man, the more I stand in awe of the gigantic evolutionary stride that our own species has taken in the last few million years. Let me very briefly enumerate a few of the characteristics that make man unique among the primates and that are highlighted by an understanding of the chimpanzee.

The chimpanzee has a wide range of calls. Each call is fairly reliably associated with some specific emotion—a scream usually signifies fear, a bark aggression, and so on. This means that the chimpanzee is able to communicate reasonably specific information about his feelings and his environment through the use of calls alone. But his repertoire of grunts and hoots cannot be compared with human language. Moreover, recent research into the speech centers of the human brain suggests that our language did not even originate from this type of primitive vocal communication.

The chimpanzee cannot convey information about the past and the future; he cannot pass on, through his vocabulary of calls, experiences of his own that might benefit his group or his species as a whole. It is true that recent work in teaching chimpanzees to communicate by means of sign language and with plastic word symbols has shown convincingly that the chimpanzee is capable of a greater sophistication of mental process than most people were formerly prepared to accept, such as the ability to appreciate abstract concepts. But the chimpanzee intellect is dwarfed by that of a species that can produce a Plato, an Einstein, a Beethoven.

When the chimpanzee Washoe, trained in sign language, was shown herself in a mirror for the first time and was asked, in sign language, "What is it?" she signed back, "Me, Washoe." This suggests that the chimpanzee may show the beginnings of self-awareness. But man not only is conscious of himself as an individual being, he also questions the reason for his existence, he searches for clues as to how he came to be the way he is, he strives to understand his own behavior and the world and the universe in which he finds himself.

The chimpanzee, especially as a young adult, does show what might be considered the beginning of an affectionate heterosexual pair bonding: a male and a female may wander off together during the female's period of oestrus and sometimes for longer. At such times they may stay on their own, away from others of their kind. But this kind of relationship cannot be compared in any way with the tenderness and passion, the understanding and compassion, the exaltation and peace, that mark human heterosexual love in its highest form.

It is of interest that man has also leapt ahead of the chimpanzee in what we might call negative as well as positive intellectual evolution. Man can use language to the advantage of himself, his group, his species. With words he can inspire others to acts of goodness and self-sacrifice. He can also use words to incite his fellowmen to deeds of evil and destruction. Despite his awareness of himself and of the individual existence of his fellowmen, and despite his understanding of the world he lives in, he can destroy not only human and animal life but the very environment that, as he alone among living creatures is capable of appreciating, enables him to live. Man, with his tremendous capacity to love, has a capacity to hate far transcending the worst of the squabbles that may break out between chimpanzees. Two neighboring communities of chimpanzees may occasionally indulge in displays of power as individuals hurl rocks and wave branches or even briefly attack one another. But they show nothing even remotely comparable to the horror of human warfare.

Today we are all aware of the tremendous need for a better understanding of our own species, yet only too often the scientist, searching for the biological basis of some human pattern of behavior, is confounded because intellect and culture have conspired to confuse the picture. We cannot afford to neglect any approach that might give us a better understanding of even a few aspects of man's evolution, his biology, his social behavior. And I believe the study of the chimpanzee is important in this respect.

And so we are continuing our intensive study at the Gombe Stream Research Center, where postdoctoral, graduate, and undergraduate students from the universities of Dar-es-Salaam, Stanford, Cambridge, and others are working as a team in our efforts to understand chimpanzee behavior more completely. Dr. David A. Hamburg is setting up a new facility at Stanford—"Gombe West"—where chimpanzees will be studied in groups in large outdoor compounds. Here we shall investigate problems that cannot be tackled in the field but that we know from our experience are relevant to the understanding of the behavior of the species. An initial area of inquiry will be the problems of adolescence; methods and goals will be closely linked not only with the research at the Gombe Stream but also with ongoing studies of human behavior. We hope that some problems arising in research with human subjects may be solved through work with the chimpanzees; in turn, work with chimpanzees may pinpoint areas of behavior that merit careful investigation in human subjects.

The study of the chimpanzee, with all his complexity of behavior and his close evolutionary relationship to man, is exciting and worthwhile in itself. If our research can help, even in the smallest way, in the better understanding of man, then our work will be even more rewarding.

11. BABOON ECOLOGY AND HUMAN EVOLUTION

IRVEN DeVORE and SHERWOOD L. WASHBURN

Reprinted from F. Clark Howell and François Bourlière (Eds.), African Ecology and Human Evolution *(Chicago: Aldine Publishing Company, 1963). Irven DeVore is an Associate Professor in the Departments of Anthropology and Social Relations at Harvard University. His long-range research interest is in the evolution of human behavior, for which he uses the results of field studies of monkeys, apes, and living hunter-gatherer peoples as aids in interpreting the archeological record. He has done field work with baboons in Africa and among the Bushmen of the Kalahari Desert. His publications include* The Primates *(with Sarel Eimerl) and* Primate Behavior: Field Studies of Monkeys and Apes. *For a biographical note on Sherwood L. Washburn, see Selection 8. This selection in its original form contains a number of photographs, maps, and tables that have been deleted here.*

■ Although man did not evolve from the baboon, study of the baboon's adaptation is important because he occupies an area that once was occupied by *Homo*. Hence the two faced similar selective pressures. In the baboon we observe an evolutionary dead end, but *Homo's* relationship to this habitat was not a dead end.

In the following selection, DeVore and Washburn provide a thorough account of baboon adaptation, which makes possible a further exploration of the common capacities of man and his nonhuman primate ancestors. For example, after we read about the ways in which baboon troop organization is designed to protect females and juveniles, can we say that such a dictum as "women and children first" is exclusively cultural? Or must we assume that the cultural rule is an expression of an inherited capacity that had its genesis in nonhuman forms? DeVore and Washburn examine several similarities and differences between baboon and human patterns of adaptation. Although they explicity deny a lineal relationship between the two, the authors suggest that our understanding of human adaptations can benefit from an examination of baboons' adaptations. Their exploration bears directly on many of the hypotheses that are suggested by Hallowell (Selection 15).

A capacity for gregariousness and organized social relations among primates is clearly established at the infrahuman level. An interesting aspect of this capacity is the process of segmentation, described below for baboons, which has remarkable echoes in many human societies. Segmentation in some groups is a process by which an original group will split into several subdivisions, each mirroring the organization of the original group. Again we can ask if human social segmentation is a process that is unique among *Homo* or an inherited capacity that is manifested under particular conditions. Whichever it is, it is evident that the inability to survive without group life descends from man's nonhuman past.

DeVore and Washburn point out that variations in baboon group structure correspond to variations in habitat. For example, they note the different group organizations that occur when the baboons are in or under trees and when they are in the open. This change is closely tied to the discussion of the relationship between territoriality and social structure, especially in intergroup relationships. Another important topic is the adaptive value of the smaller size of the females, in contrast to the males. As most male primates seem to know, it is easier to maintain a petite female than a large female.

Baboons are primarily herbivorous, and meat is only a very small proportion of their total diet. The shift from a herbivorous to a carnivorous diet was one of the major events in evolution. But first let us see what the herbivores are like.

An excellent review article has been written by S. L. Washburn, Phyllis C. Jay, and Jane B. Lancaster: "Field Studies of Old World Monkeys and Apes" (*Science,* 150 [1965]: 1541-47). Also of direct relevance are several papers in *Behavior and Evolution,* edited by Anne Roe and George Gaylord Simpson (New Haven: Yale University Press, 1958), and Niels Bolwig's "A Study of Behavior of the Chacma Baboon, *Papio ursinus*" (*Behaviour,* 14 [1959]: 136-163). ■

THE ECOLOGY OF BABOONS is of particular interest to the student of human evolution. Aside from man, these monkeys are the most successful ground-living primates, and their way of life gives some insight into the problems which confronted early man. We have been concerned with an attempt to reconstruct the evolution of human behavior by comparing the social behavior and ecology of baboons with that of living hunter-gatherer groups, and applying these comparisons to the archaeological evidence. The following description of baboon behavior and ecology is based on field data collected during 200 hours of observation by Washburn in the game reserves of Southern Rhodesia in 1955, and on more than 1,200 hours of observations by both of us in Kenya game reserves during 1959. The original study was financed by the Wenner-Gren Foundation for Anthropological Research, and the second trip was part of a study of the origin of human behavior supported by the Ford Foundation. Analysis of the field data is being completed under a National Science Foundation grant for the study of primate behavior. We wish to thank the foundations, and the numerous people who helped us in Africa—especially J. Desmond Clark, Stephen Ellis, L. S. B. Leakey, B. L. Mitchell, and B. Verdcourt.

CLASSIFICATION

This paper will primarily consider troop size, range, population density, and diet, but before discussing these topics we wish to give our views on the classification of these primates. Baboons are large, primarily ground-living monkeys of the family Cercopithecidae. As has been true of many of the primates, this group has been so divided that generic names have been applied to taxonomic groups which amount to no more than species. The most widely distributed baboon group occurs in the savanna and forest from the Tibesti Plateau in the north to Cape Town in the south, and across Central Africa from Dakar to the east coast. This group, the genus *Papio,* is usually divided into several species, including "chacma," "yellow," and "olive." There is no evidence that these forms are more than racially distinct, however, and "chacma" and "yellow" baboons occur in the same troops in the Rhodesias, although not in extreme form. It is not known whether the East and West African forms are distinct species, but it is likely that intermediate forms exist there, as they do between East and South Africa. When separated by long distances, individuals from these races appear to be quite distinct, as is the case of the "chacma" from the Cape, the "yellow" in Nyasaland, or the highland and coastal races in Kenya. But if intermediate forms exist and reproductive isolation cannot be demonstrated, these varieties are best considered races.

In West Africa there are two species of short-tailed forest baboons, the drill and the mandrill. These forest types differ no more from the widely distributed form of savanna *Papio* than the pig-tailed macaque (*Macaca nemestrina*) differs from the crab-eating macaque (*M. irus*). In North Africa and Arabia the desert baboon, *Papio hamadryas,* lives in country which is too dry and open for the other species. In summary, the genus *Papio* is divided into a number of races and into at least four species, including the savanna species with several races, two forest species, and a desert species.

The gelada baboon, *Theropithecus gelada,* is very distinct from *Papio.* Its facial skeleton is constructed differently; it jumps and uses its tail differently; and it should probably not be regarded as a baboon at all. Today *Theropithecus* is confined to the mountains of Ethiopia, in the same region in which hamadryas occupy the lowland desert and "olive" baboons the savanna between the two extremes. But *Simopithecus* is probably *Theropithecus,* indicating that the *Theropithecus* group formerly extended into East Africa. The practice of putting almost every fossil primate in a new genus nullifies the utility of the genus as a taxonomic concept. *Cercopithecoides*, for example, is a *Colobus* monkey. The increased understanding which the presence of such forms might contribute to the reconstruction of the ecology of Olduvai or Sterkfontein is lost by the multiplication of names which separate the fossils from similar living forms.

The African baboons are very similar to the Asiatic ground monkeys, the macaques. Both groups have forty-two chromosomes, and their distribution does not overlap. The newborn are usually black, changing to brown. Skulls, teeth, and general physical structures are much the same. In social life and basic habits the two groups are very similar. In contrast to all other monkeys (both New and Old World), the macaques and baboons do most of their feeding on the ground. They can cross rivers and may live in dry areas, moving far from trees. Compared to other monkeys they are more aggressive and dominance-oriented, and their average troop size is considerably larger than any other species yet studied. These characteristics have enabled the baboon-macaques to occupy a much larger area than that of any other group of monkeys. It is an area very comparable to that utilized by *Homo* before the time of the last glaciation. Ground living, ability to cross water, an eclectic, varied diet, the protective troop, and aggressive males permitted the baboon-macaques to occupy this vast area with a minimum of speciation. The contrast in the number of species between ground-living and tree-living monkeys emphasizes this point. There are more species in the genus *Cercopithecus* in the African forests than among all the baboon-macaques from Cape Town to Gibraltar to Japan. There are more species of langurs in Southeast Asia alone than species of *Cercopithecus*. Further, the most ground-living of the langurs (*Presbytis entellus*) has the widest distribution, and the same is true for the most ground-living vervet (*C. aethiops*). The taxonomic contrast between tree and ground monkeys is clearly seen in Ceylon where the island is occupied by one macaque, one dry country langur, and four forest forms. Apparently in Ceylon the rivers have been a major factor in isolating the langurs, but they do not form barriers for the macaques. The general relation between ecology and taxonomy in the monkeys appears clear: the more ground-living, the less speciation. There are many more adaptive niches in the forest than in the drier regions.

The men of the Middle Pleistocene, genus *Homo*, occupied the same range as the baboon-macaques but without speciation. Their way of life (based on tools, intelligence, walking, and hunting) was sufficiently more adaptable and effective so that a single species could occupy an area which group monkeys could occupy only by evolving into at least a dozen species. This comparison gives some measure of the effectiveness of the human way of life, even at the level of Pekin and Ternifine man. Obviously, there is nothing to be gained by being dogmatic about the number of species of Middle Pleistocene men. Perhaps when many more specimens have been found it will be convenient to recognize two or three species, but the general form of this argument will still hold. There is no suggestion that any of the known fossil men (genus *Homo*) differ in size or form as much as a chacma baboon and a drill, or a crab-eating macaque and a pig-tail macaque. Even in its most primitive form the human way of life radically alters the relation of the organisms to the environment. As early as Middle Pleistocene times man could migrate over three continents without major morphological adaptation.

Australopithecus may have occupied an adaptive position midway in effectiveness between the ground monkeys and early *Homo*. Small-brained, bipedal tool-makers probably occupied larger areas than baboons, and without speciation. It is most unlikely that the East African and South African forms of *Australopithecus* are more than racially distinct. Robinson's suggestion that the jaws from Java called "Meganthropus" are closely allied to the Australopithecoid from Swartkrans supports the notion that *Australopithecus* was already able to disperse widely with minimum biological change. The presence of small and large Australopithecoids in South

Africa at the same time suggests that their adaptation was much less effective than that of *Homo*. It may be possible to reconstruct more of this stage in human evolution with a more thorough study of the ecology of baboons, and by contrasting their mode of adaptation to that of man. With this hope in mind we will now consider the ecology of baboons in East Africa.

THE TROOP

TROOP SIZE

Careful counts of more than 2,000 baboons showed a range in troop size from 9 to 185. Estimates by the Tanganyika Game Department, for troops in the Wankie Game Reserve (Southern Rhodesia), and carefully repeated counts in the Royal Nairobi National Park all give an average troop size of 36-42. The *largest* troop in Nairobi Park numbered 87. In the Amboseli Reserve, the *average* troop size was 80 and the largest troops numbered 171 and 185. The fact that troops are twice as large at Amboseli indicates the need to study several localities before generalizing. . . .

The smallest troops we observed numbered only 9 (Tsavo Reserve), 11, 12 (Wankie Reserve), 12 (Nairobi Park), and 13 (Amboseli Reserve). Three of these troops contained two adult males, and one only a single adult male; our data do not show the number of adult males in the fifth troop. These small troops are independent, functioning societies and not temporarily detached parts of larger troops. Baboon troops are closed social systems with a high degree of inbreeding. Often, adjacent troops can be distinguished from each other by the characteristic color patterns, length of hair, and form of face or tail. During both field trips, in over 1,400 hours of observation, we saw only two individuals change from one troop to another. In these very small troops, inbreeding may be very important, and a whole generation may be the offspring of a single male.

The largest troops (103, 171, 185) were seen only at Amboseli, but for our study we selected open areas where the baboons would be visible as much of the time as possible. In areas with more rain the abundant vegetation supports more baboons, and large troops may be more common in these areas. For example, in the reserve of the north end of Lake Manyara (Tanganyika), and in the forested areas adjacent to the Athi River near Kibwezi (Kenya), we saw approximately one large troop per mile. This suggests that there were both larger troops and a much higher population density in these areas, but under conditions where continued observation was impossible.

The large troops may temporarily subdivide, and the troops of 88, 94, and 103 at Amboseli, and the troop of 77 at Nairobi, frequently split. When the troops of 77, 88, and 94 split, all the small infants and their mothers were in one section of the troop with the largest adult males. On one occasion troop 171 (Amboseli) was also seen dividing in this way. When all the individuals in a troop are together, there is a clear distinction between the large, dominant adult males, mothers, and infants occupying the center of the troop, and the other, peripheral, troop members around them (as described below under "Troop Structure"). The temporary divisions seen in these large troops are divided along these lines. Such a subdivision lasts for only part of the day and the troop reunites before nightfall. Another type of splitting, in which the troop divides into two sections with a normal distribution of males, females, and juveniles in each section, also occurs in some large troops. Troop 103 sometimes split into two troops of 66 and 37, each troop having a center, a periphery, and all the characteristics of a normal, independent troop. It seems likely that this kind of splitting represents the first stage in the formation of a new troop. Observations on troops 51 and 66 support this. These two troops stayed very close together; if one of the troops arrived at a water hole, the other was likely to appear, and, after using adjacent sleeping trees at night, they often followed the same route away from the trees the next morning. It is tempting to regard this situation as representing a large troop divided one stage further than was the case of troop 103. The reason for regarding 51 and 66 as two troops is that individuals within them did not shift; repeated counts showed that the membership of these two troops was constant, and sometimes they were entirely separate from each other, once for a period of days. It appears that large troops may become unstable, and that divisions occur in troops larger than 70 individuals which are not seen in the small troops. If this division persists, and if the division contains a normal age-sex distribution, a new troop may result.

The division of troops, their large size, and the fact that they met at water holes made counting of troops at Amboseli very difficult. Troop 103, for example, was originally counted as three troops (one of 103, one of 66, and one of 37). Similarly, troops 51 and 66 seemed to form a troop of 117 on some days. A single count made at an Amboseli water hole might include only part of a troop, or a cluster of 400 baboons representing three adjacent troops.

The density of baboons in an area is related to the food supply, but the size of the troop itself bears no such simple relationship. The ranges of the smallest (13) and the largest (185) troops at Amboseli overlapped, and the size of Nairobi troops did not correlate with the different vegetation zones in the park. Social behavior, rather than ecology, seems to determine troop size. Because adult males defend the troop from other animals, troop size is important to the individual's survival. Troop 185 contained over thirty large, adult males, compared to only two in troop 13 and one in the troop of 12 (Nairobi). Like troops of the much smaller vervet monkey (*Cercopithecus aethiops*), a small baboon troop yields to a large troop when they meet at water holes. When food supplies are limited, this gives a large troop an advantage over smaller ones.

TROOP STRUCTURE

A detailed description of the social relationships within baboon troops is given elsewhere. Here we have emphasized those aspects of troop life which are adaptations to life on the ground. Baboons are intensely social, and membership in a troop is a prerequisite for survival. Most of a baboon's life is spent within a few feet of other baboons. Baboon troops are closed social systems, individuals very rarely change to a new troop, and the troop regards any strange baboon with suspicion and hostility.

Within the troop, subgroups are based on age, sex, personal preferences, and dominance. When a troop is resting or feeding quietly, most of the adult members gather into small clusters, grooming each other or just sitting. Juveniles gather into groups of the same age and spend the day in these "play groups," eating, resting, and playing together. The most dominant adult males occupy the center of the troop, with the mothers and their young infants gathered around them, and the groups of young juveniles playing close by. These dominant males, and the small black infants near them, seem to be greatly attractive to the other troop members. During quiet periods the other troop members approach the adult males and the mothers, grooming them or sitting beside them. It is unnecessary for male baboons to herd the troop together; their presence alone insures that the other troop members will not be far away.

Around this nucleus of adult males, mothers, and young juveniles are the most peripheral members of the troop—the less dominant adult males, older juveniles, and pregnant or estrus females. Estrus females and their consorts usually stay at the periphery of the troop. Although the juvenile play groups will not wander far from the troop's center, peripheral adults may leave the troop for short periods. While the center of the troop moves slowly along, the adult and older juvenile (subadult) males and adult females sometimes move rapidly ahead to a new feeding spot. This may separate them from the rest of the troop by a quarter of a mile or more, and they may not rejoin the troop for thirty minutes or an hour. Although peripheral adult males may make such a side trip alone, or in small groups, other troop members will not leave the troop unless accompanied by the males. Healthy "solitary males" observed during the early part of our study later proved to be troop members who had left the troop for a short while.

A baboon troop that is in or under trees seems to have no particular organization, but when the troop moves out onto the open plains a clear order of progression appears. Out in front of the troop move the boldest troop members—the less dominant adult males and the older juvenile males. Following them are other members of the troop's periphery, pregnant and estrus adult females and juveniles. Next, in the center, comes the nucleus of dominant adult males, females with infants, and young juveniles. The rear of the troop is a mirror image of its front, with adults and older juveniles following the nucleus and more adult males at the end. This order of progression is invariably followed when the troop is moving rapidly from one feeding area to another during the day, and to its sleeping trees at dusk. A troop which is coming toward trees from the open plains approaches with particular caution. The tall trees in which baboons sleep are found only where the water table is near the surface, usually along a river or beside a pond. Vegetation is usually dense at the base of these trees, and it is in this undergrowth that predators often spend the day. The arrangement of the troop members when they are moving insures maximum protection for the infants and juveniles in the center of the troop. An approaching predator would first encounter the adult males on the troop's periphery, and then the adult males in the center, before it could reach defenseless troop members in the center.

Because they are in front of the troop by twenty to forty yards, the peripheral adult males are usually the first troop members to encounter a predator and give alarm calls. If a predator is sighted, all the adult males actively defend the troop. On one occasion we saw two dogs run up behind a troop, barking. The females and juveniles hurried ahead, but the males continued walking slowly. After a moment an irregular group of some twenty adult males was between the dogs and the rest of the troop. When a male turned on the dogs, they ran off. On another day we saw three cheetahs approach a troop of baboons. A single adult male stepped toward the cheetahs, gave a loud, defiant bark, and displayed his canine teeth; the cheetahs trotted away. If baboons come upon predators while en route to their sleeping trees, the troop stops and waits while the males in the center move ahead and find an alternate route (the young juveniles and mothers with infants stay behind with the peripheral adult males). Eventually the dominant males return, the original order of progression is re-established, and the troop proceeds along the new route. These behavior patterns assure that the females and young are protected in the troop's center.

The ultimate safety of a baboon troop is in the trees. When the troop is away from trees, the adult males are very important in troop defense. We saw baboons near such predators as cheetahs, dogs, hyenas, and jackals, and usually the baboons seemed unconcerned—the other animals kept well away. Lions, however, will put a baboon troop to flight. From the safety of trees baboons bark and threaten lions but make no resistance to them on the ground. The behavior of baboons when near trees contrasts strikingly with their behavior on the open plains. If the troop is under trees, it will feed on the ground within thirty yards of predators, including lions.

ECOLOGY AND SEX DIFFERENCES

The role of the adult male baboons as defenders of the troop has been described. This behavior is vital to the survival of the troop, and especially to the survival of the most helpless animals—females with new babies, small juveniles, and temporarily sick or injured individuals. Selection has favored the evolution of males which weigh more than twice as much as females, and the advantage to the troop of these large animals is clear, but it is not obvious why it is advantageous for the females to be small. The answer to the degree of sex differences appears to be that this is the optimum distribution of the biomass of the species. If the average adult male weighs approximately 75 pounds and the average adult female 30 pounds, each adult male requires more than twice the food of a female. If the food supply is a major factor in limiting the number of baboons, and if survival is more likely if there are many individuals, and if the roles of male and female are different—then selection will favor a sex difference in average body size which allows the largest number of animals compatible with the different social roles in the troop.

If selection favors males averaging 75 pounds, then it will favor females which are as much smaller as is compatible with their social roles. Since the females must travel the same distances, carry young, engage in sexual and competitive activities, there are limits to the degree of sexual differentiation, but the adaptive value of the difference is clear. For example, a troop of 36 baboons composed of 6 adult males and 12 adult females and their young (18 juveniles and infants) has a biomass of some 1,000 pounds. If the females also weighed 75 pounds each, 6 adult males and 6 adult females would alone total 900 pounds and have only one-half the reproductive potential of 6 adult males and 12 adult females. Because this would halve the number of young, it would greatly reduce the troop's chances of survival. Our data are not sufficiently detailed to analyze the actual distribution of biomass in the troops we observed, but our observations are compatible with the limited data on weights and the numbers of adult animals we saw. Viewing sexual differentiation in size as a function of the optimum distribution of biomass of the troop offers a way of understanding sexual dimorphism fundamentally different from the view which considers only sexual selection, dominance, and intratroop factors. Obviously, all factors should be considered. Adaptation is a complex process and results in compromises between the different selective pressures, but a distribution of biomass which doubles the reproductive potential of a species is so important that other factors may be minimized.

The importance of sex difference in body size is reinforced by social behavior and the structure of the troop. As described earlier, some subadult and adult males are peripheral in the structure of the troop. They tend to be first, or last, when the troop moves. They are the most exposed to predators and are, biologically, the most expendable members of the troop. Interadult male antagonism results in a social order which both protects females and young and reduces feeding competition with females and young. Without altruism, the dominance behavior of a small number of males keeps a feeding space available to subordinate animals.

Juvenile play prepares the adults for their differential roles. Older juvenile females do not engage in the serious mock fighting which characterizes the play of older juvenile males. In this "play" the males learn to fight, and by the time the canine teeth have erupted and the temporal muscles grown to adult size they have had years of fighting practice. Play, social arrangement, and structural sexual dimorphism all supplement each other, producing a pattern in which the females and young are relatively more protected than the large males. Sexual differentiation must be seen as a part of this whole complex social pattern which leads to the survival of troops of baboons.

RANGE

On an average day a baboon troop leaves its sleeping trees at full daylight and moves rapidly to a spot where the animals feed intensively for two or three hours. In Nairobi Park, this morning feeding period is often spent in a fig tree (if these are in fruit), along a watercourse, or out on the open plains. During the dry season in the Amboseli Reserve, feeding areas were usually at the edges of water holes. During the middle of the day baboons rest in the shade of bushes or trees, not far from the feeding place of the morning. The late afternoon is another period of relatively intensive feeding. It is often some distance away from the feeding area of the morning, and a different kind of food is usually eaten. If the morning was spent in a fig tree, the afternoon is usually spent eating grass on the plains; if the morning was spent on the plains, the afternoon meal often consists of the pods, buds, and blossoms of acacia trees. During such a day the troop completes an average circuit of about three miles in Nairobi Park, but this distance varies from a few yards on some days to six or seven miles on others. These figures refer to the distance between points on a map. As a troop meanders across a plain, however, the individuals actually walk twice as far as these figures indicate.

During the year a baboon troop moves over an area which probably averages about fifteen square miles in open savanna country, but which may be much smaller (for a small troop or a troop living in forest country). Even where the total animal range is as large as fifteen square miles, only parts of it are used frequently. These areas of frequent use may be called "core areas." . . . Although this study was largely confined to the forty square miles within the boundaries of the park, all but one of the nine troops in the park included the Ngong Reserve (to the south) in its range. The southern extent of these ranges into the Ngong Reserve is not known.

. . . Only the annual ranges for the Lone Tree and Songora Ridge troops are complete, but range sizes shown for the other troops probably include about 80 per cent of their annual range. In general, a large troop contains more adult males and covers a wider range. Although both the Athi River and the Songora Ridge troops numbered twenty-eight, the range of the latter is about five times as large as the former. The Athi River troop had only one large adult male; the Songora Ridge troop had six. This gives an approximate measure of the additional range required by a troop with more large males. It also indicates that troops with more males can control a larger range.

Daily routines tend to keep baboon troops apart. Although annual ranges overlap extensively, there is very little overlap of the core areas of adjacent troops. We saw no evidence that troops defend a part of their range as "territory," but in Nairobi Park one troop is seldom seen in the core area of another. The core area or areas of a troop contain sleeping trees, water, resting places, and food sources. A troop uses one core area and one grove of sleeping trees for many weeks at a time but may then shift suddenly to a new area. In 1959 throughout the dry season there were only two sources of water in Nairobi Park, which also contained tall trees: the Athi River, which forms the southern boundary of the Park, and a water hole in the core area of the Lone Tree troop. All the troops except Lone Tree had at least one core area along the Athi River, which is the boundary between the park and the Ngong Reserve. At both Nairobi Park and Amboseli baboon troops usually slept in the tall fever trees (*Acacia xanthophloea*) which grow only where the water table is high. Since the plants and fruit trees which baboons use for food also tend to be concentrated near water holes or along rivers, the core areas of a troop include food, water, refuge sites, and sleeping trees. Although a troop usually returns to the same sleeping place after its daily circuit, it also shifts from one core area to another over a period of weeks or months. The existence of alternative core areas serves to reduce contact between adjoining troops, and behavior patterns reinforce this distance. These spacing mechanisms, rather than defense of territorial boundaries, disperse baboon troops in an area.

The population density of baboons in Nairobi Park is about ten per square mile. As a result of social factors and low population density, baboon troops are seldom within sight of each other. When a troop is living on one edge of its range, for example, its neighbor on that side tends to move to a portion of its range well away from other troops. However, where core areas of adjacent troops overlap (along the Athi River), troops may sleep only fifty yards apart without any display of aggression. Only one incident of intertroop aggression was seen when human intervention was not involved. Deliberately bringing two troops together by artificial feeding, however, can cause intertroop threats. The troop with the greatest number of adult males always won in these encounters, although a troop in its core area seemed to be more aggressive than one at the edge of its range.

By contrast to the infrequent encounters between baboon troops in Nairobi Park, troops at Amboseli were seen in close proximity every day. At the end of the dry season at Amboseli (September and October), baboon troops were tightly clustered around water holes. . . . The double water hole was used by troops 51, 66, 171, 88, 70, and 57. At this place a 100-yard crescent of vegetation contained two pools 50 yards apart, and it was here that more than 400 baboons from these troops might be seen together. We never saw any fighting between troops, and it soon became clear that there was a pattern to the various troops' use of the surrounding area. Troops 51, 66, and 171 used only the northern pool, and went north and west from the pools during the day. The southern pool was used by troops 88 and 70, and probably by 57, and these troops ranged south and east from the pools. The area between the two pools was used by all, except that we have no record of troop 70 there. . . .

Ordinarily this pattern of range utilization segregates the troops into clusters which recognize and tolerate each other at short distances without any sign of nervousness or tension. If a small troop is at a water hole and a large troop which also uses that water hole arrives, the smaller troop feeds slowly away. When troop 171 came once to the water hole usually frequented by 185, however, both troops paid close attention to the other. Adult males clustered where the troops were closest. The gestures, noises, and indications of nervousness were very different from the apparent lack of attention which is characteristic of troops normally frequenting the same water hole. By comparison with the behavior of troops in Nairobi Park, troop 171 was probably at the edge of its range, and troop 185 was occupying its core area.

As in Nairobi Park, the baboons at Amboseli slept in the tall fever trees around the water holes and marshes. The importance of trees as refuge from predators is illustrated by the ranges of the Amboseli troops. The marshy area north of the causeway contained water and plenty of food, but there were no trees there. Despite the heavy competition for the limited food resources during this season, no baboon troop included this treeless area in its range. Lions were often seen near the marsh, but south of the causeway, where lions were seen even more frequently, there were trees, and three large troops lived in this area. Normally the adult males protect the troop, but against the largest carnivores the only safety is flight into the trees. In areas where there are lions, trees limit the distribution of baboons as much as does the availability of food or water.

In summary, baboon range is based on the existence of refuge sites as well as sources of food and water. These ecological factors control population density, but the interrelations of troops are based on behavior. Troop size, number of adult males in the troop, and frequency of contact between troops determine the outcome of intertroop relations. Territorial defense is not seen, but core area of different troops tend to space troops apart within ranges which may overlap extensively.

In the evolution of human behavior, hunting is the best clue to the size of the range and the area which is defended from strangers. The pattern of core areas around water, within a large range, described here for baboons, is analogous to the pattern of land use by primitive hunter-gatherers in savanna country today. The major difference between baboon range and that of human hunters is the vastly larger area which humans, like the other large carnivores, must control. The aggressive protection of the hunting territory by humans also contrasts with the behavior which spaces baboon troops apart. African bushmen and Australian aborigines range over a hunting territory of from 100 to 1,200 square miles. A range of this size is far more comparable to the ranges of wolves, wild dogs, and large felines than to the small ranges of the nonhuman primates. Within these large ranges, camp sites near water sources correspond to the core areas of baboon troops. Access to the resources within the core areas of these hunters is rigidly controlled by social custom, religious sanction, and the force of arms. Interband relations between human hunters distinguish between "friendly neighbors" and strangers, a distinction which has an ancient, prelinguistic basis in primate behavior. The most striking difference between the social organization of baboons and human hunter-gatherers is the closed social system of the former and the rules of local exogamy which are usually found in the latter. Although formal rules of exogamy depend upon the presence of language, the exogamous pattern itself may have arisen during the shift to a hunting economy by men of the early and middle Pleistocene.

DIET

A more detailed description of baboon diet is in preparation; the following discussion outlines the range of baboon foods and the relation between food supply and the troop. Baboons, like the macaques of Asia, eat a wide variety of foods. Although the bulk of their diet is vegetable food, they will also eat insects, eggs, and an occasional small mammal. Most of the Nairobi Park, where baboon foods were collected, is grassland with some scattered trees. *The Nairobi Royal National Park Guide Book* contains an excellent brief description of the flora and fauna of the park. Although the western edge of the park is dry semievergreen forest verging on woodland, . . . observations were largely confined to the open grassland country, and the description of baboon diet which follows includes few food items from the forest habitat. A study of the foods eaten by baboons in forest areas would be necessary before the full range of baboon diet in this area could be known.

VEGETABLE FOODS

The diet of baboons living in the savanna of Nairobi Park can be divided into: the vegetable foods which provide forage for them throughout the year, seasonal fruits, insects, and the live animals which they occasionally catch and eat. Grass is the baboon's single most important food. In ten months of observations, not a single day passed in which baboons were not observed eating grass, and for many weeks during the dry season, grasses composed an estimated 90 percent of their diet. The portion of the grass eaten varies with the season. When the tassels contain seeds, these are "harvested" by pulling the tassel through the closed palm or clenched teeth. Most often, however, baboons pull up the grass shoots in order to eat the thick, lower stem at the base of the culm. Before eating the shoot, the dirt in the root system is carefully brushed away, and the roots themselves bitten off and discarded. By the middle of the dry season, when grass shoots are rare, baboons concentrate on digging up rhizomes—the thick, rootlike runners of the grasses which lie from two to four inches beneath the surface. Even after many weeks or months without rain, these rhizomes are still juicy, providing baboons with considerable water. The ability of baboons to shift to subsurface rhizomes and roots when surface vegetation is dry and sparse is one of their most important adaptations to the grasslands. It enables them to feed in an area which has been denuded of surface vegetation by the many ungulates with whom they share this habitat, and to find sufficient forage during long dry seasons. Digging these rhizomes out of the hard, dry soil with the fingers is a laborious task, and in the dry season baboons spend longer hours getting their food than they do during the rest of the year. The use of a simple digging stick or sharp stone would enormously increase their efficiency in extracting this food from the ground, but no baboon was ever seen trying to use a tool in this or any other way.

There are numerous plants on the Nairobi plains which have large, tuberous roots or bulbs, and the baboons are very adept at finding the tiny stem or leaf which indicates that such a root lies below. It may take as long as twenty minutes for a baboon to uncover a large root, and require a hole as large as 24 inches long, 8 inches wide, and 15 inches deep. Where the water table is high, along the rivers in Nairobi Park and around the water holes at Amboseli Reserve, the lush grasses attract many animal species, including baboons. Not only is the grass more plentiful here during the dry season, but also the earth is softer and more easily dug, and many water plants are found which grow nowhere else in the area. Baboons spend the majority of their time feeding in the grass near the water, but they will also wade into the shallow water to eat such plants as rushes and the buds of water lilies.

The baboon's usual diet is further extended by the various bushes, flowering plants, and shrubs of the savanna. In Nairobi Park they were seen eating the berries, buds, blossoms, and seed pods of such plants. Another very important source of food throughout the year is provided by the acacia trees. Probably the buds, blossoms, and beanlike seed pods of all acacias are eaten, but those of the fever trees (*A. xanthophloea*) are particularly important. Not only is this species used almost exclusively as sleeping trees, but when they are in the height of their bloom the baboons also usually feed in them for one or more hours before starting their morning round, returning in the afternoon for another heavy feeding period at dusk. Out on the plains the ant galls on the short whistling-thorn trees (*A. drepanolobium*) are constantly plucked for the ants inside, and extrusions of its sap are eaten as well. Some edible portion—bud, flower, seed pod, sap—of one of the types of acacia tree will be available within a troop's range at almost any time of year, and acacias are second only to grasses in the quantity of food they provide for Nairobi Park baboons. In addition to the plants and trees which provide forage for baboons all year, certain seasonal foods may constitute the bulk of their diet for short periods. The most important source of these seasonal foods in Nairobi Park are fig trees. When large fig trees are in fruit, the baboons may also use them as sleeping trees.

The most important food sources in the park are the grasses, acacia, and fig trees, but despite the frequency with which they feed in these trees, baboons were never seen eating tree leaves. On the southeastern slope of Mt. Kilimanjaro, baboons were observed feeding on the forest floor, while vervets (*Cercopithecus aethiops* and *C. mitis*) fed in the lower branches of adjacent trees. Leaf-eating *Colobus* monkeys occupied the canopy of the same forest. Their ability to find food both on the open plain and in the trees is a distinct advantage for the baboons. Although they compete with a wide variety of ungulates for their food on the plains at Nairobi Park, their only close competitors in the trees are the vervets. Vervets and baboons are commonly seen feeding in adjoining trees in the park and occasionally they occupy the same tree—the baboons on the lower branches and the vervets in the canopy.

In addition to the staple diet, other vegetable foods were frequently eaten when they were available. These included "kei-apples," croton nuts, sisal plants, mushrooms, and the produce of native gardens (potatoes, yams, bananas, beans, maize, peanuts, sugar cane, etc.). Since almost all cultivated plants in this area have been imported from the New World, it is clear that baboons are very eclectic in their food habits.

INSECTS

Baboons eat many types of insects when they can find them, but the climate of Nairobi Park, with its dry season, its hot days and cool nights, does not support a very heavy insect population. The most common insect eaten in the park is the ant living in the galls of the *Acacia drepanolobium* trees. The amount of ants eaten in this way, however, is very small compared to the grasses and plants eaten during the same feeding period. If the troop is walking slowly through an area strewn with large stones, some of these may be turned over and the ground beneath them examined carefully. Under such stones an occasional beetle, slug, or cricket will be found and is quickly eaten. Rarely, an ant nest is uncovered, and the baboon bends over and licks up the contents of the nest from the earth, licking additional ants from its hands and arms afterwards. But the baboon's attitude toward insects is one of mild interest, and no troop was ever seen moving from its pathway to systematically turn over the stones in an area.

Besides the ants in acacia galls, a baboon most frequently eats the grasshoppers which it finds on the branches of the bushes or blades of grass where it is feeding. Young baboons are seldom able to capture grasshoppers, but an adult will move the hand cautiously and deliberately to within one or two feet of the insect, then grasp it very quickly in a movement which is usually successful. Not all insects encountered are eaten. When a rock is overturned, some beetles and centipedes are ignored while others are carefully selected. Too few instances were observed to be able to say whether such selection was by individual preference, or whether these insects were avoided by all baboons in the park.

Although insect food is minor in the overall baboon diet, a very heavy infestation of "army worm" caterpillars in the park showed that for short periods insects can become the baboons' most important food. Beginning in early April, during the rainy season, army worms appeared in the park in large numbers. For about ten days the baboons ate little else. Feeding on the worms in a small area were: three baboon troops, totaling 188 animals; several troops of vervet monkeys, perhaps 75 in all; and a group of about 300 Marabou stork (*Leptoptilos crumeniferus*). The different baboon troops fed very near each other, and the other animals, without incident. All were gorging themselves on the caterpillars; several baboons were timed picking up 100 army worms per minute, and continuing at this rate for from 10 to 15 minutes without a break. The eating of insects, in addition to the extensive inventory of vegetable foods, further increases the dietary adaptability of the baboon.

LIVE ANIMALS

On six, perhaps seven, occasions during the twelve months of study in Kenya and the Rhodesias, we saw baboons eating freshly killed animals. Twice they caught and ate half-grown African hares (*Lepus capensis crawshayi*). On the first occasion the male in possession of the hare was being harried not only by two more dominant males in this troop, but by a pair of tawny eagles (*Aquila repax raptor*) as well. The male in possession eluded his harassers and managed to consume most of the hare, the eagles retrieving scraps of viscera and skin. In his haste the baboon dropped the rib cage and a foreleg of the hare, with most of the flesh still attached, *but these pieces were ignored* by the other two baboons chasing him, despite their desire to obtain his catch.

Two or three times baboons were seen eating fledgling birds of some ground-nesting species, probably the crowned plover (*Stephanibyx coronatus*). On several occasions they chased fledglings some yards through the grass without catching them. We never saw baboons finding and eating eggs, but, when offered a dozen guinea fowl eggs, they ate these without hesitation. Entire eggs were stuffed into the cheek pouches and the shell broken by the hand pressing the cheek against the teeth and jaws. More significant than the few instances of baboons eating fledglings are the numerous times when baboons were seen feeding across a plain covered by bird nests without discovering the contents of a single nest. The same animals which are able to detect an underground root from only a tiny dried shoot on the surface will walk beside a bird nest six inches in diameter without noticing it. Furthermore, four species of weaver bird inhabit the park, and their nests are frequently clustered in the acacia branches where the baboons are eating, but no baboon was ever seen investigating such a nest, much less eating its contents. The baboon's attitude toward food is clearly vegetarian. It is common to see a baboon troop completely mingled with a flock of guinea fowl without incident. The only eggs or fledglings which they seem to recognize as food are those which are literally stepped on as the troop searches for vegetable foods on the plains.

On December 14, near the close of the study, two very young Thomson's gazelle (*Gazella t. thomsonii*) were caught and eaten by the adult males of a troop. The actual capture of the second gazelle was seen. An adult male baboon grabbed it, brought it above his head, and slammed it to the ground. He immediately tore into the stomach of the gazelle and began eating. Beginning with the most dominant males, five of the six adult males in the troop participated in eating this gazelle, and two hours later only skin, teeth and large bones remained. The viscera were eaten first, followed by the flesh, and finally the thin brain case was bitten open and the contents, carefully scooped out with the fingers—bits of skull being pulled through the teeth and licked clean. The incisors, not the canines, were used in biting and tearing at the flesh.

These two Thomson's gazelle were apparently only a few days old and were hiding in the grass some 150 yards from the herd of 38 with which they were no doubt associated. After the baboon troop moved on, two females from the herd of gazelle (of 35 females, two young, and one adult male) came over and paced nervously around the remains of the carcasses. It seems reasonable to assume that the discovery of these two young gazelle took place under circumstances very similar to those involved in the eating of the young hares, that is, that they were discovered accidentally in the grass. In fact, after the first gazelle had been found, and four of the males were pressing its possessor closely, the males passed within five yards of an African hare sitting in plain view. They clearly saw the hare but did not even walk over toward it.

All these cases of flesh eating have one thing in common—they involve the eating of immature animals whose defense is to hide "frozen" in the grass, and in each case their discovery by the baboons seemed fortuitous. Nothing resembling a systematic search of an area or the stalking of prey was ever observed, nor was fresh meat eaten except when it was found alive or taken up immediately by a waiting baboon. Since baboons avoid lion kills when they are away from trees and other carrion is not eaten, the lack of interest shown by the male in the portion of hare which had been dropped (described above) may be due to their avoidance of carrion. It is also possible that baboons do not recognize as edible any meat which is not alive and easily caught. In either case it seems clear that their attitude toward other animals is not that of a predator, nor do the scores of other species with which they live peacefully so regard them.

The final instance of meat eating was observed in Amboseli Reserve. While watching baboons in an open area, we heard loud screeches and chattering in a tree where baboons and vervets had been feeding peacefully for the previous hour. When we approached the tree we saw an adult male baboon walking through the branches with a juvenile vervet dangling from his mouth, and the vervet troop had left the tree. The baboon consumed most of the vervet, carrying the carcass in his mouth as he walked toward the troop's sleeping tree at dusk. This observation is in striking contrast to the many occasions when the two types of monkey were seen feeding peacefully together. During a brief aggressive interaction between the two species in Nairobi Park, DeVore saw an angry adult male baboon put a troop of vervets to rapid flight, and this case of meat eating may have been the incidental result of such a situation in the tree at Amboseli. Although Washburn saw baboons chase vervets quite frequently near Victoria Falls, he only once saw a baboon catch one. This was held in the mouth by the female who caught it. She apparently was bewildered by the situation and soon released it unharmed. In much the same way, one of the fledglings DeVore saw eaten was actually caught by a juvenile baboon, who seemed puzzled by the object and quickly relinquished it to an adult male (who promptly ate it).

In summary, baboons may be described as very inefficient predators. Meat eating, to judge by the bewildered state of the female baboon who caught a vervet and of the young juvenile who caught a bird, would appear to be learned by each generation, and meat never becomes an important source of food for the whole troop. Only one baboon other than adult males (an adult female) participated in the eating of meat in any of the instances observed during the study. Accounts of meat eating in captive baboons are contradictory. Kenya baboons kept near Nairobi Park ate meat readily, but Bolwig found that his captives refused it. In South Africa, where most reports of carnivorous baboons have originated, baboons are only now being systematically studied, and we feel that the importance of meat in the baboon diet has been considerably overstressed. The usual reason given for the habit of meat eating in South African baboons is that the hardship of drought creates the conditions under which it flourishes, but when the two Thomson's gazelle were eaten in December, the park was well into the rainy season, and the vegetable foods baboons ordinarily eat were more abundant than at any other time of year.

It would seem more reasonable to us, on the present evidence, to assume that meat has been a consistent but very minor part of the baboon diet throughout their evolutionary history. In localities where sources of animal protein can be obtained without danger, baboons apparently include these in their regular diet. At Murchison Falls, baboons are often seen digging out and eating crocodile eggs. Hall's description of the foods eaten by baboons along the coast of South Africa is very similar to the inventory of vegetable and insect foods discussed here, except that the South African baboons also eat marine foods much as mussels, crabs, and sand hoppers found along the beach. But baboons are ill fitted anatomically to be carnivores, and too great a dependence on meat eating could have been detrimental to their wide exploitation of the vegetable foods they depend upon today. By their utilization of a wide variety of plant and tree products, baboons have been able to spread over the African continent and, together with the macaques, to cover most of the tropical Old World.

In the evolution of the human species, meat eating played a very different role. We have suggested that the earliest hominids may have been living on a diet very like that of the baboons; that is, vegetable foods supplemented by an occasional small animal. The freedom to carry a simple digging implement in the hands would greatly enhance this adaptation. During the dry season in Africa, human hunter-gatherers also are very dependent on the subsurface roots and tubers sought by baboons. A digging stick greatly improves the human's chance for survival during this period of food shortage, and it may be that the presence of baboon skeletons at Olorgesaille indicates the result of competition between baboons and humans over a limited food supply. It would be an easy step from killing baboons to protect a source of vegetable foods, to killing them for meat.

SCAVENGING

Scavenging has been regarded as an important phase in the evolution of man's carnivorous habits. It seems reasonable that a primate liking eggs, nestling birds, insects, and an occasional small mammal might add to this diet and develop more carnivorous tastes and habits by gleaning meat from kills. This theory seemed reasonable, and we made a particular effort to examine kills and to observe the relations of the baboons to them. Although we saw over a dozen recent kills (including gnu, giraffe, zebra, waterbuck, impala, Grant's gazelle, warthog, Masai cattle, and goat) and have thorough records on some, we were primarily looking at baboons. The subject of scavenging is so important, especially in the interpretation of the deposits in which *Australopithecus* is found, that a much more comprehensive study is needed. However, here are our tentative conclusions.

The scavenging theory is not supported by the evidence, and primates with habits similar to those of baboons could get meat by hunting far more easily than by scavenging. There are several reasons for this. The first is that most kills are made at night and are rapidly and thoroughly eaten. When the hyenas leave at dawn, the vultures locate the remains and clean the last meat from the bones. Some kills are made by day. We saw the remains of a gnu which a pride of ten lions finished in an hour. A pride of four lions (two not fully grown) killed a gnu one afternoon and ate almost all of it in one night. The vultures finished the rest, and the bones were undisturbed for three days. Many bones disappeared on the fourth night. Similarly, we saw two lions eat a warthog, three lions eat a Grant's gazelle, and five cheetahs kill and eat an impala. Only the meat of very large animals is left for long, and Africa is well supplied with highly efficient scavengers which leave little meat to tempt a primate.

Actually there are far fewer kills than might be expected from discussions of scavenging. In the part of the Amboseli Reserve which we studied intensively there were on the order of 100 baboons to one lion. The lions move over large areas, and the chances of a troop coming on a "kill" are very few. We saw a troop around a kill left from the previous night only once in Amboseli. It had been largely eaten, and the baboons appeared to take no interest in it. During nine months of observation in Nairobi Park, baboons were seen to pass near four kills and paid no attention to the few scraps of meat left on them. A Grant's gazelle carcass, presumably a leopard kill, hung in a fig tree where baboons ate and slept, but the baboons apparently ignored it. In addition, they did not attempt to eat fresh carrion when this was found. A further complicating factor is that, when much meat is left, the lions usually stay nearby, and the neighborhood of the kill is very dangerous.

In summary, the chances of a kill within the range of a baboon troop are very small; little meat is likely to be left; and the vicinity of the kill is dangerous. Most of the killing and eating is at night, and primates have neither the sense of smell of the hyenas nor the eyes of the vultures to locate the kill. As noted earlier, the baboons seem uninterested in dead animals. A slight increase in predatory activity against young animals would yield a far greater reward than scavenging, would be much less dangerous, and would represent a smaller change in habit. The use of a stick or stone for digging would increase the baboons' food supply more than any other simple invention. Perhaps in *Australopithecus* we see a form which had such a tool to exploit vegetable foods and which also used this tool as a weapon. If tools were being used at all, their use in the deliberate killing of small animals would be only a small change from the behavior observed in baboons. Once man had become a skilled tool-user in these ways, he could extend tool use to the hunting of large animals, to defense, and to driving carnivores from their kills. Scavenging may have become a source of meat when man had become sufficiently skilled to take the meat away from carnivores, but the hunting of small animals and defenseless young is much more likely to lie at the root of the human hunting habit.

DISCUSSION

In this paper we have tried to stress those aspects of baboon ecology which are of the greatest help in understanding human evolution. Obviously, man is not descended from a baboon, and the behavior of our ancestors may have been very different from that of living baboons. But we think that in a general way the problems faced by the baboon troop may be very similar to those which confronted our ancestors. At the least, comparison of human behavior with that of baboons emphasizes the differences. At the most, such a comparison may give new insights. Many topics have been summarized above, and in this discussion we will call attention only to a few major points.

The size of baboon troops may exceed that of hunter-gatherers, and their population density far exceeds that of primitive man. The human group differs in being exogamous, so that many local groups form the breeding population. We believe that this radically different breeding structure has exerted a profound effect on the later phases of human evolution and has long been a factor in preventing speciation in man.

The social structure of the baboon troop is important to the survival of the species. Survival depends on the adult males being constantly close to the other troop members. Roles in the troop are divided between the sexes, but these are in the context of a compact troop. With man, the hunters leave the local group, sometimes for days, and then return to their home base. Such a pattern is radically different from anything known in monkeys or apes. Hunting with tools basically changed the social structure of the band, the interrelations of bands, the size and utilization of range, and the relation of man to other animals.

Diet has already been discussed and we will not repeat here, except to point out that our opinion of the importance of scavenging has changed through observation of the actual situation at the kills. It is not enough to speculate that scavenging might have been important. One must estimate how much meat is actually available to a vegetarian and how dangerous it is to get meat by scavenging.

Finally, we would stress that survival is a complex process, and that all the factors which lead to reproductive success must ultimately be considered. Varied diet, social structure, and anatomy, all are important, but their meaning only becomes clear as they are seen making possible the behavior of a population. Sex differences, peripheral animals, and range—each of these has meaning only in terms of the survival of groups. With the coming of man, every major category is fundamentally altered and evolution begins to be dominated by new selection pressures. Some measure of how different the new directions are may be gained from the study of the ecology of baboons.

12. SOCIAL UNITS OF A FREE-LIVING POPULATION OF HAMADRYAS BABOONS

HANS KUMMER and FRED KURT

Reprinted from Folia Primatologica, *1 (1963): 4-19 (Basel: S. Karger). Fred Kurt, after spending a year with Hans Kummer in Ethiopia doing field work on the baboon, returned to Switzerland to resume his studies at Zurich University. For his Ph.D., he made an intensive three-year field study of the sociology of roe deer populations in the Swiss midlands and Alps. Kurt's main interests are in the sociology and management of large mammals. Besides his university studies, he has undergone the discipline of a game warden's training. The figures and tables that appeared in the selection as originally published have been deleted here. For biographical information on Hans Kummer see Selection 8.*

■ The near-universality of the human family leads us to consider the possibility that *Homo sapiens* is genetically programmed to respond to various types of challenges with the organization of social relations that we call the family. As will be seen below, single-male reproductive units are not confined to man but are also found under various conditions among nonhuman primates. Recent research, moreover, suggests that the family also exists among nonprimate mammals, and even among some birds.

What has to be stressed in this connection is that man is not genetically programmed for a particular form of marriage and family living; if he were, *Homo* would not exhibit the several varieties of marriage and family organizations that anthropologists have documented. Instead, we must postulate, at least as a start, that *Homo* responded to various challenges early in his evolutionary career with the capacity if forming family groups. The further assumption can then be made that this capacity was genetically programmed.

How do we deduce the challenges that evoked this response? The best hint comes from the work of cultural anthropologists and sociologists who have consistently found that marriage and family relationships respond most sensitively to changing economic conditions. This is not to say that the family does not change in response to other pressures in society, such as political factors, but only that the economic variable is most consistent in its effect.

Such anthropological and sociological findings underline the importance of the following paper by Kummer and Kurt. The hamadryas baboons whom they studied sleep at night in groups as large as 700. When they awake they break up into groups of about 50, and then further subdivide into single-male family groups. These are stable groups, and they sleep alongside each other at night. The adult males are "faithful" to their wives, even when other receptive females are available; only the young males tend to "sow wild oats." More significantly, the adult male is the one who provides food for his own family group. The apparent key to this puzzle is that these baboons live in a semidesert region in which food is scarce.

This suggests that single-male family groups are an adaptive response to food shortage. If it is true that *Homo* made his first appearance in environments in which food gathering was difficult, as in the open plains, the fragmentation of the larger band into small family units was an efficient adaptation to diffused food supplies. Our similarity to earlier living forms cannot be attributed to mere coincidence.

This interpretation of the material by Kummer and Kurt was originally suggested by Bernard Campbell in *Human Evolution* (Chicago, Aldine, 1966). Further material in this connection is provided by K. R. Hall in "Variations in the Ecology of the Chacma Baboon, *Papio ursinus*" (*Symposia of the Zoological Society of London*, 10 [1963]: 1-28). The reader who wishes to go still further should consult Ernst Mayr's *Animal Species and Evolution* (Cambridge, Mass.: Belknap Press of Harvard University Press, 1963). ■

INTRODUCTION

THE SOCIAL ORGANIZATION of baboons has been discussed repeatedly in recent years in connection with various other topics. In a symposium on "The Social Life of Early Man" Washburn and DeVore have used their observations on East African baboons as the basis for speculations on the most probable social conditions of fossil men since both these types of primates are terrestrial. In another symposium of 1962 it is attempted to consider behavior intensively for a review of primate taxonomy. Such a utilization of movements specific to a species, genus or family has produced some very significant improvements in the classification of various groups of insects, birds, etc. For taxonomic purposes it has been the quantitative study of the most simple elements of behavior which has proved to be most useful, such as threatening or courtship movements in ducks. It would be reasonable to apply this experience to primatology, i.e., to start criticism of morphological taxonomy by a comparative study of behavior-elements. Unfortunately, however, a systematic description of such elements has not been accomplished in the field even for one species, since the interest of field workers concentrated on types of social organization, i.e. the composition and structure of groups. It is still quite uncertain that the social organization of primates will be of any significant help for taxonomic purposes. The problem remains unsolved how far ecological factors can change the composition of groups in one and the same species. The female leadership on the "Minoo-B-group" of Japanese macaques, observed by Kawamura and Kasai and Imanishi, is a further warning against rash generalizations. This case deviates so far from the conditions in neighboring groups that ecological factors are inadequate for explanation. It seems not impossible that small groups of primates can respond to exceptional behavior of *one* dominant animal with a social organization which also deviates from an average.

By means of the scant observations available today such further questions have been discussed as how far there seem to exist units resembling a family among nonhuman primates and how the social relations within and between such units are regulated. Imanishi assumes that among monkeys there occur no firmly united pairs, whereas in the man-like apes such ties do exist in their "familioids," though probably they do not fulfill all the criteria of human families, such as incest taboos, exogamy, relations between adjoining groups, and the division of duties.

In all three of these problems, briefly indicated here, the discussions seem to be in danger of transgressing our present, relevant knowledge. This paper, therefore, will deal very little with interpretation and rather emphasize the factual report of findings which deviate considerably from recently described social organizations in other monkeys and, hence, have already been doubted. This refers chiefly to our finding of family-like units in a lower primate such as Zuckerman had reported for captive hamadryas baboons and had assumed for wild chacma baboons.

In the savanna-baboons of Kenya, Washburn and DeVore found highly constant, mainly inbreeding troops which live in overlapping areas but never mix. Subgroups within the troops are formed, e.g., by two females and their young offspring or by an adult male with one or more females, but none of these subgroups persist for longer periods. An estrous female will leave her subgroup and mate with different males. For the chacma baboon of the Cape Reserve, Hall has reported two constant troops. One of them contained subgroups consisting of one adult male and two to four females. It is not known, however, whether these units last for weeks or months.

ETHIOPIAN FINDINGS

Our one year field study on *Papio hamadryas* was carried out in Eastern Ethiopia from November 1960 to October 1961. We first made a preliminary survey of 23 population samples along a line from Sendafa northeast of Addis Ababa to Harar in the east of the country. These samples rather constantly contained about 18 percent adult males and 33 percent adult females. The parties observed numbered from 12 up to 750 individuals. From these populations we chose one living in the relatively open savanna between Dire Dawa and Erer-Gota for detailed investigation. These baboons lived far from any cultivated crop land and, therefore, were not hunted. The mean size of the local "troops" or sleeping-parties was about 100 and the percentage of its sex-age-classes came close to the above-mentioned values.

All hamadryas baboons in the visited regions sleep on vertical rocks varying from 12 to 20, rarely up to 30 meters in height. It soon became evident that the number of baboons sleeping on one cliff changed almost every night. In the morning the party leaves as a whole and in one direction, but will soon split into smaller groups of anywhere from 5 to 40 individuals. During the day they often move out of contact with each other and may seek out different rocks for the next night. There they will meet other groups of baboons which during the previous night may have slept on different rocks. In succeeding nights we counted on a rock near Erer-Gota 92, 44, 62, 74, 53, 0, 68, 49, 68 and 62 baboons. The numbers 62 and 68 occur twice each in this series. On the same rock an identified party of only 12, consisting of two adult males with their groups was observed for two succeeding days after it had slept there for several nights in a large party which gradually moved away. We also checked the size of parties at 8 neighboring rocks regularly during the same nights and found repetitions of the same numbers to be extremely rare. These rocks were situated several miles apart.

The sleeping-party, therefore, is not a constant social unit. Nevertheless, some known individuals are encountered with considerable regularity on a given rock. The sleeping-party possibly consists of semiconstant associations which freely move and recollect for the night in different combinations. Possibly due to the open character of the sleeping-parties, solitary individuals seem to be rare, since only once did we see a solitary adult male sitting on a sleeping-rock in the early morning.

THE ONE-MALE-GROUP

The constant social unit of our baboons is a small group following one adult male, corresponding to the "harems" reported by Zuckerman and the "families" observed by Starck and Frick. Because a systematic terminology for social units is still lacking we have avoided the terms proposed so far and tentatively chose the descriptive name "one-male-group." Imanishi's term "oikia" for the minimum unit of social life would possibly fit our observations, but seems to lack a definition of the "social unit' in terms of constancy during the daily cycle and of persistence over longer periods: In the hamadryas, for instance, an association of two adult males is smaller but less constant than a one-male group. The mother-child association is smaller and its social contact is nearly permanent, but very probably it does not persist as long as the one-male group.

The one-male group consists of one to four, rarely up to nine, females following a single adult male, and the offspring of the females up to one or one and a half years of age. Outside such groups live the juvenile and subadult males and some adult males without females.

The existence of subgroups in a large association of animals can become clear in three ways:

1. In the spatial distribution of individuals,
2. in the frequencies and types of social behavior among the members differing from their contacts with other individuals of the association, and
3. in the time for which these two conditions persist.

In this paper we will consider a number of baboons as one group if they always were observed together within a party, if the frequency of social behavior among members was significantly higher than between members and non-members, and if these conditions remained unchanged for many weeks or months. The following results about these three qualities were collected on the groups described during mornings and late afternoons at the sleeping-rocks, when social life is most intense and all members can be seen continuously.

1. THE SPATIAL COHERENCE

To obtain a measure of the group's coherence we estimated the distances kept by the females from their male every 5 minutes during the periods of observation. Using as a scale the known size of the sitting male, this is possible with a maximum error of about 15 percent. From 10 adult females we obtained a total of 256 estimates from which the mean distance could be found as 0,65 ± 0,04 meters. The highest distance observed in this sample was 3.2 meters in a female that carried an infant a few days old. A female having increased its distance up to 5 meters was attacked by the male immediately after the estimate. The sleeping-rock covers about 400 square meters. On this available space, a one-male group rarely spreads over more than 5 square meters.

This strict coherence is brought about by the response of the females to follow the male whenever he moves more than a few steps. Especially young males carefully watch the distances of their females, looking back as they walk every few meters. When a female drops behind or moves on her own account, the male's staring at her with outstretched head will usually make her run towards him. This flight of a frightened individual towards the threatening male is accompanied by screaming and raising the hair on the back. It was frequently observed in a group of hamadryas baboons at the Zurich Zoo, but its function was much more evident in the field. Chance gave a discussion on this behavior in connection with the nature of the social bond in primates.

Sometimes, a female will move far apart with nonmembers staying between her and the male. In this case the latter will even attack her and bite her into the back of the neck. Several times a male could be seen running along the walking column in the morning, applying the neck-bite to a female and walking on, the screaming female following him closely. A male's search for his lost female may continue for 15 minutes with unchanged energy throughout the party. It is interesting to notice that the baboon males of Kenya, where no one-male groups are found, never force another individual to follow them. "Their presence alone ensures that the troop will stay with them at all times." DeVore's catalogue of the main threat gestures of these baboons does not contain a neck-bite. In our own observations on savanna-baboons in the colony of Paris-Vincennes the neck-bite appeared as only one of a series of relatively unspecific bites. In the one-male group organization of *Papio hamadryas* the neck-bite obviously was selectively developed as the male's instrument to herd his group together in the crowd of the party.

On a crowded sleeping-rock or in a moving column the different one-male groups stay close together, and it is usually impossible to decide whether a female belongs to one male or the other. Sometimes, however, the groups segregate clearly, e.g., if a party is offered plenty of desired food on a very small spot. The extreme crowding following the attraction of the food compresses and breaks the normal framework of social distances. Different one-male groups get mixed, males start neck-biting their females, then threatening each other. In such situations the females may collect closely in the "attack shadow" of their respective males.

2. PREFERENCE OF PARTNERSHIP AMONG MEMBERS

A second criterion of the subgroup's significance was found in the high frequency of social intercourse among members. Following a partner or exchange of a glance, as well as bodily contact of individuals sitting close together, e.g., mother and newborn, has not been included under "social contact."

The table [not reprinted] shows a marked difference between young and adult baboons.

By using the criterion of partners' preference, the young offspring are not real members of the maternal group. They spend considerable time with their mothers and other infants of the group, but have free leave to play with infants of non-members, sometimes far from their mothers. The overt social behavior of adult members, on the other hand, is mainly restricted to their one-male groups. This almost complete restriction is no consequence of the groups' spatial isolation. On the sleeping-rock, every individual will find close to it as many "strangers" as members of its own group. The difference of choice by adults between members and nonmembers is highly significant ($t_{25} = 8.5$; $P < 0.001$). If, however, we consider every baboon of a sleeping-party, i.e., about 100 individuals, as a potential partner, we find that a baboon living in a group of 6 will choose a member of it about 400 times more often than any other baboon on the rock.

3. PERSISTENCE OF GROUPS

Unfortunately, our observations did not outlast any of the 8 identified groups. Two of them were observed for 103 and 124 days, respectively, without any significant change in their structure. The mean period of observation per group was 61 days. In this period, the number of the original members increased by two newborns. Only one member, a subadult female, disappeared; whether it died or entered another group is unknown.

4. CHOICE-EXPERIMENTS BETWEEN GROUP AND SLEEPING-PARTY

Since the 3 criteria of true subgroups could all be found in the observed one-male groups, the following paragraphs merely give some details about their role in the social life of our population. For special purposes, individuals were captured near distant rocks and transported to the parties under close observation where they were liberated to study the reaction to this new situation, as will be described in a later paper. When an adult hamadryas baboon was thus given the choice to remain with its usual or its freshly established one-male group *or* with the mass of a known or unknown sleeping-party, it invariably chose the group, as will become clear from the following summary of the results.

A young female raised in captivity and set free at the rock in the evening was at once seized by an adult male in early prime which previously had no females. The female fell asleep in his arms, but when the party left in the morning it became evident that she had no adequate following-response. After much hastening back and forth, accompanied by displacement activities, the male chose to stay at the rock. He remained there for the whole day, alone with his new consort, while the rest of the party had left without paying much attention.

Single hamadryas females from a very distant rock were readily accepted by males of our rock and followed them from the first moment. If, however, a whole one-male group was set free, its male ran away from the strange sleeping-party. In this case, his females followed him into the unknown savanna for the night, one of them even leaving her infant which was then "adopted" by a male of the party.

5. A SOCIOSEXUAL AND PHYSIOLOGICAL UNIT

In early June, the following, striking observation was made. Out of 87 well isolated one-male groups, the females of 72 were in the same sexual state since either all or none of them had sexual swellings. Newborns were found only in groups where all the females lacked swellings. The data concerning this phenomenon are not yet ready for interpretation, but they indicate that the one-male group, besides being a social unit, may also be physiologically homogeneous at some times of the seasonal cycle.

Sexual activity of adult males is restricted to the females of their own groups, as could be observed in the identified males. We never saw adult males do more than merely look at estrous females of other groups, even if their own females were not receptive during two to six months of the observation period. Mating pairs were never threatened or attacked by other males. On the other hand, estrous females may sometimes copulate with 2- to 3-year-old males if not observed by their group leader, but they never leave him to form temporary consort-pairs with other males. Before, during and after the birth of their infants, the females also stay within their own groups. The one-male group, therefore, is the social frame for sexual activity of adults as it is for the birth and the first months of the infant's life. These are marked differences between our hamadryas population and the baboons in Kenya, since the latter show temporary pairing during the estrous like the rhesus macaques.

Some preliminary remarks may be added here about the male dominance hierarchy. Washburn and DeVore, Bolwig and Hall all report a dominance order among males which is relatively easy to recognize. A similar system *might* exist among hamadryas males, too. However, in several months of study, devoted especially to the mechanism of "troop" movements, we could not find a dominance order nor any leaders, dominant over all others. This may be due to the difficulty of analysing the problem in our large sleeping-parties and to the rare social contact between fully adult males, which mainly mind the affairs of their direct followers. The main reason, however, is probably found in the fact that a pronounced hierarchy of dominance can not develop in an open party which frequently changes its composition.

DISCUSSION

The observed hamadryas baboons of Eastern Ethiopia live in a social organization which differs from that of other Cercopithecoidea known so far. In the savanna-baboon of Kenya, the largest social unit is also the most stable. In the hamadryas, we find the small one-male groups to be the solid unit, the larger associations becoming less constant with increasing number of individuals. The system of social and sexual relations in a troop of Kenya baboons closely resembles that of some species of *Macaca*, but is strikingly different from that of the hamadryas belonging to the same genus *Papio*. It is evident that our present knowledge does not yet allow much generalization even on the genus level.

For technical reasons, we chose a population living in a relatively flat and open country. It is reasonable to object that its organization might be an easily reversible effect of the environment, and not be representative of the species. Thus, the scarcity of sleeping possibilities in our area might be one condition for the development of the loose organization of its hamadryas population. The single one-male groups sometimes observed during the day are very probably less successful in the defense against predators than a large troop, as is suggested by Washburn's and DeVore's results. Even such isolated one-male groups, however, will reenter a large association with high probability the same day, if only to find a suitable rock in the evening, while all such rocks are almost regularly occupied. The few rocks of the region play a very important role as meeting-points for the population which scattered during the day. This is different in a species sleeping on types of trees or rocks in a region where this type is frequent. It could—in terms of adaptive behavior—be forced to stick together in closed troops so as not to become scattered. The subgroups could also reassemble in the evening at a sleeping-place fixed by tradition, but would then lose the full utilization of food resources which is provided by moving from one sleeping-quarter to another. The closed troop, however, can afford to sleep in every suitable place where it is overcome by the dusk, as is shown by the group of howling monkeys described by Altmann. The chacma baboons in the Kruger National Park seem to show the same correlation, since they live in closed troops but have no fixed sleeping quarters.

For the hamadryas baboons, an observation by Zuckerman throws some light on the question of ecological influence. In 1925, more than 100 hamadryas baboons were liberated on "Monkey Hill" in the London Zoo. In spite of being forced to form a "closed troop" in the enclosure and in spite of the unnaturally low number of females, these baboons developed the same type of social organization that is present in Eastern Ethiopia. "A family group consists of a male overlord, his female or females, together with their young, and may sometimes include one or more 'bachelors' or unmated males." This pattern may be brought about by genetically fixed types of behavior and releasing mechanisms, or else it has been individually acquired by the London baboons before they were captured. The possibility that entire one-male groups had been imported from the wild is excluded by the fact that most of the females were added at least two years after the Hill colony had been founded. Thus, the London experience proves at least that the one-male group organization is not easily destroyed by even a radical change of environment. It is likely to be more than a temporary condition of populations with a certain sex ratio or under the influence of special ecological factors. Persistent consort relations, therefore, are not a feature specific to the "man-like apes." According to recent field work by Petter, they may even exist among certain Lemuroidea.

SUMMARY

This paper is a preliminary survey of the social structure observed in a natural population of *Papio hamadryas* in Ethiopia. Unlike all monkeys studied so far in the wild the hamadryas, at least in the described region, does not form closed troops but lives in parties consisting of highly constant "one-male groups." Such a group includes 1 adult male and 1 to 9 females with their young. The parties are unstable associations of groups sleeping together on the same rock-face and numbering anywhere from 12 to 750 baboons. They usually split up during the day, and different combinations of one-male groups may settle down on the rocks every evening. However, the groups of a party sleep together for most of the nights. The significance of the one-male groups is shown by the following observations:

1. A one-male group never splits up. Its spatial coherence is constantly watched by the male who, if necessary, enforces the following-reaction of his females by a specific bite into the nape of the neck, not known in other species of baboons.
2. Quantitative studies over several months demonstrate that nearly all the social activities of adult animals are restricted to partners of their own one-male group.
3. Eight groups were observed for periods up to 6 months. During these periods, one female disappeared and two infants were born, but none of the groups disintegrated even temporarily.
4. An adult hamadryas may follow his or her one-male group even if this means leaving the entire rest of the party, their own infant or a sleeping-rock in late evening.
5. The identified adults exclusively copulated with members of their own one-male group.

The evolution of the difference in social structures between hamadryas baboons and other species of *Papio* and *Macaca* is discussed in connection with the number of potential sleeping-places available in their habitats. The organization built up by a Zoo colony of hamadryas baboons was strikingly similar to our findings in the wild. Ecological factors, therefore, are not likely to modify the basic structure of hamadryas populations.

13. KINSHIP AND THE FAMILY IN MONKEYS, APES, AND MAN

V. REYNOLDS

Reprinted from Man, *Vol. 3, pp. 209-23 with permission from the Royal Anthropological Institute of Great Britain and Ireland. Dr. Reynolds is presently a member of the Department of Physical Anthropology at the University of Oxford, England. His primary interests are the social organization of human and prehuman primates and he has been concerned with bridging the two fields of study. Among his publications are* The Apes *and* Budongo: A Forest and Its Chimpanzees.

■ The precepts of evolution suggest that each species reflects the adaptations of its earlier ancestors in addition to its own. *Homo sapiens* is no exception. Until recently, family and kinship organizations were regarded as uniquely human characteristics, representing a sharp break between human "culture" and nonhuman "nature." In this selection, Reynolds synthesizes data from the study of monkeys and apes, hypotheses about proto-hominids, and data from the study of hunter-gatherers (the earliest known form of human society) in an attempt to explain the evolution of family and kinship organizations.

Evolution entails both continuity and change. Since the former process often tends to be overlooked when discussing the place of humans in evolution, Reynolds devotes most of his paper to the common threads in social relations among monkeys, apes, and man. He concludes that the flexibility of social organization that developed in monkeys and apes may have provided the basis for the successful evolution of the proto-hominids and, therefore, of *Homo sapiens.*

Such questions as the nature of kinship and family in the evolution of our species draw us into conjecture, for there are neither bones nor stones, neither tools nor records, that enable us to verify or disprove these speculations. The greatest danger is that metaphors and analogies may be mistaken for data. Two popular and controversial books that speculate along these lines are *Men in Groups,* by Lionel Tiger (New York: Random House, 1969) and *The Imperial Animal,* by Lionel Tiger and Robin Fox (New York: Holt, Rinehart and Winston, 1971). The reader who wants to compare family and kinship relations among nonhuman primates with those of hunter-gatherers will find many valuable empirical studies in *Man the Hunter,* edited by R. B. Lee and I. DeVore (Chicago: Aldine, 1968). There are several excellent descriptions of family and kinship organizations at more advanced levels of cultural development in *Man in Adaptation: The Cultural Present* and *Man in Adaptation: The Institutional Framework.* ■

INTRODUCTION

IN AN EARLIER ARTICLE in this journal an attempt was made to show that certain characteristics of the social behavior of the large apes, chimpanzees, orangs and gorillas, may have been present in the common ancestor of the large apes and man about 25 million years ago; this led to a new perspective on the possible evolution of some typical and universal patterns of human behavior. This article presents a more detailed examination of the system of kinship recognition often thought to be unique in man, and suggests how such behavior might have evolved.

The recognition of kinship relationships between individuals is an important factor in determining behavior in all human societies. Kinship terms relate to an individual or category of persons, of which certain types of behavior are expected, and to which certain types of behavior are directed. During the evolution of hominid social patterns, many kinds of behavior came first and terminology afterwards. In the evolution of language, as distinctive sounds became attached to specific things and individuals, names could be used to reinforce relationships already recognized by differential behavior. Is there any recognition of kinship ties among nonhuman primates? In particular are there any comparable patterns found among the large apes, our nearest relatives? If so, it could be that the proto-hominids shared the same patterns at their emergence; and that subsequent adaptations to the pressures of terrestrialism and savannah ecology worked on this substructure of inherited behavior patterns.

THE JAPANESE MONKEYS

While monkeys are not so closely related to man as are apes, the only studies specifically of kinship among primates have been made on some troops of Japanese macaques by research primatologists at the Japan Monkey Centre. A number of macaque troops have been subjected to long-term observation since the 1950's. The monkeys are in fact wild, but have become accustomed to being "provisioned" at the observation centers set up on part of their range. Several troops now spend some hours at the food hoppers daily and, as they are never interfered with or frightened, they behave confidently and can be identified and watched closely. For two troops named the "Koshima" troop and the "Minoo B" troop, each individual has been identified and its genealogy recorded over three monkey generations. One striking fact to emerge is the importance of the ties between matrilineally related females. This is reflected in ranking among adult females. For example, in the Minoo B troop there were four senior females in a rank order. When their daughters reached maturity each assumed the rank of its mother.

The process by which this comes about begins, we are told, when the young monkey behaves with superiority towards an adult of lower status than its mother, while the mother supports it with her presence. Dependent ranking of this kind has been observed in juvenile monkeys due to the presence and support of a high ranking mother, elder sister, mother's sister, and on one occasion, mother's brother. A young male monkey receives this sort of protection, allowing him to act according to the rank of his mother, until at the approach of adolescence he normally leaves the central part of the troop and joins the other young males on the periphery. Here he must fight out his own basic status on his own. On one or two rare occasions, however, the Japanese observers have recorded cases where a young male with a very high ranking mother to protect him has been allowed to remain throughout adolescence in the privileged central part of the troop, and has achieved the status of a dominant male without ever having fought a status battle in his own right. This is also reported in Koford from his studies of rhesus monkeys on Cayo Santiago.

These examples of how status within a Japanese macaque troop may be influenced by blood relationships are only one measure of the way kinship affects behavior. The incidence of sitting together and forming subgroups for grooming is highest between mother and offspring even into adulthood, and next highest between friendly females who are often siblings or cousins. Animosity is marked when monkeys meet at the food hoppers but there is a high degree of mutual tolerance between mothers and offspring and between siblings into adulthood. Another interesting observation is that a sudden appearance of a new item of behavior, "potato-washing," spread quickly among the members of the Koshima troop's solidary matrilineage. One final point is that the Japanese have not seen an instance of sex relations between mother and son. These reports are to be found in the translated edition of some of the key Japanese monkey papers. It thus appears that among these monkeys primary kinship ties are recognized by the individual throughout life, and these affect the behavior of the individual and often of the rest of the troop towards related individuals. Kinship ties, in particular those between females, appear to affect the placement of individuals in the social structure of the troop. Without knowing whether kinship recognition of this order is typical of the higher primates in general, the time is ripe for long-term studies on other species, to assess how it affects social organization according to the species-specific way of life.

THE LARGE APES

Chimpanzees live in what I have termed an "open community." A study area of six to eight square miles of forest contained a local population of sixty to eighty chimpanzees who were frequently to be found feeding somewhere in the area. It was not a troop of permanent membership, but composed of temporary bands roaming the area, meeting, joining and splitting. At some seasons when abundant fruit was concentrated in one part of the forest, large noisy gatherings occurred. At other times there were three or four smaller subgroups feeding in different places, and if fruit was scarce or widely scattered, individuals foraged in ones and twos. There seemed to be no boundaries separating the populations of different places, indeed there was a continuous traffic of small parties between all neighboring communities; but at any one time there was a core population of familiar inhabitants to be found somewhere among the favorite feeding places in each area. Ecologically the social system of chimpanzees can be viewed as adaptive to the needs of large fruit-eaters in a forest environment, where there is little predator pressure; for the flexible system allows dispersion and aggregation to exploit seasonal variations in availability and distribution of fruit. While there are no permanent groups, certain types of bands are frequently forming. From an analysis of my own field data, small groups of two or three mothers with their offspring were among the commonest groups found feeding together in a tree or traveling through the forest. Such a group of mothers might remain in the vicinity of a fruiting tree for several days while the more energetic and mobile members of the community, the adult males, the childless females and the adolescents, would visit several different feeding places in the course of a day.

Another frequently observed grouping consisted of two to five adult males in a band, often miles from any other chimpanzees. Elderly adult males were sometimes seen accompanying a band of mothers, never acting as leaders, but just as companions, as though they preferred the peaceful life. Sex relations were promiscuous and not controlled by dominance. It was not possible in my observations of a population of wild forest chimpanzees to determine what part kinship relationships play in the formation of bands and association patterns within the community, or between communities. It is possible, inferring from the evidence of the macaques, that ties between mother and offspring, and between siblings, are recognized throughout life, and that these determine to some extent habitual associations, and visiting patterns. Some indication that this is indeed so comes from the long-term studies of a chimpanzee population in a Tanzanian reserve by Jane van Lawick-Goodall. She has found from her detailed observations of a community of chimpanzees over seven years, that the relationships between a mother and offspring persist into adult life. The mother and her infant and preadolescent juveniles stay together all the time. At adolescence both males and females tend to spend less time with the mother, but frequently return to greet her affectionately, and to spend some time with her and their younger siblings. The male starts to join in the activities of the older males in the community, to form friendships and enmities and to establish himself as having a particular status.

Bonds between siblings seem to be strong, and Goodall reports an instance where a sister adopted an infant, when their mother died, and cared for it in her place. Among the adults, Goodall observed many instances of permanent friendships and alliances, and although she cannot know for certain the origins of these attachments, she speculates in the light of her findings concerning family ties among chimpanzees that "perhaps the companionship between young siblings develops into the casual but persistent friendships that we see between some of the adult chimpanzees."

Gorillas differ in some important ways from chimpanzees in their behavior, but there are also important similarities. The chief differences are probably correlated with the gorilla's specialization to a predominantly terrestrial life as a forest-dwelling leaf, stem and pith eater. There are established groups in gorillas, usually consisting of one or two fully adult males and a number of females with their young. Such a group appears to live in an area of about fifteen square miles within which it wanders more or less at random. The interesting characteristic that gorillas share with chimpanzees is the "open-group" system. There is no exclusive use of any range or even "core area" by any one group. Several groups use the same stretches of forest, and when two groups meet they may even join up and forage together for a few days; or they may take a long look at each other and go off in opposite directions. Like chimpanzees, gorillas have wandering males, though they are usually alone in the case of gorillas. These lone males often attach themselves to one group or another for varying lengths of time, and may mate with a female in the group without provoking any anger or even interest from its male leader. Although we lack data on the formation of gorilla groups, it is possible that some or all males leave their natal groups at maturity and may for some time lead a wandering life attached to different groups. In the course of time they may form a new group from part of an overlarge established one, or take over one where the leader is ailing or dies. This pattern would explain the various relationships existing between members of different groups and the tolerance for visiting males as though they were known to the group, and would be consistent with the widespread primate tendency for males to leave the mother's circle at adolescence or maturity. From what we know of monkeys and chimpanzees, it seems possible that the females making up the groups are related by ties of primary kinship, and that they tend to remain in their natal group.

The third large ape, the *orang-utan*, has so far yielded less information to observers because of its very low population density. But from early reports and recent observations, it seems likely that orang society resembles very closely the flexible group system of chimpanzees, with allowance for the drastically fewer numbers.

HYPOTHESES CONCERNING THE PROTO-HOMINIDS

I have briefly sketched the social organizations of the large apes with special reference to the probable or possible parts played by ties of kinship between mother and offspring and between siblings. Although each of the large apes has evolved certain characteristics of its own in adaptation to its particular ecological situation, they have all three remained within a tropical forest or woodland habitat such as was the scene of primate evolution for millions of years. Their organization has therefore probably changed much less in the 25 million or so years since they had a common ancestor with the hominid line, than has the hominid line itself, which took to savannah-dwelling. What interests us in this article is the social behavior, especially that part which has to do with kinship recognition, which the three large apes seem to share with each other. For these characteristic forms of behavior may well be the ones the earliest proto-hominids took with them when they first emerged onto the savannahs. It may be these characteristics, modified by the new pressures for survival on the ground in open country, which determined the direction and success of hominid evolution.

The behavioral characteristics which may have preceded the development of hominid kinship and family patterns include:

1. An *open community* social system whereby bands within a local regional population gathered or dispersed according to food distribution and inclination, such bands being connected by relationships between them and,

where groups were nonexclusive, by interaction and visiting between neighboring communities.

2. The *matrifocal* family of mother and immature offspring as the basic and independent unit of the community, such units being commonly attracted to each other where primary kinship ties joined the adult females involved.

3. *Adult male bands* for exploratory and food-finding purposes, composed of males who left their family at adolescence and joined in the activities of the dominant males of the community.

4. The importance of *ties of primary kinship* (defined as relationships between ego and his/her mother, father, siblings, offspring, and spouse) in determining temporary groupings, such as, for example, the association of males with their mother's group, or young females with a brother's band. In a community of the type described for the chimpanzee, any individual is a part of a number of different groupings at different times, but retains his relationship with the other groups too. The main point here is that, descriptively, chimpanzees and, to a lesser extent, the more specialized gorillas behave in this way, and from indirect evidence from chimpanzees and other primates, it is at least likely that one of the chief factors in the formation of the various temporary associations may be primary kinship linkage.

5. The probable predominance of *mating outside the family*. Such a system would tend to promote sexual relationships over the whole community, as it probably does in chimpanzees and gorillas. Although observations are not detailed enough in these species to state definitely that mating does not occur between mother and son (as was reported for Japanese macaques) or between siblings, the prevalent social patterns would seem to foster breeding outside the family group. For example, there are the tendencies for young males to spend less time with their families at adolescence and join with the other males in the community, or to go off on their own to attach themselves to other bands in gorillas. However, we know also that a special and permanent relationship between family members is recognized throughout life. It has been suggested by Fox that the existence of a familial relationship precludes the circumstances necessary for the initiation of a sexual relationship, as in some human societies. Perhaps the same is true for chimpanzees. When the population is gathered together at times of abundant food, social excitement is high and mating frequent, as has been noted by Goodall. This pattern of sexual interaction is in marked contrast to that of some other primates such as macaques or baboons where the permanent group is also the breeding unit, but it bears some resemblance to some human behavior patterns.

EMERGENCE AS A FOREST FRINGE SPECIES

Plasticity of social organization could have determined much of the success of the proto-hominids. For while bands of exploratory males ventured onto the savannahs opportunistically scavenging, catching small prey, or frightening away carnivores from a kill with noisy displays, the female groups could forage in the forest edges. Sometimes the matrifocal family might forage alone, at other times in small bands of sisters, mothers and offspring. An elderly male or two might accompany them as a benign protector. The male bands would be drawn from the community as a whole and, by their various kinship ties and visits to their mothers and siblings, link up the different family groups in the community.

Millions of years may have passed like this before the proto-hominids, increasing in skill as hunters and scavengers, became a predominantly savannah-dwelling species. Then, the very different ecological situation, the necessity for a much lower population density, the need for protection for mothers and young, the seasonal changes in climate and the dependence on water sources, must have exerted many pressures on their behavioral evolution. But evolutionary modification occurs on the basis of what was there before, and the present hypothesis of the behavioral origins of man will find support if it is possible to detect a logical continuity between what has been hypothesised for the proto-hominids and what we know to be the case for hunting and gathering nomads living today.

HUNTER-GATHERER BANDS

At the recent Wenner Gren conference on "Man the Hunter" in 1966 much new material was presented on existing hunter-gatherer societies, a few of which had been newly studied, and many others of which had been studied and restudied for generations. The emphasis in the reports was on the observed actualities of everyday behavior, as opposed to the construction of models of social institutions. In almost every society described the picture which emerged was of a shifting band composition with nuclear families as the basic and relatively independent units.

Helm's terminology in her description of northeastern Athapaskan Indians has wide relevance, not only to other hunter-gatherer societies but also to what I have been inferring for chimpanzees and proto-hominids. She defined three types of band, the regional band, the local band and the task force. The regional band was a social identity only; an individual thought of himself as belonging to the community of a particular region, although he might not have lived there for some time. In any region, moreover, individuals or families from other regional bands might be living permanently or temporarily. This term "regional band" is somewhat analogous to my use of the term "open community" with respect to chimpanzees and proto-hominids, although my term specifically refers to the *actual* population of a region at any one time rather than a social identity. The Athapaskan regional band consisted on average of twelve to fifty nuclear families in shifting and intermittent communication, which even in its size is strikingly comparable to my estimate of sixty to eighty chimpanzees in a local community.

Within the loosely structured regional band is Helm's "local band," consisting of between two and twelve nuclear families which tended to remain together most of the time. Families in such a band would always be related by close kinship ties, consanguineal or affinal. Often the band comprised a set of brothers and sisters with their husbands and wives. The socially independent unit in Athapaskan society was the nuclear family, which could go to visit or to live in any local or regional band where either spouse had a primary kinship tie. Helm's local band is in many ways comparable to my group of related mothers amongst chimpanzees and proto-hominids, although the latter lacks permanent males; while amongst the Athapaskans it is the nuclear family which is the basic unit, in my analysis it is the matrifocal family of mother and offspring without a permanent adult male.

Lastly, Helm's "task force"—a purposive *ad hoc* group formed of men or men and women—is very analogous to the all-male chimpanzee band. While for chimpanzees I have suggested that the flexible social system was a good adaptation to the optimum exploitation of a seasonal fruit distribution, it has been suggested that for the Athapaskans the not dissimilar system was adaptive for "environmental instability and recurrent disaster." In other words, the recognition of primary kinship ties ensuring certain hospitality over whole populations outside the region or residence group, and the independent mobility of the individual family unit, prepared the Athapaskans to survive many eventualities. These features enabled the maximum value to be obtained from any particularly abundant food source by the joining of groups; it enabled maximum dispersion in hard times; and should one local group suffer decimation, then the survivors could join up with any other group in which they had relatives.

Although Helm's useful terminology was used in her description of the Athapaskans, it seems to have a wide relevance for other hunter-gatherer tribes. I shall give a few references from other reports to the Wenner Gren conference to show how world-wide these shifting and flexible social systems appear to be.

Steward said of the Great Basin Shoshoni that there were no structures of permanent membership or lasting sociocultural integration larger than the nuclear family. He went on to mention that Shoshoni nuclear families were connected to one another through intermarriage in a network extending from Death Valley to northern Utah and southern Idaho. And he made an additional more general point that many so-called hunting bands were little more than statistical norm-groups that assembled for varying lengths of time in certain circumstances and for particular purposes, but that readily fell apart into their component families.

Balikci, in an ecological study of the Netsilik Eskimos, found a system similar to that of the Athapaskans. Igloo communities comparable to Helm's regional bands consisted of sixty or more individuals, comprising a number of distantly related extended families. The latter are similar to Helm's local bands. There was fusion and fission of the community at different times of the year; nuclear families went off alone on separate trips, and there was great flexibility in residence. Balikci suggests that the pattern was adaptive to constantly changing opportunities and pressures.

Damas summarized work on three major groups of Eskimos, the Netsilik already mentioned, the Iglulik, and the Copper Eskimos. Again there were the regional groupings, with a shifting membership, among which sixty to seventy percent returned to the same regions from one year to the next. Any local group identified itself with a particular region but this was not a defended hunting area. Extended families and nuclear families were the basic independently acting units in a community, and again it appears that primary kinship ties were the bonds linking any nuclear family with others in the local and the regional group. A nuclear family could go to live with primary kin on either the wife's or the husband's side.

The Hadza of east Africa described by Woodburn have a similar shifting social system. They have "three roughly defined localities," in which live the equivalents of regional bands, which at any one time are likely to contain about 100 individuals each. Persons and nuclear families pass freely from group to group and from region to region. Within any local group the bonds are based on primary kinship ties on either the husband's or wife's side. Though residence with the wife's mother is more frequent than with the husband's mother in this tribe, many families are to be found with the husband's mother or other relatives on extended visits.

The !Kung Bushmen, described by Lee and previously by Marshall present a similar picture once again. In this case while they averred themselves to be virilocal, Marshall found that over half of the !Kung adult males were living with their wives' bands.

Another African hunter-gatherer society is that of the BaMbuti pygmies studied by Turnbull. They show a very similar pattern of regional bands with a social identity, local groups with a shifting membership, independent nuclear families, and the preeminence of primary kinship ties. A girl may marry within or outside her own band provided she and her partner have no common grandparent that anyone can recall. When married, the couple may go to live with the parents of either of them, according to where they are most needed.

From India we have data on the Birhor hunter-gatherers, studied by Williams. The band size fluctuates from a large aggregate in the wet season to as little as a single nuclear family in the dry. The Birhor aver themselves patrilineal, but the coresidential unit which we would term local group or band, is not a patrilineally defined unit. If a man's father is dead, he is more likely to be found living with his wife's kin than with his own patrikin. Again the nuclear family visits independently wherever it has primary kinship links.

Australia is traditionally associated with territorial patriclans or hordes, consisting of patrilineally related males, their wives and unmarried female offspring. These generalizations about the nature of Australian aboriginal social systems were based on Radcliffe-Brown, but lately they have been questioned by four fieldworkers, independently of one another: Hiatt working among the Gidjingali of Arnhem Land, Pilling in south-eastern Australia, Meggitt among the Walbiri, and Rose in a general survey.

Hiatt maintains that while it is true that an individual identifies himself with a particular patriclan and a particular stretch of territory, the actual observed behavior concerned with residence, group movements and exploitation of resources is not based on the patriclan. There are many reports of dispersal and reunion of groups according to season, and joint participation of many different communities in ceremonies. Many observers have reported unrestricted movements of food-seekers over regions belonging to many different patriclans. Boundaries are found to be uncertain and ill-defined. Hiatt found that the Mara patriclan, for example, lived in a community called the Anbara, which comprised at least six patriclans, all the members of which moved, camped, and fished over the whole area, although each had a *ritual* connexion with a piece of estate belonging to it where its totems dwelt. In fact the "patriclan" turns out to be rather similar to our previously defined local band; Hiatt found local groups containing nonagnatic kinsmen, and Pilling supports this with many examples from earlier accounts, including cases where the brother's son of a leader of one group was the leader of a different group. Rose, in a very thorough study of the people of Groote Eylandt, where conditions were near-traditional in the recent past, has the same view. He found that while the patrilineage was the land-owning group and existed as a political unit, it hardly ever came together as a foraging or residential unit. A local band could be a single nuclear family, or it could comprise 200 persons from many different patriclans, entirely according to availability of food.

CONTINUITIES AND DISCONTINUITIES BETWEEN APE-LIKE ANCESTORS AND MAN

Such accounts suggest very strongly that there are underlying patterns of social organization among many hunter-gatherer societies, and that these in some respects bear striking similarity to those found in man's closest relatives, the large apes. This supports the probability of a *continuity* of behavioral evolution from a common ancestor.

Hitherto it has been man's pattern of mating beyond the family or band with the resultant network of kinship ties linking group with group and community with community, and his division of labor into groups of males or females for different purposive activities which have seemed to be unique to man. These have seemed to many, such as Lévi-Strauss, to be rational modes of behavior which marked a qualitative difference between the societies of men and those of animals. From the preceding comparison of characteristics of ape societies and hunter-gatherers it can be seen that the *observable* patterns of actual social behavior are in many respects similar, even though the apes are forest-dwelling vegetarians and man a hunter on the open savannahs.

Let us in our search for continuities and discontinuities, refer back briefly to some characteristics of the large ape societies which, it was suggested, might have been present in the first proto-hominids and have preceded later adaptations to savannah ecology.

First there was the open community where groups are nonexclusive, with a network of interaction within and beyond the community, where the local population gathers or disperses according to food distribution. This form of social organization is seen in the hunter-gatherer societies in their regional and local bands. It is suggested that it was the inheritance of such behavioral tendencies which pre-adapted the early hominids for their successful colonization of the savannahs.

Second, there was the matrifocal family of mother and offspring as the basic and independent unit of society—this is comparable to, but in one respect different from, the nuclear family in hunter-gatherer societies. This discontinuity, the permanent incorporation of one adult male into the matrifocal family, will be discussed in the final section of this article. The tendency for matrifocal families to be attracted to each other on the basis of primary kinship ties is paralleled by the attraction between nuclear families in the local bands among the hunter-gatherers described. The only big difference is that at the ape-like stage relationships in the female line only are recognized, whereas in the hunters with the male as part of the family, kinship ties cause the attraction of families to be matrilineal, patrilineal, or, most often, bilateral.

Third, the males leaving their family unit at the onset of maturity, to join in the activities of the other adult males, and to join with them in small bands on explorations, on forages, and on visits to other groups and communities—this descriptively is exactly the same in the hunter-gatherers as in the apes. The banding of adult males for joint forays may have been a pre-adaptation for the hunting bands which made life on the savannahs possible.

Fourth, the importance of primary kinship ties in the formation of groups and in visiting patterns is the same in hunter-gatherers as in apes, except that the ties recognized by the former are in the male as well as female line, and are affinal as well as consanguineal. But the essential characteristic which makes human society as complex as it is—the playing by the individual of different roles in different groups at different times, and the recognition of a nexus of relationships with other individuals over time and space—this characteristic is present in chimpanzees and to some extent gorillas, and was, it is suggested, present in the common ancestor of apes and man.

Finally, mating outside the nuclear family —this is universally characteristic of human societies. There is a correlation between the number and differentiation of degrees of kinship in the terminology and the complexity of regulations determining mating possibilities. For example, among the BaMbuti pygmies a couple should not marry only if someone remembers they have a common grandparent, but people do not remember well. Among the Australian aborigines, who record descent in great detail because the succession to the ritual estate must be determined, complex regulations govern the choice of marriage partner.

THE PAIR BOND: AN EVOLUTIONARY ADAPTATION TO THE SAVANNAHS

In my comparison of the social organization of the large apes and primitive man we have seen that the chief and basic difference is the attachment, in man, of one adult male to one, or sometimes more, females and offspring. This male is a permanent sexual partner to the female or females and the offspring are his. Among students of animal ethology, such an arrangement may be called a "pair bond" and is a common pattern of social behavior in many birds, fishes, and mammals. The formation of such a bond between a male and a female is typically preceded by a ritual form of courtship, which in many species has a seasonal onset. Bonding may be for life, as in some geese or gibbons, for example, or may last for a season only, as in robins. In many species sexual attraction is the basis for the formation of a pair bond, but this is not always the case. In the case of the greylag goose, for example, described by Lorenz, the young male pushed out of his family on the birth of new offspring looks for a mate to perform his "triumph ceremony" dance with. Normally he chooses a female, but by mistake he may pick another male. In either case the ties so formed by the performance of the ritual bind the pair for life. If he pairs with a female, they will not become interested in mating until the next season, but will dance together all winter long. Ardrey reviews a large amount of ethological literature, showing how in many species bond formation occurs when the female is attracted to the *territory* of the male.

Pair bonding as a survival mechanism has evolved, as Ardrey says, for the welfare of the offspring where they are "too numerous, too complicated or too long in their growing up." Such a pair relationship has to have behavioral means of expression and reinforcement for its maintenance and this is provided in a variety of ways. The caring for the young itself reinforces the pair bond, but there are other ways as well which operate outside the breeding season. In a species such as the gibbon, sexual activity is nonseasonal; in some birds duets of singing, in some cases songs specific to the pair, reinforce their relationship; the triumph ceremony of the geese has already been mentioned; and in numerous species the joint attachment of the pair to the territory and its defense is the major factor uniting them. Food sharing and home-building also act in the same way. But the evolutionary justification of the pair bond is the linking of a male and female in a relationship which is to the advantage of the offspring, and which normally lasts long enough for the young to be reared to independence.

There are of course other ways of ensuring the survival of the young besides the pair bond. The troop system of baboons is one example, where the vulnerable mothers and infants are protected in the centre of the troop by a circle of adult males. Again, chimpanzees get along without pair bonds. In their case, with little to fear from predators inside the forest, and with a good communication system of calls from one individual or group to another, the banding together of a number of mothers and young in temporary associations is sufficient to allow the opportunity for play and interaction with other juveniles which is so important in the development of the young primate.

One difference between the pair bond in hunter-gatherer societies and the pair bond as it is found in most other species, is that in the case of man, the pair operates as a unit within a larger network, i.e., both the male and the female maintain other relationships of friendship, kinship, status, etc. In most pair bonding species, the pair lives (analytically) in opposition to others, as an independent unit, or within a community of similar units but without joint interaction with other pairs. This is the most interesting aspect of the pair bond or the nuclear family in man, for it parallels, as we have seen, the independent mobility of the matrifocal family unit in the ape-like system. We must therefore look for those selection pressures in the evolution of the hominids, which favored the development of pair bonding *within the existing social system.*

We saw that the independent mobility of matrifocal family units, and the nexus of recognized relationships between communities and between individuals, was adaptive to seasonal dispersion and aggregation according to a fluctuating food supply. We saw too that the food-finding bands of exploratory males may have been preadapted to becoming the scavengers and hunters of the emergent hominid species. During a stage of being forest-fringe dwellers, the females may have congregated into extended matrilineal groups which split up to forage but collected together for sleeping, companionship and for the children to play, while the males were out on the savannahs. I have suggested that when the proto-hominids, male and female, began to exploit the savannah more and more and became a predominantly savannah dwelling species, the flexible social organization and sexual division of labor may have determined its success.

Certain new pressures would have then begun to operate. While small social units would not need a male in the forest, a mother and offspring alone on the savannah would be in need of an adult male's protection from predators. Food sharing and food storage would become economically necessary. Those females who attracted meat-possessing males to share their family shelter would gain much needed nourishment, and thus selection would favor the lengthening of the period of sexual receptivity. Conversely, those females who had stored vegetable foods to give the males in exchange for meat would have an advantage, as would the sharing males.

In this new ecological setting the pair bond could well have evolved because the male and female who had formed a successful and permanent relationship would gain selective advantages in terms of the survival of each of them and of their young. They had achieved the smallest hunting and gathering unit possible in terms of efficiency and safety. It was efficient economically for the male to go to hunt, while the female stayed at "home" in a protected place, foraged near it, and cared for the offspring; the male could protect his family at night, being thereby also able to have year-round sexual relations with his female. While such a family would form a food-sharing unit within a larger community, it could retain an independent mobility to allow the dispersion and aggregation of groups according to resources. And as each partner recognized kinship ties to all near relatives throughout and beyond the community, the family could go peacefully into many different localities, in which each partner could enter into cooperative daytime activities with members of other families, in groups of mothers and offspring for foraging and infant care, or in large hunting bands.

In this way, the evolution of pair bonding within a community system inherited from an ape-like ancestor is probably one of the most important adaptations which the hominids underwent and which helped to determine their success. It was the most efficient and yet enormously flexible social structure ever to attempt to exploit the savannahs, and it was because so many of what have always been thought to be typically and rationally human sorts of behavior were *already there* at the first emergence of the proto-hominids that success was assured.

HUMAN SPECIES-SPECIFIC CHARACTERISTICS

In this article the main concern has been with the typicalities and universalities of the hunting and gathering way of life, and its likely behavioral derivations from a similar social organization among the large apes, our closest evolutionary relatives. This kind of argument involves making certain assumptions about the nature of social patterns.

Crook has reminded us of the numerous interrelated factors which may determine the form a social organization will take, and in particular has stressed intra-species variability according to different ecological conditions. Specifically, primate fieldworkers have found differences among groups of the same species living in different environmental conditions, and at different population densities; differences in the frequency and intensity of social interactions such as intra- and inter-group aggression, territoriality, dominance and sexual jealousy.

However, the demonstrated flexibility in social organization involves *quantitative* differences. The basic "social structure phenotype" of a species is not *determined* by prevailing environmental conditions. For example a species such as the gibbon, which lives in territorial pairs, cannot be persuaded under any circumstances to form a troop organization based on a dominant male caucus such as is typical of baboons, nor a shifting band society as found among chimpanzees, nor an aggregation of harem groups such as that of the hamadryas or gelada. A good example of such inflexibility occurred many years ago at the London Zoo, when a colony of hamadryas baboons was set up. This species lives in harem groups in the wild, and unfortunately males outnumbered females in the new colony. Rather than set up a society based on either promiscuity or "polyandry," the males killed each other off, each one no doubt determined (in the genetical sense) to establish his own harem.

The form of the "social structure phenotype" depends on the patterns of bonding and dominance between individuals in the group according to age and sex. These normally develop during the maturation of each individual provided he experiences a minimum of the social experience normal for his species. Thus Harlow and Harlow in the well-known work on the development of rhesus monkeys, found that if an infant's emotional needs for comfort and security were met by its mother, and it had opportunity to play with other juveniles, species-typical patterns of sexual behavior, maternal care of offspring, self-assertion of the male, submission or subterfuge of the female, etc., occurred spontaneously. The normal maturation of species-specific patterns of dominance relations and bonding is genetically controlled in the individual, while dependent on the right environmental input at critical stages of development; if the social environment is wrong, abnormality and neurosis result, rather than an alternative form of social adjustment. This is as true for man as for other species of primates.

Hominids have lived for perhaps ten million years as hunters and gatherers, and it is to the hunters and gatherers remaining today that we must look for the clearest expression of the underlying species-specific characteristics of man; for in ten million years of evolution, forms of behavior conferring advantages on the human hunter or his band have been selected for, and those that were disadvantageous, selected against, at the genetic level. In the few tens of thousands of years since "mainstream" man made sweeping changes in his economy and became pastoral or agricultural, and in the few thousands in which he has been urbanized, modifications of behavior have been culturally innovated and transmitted. Institutions and sanctions have developed to promote some sorts of behavior adaptive to settled societies, and to discourage others. But the extent of cultural variation is more apparent than real, and the veneer is relatively thin. The same patterns of emotional involvement (bonding) and the same sex and age differences in dominance concern all humans, and these determine limits to the form of social institutions everywhere. For example, while there are enormous technological and economic differences between modern Britain and a hunting and gathering population, an objective observer would note many similarities in the patterns of social behavior. He would see a society divided into regional communities, with nuclear families moving about between regions, each family linked by a complex system of visits (or telephone calls) to a nexus of primary kin. And he would see that while the women and children remained in the residential area or camp, the adult males, alone or in organizations, competed, explored and tested their skill on the wide savannahs and in the attractive jungle of the modern city.

14. STATUS

ALISON JOLLY

Reprinted with permission of Macmillan Publishing Co., Inc. from Evolution of Primate Behavior *by Alison Jolly (Copyright © 1972 by Alison Jolly). The author received her B.A. from Cornell University in 1958 and her Ph.D. from Yale in 1962; she has been a Research Associate with the New York Zoological Society and is presently teaching at the University of Sussex. Her main field of study is primate behavior and its relevance to the evolution of human behavior.*

■ I suggested in my introduction to Van Lawick-Goodall's paper on chimpanzees (Selection 10) that it is necessary to keep the study of the evolution of individual behavior separate from that of social organization. In the previous selection, Reynolds concentrated on the possible continuities and discontinuities in the evolution of social relations, though these are not without behavioral significance for individuals. In this selection, Alison Jolly focuses on the evolution of behavior that is not without its components of social organization.

The task she sets herself is to understand anger, aggression, territoriality, predation, and status. Unfortunately, as she observes in her opening sentence, these terms are often erroneously lumped together; each must be dealt with separately. For example, anger is an emotion; status, territoriality, and predation, however, are relations. Status and territoriality refer to relations within a species, while predation refers to relations between species and entail selective pressures that may not be operating in connection with the other relations.

But even this amount of caution may be insufficient to avoid misunderstanding, because these categories are not so simple as one may imagine. For instance, consider aggression. Jolly observes that each species has its own ways of expressing aggression, its own forms and levels of aggressive behavior, different from other species. The reason for this is that aggression is not the expression of a single drive; instead, the aggressive behavior of each species expresses its adaptation to the conditions under which it lives. Depending on circumstances, there may be considerable variation in aggression even among the members of a single species.

In "The Origin of Society" (*Scientific American*, 204 [September 1960): 76-87], Marshall Sahlins discusses the proposition that the common capacities of humans and nonhuman primates notwithstanding, there are overriding discontinuities between them. *Social Groups of Monkeys, Apes and Men*, by Wolfgang Wickler (Garden City, N. Y.: Doubleday, 1972) concentrates on mating behavior, procreation, pair-bonding, and parental behavior. ■

DEFINITIONS: ANGER, INTERSPECIFIC CONFLICT, TERRITORIAL CONFLICT, AND STATUS

ANGER, STATUS, *territoriality*, and *predation* are four terms that often, wrongly, are lumped together. They are not even the same *kind* of word. Anger is an emotion, the other three are relations. Status and territoriality are intraspecific relationships and subject to quite different selective pressures than interspecific predatory relations.

This chapter is chiefly concerned with status, which is a set of long-term relationships among individuals who know each other. Although status is often connected with aggressive acts, the connection may be fairly remote in the history of either individual or species.

In ethology, agonistic situations are situations of conflict between conspecifics. We recognize agonistic situations, at one extreme, where animals actually damage each other and, at the other extreme, where one animal approaches and the other avoids. We learn to recognize ritualized antagonism in situations where the animals' gestures bear a close temporal relationship or a formal similarity to either physical conflict or approach-avoidance. The rhesus head-bobs in threat, in a clear intention movement of lunging. Conflict, as noted earlier, is often more difficult to recognize—for instance, Ellefson thought the morning display of male gibbons a sequence of joyous acrobatics until he saw one territorial male actually pursue and bite his rival. Similarly, the territorial "battles" of the sifaka seem like nothing at all to the unfamiliar observer—the troops leap toward or through each other in close formation, each side facing outward from its own territory. Only when one knows the animals is it clear what is relatively fast or slow movement, tight or loose grouping, and which animals belong to which territory.

Granted that we can recognize conflict situations, what is *anger* or feeling aggressive? It is emotion, motivation. We do, in fact, know what we mean by feeling aggressive ourselves. It seems to me perfectly legitimate to say that animals in many conflict situations also feel aggressive.

Aggression, in the sense of a propensity to intraspecific conflict, is common in the majority of vertebrate species. That is, most individuals of most species, brought up under most normal conditions and a variety of abnormal conditions, will show a range of agonistic behavior. In this sense, Lorenz' recent book *On Aggression,* is salutary precisely because it points out, and brings home, the fact that man as well has a strong tendency to aggressive behavior. This is not a return to the doctrine of original sin—original sin was an absolute, visited on all under all circumstances. Aggressive behavior is a product of circumstances, and it is shaped in man or the swinging gibbon through evolution and through learning. A few rare tribes of men, the Arapesh, the Lepchas, the Congolese pygmies, do not hurt or kill their neighbors. Having stated that aggression is at the innate end of our behavioral spectrum, one is then left with the interesting questions: in what situations does it appear, in what ways is it ritualized, what is its normal ontogeny? If the Arapesh are "abnormal," how could the rest of us learn to be abnormal like them?

The other three terms describe types of situations in which conflict behavior occurs.

Interspecific conflict, or competition and predation, is not usually classed with intraspecific conflict. It is related mainly because the weapons of predation—claws, canine teeth, hand axes—can be used on one's neighbors as well as on one's prey. There is no need to attribute angry feelings to the predator, tiger with sambhur or man shooting pigeon. Perhaps we are angry at a member of another species when it becomes competitor, not prey. It is not the target fowl or the squab on toast we hate, but the pigeon gobbling lettuce seedlings out of the back garden.

In *territorial conflict* aggressive behavior has a geographic aspect. Animals defend particular areas. This may result in a kind of community. Territorial animals typically recognize neighboring individuals and know how far they can go, in both literal and social senses. Territorial defense has arisen over and over by convergent evolution in unrelated lines of primates as an adaptation to particular ecological niches, whereas even closely related species may differ in their territorial behavior. Territorial defense also shades into "status."

In troop-living primates, it is usually fairly easy to divide "dominance" and "territorial" relations—neighboring troops space themselves out, and this spacing generally has some reference to geography, whereas members of the same troop move about close together and are therefore said to stand in a dominance relationship to each other. However, when animals change troop frequently, or troops join and divide, the distinction may become arbitrary or useless.

The idea of *status* or dominance has a checkered history. Schjelderup-Ebbe, who discovered the pecking-order of hens, enlarged his findings to a Teutonic theory of despotism in the structure of the universe. For instance, water eroding stone was "dominant." Schjelderup-Ebbe called animals' ranking "dominance," and many workers, with an "aha," recognized dominance hierarchies in many vertebrate groups.

Among the primates, dominance soon seemed even more baroque. Baboons and macaques mount each other in a gesture of assertion or present their rumps to be mounted as a gesture of submission. Rarer variants are the stronger animal backing up to the weaker and forcing the weaker to mount, which happens among bonnet macaques, or else juvenile males (playfully?) mounting adult males, as in hanuman langurs. This sexual gesture, transferred into threat-submission situations, combined with Zuckerman's description of zoo hamadryas baboons fighting to the death over females and the general influence of Freudian theory, all led to a view of aggressive dominance as a universal primate ranking, inseparably bound to sexual priorities. Of course, Carpenter's howlers rarely threatened each other at all and mated promiscuously within the troop, but people then said New World primates might best be ignored.

At last, DeVore and Washburn pointed out that in a savanna baboon troop a recognized hierarchy stabilizes the society. Far from a continuous pecking down the line, the hierarchy makes it possible to avoid fights in most situations, for each animal knows the other's strength and respects his rights.

Finally, with much more analysis of differing species, dominance has been sorted out into component parts. It is perhaps better called "status" ranking or even "role playing," as the word "dominance" is so contaminated with the notion of a single rank-order established and maintained by threat.

FREQUENCY OF AGGRESSION: DIFFERING SPECIES

Threat behavior differs among species both in absolute frequency and in the form of the resulting hierarchies. Whereas it is clear that frequency differences are important, field techniques are such that one can only compare species by loose subjective categories—those of "none," "rare," "common."

Davis *et al.* have made an attempt to directly compare the frequency of aggressive behavior among species in the laboratory. They released adolescent animals into the center of a runway 9 feet by 1 foot by 2 feet high, with an animal of the same species in a carrying cage at either end and nothing to manipulate except the iron bars of the runway. They then scored visual survey, cage manipulation, social behavior, rapid energy expenditure, and self-directed behavior. Their results were highly consistent, with five to nine individuals of six different primate species. In many measures, the scores of different species did not overlap at all, and in a majority the scores differed at the 5 per cent level. However unnatural the situation, it is a first attempt at standardized comparison.

The ringtailed lemur showed the lowest proportion of social behavior, and a large amount of this was only social observation. However, before jumping to phylogenetic conclusions, we should note that differences among the three species of macaque were as great as among the more distantly related primates. Stumptailed macaques were particularly affiliative, spending nearly all the time in social grooming.

What emerges most clearly is that each species has its own profile—its own quantitative preferences for various sorts of activities, at least in this very restricted situation.

Turning to field studies, can one make any generalizations about frequency of aggressive interaction? Two points seem to emerge. . . . First, the macaques and baboons, as a group, are by far the most aggressive. Macaques and baboons are adapted to semiterrestrial life in more open country than other primates. It has often been argued that man's aggressiveness is also related to life in the open. However, as there are data from only the one phylogenetic group, the case is hardly convincing. Forest-living baboons, the drills and mandrills, are only now being studied, and preliminary results indicate that they have similar group size and dispersion to their savanna cousins. The guenon-patas group, which also ranges from forest to savanna, tends to have a low level of intragroup aggression in either place, and so do chimpanzees. Although some differences in status hierarchy occur *within* species in differing ecologies (see following), it does not seem clear whether one can really attribute macaque-baboon aggression to savanna life.

Second, among forms with ritualized long-distance spacing behavior, the level of aggression is very low within the group. This is more convincing as such species occur in every line of primates: sifakas among the lemurs, titis and howlers in the New World, leaf monkeys in the Old World, and gibbons among the apes. Three of these forms are leaf eaters, and all live in smaller groups than other primates, or even in monogamous pairs. It is not clear whether one should argue that the *general* level of aggression is low in these animals because many of their ritualized battles seem perfectly calm to the naive observer, or whether one should argue that the battles are highly aggressive and that they somehow "take it out on" their neighbors instead of behaving aggressively within the troop. The question is probably not useful to ask, because we can have little insight into the feelings of a langur or sifaka as he whoops or scent-marks. It is more useful to ask why we ask it. Many people do, in fact, still often think of aggression as a unitary drive—a fixed quantity of energy that could somehow be channeled into either territorial display or intratroop bickering—whereas the whole effort of this book is to show how each species' pattern of behavior is complexly adapted to its own particular pattern of life. Indian rhesus macaques are probably aggressive toward either troop members or troop neighbors with the same feelings, as they literally act the same way toward both, with the same threat gestures and expressions. But one can hardly conclude that the leaping, great-calling gibbon, displaying in a tree top at his male neighbor, is aggressive in anything like the same sense as when he glances at his mate and reaches first for a fig —his feelings quite likely differ not only in intensity but also in kind.

FREQUENCY OF AGGRESSION: DIFFERING ENVIRONMENTS

Even within a species, the quantity and kind of aggression vary enormously with circumstances. These can be considered under two headings: changes under experimental manipulation and differences among wild populations in different habitats.

The most naturalistic experiment on primate aggression has been that of Southwick, who caged a group of rhesus monkeys as large as many wild troops: three males, fourteen females, and ten juveniles. The animals were released into a cage 40 by 30 feet and allowed 15 weeks to settle down to a steady level of aggressive encounters. Of course, this could never really approximate a natural troop whose relationships had been built up over a lifetime, but as with humans put in prison or the random encounters of a holiday camp, a fairly constant social structure emerged rapidly. After this initial baseline period, the group was subjected to various stresses.

Far and away the most extreme procedure was the introduction of strange rhesus the group had not known before. The group attacked the strangers. Interestingly, group members were most aggressive toward strangers of their own sex and age—males fought males, females fought females, young fought juveniles. Two juveniles, two males, and four females were introduced, two at a time. The level of agonistic behavior dropped in each case by the third day, but two of the females had been badly bitten and had not eaten in 3 days, so that the experimenters had to nurse them back to health.

Other experiments varied the amount and distribution of food. When all food was concentrated in one place, the number of threats rose—such as happens at any winter bird table. However, when the quantity was simply cut by half, the threats decreased, as did all social interaction. The starving animals moved about in apathetic lethargy. In fact, this parallels reports from prison camps and a Wisconsin experiment on starving humans.

The hierarchy really did determine priority to food. Dominant animals ate their fill, much as usual. After 5 days of half rations, the two lowest-ranking subadults almost died from starvation and had to be removed and hand-fed to save them. Southwick's results are confirmed by a study of provisionized rhesus macaques in Cayo Santiago Island, when a monkey chow shipment was 3 weeks delayed. Not only fighting but all social behavior lessened dramatically while the monkeys foraged for natural food.

However, changing the environment had little effect: either scattering rice to be hand-picked from the dust, introducing a paddling pool for play, or erecting a partition nearly across the cage. But reducing total cage space by half resulted in a highly significant rise in aggressive acts.

Other experiments have tested different variables. For example, Plotnik *et al.* tried squirrel and cebus monkeys in a competition situation, wherein the group had to pass one at a time through a narrow tunnel either to obtain food or to avoid shock. In both cases and both species, the frequency of aggression rose. Plotnik *et al.* pointed out that in contrast to Bernstein's study of cebus in a large open cage, their study clearly identified a boss cebus with priority to everything, although the rest of the hierarchy was unstable. They finished with the important point: "It appears there is a continuum of aggressive behavior [in different circumstances]. The kind of aggressive pattern observed depends on what part of the continuum is selected for measurement."

This is clear in the study of wild populations as well. Many of the formative ideas of primatology have come from the study of savanna baboons in the harsh conditions of thorn savanna. Rowell observed baboons living in riverine gallery forest. Although they foraged during part of the day in adjacent grassland, the troops could retreat for safety to the trees and found most of their food among the lush, if bitter, fruits of the forest. The adult males formed a coherent cohort, constantly aware of each other's movements—but with scarcely any aggressive interactions.

Among vervets on Lolui Island in Lake Victoria, males changed troops fairly frequently. One of the most consistent roles of the adult males was loud "territorial" or "spacing" vocalization at neighboring troops. There, it seemed unreasonable to talk of a hierarchy within the troops—much better to calculate the frequency with which each animal gave certain threats and gestures, such as the spacing call, characteristic of adult males. In savanna, however, there was a perfectly clear-cut threat hierarchy among troop males and less frequent, unritualized bickering among all the members of adjoining troops.

The most dramatic case is the hanuman langurs. In the troop Jay watched in Central India, six males lived in perfect harmony, rankable only by the slightest gestures. When the beta male became alpha, he marked the change by whooping at the far end of the grove. In Sugiyama's crowded South India population, all-male bands attacked and fought the harem males for possession of harems, then fought each other, exiling the weakest of the male band as well as male juveniles and children of the harem.

In Sugiyama's case, there is perhaps the clearest sexual selection for male dominance. The new harem leader gained exclusive right to his females. Then he commonly bit young infants after the takeover. Their mothers hung them on a branch to die and came into estrus, ensuring an even higher proportion of the new lord's children. One of the few parallels in other mammals is the Bruce effect in mice. Pregnant female mice exposed to the presence or even the odor of a strange male abort their litter and come promptly into estrus. Slater suggests that this may conceivably be of advantage to the female as well as the male, if it reduces inbreeding. It might also be of advantage if the male mouse or langur tends to be antagonistic to a female he has not mated, to such an extent that she would be likely to lose her infant or litter anyway, after putting still more time or effort into it. Some such advantage to the female might at least make it more understandable in mice, where the female's own physiology must play some role in the abortion.

A final, less gruesome, but still more important example is the town-living rhesus monkeys of India. Singh paired urban rhesus monkeys with their counterparts from forest habitats. The town monkeys were far more aggressive than the rural ones, both among each other and toward the strange forest monkeys.

Thus, any description of "the" level of aggression in a primate individual or species must specify which level in which circumstances.

MEASURES OF STATUS

Status may be measured first by an animal's priority to a desired object, usually either food or sex. This, after all, is the selective advantage to the individual of being dominant: to survive when resources are limited and to have a high percentage of offspring.

However, such priorities are only one dimension of status. A second criterion is frequency of threats—of conflict situations and who won them. Measuring threats as such tends to emphasize the role of the argumentative animal; furthermore, different threat gestures are not perfectly correlated with each other. A more telling category is "approach-retreat" situations. Rowell points out that it is really the subordinate's reaction that shows whether the situation contains potential conflict. If one animal just walks toward another, this means little. The second may stay put and groom the first, making it a friendly interaction, or the second may choose to step aside or run away, thus defining a status interaction. Eisenberg and Kuehn, studying captive spider monkeys, arrived at different rank orders in the two calculations of percent of times an animal retreated and percent of times the animal was retreated from. This complexity is not confined to primates. For instance, Baenniger found no correlation in rats between priority to food or water and spontaneous threats, and even chickens have no single, simple pecking order if given space to express themselves. Presumably, the threats, or prevalence and stability of approach-avoidance situations, have evolved away from the original priorities. Instead of fighting for food or mates, social animals have evolved ritualized means of self-assertion and self-abasement. Instead of even self-assertion in each new situation, they have become capable of recognizing their station in life and have evolved self-assurance and deference.

OTHER ROLES OF THE DOMINANT

Males of nearly every primate species protect the troop. They threaten predators, particularly predators holding or approaching an infant. This is true even in species whose males do not pay much other attention to infants. Bernstein, in a series of tests with a confined troop of pigtailed macaques, showed that the males differentiated between their own troop members and other monkeys. They attacked an experimenter who held a member of their own troop, but if another monkey was held, in the same circumstances, they sometimes attacked the strange monkey instead. Male defense is not universal. Among Rowell's forest baboons, the males galumphed fastest to the trees. Spider monkey males are often not with the troop, and lemur and sifaka females mob as energetically as their consorts. However, in general where there is real defense it is the function of males, and particularly of the dominant males in the internal threat hierarchy.

Territorial defense against other troops is also frequently a male role and in many species particularly that of the dominant male.

Another common pattern is stopping fights among other members of the troop, even literally punishing one or both of the combatants. This appears in many species and is consistently directed down the threat hierarchy, never up.

Leading the troop in geographic movements is often harder to analyze. It certainly occurs in many distantly related species. In macaques and langurs the alpha male stalks off first. In gorillas, however, the troop ambles off amoebically. If the main silverback male comes along, the troop makes progress, if not, they all subside again. One male of the fifty to one-hundred talapoins in a troop, scattered green and invisible through the obscuring forest, gives a "leader call"—and the others move accordingly. A gorilla troop changed its range when a new male took control, and the Japanese monkeys of the Takasakiyama troop abruptly changed their ranging times when Titan took over from Jupiter.

In this last case, there is a suggestion of more subtle influences. The provisionized Takasakiyama troop contained five-hundred animals at the time, with twenty-four leading males. They all recognized a single male hierarchy, with Jupiter as alpha—an animal Itani called "violent to the point of cruelty." However, Jupiter weakened and died, and Titan took his place. With the gentler Titan in control, the whole troop became more gregarious, with much closer spacing and more frequent contact among animals. Although other factors might have operated as well, it is possible that here the personality of the leader affected the whole level of social interaction within the troop. . . .

III.
EXPLORATIONS IN THE NATURAL HISTORY OF HUMAN ADAPTATIONS

MAN DID NOT emerge *de novo* on the evolutionary scene; he inherited many adaptive capacities from his nonhuman past. To study man's physical evolution, the many features of his adaptations must be studied one at a time and in relation to the others. To appreciate the uniqueness of man we must take a very close look at his hands, at his shoulders and elbows, at the hemoglobin that constitutes and colors his blood, at his skin color and the like. But each feature must be looked at in the context of the other features.

The gaps in our knowledge about human physical evolution and adaptation are greater than the solid information we possess. For example, we know almost nothing about the processes by which such biological variations as skin color, hair texture, and eyelid structure become fixed. An important qualification in the study of physical adaptation is that it probably is not possible to measure the effectiveness of adaptation until after the fact. If a population has been able to survive over many generations, we can assume that its adaptation has been effective; if a population became extinct, we can assume that its adaptations for its particular habitat were ineffective.

One of the concepts that is necessary for understanding specific physical features in the evolutionary perspective is differentiation. The amount of differentiation in an organism's architecture is what biologists usually mean when they speak of the role played by "organization" in the adaptive process—the amount and kind of specialization that characterizes the organic systems. When we look at the total span of physical evolution we observe a process of increasing differentiation within organic architectures; indeed, the degree of an organism's ability to adapt to its habitat is determined by the extent of differentiation in its architecture.

For example, both a primitive metazoan and a human being carry on activities such as metabolism, growth, responses to irritability, and reproduction. In a one-cell organism, all of these activities are carried out in the protoplasm of its cell; in a human being, as in other complex organisms, there is not only cellular differentiation but each activity is carried out by different organs and organ systems. One of the clearest examples of differentiation as an aspect of adaptation is described by Hockett and Ascher in Selection 29 in their discussion of the development of human speech. They observe that when the mouth was no longer used for carrying objects, it was freed for chattering.

An important correlate of differentiation in human architecture is that man has become extremely generalized. Generalization—in contrast to specialization—means that an animal, organ, or organ system is not specifically tied to a particular habitat or pattern of social life. No animal, of course, is entirely specialized or entirely generalized; both characteristics are found in all animals. The important point, however, is that some animals are more generalized than others.

Specialization and generalization have adaptive advantages and disadvantages. Specialization's clearest advantage is that it enables an organism to meet specific habitational challenges; its principal disadvantage is that it limits change; and a given specialization may represent an evolutionary dead end, a one-way path. A generalized form, however, is more plastic and can develop more varieties on basic biological themes. Many more alternatives are open to it when it is confronted with challenges that require modification.

Neither man's architecture nor his culture alone embodies his adaptive and evolutionary potential; both are required, and there is constant feedback between the two. Man's physical equipment could not have developed to its present level without culture, nor could man have achieved his capacity for culture without his physical evolution. Culture enables man to maximize his physiological potentials and to adapt to a variety of habitational conditions; it provides man with the potential for freeing himself from the limitations of his biology and habitat. With every major advance in his mastery over his natural milieu, man has capitalized on and exploited his ability to adapt, and in these terms *Homo sapiens* is the most advanced living form.

Also in these terms adaptation is the mechanism of evolution, which is the continuous succession of adapted forms that is observable in nature. When we speak of evolution, whether in physiology or in culture, we imply that *B* grows out of *A* and precedes *C*, because *A* contained the necessary germ for the emergence of *B* under the proper habitational conditions, as did *B* for *C*. The notion of the succession of surviving forms obviously assumes that each has more or less successfully adapted to its habitat. *A* cannot lead to *B* if *A*'s failure to adapt successfully led to its extinction; when this occurs, the succession of forms in a particular lineage comes to an abrupt halt.

The greatest of all adaptations is the ability to adapt to changing habitational conditions. There are many examples of this aspect of evolution, and man in particular can adapt to a great variety of habitats. Evidence of man's highly developed adaptability will be seen below in the emergence of the human hand; in varieties of stature; in alterations in the blood flow of the hand in response to cold;

and in changes in respiration, size of the chest cavity, and hemoglobin in response to the conditions of high altitudes.

This section begins with a consideration of man as the most generalized animal. One of the clearest examples of his generalization is his hand (Selection 16), whose evolution had profound consequences for his ability to make and use tools (Selection 17). However, although all men have emerged with the same basic architecture, there are important variations within the species, and one of these is the subspecies diversity that is commonly referred to as racial differences (Selections 18 and 23). It cannot be said that these particular variations are tied to differences in habitat, but other physiological variations within the species are closely related to habitational factors. One of the most significant examples of variation for the study of adaptation and evolution was the emergence and temporary dominance of *Australopithecus,* who was one of our most important ancestors.

We will further observe in Part IV comparable processes of adaptation in contemporary populations that can be studied first hand: adaptations to cold among Eskimos (Selection 24). Other variations within the species arise in connection with technological (cultural) factors, such as weaponry, and Selection 27 deals with the hypothesis that different body builds are adaptive to different types of weapons, such as bows and spears. Finally, as will be seen in the analysis of some physiological characteristics (steatopygia, large deposits of fat over the buttocks) of residents of the Kalahari Desert in South Africa (Selection 28), there are variations within the species that continue to defy our understanding.

In concluding this introduction I want to raise an issue in the study of human adaptation that, unfortunately, has been seriously neglected: the consequences of overcrowding on *Homo sapiens.* There is very little that man can do about his physiological adaptations but there is much that he could do if it is found that the environments he has created, such as modern metropolises, are out of keeping with his biological constitution.

If we compare the kinds of groups in which modern industrial man lives with those in which he lived shortly after his emergence on the evolutionary scene, we come to a very disturbing conclusion about the relationship of man's physiology to the sociocultural component of his environment. Man's genetic constitution and his social organization must have been in equilibrium during his early history; that is, we can assume there was a close and good fit between *Homo's* genetic constitution and life in very small groups. As far as can be determined, man's genetic constitution has not changed appreciably during the last few million years, but now he is living in groups whose sizes would have defied the imagination of early man. Has discord arisen, in the course of evolution, between man's genetic capacities and his social organization? Or are modern settlements further evidence that man has freed himself from the limitations of his genetic constitution?

This question cannot be answered at the present stage of theory. Whatever evidence exists—and it is inferential at this time—points to the hypothesis that there is indeed a widening gap between man's genetic capacities and his modern social systems, that man is not programmed for the kinds of groups in which he tends more and more to live.

Recent research by a variety of investigators who study nonhuman animals (such as rats) suggests that overcrowding *per se* produces deleterious effects, such as a sharp rise in mortality, sterility, and susceptibility to infection, as well as maternal neglect and cannibalism. Was this, then, the reason that during pre-industrial stages human groups regularly and recurrently subdivided after they reached what appears to have been optimal size? Was pressure on their available food resources the only reason such groups recurrently split? Is the failure of modern social systems to provide for such splitting one of the reasons for disorganization, maternal neglect, and short life expectancy among overcrowded people in urban slums? Is this a justification for limiting the size of populations? Can man continue to increase the population density in his cities without heeding the analogy with the mechanisms that govern the control of population density among nonhuman animals and primitive groups? Some students of modern social organization do not agree that overcrowding produces deleterious effects in man as it does among nonhuman animals; however, other students feel that it might and that there is urgent need to investigate the problem.

It has been suggested that overcrowding produces such strong pressures on available resources—including space—that there would be insufficient resources for all the members of a group if they continued to multiply in the same territory; therefore, group subdivision, as well as the deleterious effects of overcrowding, might be important adaptive responses that allow some members of the species to survive instead of having all of the members starve or die from other causes. Must we, however, continue primordial patterns of adaptation? Can we continue to disregard or fail to investigate, what is being learned about the relationship between man's genetic capacities and social organization? Or is it an adaptive imperative that we incorporate this knowledge into social planning as actively as we apply what is known about, say, oral hygiene and the planning of traffic flow?

Because new additions and revisions are made every few months in the record of man's emergence on the evolutionary scene, it is almost impossible for works in print to keep abreast of the fossil record, but there are several general works that provide a schematic outline of that record. A beautifully illustrated and well-written introductory work is *Early Man,* by F. Clark Howell and the editors of *Life* (in the *Life Science Series* [New York: Time, Inc., 1965]). One of the best introductory works is Robert J. Braidwood's *Prehistoric Men* (Chicago Natural History Museum, Popular Series, Anthropology, No. 51, 1959), which, although originally published more than two decades ago, provides a good introduction to the relationship between physical and cultural evolution. The advanced reader can consult *The Evolution of Life: Its Origin, History, and Future* (edited by Sol Tax), which is Volume I of *Evolution after Darwin: The University of Chicago Centennial* (Chicago: University of Chicago Press, 1960). In Volume II, *The Evolution of Man: Mind, Culture, and Society* (also edited by Tax), the first four papers are especially relevant to our general area: "Evolution, History, and Culture," by A. L. Kroeber; "The Origin of the Genus Homo," by L. S. B. Leakey; "Human

Evolution and Culture," by S. L. Washburn and F. C. Howell; and "Dating Human Evolution," by S. Emiliani.

Several works provide a good introduction to the fossil record: *The Ascent of Man: An Introduction to Human Evolution,* by David Pilbeam (N. Y.: Macmillan, 1972); *The Emergence of Man,* by John E. Pfeiffer (N. Y.: Harper and Row, 1972); *Human Evolution,* by Bernard Campbell (Chicago: Aldine, second edition, 1974); F. C. Howell, "Recent Advances in Human Evolutionary Studies," *Quarterly Review of Biology* 42 (1967): 471-513; *The Prehistory of Africa,* by J. D. Clark (N. Y.: Praeger, 1970); and the first two volumes of the Time-Life series *The Emergence of Man: Life Before Man* and *The Missing Link* (N. Y.: Time-Life, 1972).

Almost all of the work on the consequences of overcrowding has been conducted with nonhuman animals, but one of the most suggestive inquiries has been conducted by John J. Christian, who concluded that these consequences are basically sociopsychological. His paper, "Phenomena Associated with Population Density" (*Proceedings of the National Academy of Sciences* [April, 1961], 47 [4]: 428-49), is an excellent starting point. John B. Calhoun's "Population Density and Social Pathology" (*Scientific American,* 206 [February, 1962]: 139-148), deals with rats in experimental situations. A thorough review of the relevant literature is D. D. Thiessen's "Population Density and Behavior" (*Texas Reports on Biology and Medicine* [Summer, 1964], 22 [2]: 266-314).

In addition, the reader who is interested in the consequences of overcrowding should consult the following papers: "Ecology and the Proto-Hominids," by George A. Bartholomew and Joseph B. Birdsell (*American Anthropologist,* 55 [1953]: 481-98); "Ecology and the Population Problem," by LaMont C. Cole (*Science,* 122 [1955]: 831-32); and "Increase of Settlement Size and Population since the Inception of Agriculture," by Max Petterson (*Nature,* 186 [1960]: 870-72.

Although we usually think of physiological adaptations to environmental pressures without reference to their possibly maladaptive consequences, the latter should not be overlooked. In "Problems in Bioclimatology" (*Proceedings of the National Academy of Sciences,* 45 [1959]: 1687-96), Rene J. Dubos raises some interesting questions about the possibly maladaptive consequences of air conditioning.

15. THE PROTOCULTURAL FOUNDATIONS OF HUMAN ADAPTATION

A. IRVING HALLOWELL

Reprinted from Sherwood L. Washburn (Ed.), Social Life of Early Man *(Chicago: Aldine Publishing Company, 1961). A. Irving Hallowell is Emeritus Professor of Anthropology at the University of Pennsylvania. In his field work he has been concerned mainly with the Northern Ojibwa in Canada (see his book* Culture and Experience*), but for many years he also has been interested in the problem of hominid evolution, particularly in its psychological dimension. He links this topic with the study of culture and personality, and thus with human adaptation generally.*

■ One of the most important early contributions of anthropology to the understanding of man was the idea that one of the ways by which a culture is perpetuated from generation to generation is through its techniques for shaping the minds of the young members of the society. A. Irving Hallowell, the author of this selection, was among those most responsible for giving substance to this principle. He points out in the following paper that man's capacity for the transmission of culture was established at the prehuman level of evolution. As we have noted, this is an important aspect of man's kinship with his ancestors, and it is a key to an understanding of man himself. *Homo's* inheritances have not only given him potentialities for the development of culture, they also set limits for his modes of adaptation.

It should be reiterated that the concept of man's inheritance of such capacities from his prehuman ancestors in no way suggests that *Homo* inherited specific modes of behavior from his nonhuman forebears, any more than one man's kinship with another is an index of their similarity. One way of viewing man's uniqueness—his humanity—is by examining what has happened to the capacities that were established in his nonhuman past. This means, as Hallowell maintains, that the study of evolution and of adaptation cannot disregard the evolution of mind and of psychological functions, nor can the study of mental functions disregard evolution and the processes of adaptation. To study man's mind in an evolutionary vacuum is to assume that it developed at some critical moment in evolutionary time, and this implies that man has no roots in earlier modes of adaptation.

In addition to the ability to learn, accumulate, and transmit experience, Hallowell discusses three other levels of functioning that man shares with infrahuman primates: group life or gregariousness, terrestrial rather than arboreal living, and patterns of communication. The four cannot be separated one from the other, in their development or their functioning.

Hallowell's analysis is conjectural, but any statement about the evolution of the first human social and mental processes must be conjectural because it is impossible to observe the psychosocial lives of populations that are represented only by fossil remains. Are there ways of evaluating such hypotheses? There are, although the evaluations must be based on deduction rather than on direct observation. Comparisons of primate and rudimentary human social relations, such as those of Reynolds and Jolly, and the analysis of primate group structure in relation to habitational pressures, as in the paper by Kummer and Kurt (Selection 12), are of the kind from which very suggestive insights can be gleaned for the evaluation of Hallowell's hypothesis.

Hallowell has been almost alone among anthropologists in his concern for the psychological dimensions of human adaptation; this paper is an excellent introduction to much of his other work, which is represented best in a collection of many of his papers, *Culture and Experience* (Philadelphia: University of Pennsylvania Press, 1955). Also relevant is a chapter in Anthony F. C. Wallace's *Culture and Personality* (New York: Random House, 1961): "The Evolution of Culture and the Evolution of Brain" (Wallace was a student of Hallowell). The last three chapters of Campbell's *Human Evolution* (Chicago: Aldine, 1974) are especially relevant to the issues being explored by Hallowell; they contain some ingenious hypotheses about the continuity of man's mental functioning since the nonhuman past. A highly technical and pertinent paper is Clifford Geertz's "The Growth of Culture and the Evolution of Mind" (in *Theories of the Mind,* edited by Jordan M. Scher [New York: Free Press of Glencoe, 1962]).

A current debate among scholars about organizations of social relations and the use of tools is very relevant to the study of the continuity between nonhuman and human primates. One of the recent exchanges in this debate is that between S. L. Washburn and Phyllis C. Jay on one side and Adriaan Kortlandt on the other. The antecedents of this debate are listed in the bibliography of this exchange *(Current Anthropology,* 8 [1967]: 253-57). ■

IN A PAPER contributed to the centennial celebration in Chicago of the publication of Darwin's *Origin of Species,* I pointed out that vital evolutionary questions of a sociopsychological order need reconsideration in the light of twentieth-century knowledge. While questions of this kind were among those originally broached in the nineteenth century, along with others essentially biological in nature, evolutionary thinking among psychologists and cultural anthropologists fell into abeyance, as time went on, with the rejection of the recapitulation theory in its classical form and of unilinear theories of cultural evolution. The study of human evolution became more and more restricted to biological problems dealt with by physical anthropologists. But, with the discovery of new types of early hominids (small brained but bipedal in locomotion), the accumulation of observations on the social behavior of nonhominid primates in their natural state, the development of psychoanalytic theories, culture and personality studies, and the conceptualization of the nature of culture provided by twentieth-century cultural anthropologists, we now have a more fruitful point of departure for enlarging the boundaries of evolutionary thinking beyond a morphological frame of reference. What appears to be indicated is a conjunctive approach to problems of hominid evolution in which relevant data from various specialized disciplines can be integrated and major categories of variables defined in the general framework of behavioral evolution.

Whether we consider hominid evolution in an ecological, a social, a psychological, or a linguistic frame of reference, behavior is the unifying center to which we must constantly return at any adaptive level. As we proceed to new levels, we must consider novel integrations of determinants brought about by potentialities for behavioral adaptations that did not previously exist. In the evolutionary process, differential behavior patterns provide major clues to significant variables. The social behavior characteristic of the mode of cultural adaptation that eventually became the most distinctive feature of hominid development could not have arisen *de novo.* It must have complex roots in the evolutionary process. It could not have emerged suddenly as a saltatory configuration. Unique as a cultural mode of adjustment appears to be when observed in *Homo sapiens,* there are behavioral continuities as well as discontinuities to be observed when man is considered in the total setting of his primate heritage. In the paper referred to above, as well as in a previous publication, I suggested that the level of development represented by cultural adaptation can be focused more sharply in evolutionary perspective if we hypothecate a *protocultural* phase in hominid evolution and attempt to define its characteristic features. This earlier stage in development, deductively conceived, should embody some of the necessary, but not all of the sufficient, conditions for a fully developed human level of existence. On the one hand, it must have constituted a behavioral link between early hominids and other primates. On the other hand, it must have provided a preadaptive stage necessary for the later full-blown mode of cultural adaptation with which we are familiar in *Homo sapiens* by direct observation and experience.

What are the earmarks of a protocultural stage, and how may we identify them? We can best proceed, it seems to me, by selecting very broad categories for the purpose of comparing man with other hominids and infrahominid primates that, in addition to being relevant to all species, likewise bring into focus behavorial dimensions in which wide changes must have occurred in the course of the evolutionary process. What the selective pressures may have been that initiated such changes is not our present concern. The categories chosen here for brief discussion are: (1) social behavior and social structure, (2) ecological relations, (3) modes of communication and their properties, and (4) psychological capacities and organization. Observed behavioral similarities and differences, when considered with reference to the evolutionary process, indicate continuities and discontinuities in such behavioral categories and suggest some of the crucial features that, in combination, distinguish a protocultural phase in hominid evolution from a later and more fully realized level of cultural adaptation.

SOCIAL BEHAVIOR AND SOCIAL STRUCTURE

Perhaps the major clue to the basic continuity that links the Hominidae to the other primate groups, and thus makes comparisons of similarities and differences in this category of behavioral evolution significant, is the fact that we are dealing with gregarious animals. Whatever the ultimate determinants of sociality in the primates may be shown to be, all forms of cultural adaptation, as we know them in their fully developed stage, are based on some system of social action. But systems of social action are not unique in man. They also occur in infrahuman primates, and, structurally varied as they may be in different species, they constitute, nevertheless, a generic and characteristic mode of adaptation. Consequently, we may infer that social structure long antedated any form of cultural superstructure that, when eventually built into an organized system of social action in the course of hominid evolution, established the foundation of a new level of social living with the inherent potentialities that led to the emergence of various types of socio*cultural* systems. Cultural adaptation, then, is a mode of social existence deeply rooted in the behavioral evolution of the primates, where systems of social action were an ancient and typical feature of primate life. More detailed analysis shows basic similarities, as well as differences, in mating patterns and principles of organization that are meaningful in evolutionary perspective.

Mating patterns, of course, have suggested the closest human analogies. We see these analogies in types of mateship and in the range of their variation. Since lar gibbons, for example, live in groups that consist of one male and one female and their young, we have a close analogy to the "nuclear family" in man, which likewise represents a monogamous type of mateship. Some biological writers have applied the term "family" exclusively to this kind of primate social unit, despite the fact that in anthropological writing the term "family" is never limited to the nuclear family. The gibbon type of mateship, in which the sexual drive of the male appears to be low, would seem to be a limiting case in the total range of social units found among infrahuman primates and without evolutionary implications. In *Homo sapiens* we find two types of polygamous mateships, polygyny and polyandry, and social structures based on these are ordinarily called "families." Relatively rare in man in an institutionalized form, polyandrous mateships appear to be absent in infrahuman primates. On the other hand, polygynous mateships are common in both monkeys and apes. In the chimpanzee and gorilla this type of mateship seems to furnish the basis for independent social groups. In some monkeys, as, for instance, the baboon, "harems" occur as subgroups within the larger "troops" or "bands" found in these animals. Monogamous mateships, on the other hand, do not occur in groups of larger size because females in heat mate with more than one male.

Past attempts to establish any regular evolutionary sequence of mateship within *Homo sapiens* have failed, as well as have attempts to link any particular type of mateship in the nonhominid primates with early man. Perhaps it might be better to recognize that, since there are only a limited number of possibilities in mateships, it is not surprising to find them recurring at both the nonhominid and hominid levels of evolutionary development in the primates and in social units of varying size and composition. Whatever form they take, all these mateships serve the same reproductive ends. Their importance lies in this constancy in biological function rather than in any direct relation that can be shown to the evolution of group organization. They all lie close to biologically rooted central tendencies and continuities in behavioral evolution that link *Homo sapiens* to his precursors. For what we find as the common social core of all but the lowest primate groups, despite their variation, is the continuous association of adults of both sexes with their offspring during the portion of the latter's life cycle that covers the period from birth to the threshold of maturity. This core pattern of associated individuals, when considered with reference to their interrelated roles, is linked with the fact that basic functions are involved, that is, the procreation, protection, and nurture of offspring—born singly, relatively helpless at birth, and dependent for a period thereafter. Variations in mateship or size of group may occur without affecting these functions. In addition, the sex needs of adults and the food needs of all members of the group can be taken care of. The role of the female in relation to her young does not seem to vary widely, nor does the behavior of infants and juveniles. The protective role of the male in relation to infants and juveniles is similar in gibbon and howler, even though the young of the group in the latter genus are not all his own offspring, and the actual biological relationship between these two species is remote. Among monkeys and apes the adult males never provide food for juveniles or females. After being weaned, the juveniles always forage for themselves. Whether we call nonhominid primate groups "families," "clans," "troops," or "bands," their basic social composition can be expressed by the same general formula: X males + X females + X infants + X juveniles.

Whatever the mating types or size of early hominid groups may have been, their social composition must have conformed to this fundamental pattern. This generic type of social structure, associated with territorialism, must have persisted throughout the extremely long temporal period during which major morphological changes occurred in the species of the primate order, including those that ultimately differentiated the Hominidae from the Pongidae and later hominids from earlier ones. Underlying it, physiologically, was the type of ovarian cycle characteristic of practically all the primates. In contrast to some mammalian species, in which females have only one oestrus period a year, primate females, along with those of a limited number of other mammalian species, are characterized by successive oestrus cycles in the course of a year. Breeding is not seasonal but continuous.

The evolutionary significance of the social organization of primate groups cannot be fully appreciated, however, without considering behavioral patterns other than those directly connected with reproduction. For the structuralization of infrahuman societies is by no means a simple function of differential roles determined by sex and age. Of central importance in many of the groups so far investigated is the existence of interindividual behavior influences by an order of social ranking in the group, a dominance gradient. Males are, quite generally, dominant over females, and the females associated with them may sometimes outrank other females. While it appears that in different species the "slope" of the dominance gradient varies considerably, some kind of rank order occurs. This factor in the operation of the social structure is important because it reduces aggression between males, determines priorities to mates and food, influences the spatial disposition of individuals within the group, affects the socialization of group habits, and may determine the relations of groups adjacent to one another. Nevertheless, the ranking position of individuals is not fully determined once and for all; an individual's role in the dominance hierarchy may change. There are psychological factors that must be taken into account.

Enough has been said to indicate that in evolutionary perspective a necessary locus and an indispensable condition for a cultural system is an organized system of social action in which social behavior is patterned by role differentiation. Role differentiation in the non-hominid primates, in other gregarious animals, and in man exemplifies a basic principle in the organization of social relations, whether the determinants be innate or learned or a combination of both. A social structure, therefore, can be identified as one of the characteristic features of a protocultural stage in hominid evolution. Once this is recognized, I believe that the emergence of a cultural system is made intelligible if we assume that any adaptive genetic changes that took place inevitably became of vital importance to the social order. The interplay and cumulative effects of changes of all sorts must have been fed back into the system of social action that prevailed and led to modifications in its operation. Cultural adaptation, indeed, may be viewed as the culmination of *social* evolution in the primates. It could not have occurred if there had not been changes in ecological relations, psychological capacities, and codes of communication that directly affected both the behavior of individuals and the social structure. Overemphasis sometimes has been given to the brain as such, in relation to the development of culture. We now know that it was bipedal locomotion rather than brain size that gave initial morphological impetus to the hominid radiation. While no one would wish to minimize the importance of the later expansion of the brain in behavioral evolution, cause-and-effect relations are oversimplified if we do not take into account the continuing social context of behavior, the potentialities for change in the patterns of interindividual relations and, consequently, in the attributes of the social order considered as an evolving system. Whatever new potentialities may be attributed to the acquisition of additional neurones in the brain, and their organization, the resulting behavior must have become functionally manifest in a system of social action already in existence.

ECOLOGICAL RELATIONS

A cultural level of adaptation, in addition to requiring a preadaptive base in a system of social action, also required an environmental setting in which ecological relations at a protocultural stage provided the foundation for later developments. Whatever part arboreal adaptation may have played in the earlier evolution of the primates, including the development of distinctive psychological capacities and behavioral patterns, it is difficult, if not impossible, to imagine an arboreal niche as the basic ecological matrix of the hominid line of evolution that eventuated in a cultural mode of adaptation. It was terrestrial living that provided the ecological framework of this development and, when the necessary psychological capacities, experience, and technological traditions had been developed, enabled the hominids to accelerate the behavioral differences between themselves and other primates by exploiting the resources of their environment through knowledge of it and a succession of discoveries and inventions.

Even if capacities for *tool-using* were present in arboreal primates, how could the properties of stone have been discovered, exploited, and developed through shaping techniques into the lithic industries of a *tool-making* tradition by creatures who spent relatively little time on the ground? How could fire have become of importance in the life of primates confined to an arboreal niche? It was terrestrial living that provided the opportunity for the discovery of new food resources and made possible the shift to a carnivorous diet and the cooking of food, which ultimately led, through a scavenging stage perhaps, to the hunting of large mammals. If an upright posture with bipedal locomotion be taken as crucial generic features in hominid structural and behavioral differentiation, the terrestrial adaptation that accompanied them led to radical changes in the ecological relations of evolving hominids as compared with their primate forebears and the arboreally adapted monkeys and pongids. Motor functions already present, like grasping, were freed for new uses, and the discriminatory functions of binocular stereoscopic vision facilitated new developments in tactile skills, in manual dexterity, and probably in visual imagery, which ultimately became increasingly mediated through the more complex level of cortical organization made possible by the expansion of the brain. A new ecological niche provided the opportunity for the exercise, at a new level of behavioral organization, of behavioral potentialities already present, as well as for the development of new behavioral patterns. From an ecological point of view, a terrestrial habitat was a necessary setting for the protocultural stage in hominid adaptation that established the behavioral foundation for subsequent cultural adaptation.

One of the most characteristic features of the adaptation of infrahuman primates is territoriality. The locus of the social structures already discussed is a bounded area defined by the spatial range of the daily activities of members of each group. Ecologically, territoriality is the means by which the dispersal of the total primate population of a given region is spatially ordered and the independence of these breeding and nurturing groups maintained as distinguishable social units. Ordinarily, members of the different groups in a given region do not freely mix, nor do adults of different groups interbreed. The strong avoidance behavior that prevails between different groups is complemented by the factors that promote in-group integration. Territoriality, as observed among living primates is, therefore, a fundamental ecological adaptation that, at the same time, functions as a barrier to social integration of a higher order and to more complex social composition and role differentiation.

If we assume that territoriality persisted among the earliest hominids, some interesting questions arise. What was the size and range of these groups, and at what point in hominid development and under what conditions were groups of a higher order of complexity formed? For in men of the historic period, at least, we always find types of social organization that transcend in composition and role differentiation, if not always in size, what we find at the infrahuman primate level. The later, more evolved, forms of social organization incorporate the nuclear family, as well as other types of family structure, in a larger whole that includes individuals of all ages, as well as both sexes, and three or more generations. At this level, of course, sexual differentiation of roles in the performance of economic tasks has emerged, a phenomenon unknown in the nonhominid primates. We can only assume that, in the course of hominid evolution, factors must have come into play that made possible the functional integration of groups with radically different social composition and role differentiation from those that existed at the earliest stage of hominid development. At the same time it also became possible to transcend, through the development of new patterns of ecological relationships, the older form of ecological adaptation that formerly prevailed. Although social organization and ecological adjustment of the kind just mentioned must have considerable historical depth, for which there is some archeological as well as ethnographical evidence among hunting and gathering peoples, one must associate such developments with euhominids, who already had arrived at a cultural level of adaptation. As Sahlins has said:

> Primate territorial relations are altered by the development of culture in the human species. Territoriality among hunters and gatherers is never exclusive, and group membership is apt to shift and change according to the variability of food resources in space and time. Savage society is open and, corresponding to ecological variations, there are degrees of openness.

At an earlier protocultural stage the size, composition, structure, and behavioral range of social groups was determined by the same basic factors of ecological adaptation generally characteristic of nonhominid primates. At the same time, behavior was limited by psychological factors that made it impossible for systems of social action of a higher order to arise. Washburn and Avis point out:

> The acquisition of hunting habits must have been accompanied by a great enlargement of territory, since the source of food was now more erratic and mobile. . . . Whether early man scavenged from the kills of the big carnivores, followed herds looking for a chance to kill, drove game, or followed a wounded animal, his range of operations must have been greatly increased over that of arboreal apes. The world view of the early human carnivore must have been very different from that of his vegetarian cousins. The interests of the latter could be satisfied in a small area, and other animals were of little moment except for the few which threatened attack. But the desire for meat leads animals to know a wider range and to learn the habits of many animals. Human territorial habits and psychology are fundamentally different from those of apes and monkeys . . . this carnivorous psychology was fully formed by the middle Pleistocene and it may have had its beginnings in the depredations of the Australopithecines.

MODES OF COMMUNICATION AND THEIR PROPERTIES

The prevailing sensory modes of communication among primates are visual and acoustic. Both appear to be extremely important. Schultz speaks of the intricate "silent vocabulary" of the nonhominid primate.

> Crouching down, presenting buttocks, extending hands in pronation, exposing teeth partly or fully, raising eyebrows, protruding lips, shaking branches, pounding chest, dancing in one place, etc., all are actions full of definite meaning. . . . [Although] the long lists of different postures, gestures, and facial movements characteristic of monkeys and apes have not yet been compiled, . . . any careful observer realizes that they represent an intricate "silent vocabulary" of great aid in social intercourse.
>
> In the perfectly adapted arboreal life of monkeys and apes the limited variety of sounds, together with the great variety of meaningful gestures and facial expressions, is fully adequate for all social life within such close contact as permits seeing and hearing these detailed means of communication.

The utterance of sounds, Schultz says, "is the essence of primate life . . . , the simian primates are by far the noisiest of all mammals." In species that have been closely investigated, like the howling monkeys of Panama and the lar gibbon, differentiated vocalizations have been shown to have functional significance in the social co-ordination of the individuals belonging to a group. Schultz says the primatologist

> regards *language* not as the result of something radically new and exclusively human, but rather as a quantitative perfection of the highly specialized development of man's central nervous control of the anatomical speech apparatus in the larynx, tongue, and lips, the latter being as good in an ape as in man. . . . As soon as the early hominids had ventured into open spaces, had begun to use and even make tools, and had co-operated in hunting, the total variety of all means of expression needed additions, which could come only from an increase in sounds, since the comparatively little changed anatomy had already been fully used for all possible gestures, etc. . . . . Gestures have always persisted in human evolution, but they have become overshadowed by an infinitely greater variety of sounds in increasing numbers of combinations.

Oakley and others have suggested that early hominids may have depended primarily on gestures "mainly of mouth and hands, accompanied by cries and grunts to attract attention" and that speech may have been a comparatively late development. If so, a nonhominid mode of communication would have persisted in the protocultural phase of hominid evolution. Unfortunately, this interpretation must remain speculative. It is difficult to imagine, however, how a fully developed cultural mode of adaptation could operate without speech. However, if one of the necessary conditions for the functioning of a typically human system of communication is a speech community, an organized social system is as necessary for human language as it is for a cultural mode of adaptation. This condition was present even at the nonhominid level. So what we can discern in primate evolution is a behavioral plateau that provided the necessary context, but, at first, not all the sufficient conditions for either speech or culture.

Hockett has recently pointed out that "part of the problem of differentiating man from the other animals is the problem of describing how human language differs from any kind of communicative behavior carried on by nonhuman or prehuman species. Until we have done this, we cannot know how much it means to assert that only man has the power of speech." He has approached the problem by identifying seven "key properties" of the speech of *Homo sapiens* and comparing them with the available data on nonhuman systems of communication, discovering that there was considerable overlapping in the properties selected, although they did "not recur, as a whole set, in any known nonhuman communicative system." This suggested that the combination of properties that characterize speech, "those design-features . . . which seem to be of crucial importance in making it possible for language to do what it does," did not arise full blown. Hockett argues that this assemblage of properties, considered with reference to man's lineage, "could not have emerged in just any temporal sequence. Some of them either unquestionably or with high likelihood imply the prior existence of some of the others." Consequently, he is led to suggest a tentative evolutionary reconstruction. Since one of the key properties of a human system of communication is "cultural transmission," a property absent in the communication systems of primates and other animals, this factor becomes highly significant chronologically and, I think, has wider implications than those developed by Hockett, who suggests, in effect, that, although learning and the social transmission of habits, or what he calls "culture of a rather thin sort," may have existed at a very early stage in the development of the higher primates, the associated system of communication that prevailed may have operated without "cultural transmission." The significance of the fact that these earlier codes of communication did not function through learning and social transmission lies in the limitations this imposed upon the systems of social action developed in nonhominid and, perhaps, the earliest hominid groups. At the same time, these codes of communication, operating through the same sensory modes that appear at a later level may be considered prerequisite for the evolutionary development of a communication system characterized by the total assemblage of properties discussed by Hockett.

This kind of evolutionary inquiry is, of course, a far cry from earlier approaches, particularly those that began by concentrating on the problem of "primitive" languages spoken by *Homo sapiens*. Hockett's approach does permit us to have a fresh look at speech in greater evolutionary depth. And by direct observation we know that, whereas some of the great apes have been able to acquire a "thin sort" of human culture when closely associated with members of our species, they do not have the capacity to acquire and use our distinctive form of symbolic communication, even when systematically motivated. There seems little reason to doubt that in the course of behavioral evolution psychological capacities of crucial importance lay back of the ultimate emergence among the hominids of a characteristic system of linguistic communication. While this system shared some "design features" with that of nonhominid primates, capacities that transcended those of the other primates permitted the development and integration of novel features. These, in turn, resulted in the functional potentialities of speech as we know it in *Homo sapiens*.

PSYCHOLOGICAL CAPACITIES AND ORGANIZATION

Far down the evolutionary scale we have evidence that indicates that some activities of animals may originate, or be changed, through experience and affect subsequent behavior. When such responses cannot be reduced to innate determinants, or maturational processes, they are ordinarily referred to as "learned," although the conceptualization is loose. Harlow maintains that "there is no evidence that any sharp break ever appeared in the evolutionary development of the learning process" while, at the same time, "it is quite clear that evolution has resulted in the development of animals of progressively greater potentialities for learning and for solving problems of increasing complexity." My principal concern here is with the relevance of learning to the question of a protocultural platform in hominid evolution. While, as Nissen once said, "experience will not make a man out of a monkey," nevertheless, the extent to which learning is an integral part of the systems of social action, ecological relations, and modes of communication in monkeys and apes is relevant for an understanding of hominid evolution in an inclusive evolutionary perspective. What needs particular emphasis is what is learned and what is not, and the fact that what is individually learned by one animal may directly influence the behavior of other animals. It is not learning as such that requires consideration as a diagnostic characteristic of a protocultural stage. What is significant is that the part that learning plays in the life history and social relations of the nonhominid primates closely parallels, at so many points, the part that it plays in human sociocultural systems.

Beach, for example, says: "Descriptions of mother-infant relations in monkeys and chimpanzees leave no doubt as to the importance of learning in the filial responses of immature primates. The infant learns to obey gestures and vocal communications given by the mother and derives considerable advantage from her tuition and guidance." Socialization of the young, moreover, is an important factor in the formation and maintenance of infrahuman primate groups, as Carpenter pointed out long ago. And Collias, considering socialization in the wider perspective of behavioral evolution, points out: "In both insect and vertebrate societies, maintenance of cooperative relations depends to a large extent on socialization of the young. Among vertebrates, this trend reaches its climax in the primates." The formation of dominance gradients likewise involves learning, even in lower mammals and the phenomenon of territoriality in the primates also requires learning. It seems reasonable to assure, therefore, that the intimate relations between learning, social structure, and ecological adaptation, so fundamental in the functioning of culture, were well established in the nonhominid primates prior to the anatomical changes that led to both erect posture and the expansion of the brain.

Even more important, perhaps, is that fact that, at this same stage in both monkeys and apes, learned habits might not only be acquired by individuals of various ages but also could be transmitted through social interaction to other individuals in the group. The most striking cases have been reported by observers who have been studying *Macaca fuscata* at the Japanese Monkey Center. These primates have been lured from their forest habitat into open feeding places, where, among other things, they have been offered new foods. Systematic observation has shown that newly acquired food habits, such as eating candies, become quite readily socialized. Imanishi points out, moreover, that young macaques acquire the candy-eating habit more quickly than do the adults and that some mothers learned to eat candies from their offspring, rather than the other way round. It has likewise been observed that the spread of a new food habit may be directly related to the dominance gradient that is a central feature of their social structure. Adult females of high rank were observed to imitate the wheat-eating of a dominant male very quickly, and the habit was passed on to their offspring. Females of lower rank, in a more peripheral position in the group, only later acquired the habit from their offspring, who, in turn, had picked it up through association with their playmates. The rate of transmission was extremely rapid in this case, the entire process occurring within two days. In another instance, a young female initiated the habit of washing sweet potatoes before eating them. This habit, having been transmitted to her playmates, as well as to her mother, was slowly transmitted to a number of groups during the next three years. The same class of phenomenon in the anthropoid apes is illustrated by nest-building in chimpanzees and the transmission of the technique of working the drinking fountain at Orange Park, which chimpanzees learned from each other.

In the past, the social transmission of acquired behavior patterns has sometimes been stressed as one of the distinctive characteristics of culture. But in the light of our present knowledge of primate behavior it is better to consider it as one of the conditions necessary for cultural adaptation rather than as the distinguishing feature of it. Social transmission of acquired behavior patterns is, rather, a prerequisite of culture and an earmark of an earlier protocultural behavior plateau. The fact that even some animals other than primates may learn from one another or that some chimpanzees in social interaction with members of our species have acquired "culture traits," is no indication that a full-fledged level of cultural adaptation has been reached in these species. It only confounds the conceptualization and the investigation of hominid evolution if the term "culture" is applied, without qualification, to the phenomena of social transmission of simple habits in infrahuman species. J. P. Scott, for example, writes:

> The more the capacities for learning and for variable organization of behavior are present, the more it is possible for an animal to learn from its parents and pass the information along to the next generation. As we accumulate greater knowledge of natural animal behavior, we find more and more evidence that many animals possess the rudiments of this new ability, which we can call cultural inheritance. The migration trails of mountain sheep and the learned fears of wild birds are two of many examples. . . . . At the present time all our evidence indicates that cultural inheritance exists only in quite simple form in animals other than man, but future research may show that it is more common and complex than we now suspect.

While it is true that a variety of gregarious animals possess the *rudiments* of an ability to be influenced by the behavior of other individuals of their species, the part this ability plays in their total life history and social relations is what needs precise analysis. In phylogenetic perspective it is only in the primates that capacities and conditions arose which led to the transcendence of a rudimentary stage. And at this stage the primates are distinguished from other animals by a higher capacity for observational learning. Munn concludes that: "it is only in monkeys and apes that anything clearly approximating such observational learning can be demonstrated and even at this level the problems solved by imitation are relatively simple." If we use the term culture to refer to different levels of behavioral evolution, our vocabulary fails to discriminate the quantitative and qualitative differences between cultural adaptation in man and the very rudimentary "cultural" manifestations found in infrahuman animals, to say nothing of possible differences between primates and nonprimates. Dobzhansky, in a brief discussion of the "Rudiments of Cultural Transmission among Animals," has pointed out one essential difference between a protocultural and a cultural level of behavior, although he does not analyze specific cases in detail and his chief citations refer to birds rather than to primates. He says:

> In animals the individuals of one generation transmit to those of the next what they themselves learned from their parents—not more and not less. Every generation learns the same thing which its parents have learned. In only very few instances the evidence is conclusive that the learned behavior can be modified or added to and that the modifications and additions are transmitted to subsequent generations.

Simple conditioning and possibly observational learning account for these facts. The greater capacity for observational learning in primates also accounts for the socialization of nest-building habits in chimpanzees and the spread of the habit of washing sweet potatoes observed in the macaque group already referred to. But, so long as social transmission was dependent on capacities for observational learning, this fact limited the kind of acquired habits or innovations that could become significant in the adaptation of the group. Intervening factors were required before quantitative and qualitative differences in the kind of innovations possible at this level could be modified or changed and become effective through other mechanisms of socialization. It is difficult, for example, to imagine how the manufacture of tools, and the development of tool-making traditions could have arisen at a protocultural stage at which the mechanism of social transmission was exclusively observational learning and at which communication was mediated through signs rather than through any form of symbolic representation. Washburn and Avis, moreover, have expressed the opinion that tool-using may require

> much less brain than does speech and might have started as soon as the hands were freed from locomotor functions. Oral traditions essential for complicated human society probably were not possible with less than 700 or 800 cc. of brain [as contrasted with a range of about 450 to 600 cc. in the Australopithecines], and there is no likelihood that elaborate traditions of tool making are possible at lesser capacities, although simple pebble tools might well be.

Among other things, too, tool-making must have involved a whole series of discoveries and the accumulation of information necessary in a discriminating search for and selection of lithic materials with particular properties, in addition to the development and application of skilled methods of chipping. Even if we assume that there must have been successive stages in the development of tool-making traditions, these cannot be envisaged in a social and psychological vacuum, so that questions about a capacity for temporal orientation toward the future, the existence of property rights, as well as the kind of communication system to be assumed are relevant to the problem. It has often been said, for example, that a fully developed tool-making tradition is difficult to conceive in the absence of speech. It becomes all the more significant, then, that, despite the part that learning plays in the life of living primates at a protocultural level, they can be negatively characterized by the fact that no code of communication exists with the assemblage of properties appearing in speech and that whatever sublinguistic codes prevail appear to be transmitted genetically rather than through learning. We also know that chimpanzees cannot be taught to speak, despite the fact that, when closely associated with members of our species, it is possible for them to acquire many human habits through learning. Whatever inferences we make about the transition from a protocultural stage to a level of cultural adaptation, we must consider what habits were socially transmitted and what were not. And, quite aside from the properties of any code of communication, the question of its social transmittal must be taken into account.

Thus, while we may say that in the course of hominid evolution all the characteristic features of a protocultural stage were incorporated at a subsequent level of cultural adaptation, at the same time we must account for the differences observed. Here, organic changes, considered as intervening factors, must be taken into account. At the protocultural stage the psychological capacities of the actors determined the limiting framework of social and ecological adaptation. Although we cannot now observe the behavioral characteristics of the protohominids themselves, subsequent hominid developments in social structure, ecological relations, and modes of communication can hardly be dissociated from the known organic changes in the central nervous system after prior morphological changes in posture and locomotion have distinguished the hominid radiation. Account must be taken also of the sociopsychological effects produced by biological factors that prolonged dependency of the young, delayed reproduction, and increased the life span in an already well-advanced hominid whose psychological functioning was, at the same time, being greatly enhanced and restructured. All the distinctive features of a protocultural stage were being raised to a new level of sociopsychological integration through the increasing part that cortical processes came to play. In time, this new level of psychological organization affected every aspect of the earlier mode of protocultural adaptation. It led to the transformation of provincial social structures, through a change in their underlying dynamics, into the more inclusive, complex, and diversified socio*cultural* systems of the euhominids. At this more evolved stage a normative orientation became an inherent and distinctive feature of these systems of social action. Psychological factors became paramount in the functioning of these systems because the socially sanctioned values that characterized them were linked with the cognitive processes, motivations, and need satisfactions of individuals through the formation of a new and distinctive type of personality organization molded in the socialization process. What was learned in this process, beginning in infancy, not only included habits, roles, and adjustment to a physical environment, but also speech and a sense of values that pervaded every phase of personal adjustment and behavior. Conduct was evaluated in relation to socially sanctioned ethical standards. Food and material objects were not merely possessed; possession was regulated by a system of property rights. Skills and techniques used in the manufacture of material objects were also appraised in relation to recognized standards. Knowledge and beliefs were judged true or false, and art forms and linguistic expression were brought within the sphere of a normative orientation. All sociocultural systems became infused with appraisals that involved cognitive, appreciative, and moral values. If the total ramifications of the normative orientation of human societies are taken into account, we have a major clue to the kind of radical psychological transformation that must have occurred in hominid evolution and a measure of its depth and significance for an understanding of the dynamics of a cultural mode of adaptation as compared with what we find at a protocultural level.

Psychologically, a normatively oriented social order requires a capacity for self-objectification on the part of the individual actors. This makes possible self-identification over time, and an appraisal of one's own personal conduct and that of others in a common framework of socially transmitted and sanctioned values. Without the capacity for a psychological level of organization that permits the exercise of these and other functions, the social system could not function at the level of normative orientation nor could moral responsibility for conduct exist. Learning remains important, of course, but it operates at a higher level of sociopsychological integration than was possible at a protocultural level. The relations between needs, motivation, socially recognized goals, and learning are more complex because cortical processes have become increasingly important. It is impossible to attribute an equivalent level of psychological functioning to the earliest hominids.

What occurred in the psychological dimension of hominid evolution was the development of a human personality structure in which the capacity for self-awareness, based on ego functions, became of central importance. The functioning of ego processes contributed new qualities to the psychological adjustment of individuals in the socialization process. Ego functions became integral factors in determining responses to the outer world in the interests of inner needs, particularly when delay or postponement of action is required. They became intimately connected with such cognitive processes as attention, perception, thinking, and judgment. Considered in evolutionary perspective, ego may be said to be the major "psychological organ" that structurally differentiates the most highly evolved members of the Hominidae from infrahuman primates. At the same time, there is some evidence that suggests that rudimentary ego functions may be present in some of the higher apes, so it is possible that equivalent functions may have been present in the early hominids.

In ontogenetic development, as observed in *Homo sapiens*, ego processes can be identified in the first half-year of life, but a fully developed sense of self-awareness represents a psychological level of functional integration that is only manifest later. The initial development of the ego process does not appear to be dependent upon the prior existence of speech or culture, whereas self-awareness, on the other hand, requires socialization, a normative orientation, and the manipulation of what I have called extrinsic forms of symbolization. In other words, self-awareness is an integral psychological factor in cultural adaptation itself. It is rooted not only in ego functions but also in an already existent psychological capacity to abstract significant bits from the flow of experience and to represent their content in a meaningful form of expression extrinsic to the experience itself. This capacity to *project and objectify* significant aspects of experience may be contrasted with the evidence for *intrinsic* symbolic processes that occur in nonhominid primates and even lower mammals, that is, central processes that function as substitutes for, or representatives of, sensory cues or events that are not present in the immediate perceptual field. In the evolved hominid, processes of this kind can become socially significant by objectification in a variety of extrinsic symbolic forms. In nonhominid primates, on the other hand, only outward behavior in its concrete forms can become meaningful through perception. And the response, as has been indicated, may be observational learning. But what is privately sensed, imaged, "conceptualized," or "thought" cannot be responded to without an overt sign that represents it, but is, at the same time, extrinsic to the experience itself. In the evolved hominid, extrinsic symbolic forms, functioning through vocal, graphic, plastic, or gestural media, make it possible for groups of human beings to participate in a common world of meanings and values that is no longer confined to the perception of outward behavior alone or to concrete objects or events immediately given in perception. Both art and speech exploit this novel capacity for extrinsic symbolization. The artists of the Upper Paleolithic were capable of invoking intrinsic symbolic processes (memory images of animals), abstracting significant features, and representing these animals in a graphic form. In principal, the same capacity, expressed in arbitrary sound clusters that have no iconic relation to the objects and events represented, is one of the characteristics that distinguishes speech from infrahuman forms of communication in which signs, without symbolic value, are found.

The capacity for individual and social adaptation through the integral functioning of intrinsic symbolic processes and extrinsic symbolic forms enabled an evolving hominid to enlarge and transform his world and to become, at the same time, an object to himself. Means now became available whereby inwardly as well as outwardly directed references to an individual's own experience and that of others, and to objects and events in his world other than self, could find common ground through symbolic mediation. The immediate, local, and time-and-space-bound world of other primates who could not deal effectively with objects and events outside the field of direct perception was transcended. Speech, through the use of kinship terms, made it possible, among other things, for an individual to symbolize, and thus objectify himself, in systems of social action. And, as Professor Grace A. De Laguna has pointed out, becoming an object to one's self "carries with it the awareness of other persons not only as *objects*, but as *fellow-subjects*. An 'other' person is not only 'him' *of* whom I speak, but you *to* whom I speak and in turn an 'I' who speaks to me." As a consequence of self-objectification, sociocultural systems could function through the commonly shared value orientations of *persons*, self-conscious individuals in contrast to the societies of nonhominid and early hominid primates where ego-centered processes, even though they existed in a rudimentary form, had not yet become salient at the psychological level of self-awareness. In fact, when viewed from the standpoint of this peculiarity of man, culture may be said to be an elaborated and socially transmitted system of meanings and values that, in an animal capable of self-awareness, implemented a type of adaptation that made the roles of the human being intelligible to himself, with reference both to an articulated universe and to his fellow men.

In anthropological writing prior to the culture and personality movement, the connection between learning and culture remained vague because it had not been carefully analyzed in relation to the development of personality structure, cognitive orientation, motivation, etc. The fact had been overlooked that the only way in which a sociocultural system can be perpetuated is through the characteristic psychological structuralization of individuals in an organized system of social action. In the perspective of hominid evolution it is significant that the foundation for this later development was laid at the protocultural level, where learning was also intimately linked with the functioning of social structures, dominance gradients, and with the social transmission of habits. But at this protocultural stage what was learned was greatly restricted by the psychological capacities of the nonhominids. In *Homo sapiens,* on the other hand, we see the quantitative maximization of learning that, because of expanded psychological capacities, has led to qualitatively distinctive consequences. Among other things, we find cognitive processes raised to a higher level of functioning by means of symbolic forms, which can be manipulated creatively through reflective thought and experience. Cultural modes of adaptation, or certain aspects of them, learned and transmitted as they may be, can be objectified, thought about, analyzed, judged, and even remodeled. Man has never been completely enslaved by this traditional cultural heritage. The great novelty, then, in the behavioral evolution of the primates was not simply the development of a cultural mode of adaptation as such. It was, rather, the psychological restructuralization that, occurring in a primate where a system of organized social action already was present, not only made possible a more advanced level of social existence but laid the foundation for subsequent cultural readjustment and change. The psychological basis of culture does not lie only in a capacity for highly complex forms of learning and personality organization. What should not be overlooked is the potentiality that exists for transcending what is learned—a capacity for innovation, creativity, reorganization, and change in sociocultural systems themselves.

16. THE EVOLUTION OF THE HUMAN HAND

BERNARD CAMPBELL

Reprinted from Human Evolution: An Introduction to Man's Adaptations *(Chicago: Aldine Publishing Company, 1966; second edition, 1974). Bernard Campbell took his undergraduate degree and his Ph.D. at Cambridge University in natural sciences and physical anthropology. Now Professor of Anthropology at the University of California, Los Angeles, he has also taught at Cambridge and Harvard University. His main interests are in fossil man and the roots of human behavior. He is co-editor (with Kenneth P. Oakley) of the* Catalogue of Fossil Man. *He is also editor of* Sexual Selection and the Descent of Man *and the author of numerous articles.*

■ Following the methodological requirement that the features of man's adaptation and evolution must be studied one at a time, we begin this selection with some of the characteristics of the human hand. This organ is in some respects unique, especially in its ability to perform accurate and delicate movements, such as holding a pen or a screwdriver between the pulps of two fingers in what Napier has called a "precision grip." (But our hand is not in every respect unique; it retains many movements and postures of the foot-hand of nonhuman primates.) In addition to freeing the hand from brachiation for the manipulation of tools (see Selection 17)—an event of momentous importance in evolution—the emergence of man involved a crucial addition to the many movements that primate hands already possessed.

The hand is one of the most generalized organs in man. It is not designed to meet specific habitational challenges, but the possibilities of doing things with it have been greatly increased over the hand of any previous animal. The potential advantages of the generalized human hand, however—the adaptability inherent in it—would be very limited without tools, weapons, and other cultural implements. Man's ability to adapt resides largely in his generalized hand, and in other generalized features of his architecture, such as his central nervous system.

In this selection Campbell shows that monkeys and apes are able to perform most of the hand movements of man. He suggests that this ability underlies the development of perception in primates and, in addition, that it has important implications for their locomotor efficiency. But a hand is not sufficient for performing such tasks as flaking stone for tools; apparently, on the basis of fossil remains discovered thus far, a relatively large brain also is necessary.

The reader can pursue this particular line of study in "The Evolution of the Hand" by John Napier (*Scientific American* [December, 1962], 207: 56-62). ■

WE HAVE SEEN that monkeys use their hands for a wide variety of purposes, and one of the most important, which we must now consider, is their contribution to the satisfaction of effectance motivation: The hands make possible a detailed examination of parts of the environment that can be manipulated. Monkeys, apes, and man are almost the only animals that fiddle about with things, that turn them over and examine their form and texture. This manipulation of objects is not necessarily directly related to the procurement of food but is simply a process of investigation, equivalent to the olfactory investigation of the environment so characteristic of a dog. But in manipulation a monkey is investigating not the whole environment but only one particular part of it—and frequently a part that can be separated physically from the rest. This ability of higher primates to extract an object from its setting and examine it visually and three-dimensionally from all sides is a development of the utmost importance in human evolution. A carnivore will examine the olfactory nature of objects but cannot at the same time see them as part of the visual pattern of the environment because during examination by the nose the objects are more or less out of sight. Admittedly, a ball or bone can be manipulated with the paws and tossed with the mouth, but the range of objects examined is limited, and, compared with the primate hand, the paws and mouth give only a rough indication of shape and texture.

The detachment of objects from the environment appears to be a most important prerequisite for the evolution of primate perception. This examination of things as objects we owe to the evolution of the primate hand and the opposable thumb. The recognition of different kinds of objects we owe to primate visual and tactile examination. Only a primate can, as it were, extract an object from the environment, examine it by smell, touch, and sight, and then return it to its place in its surroundings. In this way the higher primates have come to see the environment not as a continuum of events in a world of pattern but as an encounter with objects that proved to make up these events and this pattern.

This all-important development may have come quite early in primate evolution, with the appearance of the opposable thumb. In man, a further advance in opposability has perfected this remarkable organ of manipulation. Napier has examined the thumb of the primate in order to analyze the development of opposability in hominid evolution. It reaches its most evolved condition in modern man and involves "a compound movement of abduction, flexion and medial rotation at the carpal-metacarpal articulation of the pollex." That is to say, a saddle joint evolves at the base of the thumb that allows a 45-degree rotation as well as movement in two planes.

Though monkeys have an opposable thumb, it is only in man that the thumb is long enough and divergent enough to carry a heavy musculature. This length and strength make possible a precision grip, strong yet delicate. At the same time, with the hand no longer functioning to grasp branches, the *metacarpals* and phalanges straighten out in the course of human evolution, and the terminal segments broaden.

When apes and monkeys carry objects about, which they often do, their locomotor efficiency is impaired. Tool-carrying, food-carrying, and food-sharing (which is possible only if the food is first carried) all have been reported to a limited extent among chimpanzees, but they cannot move fast with a handful of objects and do so only for very limited distances. In early man the survival value of being able to carry objects without any disadvantageous effect on locomotion is of the greatest importance. Not only is it most advantageous to be able to carry food when surprised by a predator, but it can lead eventually to food sharing, and in particular to hunting. A weapon at the ready is also of great survival value. The survival value of carrying also serves in turn to promote a physical structure adapted to bipedal locomotion.

Tool-using is not so rare as might be supposed in the animal kingdom. The use of sticks and rocks in agonistic display by chimpanzees and baboons is familiar. Tool modification—the next stage in the evolution of material culture—has been reported among chimpanzees, which prepare and collect sticks of a certain length for extracting termites from their nests. At Olduvai, stone tools have been found at an early level dated nearly 1.8 million years B.P (Before Present) in direct association with one or more species of *Australopithecus.* Here we see pebbles obviously used for pounding roots and other vegetable matter, preserved, after flaking, as cutting tools for the preparation of meat. From South Africa we have evidence of bones modified as tools from a site and stratum inhabited by *Australopithecus africanus.* The transition from tool-modifying to tool-making is slow and continuous, but there is no doubt that a cutting tool was a great asset and is particularly necessary in the evolution of a carnivorous diet.

But the tools we have found associated with the man-ape *Australopithecus* are relatively simple, and it is relevant that this creature had a brain not really much bigger in absolute size than that of a gorilla. Fragments of a hand are known, and they are reported to be somewhat less than fully human. Finely flaked stone tools are found—to date—only in association with the bones of hominids that had larger brains. Fine Acheulian hand-axes are known from Ternifine as well as Olduvai, where they are associated with remains of *Homo erectus;* in this hominid the brain may well have amounted to nearly 1,000 cc. Clearly, the final perfection of the hand was a perfection not only in anatomy but in sensory perception and motor control, with a highly developed brain. The most clear-cut neural correlate of the evolution of the human hand is the increase in representation that this organ carries on the motor and somatic sensory cortexes of the brain. We have also noted significant improvement in the direct nervous pathways to the hand of the ascending and descending fibers.

We can summarize the evolution of the human hand by marking two stages in its development. First, the primate terrestrial environment selects the opposable thumb, for grasping branches, and this thumb appears among the prosimians and in turn makes possible manipulation among Old World monkeys. Second, the ecological move to the open plains correlated with bipedalism frees the hand from all locomotor function and allows the perfection of the precision grip by some minor modifications of the thumb. The hand has contributed as much as the eye to the making of man; together they gave him a new perception of his environment and, with his material culture, a new control of it.

17. MAN AS TOOL-MAKER

KENNETH P. OAKLEY

Reprinted from "On Man's Use of Fire, with Comments on Tool-Making and Hunting," in Sherwood L. Washburn (Ed.), Social Life of Early Man *(Chicago: Aldine Publishing Company, 1961). Kenneth P. Oakley is head of the Anthropology Sub-Department of the British Museum (Natural History). Trained as a geologist and anthropologist, he has brought the two disciplines to bear on his principal scientific interest, the dating of fossil human remains. By applying the fluorine dating technique to Piltdown, he began the demolition that led to this famous skull's being proved fraudulent. Another anthropological matter that has engaged his attention is the importance of the use of fire in man's cultural evolution. He is the author of* Frameworks for Dating Fossil Man *and* Man the Tool-Maker *and co-editor (with Bernard Campbell) of the* Catalogue of Fossil Man.

■ To show that man could not survive without tools requires no elaborate demonstration. It has been suggested that *Australopithecus* was a tool-user, and most authorities agree, but this is not surprising because tool use has been established in nonhuman primates and even in birds. One of the most important questions in this regard, as Oakley observes in this selection, is what evolutionary process was involved in the development of the capacity to make tools with which to make other tools?

There can be no doubt that the use of tools was an adaptation to life in open country, away from forests. It is also clear, as Oakley observes, that the use of tools arose in conjunction with a reduction of tooth size, the disuse of teeth as a weapon, an increase of meat in the diet, and an increase in intelligence. Also, Oakley suggests that the development of tools accompanied the development of patterns of cooperation.

In this highly imaginative paper, Oakley hypothesizes that tool making was one of the cornerstones for all human adaptation. Certainly, man could never have transcended the limitations of his habitats and his genetic constitution without tools. But Oakley also shows, especially in tracing out the ways in which one habit leads to another, that the capacity for tool use and manufacture depended on many other evolutionary and adaptive developments.

The reader should consult Oakley's *Man the Tool-Maker* (Chicago: University of Chicago Press, 1960) and Sherwood L. Washburn's "Tools and Human Evolution" *(Scientific American* [September, 1960], Vol. 203). ■

THE ANCESTRAL STOCK of apes and man was probably not highly adapted to forest life. Some of the monkey-like apes, such as *Proconsul,* lived in woodlands and could brachiate but could also probably run short distances bipedally on the ground. During Miocene times in East Africa they inhabited a mixed environment of corridor forests along streams and lakes, separated by bushy grassland. It has been suggested that members of the group were sometimes forced to move through the tall grass between forest areas and that, since this necessitated raising their level of vision, bipedal abilities would have had selective advantages. Supposedly, some members of the group moved into permanent forest and became solely arboreal, with hands specialized for hooking onto or hanging from boughs, while other members became adapted to more open country and evolved into bipeds.

In these latter—one might call them protohominids—the feet became rigid supporting organs, with pelvic changes accompanying bipedalism. The hands were freed as manipulative organs. It seems a mistake to think that tool-making depended upon any evolution of the hand; probably a generalized pongid hand could make tools if enough brain capacity were present. As Wood-Jones has indicated, manual skill depends upon initiation and coordination in the cerebral system. Men have developed manual skill even when their hands and limbs were maimed. Refined stereoscopic-color vision and erect posture with a vertical position of the skull are probably important. These allow close visual concentration over a wide field. The earliest hominids would have been anatomically equipped to use tools. It is a question of when they did, and why.

Tools are additions to the body that supplement the hands and teeth. In the arboreal prosimians, hands are climbing and feeding organs. These hands could have been used for other activities whenever such animals became ground-dwellers. Even baboons, partly ground-dwelling but adapted for quadrupedal progression, occasionally use a pebble to kill a scorpion needed as food (observation by Professor D. M. S. Watson). Chimpanzees are rarely known to use tools in the wild but will do so in captivity when they are forced to live on the ground.

When did tool-using begin? If the protohominids inhabiting the edge of the forest spent much time on the open ground, it is conceivable that they began picking up things to use. Life in the open is more precarious than life in the trees, and tool-using would offer a selective advantage. An increased period of learning would also be an important factor; the mother would assist and teach the child for a protracted period if adult status was retarded (as it may well have been) through the acquisition of bipedal habit and through continuing operation of the principle of pedomorphism.

Tool-making requires a higher order of intelligence than does tool-using. Chimpanzees are the only reported animals that *make* tools, and then only in captivity. Sultan, the chimpanzee observed by Kohler, was capable of improvising tools in certain situations. Tool-making occurred only in the presence of a visible reward, and never without it. In the chimpanzee the mental range seems to be limited to present situations, with little conception of past or future. The power of conceptual thought is basic to tool-making but is only "incipient" in apes. For instance, a chimpanzee can learn to make fine discriminations in color shading but retains the learning for only a short time; a chimpanzee, through constant exploratory activity, will see possible single sticks to use as tools in a broken box but not in a whole box. Man can see a tool in a formless lump of stone. This ability of conceptual thought may have been present in a few individuals in a group, becoming extended by selection when conditions demanded. The range of tool types already present in the oldest industries includes *tools for making other tools* (e.g., hammerstones), illustrating that what we regard as the unique foresight of man was present at a very early stage in his evolution.

By the end of Lower Pleistocene times, the human level of cerebral development had certainly been reached, and stone artifacts of crudely standardized types were being made (e.g., Oldowan pebble tools). The incipient standardization shows that the manufacture of such tools (e.g., at Sterkfontein) was not an isolated occurrence but was already a tradition that served certain permanent needs of the earliest human beings. What were these needs? The use of tools and weapons was surely the means whereby the Hominidae kept themselves alive after they had abandoned the protection and sustenance provided by forests.

The tree-climbing primates had no use for tools. Tool-using arose in connection with adaptation to life on open ground away from forests. In the evolution of the primates the forelimbs have continually showed a tendency to take on functions performed in their ancestors by the teeth. The use of tools is evidently an extension of this trend, and we may suppose that tools were largely substitutes for teeth.

The apes of today are forest creatures, subsisting almost exclusively on fruits, leaves, shoots, and insects. All known races of man, on the other hand, include a substantial proportion of mammalian flesh in their diet. We have ample evidence that Peking man, Neanderthal man, and Late Paleolithic races of *Homo sapiens* were meat-eaters. I suggest that meat-eating is as old as man the toolmaker, that with adaptation to partly open forest margins the diet of proto-men inevitably became more varied, and that they changed from being eaters largely of plants and the fruits of plants to being in part meat-eaters.

It seems probable, on the important analogy of the baboons, that early Hominidae living in such country (savanna) may have become increasingly addicted to flesh-eating as a result of the intensification of the struggle for existence by excessive drought. It may be recalled that baboons, which in daytime range into grassland away from trees, occasionally prey on lambs and other animals of similar size, using their powerful canine teeth as offensive weapons, and, moreover, that this habit is likely to become more prevalent when conditions of existence are hard. Owing to the extensive folklore associated with baboons, reports of the carnivorous habits of those in South Africa have been discounted by some zoologists, but information from many different observers, collected by my friend Mr. F. E. Hiley in 1950, leaves no doubt that such reports are substantially true. A report from Captain H. B. Potter, with long experience as a game conservator in Zululand, is typical of those received:

> The following are my personal observations over a period of twenty years' wardenship in the Hluhluwe Game Reserve: I have seen full-grown poultry killed and actually eaten by baboons, mostly however by aged individuals. Eggs and chickens are taken by the dozen, by old and young baboons. I have on many occasions actually witnessed apparently organized hunts which often result in the death of the intended victim. The baboons, usually led by a veteran of the troop, surround an unsuspecting three-parts-grown Mountain Reedbuck, or Duiker, as the case may be, and on one occasion a young Reedbuck doe was the victim. It would appear that on a given signal the baboons close in on their quarry, catch it and tear it asunder. As a matter of interest I have refrained from interference in these grim encounters so that I would be in a position authentically to record the results. In nine cases out of ten the game animal is devoured limb by limb and after the affair is over all that is to be found are the skull and leg bones.

Baboons, like some other monkeys and apes too, have powerful canine teeth, serving mainly for defense against carnivores and, in males, for gaining dominance. It has been suggested that the reduction in the size of the canine teeth in hominids was an outcome of functional replacement by hand weapons. The canine teeth may have been reduced in the Hominidae at an evolutionary stage below that of systematic tool-making. That australopithecines, with small canines, must surely have been at least tool-*users*. Certainly the protohominids would have needed some means of defending themselves in the open and, having their hands free, may well have used stones as missiles and sticks or animal long bones as clubs.

In dry, open country the protohominids, like baboons, might readily have taken to eating flesh, particularly in times of drought or scarcity of food. Although they lacked teeth suited to carnivorous habits, they could easily have killed small mammals. Life in the open set a premium on co-operation. Drawing on our knowledge of the mentality and social life of other primates, particularly baboons, it seems not unreasonable to suppose that, hunting in groups, the protohominids could have killed medium-sized mammals, say, by cornering them and using improvised hand weapons such as they might earlier have *learned to improvise in the first instance for their own defense*. The protohominids could certainly have killed small mammals.

It should not be forgotten that in the wild, after a kill by one of the larger carnivores, there is a scavenging food queue; when the lions, for instance, have had their fill, the hyenas and then the vultures enter the scene of slaughter. The protohominids may have first obtained the meat of larger mammals by entering this queue at an early stage. It has been reported that African children have been known to drive lions from their kill by beating tins. It is certainly conceivable that the protohominids used tactics of intimidation to facilitate their scavenging and that this preceded the hunting of larger wild game.

This is frankly speculation, for there is still no direct evidence that the earliest Hominidae were semicarnivorous or that they passed through a tool-using stage before becoming tool-makers. It is true that Dart has claimed that the quantities of broken animal bones found at Makapan in association with *Australopithecus* represent the food refuse of this creature and that some of the bones had been *used* as weapons, but other authorities are unconvinced and consider that the Makapan "bone statistics" can be satisfactorily explained as representing the product of *several* selective agencies, notably hyenas. The few pieces of mammalian bone found in the tool-bearing layer at Sterkfontein may yield some important clues bearing on this problem. Although there is a strong probability that all australopithecines were tool-users, and that some at least were tool-makers, the evidence remains *sub judice*.

By the time the Hominidae had evolved into tool-makers, they were evidently largely carnivorous—quantities of meat bones were associated with the remains of *Pithecanthropus pekinensis*. It is easy to see how the one habit led to the other. Although the killing of game may have been accomplished easily enough in some such way as that suggested above, the early hominids must often have encountered difficulty in removing skin and fur and in dividing the flesh. In the absence of strong canine teeth, the solution would have been overcome most readily by using sharp pieces of stone. Here, surely, was the origin of the tradition of tool-making. Where no naturally sharp pieces of stone lay readily to hand, some of the more intelligent individuals saw that pebbles, which broke on the ground when thrown, provided the solution. By breaking pebbles, fresh sharp edges were produced. Once the tradition of tool-making had begun, the manifold uses of chipped stones became obvious. They were useful for sharpening sticks for digging out burrowing mammals; for making spears sharp enough to be effective weapons in hunting larger game; for scraping meat from bones, splitting them to get at the marrow; and for chopping the meat into convenient mouthfuls. All the main uses of stone tools were, I suggest, connected in the first place with adoption of semicarnivorous habits.

From the endowment of nature we should be vegetarians. We lack the teeth evolved by true carnivores, and we have the long gut associated with a herbivorous diet. Furthermore, we are the only members of the Hominoidea accustomed to eating meat on any considerable scale. It is true that anthropoid apes, like most herbivores, consume small quantities of animal protein; some of them occasionally rob birds' nests of eggs and fledglings, but by and large they are fruit and plant eaters.

One can well imagine that a changing environment, for instance during a period of desiccation, may have produced an abnormal appetite in the early hominids. Gorillas in captivity quickly develop a liking for meat, and this appears to be due to a change in the flora and fauna of their intestines. Normally, their intestines are richly supplied with ciliate protozoa (Infusoria), which serve to digest cellulose. According to Reichenow, under the abnormal conditions of captivity the Infusoria are ingested, and, with their disappearance from the intestines, the animal develops an abnormal appetite and readily takes to eating meat—and may even prefer meat to its normal fare.

By widening his diet and becoming a tool-maker, man became the most adaptable of all primates. The change from herbivorous to semicarnivorous habits was important from the point of view of the use of energy. To obtain a given amount of energy, a carnivore subsists on a smaller quantity of food than does a herbivore. Instead of eating almost continually, like their ancestors, the Hominidae spent much of their time hunting. This led to increased interdependence. New skills and aptitudes were developed through this new way of life, and with increasing control of environment through the use of manufactured tools, man became the most adaptable of all creatures, free to spread into every climatic zone.

An important step in the control of environment in northern climes was the making of fire. This could have been discovered only as an outcome of tool-making.

To sum up: I think it may fairly be claimed that tool-making is one of man's fundamental characteristics from a biological point of view. But the definition of man as the tool-*making* primate carries the implication that the term "human" should be applied only to the later members of the family Hominidae.

18. GENETICS AND THE RACES OF MAN

THEODOSIUS DOBZHANSKY

Reprinted from Sexual Selection and the Descent of Man, 1871-1971 *edited by Bernard Campbell, pp. 59-84. Copyright © 1972 by Aldine Publishing Company. For biographical information on Theodosius Dobzhansky, please see Selection 2.*

■ The concept of race has long proved to be scientifically useful but socially misleading. Scientifically, the value of the concept is that it directs our attention to the adaptation to different habitats of groups within a species. It is a socially dangerous concept because it is often used to suggest that racial differences imply inequality, and even superiority and inferiority. But this misapplication of the concept of race does not mean that we should discard the idea; the social use of the concept does not invalidate its biological value. Elaborate and often cruel forms of social stratification—such as caste restrictions in India and Japan—have been devised and maintained without even the slightest hint of racial distinction. Thus race may serve as a basis for caste divisions, but it need not; similarly, notions of superiority and inferiority may be developed without the concept of race.

The distinction between generalized and specialized adaptation within the human species is one of the principal themes of this selection. Paralleling this distinction are issues of human uniformity and diversity, which Dobzhansky discusses within the framework of genetics and racial differences. Dobzhansky covers a wide range of questions surrounding the problem of race, succeeds in demolishing many commonly held misunderstandings, and places the concept in a scientifically useful perspective.

This selection concludes with the socially sensitive topic of mental development and its political and ethical implications. It is abundantly clear that neither individuals (except for identical twins) nor populations are identical; thus to equate diversity with social inequality is logically absurd. By nature, *Homo sapiens* is a diverse species with a universal capacity for educability. But what must always be remembered is that equality and inequality are socially—not genetically—determined.

Two books with stimulating collections of papers can be consulted by the reader who wants to explore these issues further. Most useful for the anthropologist is *Science and the Concept of Race,* edited by Margaret Mead and others (New York: Columbia University Press, 1968). For papers that suggest that the concept of race has no scientific usefulness, see *The Concept of Race,* edited by M. F. Ashley Montagu (New York: Free Press, 1964). ■

INTRODUCTION

"IT IS NOT MY intention here to describe the several so-called races of men; but to inquire what is the value of the differences between them under a classificatory point of view and how they have originated." The above is the opening sentence of Darwin's chapter "On The Races of Man" in *The Descent of Man.* A century later, it remains appropriate to open with the same sentence a discussion of the races of man in the light of genetics. For despite the enormous growth of the information concerning racial variation both in man and in other organisms, the problems that occupied Darwin are still at issue. In a sense, the uncertainties have increased. It is the contention of a small but vociferous group of students that mankind is not differentiated into races. But even if this contention were justified, our discussion would not lose its point. If mankind has no races, it is surely not homogeneous or uniform. The diversity would still have to be described, studied, and explained. Two kinds of diversity can be distinguished: that of individuals and that of populations.

DIVERSITY OF INDIVIDUALS

"What is man," asks the Psalmist, "that Thou art mindful of him, and the son of man, that Thou visitest him?" The contribution of biology toward answering this question is the assertion of irreducible singularity of every man as individual, unprecedented, and nonrecurring. Monozygotic,[1] or so-called identical, twins bear the greatest resemblance to one another; barring mutations, they have the same genetic endowment; and yet even they have nonidentical life experiences, which make them different individuals. The appearance and the behavior of an individual is called his phenotype. Though changing with age and even from one moment to the next, the phenotype is always a product of the interactions of the heredity, the genotype,[2] and the environment. What we observe directly, or with the aid of instruments and laboratory tests, are the phenotypes of individuals.

Good or bad health, beauty or ugliness, goodness and honesty or wickedness and evil, are phenotype characteristics. It is the phenotype which lets some individuals survive and others die, some to leave offspring and others to remain childless. The statement found in some biological writings, that only the genetic endowment, the genotype, is important is wide of the mark. The genotype is important inasmuch as it engenders some phenotypes and not others. The quality of a genotype is appraised through the phenotypes which it produces in the environments that exist or can be devised by our technology. Analysis of the evolutionary process deals nevertheless largely with genotypes. Natural and artificial selection are powerless if the materials at their disposal are genotypically uniform. Biological evolution can be defined as sustained genetic change. Without genetic changes the phenotype would be at the mercy of environmental fluctuations. Genetic change confers at least a relative stability and irreversibility on evolutionary alterations.

It is now logical to inquire how extensive are the genetic differences between individuals of the same species, particularly of the human species. This problem has long been, and to some extent still is, a subject of controversy. Classical geneticists, beginning with De Vries and Morgan, assumed, as often implicitly as explicitly, that most genes are held in homozygous[3] condition in most individuals of a species. De Vries believed that new species arise through single mutational events. *Drosophila* geneticists were bedeviled by the concept of the normal or wild type. A vast majority of the flies of a given species of *Drosophila* found in their natural habitats are rather uniform in appearance. Bred in laboratories, these wild-type flies throw occasional mutants, some of them sharply, even dramatically, differing from the wild type in structure, color, or other characteristics. Mutant and wild-type flies can be crossed, and show segregation of normal and mutant traits as single entities.

Further and more penetrating genetic studies gradually disclosed a more complex but interesting situation. In *Drosophila,* it was found rather early that, in addition to the major mutants, there exist also less spectacular modifiers, polygenes in modern terminology, which are responsible for minor variations in sundry traits of the flies. In many species, the wild-type concept is patently inapplicable. What would one choose to designate the "normal" or wild-type man? More and more species, including man, were shown to be polymorphic, that is, to have two or more clearly distinguishable genetic forms in their populations. The many blood group polymorphisms, in addition to such traits as eye colors, are examples of human polymorphisms. The external uniformity of natural populations of *Drosophila* proved to be deceptive; it conceals polymorphisms in the chromosome structures,[4] owing chiefly to the occurrence of inversions of blocks of genes. As stated above, the term "poly-

1. Monozygotic: offspring derived from the same fertilized egg. (This and the definitional footnotes that follow have been added by the editor.)
2. Genotype: the genetic constitution of an organism.
3. Homozygous: a genetic characteristic received from both parents.
4. Chromosome: structures within the nucleus of a cell which is made up of genes.

morphism" is usually applied to situations where the genetic variants in the populations are clearly distinguishable by phenotype. Distinguishability is, however, a function of the refinement of the methods of study. The genotypes giving rise to different blood groups are accurately indentifiable, but those responsible for the infinite variety of facial features are not. Yet at the gene level, human populations are doubtless polymorphic for many more genetic variants than we are able to diagnose precisely. . . .

The estimates obtained are that about 40 percent of genes are polymorphic in the populations of *Drosophila pseudoobscura*, and that an average individual is heterozygous[5] for some 12 percent of its genes. The estimates for man are surprisingly similar: 40 percent of the genes are polymorphic, and the average heterozygosity per individual is about 16 percent. It should be noted that these figures are low since the electrophoretic[6] techniques distinguish only those enzyme[7] variants which differ in the electric charges on their molecules. But even with these underestimates, the genetic variety present in human and other populations appears to be prodigious. A minimum estimate of the number of genes in a human sex cell is something like 20,000. With this number, at least 8000 genes will be polymorphic in human populations and an individual will be heterozygous on the average for some 3200 genes. With these figures, the potentially possible human diversity is immense. A heterozygote for n genes can form 2^n kinds of gametes with different gene complements; two parents heterozygous for the same n genes can give 3^n different genotypes in the progeny, and if heterozygous for n different genes can give 4^n different genotypes. With n being 3200 or greater, the potentially possible number of human genotypes is between 3^{3200} and 4^{3200}. Either figure is vastly greater than the number of atomic particles in the universe. What these figures mean is only that the variety of human constitutions that can ever be realized is vanishingly small compared to the potentially possible variety.

Taking into consideration any two variable genes, the genotype of an individual can be represented by a point in two-dimensional space; with three genes, the space is three-dimensional. With n variable genes, every genotype is symbolically represented as a point in an n-dimensional space. Except for monozygotic twins, every individual now alive, having lived, and to live in the future is symbolized by a point different from all others. The array of points for all these individuals comes nowhere near filling the space: a vast majority of possible points have no individuals corresponding to them.

Is any one of the virtual infinity of actually existing or potentially possible human genotypes in any meaningful sense normal or optimal? It is doubtful, to say the least. It is no longer reasonable to maintain that, for every one of the thousands of variable genes, one allele is always beneficial and the rest belong to the genetic load. The partisans of the classical theory of genetic population structure have retreated to a less easily assailable position: most genetic variants found in populations of the human and other species are neither useful nor harmful. They are adaptively neutral. Natural selection neither promotes nor inhibits their spread. We shall consider this possibility below, together with the problem of adaptive significance of human racial traits. Here we may conclude our discussion of the diversity of human individuals by a negative statement: the Normal Man does not exist. Normal men, if one wishes to retain this rather equivocal designation, are a great array of individuals, no two of whom (excepting monozygotic twins) are either phenotypically or genotypically alike. The sole, though obviously important, characteristic which these persons share in common is satisfactory physical and mental health, which permits them to survive and to function in their environments. In man, the key features of the environments are those created by human cultures. The array of genotypes which compose the adaptive norm of the human species allow

5. Heterozygous: a genetic characteristic received from only one parent.
6. Electrophoretic technique: the process of separating charged molecular substances by migration toward oppositely charged poles.
7. Enzyme: a macromolecule which plays an important role in catalyzing chemical reactions within the body.

their carriers to function at least passably well as members of human societies.

DIVERSITY OF POPULATIONS

Immanuel Kant, who was a naturalist before he became the prince of philosophers, wrote in 1775 the following remarkably perceptive lines:

> Negroes and whites are not different species of humans (they belong presumably to one stock), but they are different races, for each perpetuates itself in every area, and they generate between them children that are necessarily hybrid, or blending (mulattoes). On the other hand, blonds and brunettes are not different races of whites, for a blond man can also get from a brunette woman, altogether blond children, even though each of these deviations maintains itself throughout protracted generations under any and all transplantations.

Kant understood the distinction between individual (intrapopulational) and group (interpopulational) variability better than do some modern authors.

The make-up of group variability is a function of the reproductive biology of the organism. It is quite distinct under sexual and asexual reproduction. Consider first organisms, chiefly though not exclusively plants, in which the prevalent or obligatory method of reproduction is asexual fission, budding, diploid parthenogenesis, or self-pollination. Thousands or even millions of individuals of such organisms belong to the same clone or pure line, and have the same genotype. These are, indeed, "pure races." The inhabitants of a territory can be described by listing which pure races occur there, and specifying their relative frequencies. To cite a single example, Allard and Kannenberg found in central California at least eight pure lines of the grass *Festuca microstachys;* the relative frequencies of these lines differ from locality to locality, apparently depending on which lines are best adapted to the environments of the particular locality.

Mankind is a sexually reproducing species. Excepting monozygotic multiple births, it has no clones, pure lines, or pure races. And yet anthropologists have long been under the spell of the notion of pure race. This notion is, in turn, one of the protean manifestations of the typological way of thought, the roots of which go back at least to Plato's eternal Ideas. Human populations are polymorphic for eye and hair colors, blood groups, and many other traits. Could one declare blue-eyed persons to constitute one race and brown-eyed another? Or may one race include the possessers of blood group O, another of A, and the third of B? These divisions would be ludicrous; parents and children, as well as siblings, would often find themselves belonging to different races. Nevertheless, "races," almost equally meaningless, the dolichocephalics[8] and the brachycephalics,[9] have been copiously discussed and written about. The genetics of the human head shape is not well known; several pairs of genes are probably involved, and all gradations from extreme longheadedness to roundheadedness can be observed.

The notions of pure races and of Platonic types or ideas are lurking in the background of various taxonomies of constitutional types. Kretschmer's pyknic,[10] athletic and asthenic[11] body builds and associated psychological types are admittedly rare in their extreme or pure forms. Nevertheless, they can be somehow perceived among the manifold products of the general miscegenation in which mankind has been engaged for millennia. Sheldon's 88 somatotypes[12] are combinations of graded series of three supposedly independent variables: endomorphy,[13] mesomorphy,[14] and ectomorphy.[15] The genetics of these variables is obscure. If there existed series of multiple alleles[16] in each of the three genes, for endomorphy, mesomorphy and ectomorphy, the 88 somatotypes could be interpreted as an elaborate instance of intrapopulational polymorphism.

8. Dolichocephalic: longheaded.
9. Brachycephalic: short- or broadheaded.
10. Pyknic: fat or stout.
11. Asthenic: slender; slight muscular development.
12. Somatotype: the structure, function and development of the body.
13. Endomorphy: fat or stout body build (pyknic).
14. Mesomorphy: athletic body build.
15. Ectomorphy: slender, slight body build (asthenic).
16. Allele: either of a pair of different genetic characteristics. When more than one allele contributes to a characteristic, it is multiple allelic.

This hypothesis, however, remains to be demonstrated.

For more than half a century, the Polish school of anthropology consistently adhered to strictly typological assumptions. Populations are described in terms of the incidence in them of racial "types," which are "distinguished by diagnosing the racial affinities of individuals independently of their ethnic origins." European populations are composed of Nordic, Mediterranean, Armenoid and Lapponoid races or racial types. The frequencies of these types in each population are given with precision to one-tenth of one percent, as though men belonged to one or another clone or pure line, like *Festuca microstachys* grasses. Mediterranean individuals in the Polish, Swiss, and Italian populations are assumed to be more alike than any of these are to their Nordic or Armenoid neighbors or brothers. Michalski and Wiercinski can identify as many as 16 "racial elements" of which mankind is a composite. Each of these races or types or elements is recognized by a constellation of chiefly morphological[17] traits, such as stature, eye and hair colors, hair form, cephalic, orbital, facial and nasal indices.[18] A set of mathematical techniques has been devised to identify to which racial type every individual belongs. Can this identification really be accomplished? As with biological mathematics generally, the results of even the most precise and elaborate calculations do no more than give numerical expressions of the biological assumptions put at the base of the mathematical model. The assumptions used in the present case, insofar as they have been stated at all, are untenable. The crucial assumption is that the trait constellations which supposedly identify the racial types, such as Nordic, Armenoid, Lapponoid, are inherited as alleles of a single gene, somewhat like those giving O, A, and B blood groups. This assumption is in flat contradiction with all that is known about the genetics of these traits, scanty as this knowledge admittedly is. Wiercinski gives an example of a family in which the father is diagnosed as Alpine, the mother Nordic, and their two children Lapponoid and Nordic. This classification stretches one's credulity to the breaking point.

17. Morphological: physical form.
18. Cephalic, orbital, facial, nasal indices: four different ways of measuring the head to accurately define its structure.

MENDELIAN POPULATIONS

A person has two parents, four grandparents, eight greatgrandparents, and so on. Continued for some 32 generations, the number of ancestors would turn out greater than the total world population. Of course, this description is not so. Notwithstanding the universality of incest taboos, all our ancestors were more or less distant relatives. Though it cannot be documented, all humans are relatives. If one could construct a complete pedigree of all mankind, it would be a complex network in which every individual is multiply related to every other. Mankind is a Mendelian population, a reproductive community all members of which are connected by ties of mating and parentage. A Mendelian population may be said to have a common gene pool. The genes of every individual are derived from, and unless he dies childless, some of them return to this pool.

In theory, mankind could be described by listing its gene loci, and indicating the frequencies in the gene pool of the different alleles at each locus. Such a description, even if it were possible to carry out in practice, would not be entirely satisfactory. Mankind is not a panmictic population, in which every individual would have an equal probability of mating with every individual of the opposite sex and of the appropriate age. The chance that a boy born in Canada will marry a Canadian girl is greater than that he will marry a girl from China or Uganda. In common with many sexually reproducing animal and plant species, mankind is differentiated geographically into subordinate Mendelian populations; the intermarriage within these subordinate populations is more frequent than between them.

There are also specifically human agencies which cause further discontinuities in the intermarriage patterns: these agencies are economic, social class, linguistic, religious, and other subdivisions.

Mankind, the biological species, is the inclusive Mendelian population. Within it there is a hierarchy of subordinate Mendelian populations, geographically or socially partially isolated from each other. Only the smallest subdivisions, inhabitants of some villages, groups of equal social status in small towns and the like, may be regarded as approximately panmictic. Races are Mendelian populations within a species. They are not Platonic or statistical types, not collections of genetically identical individuals, and not subdivisions of primordial mankind submerged by long-continued miscegenation. They are Mendelian populations which differ in the incidence of some genes in their gene pools.

The delimitation of the Mendelian population of the human species presents no difficulty. At our time level there is no gene exchange between the gene pool of the human species and those of even its closest biological relatives, pongids.[19] It is sometimes asked how one defines a human being. Biologically, the answer is simple: any individual is human whose genes are derived from the gene pool of the human species. By contrast, the delimitation of the Mendelian populations which are called races is always to some extent vague, because their gene pools are not wholly disjunct. This is a restatement in modern terms of Darwin's conclusion that "The most weighty of all arguments against treating the races of man as distinct species, is that they graduate into each other, independently in many cases, as far as we can judge, of their having intercrossed."

The subordinate, intraspecific, Mendelian populations, in man as well as in other sexual organisms, are as a rule not fully discrete. Because of gene exchange, they merge into each other. Very often one cannot tell where one ends and the other begins. This fact is disconcerting to some orderly minds. How can one make something so ill-defined the basic unit of biological and anthropological study? Two observations can be made in this connection. First, the complexities of nature should not be evaded. Second, the only way to simplify nature is to study it as it is, not as we would have liked it to be. . . .

The genetic nature of race differences is only beginning to be understood. The classical race concept, in anthropology as well as in biology, was typological. Every individual of the negroid, mongoloid, caucasoid, and even Jewish and Nordic "races" was a variant of a quasi-mythical type of his race, and "never the twain shall meet." Operationally, this concept led to characterization of the racial types by systems of average values of measurements and observations made on samples of populations living in or descended from ancestors who resided in different territories, belonging to different castes, speaking different languages, and the like. The more separate measurements or traits went into the construction of a racial type, the more valid and reliable it was supposed to be. This typological approach reached its extreme, almost a reductio ad absurdum, in the above discussed attempts to find different racial types among individuals of the same population, and even among members of the same family.

Genetics has gradually made the ineptitude of typological approaches evident to an increasing majority of anthropologists. Mendelian populations should be described in terms of the incidence in them of separate characteristics, and ideally of the alleles of variant genes. Boyd's pioneering attempt to carry out such a description using the then known blood genes met a sceptical reception, although the race classification he arrived at was little different from some current typological ones. The two decades since then greatly increased the information available, and also revealed complexities that were not so clearly apparent before.

When the frequencies of gene alleles or of separate phenotypic traits are plotted on

19. Pongids: apes.

maps, one finds as a rule gradients (clines) of increasing or decreasing frequencies toward or away from some centers. Thus, the allele I^B of the OAB blood groups system reaches frequencies between 25 and 30 percent in Central Asia and northern India. Its frequencies decline westward to 15-20 percent in European Russia, 5-10 percent in western Europe, and even lower in parts of Spain and France. The frequencies decline also southeastward, practically to zero among Australian Aborigines, and northeastward, to below 10 percent among the Eskimos and to zero in unmixed Amerindians. The center of light skin and eye pigmentation is northwestern Europe; the pigmentation becomes darker eastward and especially southward, reaching maximum in subsaharan Africa, southern India, and Melanesia. Rohrer's index (body weight divided by the height cubed) reaches its highest values among the Eskimos, and its lowest in southern Asia, Australia, and Africa.

Are any genetic differences between human populations qualitative, in the sense that some gene alleles are absent in some and reach one hundred percent frequencies in other populations? As noted by Darwin, "Of all the differences between the races of man, the color of the skin is the most conspicuous and one of the best marked." Indeed, cases of albinism excepted, no native of subsaharan Africa is born as lightly pigmented as the natives of Europe, and no European develops as dark a pigmentation as an African. However, this skin color difference is due to the additive effects of at least three pairs of genes without dominance. Heritable color variations are found among Europeans as well as among Africans. One can well imagine that skin color could be darkened considerably by selective breeding in a population of European descent and lightened in a population of African descent. Whether or not the intraracial pigment variations are due to the same gene loci which are responsible for the interracial differences is uncertain. One of the alleles of the Rh system (cDe) reaches frequencies above 50 percent in African populations, but it occurs with low frequencies, generally below 5 percent, in individuals without known African ancestry elsewhere in the world. An allele of the Diego locus seems to be lacking among Europeans and frequent among Amerindians, not reaching however one hundred percent frequencies among the latter. An allele of the Duffy system has frequencies above 90 per cent among Negroes in western Africa, and is also found but with low frequencies among Europeans.

Our tentative conclusion, subject to modification by future findings, must be that qualitative differences, in the above defined sense, are absent among human populations. This conclusion is in no way contradicted by our ability to distinguish any individual native of (for example) Congo or Ghana from any Scandinavian, and both of these from any native of eastern Asia. The reason is, of course, that the populations native in these countries differ in frequencies not of a single but of many genes. What is important is that typological race concepts must be replaced by populational ones. Individuals are not accidental departures from their racial types. On the contrary, the interpopulational, racial, differences are compounded of the same genetic variants which are responsible for the genetic differences among individuals within a population.

RACIAL DIFFERENCES AND NOMENCLATURE

The apparently endless variety of living beings is as fascinating as it is perplexing. There are no two identical humans, as there are no two identical pine trees or *Drosophila* flies or infusoria.[20] The runaway diversity of our perceptions is made manageable by means of human language. Classifying and giving names to classes of things is perhaps the "primordial" scientific activity. It may antedate the appearance of *Homo sapiens,* and it is bound to continue as long as symbol-forming animals exist. Biologists and anthropologists describe and name the complexes of organisms which they study in

20. Infusoria: protozoan; single cell organism.

order to identify for themselves, and let others know, what they are talking and writing about.

Human beings whom we meet, and about whose existence we learn from others, are numerous and diversified. We have to classify them and attach recognition labels to the classes. So, we distinguish the speakers of English, Russian, Swahili, and other languages; college students, industrial workers, and farmers; intellectuals and the "silent majority." Those who study physical, physiological, and genetic variations among men, find it convenient to name races. Races are arrays of Mendelian populations which belong to the same biological species, but which differ from each other in the incidence of some genetic variants.

The question is often posed: Are races objectively ascertainable phenomena of nature, or are they mere group concepts invented by biologists and anthropologists for their convenience? Here we must make unequivocally clear the duality of the race concept. First, it refers to objectively ascertainable genetic differences between Mendelian populations. Second, it is a category of classification which must serve the pragmatic function of facilitating communication. One can specify the operational procedures whereby any two populations can be shown to be racially different or racially identical. The populations may contain different arrays of genotypes or similar arrays.

Racial differences exist between populations "out there," regardless of whether or not somebody is studying them. Yet this does not mean that any two genetically different populations must receive different race names. For example, Cavalli-Sforza has found no significant genetic differentiation between inhabitants of the villages on the densely settled Parma plain; he did find such a differentiation between the villages in the more sparsely settled mountains. Racial differences are, therefore, ascertained among the latter but not among the former. It would nevertheless not occur to anyone to give race names to the populations of every mountain village. This nomenclature is uncalled for, because the village names are adequate labels for the populations that live in them.

How many arrays of populations in the human species should be provided with race names is a matter of expediency. Already Darwin noted that "Man has been studied more carefully than any other organic being, and yet there is the greatest possible diversity among capable judges" concerning the number of races recognized and named. Different authors referred to by Darwin named from two to as many as sixty-three races. The incertitude is undiminished today. Hardly any two independently working classifiers have proposed identical sets of races. This lack of unanimity has driven some modern "capable judges" to desperation. They claim that mankind has no races. Furthermore, they say the very word "race" should be expunged from the lexicon. This proposal is often motivated by a laudable desire to counteract racist propaganda. But will this be achieved by denying the existence of races? Or will such denials only impair the credibility of the scientists making them? Is it not better to make people understand the nature of race differences rather than pretend that such differences are nonexistent?

To give an example of a race classification by an author fully conversant with modern biology and anthropology, Garn recognizes 9 "geographical races" and 32 "local races," some of the latter being subdivisions of the former. The geographical races are as follow:

1. Amerindian
2. Polynesian
3. Micronesian
4. Australian
5. Melanesian-Papuan
6. Asiatic
7. Indian
8. European
9. African

Among the local races not included in his major geographical groups, Garn distinguishes three interesting categories: Ainu and Bushmen are "long-isolated marginal;" Lapps, Pacific Negritos, African Pygmies, and Eskimos are "puzzling, isolated, numerically small"; and American Negroes, Cape

Colored, Ladinos, and Neo-Hawaiians are "hybrid local races of recent origin." On the other hand, Lundman recognizes only 4 main races: white, yellow, red (Amerindian), and black; his 16 subraces correspond in part, but only in part, to Garn's local races. If one of these classifications be accepted as correct, must the other necessarily be incorrect? This, I believe, is a wrong way to judge race classifications; we should rather ask which classification is more convenient, and for what purpose.

In sexually reproducing and outbreeding organisms, every individual can usually be recognized as a member of one and only one species. Adherents of the typological race concepts believed that the same should be true of races, that every individual, excepting only the progenies of inter-racial crosses, should be classifiable as belonging to a certain race. This classification is not possible. Species are genetically closed, while races are genetically open systems. There is no individual whose membership in the species man or the species chimpanzee could be called in question. These species do not exchange genes. There are however many local populations in northwestern Asia intermediate between the white and the yellow, and in northern Africa intermediate between the white and the black races. Moreover, individual members of these intermediate populations do not necessarily have two parents belonging to "pure" white, black, or yellow races. Whole populations are intermediate. Sometimes this situation is due to secondary intergradation, that is, to interbreeding of genetically distinct populations in the near or remote past. The gene exchange is, in fact, the origin of Garn's "hybrid local races." More often the intermediate populations are autochthonous;[21] primary intergradation is a result of gene diffusion taking place while the racial divergence of the populations is increasing, as well as after the populations have diverged. The gene gradients or clines result from both primary and secondary intergradation.

Gene gradients make it only rarely possible to draw a line on the map on the two sides of which live different races. Race boundaries are more often blurred than sharp. Worse still, the gradients of the frequencies of different genes and traits may be only weakly or not at all correlated. This can easily be seen on maps that show the frequencies of various traits in human populations, such as different blood antigens, pigmentation, stature, etc. Human races are not the discrete units imagined by typologists. Some disappointed typologists have seen fit to draw the radical conclusion: races do not exist.

But let us take a closer look at the situation. If gene or character gradients are uniform, the gene frequencies increase or decrease regularly by so many percentages per so many miles travelled in a given direction. With uniform gradients, race boundaries can only be arbitrary. However, often the gradients are steeper in some places and are more gentle or absent elsewhere. Consider two gene alleles, A_1 and A_2, in a species with a distribution area 2100 miles across. Suppose that for 1000 miles the frequency of A_1 declines from 100 to 90 per cent; for the next 100 miles from 90 to 10 per cent; and for the remaining 1000 miles from 10 to 0 per cent. It is then reasonable and convenient to divide the species in two races, characterized by the predominance of A_1 and A_2 respectively, and to draw the geographic boundary between the races where the cline is steep.

Why are the gene frequency gradients gentle in some places and steep in others? The steepening of the gradients usually coincides with geographic and environmental barriers that make travel difficult. Barriers to travel are also barriers to gene diffusion. Newman has analyzed human racial taxonomy in a very thoughtful article. His general conclusion is "that there are valid races among men, but that biology is only beginning to properly discover and define them. . . . I consider some of Garn's races probably valid, others probably invalid, with still others in the 'suspense' category for lack of adequate data." He validates Garn's Asiatic,

21. Autochthonous: indigenous; native.

African, and Amerindian races as showing good trait correlations in such visible traits as pigmentation, hair form and quantity, nose and lip form, cheek bone prominence, eyelid form, and general body shape. European, Indian, and Australian are "unwarranted abstractions," on account of the high variability and discordance (lack of correlation) in the geographic distribution of many traits. Melanesian, Polynesian, and Micronesian are in the "suspense" category.

The adherents of the "no races" school argue that one should study the geographic distributions of genes and character frequencies, rather than attempt to delimit races. The truth is that both kinds of studies are necessary. Gene and character geography is the basis of the biological phenomenon of racial variation; classification and naming are indispensable for information storage and communication. The fact that races are not always or even not usually discrete, and that they are connected by transitional populations, is in itself biologically meaningful. It is evidence that gene flow between races is not only potentially possible but actually taking place. Gene flow between species, however, is limited or prevented altogether. To hold that because races are not rigidly fixed units they do not exist is a throwback to typological thinking of the most misleading kind. It is about as logical as to say that towns and cities do not exist because the country intervening between them is not totally uninhabited.

RACE DIFFERENCES AS PRODUCTS OF NATURAL AND SEXUAL SELECTION

A century ago Darwin felt "baffled in all our attempts to account for the differences between the races of man." In particular natural selection could hardly be invoked, because "we are at once met by the objection that beneficial variations alone can be thus preserved; and as far as we are enabled to judge (although always liable to error on this head) not one of the external differences between the races of man are of any direct or special services to him." He put more faith in sexual selection: "For my own part I conclude that of all the causes which have led to the differences in external appearance between the races of man, and to a certain extent between man and the lower animals, sexual selection has been by far the most efficient." On the 828 pages of the first edition of *The Descent of Man and Selection in Relation to Sex,* Part I, "On the Descent of Man," takes 250 pages, and part II, "Sexual Selection," more than twice as many.

It is almost incredible that, a century after Darwin, the problem of the origin of racial differences in the human species remains about as baffling as it was in his time. Several circumstances have conspired to make it so. The chief one was that until less than a generation ago, the leading anthropologists believed race differences to be mostly adaptively neutral, and consequently made little effort to discover their selective values. Radical changes in human environments brought about by cultural developments made the problem particularly difficult to approach; a genetic trait may have played a role a million years ago which was quite different from its role ten thousand years ago, and that again different from what it is at present. And finally, by a curious twist of reasoning, the doctrine of human equality seemed to exclude the possibility of differential genetic adaptedness.

The adaptive significance of even so obvious a trait as skin pigmentation has not been fully clarified. The notion that dark pigmentation is protective against sunburn is very old. It is made plausible by the fact that dark-skinned races are (or were) inhabitants of the tropics, and light-skinned ones of temperate and cold countries. This rule is however not free of exceptions; the Indians of equatorial South America are not particularly dark, and some of the natives of northeastern Siberia are at least as dark as those of Mediterranean Europe. These exceptions have been "explained" by assuming that the relatively light people in the hot countries and the relatively dark ones in cold countries are recent immigrants, or that they live mostly in forest shade rather than in the open. It has also been supposed that light skins facilitate the synthesis of vitamin D in countries with little sunshine, while dark pigmentation protects against excessive amounts of this vitamin where sunshine is abundant. There is good evidence that light skins are more prone to develop skin cancers owing to sun exposure than are dark skins. Still another surmise is that dark skin pigmentation may facilitate absorption or solar radiation "where energy must be expended to maintain body temperature, as at dawn and dusk in otherwise hot climates." A dark skin may give protective coloration to a hunter stalking game or escaping from predators.

The above hypotheses concerning the adaptive significance of pigmentation are not mutually contradictory or exclusive. And yet their multiplicity attests to the inadequacy of our understanding of the most conspicuous of all human racial differences. A considerable amount of careful study has been devoted to the physiology of human populations adapted to certain particularly rigorous environments, such as Indians of the Andean Altiplano (cold, low oxygen supply) and Eskimos of the Arctic. Riggs and Sargent and others compared the reactions of young negro and white males to exertion under humid heat condition. Some statistically assured differences in the expected directions have been found, but it is not ruled out that part of these differences may be the product of physiological adaptation to the environments in which the persons grew up.

The racial differences in the incidence of various blood groups have long been a challenge to those who believe that all racial differences must be established by natural selection. There is no doubt that certain pathological conditions (for example, duodenal ulcers) occur more often in carriers of some blood antigens than in others, but it is questionable whether these correlations are even in small part responsible for the racial differences. Attempts to correlate the blood groups with resistance to some infectious diseases, such as plague, smallpox, and syphilis, have thus far been unconvincing.

Sexual selection "depends on the advantage which certain individuals have over other individuals of the same sex and species, in exclusive relation to reproduction." In our present view, the difference between natural and sexual selection is not fundamental. The selection coefficient, that is, the difference between the Darwinian fitnesses of different genotypes, measures the relative rates of transmission of certain components of these genotypes from generation to generation. It is of lesser consequence, though certainly not immaterial, that differential gene transmission is in some instances due to greater success in mating, while in others it is caused by differential mortality, or fertility, or greater speed of development, or anything else. Genetic variants which are favored by the balance of all these causes will increase, and those disfavored will diminish in frequency in the populations. Lessened success in mating may be compensated by a greater viability or fertility, or vice versa.

That sexual selection, in the classical Darwinian sense, occurs in man is clear enough. Although almost everybody in tribal societies has an opportunity to mate and produce offspring, socially more influential and more prosperous individuals may not only have access to more mates but be able to provide better conditions, which increase the probability of survival of their offspring to maturity. Good evidence of this has been provided, for example, by Salzano, Neel and Maybury-Lewis and Chagnon et al. for Xavantes and Yanomamas, two of the surviving primitive tribes of South American Indians. In both tribes, "Whereas women are uniformly exposed to the risk of pregnancy and rarely fail to reproduce, men, on the other hand, are characterized by an appreciably higher variance in their reproductive performance." In one of the villages studied two head men sired approximately one-fourth of the total population.

Some forms of physiological sterility of females as well as of males are genetic. The genetic basis of the psychological variables which predispose individuals to spinsterhood, bachelorhood, or prolificity in technologically advanced societies is another matter, reliable data on which are almost totally lacking. It is, for example, an open question whether homosexuality has an appreciable genetic component. However, if the existence of such a genetic predisposition were proven, its bearing on race differentiation would still be in doubt. Genetically caused partial or complete sterility, better known in *Drosophila* than in human populations, is part of the concealed genetic load; when the components of this load become overt, they come under control of normalizing natural selection. Normalizing selection can hardly bring about appreciable racial differentiation. What interests us rather is to what extent the racial divergence of populations is brought about by directional selection, of either the sexual or the natural kind. In other words, one wishes to know not only whether a given genetic trait is influenced by selection, but also, and this is a more difficult problem, why different variants of this trait are favored in different populations. The present state of knowledge in this field is quite unsatisfactory. . . .

DO RACES DIFFER IN GENETIC CAPACITY FOR MENTAL DEVELOPMENT?

This is a "sensitive" issue, arousing violent emotional reactions in many people. Yet a discussion of the genetics of race would be gravely incomplete if this issue were sidestepped, as it so often is. In my previous writings, I have tried to point out that some of the emotional reactions are due to sheer misunderstanding. Equality is confused with identity, and diversity with inequality. Even some eminent biologists who should have known better wrote that biology showed that men are unequal! Human equality and inequality are not statements of observable biological conditions. They are policies adopted by societies, ethical principles, and religious commandments.

People can be made equal before the law, equality of opportunity may be promoted or guaranteed, human dignity equally recognized, and human beings can be regarded as equally God's sons and daughters. To have equality, people need not be identical twins; they need not be genetically alike. And vice versa: Individuals can be treated unequally, as social superiors and inferiors, masters and servants, elite and plebian. There is no reason why monozygotic twins must necessarily be social equals, even though they are genetically as nearly identical as two individuals can be. People can be made equal or unequal by the societies in which they live; they cannot be made genetically or biologically identical, even if this were desirable, which it surely is not. In principle, human diversity is compatible with equality and with inequality.

Any two human individuals, identical twins excepted, carry different sets of genes. This fact has traditionally been stressed by partisans of social inequality and softpedaled by champions of equality. It should be the other way round: equality is meaningful only because people are not identical. Like illness, the genetic lottery is no respecter of the social position or rank of the parents. Owing to gene recombination in the progeny of highly heterozygous individuals, genetically well- and poorly-endowed children can be born to parents of either kind. This does not deny the existence of some positive correlations between the genetic endowments of parents and offspring. The fact that these correlations are far from complete is socially and ethically no less important than that the correlations are there.

The above is a necessary preamble to the substantive issue: How great is the genetic diversity among individuals and among populations, such as races, in their capacities for mental development? This issue is entangled with the perennial arguments about nature vs. nurture, heredity vs. environment. Some sophistication has been achieved in the approach to this problem: no informed person now dichotomizes human mental abilities into those due to heredity and those due to environment. It is rather a question of heritability, of partitioning the observed variance into its genetic (additive, dominance, epistatic) and environmental components. Most extensive, though far from sufficient, data have been accumulated on the heritability of the intelligence quotient. Despite some weaknesses, the technique of comparison of monozygotic and dizygotic twins remains the best source of information. Having reviewed all twin data reported in the literature, Jensen has arrived at an average heritability figure of 80 per cent for the intelligence test scores. Taking into consideration also the correlations between relatives other than twins, Jensen obtains an only slightly lower estimate of 77 per cent.

The meaning of the intelligence quotient continues to be in dispute. It is fairly generally admitted that the IQ does not measure a single faculty, but rather a battery of at least six aptitudes: verbal ability, word fluency, numerical, spatial, and reasoning abilities and memory. These six abilities are not very strongly correlated, and may well be independent in their genetic conditioning. Some progress has been achieved toward their independent measurability. A person who is top-flight in one or more of them may be mediocre or deficient in others. On the other hand, defenders of IQ usability point out that the IQ ratings have vindicated themselves by their predictive power of probable scholastic achievement. There is an appreciable correlation, between 0.4 and 0.7, of IQ rating and the occupational status achieved or likely to be achieved.

Statistical differences in mean IQ's have been demonstrated again and again between representatives of different racial groups, especially in the United States. The means are highest for persons of European descent and lowest for those of African and Amerindian origin. But what is the significance of these means? Racists have seized upon them as proof of racial superiorities and inferiorities. By egregious miscomprehension or deceitfulness, they argue that since the heritability of the IQ has been shown to be high, race differences in the IQ averages are genetically fixed and irremediable. This argument is certainly unproven and unconvincing. It is obvious that the environments in which the racial groups live in the same country are appreciably and often drastically different, and furthermore the environmental differences are most pronounced precisely in those aspects which are most relevant to intellectual development. Heritability is not a fixed parameter; under an ideally uniform environment it would be 100 per cent, and it is reduced more or less strongly by environmental heterogeneity; it will be zero in genotypically uniform populations, and increased by genetic heterogeneity. One can only conclude that the degree to which the differences in the IQ arrays between races are due to genetic predisposition is at present an unsolved problem.

The differences between the mean IQ's of white and black populations in the United States amount to about one standard deviation in the white population, or less. In other words, the distributions of the IQ scores of these two races or castes broadly overlap. There are numerous, not merely exceptional, blacks whose IQ's are higher than the white mean, and numerous whites below the black mean. There is every reason to think, although the data are scanty and unreliable, that the same is generally true of all race differences within the human species. The practical consequences of this are evident. Individuals should be judged by what they are, not by what race or subrace they come from. Equality of opportunity means that everybody must be given equal chance to develop and demonstrate his abilities; it does not mean that everybody should be forced into the same position or profession, or given the same role to play in economic or intellectual life. Moreover, however excellent may be the genetic potential, it can be frustrated and defeated by lack of environmental and educational opportunity. Conversely, nobody, short of gross pathology, can be wholly impervious to environmental improvement.

Finally, consideration must be given to an argument which is the more misleading since it is superficially so plausible. It runs approximately as follows. Races of animal and plant species develop adaptedness to their respective environments. Breeds of domestic animals and plants are made to serve different purposes or to live under different conditions. These racial characteristics are very largely genetically fixed. Why, then, suppose that man's races are a singular exception? Modern technologically advanced societies have been constructed chiefly by a minority of human breeds. It is likely that at least some of the "lesser" breeds do not possess aptitude sufficient to create, manage, or even to live in these societies.

The above argument must be honestly and squarely faced. Yes, mankind is a product of a unique evolutionary pattern. The cardinal distinction between mankind and all other forms of life is that man's adaptedness depends more on his cultural than on his genetic inheritance. Culture is acquired by each individual by learning, and transmitted by instruction, chiefly, though not exclusively, by means of a language consisting of socially agreed-upon symbols. To adapt to new environments, mankind changes mainly its cultural inheritance, rather than its genes, as other organisms do. Genes and culture are not independent but interdependent. It is man's genetic endowment which makes him able to think in symbols, abstractions, and generalizations. The potentiality of cultural evolution, which is uniquely human, has developed through the evolution of his gene pool. But the contents of his gene pool do not determine the contents of his culture. The genes give man his ability to speak, but do not decide what he shall say.

The basic and unique capacity of man is his genetically established educability by means of symbolic language. This educability is a species trait, common to all races and all nonpathological individuals. Its universality is no more surprising than that all people have a body temperature and blood pH which varies within narrow limits. Educability and symbolic language became universal human traits because existence in manmade environments depended on the possession of these traits. Nothing of the sort happened to any other animal species, wild or domesticated. Genetically fixed specializations in both morphology and behavior have often been deliberately built into different domestic breeds.

The foregoing remarks should not be taken as denying that the variability of human behavior in general, and of educability in particular, has a considerable genetic component. Suppose, for the sake of argument, that it can be established that some fraction of the mean difference in the IQ scores of whites and blacks is indeed genetic. Would that be a vindication of the racists in Alabama, South Africa, and elsewhere? Certainly not. Two basic facts refute the racists; first, the broad overlap of the variation curves for IQ's and other human abilities; second, universal educability, and hence a capacity for "improvement," however this last term be defined. We may accordingly agree with Darwin that

> Although the existing races of man differ in many respects. . . . yet if their whole organization be taken into consideration they are found to resemble each other closely in a multitude of points. . . . The same remark holds good with equal or greater force with respect to the numerous points of mental similarity between the most distinct races of man. The American aborigines, negroes and europeans differ as much from each other in mind as any three races that can be named; yet I was incessantly struck, whilst living with the Fuegians on board the "Beagle," with the many little traits of character, shewing how similar their minds were to ours; and so it was with a full-blooded negro with whom I happened once to be intimate.

19. MAN FOR ALL SEASONS

BERNARD CAMPBELL

Reprinted from Sexual Selection and the Descent of Man, 1871-1971 *edited by Bernard Campbell, pp. 40-57. Copyright © 1972 by Aldine Publishing Company. For biographical information on Bernard Campbell, please see Selection 16.*

■ We have seen in previous selections that the course of human evolution may be reconstructed by a variety of means. For example, Washburn (Selection 7) pointed to the approaches associated with general evolutionary theory, the study of the fossil record, and knowledge of living primates. In this selection, Campbell focuses on the classical ecological approach, examining the relationships of our human ancestors to the tropical climate in which the species originated and the temperate and arctic climates into which different populations moved.

One of the central features of an ecological frame of reference is to determine a population's ability to extract energy from its habitat. Energy is what makes work possible; a population's success in adapting to a particular habitat depends largely on its ability to convert potential energy—food, minerals, other natural resources—to productive ends. Within this framework, Campbell reviews the successive adaptations that the species has undergone, beginning with man's separation several million years ago from the line of apes and concluding with the contemporary industrialized ecosystem. His goal is to learn how individuals in succeeding populations have increased the rate of energy-extraction from the habitats they have exploited, thus facilitating their breeding success.

In the concluding portion of this selection, Campbell turns to some of the possible implications of his findings for modern life. An important aspect of human adaptations during the past 5,000 years is that the local ecosystems in many parts of the world have been altered, and much of their natural balance has been destroyed. Campbell suggests that with increasing urbanization and a growing dependence on resources from other habitats, *Homo sapiens* has created a single worldwide ecosystem. The social and political implications of this economic phenomenon are far-reaching.

The reader who wants to explore further along these lines may profitably consult *Environment and Archeology: An Ecological Approach* by Karl W. Butzer (Chicago: Aldine, 1973, second edition); this book also stresses the methods used in these kinds of explorations. Also of great relevance are many of the chapters in *African Ecology and Human Evolution,* edited by F. Clark Howell and Francois Bourliere (Chicago: Aldine, 1963). ■

In each great region of the world the living mammals are closely related to the extinct species of the same region. It is, therefore, probable that Africa was formerly inhabited by extinct apes closely allied to the gorilla and chimpanzee; and as these two species are now man's nearest allies, it is somewhat more probable that our early progenitors lived on the African continent than elsewhere.

Charles Darwin, *The Descent of Man,* 1871

INTRODUCTION

WHEN DARWIN WROTE *The Descent of Man,* his object was to consider first "whether man, like every other species, is descended from some pre-existing form; secondly, the manner of his development; and thirdly, the value of the differences between the so-called races of man." Today, there is no need to argue that man is descended from some pre-existing form; this is a well-founded theory that finds wide acceptance.

The manner of man's evolution is the aspect of the descent of man to which Darwin was least able to address himself. In 1871 he was not in a position to cite any fossil men which could be considered intermediate links in the lineage of man's ancestors. The Neanderthal skull, one of the first found and most famous fossil men of the time, had been described by Huxley as "the extreme term of a series leading gradually from it to the highest and best developed of human crania." The large cranial capacity, among other features, led Huxley to conclude that the skull did not fall far enough outside the normal range of variation of modern man to justify regarding it as a "missing link." Darwin followed Huxley in this conclusion, and it was in a sense correct, insofar as we now recognize that West European Neanderthal man was in the relationship of cousin rather than ancestor to modern man.

Darwin did not appear perturbed by the absence of fossil evidence. Breaks in the fossil record were common, and the discovery of vertebrate fossils had been, until that time, a very slow and fortuitous process. Darwin added that "our progenitors, no doubt, were arboreal in their habits, and frequented some warm, forest-clad land" suggesting Africa to be the more probable. Thus the regions most likely to afford fossils had not at that time been searched by geologists. During the past one hundred years, however, the fossil evidence for human evolution has accumulated at an accelerating pace. We now have not only the direct evidence of the evolutionary process, but enough detail to consider in a limited way the manner of man's descent: his successive adaptations.

Following Simpson, we may define adaptations as those characteristics of an organism advantageous to it or to the conspecific group in which it lives. Obviously such characteristics are evolved and accumulated during evolution, and the word could be applied to any characteristic of a successful species. What we are concerned with here are those characteristics that man evolved since his separation from a common lineage with the apes and that gave him success in the new environments he entered during his evolution. They include a vast range of anatomical, behavioral, and cultural structures and traits that could not be described in detail in a large book, let alone a short paper. We therefore need to consider them as a whole and in limited aspect.

The aspect I have chosen, classical ecology, is one that would have appealed to Darwin. In short, I wish to consider how man managed to tap an increasing proportion of energy from his ecosystem and how this was done in the different climatic zones he successively entered. Since all organic life depends on solar energy as the ultimate source of its food, evolutionary success can be seen as the proportion of the total solar energy that any species can tap and incorporate into its own biomass. Since the quantity of energy is finite, and the conversion rate more or less fixed by the process of photosynthesis (in the region of 1 per cent), gains by one organism will usually be at the expense of another—a generalization that is particularly true in the case of man.

My view of man, then, is to be economic. Economics seem always to have controlled our destiny, as they do today. I will review the series of adaptations which man has undergone and devised to increase the extraction rate from his environment: the extraction of energy in the form of food, and later fuel and power. To this end I shall briefly review man's biological and cultural adaptations to the tropical, temperate and arctic biomes, and attempt to summarize an overall pattern of successive adaptation in human evolution.

FOREST AND SAVANNA: THE FIRST PHASE

The detailed fossil evidence for human evolution has reduced theories of the origin of the Hominidae from a large number to just two, and these two have a great deal in common. Both state that the earliest Hominids split from the apes in tropical Africa, and both would accept that this split occurred about the same time as the apes themselves divided into the two species chimpanzee and gorilla. In one theory the split is believed to have occurred more than 15 and probably nearer 25 million years ago. In the second the split is believed to have occurred less than 10 million years ago.

The differences involved in these two theories are of degree rather than of kind, and the degree is mainly a matter of timing. The evolutionary process would have been broadly the same in each case, but while in the first theory the split would have occurred among apes which were primarily adapted for arm swinging as a mode of locomotion, in the second theory the split would have occurred among knuckle-walking apes. There is therefore an important difference in the locomotor adaptation that we should attribute to the earliest Hominids. The difference of opinion is based on a wide range of indirect evidence, but in the end we should be able to elucidate which hypothesis is correct by the collection and evaluation of the fossil evidence. In particular, we should ask if there are any fossil Hominidae of 10 or more million years of age. The record of *Australopithecus* goes back in a very clear and satisfactory series to 5.5 million years, but before that time we have no *certain* fossil evidence.

From an earlier date we have specimens of the genus *Ramapithecus* (= *Kenyapithecus*) from India and East Africa which have been described as early hominids. The Indian fossils are more than 8 millions years of age, and the Fort Ternan fossils (Kenya) carry a date of 14 million years BP. The Kenyan specimen is of great interest: in its anatomy it seems to fall neatly between the later hominids and the early Dryopithecines. Particularly striking is the first lower premolar tooth, which is halfway between the sectorial ape tooth and the bicuspid human tooth. Anatomically these fragments fulfill our expectations to a remarkable degree.

The status of these fossils is, however, not a foregone conclusion. As we pass back in time, the fossil record becomes less well documented, and bones of *Ramapithecus* are rare and fragmentary. At the same time, the nearer we come to the split, the more difficult it will be to distinguish the bases of the two diverging families; the fossils will carry more characters of common inheritance and fewer characters of independent acquisition. It may therefore take some time before the matter is resolved. We urgently need evidence of the skull and limb bones of *Ramapithecus*. Extensive fossil evidence may well prove necessary to elucidate the status of these early and closely related groups of hominoid primates.

The faunal context of *Ramapithecus,* in both India and Kenya, implies riverine forest and possibly more open woodland (perhaps bordering on savanna). The climate was warm and moist, and it looks as though ecologically, *Ramapithecus* was not very different from the chimpanzee. The evidence further indicates that there was an ecological corridor between Eurasia and East Africa, and many of the fauna of the Fort Ternan site are from Eurasia. Therefore it is possible that *Ramapithecus* came from Eurasia and evolved from an early Eurasian *Dryopithecus* ape. It is equally possible that the movement of *Ramapithecus* was in the opposite direction, and in this case it is probable that the ancestral form was chimpanzeelike, perhaps close to *Dryopithecus* (*Proconsul*) *africanus.* Alternatively, the early hominids may have evolved over a wide area of the tropical and subtropical regions of the Old World. But the similarity of man to the chimpanzee, which has been demonstrated by Wilson and Sarich through biochemical traits as well as by classical comparative anatomy, strongly suggests that we are right to look for man's origin in Africa.

If *Ramapithecus* is a hominid in its total morphological pattern, then we can be fairly sure that the split occurred sometime earlier, at least 15 million years BP. If *Ramapithecus* is not an early hominid, the opinion of Sarich and Washburn that the split occurred less than 10 million years ago may hold. It will still be necessary to find intermediate fossils from the Pliocene as positive evidence for this theory. Whatever the date of the split, the evidence at present strongly supports Darwin's prediction that our progenitors lived in "some warm, forest-clad land," and it is likely that this land was indeed Africa.

Evidence of the changes in climate in prehistoric Africa are not yet so well understood as those of prehistoric Europe. The faunal evidence suggests a reduction in rainfall toward the end of the Miocene and an accompanying reduction in the area of tropical rain forest. The forest was in part replaced by woodland, tree savanna, and open grassland savanna, though the actual area of ecotone between forest and savanna may not have been materially altered. By the early Pleistocene, evidence from the fauna and sediments at the base of Olduvai Gorge and in the Transvaal cave deposits suggests that the climate was not very dissimilar to that of today, though some variation in rainfall was not uncommon.

A visitor to the Serengeti and Kenya Rift Valley today, however, will find a great predominance of open grassland, which in some places has degenerated to a scrubby woody vegetation as a result of overgrazing. The grassland savanna is largely maintained by periodic fires set by pastoral tribes such as the Masai to encourage grazing for their cattle and to kill tsetse fly and ticks. Fire allows the growth of grasses but kills most trees and tree seedlings. Such open conditions are therefore recent. It seems probable that in Pleistocene times the extensive plains of East Africa that lie at an elevation of between 4,000 and 6,000 feet carried perennial grassland, sometimes open, but more usually with scattered trees (such as species of *Acacia*) in most areas. There are still today limited regions of montane and riverine forest, and in the past these were probably more extensive. The use of fire to maintain pasture in East Africa has certainly changed the ecology of the tropical savanna.

These considerations are important in developing our ideas about the adaptations of the early hominids. They evolved from an arboreal apelike species and would have depended on trees as an escape from predators and a safe sleeping place. The first and most significant phase in human evolution almost certainly took place in forest fringe woodland where water was available, an ecotone that offered the safety and familiarity of the forest together with the richness of the savanna fauna. This development was not so much a change in habitat as an expansion into the exploitation of an ecotone with greater diversity of species. Undefended by size or large canines, these small hominids must have survived by cunning and an effective escape route from predators.

Reliance on meat would have developed slowly. From a diet containing very little meat (such as that of a chimpanzee), the proportion would have increased, especially in the dry season, or when vegetable foods were scarce. That meat would ever have composed more than 50 per cent of the diet of early hominids seems extremely improbable; a far smaller percentage is a more likely figure. Today the Bushmen of the Central Kalahari desert survive in much more arid regions, and when game is absent during the dry seasons they survive for long periods on vegetable foods. High meat diets are a recent adaptation to winter arctic desert conditions where there is no vegetation whatsoever. The dentition of the Hominids at all known stages is oriented toward chewing and grinding rather than cutting and slicing and carries clear evidence of a preponderantly vegetable diet.

We can therefore suppose that the early hominids came to exploit a range of fauna as a subsidiary food resource, with the adaptations of the bipedal stance and run and, later, the use of the cutting tool. There is no convincing evidence at this early stage of the use of weapons to kill game, but it is quite possible that sticks and bones would have been used as clubs. Sticks are thrown in agonistic display by apes and would almost certainly have been used as clubs by hominids. Techniques of killing must have been developed that did not involve the use of sharp canine teeth.

There is little doubt that during this period of adaptation the social structure of the hominids underwent some modification. Even among living chimpanzees we see a tighter social structure in savanna woodland than in forest areas, and this pattern was undoubtedly followed by early hominids. The relationship between social structure and ecology has been clearly demonstrated among primates. This relationship is important for our consideration of early hominid adaptations, because the social group sets the scene for social hunting.

By the time of the deposition of Bed I, Olduvai Gorge, above the 1.8 million-year-old lava flow, a stable adaptation to woodland savanna was almost certainly achieved. Food remains on the living floors at many sites testify to successful hunting as well as scavenging. The move to the savanna was completed. We know, however, that the Olduvai sites were on a lake shore, and it seems certain that the hominids were dependent on both a constant supply of fresh water and the presence of trees. Arid areas without fresh water but with saline lakes would not have been satisfactory sites for hominid occupation, as Bed III, Olduvai, testifies.

Further adaptations to arid savanna regions must have appeared by Olduvai times. We know that the stone tool kit was by now quite extensive, and probably included tools for making implements of wood and bone. Large stones suitable for tool making were collected from point sources as much as eight miles from the lake shore. Animal products such as skins and ligaments would probably have been used and prepared with stone tools. The site at DK carries evidence that has been interpreted as indicating some sort of shelter from rain or wind. The relative size of the brain compared with body size of the australopithecines had increased over that of the apes, and the pelvis was manlike in many features. Their hunting skills were quite well developed. Their survival indicates that these hominids were successful until more advanced hominids had evolved by the Middle Pleistocene. We do not yet know whether this took place in Africa or further north, but the fossil and archeological remains suggest that adaptation to cold winters was an important factor in the evolution of man.

TEMPERATE ADAPTATIONS: A CRITICAL ADVANCE

It now seems possible that the expansion of the hominid populations into temperate zones was the most significant step in the evolution of *Homo.* While it was previously thought that the appearance of *Homo* coincided with the successful hunting of big game, this now seems to have been an art mastered by *Australopithecus,* and we have to look elsewhere for the hallmark of *Homo.* The expansion into temperate zones was more than a shift between adjacent tropical biomes such as occurred among earlier hominids; it involved a major climatic change and was accompanied by many important new adaptations. The most striking adaptations, however, were cultural rather than biological.

Unlike the tropical biomes, the temperate regions are subject to quite extensive seasonal fluctuations in temperature: there are usually two or three months of frosty weather in winter when plant growth ceases. There is also abundant fauna and a richly diverse flora, and the temperate woodland is second only to the tropical rain forest in diversity. The rainfall is evenly distributed, and in the woodland regions lakes, permanent streams, and rivers are common.

Adaptations to this biome would have opened up very extensive food resources to early man. While geographical (mountain) barriers may well have delayed expansion into this environment, cultural adaptations to winter were undoubtedly the most important factor enabling man to move north.

From the period from two to one million years BP we have tropical fossil hominids from Java which seem to be somewhat similar to the later *Australopithecus* (= *Homo habilis*) fossils from East Africa. These populations could have entered Southeast Asia via a tropical woodland and savanna corridor at this time if not earlier. They are usually classified as *Homo erectus.*

Perhaps the earliest occupation of a temperate zone is that recorded at the Vallonet Cave in southern France (Alpes-Maritime), which is believed to be of Günz age. There is evidence of hearths in the Escale Cave in the nearby region of Bouches-de-Rhone; these deposits date from a period which is at least early Mindel. Neither of these sites, however, is so convincing as the inter-Mindel site at Vértesszollos in Hungary, where there are hearths with burnt bones and numerous broken and split bones of a substantial mammalian fauna, particularly rodent bones but also those of bear, deer, rhino, lion, and canid. There are several hundred artifacts of chert and quartz, mainly choppers and chopping tools, flake tools, and side scrapers.

The great cave of Choukoutien, near Peking, is another important site of about the same age. It has very extensive deposits containing numerous hearths with food remains of 45 different species, including sheep, zebra, pigs, buffalo and rhinoceros, together with deer, which form about 70 percent of the total. The tool kit has much in common with that of Vértesszollos and indeed with the "developed Oldowan" from Olduvai Gorge.

Dating from the late Mindel we have two other important European sites, and both show the use of fire. At Torralba and Ambrona in Spain (two contemporary sites, three kilometers apart) we have evidence of extensive butchery by bands of *Homo erectus* who appear to have been trapping elephants in a bog and butchering them on the spot. Bones of more than 20 elephants have been found in a small area together with remains of horses, cervids, aurochs, rhinoceros, and smaller animals: evidently extensive slaughter took place at this site from time to time. There is also evidence of shaped and polished tools of wood and bone and there is a rich Early Acheulian industry.

At Terra Amata, in present-day Nice, we have evidence of seasonal habitations on coastal dunes. There are ovoid arrangements of stones with regularly spaced post holes. Within the shelters these represent, the floors were covered with pebbles or animal hides (imprints are preserved). Hearths occur in holes or on stone slabs sheltered by low stone walls. Food residues include elephant, deer, boar, ibex, rhinoceros, small mammals, and marine shells and fish. The industry is of Early Acheulian type, and includes a few bone artifacts.

Although he probably entered this latitude during the preceding warm interglacial, on the basis of the present evidence man seems to have become well established in north temperate zones during the Mindel glaciation. By Mindel times (variously dated between 700,000 and 400,000 B.P.), highly efficient and productive hunting techniques were employed; and it appears that no animal was too large or too dangerous to be killed by hunting bands. The product of the hunt would have served to support an increased population through an improved food supply and other animal products, such as skins, that could be used in an advancing technology.

Because at most times of the year vegetable foods in the temperate regions would have been fairly plentiful, systematic hunting was primarily a means of supplementing a diverse vegetable diet. Mammal meat would have become a primary food resource only during late winter and early spring. Berries and seeds would have been eaten by this time, and the new succulent vegetation would not yet be grown. Even the game would have dispersed to find food. Because this limiting period of the temperate year would have kept hunting populations fairly sparse, both gene flow and the transmission of cultural traits would have been restricted. New developments in human evolution were in future to be tied to adaptations which increased man's extraction rate by the exploitation of the additional food resources by food resources and by food storage techniques.

THE SIGNIFICANCE OF WINTER

There is very limited evidence of biological adaptation to cold in modern man. Biologically suited to the tropics as he was, his survival through the cold winters of the temperate zone required extensive cultural adaptations unmatched by the evidence from Africa.

Whether in a grass hut shelter (such as those found at Terra Amata) or cave site, fire would have been maintained not only for warmth but as essential protection against giant carnivores such as cave bears and brown bears which were dangerous competitors for life space in cave sites. Archeological evidence strongly supports the notion that fire was used in temperate biomes for a very long period in human evolution. The earliest evidence of fire in Africa is from the Cave of Hearths and is dated about 55,000 B.P.

Cold winters also necessitated considerable development in social behavior. It seems inescapable that there would have been a fairly complete division of labor by this time: the men hunting and the women minding the babies and gathering vegetable foods, water (in skins), and fuel. Perhaps for the first time babies were put down and left in charge of siblings or aunts at the base camp. The division of labor and separation of the sexes must have increased the need to communicate abstract ideas by the development of language, and the vocabulary expanded. Perhaps the expression of the emotions (which language could replace) was first inhibited in a closely knit cave-dwelling band, and this emotional inhibition was to become increasingly important. It may prove to have been one of the most fundamental social developments which has shaped the psychology of modern man.

From the skeletal evidence at Choukoutien we can deduce that more than 50 per cent of the population died before the age of 14, before they reached full reproductive age. This suggests that it would have been necessary to produce four children per family simply to maintain the population level, and it is probable that more than four children were born in each family. The impossibility for hunter-gatherers of carrying and nursing more than one child at a time indicates that a cultural adaptation such as infanticide was possibly quite common.

The cold made demands on man's ingenuity to devise protective facilities such as clothing and tents. It was surely an important factor in the evolution of human intelligence. All the necessary adaptations had their anatomical correlates in the brain, the skeleton, and the soft parts of the body. The people of Choukoutien show a great advance over *Australopithecus.* Their endocranial capacity is twice that of *Australopithecus* and falls into the range of modern man. At Choukoutien the cranial capacity varies between 915 and 1,225 cc, while at the Vértesszollos site in Hungary (which is at least half a million years old), the cranial capacity has been estimated to be about 1,550 cc. This is three times the mean for *Australopithecus* and above the mean for modern man (1,325 cc). The only well preserved cranium from Africa of *Homo erectus* is the skull from site LLK II at Olduvai Gorge, which has a cranial capacity of 1,000 cc. Such is the variability of cranial capacity in man that samples of one individual have little significance, but they are suggestive.

The people at Choukoutien were also anatomically more advanced than *Australopithecus;* they were of greater stature and more sturdily built, with a well-developed bipedalism. Yet they still carried a heavily built masticatory apparatus that clearly distinguished them from modern man.

The cultural adaptations man developed to survive in temperate and arctic biomes were an extraordinary achievement, and northern winters were undoubtedly a factor of great importance in the evolution of *Homo sapiens.*

ARCTIC ADAPTATIONS

None of the sites discussed so far carry evidence of permanent cold, but only of seasonal frost. The fauna and pollen data suggest a cool climate becoming either colder (as at Vértesszollos) or warmer (as at Torralba). The climate at Terra Amata near the sea was certainly mild, and that at Choukoutien was of interstadial or interglacial type.

Man at this time survived cold temperate winters, but there is no evidence that he had yet adapted to the arctic conditions of the Mindel glaciation. Following the Mindel period, we have more northerly fossils of late

Riss-Würm interglacial date (Swanscombe and Steinheim) which represent the expansion of human populations northward during warm temperate spells. In view of the extreme difficulty of survival in northern coniferous forest and arctic tundra, it is not to be expected that man would have entered these zones at a very early date. Today it is hard to see how early man could possibly have adapted to arctic conditions without domesticated animals such as reindeer or dogs. The presence of Neanderthal man in the first Würm glaciation of western Europe is a surprising fact. It suggests an advanced use of both tools and facilities.

Though the present evidence suggests that Neanderthal man did not survive throughout the first major advance of the Würm glaciation, it does clearly demonstrate that he could survive extreme cold and must have lived some thousands of years under arctic conditions. This Würm advance of the northern ice sheets brought a cold moist climate characterized by animals such as mammoth, woolly rhinoceros, reindeer, musk-ox, ibex, blue fox and marmot. All these were hunted, together with the formidable cave bear.

Neanderthal man was well established in southern and central Europe before the colder weather descended, and he survived the cold to a great extent by using caves and rock shelters. Judging by the extent of their cultural remains the Neanderthal people adapted successfully to the climate and were able to exploit the huge herds of reindeer and other animals. In some areas such as the Dordogne, the local topography must have offset the very extreme conditions. The Dordogne river and its tributaries dissect deeply into a limestone plateau and offer a number of sheltered valleys. Possibly the vast herds of animals which must have undertaken regular migrations used these valleys as migratory routes. It seems possible that here, as in southwest Asia, these people came to rely on harvesting migratory animals. Many temperate animals, and more particularly arctic species, migrate regularly in the spring and autumn between coastal plain and mountain pasture. These people were thus not only able to harvest "earned" resources (which gain their food within the local habitat where they live), but to tap "unearned" resources—animals which pass through or spend some portion of their annual life cycle in one biome and yet gain most of their food (energy) in another biome. To settle along the migration routes of herd mammals such as reindeer, musk-ox or ibex and intercept them between their summer and winter feeding grounds is a sophisticated adaptation that we can fairly safely attribute to late Neanderthal man. It was a simple step to allow autumn-killed meat to dry and freeze for use during the winter, as many Eskimo do today. We can also deduce that the game was sufficient for their needs, for they must have relied to a great extent on meat during the winter.

An interesting clue is provided by the teeth of the man from La Ferrassie. They show a particular type of extreme wear also found today among the Eskimo and some other hunters caused by chewing animal skins to soften them for clothing. It is indeed highly probable that Neanderthal man had exploited the whole range of animal products, and especially skins, to develop a well-differentiated material culture. He could well have made the kind of clothing that we find among the Eskimo, though the ready-made shelter of rocks and caves would probably have stood in place of the warm and intimate family igloo.

Probably the most difficult problem facing these arctic people was transport. Without dogs or sleds, they would be confined to a small area during the winter months, and their movements would have been limited to the valleys in which they lived. This restriction shows how successfully they had been able to exploit the local food resources of the region.

The successors of Neanderthal man in western Europe, the Aurignacian and Magdelenian peoples, have left us a far more detailed picture of their adaptations than their predecessors. Migratory herds of reindeer were harvested in very large numbers, and often formed 85 to 90 per cent of the faunal assemblage. (Other animals hunted include mammoth, bison and horse.) When climatic conditions became severe again, as they did toward the end of this period, the later Magdelenians (14,000-12,000 B.P.) began the systematic hunting of new kinds of unearned resources: migratory birds, aquatic mammals, and fish. The significance of these additions to their food supply can scarcely be overstated. Migratory fish and fowl appear in early spring, a time of maximum food shortage, and make possible the survival of a much larger population throughout the year. At the same time, it is probably important that fish oils (unlike terrestrial animal fats) contain vitamin D, and this may have been an essential vitamin source to people living in areas where insolation was low and clothing essential at all times. Evidence of rickets is present among Neanderthal skeletal remains; as far as we know, these people were unable to catch fish and complement their own low vitamin D production. Shortage of this vitamin may indeed have been one of the factors that mitigated against their survival.

This extensive exploitation of migratory animals, which must have arisen slowly through the later stages of human evolution, was to have a profound effect on the evolution of man's social life. The most obvious result was that it allowed a more sedentary way of life to develop. Home bases could be occupied for longer periods of time during migrations and the ensuing winter, and there was no longer such a premium on the mobility of hunting bands. During the early phases of hunting and gathering (a way of life which Bushmen and Australians still follow) possessions and infants were limited by the need for mobility. A mother with a baby or a small child who must break camp frequently and transport her baby as well as her household gear will not welcome a second infant to care for and carry or a mass of material possessions. She will have few compunctions about taking any means necessary to limit family size. Today, hunter-gatherers practice infanticide, abortion, and other means of birth control to retain their essential mobility.

Sedentism changed all this. As soon as a band could remain more or less permanently in one place, an increase in possessions and in population densities was possible. The limitation on numbers was removed, and population could now expand to a level related to that dictated by the increased food supply. When we compare the sites of the earlier Magdelenian to those of the later period (since 14,000 B.P.), we find that the living sites are more numerous, larger, and more often situated low on river banks, frequently at places where the river narrows. Many of these sites have yielded evidence that they were inhabited throughout the year. Thus we find these permanent settlements associated with an increase in density and in group size. These profound changes have been discussed elsewhere. It is also clear that these developments may have required a much greater complexity of social structure, compared with the essentially egalitarian local bands that characterize most hunter-gatherer groups. It opened the way to the developments that characterize the Neolithic period.

ALTERATION OF THE ECOSYSTEM

In his adaptations as a hunter-gatherer, man remained part of a more or less stable natural ecosystem, as he adapted to the particular conditions of each biome and each region. Because food supplies were not the limiting factor in population growth, except perhaps at certain very critical periods, we do not find evidence of overkill or of any serious instability following the appearance of man in a region. As a hunter he was competing with carnivores for herbivores, but the ecosystem is characterized by a functional dynamism that allows it to equilibrate in the face of climatic and other minor changes, especially if it is diverse in species.

Farming is the protection of food plants and animals at the expense of wild forms of less nutritive value to man. It also involves domestication, which is the selective breeding of certain species for their tameness and their value as food. The overall effect of the practice is a reduction in the diversity of organisms in an area, which is balanced by an increase in the domestic species, and a larger proportion of the solar energy in the area is turned into human food, either as plants or meat. Pastoralism can be a surprisingly effective adaptation in this respect, especially when more than one animal species is herded is a single area. Where field agriculture results in whole areas of the ground being covered in one or two food species, the conversion rate of solar energy into human food is even higher. Although we only cultivate 10 percent of the earth's land surface today, it has been estimated that the human population has increased from a potential maximum of about 10 million hunter-gatherers to its present size: some 3½ billion (many of whom suffer from malnutrition and starvation). But the ecological cost of the introduction of pastoralism and agriculture is high; it implies the destruction of the natural ecosystem and of the diversity of species which assure its stability. The Neolithic was the start of the destruction of man's natural environment, and as the rate of population increase grew, the rate of destruction increased.

In the past 5,000 years man has altered the ecosystem in many parts of the world and destroyed the natural balance. Pastoralism itself has been one of the most destructive forces; it is clear that wherever it has been carried out in semi-arid regions, whether in Australia, Asia, Africa, the Americas, or in limited areas in the Mediterranean regions of Europe, there has been degradation of the grasslands and the threat or reality of soil erosion. The local fauna has been destroyed and the existing ecosystem degraded beyond the point where it can naturally equilibrate. Where soils are eroded, the loss is irrevocable. The displacement of game and the destruction of their natural environment has done far more damage to natural life than all the hunting of the Pleistocene. In the same way agriculture and deforestation for timber have involved the destruction of vast areas of forest (areas of naturally high rainfall), and we have lost both the forest with its associated flora and the forest animals (which often have a very limited distribution). All these developments, though they may eventually prove to endanger his survival, have enabled man to increase his extraction rate from his environment and place himself increasingly at the top of the energy food chain of the biosphere.

TRENDS OF HUMAN ECOLOGY AND ADAPTATION

During his evolution man has occupied an increasing series of subsystems within the world ecosystem. From his beginning in the forest/savanna ecotone of Africa (and possibly Asia) he spread northward into temperate and arctic biomes. The evolution of man through this climatic spectrum shows a number of trends.

FLEXIBILITY AND RANGE OF BEHAVIOR

Man's relatively large brain is a reflection of a particular kind of progressive evolution that has been described by Herrick as "change in the direction of increase in range and variety of adjustments of the organism to its environment." An increased dependence on learning as a basis for appropriate response to environmental stimuli has allowed a much greater flexibility in that response. Innate mechanisms form the basis for learning, rather than for programming, the actual behavior. Behavioral flexibility allows the evolution of complex social behavior, which is facilitated by language, a novel behavior pattern unique to man. The flexibility and range of human behavior is man's primary biological adaptation, which has made possible his unique mode of cultural adaptation. The evolution of increasing cranial capacity began to accelerate about one million years ago and has continued throughout man's adaptation to temperate regions. The correlation may well be significant.

CULTURAL ADAPTATIONS

Throughout his evolution man's behavioral development takes the form of a material culture. From an early use of very simple tools, complexity developed. Adaptations to temperate climates in particular involved dependence not only on tools but on facilities, defined by Wagner as objects that restrict or prevent motion or energy exchanges (such as dams or insulation), so that anything that retains heat is included (tents, houses, or clothing). Containers of various sorts—skins for carrying water, pots or boats, fences or even cords—fall into this category. Temperate adaptation requires far more facilities than tropical adaptation; and their most important aspect is the extent to which they enabled man to become even marginally independent from certain limiting factors in the environment.

Much later in human evolution during the Neolithic, the first machines were invented. The atl-atl, bow and arrow, firedrill, and eventually the wheeled cart enabled man to carry out tasks previously beyond his power. His advancing technology gave him enormous power over his environment and the food resources it contained.

INCREASE IN TROPHIC SPAN

Advances in material culture allowed man to increase his trophic span. From a mainly vegetarian diet in the forest, man spread onto the savanna to become a part-time hunter. In this way he came to tap a second trophic level in a very diverse environment. Evidence was found at Choukoutien that man was obtaining food from three trophic levels by killing and eating carnivores as well as herbivores. The Eskimo today (as perhaps the Magdelenians in their time) kill carnivorous sea mammals in the next trophic level, and this gives them an even greater trophic span. As man moved north, he moved up the food chain; his original dependence on a single trophic level was spread to four trophic levels. This gave him a much more reliable food supply in northern regions of low diversity, where oscillations in populations of animals are common and where vegetable foods may be almost unobtainable during the winter and early spring. This adaptation to a very varied diet is probably unique to man and allows his survival in a wide range of habitats. However, owing to the inefficiency of energy transfer from one trophic level to the next (which lies in the region of 10 per cent), the transfer of solar energy into human beings is most efficiently achieved at the first trophic level. During the present period of overpopulation more and more people are having to live on a vegetable diet as meat becomes a scarcity. In this way the world population is fed, though limitations in variety of vegetable foods may result in serious protein deficiency.

EXPLOITATION OF UNEARNED RESOURCES

The use of unearned resources of food (and later fuel and minerals) had a far-reaching effect on the balance of man's ecosystem: sedentism followed and then the expansion of populations. Sites that were favorable for the exploitation of terrestrial and aquatic unearned resources were settled, and social organization became more complex. Many groups budded off from parent populations and were forced into less favored semi-arid regions. It was under these circumstances that the harvesting of seeds became adaptive. The first documented attempts at agriculture in both Old and New Worlds appear in semi-arid zones bordering established riverine settlements. With the development of agriculture and its associated stored surplus, there was a further expansion of population and an increased dependence on other kinds of unearned resources, such as water for irrigation, and the grain crops themselves, as they were transported over long distances into growing villages and towns.

EXPLOITATION OF ENERGY SOURCES OTHER THAN FOOD

Nearly a million years ago man may have learned to capture and maintain fire both for protection and warmth. The use of timber for fuel and eventually for smelting made possible important cultural developments and resulted in the deforestation of vast areas. The use of fossil fuels in the past 300 years has in turn made possible our western civilization, and these limited energy sources will need replacement by nuclear, solar, or tidal energy in the near future. Following the development of field agriculture the use of draft animals was another important energy source: ploughing was developed and heavy transportation was facilitated. Eventually man learned to harvest the energy of the wind and rivers.

URBANIZATION

The appearance of agricultural surplus brought technological specialization and more complex social organization. The development of storage and transport facilities that made possible trade was a basic adaptation that brought the interdependence of human groups. As settlements increased in size and density, the effects of disease (which perhaps were not so important in the earlier stages of human evolution) would have become evident and a significant factor in population growth. After the development of agriculture, when population levels again came to stability, far more people would have died prematurely of disease (including malnutrition) than were ever killed by hunter-gatherers as newborn infants.

THE MEDICAL REVOLUTION: REDUCING THE DEATH RATE

As a constant factor in man's evolution, endemic disease would have had little impact on human consciousness. But with increasing populaton density and mobility, epidemics would have become significant factors in human selection and survival. The horror of this natural check on population growth was well-known in cities such as Paris and London that suffered the plague and black death. Toward the end of the nineteenth century, new discoveries in medicine (such as Pasteur's discovery of germs) brought about a further increase in human populations (and later in domestic herds), not by increasing the effective birth rate but by reducing the death rate. Today lives are preserved to suffer malnutrition and starvation, which the doctors cannot treat.

ONE ECOSYSTEM

All these trends have interacted in such a way as to produce an accelerating development of human biomass and culture. From a series of self-supporting social units of perhaps 25 to 50 individuals, each adapted to a particular region, we now find cities of many millions of people living under artificial circumstances in the sense that they are totally dependent on others for their survival, and their environment is totally manmade. Dependence on unearned resources has reached a new level. The city of Los Angeles is an extreme example: it receives 90 per cent of its water from watersheds other than its own, some sources hundreds of miles away where otherwise fertile valleys have been robbed. The United Kingdom imports approximately half its food requirements from all parts of the world. Such dependence on foreign resources and the necessary transport of materials and food from one biome to another have brought the different ecosystems of early man together, so that they form today one great system with much-reduced differentiation. Although outposts of wilderness (the name we give the natural world) do survive, all are becoming caught in man's unbalanced ecosystem.

I thought, We have geared the machines and locked all
together into interdependence; we have built the
great cities; now
There is no escape. We have gathered vast populations
incapable of free survival, insulated
From the strong earth, each person in himself helpless, on
all dependent. The circle is closed, and the net
Is being hauled in. . . .

—Robinson Jeffers, "The Purse-Seine", 1937

20. HOMINID ADAPTATIONS

DAVID PILBEAM

Reprinted with permission of Macmillan Publishing Co., Inc. from The Ascent of Man *by David Pilbeam. (Copyright © 1972 by David R. Pilbeam.) David Pilbeam is Associate Professor of Anthropology at Yale University, where he has been since 1968. Before that he taught at Cambridge University. His principal interests are in human evolution and the relationships between structure, function, behavior, and ecology.*

■ In the previous selections, Oakley focused on man as a tool-maker, while Campbell presented a wide-ranging picture of human evolution within the framework of a changing climatic habitat. In this selection, Pilbeam turns to a synthesis of hominid evolution by focusing on the evolution of the brain. He also incorporates the evidence from ecology, behavior, and anatomical function to illuminate the agencies for natural selection that ultimately produced *Homo sapiens.*

The two principal tools at a scientist's disposal are analysis and synthesis; Pilbeam makes excellent use of both. The only feasible way of deciphering the complex process of evolution is the analysis of its component parts. Pilbeam concentrates here on the brain. But overanalysis of a problem may obscure the structure of the total system in which a component functions, and Pilbeam avoids this pitfall by carefully resynthesizing the components he analyzes into a coherent whole.

Pilbeam's analysis strongly emphasizes behavior, showing that it is behavior on which the agencies of natural selection operate. With this in mind, it is clear that the adaptation of hominid hunting probably provided the most significant behavior on which natural selection acted. Hunting required foresight; communication; the division of labor; pair-bonding; efficient means of body temperature regulation; emotional inhibition; and other physical, behavioral, and mental adaptations, many of which are characteristic of modern *Homo sapiens.*

An excellent overview of physical evolution is given in *The Emergence of Man,* by John E. Pfeiffer (New York: Harper & Row, 1972). *Human Evolution: Readings in Physical Anthropology,* edited by Noel Korn and Fred Thompson (New York: Holt, Rinehart and Winston, 1967) and *Evolutionary Anthropology: A Reader in Human Biology,* edited by Hermann K. Bleitbrau (Boston: Allyn and Bacon, 1969) contain many good articles that can be consulted with great profit. ■

THE BRAIN AND BEHAVIOR

PERHAPS THE MOST profound differences between man and other animals are behavioral. Complexity and organization of behavior are directly related to the fact that the brain of *H. sapiens* is greatly enlarged compared with the brains of other primates. In considering the evolution of the hominid brain, we have to rely on indirect evidence, for there are unfortunately no brains fossilized so that their internal structure is preserved. Available are external casts of brains or internal molds of brain cases, neither of which are particularly satisfactory, for they can provide information only about the grosser aspects of external brain morphology. Another source of information is study of the internal structure and organization of brains in living primates. Thus, ape brains can be used as "models" for early stages in hominid brain evolution, although this can be done only with extreme care, for the living nonhuman primates are themselves products of long periods of independent evolution. For example, the chimpanzee brain may not resemble particularly closely the brain of the common ancestor of the African apes and man. The final type of behavioral data available is the information that can be gleaned from studying teeth, jaws, postcranial bones, stone tools, associated animal remains, settlement patterns, burials, and cave paintings. In a sense this is all fossilized human behavior.

The human brain is approximately three times as large as that of the largest nonhuman primate. In modern man the average brain volume is around 1,400 cm^3, whereas that of the gorilla is a little over 500 cm^3; the chimpanzee mean volume is around 400 cm^3. In particular, the cerebral hemispheres are greatly expanded in man, but it must not be forgotten that other important parts of the brain have expanded, too. The increase has been differential, not all parts expanding equally. For example, the association areas of the cerebral cortex, together with their related subcortical regions, have become greatly enlarged, especially in the parietal region.

This threefold expansion in human brain size has not been accompanied by an equivalent increase in the number of brain cells, for there are only about 25 per cent more cells in the human cerebral cortex than in that of the chimpanzee (this amounts to around 1.4 billion extra cells). However, human brain cells (neurons) are larger, more complex, and spaced further apart than the ape's. Neuroglial cells—cells in close physical and metabolic contact with neurons—also increase in number with increasing brain size. Although it would seem likely that the more complexly branched a neuron is the more interconnections it can make with other neurons and the richer and more varied behavior can be, the specificities of human behavior cannot be explained in terms of any one of these parameters. Much more likely, it is the interaction of these factors with changes in the internal organization and interrelationships of the brain that produce the uniquely human brain output. That organization is important is clear from the fact that human microcephalics (very small-brained people) of various sorts behave in specifically human ways, although they may have brains no larger than those of apes, brains that probably contain fewer cells than those of average healthy gorillas or chimpanzees.

These considerations point up the fact that comparing 1 cm^3 of human brain with 1 cm^3 of ape brain is not a comparison of equivalents. It seems equally clear that it is impossible to draw any sort of "rubicon" of brain volume beyond which a creature can be considered human, for the brain did not simply expand but became restructured. Indeed, it can be plausibly argued that the restructuring of the brain probably came first and the expansion (correlated with changes in neuron size, branching, and density) followed later.

The cortex in higher primates can be divided into a number of regions, differentiated in structure and function. These regions mature at different times, maturation being defined as the acquisition by neurons of a myelin sheath, a kind of insulating coat. The areas that mature first are described as "primordial zones" and consist of the motor cortex, the three primary sensory regions for visual, auditory, and somesthetic (tactile) input, and the limbic system. The motor cortex forms the back of the frontal lobe and the somesthetic cortex forms the front of the parietal lobe; visual and auditory areas are located in the occipital and temporal lobes, respectively. Deep in the cerebrum, surrounding the brain stem, is the limbic lobe (the term refers to the older—in an evolutionary sense—parts of the cortex and its associated nuclei). The limbic region is closely connected with the lower parts of the brain important in the control behavior vital to reproduction and self-preservation and apparently is intimately involved with all those basically emotional behaviors necessary for species survival.

In the most primitive mammals the limbic region makes up most of the cerebral cortex. In more advanced types the other primordial zones contribute relatively more. In higher primates phylogenetically newer areas of the cortex known as "association areas," which mature later than the primordial areas, have differentially expanded. Some of these association areas receive input from only one of the primordial zones, and it is generally agreed that there is an association area for each primordial zone. The association areas are involved with the integration of behavior. At this point, a little will be said concerning what is known about the functioning—in very broad terms—of various association areas.

At least part of the frontal lobe association areas is involved in drive inhibition. Damage to the human frontal region often produces profound changes in mood and personality. Experimental work with rhesus monkeys has shown that removal of the frontal lobes alters the dominance status of certain animals, principally because previously subordinate animals appear less inhibited, less cowed, in their interactions with animals that are normally more dominant. Other experiments have demonstrated that the neural pathways that mediate these behaviors involve parts of the limbic system as well as areas in the midbrain and brain stem.

The temporal lobes are involved in the storage of auditory and visual information and also in the performance of sequential activities. Monkeys can be trained to respond to stimuli given in a specific sequence. For instance, one response is required when a stimulus of type A follows one of type B, whereas another is necessary with the reversed sequence. The stimuli can be visual or auditory—flashes of light or clicks, for example. If certain parts of the temporal lobes are damaged experimentally, this type of sequential behavior becomes difficult or impossible. Once again, limbic interconnections are important in these types of behavior. The temporal lobes and their connections are clearly very important in higher primates, highly emotional and often aggressive animals that live in structured social groups where they are more or less constantly surrounded by their fellows and where continual adjustments in individual behavior have to be made. It is vital in a social context such as this that subtle changes in an animal's emotional state be conveyed to others in the group and that the messages be understood. The social communication of primates is extremely complex, involving postures, gestures, facial expressions, and vocalizations, and utilizes a number of channels—visual, auditory, olfactory, and somesthetic. All these sensory inputs have to be related to limbic (emotional) responses, and they must also be related to each other. A great deal of such social information is conveyed in a sequentially significant manner; the order in which the bits of data come affects the meaning of the whole message.

There are internal connections from each of the three sensory association areas to other parts of the brain. For example, the visual association area in one side of the brain has interconnections with the visual association area in the other side, with the motor association cortex, and with the association region of the limbic system. Other sensory modalities appear to have the same types of links. Almost all monkey and ape behavior depends upon sensory-limbic interconnections. For example, learning generally requires such associations, whether the behavior is learned by reward or punishment. The temporal lobe is particularly important in forming interconnections between auditory and visual signals and the limbic system.

Another type of connection is said to be at a rudimentary level in apes. This involves the development of a new association area in the parietal region, one having connections only with other association areas and not with primary sensory or motor regions or with the limbic system. It is this part of the brain, the so-called association area of association areas, that has become so greatly expanded in man. Only man is capable of forming, to any great extent, nonlimbic-nonlimbic-limbic associative chains; this is of great importance in any discussion of language or tool making.

There is some debate as to whether or not frontal and temporal lobes have become enlarged "relatively" in the evolution of the hominid brain, and the matter has yet to be resolved. However, the parietal lobe has expanded enormously. It should not be thought, though, that human behavioral specificity lies in the parietal lobe, or indeed that it lies in any particular place. It is clear that the human brain has become reorganized internally, so that although it is still undoubtedly a primate brain its output is qualitatively different from that of other primate brains. We cannot define here precisely how and why human behavior differs from that of our nonhuman relatives. However, it is possible to list some of the more obvious human characteristics.

In discussing the behavior of a mammal with a nervous system as complex as that of higher primates, it is necessary to remember that a great deal of behavior is learned. The old dichotomy between learned and instinctive behavior (and also between environmentally and genetically determined traits) is widely regarded now as being of little or no help. All behaviors are at the most basic level genetically determined, but some of them require enormous amounts of learning and, being largely learned, are subject to considerable variability of expression. Thus, language abilities in man are genetically based; it is almost impossible to suppress language acquisition in a normal human, and the way in which speech develops in children tends to follow a rather regular pattern. However, we do not possess the ability to speak *a* language, but the capacity for the acquisition of language. Individuals within a species generally learn most easily to perform those behaviors that are necessary for survival, and the learning of these behaviors is, in general, in some way pleasurable.

Human behavior is characterized by at least some of the following: the ability to behave cooperatively, to suppress or channel rage and aggression, to sustain motivation or drive over considerable periods of time, to form close affectional ties with other adults of the opposite sex, to make and use tools, and to communicate linguistically. It can be argued, and has been, that these and other attributes are in some way bound up in a behavioral package that developed as the hominids became hunters and gatherers.

Modern hunters live in small bands of around fifty individuals, the band being composed of subunits, the so-called nuclear family, containing an adult male and female united in a formal, legal marriage and their offspring. To what extent the pair bond between male and female is genetically programmed is unknown. Possibly, it is relatively easy for humans to form close affectional relationships with one other adult of the opposite sex, these ties being sometimes intense and frequently of relatively long duration. But in hunting society other factors are involved, and it can be argued that economic and political factors are what maintain the relationship by reinforcing such biological determinants as there are. There is division of labor between sexes, and it can be argued that the smallest reciprocal economic unit is the pair bond, the hunting and gathering roles complementing each other. Marriage in hunting society is exogamous—mates are found outside the group into which an individual is born—and so marriage also performs a political service in that it spreads the nexus of kinship ties between groups, not just within them. This tends to result in friendly relationships between bands in adjacent hunting ranges, very important during harsh conditions when resources must be shared widely between groups.

Cooperative behavior is strongly developed in hunting society, perhaps more so in males, who need to plan hunting activities. It should be emphasized that this type of behavior in man is no more an instinct than is the pair bond, although it probably does have genetic and neurological foundations that make it relatively easier, given the appropriate circumstances, to learn this type of behavior. Aggressive and dominance behaviors are generally less frequent in hunting society than in societies of other primates. This is probably because of constraints imposed by the economics of the situation, for hunters cannot afford not to be cooperative sharers, and there is no place for dominance hierarchies in such societies. This is not to say that man is not aggressive, or that his aggressive behavior does not have biological roots, but that its control or expression generally depends upon social factors.

The nuclear family contains a number of offspring, often ranging considerably in age. Ties between young and their parents, especially the mother, are very long-lasting and extremely important for the learning of adult skills. Socialization of the young is also profoundly influenced by play with peers. Learning in man differs from that in other primates in duration and intensity and is also greatly facilitated by a system of social rewards; these depend ultimately upon language for their transmission.

Of all the behavioral differences between man and other animals, language and tool making stand out as most distinctive and most important. In general, it can be said that the communication systems of nonhumans convey information about an animal's motivational state and depend directly upon limbic connections, whereas language conveys another type of information and makes statements that do not reflect the internal emotional state. This can be tied in with the great enlargement of the human neocortex, particularly those parts that do not have direct connections with the limbic system.

There are many "design features" that distinguish human language from other types of vocal communication, and only a few of them shall be briefly mentioned here. Language is a vocal sequential sign system; more important, the order of the elements within the sequence controls meaning. Thus it is a "relational" system rather than a purely "combinational" one. As we have seen already, higher primate communication also involves sequential signs, some of which may be relational, and it is possible to train monkeys to respond appropriately to relational sequences. However, these abilities are apparently quite limited, and the information conveyed in human language differs from that carried by monkey and ape communication. Part of the human temporal lobe, the so-called Wernicke's area, is important in human language production. As we have noted, this brain area apparently controls sequential signals in nonhuman primates.

Language is an open system; that is, a theoretically infinite number of meaningful messages can be generated. This is possible for a number of reasons, one of which is that language is referential, it can be used to refer to objects (including ideas, qualities, and feelings as well as noun-type objects). It has been suggested that this process of object naming involves the linking of an auditory cue with a cue in some other sensory modality, without directly involving the limbic system. Such cross-modal transfers are easily performed in man but (it is said) can be completed only with great difficulty in nonhuman primates, because the development of the parietal superassociation area in man provides a way of interconnecting sensory modes without utilizing the limbic system. Language also has the property of displacement; it can be used to talk about events in the past or future, or objects out of sight. Once again, emotional state need have nothing to do with such language usage. Finally, language is hierarchic. Meaningful elements (words) are composed of essentially meaningless sounds (this is termed "duality of patterning") and can be further combined into more meaningful groupings. Language can be said to have a nested structure.

Thus, language is the product of a brain that can, among other things, generate and understand relational linguistic codes that are hierarchic, that name objects (in a broad sense), and that permit some degree of displacement. All of these behaviors (although not their linguistic manifestations) are present at least on a rudimentary level in higher primates.

Some effort has been expended in attempts to teach chimpanzees to talk. The Hayeses had tremendous difficulty training Viki, their young female chimpanzee, to utter a very few words, although she possessed good memory, intelligence, and imitative skills. She would vocalize in this way only in a highly specific motivational context and evidently experienced considerable distress in doing so. Chimpanzee brains are not built to generate vocal language. (It should be remembered that some human microcephalics with brains no larger than those of chimpanzees are able to talk.)

More recently, the Gardners have been attempting with some success to teach a young chimpanzee, Washoe, American Sign Language. Washoe has already accumulated a large store of "names," most of which involve visual-visual interconnections, but some of which appear to involve cross-modal transfers. For example, she will use the sign for "dog" upon hearing a dog bark. She has also spontaneously invented new signs and produced sequences of signs, although none of these have yet been relational.

If we assume that the brains of protohominids were capable of producing behavior that, in the appropriate combination, would allow the production of a rudimentary language, what selection pressures might have caused such a change? For want of any better hypothesis, the best suggestion might be that it was the development of hunting as a way of life, for it would certainly seem that planning and cooperation among males, organization and coordination of economic activities in both sexes, and many other activities would require language.

Stone tools first appear in the fossil record of a little more than 2½ million years ago and probably were being manufactured even earlier. These tools consist of a number of types, presumably each one serving several functions. Unlike tool modification in chimpanzees, human use of a tool is not limited to one specific function; human tools are more complex—more activities are involved in their manufacture and their form is governed by (arbitrary) norms or rules. Tool making, like language, is a hierarchical activity, and this factor, plus the standardization of tools and the fact that they are the

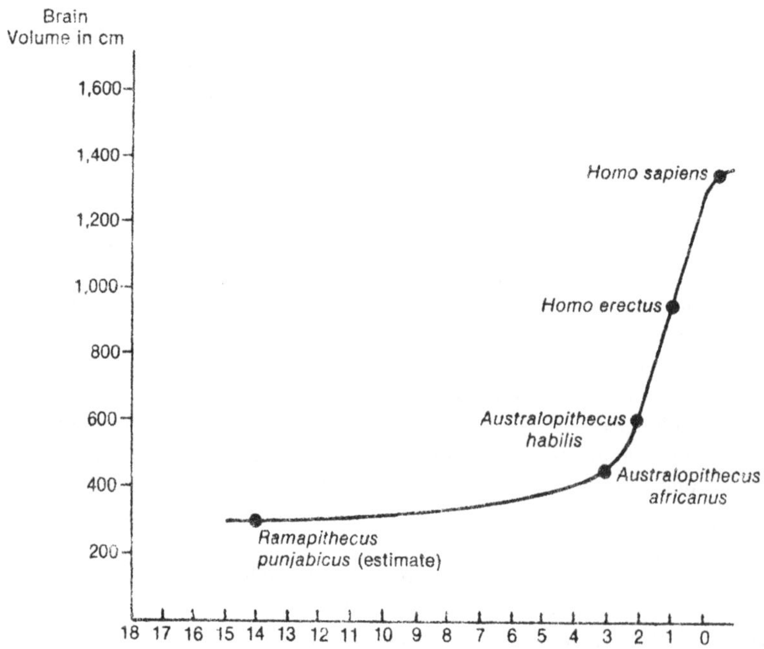

FIGURE 20.1

Relationship between brain volume and time in hominid evolution. The capacity of *Ramapithecus* is estimated; the temporal positions for the later species are approximate.

expression of arbitrary imposition of form upon the environment, suggests that the type of brain that was capable of tool making could also generate at least rudimentary language.

Hominids, unlike other primates, are dependent on tools for survival, and tool making is intimately bound up with the hunting way of life. It can be argued that hunting, language, tool making, division of labor between sexes, cooperation, formation of the nuclear family, incest prohibitions, and rules of exogamy are parts of a multifaceted behavioral complex, and, if this is indeed so, then we can trace back at least the rudiments of these behaviors to the time of the earliest stone tools, some 2½ million to 2¾ million years ago.

Something can be said concerning the evolution of brain and behavior during the late Pliocene and Pleistocene from a study of natural brain casts and endocranial molds, stone tools, and from associated remains other than the hominids themselves. If we estimate that the hominoids ancestral to hominids weighed around 40 to 50 pounds on average and if we further assume that the relationship between body weight and brain size was the same in these hypothetical protohominids as it is in pygmy chimpanzees, we can give a tentative value of around 300 cm^3 for their mean brain volume. We know nothing of the brain in Miocene *Ramapithecus,* but it is known that late Pliocene *Australopithecus* species had bigger brains than that. The small species, *A. africanus,* weighing around 50 pounds, had an average brain volume of a little less than 450 cm^3, and the larger forms *(A. boisei,* for example), weighing 150 pounds or more, had brains between 500 and 550 cm^3 in volume. In *A. africanus* there is evidence to indicate parietal lobe expansion, and the cerebellum in the more robust species was —at least in external morphology—more like that of man than of the apes (Holloway, personal communication). We know too that at least one of these early species was a tool maker, and both were bipeds. Therefore, all the evidence points to brain reorganization and also to some expansion, due perhaps to internal changes such as increased neuron size and reduced cell density. These forms may well have communicated linguistically with each other and lived in social groups more like those of man than of apes. By the early Pleistocene, *A. habilis,* the "descendant" of *A. africanus,* had a brain volume of around 600 cm^3 or more, although average body size was small, perhaps still no more than 50 or 60 pounds.

After the early Pleistocene, the brain increased rapidly in size, possibly involving some extra differential expansion of the temporal and parietal lobes. Stone tools became more complex, and the incidence of hunting big game also rose. The most reasonable assumption is that social organization was becoming more complex as language grew more efficient and individuals became more intelligent. Presumably, hominid social life was much as it was in later times. The increase in brain size during the last half of the Pleistocene is likely to have been due mostly to an increase in size and complexity of brain cells, rather than to a growth in numbers, and also perhaps to changes in biochemical efficiency. The brain reached its present size at least 100,000 or more years ago, and there were probably few dramatic changes in the biological bases of social organization and social behavior after that time.

21. NEANDERTHALERS

BJORN KURTÉN

Reprinted from Not from the Apes *(Copyright © 1972 by Bjorn Kurtén) by permission of Pantheon Books, A Division of Random House, Inc. Bjorn Kurtén is well known for his studies of Pleistocene Carnivores and of human evolution. He is the author of* The Age of the Dinosaurs *and* Pleistocene Mammals of Europe. *His present position is Lecturer in Palaeontology at the University of Helsinki.*

■ Now that we have examined some of the details of hominid evolution and adaptation, we may turn to a particular fossil example among our ancestors: Neanderthal man. Evolution may take any number of courses, and there is still controversy among anthropologists about Neanderthal's place in the evolutionary scheme. Some observers regard him as a direct precursor of modern man, and therefore as a subspecies; among the proponents of this point of view are Bernard Campbell and C. Loring Brace. Others, like John Napier and the late Louis Leakey, regard Neanderthal as a separate line in hominid evolution that branched off from the species that eventually evolved into *Homo sapiens.* In this selection, Kurtén subscribes to the latter point of view. Though Neanderthal had a past, Kurtén implies that he had no future.

An important question is why Neanderthal disappeared between 35,000 and 40,000 years ago, but that remains unanswered. Kurtén uses data similar to those Pilbeam describes in the previous selection: dentition, the postcranial skeleton, and the brain. Neanderthal's dentition suggests that he was omnivorous. Hence, he must have been a hunter and was characterized by a social organization designed for this. Neanderthal's postcranial skeleton reveals a short but robust and powerful build, well adapted to a dense scrub forest habitat. Most important, the Neanderthal brain was larger than that of modern man—but the significance of this is not clear. In any event, as Kurtén notes, brain size alone is not the sole determinant of intellectual and cognitive ability.

A highly readable account that treats Neanderthal as a fellow *Homo sapiens* is offered in Chapter 8 of *The Emergence of Man,* by John E. Pfeiffer (New York: Harper & Row, 1972). One of the most dramatic finds in the evolutionary reconstruction of Neanderthal will be found in *Shanidar: The First Flower People,* by Ralph Solecki (New York: Knopf, 1971). ■

THE SCENIC NEANDER VALLEY near Dusseldorf in western Germany was named for the clergyman and hymn writer, Joachim Neander. The psalmist would have been astonished to learn what the name Neanderthal would mean to later generations. The sides of the valley are formed by limestone bluffs containing many caves, and in 1856 a quarryman discovered a number of fossil bones in one of the caves. The bones were sent to J. C. Fuhlrott, a high school teacher in nearby Elberfeld.

One can almost imagine Fuhlrott's surprise upon seeing those fossils. The skullcap was long and low with immense eyebrow ridges, and the thighbone was heavy and peculiarly crooked. Nothing like this had previously been known to science, although it was later found that a similar skull had been unearthed in a Gibraltar cave as early as 1848. But nobody had realized the importance of the Gibraltar find, whereas Fuhlrott was quite clear about the significance of the man from Neanderthal. And so 1856 became the year in which paleoanthropological science was born—three years before the appearance in print of Darwin's *Origin of Species.*

Fuhlrott found a collaborator in H. Schaafhausen, an evolutionist and anatomy professor in Bonn who saw the Neanderthaler as an extinct, primitive type of man.

The find was presented at a congress in Kassel in 1857, but, as usual, the discovery and the theories were too novel and foreboding to be accepted immediately. Instead, the Neanderthaler was explained away as a microcephalic idiot, a Cossack from the war of 1814, or an ancient Dutchman or Celt. Virchow regarded him as a pathological specimen, crippled by rickets and other maladies, beaten over the head to receive the unnatural swellings over the eyes, and finally suffering from gout in his old age.

Fuhlrott and Schaafhausen remained undaunted, and the victim of all these imaginary misfortunes, the Neanderthaler himself, soon became acceptable in anthropological circles. Although Virchow's authority dominated in Germany, C. Lyell from England visited the site with Fuhlrott and found evidence of the great antiquity of the find, and W. King described the fossil man as a distinct species, *Homo neanderthalensis.*

Then the Gibraltar skull was remembered and was duly presented at a congress in 1864. Two years later E. Dupont found a jaw fragment of a Neanderthaler in a cave at La Naulette near Dinant, Belgium. Other finds came from Bohemia, and in 1882 K. J. Maska discovered the jaw of a child in the Sipka cave in Moravia. Two Neanderthal skeletons were excavated by J. Fraipont and M. Lohest in definitely Pleistocene deposits in the cave at Spy d'Orneau near Namur. Since then many more discoveries have been made.

Gradually it could be shown that the Neanderthalers were comparatively late in time. Most date from the last, or Weichselian, glaciation, or more precisely, its early part up to about 35,000 years ago when this type of man apparently vanished. A few date back to the preceding warm phase, the Eemian interglacial.

But the Neanderthalers have antecedents in Europe, and it might even be said that there was a special European, or perhaps more properly northern Eurasian, line of evolution that got onto a somewhat different tack than that of widespread *Homo erectus* of the south and east.

The earliest known member of this line is Heidelberg man, whose jaw was found in 1907 in the Grafenrain sand pit at Mauer near Heidelberg. The site had long been studied by O. Schoetensack, who alerted the owner and workmen to the possibility of finding early human remains among the animal bones that were constantly being unearthed there. So nobody was surprised when the jaw came to light, and a notary public was called in to testify to the circumstances.

And its age? Fortunately there is no difficulty about that. The great mammalian fauna from Mauer, as well as the stratigraphy of the site, place it firmly in the Cromerian interglacial, meaning about half a million years ago—well within the time span of *Homo erectus* in Asia and Africa.

But the jaw differs in various ways from that of the typical *Homo erectus*. The teeth, for instance, are relatively small although the jaw itself is quite massive, with a very low, broad, and shallowly notched ascending branch. Such differences from *Homo erectus,* as well as Neanderthal-like characteristics, have been pointed out by many students, most recently by F. C. Howell.

Further proof of special traits in the Europeans of the middle Pleistocene emerged in the sixties when human remains were unearthed at Verteszollos in Hungary from deposits that are definitely datable as Elster interstadial—a slightly milder phase within the great glaciation that succeeded the Cromerian interglacial. These fossils probably are about 400,000 years old. In the same strata, lots of burned bones were found. These bones represent the earliest definite evidence known to date of the use of fire.

The most important specimen is the back part of a skull. It resembles *Homo erectus* in the development of big ridges for the neck muscles, indicating that this human must have been just as bull-necked as the contemporary Asiatic and African forms. But the angle formed by the roof and floor of the skull is much wider than in *Homo erectus,* indicating that the volume of the brain case was considerably larger. In fact, A. Thomas, the Hungarian anthropologist, estimates it at about 1400 cc., which is well above the range of *Homo erectus* and large even by present-day standards. So the deductions based on the Mauer jaw turned out to be right: these ancient Central European peoples were evolving in their own direction.

At the end of the Elsterian glaciation, conditions again became warmer, and we enter the Holsteinian interglacial, when the Acheulian hand-axe culture flourished in Europe. There are also flake-tool cultures of the so-called Clactonian type. In the thirties, the discoveries of Steinheim man by F. Berckhemer in Germany, and Swanscombe man by T. Marston in England, identified the creators of the Acheulian hand-axe culture. The age of these fossils is about 250,000 years. Both are associated with Acheulian hand axes much like those produced by *Homo erectus* in Africa, but the skulls themselves are very different from *Homo erectus.*

The Steinheim skull is well preserved except for some damage to the face and the loss of the lower jaw. Here, too, the base had been broken up, showing that the brain had been taken out. It was found in river gravels at Steinheim near Marbach in Württemberg, at the same level as the rich mammalian fauna of Holsteinian date.

The skull is a rather small, gracile head that probably belonged to a woman. The volume of the braincase is relatively small, some 1150 cc., but the head is rounded with a well-developed forehead. There is no trace of the constriction of the skull behind the eyes seen in *Homo erectus,* and although the eyebrow ridges are well developed, they are much smaller than in contemporary Peking man. The back of the head is also rounded and lacks the strong crests. Many details are fairly similar to those of modern man: for example, the shape of the cheekbone and that of the mastoid process just behind the ear as well as the high skull vault. Other details may point to Neanderthal.

The Swanscombe skull, found in river gravels of the Thames' 100-foot terrace and positively dated as Holsteinian in age, belonged to a somewhat larger individual and so had a larger braincase—it has been estimated at about 1300 cc. The variation found in the Vertesszollos, Swanscombe, and Steinheim skulls—1400 cc., 1300 cc., and 1150 cc. respectively—would be a normal one in a population of this type, with the Swanscombe value as an approximate average.

Unfortunately, the Swanscombe skull is far from complete. Originally, two bones were found, the left parietal and the occipital; the right parietal of the same skull was discovered after the war by J. Wymer. The resemblance to Steinheim man is striking, although the skull bones are somewhat thicker (this was presumably a male). As a detail, it may be noted that the ramification pattern of the meningeal artery, which may be studied from the impressions on the inside of the skull bones, is very advanced in comparison with that of *Homo erectus.* This artery supplies the brain with blood, and its complexity reflects that of the brain. The pattern seen in Swanscombe man is of a type encountered even in some modern men.

According to these finds, then, Holsteinian people in Europe had lost some of the ruggedness that characterized the head of Elsterian man, but continued to be relatively large-brained. There is little difficulty in deriving modern man from something like Steinheim-Swanscombe. But they may also be ancestral to Neanderthal man.

There is still quite a step in time from 200,000-250,000 years of the Holsteinians up to the early true Neanderthalers of the Eemian, some 100,000 years ago. It is bridged by a find from the cave of Montmaurin near Saint-Gaudens in France, which dates either from the end of the Holsteinian or from an interstadial in the Saalian glaciation. Its age may be roughly set at 150,000-

200,000 years. As H. V. Vallois has noted, the jaw would fit the Steinheim skull very well. At the same time it has some characteristics—such as the robust build and small teeth—that are reminiscent of the even older Mauer jaw.

Although physically quite unlike *Homo erectus,* these Europeans had the same type of culture, which again shows that the concepts of race and culture should never be confused.

The Acheulians of Swanscombe made their camps along the banks of the Thames, and finds of charcoal in the area are thought to have come from their hearths. At Hoxne (another site of the same age), there is evidence suggesting that the forest was deliberately fired. Pollen diagrams show a sudden expansion of grasslands at the same level as lumps of charcoal and Acheulian implements.

During most of the later Pleistocene Ice Age—the Eemian interglacial and the early part of the Weichselian glaciation, or roughly from 100,000 to 35,000 years B.P. (before the present)—Europe and neighboring areas were ruled by Neanderthalers. The "classical" or extreme Neanderthal man, exemplified by Fuhlrott's original find, turned out to be the typical European of the earlier Weichselian glaciation. There are many finds, mostly from cave deposits in Europe, for example, the complete skeleton from La Chapelle-aux-Saints in southwestern France. This was an old, almost toothless man, buried in sleeping position in a shallow grave in the cave floor with a bison leg to sustain him during his journey in the unknown.

The physical characteristics of these people were peculiar. They were short—men on an average were slightly over and women slightly under five feet tall—but they were exceedingly robust and powerful. The size of the head is remarkable. It is enormously long, much longer than the head of a normal modern man, and the skull looks low and narrow because of its length, although in fact it is just as high and broad as that of most modern men. From this it may be concluded that the braincase of the Neanderthalers was exceptionally large, and this is true even though the skull bone is very thick. The brain averages some 1450 cc., which is somewhat more than that of modern man. For his size, at any rate, the Neanderthaler had a bigger brain than we do.

Sheer size, however, does not necessarily indicate superior intelligence, for the structure of the brain is important too. Much larger than that of man, the elephant brain is one of the largest brains in existence. Yet clearly, many other animals are more intelligent than elephants.

In the Neanderthaler, the excess size falls mainly in the back part of the brain, whereas the most important brain centers for intelligence are in the fore part. The latter may have been somewhat more primitive than in modern man, but the true meaning of such differences is very difficult to assess. The level of intelligence varies greatly in modern man, and the same was probably true for Neanderthal man. The best might have been on a par with geniuses in modern times, given the same advantages. The Neanderthaler's world, however, was not one in which geniuses would have flourished.

Like the braincase, the face is big, with well-developed eye sockets and nasal cavities. The eye sockets are almost circular in shape, and the upper jaw seems inflated. The cheekbones and lower jaws are powerful and, like the temporal lines, indicate that the chewing muscles were very strong.

Because of the heaviness of the face, the neck muscles had to be strong in order to balance the head; thus we find familiar ridges in the back corner of the skull that indicate a bull neck. The hands and feet are big, but the arms and legs are rather short. Most long bones are slightly crooked, and the two bones of the forearm are more widely spaced than in ourselves. Obviously the Neanderthaler had great physical strength. His was a compact, robust body build, probably excellently suited for bursting forth through the dense scrub forest.

The special culture of the Neanderthaler is the Mousterian, a flake culture (i.e., a culture in which many implements were made from stone flakes) typical of a hunting people, and the animal bones left in the caves where he lived also testify to his hunting habits. At the same time, the very robust dentition may suggest that, like the robust Dartians, the Neanderthalers were dependent on vegetables for part of the year.

The Neanderthalers knew how to use fire—fossil hearths and burned bones are common among the remains. They were also cannibals at times. At Krapina in Yugoslavia, many Neanderthal bones have been found, some of them split open and charred. In a cave in Monte Circio near Rome, a Neanderthal skull was found with a broken-up base, suggesting that it may have been used as a ritual drinking vessel.

Neanderthalian table manners are revealed by a microscopic study of front teeth, discussed by F. E. Koby. The enamel is defaced by innumerable parallel scratches, which were produced when the food was stuffed into the mouth and the part that didn't fit was cut off with a flint knife. The knife would touch the teeth and leave a scratch. This phenomenon is also known among the Eskimos and other peoples. The direction of the scratches shows that the Neanderthalers were right-handed. A vivid picture is suddenly called forth, showing these savage cave-dwellers congregating in a hungry circle around a recent kill. There are cries and talk, the booming and flickering light of the fire, the sizzling of burning fat, and the smell of roasting, smoke, and drying skins.

All this deals mainly with the glacial Neanderthalers who lived in the earlier half of the last glaciation. Their predecessors in the Eemian interglacial are in some ways less extreme and somewhat reminiscent of the even earlier Steinheim-Swanscombe group.

In the area around Weimar in eastern Germany, there are thick deposits of a calcareous tufa called travertine, which was laid down by springs in interglacial times. Fossil bones are very common in the travertines, and among them are a number of human skull and jaw remains. Animals include red deer, roe deer, fallow deer, lynx, wild cat, straight-tusked elephant, and other typical members of the interglacial forest fauna of the Eemian.

Other finds come from the Eemian deposits at Saccopastore in Italy. Like the German fossils, Saccopastore man is less extreme than the later Neanderthalers. The skulls are more rounded, eyebrow ridges less protruding, and faces somewhat less inflated. A most remarkable pair of specimens comes from Eemian deposits in the cave of Fontéchevade in France. Although relatively flat-headed, Fontéchevade man has a steep and modern-looking forehead, and one of the two fragmentary skulls shows complete absence of the eyebrow ridge so typical of Neanderthal man. These skulls are obviously quite close to modern man.

All of these specimens of Eemian date are 80,000-100,000 years old and so indicate a carry over from the even older Holsteinians of Steinheim and Swanscombe. There seems to be a line of evolution leading by gradual changes from early immigrants of the Heidelberg type to the characteristic extreme Neanderthalers of the last glaciation. But at that point the line comes to an end. After about 35,000 B.P. the Neanderthalers are gone. There has been much speculation about their fate, but more about that later.

Remains of peoples closely allied to Neanderthal man are known in North Africa, the Levant, and eastwards through Iraq and Uzbekistan to Mongolia and China. The Chinese Neanderthaler is represented by a skull from Mapa in Kwantung and may be somewhat older than the Eemian, though not as old as Peking man. The Mongolian find, from Ordos, is just a tooth. In Uzbekistan, the skeleton of a buried child was dug up in the cave of Teshik Tash. The big Shanidar cave of northern Iraq has yielded several skeletons of Neanderthalers, adult as well as children. Some were found lying where they were crushed under rocks that had fallen from the roof.

In the caves of Mount Carmel, Shukbah, and Zuttiyeh in Israel, and at Ksar Akil in Lebanon, other Neanderthal-like men have been unearthed. These eastern Neanderthalers, though contemporary with those in Europe, are slightly less extreme on various accounts, such as the development of the eyebrow ridges—in the Europeans these tend to form a single bar across the whole face, in the eastern ones the ridges over the two eyes are separate.

IV.
ADAPTATIONS IN LOCAL POPULATIONS

ONE OF THE PUZZLES that have faced students of human beings, even before Darwin's revolutionary work, has been the fact that all humans may interbreed although they are clustered in groups with marked observable differences. Thus blond and blue-eyed northern Europeans may (and do) mate with dark-skinned people from central Africa and produce offspring, as do lanky coastal dwellers and barrel-chested alpine residents in Peru. It is an unfortunate characteristic of *Homo sapiens* to assume very often that differences in appearance entail mental and physical superiority and inferiority. The greatest tragedies that have ensued from such notions have, of course, been social; there is no need to document them here. But they have also had scientific consequences. It must be remembered that scientists are ordinary people who, by the time they choose their vocations, have been strongly affected by the ideas about social groups that are imposed on them in the course of growing up. This is especially important in connection with the study of race.

The concepts of natural selection, adaptation, and evolution suggest that the characteristics that distinguish different populations developed and became fixed for adaptive reasons, that is, to enable the members of these groups to survive in their habitats. More particularly, it may be assumed that those mutations that facilitated adaptation were selected for survival and transmission by habitational pressures, that the people who did not have those genetic predispositions died before they could reproduce. For this reason many students of evolution assume that, for instance, skin color—one of the most widely used criteria for racial classification—has adaptive value. But the test of the assumption has eluded scientists for many years. Could observable physical differences have come about by accident?

One of the most serious obstacles to an answer to this question, as suggested by Dobszhansky in Selection 18, is that scientists have tended to focus on the centers of populations, in which particular traits predominate, and have often lost sight of their edges, the points at which such characteristics as skin color vary from the local norm and start to approach the norms of other populations with different centers. These contrasting areas provide the most important clues in examining the adaptive value of physical traits in local populations, as seen in Weiner's analysis of pigmentation, stature, and body shape. (Selection 23).

Advances in the study of blood groups and serum proteins (discussed by Penrose in Selection 5) also contribute to an answer. Using the notion that a genetic trait may be both adaptive and maladaptive, it has become possible to inquire into the possible adaptive value of blood characteristics like the sickle cell trait, which now poses a serious medical problem for blacks in the United States and other western countries. Livingstone employs all these concepts and techniques in his investigation of the sickle cell trait in Selection 22.

In Selection 23, Weiner points to the significance of the need to adapt to pressures of heat and cold, evoking Campbell's analysis of adaptations to different climates (Selection 19). In Selection 24, Brown and Page discuss an example of specific adaptation to cold among the Eskimo that has long intrigued students of physical and cultural adaptation. One of the contributions of Katz and Foulks in Selection 26 is their demonstration that no physical adaptation can be understood by itself; instead physiological adaptations to habitational stress must be evaluated with an understanding of the cultural practices of the group in which it appears. This is also a theme in Baker's investigation of physiological adaptations to extremely high altitudes among the inhabitants of the Peruvian Andes, (Selection 25) and in Livingstone's paper on the sickle cell trait.

Together, these selections point to a still broader problem: the nature of health and illness. In Western societies, health is usually referred to as the absence of symptoms that indicate illness; conversely, the presence of such indicators are used to suggest that a person is ill. In recent years this concept has been questioned by a growing number of scientists. One of the difficulties in the conventional view of health and illness is that it does not take into account the pressures to which people must adapt; in this view, what indicates illness in one person or population may not be symptomatic of illness in another. Thus physicians who treat the same symptom in the same way in all people under all circumstances—representing the standards of mass production, as it were—may be negligent, to put it mildly.

The evaluation of what is often referred to as mental illness is as important as the evaluation of physical illness. So-called mental illnesses such as depression or schizophrenia may be adaptive means of coping with stress. There cannot be any doubt that genetic predispositions to schizophrenia, depression, hysteria, and the like, exist, and some forms of behavior such as those now loosely defined as schizophrenia may have been maladaptive in pre-modern societies, and there may have been a tendency for their bearers to be removed from breeding populations through the action of natural selection. This may no longer occur in modern industrial societies, since modern habitats are more highly protective of deviant individuals and medical means are available to control behavior that was maladaptive under earlier conditions.

As I suggested above, so-called mental illness may not be simply a psychological phenomenon, as it is usually regarded by the popular wisdom. Evidence is growing to support the idea that much of what is referred to as mental illness may have its source in nutritional deficiencies and other pressures on the individual that exacerbate genetic predispositions; the basis of one such relationship is illustrated by Katz and Foulks in Selection 26.

My speculations about mental illness are intended as provocations rather than as conclusions. I believe that physical anthropologists have an important contribution to make in this area, which traditionally has been regarded as the province of psychologically oriented cultural anthropologists. By carefully combining genetic study and investigations of nutritional standards and evaluating patterns of behavior in different societies, physical anthropologists may be more successful than other behavioral scientists in directing attention to the sources of aberrant behavior and, by implication, to its amelioration.

The reader who wants to explore further along some of the lines suggested in this essay can consult several works. In addition to the works of Theodosius Dobzhansky cited earlier, his concern with the relationship between biology and social questions is spelled out in *The Biology of Ultimate Concern* (New York: World Book, 1969) and *Genetic Diversity and Human Equality* (New York: Basic Books, 1973). A popular and stimulating discussion of the organization of the brain is *The Human Brain*, by John Pfeiffer (New York: Harper, 1955). Also of great relevance is the now classic book by N. Tinbergen, *The Study of Instinct* (Oxford: Clarendon Press, 1952). Many of the ideas about health adumbrated above owe much to the work of J. Ralph Audy. His most recent views, together with suggestions for future research, are succinctly spelled out in "Measurement and Diagnosis of Health" (in *Environ/Mental. Essays on the Planet as a Home*, edited by P. Shepard and D. McKnley, pp. 140-62 [Boston: Houghton-Mifflin, 1971]).

22. ANTHROPOLOGICAL IMPLICATIONS OF SICKLE CELL GENE DISTRIBUTION IN AFRICA

FRANK B. LIVINGSTONE

Reprinted from American Anthropologist, *Vol. 60, pp. 533-557. Copyright © 1958 by the American Anthropological Association. Frank B. Livingstone has spent his entire academic career at Michigan after obtaining a doctorate there in 1957. The present article is based on his dissertation. His main research interest has continued to be the abnormal hemoglobins and the action of selection on them. He is the author of a major work on the subject,* Abnormal Hemoglobins in Human Populations *(Chicago: Aldine, 1967).*

■ In Selection 4, Medawar discussed the sickle-cell trait as an example of a genetic predisposition that has both adaptive and maladaptive qualities. When inherited from only one parent, it provides a defense against malaria; when inherited from both parents, it is usually lethal because it results in "sickle-cell anemia." The sickle cell trait is confined to black populations; this has led many observers to assume that it evolved as a specialized adaptation to particular habitational conditions in Africa. Moreover, this trait may also be regarded as reflecting the human potential for diversity, discussed by Dobzhansky in Selection 18.

Livingstone analyzes the agent for natural selection of the sickle-cell trait. Since those who inherit the gene from both parents—who are homozygous for the trait—usually die early and before they reproduce, the gene might have disappeared from the population. However, since those who inherit the gene from only one parent—who are heterozygous for the trait—have increased resistance to malaria, the gene has remained in the population because it confers a selective advantage to this significant habitational pressure. We may therefore assume malaria operates as a selective agent and accounts for the prevalence and spread of the trait.

But this is a tautology that suggests, in effect, that the genetic trait persists because it persists. By plotting the geographical distribution of endemic forms of malaria and the distribution of the sickle-cell trait, Livingstone shows that malaria alone is insufficient to account for the persistence of the trait. He maintains that the principle of gene flow must also be considered as an operating influence. Drawing on the evidence provided by linguistic distribution, archeology, plant domestication, and culture history in general, Livingstone establishes a clear relationship between the sickle-cell gene and the spread of agriculture in West Africa.

Thus, this selection illustrates a complex set of adaptive relationships in which *Homo sapiens* is only one element. A genetic predisposition must be evaluated in terms of the particular habitational pressures in relation to which it emerges. Both malaria and the sickle-cell trait could remain relatively infrequent in pre-agricultural societies; with the advent of agriculture, however, both gained in frequency. Now that industrialization has spread into many parts of the world, the sickle-cell trait has been brought into this new social milieu, although what had been adaptive for many people in agricultural societies is clearly maladaptive in industrial societies.

A further analysis of the problems discussed in this paper is offered by Livingstone in his paper "Anthropological Implications of Sickle Cell Gene Distribution in West Africa" (*American Anthropologist,* 60 [1958]: 533-62). An important and well known paper in this field is "Natural Selection in Man: The ABO(H) Blood Group System," by John Buettner-Janusch (*American Anthropologist,* 61 [1959]: 437-56). A somewhat technical but highly rewarding paper is "Genetic Changes in Human Populations, Especially Those Due to Gene Flow and Genetic Draft," by Bentley Glass (*Advances in Genetics,* 6 [1954]: 95-139). ■

DURING THE PAST fifteen years, data on the frequency of the sickle cell gene have accumulated to such an extent that its world distribution can now be outlined in considerable detail. Frequencies of more than 20 percent of the sickle cell trait have been found in populations across a broad belt of tropical Africa from the Gambia to Mozambique. Similar high frequencies have been found in Greece, South Turkey, and India. At first it appeared that there were isolated "pockets" of high frequencies in India and Greece, but more recently the sickle cell gene has been found to be widely distributed in both countries. Moreover, between these countries where high frequencies are found, there are intermediate frequencies, in Sicily, Algeria, Tunisia, Yemen, Palestine, and Kuwait. Thus, the sickle cell gene is found in a large and rather continuous region of the Old World and in populations which have recently emigrated from this region, while it is almost completely absent from an even larger region of the Old World which stretches from Northern Europe to Australia.

When the broad outlines of the distribution of the sickle cell gene first began to emerge, several investigators attempted to explain various aspects of this distribution by migration and mixture. Lehmann and Raper attempted to show that the differences in the frequency of the sickle cell gene among the Bantu tribes of Uganda were due to varying degrees of Hamitic admixture; Brain and Lehmann postulated migrations from Asia to account for the distribution of the sickle cell gene in Africa; and Singer, using an age-area type of argument, postulated that the sickle cell gene arose by mutation near Mt. Ruwenzori and diffused from there. However, it was recognized early in the development of the sickle cell problem that regardless of the extent to which migration and mixture explained the distribution pattern of the sickle cell gene, its high frequencies in various widely scattered areas raised some additional and striking problems in human population genetics.

Since persons who are homozygous for the sickle cell gene very rarely reproduce, there is a constant loss of sickle cell genes in each generation. In order for the gene to attain frequencies of .1 to .2, which are equivalent to about 20 to 40 percent of the sickle cell trait, there must be some mechanism which is compensating for this loss. In other words, there must be some factor which is tending to increase the number of sickle cell genes in the population. Neel first pointed out that there are two outstanding possibilities; either the sickle cell gene is arising frequently by mutation, or the heterozygote for the sickle cell gene possesses a selective advantage over the normal homozygote which offsets the selective disadvantage of the sickle cell homozygote (balanced polymorphism). Since the evidence indicated that the mutation rate was not sufficient to maintain the high frequencies, selection in favor of individuals with the sickle cell trait seemed to be implicated as the factor which was maintaining them.

When Allison advanced the hypothesis that the heterozygote for the sickle cell gene possessed a relative immunity to falciparum malaria, he marshalled the first clear evidence for the mechanism by which selection maintained the observed high frequencies. In addition to experiments on sicklers and nonsicklers which seemed to show that the sicklers could cope more easily with a malarial infection, Allison also showed that the tribal frequencies of the sickle cell gene in Uganda and other parts of East Africa could be explained as well by his malaria hypothesis as by varying degrees of Hamitic admixture. Thus, Allison's work showed that selection must be taken into consideration in any attempt to explain the distribution of the sickle cell gene.

Although selection has undoubtedly played a major role in determining the frequencies of the sickle cell gene in the populations of the world, in many areas other factors in addition to selection may well be involved. Allison has shown that most of the tribes of East Africa seem to have frequencies of the sickle cell trait which are in approximate equilibrium with the amount of malaria present, but there appear to be many populations in West Africa and elsewhere for which this is not so. It will be the purpose of this paper to show how the distribution of the sickle cell gene in West Africa is the result of the interaction of two factors, selection and gene flow. Gene flow will be used here to include both migration and mixture; the term migration is used where the gene flow involves the movement of breeding populations or large segments of them, and mixture where the breeding populations remain rather stationary and the gene flow involves the exchange of individuals between them. Of course, any actual situation is usually a combination of these two "polar" concepts.

According to modern genetic theory as developed by Wright and others, there are five factors which can contribute to gene frequency change: selection, mutation, gene drift, gene flow, and selective mating. Strictly speaking, an attempt to explain the distribution of any gene must take into consideration all five. However, three of these factors —mutation, gene drift, and selective mating —are thought to have had relatively little effect on the features of the distribution of the sickle cell gene in West Africa which this paper will attempt to explain, and thus will not be discussed at any length in this paper.

The general plan of the paper will be as follows. First, the distribution of the sickle cell gene in West Africa will be plotted; then an attempt will be made to correlate this distribution with that of falciparum malaria in West Africa. It will be assumed that the high frequencies of the sickle cell gene are in equilibrium with the particular endemicity of malaria in which they are found. Thus, by comparing these two distributions we can determine where the frequencies of the sickle cell gene appear to be explained by selection (i.e., are in equilibrium), and we can also determine where the frequencies appear to be very far from equilibrium and hence where other factors in addition to selection appear to be involved. The rest of the paper will then be concerned with the populations which do not appear to be in equilibrium. In order to explain why the frequencies of the sickle cell gene in these populations are not in equilibrium with the present-day endemicity of malaria, it is necessary to have some idea of the ethnic and culture history of West Africa. The literature on the culture history of West Africa is rather sparse, so the major part of this paper will be an attempt to infer its broad outlines from the distribution of language and of certain domesticated plants in West Africa.

THE DISTRIBUTION OF THE SICKLE CELL GENE IN WEST AFRICA

In the following compilation of data on the distribution of the sickle cell gene in West Africa, several early publications of surveys have been omitted. In all of these reports, the tribe of the persons tested is not given, and the reports could thus contain subjects from several breeding populations with very different frequencies of the sickle cell gene. Data by tribe are available for the areas covered by these surveys, except for part of Evans' survey. His sample from the Cameroons has been included since there are no other data from this area.

Where the same tribe has been tested by different investigators, differences in the frequency of the sickle cell trait have been tested by a chi-square test. If the differences were not significant, the results have been combined. However, for several large tribes which extend over considerable distances and into several different countries, the samples have been kept separate when they were obtained in different countries.

For the surveys in which paper electrophoresis or other biochemical tests were done on the bloods, all individuals who would have been positive for the sickle cell test were counted as positive without regard to whether they appeared to be homozygous or heterozygous for the sickle cell gene. Thus, the frequency of the sickle cell trait, as used in this paper, includes both heterozygotes and any living homozygotes for the sickle cell gene. However, recent studies indicate that homozygotes for the sickle cell gene rarely survive the first years of life, so that most likely very few homozygotes are included in the tribal samples. Throughout the discussion, sickle cell trait frequencies will be used instead of gene frequencies, since the trait frequencies are used by most investigators and hence their significance is more easily comprehended. Since very few homozygotes are included in the samples, the gene frequency would be close to one-half the trait frequency in all cases.

Except for the Ivory Coast, Dahomey, and the Cameroons, the compilation is by tribe. The Dahomey and Cameroons samples have been included in an effort to fill up large gaps in the distribution in areas where tribal investigations are nonexistent. These samples have combined several tribes and thus have probably combined data from isolates which differ significantly from one another in the frequency of the sickle cell trait. Since they are also quite small samples, this paper will not consider them in detail.

Due to the lack of investigations, and also to the multiplicity of small tribes which inhabit the Ivory Coast, the tribal samples from there are all rather small. Since the frequency of the sickle cell trait is 0 percent in Liberia to the west of the Ivory Coast and greater than 20 percent in Ghana to the east, the Ivory Coast is an area of crucial concern to this study. For this reason, the tribal samples have been combined into larger linguistic units to increase the sample sizes and thus give them more reliability. The tribes which have been combined are very closely related, since in most cases they speak the same language with only dialectic differences between them. Although the individual tribal samples are small, there is no indication that this procedure has combined tribes which have very different frequencies of the sickle cell trait.

Table 22.1 shows the frequency of the sickle cell trait for West Africa by tribe and also by country. For the purposes of further discussion, the spelling of all tribal names has been standardized. In Table 22.1 the

TABLE 22.1

THE FREQUENCIES OF THE SICKLE CELL TRAIT IN THE TRIBES OF WEST AFRICA

Country Tribe	*Investigations*	*Number Examined*	*Number Positive*	*Sickle Cell Trait (%)*
Senegal				
Wolof (Oulof)	16, 18	2277	151	6.63
Lebu (Lebou)	16, 18	522	31	5.94
Serer	16, 18	1515	50	3.30
Soce	16	70	11	15.71
Fulani (Peul)	16, 18	299	27	9.03
Tukulor (Toucouleur)	16, 18	634	60	9.46
Dyola	18, 19	39	2	5.13
Mandiago	16, 18, 19	101	1	0.99
Gambia				
Mandingo-Western Division	2	167	18	10.78
Mandingo-Keneba	2	240	15	6.25
Mandingo-Jali	2	115	7	6.09
Mandingo-Manduar	2	59	10	16.95
Mandingo-Tankular	2	132	32	24.24
Fulani (Fula)	2	127	24	18.90
Dyola (Jola)	2	312	53	16.99
Wolof (Jolloff)	2	104	18	17.31
Saracole (Serahuli)	2	96	8	8.33
Bainunka	2	90	15	16.67

TABLE 22.1 (Continued)
THE FREQUENCIES OF THE SICKLE CELL TRAIT IN
THE TRIBES OF WEST AFRICA

Country Tribe	*Investigations*	*Number Examined*	*Number Positive*	*Sickle Cell Trait (%)*
Portuguese Guinea				
Papel	24	500	15	3.00
Mandiago (Mandjaca)	15	500	16	3.20
Balante (Balanta)	15	500	25	5.00
Feloop (Felupe)	15	466	6	1.72
Baiote	15	473	6	1.27
Nalu	15	501	14	2.79
Saracole	15	286	24	8.39
Mandingo (Mandinga)	15	500	75	15.00
Biafada (Beafada)	15	505	77	15.25
Pajadinca	15	358	66	18.44
Fulani (Fula-Foro)	15	500	115	23.00
Fulani (Fula-Preto)	15	430	108	25.12
French Guinea				
Fulani (Foula)	15, 16, 18	682	109	15.98
Susu	5, 18, 19	48	15	31.25
Kissi	19	18	4	22.22
Loma-Kpelle (Toma-Guerze)	19	40	8	20.00
Sierra Leone				
Creole	2	42	10	23.81
Timne	2	52	15	28.95
Mende	2, 23	1124	330	29.36
Liberia				
Kissi	17	298	58	19.46
Mende	17	77	13	16.88
Gbandi	17	352	54	15.34
Vai	17	93	13	13.98
Kpelle	17	982	128	13.03
Loma	17	511	65	12.72
Gola	17	183	22	12.02
Belle	17	29	3	10.34
Bassa	17	811	58	7.15
Dei	17	53	2	3.77
Mano	17	709	15	2.12
Gio	17	428	9	2.10
Grebo	17	69	1	1.45
Krahn	17	154	1	0.65
Kru	17	148	1	0.68
Webbo	17	77	0	0.00
French Sudan				
Moor (Maure)	18, 19	70	4	5.71
Saracole	16, 18	196	16	9.18
Bambara	16, 18	262	27	10.31
Mandingo (Malinke)	18, 19	50	8	16.00
Fulani (Peul)	20	152	22	14.47
Songhoi	20	100	11	11.00

TABLE 22.1 (Continued)

THE FREQUENCIES OF THE SICKLE CELL TRAIT IN THE TRIBES OF WEST AFRICA

Country Tribe	*Investigations*	*Number Examined*	*Number Positive*	*Sickle Cell Trait (%)*
Ivory Coast				
Senufo	5, 19	33	8	24.24
Agni-Baule	5, 19	53	7	13.21
Dan-Gnouro	5, 19	30	0	0.00
Lagoon	5, 19	48	2	4.17
Bete	5, 19	53	1	.89
Bakwe	5, 19	63	1	1.59
Upper Volta				
Samogo	20	120	8	6.67
Bobofing	5, 19, 22	232	57	24.57
Lobi	5, 19	15	3	20.00
Mossi	5, 19, 20	207	24	11.59
Gurma	5, 19, 20	34	3	8.82
Gurunsi	5, 19	14	1	7.14
Ghana				
Mossi (Moshie)	10, 11	121	5	4.13
Dagarti	11	97	11	11.34
Dagomba	11	71	3	4.23
Ewe	2, 10, 11	232	54	23.28
Fanti	2, 10, 11	204	48	23.53
Ga	2, 10, 11	367	67	18.26
Twi	2, 10	111	24	21.62
Ashanti	2	102	23	22.55
Frafra	9	680	66	9.71
French Togoland				
Kabre	4	1104	109	9.87
Dahomey				
Dahomeans	5, 19	55	5	9.09
Niger				
Djerma (Zabrama)	2, 19	69	15	21.74
Tuareg	3	93	5	5.38
Nigeria				
Yoruba	7, 13, 25	3477	853	24.53
Igalla	25	155	28	18.06
Ibo	2	51	11	21.57
Cameroons	12	138	21	15.22
Kerikeri	14	159	17	10.69
Fulani	14	184	31	16.85
Hausa	1, 6, 14	611	107	17.51
Lake Chad				
Mohur (Mobeur)	21	273	49	17.95
Kanembu (Kanembou)	21	76	17	22.37
Mangawa	21	58	12	20.69
Sugurti (Sougourti)	21	37	6	16.22

names used by the original investigators are shown in parentheses after the standardized name.

The distribution of the frequency of the sickle cell trait in West Africa is shown on Figure 1 [not reproduced]. In order to make the general configuration of the distribution more easily visualized, the frequencies have been grouped into five categories: 0-2, 2-8, 8-15, 15-22, and greater than 22 percent. The frequency of the sickle cell trait can be seen to exhibit extreme variability, sometimes over very short distances. In many cases there are significant differences in the frequency of the trait even within the same tribe. For example, the Fulani have frequencies ranging from 8 to 25 percent, and the Mandingo in the Gambia vary from 6 to 28 percent. Although this great variability impedes generalizing about the distribution, some significant generalizations can nevertheless be made.

Generally, the higher frequencies tend to be toward the south, and, despite many exceptions, there is some indication of a north-south gradient in the frequency of the sickle cell trait. The distribution of falciparum malaria follows a similar gradient, and, in addition, all the populations which have sickle cell trait frequencies greater than 15 percent inhabit areas where malaria is either hyperendemic or holoendemic.

In an environment in which malaria is hyperendemic or holoendemic, the disease is transmitted throughout most of the year, so that the individuals are continually being reinfected. The average number of infective bites per person per year is always greater than about 5, and in some areas ranges up to 100 or more. Thus, infants are infected with malaria shortly after birth, and for about the first five years of life every child is engaged in a mortal struggle with the parasite. During these years the parasite rate (i.e., the percentage of individuals harboring malaria parasites) is close to 100 percent, and there is a considerable mortality from the disease. Those individuals who survive this struggle have a solid immunity to malaria. In later years they are being continually reinfected with malaria but are able to keep their infection at a subclinical level. The parasite rate then decreases among older children and is lowest in adults. In holoendemic malaria the adult parasite rate will be about 20 percent and the adults will almost never have any clinical symptoms of malaria, while in hyperendemic malaria the adult parasite rate will be somewhat higher and the adults will sometimes have clinical symptoms, usually chills and fever. However, in both these conditions there is seldom any adult mortality from malaria.

It is an environment in which malaria is either hyperendemic or holoendemic that the heterozygote for the sickle cell gene has been postulated to have a selective advantage over the normal homozygote. Allison and Raper have shown that, although sicklers are infected with falciparum malaria almost as readily as nonsicklers, in the younger age groups the very high densities of parasites are not found as often among sicklers. In addition, Raper has shown that the sicklers do not suffer from cerebral malaria and blackwater fever as much as nonsicklers. Since these are the complications of falciparum malaria which result in death, the sicklers had a lower mortality rate from falciparum malaria. In addition, I have postulated that if the sickling females did not have as heavy falciparum infections of the placenta as did normal females, they would have a higher net reproduction rate and hence this could be another mechanism by which malaria was maintaining the high frequencies of the sickle cell gene. Although the evidence is not conclusive, it seems for the most part favorable to this hypothesis. When the evidence for both these mechanisms is considered as a whole, it seems to be conclusive that malaria is the major cause of the high frequencies of the sickle cell gene. One would therefore expect to find high frequencies of the sickle cell trait in areas in which malaria is either hyperendemic or holoendemic.

From about the latitude of the Gambia south, West Africa is almost entirely characterized by hyperendemic or holoendemic malaria; hence, high frequencies of the sickle cell trait would be expected. However, there are many populations in this region with very low frequencies of the trait. The majority of them are found in three areas: (1) Coastal Portuguese Guinea, (2) Eastern Liberia and the Western Ivory Coast, (3) Northern Ghana. The low frequency population which are found in Northern Ghana differ from those in the other two areas by having high frequencies of the gene which is responsible for Hemoglobin C. This gene is an allele of the sickle cell gene, so that in Northern Ghana the sickle cell locus is a tri-allelic system. Since the selective values associated with the various phenotypes of this system are not known at present, the equilibrium frequencies for these populations cannot be ascertained. Thus, one cannot say whether or not these populations are in equilibrium for this locus. The rest of this paper will therefore be concerned with the two areas, Coastal Portuguese Guinea and Eastern Liberia-Western Ivory Coast, where the Hemoglobin C gene is almost completely absent.

Cambournac in Coastal Portuguese Guinea and Young and Johnson in Eastern Liberia found malaria to be either hyperendemic or holoendemic in these areas where low frequencies of the sickle cell trait are found. Thus, these frequencies appear to be very far from equilibrium, and hence do not seem to be explained by the factor of selection alone. An attempt will not be made to show how the explanation involves the two factors, selection and gene flow. More specifically, two hypotheses will be advanced to explain these low frequencies:

(1) The sickle cell gene has been present in some parts of West Africa for a considerable time, but, due to the comparative isolation of the low frequency populations in Portuguese Guinea and Eastern Liberia, is only now being introduced to them.

(2) The environmental conditions responsible for the high frequencies of the sickle cell gene have been present for a relatively short time among these populations, so that the spread of the sickle cell gene is only now following the spread of the selective advantage of the gene.

In order to demonstrate these propositions, two general types of evidence will now be considered; first, the distribution of language in West Africa, from which an attempt will be made to ascertain the general outlines of the migrations which have occurred there; then, the archeological evidence and the distributions of certain domesticated plants in West Africa, from which an attempt will be made to determine the broad outlines of the culture history of the area. From a consideration of the culture history of West Africa and the relationship between culture patterns and the endemicity of malaria, the spread of the selective advantage of the sickle cell gene will be inferred.

THE DISTRIBUTION OF LANGUAGE IN WEST AFRICA

In the following discussion Greenberg's classification of African languages will be used, since it is the most recent and also the most widely accepted. In addition, Greenberg is attempting to make a "genetic" classification of African languages. Languages are said to be genetically related when their similarities are due to their development from a common ancestral language. It is this type of linguistic relationship which is most likely to have biological significance, since the ancestors of the speakers of genetically related languages were probably once members of the same breeding population and thus biologically related. Greenberg's classification is concerned with the larger linguistic families of Africa and the larger subgroupings within these families. Since it will be necessary at times to separate the languages into smaller sub-groups, other sources will be used, but only when these agree with Greenberg's overall classification.

Except for the Songhai, Hausa, Kerikeri, Tuareg, Moor, and the tribes around Lake Chad, all the tribes listed on Table 22.1 speak languages belonging to the Niger-Congo family. The exceptions noted above speak either Songhai, Central Saharan, or Afro-Asiatic languages. These tribes are in the northern and eastern parts of West Africa and a considerable distance from the two low frequency areas of the sickle cell trait with which we are concerned. Therefore, this discussion will be concerned only with the Niger-Congo languages. . . .

The Niger-Congo family contains seven subfamilies: (1) West Atlantic, (2) Mande, (3) Gur, (4) Kwa, (5) Central Group, (6) Ijo, (7) Adamawa-Eastern. All of these subfamilies have some member languages in West Africa, but, with the exception of the Adamawa-Eastern speakers in northern Central Africa, the Niger-Congo languages in Central, East, and South Africa all belong to a single subfamily (Central Group) and even to a single subgroup (Bantu) within that subfamily.

Because of the great linguistic diversity in West Africa, this area appears to have been inhabited for a relatively long time by speakers of Niger-Congo. On the other hand, because of the similarity of language in the area inhabited by the Bantu peoples, this area has undoubtedly been peopled by a relatively recent spread of those peoples. As Greenberg states:

> If the view of the position of the Bantu languages presented here is accepted, there are certain historical conclusions of considerable significance which follow. When Sapir demonstrated that the Algonkian languages were related to the Wiyot and Yurok languages of California, it was clear that, if this demonstration was accepted, it constituted a powerful argument for the movement of the Algonkian-speaking peoples from the west to the east. Here we have not two languages, but twenty-three peoples from the west to the east. Here we have not two languages, but twenty-three separate stocks all in the same general area of Nigeria and the Cameroons. The evidence thus becomes strong for the movement of the Bantu-speaking peoples from this area southeastwards. The usual assumption has been a movement directly south from the great lake region of East Africa. It will also follow that this is a relatively recent movement, a conclusion which has generally been accepted on the basis of the wide extension of the Bantu languages and the relatively small differentiation among them.

In discussing the archeological and ethnological evidence, an attempt will be made to give reasons for the relatively recent spread of the Bantu from Nigeria, as well as to show that this other evidence seems to support the linguistic evidence.

In West Africa west of Nigeria, there are four subfamilies of Niger-Congo: West Atlantic, Mande, Gur, and Kwa. With the exception of the rather recent movement of the Fulani pastoralists across the entire length of West Africa, the West Atlantic languages are all located along the coastal fringe of West Africa. The Kwa languages are distributed along the Guinea Coast from Liberia to Central Nigeria, with the great majority of them located in the tropical rain forest. In the central part of West Africa, in two large blocks, are the Mande languages on the west and the Gur languages on the east. These languages are for the most part located in the sudan, although several Mande groups have penetrated the tropical rain forest in Sierra Leone, Liberia, and the Ivory Coast.

The tribes with low frequencies of the sickle cell trait in Portuguese Guinea speak West Atlantic languages, but some Mandingo groups in the Gambia, who speak a Mande language, also have relatively low frequencies. In Eastern Liberia and the Western Ivory Coast, the tribes with low sickling frequencies include speakers of Kwa and Mande languages. Thus, with the exception of Gur, all these subfamilies include some languages whose speakers are far from equilibrium with respect to the sickle cell gene. Since these subfamilies also include some languages whose speakers have high frequencies of the sickle cell trait and seem to be close to equilibrium, the frequency of the trait is not correlated with language. This seems to indicate that the gene has been introduced into this part of West Africa since these subfamilies of Niger-Congo began to separate. However, since there is considerable linguistic diversity within the subfamilies, their separation occurred long ago.

Although there is no correlation of the frequency of the sickle cell trait with the linguistic subfamilies in this part of West Africa, the tribes with low frequencies in both Portuguese Guinea and Eastern Liberia seem to be the indigenous inhabitants of West Africa who have been forced back into these areas by later migrants from the east. The distribution of the West Atlantic languages along the coast with some isolated pockets in the interior indicates that the speakers of these languages were once more widespread and have been forced back to the coast by more recent invaders. This retreat of the West Atlantic speakers is documented to some extent, and there is general agreement that the general trend of migration has been toward the west. Of course, the West Atlantic peoples probably occupied the coastal regions at an early time also, but their present concentration there results from their displacement from a wider area by invaders from the east.

Several authorities state that the Baga, who now inhabit the coastal regions of French Guinea, originally inhabited the Futa Djallon, which is the highland area of Central French Guinea. The Baga were forced out of there by the Susu, who were in turn forced out by the Fulani. This forcing back of the West Atlantic speakers was also noted by Beranger-Ferand in the Casamance River area of the French Senegal. He divides the populations of this region into three groups:

A. Peuplades primitives (Feloupes, Bagnouns).
B. Peuplades envahissantes (Balantes, Mandingues, Peuls).
C. Peuplades adventives (Ouolofs, Saracoles, Toucouleur, Mandiagos, Machouins, Taumas, Vachelons).

He then states that A are the indigenous inhabitants; B are the fighters who conquered; and C are the traders or farmers who infiltrated in small groups. In Gambia the same migrations have been noted by Southorn and Reeve. Reeve states:

> The only relics that are to be found today of the primitive negro race which originally occupied the forest belt between the Senegal and the Rio Grande are the Serreres on the coast, north of the Saloum River, who are pagans and were cannibals; the Feloops, Floops, or Flups, as called by early voyagers, but now, in the valley of the Gambia, known as the Jolahs, occupying the territory between the sea-coast and the headwaters of the Vintang Creek, about one hundred miles inland; the Patcharis or Pakaris in the Middle Valley, and the Bassaris including the Kunyadis, in the Upper Valley. These will be again referred to, and it is evident, from the chronicles of the different writers on the subject of slavery in this part of West Africa, that it was these Arcadians and forest dwellers, with their simple manners and customs of sustaining life from the products of the forest, field, and streams, who supplied the bulk of the trade, under the pretext that they worshipped idols, and therefore were considered to be outside the pale of humanity by the races that had adopted the Koran.

Thus, it can be seen that these writers agree that the Feloops, who have one of the lowest frequencies of the sickle cell trait, are one of the indigenous tribes. In addition, Reeve states that the Serer, who also have a low frequency of the trait, are the indigenous inhabitants in the north and in the past were hunters and gatherers and not agriculturalists. It should also be noted that Leite and Ré, who tested the tribes of the Portuguese Guinea for sickling, give a similar explanation for the differences in the frequency of the sickle cell trait which they found.

The tribes with low sickling frequencies in Eastern Liberia and the Western Ivory Coast include speakers of Mande languages and of Kwa languages. All of the speakers of Kru and Lagoon languages, which belong to the Kwa subfamily, have very low frequencies of the sickle cell trait, and the positives for the trait who do occur among these peoples are in the eastern tribes where they are in contact with the Agni, Baoule, and other Akan speakers. On the other hand, the Kwa speakers who are to the east of the Kru and Lagoon peoples all have relatively high frequencies of the sickle cell trait. Viard states that the Guere, who speak a Kru language, came from the east, and Yenou makes a similar statement for the Alladians, who speak a Lagoon language. Since the linguistic relationships point to the east, these statements are probably true. Much has been written about the migrations of the Akan, Ewe, Ga, and other Kwa speakers who are to the east of the Kru and Lagoon speakers, and most authorities agree that the general direction of migration of these tribes has been to the southwest. Since the Lagoon languages are quite similar to the Togo Remnant languages, it seems that the speakers of these languages were forced back into peripheral areas by the Akan peoples (i.e., Ashanti, Fanti, Agni, Baoule), when they migrated to southern Ghana. The movement of the Agni and Baoule into the Ivory Coast is quite recent—17th century according to most authorities. Thus, it seems that some Kwa speakers were more widespread through the tropical rain forest when the later Kwa migrants entered it and were then forced back by these later migrants. Since the later migrants have high frequencies of the sickle cell trait, it appears that they introduced the sickle cell gene into this part of West Africa.

In addition to the Kru and Lagoon-speaking peoples, there are several tribes with low sickling frequencies who speak Mande languages in Eastern Liberia and the western Ivory Coast. These are the Mano, Gio, Dan, Gouro, and other smaller groups. At the border between the Mano and the Kpelle, the frequency of the sickle cell trait increases sharply. Although these peoples both speak Mande languages, they belong to different subgroups of the Mande subfamily. Kpelle is related to Mende and Susu to the northwest in Sierra Leone, and this tribe has undoubtedly come into Liberia from that direction. However, Mano and the other Mande languages whose speakers have low frequencies of the sickle cell trait are related to several Mande languages in the Upper Volta Province of French West Africa and also to a Mande language in Nigeria. Vendeix states that the Dan, and Tauxier that the Gouro, came into their present habitats from the northeast. Donner states that the Dan came from the north into the forest and forced the Kru peoples ahead of them. It would thus appear that these Mande tribes with low sickling frequencies came into their present location by a different route than that of their Mande neighbors to the northwest in Liberia and Sierra Leone. The Bobofing, who speak a language related to these Mande languages whose speakers have low sickling frequencies, have 25 percent of the sickle cell trait and are some distance to the northeast of the Dan and Gouro; so that it seems that the sickle cell gene was introduced after the separation of these languages. The Mandingo are to the north of the Mano, Dan, and Gouro, and between them and the Bobofing. From the 12th to 15th centuries A.D. when the Mali Empire, which was ruled by the Mandingos, was at its height, these people are known to have expanded out from their original homeland. It would appear that this expansion of the Mandingo forced the Mano, Dan, and Gouro into the forest and separated them from their relatives to the northeast.

The two areas of low frequencies of the sickle cell trait thus seem to be inhabited by peoples who have been forced back into these peripheral areas by later migrants from the east and northeast. However, this does not mean to imply that all the later migrants had the sickle cell gene. It is possible that the Kwa migrants to Southern Ghana introduced the gene into this part of West Africa by migration; but along the West Atlantic coastal fringe, the sickle cell gene seems to have spread in the past by mixture, and is still spreading in this manner today.

In the Central Ivory Coast on the border of the Kru and Lagoon peoples of the west and the Akan peoples on the east, there is a sharp increase in the frequency of the sickle cell trait. Since all the Kwa peoples from the Akan east to the Ibo in Nigeria have very high frequencies of the trait, it seems that these peoples possessed the sickle cell gene when they migrated into these regions from the east and northeast. However, along the Atlantic Coast of West Africa from the Senegal to central Liberia, the gene does not seem to have been introduced by large-scale migration. The highest frequencies of the sickle cell trait in this region are found in the Gambia and in Sierra Leone, which are also the two places where Mande peoples have penetrated to the seacoast in large numbers. Since the Mande peoples were the migrants from the east, it would appear that they introduced the sickle cell gene into this part of Africa. However, the smooth gradient in the frequency of the trait in Sierra Leone and Liberia seems to indicate that the gene was introduced after the original Mande migrations. Starting with the Susu in northwest Sierra Leone who have a sickling frequency of 31 percent and proceeding southeastward, there is a smooth gradient in frequency which is not correlated with language. The speakers of Southwest Mande-fu languages, the Mende in Sierra Leone, the Mende in Liberia, the Gbandi, Loma, and Kpelle, have 29, 17, 15, 13, and 13 percent, respectively, while the West Atlantic speakers, the Timne, Kissi, and Gola, have 29, 19, and 12 percent respectively. The Vai, who speak a Mande-tan language and are the latest immigrants from the interior, have a frequency of 14 percent, which is also in agreement with this gradient. In Portuguese Guinea, where the Mande peoples have not penetrated in great numbers, there is also a smooth gradient in the frequency of the sickle cell trait. Starting on a small section of the seacoast between the Casamance River and the Rio Cacheu where the Feloop and Baiote have 1 to 2 percent, the frequency increases going inland to 5 percent among the Mandjak, and then to 15 percent among the Biafada and Mandingo. It thus seems that along the West Atlantic coastal fringe of West Africa the sickle cell gene has spread and is still spreading by mixture and not by large scale migration, while the gene appears to have spread through the tropical rain forest along the Guinea Coast by the migration of the Akan and other Kwa-speaking migrants from the east. The archeology and culture history of West Africa will now be examined in an attempt to provide some explanation for the manner by which the sickle cell gene has spread there.

THE ARCHEOLOGY OF WEST AFRICA AND ORIGIN OF THE WEST AFRICAN NEGRO

Although there has been less archeological excavation in West Africa than elsewhere in Africa, it is now beginning to appear that West Africa was inhabited during most of man's cultural development, as was most of the continent. Lower Paleolithic hand axes and Middle Paleolithic Levallois flakes have been found in scattered places throughout West Africa. However, no rich sites comparable to those in East and South Africa have been excavated for these stages. Nevertheless, the scattered finds indicate the presence of man in West Africa during these periods, which lasted up to the end of the Pleistocene. Following these periods of time, microlithic sites are documented for Ghana, French Guinea, Nigeria, and other places in West Africa. Some of these microlithic cultures seem fairly recent and perhaps attributable to the ancestors of the present Negro inhabitants. However, little skeletal material has been found.

The earliest skeletal material which is found close to West Africa is a skull from Singa in the Sudan. This find has been dated by Arkell as Upper Pleistocene and is associated with a Levallois culture. The skull is stated to be archaic Bushman and related to the Boskop skull from South Africa. From this find it appears that the Bushman was once much more widespread than today and in Upper Pleistocene times Bushman-like peoples were in the Sudan. This statement is supported by the presence of Bushman-like rock paintings and archeological cultures similar to that of the present day Bushman over most of the southern half of the African continent. The presence today in Tanganyika of the Hatsa, who speak a Khoisan language and still have a predominantly Stone Age culture, also supports it.

The first appearance of skeletal material which has Negroid affinities is in this same area of the Sudan, but apparently much later. At Esh Shaheinab, which is on the Nile near Khartoum, several skeletons with Negroid affinities have been found in association with a microlithic hunting and gathering culture, which also had pottery. Around the fringes of the Sahara there are other finds of Negroid skeletal material, all of which seem to belong to this general period. The famous Asselar skull from north of Timbuktu, which is considered to be Negro, is from this general period, and Alimen also indicates that some of the skeletal material associated with the Capsian culture in Tunisia has Negroid affinities. In addition to this skeletal material, many of the early rock paintings in the Sahara seem to depict Negroid peoples.

The Esh Shaheinab site has been dated by radiocarbon as 5200 years ago, or shortly after the beginnings of agriculture in Egypt. The radiocarbon dates on the Capsian culture are about 7500 years ago. Alimen indicates that the Neolithic of Capsian tradition is found in French Guinea, but this is probably much later than the Capsian sites which have been dated by radio-carbon. It should also be noted that in this context Neolithic does not mean food-producing, but only that the culture had polished stone artifacts.

The first archeological evidence of the Bantu in South and Central Africa is much later than the evidence from northern West Africa, and appears to be after the beginning of the Christian era. Alimen states: "Iron entered the Congo very late, by means of the Bantu invasion, which later spread to the Rhodesias in only 900 A.D." Further, Alimen states that ironworking came to the upper valley of the Orange River in the 13th century A.D. and here too is associated with the arrival of the Bantu. Previous to the expansion of the Bantu, East and South Africa were inhabited by Bushman-like peoples.

The archeological evidence thus seems to indicate that at about the time of the introduction of agriculture into Africa, Negro peoples with a microlithic culture were living around the fringes and even in the middle of the Sahara, while most of South and East Africa was inhabited by Bushman-like peoples. Since the Pygmies would seem to be indigenous to Central Africa, they were perhaps responsible for the microlithic cultures found there. For West Africa there are numerous legends of Pygmies, so it is possible that at this time Pygmies also inhabited West Africa. However, Joire thinks there is no evidence for Pygmies in West Africa and assigns the microlithic sites in French Guinea to the Baga tribe. The diffusion of agriculture through Africa, and its effect on the preceding distribution of peoples will now be considered.

THE INTRODUCTION OF AGRICULTURE AND IRON WORKING INTO AFRICA

The first evidence of a farming economy in Africa occurs in Egypt at Fayum, which dates about 4000 B.C. Because of the domesticated plants and animals associated with this culture, it is thought to be derived from Asia Minor. Seligmann shows instances of Egyptian contact with Negroes in the late predynastic period, which he dates at about 3000 B.C., and Negroes are also known to have been living in the Sudan at Esh Shaheinab at about the same time. The inhabitants of Jebel Moya in the Sudan are also stated to be Negroes, who were forced westward by the Arabs around 700 B.C. Thus, agriculture seems to have spread from Egypt to the Negro peoples who have since been forced south and west by the Arabs and by Berber peoples such as the Tuareg.

Iron working was also introduced into Africa from Asia Minor via Egypt. There was a considerable iron industry flourishing at Meroe in the Sudan in 600 B.C., about which Arkell states: "Indeed there is little doubt that it was through Meroe that knowledge of iron working spread south and west throughout Negro Africa." The next evidence for the spread of iron southwest of the Sudan is in Northern Nigeria where Fagg has discovered the Nok culture, which is dated in the second half of the first millennium B.C. by geological methods. Assuming that iron working spread here from Meroe, this is about the date which would be expected. This culture contains both iron and stone axes; but since the iron axes have the same shape as the stone ones, this appears to be a transitional culture which had only recently adopted iron working. Since Mukherjee, Rao, and Trevor found the inhabitants of Jebel Moya to be most similar in physical type to the West African Negro, the westward migration of these people in the first millennium B.C. could very likely have been the method by which iron working was introduced to West Africa. In any case, this appears to be one route by which iron working was introduced into West Africa.

In the western part of West Africa, iron working seems to be somewhat later, and the evidence seems to indicate that it was not introduced via Meroe. Corbeil, Mauny, and Charbonnier think that iron working was introduced into the Cape Verde region around Dakar by Berbers who arrived there from the north about 300 B.C. Later, Mauny states that iron working was introduced into this region by the Phoenicians in the first century A.D., since the words for iron in many of the languages of this region seem to be derived from Phoenician. Although it is possible that some peoples along the coast obtained iron from the Phoenicians, it would seem more likely that iron working was brought across the Sahara, since contact with the Phoenician ships would not seem to have been close enough for the transference of all the techniques which iron working requires. Cline states: "Within the bend of the Niger lies the only large area where iron remains have been found associated with stone-using cultures." However, Nok culture in Northern Nigeria had not been discovered at the time Cline was writing, so that there appear to be two areas with these transitional cultures. In the same area Cline describes another type of iron working site which has copper and a much richer assemblage altogether. These sites he associates with the Ghana Empire. This empire was founded about 300-400 A.D., at about the time the camel was introduced into the western Sahara, and its rise to eminence is associated with increasing trade with Mediterranean civilizations. It thus appears that iron working was introduced into the western part of West Africa shortly before this empire was founded and probably was introduced from the north across the desert.

The preceding evidence indicates that both agriculture and iron working were introduced into West Africa from Asia Minor via Egypt, although both were no doubt diffused along several different routes. Agriculture was present in Egypt centuries before iron working and probably began to spread through Africa before iron working was introduced from Asia Minor. However, this early spread of agriculture seems to have been mostly stimulus diffusion, since the basic crops of Egypt, wheat and barley, did not spread to the Sudan. Even today, millet and sorghum are the basic crops throughout the Sudan. Both millet and sorghum, or at least some species of them, are considered to have been domesticated in Africa and to have been cultivated there "since antiquity." Viguier states: "Aug. Chevalier considers the western Sudan and its saharan border as one of the centers of the origin of domesticated sorghum." Since the agricultural methods used for them are similar to those for wheat and barley, and in addition the crops are all grains, it would seem reasonable to postulate that an early spread of agriculture from Egypt involved these crops. The techniques involved in the cultivation of these grains did not entail any considerable technological change from that of a microlithic hunting and gathering culture. The tool assemblage at Fayum in Egypt is not very different from that of the Natufian in Palestine or that of the Capsian. As this early agricultural economy spread, it either drove the hunting cultures before it or perhaps was adopted by these peoples. However, one of the hypotheses of this paper is that this economy could not spread throughout tropical Africa.

Although a Neolithic millet and sorghum economy could spread through the sudan, it was not until the introduction of iron working and/or better yielding tropical crops that the Negro agriculturalists could exploit the tropical rain forest. Thus, the forest remained the home of primitive hunters until quite recently. In West Africa these hunters appear to have been Negroes whose descendants can be seen today in the low sickling frequency areas of Portuguese Guinea and Eastern Liberia; and in Central Africa they were Pygmies, whose descendants are the low sickling frequency "true" Pygmies, the Babinga of French Equatorial Africa.

A combination of three factors prevented the spread of this agricultural economy through the tropical rain forest: (1) the poor quality of the soils, which wear out after a few crops; (2) the difficulty of clearing the forest with stone tools; and (3) the low yields of millet and sorghum.

In Northern Ghana and Northern Nigeria, where millet and sorghum are still the basic crops today, in many places the same fields are cultivated year after year. On the other hand, in Sierra Leone a new field is cleared every year, and in the forest regions of Nigeria, Gourou states that it takes 30 years for the soil to recover after one crop, while Forde indicates that the fields are cultivated for three or four years before being left fallow. Some comparison of the relative yields of the various crops can be obtained from Gourou's figures of yields in the French Sudan, although this is not tropical rain forest. Millet yields 5 cwt. per acre; yams, 15 cwt. per acre; and cassava, 32 cwt. per acre. However, from a nutritional standpoint the important yield is the number of calories per unit of land. Combining data from several African countries, Brock and Autret give the following figures for the yields of various crops in thousands of calories per hectare: millet yields 1,530; sorghum, 1,854; yams, 3,554; and cassava, 7,090. Thus, when these three factors are considered together, it would seem to be difficult for a Neolithic millet and sorghum economy to exist in a tropical rain forest environment. It should be emphasized, however, that this hypothesis does not mean to imply that there were no agriculturalists in the tropical rain forest prior to the introducton of iron working and tropical root crops. There was undoubtedly some agriculture and "whittling away" at the tropical rain forest in the areas which border on the sudan. However, these innovations were a necessary prerequisite for the great explosion of the Bantu peoples out of Nigeria, which filled up half a continent in a relatively short time.

Together with iron working, the domestication of two indigenous crops opened the tropical rain forest as a habitat exploitable by the Negro agriculturalists. Chevalier states that the yam, *Dioscorea latifolia,* was domesticated in West Africa. Today the most widespread species of yam in Africa is *D. cayenensis,* which is derived from *D. latifolia.* From its distribution it would seem most probable that these yams were domesticated in Nigeria. With the yam and iron working, the Bantu peoples then spread throughout the Central African tropical rain forest from their original homeland, which Greenberg places in the central Benue River valley in Nigeria. In many places today the Bantu do not have the yam as a staple crop, but this theory only attempts to explain the original rapid spread of the Bantu. This theory is supported by linguistic evidence, by the fact that transitional iron working cultures are known in Northern Nigeria, and also by the fact that the spread of iron working in Central and South Africa is associated with the spread of the Bantu.

In addition, in several areas where yams are no longer the Bantu staple there is still ritual associated with this crop, which seems to indicate that it was previously more important. For example, among the Kpe in the Cameroons, where cocoyams are now the staple crop, Ardener states. "Although subsidiary in Kpe agriculture, this crop [i.e. yam] is remarkable for the fact that it is the only one to which some degree of ritual is attached. . . . The ritual elements in the cultivation of the yam, the present economic importance of which is quite small, suggests that this crop . . . may have been a staple food in the past history of the Kpe."

Also from Nigeria, some of the Kwa peoples spread in similar fashion through the West African tropical rain forest to the Ivory Coast and forced other Kwa peoples, the Kru and Lagoon speakers, westward into the Ivory Coast and Liberia. The Kru and Lagoon peoples were probably in the tropical rain forest as hunters and gatherers prior to this spread of agriculture. Agriculture has since been introduced to most of the Kru and Lagoon peoples, but it usually has rice as the basic crop, which comes from a different center of dispersal, or manioc, which was introduced into West Africa from the New World. Even today in the Ivory Coast, as several botanists have remarked, there is a sharp boundary of yam cultivation on the Bandama River, which is also the border between the Baoule and Kru peoples. In addition, the yam cultivators, such as the Agni, have an elaborate ritual associated with the yam harvest, which indicates great reliance on this crop. Although the Kru peoples have for the most part adopted agriculture, there is still more reliance on hunting in the Kru area, and there are some groups who are still mainly hunters. In Eastern Liberia, Schwab states: ". . . there is one clan or small tribe . . . living to the north of the Tchien near the Nipwe River who have a reputation as elephant hunters, like the pigmies of the southeastern Cameroun."

The cline in the frequency of the sickle cell trait coincides with this spread of yam cultivation. The Kru and Lagoon peoples have almost 0 percent of the sickle cell trait, except where they come in contact with the yam cultivators, while the yam cultivators in the Eastern Ivory Coast, Southern Ghana, and Nigeria all have high frequencies of the trait. Thus, it seems that the sickle cell gene was brought into this part of Africa by the migrations of the yam cultivators westward from Nigeria, and at present both agriculture and the sickle cell gene are spreading to the hunting populations, which were in the forest prior to the spread of yam cultivation.

Perhaps a little later than this spread of yam cultivation, there was another spread of agriculture through the West African tropical rain forest. Porteres has shown that somewhere around the Middle Niger River Valley, a wild African species of rice, *Oryza glaberrima,* was domesticated. He dates this domestication at about 1500 B.C., but the spread of this crop through the tropical forest seems to be much later than the postulated date, and even later than the introduction of iron. There is evidence that the first Mande peoples to enter the tropical rain forest were hunters. Little dates this migration at least 400 years ago. However, the most plausible date seems to be about 1300 A.D., when the Susu appear to have migrated to French Guinea from the Middle Niger region. Thus, it would seem that the Mande and West Atlantic peoples in the tropical forest were still hunters about 600 years ago, and that rice agriculture has since been introduced to them. Joire assigns the microlithic archeological sites which are known in French Guinea to the Baga people, who speak a West Atlantic language. These people thus were in the tropical forest prior to the immigration of the Mande peoples and to the later spread of rice agriculture.

The spread of iron working and rice cultivation through this part of the West African tropical rain forest, after the original Mande migration, does not seem to be associated with any large scale migration; it probably occurred by diffusion, since the

Mande peoples who have now adopted rice cultivation were in the same location as hunters. Thus, according to the evidence, the spread of rice agriculture by diffusion seems to coincide with the spread of the sickle cell gene by mixture. In addition, the spread of rice cultivation appears to be later than the original Mande migration, as does the spread of the sickle cell gene. Rice cultivation also diffused to the West Atlantic-speaking peoples, as did the sickle cell gene. Thus, the type of gene flow—in one case migration and in the other mixture—which was responsible for the spread of the gene in West Africa seems to be related to the manner of the spread of agriculture. However, agriculture seems to have spread farther than the gene. The Kru peoples in Eastern Liberia, and the West Atlantic peoples in coastal Portuguese Guinea, are today rice cultivators. The reason for this lag in the spread of the sickle cell gene is due first of all to the fact that it takes several generations for the gene to build up to appreciable frequencies, but it also seems to be due to the relationship of the selective advantage of the sickle cell gene to slash and burn agriculture. This relationship is due in turn to the complex epidemiology of malaria in West Africa, which we will now consider.

MAN, MALARIA, AND MOSQUITO IN WEST AFRICA

In West Africa the relationship between man, malaria, and mosquito is very highly evolved, due largely to the habits of the major vector of malaria, *Anopheles gambiae*. This mosquito is attracted to human habitations and usually rests in the thatched roofs of an African village. It bites man regularly, and breeds in a variety of places. Wilson has estimated that 75 percent of the malaria in Africa is due to *A. gambiae*. Its breeding places are so diverse that, when attempting to delimit them, entomologists usually state where it cannot breed. *A. gambaie* cannot breed in (1) very shaded water, (2) water with a strong current, (3) brackish water, (4) very alkaline or polluted water.

If we now consider the types of water which would be found in the tropical rain forest, it can be seen that there would be few places for *A. gambiae* to breed in unbroken tropical rain forest. The high emergent shade trees and the trees of the middle "story" of the forest so effectively shade the ground that there would be few, if any, areas that were unshaded. In addition, the layer of humus of the forest floor is very absorbent, so there would be few stagnant pools. It is only when man cuts down the forest that breeding places for *A. gambiae* become almost infinite. First, with continued cutting of the forest, the soil loses all of its humus and becomes laterized. At this stage it is practically impervious to water; puddles are constantly renewed by the frequent tropical rains and so persist indefinitely. Second, man's refuse and his villages provide more abundant breeding places for the mosquito. Third, the swamps become open and hence possible breeding places.

In a hunting population, which does not destroy the forest, malaria would thus not develop this complex relation with man. Malaria could still be present, but not the holoendemic malaria which characterizes most of Africa today. Hunters do not build the type of permanent habitation in which *A. gambiae* lives, and since a hunting population moves frequently the mosquito could not keep up with the human population, so to speak. Also, in the epidemiology of any disease there is a critical size for the population below which the disease cannot persist. Since hunting populations are small, they would be closer to this critical size and perhaps even below it.

The Pygmies provide an example of such a hunting population, but unfortunately no malaria surveys of hunting Pygmies are available. Schwetz, Baumann, Peel, and

Droeshant did examine three groups of Pygmies for malaria and found that they had less than the surrounding Negroes, but these Pygmies were building houses and farming, and so cannot be considered a hunting population. Putnam, who lived with the hunting Pygmies for 20 years, states that they do not suffer from malaria. His account also shows that the Pygmies do not cut down the forest and do not build their rude huts in a clearing but in the middle of the forest. These customs would appear to be the reasons for the absence of malaria among them.

If this complex relationship between parasite, host, and vector which is characteristic of holoendemic malaria could not have developed in hunting populations, then the selective advantage of the sickle cell gene would not be present in these populations. If, as has been postulated, the Feloop and other peoples in Portuguese Guinea and the Kru peoples of Eastern Liberia and the Western Ivory Coast were the last remnants of hunting populations which once were spread through the tropical forest, then the absence of the selective advantage of the sickle cell gene in these populations would have prevented it from becoming established, even if there had been some gene flow from neighboring Sudanic peoples. Although considerable areas of tropical rain forest are shown on any vegetation map of West Africa, these are greatly broken up by agricultural settlements and fields. Nevertheless, the last northern remnants of the forest are located in Portuguese Guinea near one area of low sickling frequencies, and the other area in Eastern Liberia is in the center of the largest remaining block of tropical rain forest.

The frequencies of the sickle cell trait among the Pygmies also support this theory, although the comments of several authorities might seem to contradict it. Regarding the Pygmies and Pygmoids, Hiernaux states: "They generally show a lower frequency of sicklemia than the surrounding populations, as shown in Table 2. In all cases but one, the frequency is lower in the Pygmoids. The most striking difference is between the Bondjo and Babinga, who are true Pygmies." Since most Pygmy groups have formed symbiotic relationships with their Negro neighbors, the frequencies among them can easily be explained by mixture, which is known to be occurring.

There is other evidence that *A. gambiae* has spread rather recently through the West African tropical rain forest. In the area around the Firestone Plantation in Liberia, shortly after the forest had been cut down, Barber, Rice, and Brown found that *A. gambiae* accounted for 46 percent of the mosquito population found in the native huts, while *A. funestus* accounted for 51 percent, and *A. nili* for 3 percent. However, at the present time in this same area, *A. gambiae* accounts for almost 100 percent of the mosquito population. Barber, Rice, and Brown found holoendemic malaria, which is not present today; however, this change is due to malaria control and not to changes in the mosquito population. These figures thus indicate a significant increase in *A. gambiae* when the forest is cut down. Even more significant are Barber, Rice, and Brown's comments on the effects of reforestation on the mosquito population. They state:

> We felt that it would be interesting to know what would be the condition of things when the rubber trees had grown and the unplanted ravines and swamps had become "rejunglized." We surveyed Mt. Barclay Plantation where the stream borders have grown up with brush or long grass. After a long search in the streams we found only two or three larvae, *A. mauritianius* and *A. obscurus.* In a pool near a village *A. costalis* was plentiful.

It can thus be seen that *A. gambiae* (the authors call the species *A. costalis*) was not present in natural water but only near a village. The authors also discuss "rejunglization" as a means of malaria control, but state that it would not be feasible due to the breeding places which would persist around the villages. In the absence of these villages, which are not built by hunting populations, and in the presence of unbroken tropical forest, the intensity of malaria would be much less. This seems to have been the situation in West Africa prior to the spread of slash and burn agriculture. Therefore, the spread of this agriculture is responsible for the spread of the selective advantage of the sickle cell gene, and hence for the spread of the gene itself.

SICKLE CELLS, DISEASE, AND HUMAN EVOLUTION

The preceding explanation of the distribution of the sickle cell gene and its relation to the culture history of West Africa has broad implications for the role of disease in human evolution. In considering the epidemiology of the sickle cell gene, Neel suggested that either the mutation which resulted in the sickle cell gene was very rare or else the spread of the gene was at present favored by special circumstances of relatively recent origin. The detailed arguments of this paper would seem to show that there are indeed special circumstances of recent origin, while at the same time not excluding the possibility that the mutation is quite rare. The special circumstances are considered to be the conditions necessary to maintain holoendemic malaria due to *Plasmodium falciparum*. This parasite is in fact regarded as evolutionally the most recent species of malaria to parasitize man. If, as has been proposed, a mobile hunting population in the tropical rain forest could not develop holoendemic malaria, then this high endemicity would perhaps be even later than the adaptation of the parasite to man as its host. Since the agricultural revolution occurred only about 7000 years ago and spread much later to Africa, it appears that the development of the environmental conditions which are responsible for the spread of the sickle cell gene are relatively recent, as Neel postulated they should be.

The agricultural revolution has always been considered an important event in man's cultural evolution, but it also seems to have been an important event in man's biological evolution. Prior to this revolution, the size of the human population was controlled to a large extent by the size of its food supply, and man's ecological niche was comparable to that of the large carnivores, or more closely perhaps to that of a large omnivore such as the bear. With the advent of the agricultural revolution, the food supply was no longer the major factor controlling the size of human populations. Man broke out of his ecological confinement and there was a tremendous increase in the size of the human population, an increase which was limited only by the available land. Haldane has stated that disease became the major factor controlling the size of human populations at this time, and his statement seems to be supported in one case by the spread of holoendemic malaria.

Two results of the agricultural revolution seem to account for this change in the role of disease in human evolution: (1) the great changes in the environment, and (2) the huge increase in the human population. Both of these seem to be involved in the development of holoendemic malaria. First, when man disrupts the vegetation of any area, he severely disrupts the fauna and often causes the extinction of many mammals, particularly the larger ones. When this happens, there are many known instances of the parasites of these animals adapting to man as the new host. It is thus possible that the parasitization of man by *P. falciparum* is due to man's blundering on the scene and causing the extinction of the original host. Second, concomitant with the huge increase in the human population, this population became more sedentary and man also became the most widespread large animal. Thus, he became the most available blood meal for mosquitoes and the most available host for parasites. This change resulted in the adaptation of several species of the Anopheline mosquito to human habitations and the adaptation of many parasites to man as their host. Under these conditions, holoendemic malaria and probably many other diseases developed and became important factors determining human evolution. It should be noted, however, that through domestication man has created large populations of other animals and these have influenced the epidemiology of several human diseases including malaria. The sickle cell gene thus seems to be an evolutionary response to this changed disease environment. Hence, this gene is the first known genetic response to a very important event in man's evolution when disease became a major factor determining the direction of that evolution.

23. ADAPTATION AND VARIATION AMONG HUNTER-GATHERERS

J. S. WEINER

Reprinted from Chapter 4 of The Natural History of Man *(New York: Universe Books, 1971). J. S. Weiner is Professor of Environmental Physiology at the University of London. He taught anthropology at Oxford from 1945 to 1953 and took the lead in 1953 in exposing the Piltdown fossil forgery; his book,* The Piltdown Forgery, *has received worldwide acclaim. For his field work in the Far East, southern Africa, and central Tanzania, Professor Weiner received the Pitt-Rivers Memorial Medal of the Royal Anthropological Institute.*

■ A principal theme of this selection is that human subpopulations emerge in response to regional pressures. Using nomadic hunter-gatherers as the source of his data, Weiner observes first that most human physical traits are polygenic—that is, they are representations of several genes—and second that hominid evolution has been affected by a variety of habitational factors, there being no one-to-one correspondence between a particular trait and a particular gene or habitational pressure.

Within this frame of reference, we are able to observe a convergence of the issues raised by Katz and Foulks on the role of nutrition (Selection 26), Campbell on adaptation to seasonal variations (Selection 19), Pilbeam on the relationship between behavior and the evolution of the brain (Selection 20), and Dobzhansky's analysis of race (Selection 18). Weiner's discussion helps us to understand the geographical variations in observable physical characteristics, as in the relationship between skin color and solar radiation.

In this analysis of the interplay of biological, social, and habitational factors in population control—an issue of great practical importance —we can see that there are neither simple explanations for the size of adaptive units nor easy solutions to questions of population pressure on available resources. Likewise in connection with the relationship between physique and climate. For example, as Weiner observes, people in hot regions tend to be light in weight and long-limbed; those in cool climates tend to be tall. As he explains, each of these body forms has adaptive advantages; as they appear in individuals, however, they are also affected by nutrition and genetic endowment.

This provides a good opportunity to reiterate a central point in the study of evolution and adaptation. Different subpopulations did not develop distinctive physiques so they could live in particular habitats. Instead, each local group may have begun with a variety of physiques in its membership and over time, habitational pressures selectively favored those individuals who were best suited to that locale. These individuals survived long enough and effectively enough to reproduce, that is, to transmit their genes for the appropriate physique to future generations. Those whose stature was not suited to their habitats—for example, lightweight and long-limbed individuals in cool climates—lost out in the competition for mates or died before they could reproduce themselves. It is necessary to stress that natural selection—and not purposiveness—is what determines the course of evolution.

An excellent introduction to the evolutionary study of population control is *Populations and Societies,* by Judah Matras (Englewood Cliffs, N.J.: Prentice-Hall, 1973). ■

ALMOST THE WHOLE evolutionary history of modern man, indeed of the hominids, was encompassed within the hunting and gathering of wild food. Man's biological constitution and capacities are therefore to be understood in very large part as moulded by the stresses inseparable from this precarious way of life—an existence pursued by the hominids for over a million years, by the genus *Homo* for about half a million years and by *Homo sapiens* himself for all his quarter of a million years, except for the recent 8,000 years when at last he became a settled being.

A major evolutionary fact to be kept in

mind is that the hominid period falls into two unequal parts. We must distinguish between a first period of relatively gross morphological and skeletal change when, as we have seen, the major transformation of *Australopithecus* to *Homo erectus* took place; and a second during which the latter is transformed into *Homo sapiens*. The first period is the longer, about 750,000 years at least; the second is not much more than about 250,000 years. In the first period the working of evolutionary forces is, as we have seen, prominently displayed in the fossil remains and particularly in the brain, face, dentition, hands and locomotor skeleton generally. In the second period the palaeontological record still reveals evolutionary change, but while this becomes much less obtrusive in the skeleton, there is plenty of direct and detailed evidence of the results of long continuing genetic differentiation to be found in present day populations. Throughout this period of ecological change we are witnessing a continuous process of modification of the genetic basis of hominid anatomy, physiology and behaviour, as well as in the regulation of population numbers and the utilisation of resources. The basic biological equipment of the emergent *sapiens* species was fashioned in the first long period of hominid existence, in the African equatorial savannah-bush environment where *Australopithecus* gave rise to the first forms of *Homo*. It was this biological endowment that opened the way to the era of *Homo*, when the range of biotopes widens so greatly to encompass eventually nearly the whole world.

The nature of this endowment is quite clear. It consists of an array of characteristics conferring on the species as a whole a high degree of adaptive flexibility. Specifically, this affects, as we shall see, the bodily systems making for survival—nutritional and metabolic flexibility, the capacity for sustained physical work, body temperature regulation, protection against ultra-violet radiation and immunological responses to infectious disease. The adaptability displayed in the functioning of these systems remains the common property of the *sapiens* species as a whole, continuing to provide the essential biological basis for adjustment to the many changes and stresses of environment and activity to which modern man subjects himself. Since these properties are held in common by *sapiens* populations of today, and since they clearly reveal their origin in the hunting way of life in hot, sunny climates—and must have required a long period for their successful development—we may confidently regard them as the product of the environmentally restricted but very protracted period of the early hunting hominids in Africa—the Australopithecines and first varieties of *Homo erectus*.

The biological achievement of this phase (and it involves, of course, a major development of brain capacity) made possible later on the occupancy and exploitation of ever more varied ecological niches by early and modern kinds of *Homo sapiens*, culminating in their spectacular irruption, while still in the hunting phase, into the variegated environments of the circumpolar regions of the New World and into Australasia.

As the *sapiens* hunting populations dispersed, they became subject to the action of novel micro-evolutionary forces; migration and seasonal movements are inseparable from this way of life and bands are continually forming and moving away to make contacts with new groups. The evidence from prehistory, as well as the practices of contemporary hunters, makes clear that while shifting networks of hunting bands must have been subject to virtual isolation over long periods of time, isolation of individual breeding within the "network" was never complete and there was a renewal of contact from time to time. Nevertheless, in the geographical dispersion of *Homo sapiens* genetic differentiation was inescapable in the face of such facts as migration, environmental stress, small population size, and chance breakup of isolates combined with intermittent hybridisation and interaction with other groups.

As part of this process of local differentiation, those features conferring general species adaptability have undergone only a small degree of modification; it is more pronounced in some characters but the basic adaptive property persists.

NUTRITION

A fundamental biological endowment of man, the inescapable consequence of the gathering-hunting mode of life in the Australopithecine stage, as we have noted, is the dependence on an omnivorous, largely carnivorous diet. This departure from the general anthropoid diet was not an absolute one, but one of degree. The use of vegetable nutrients, of fruits, nuts, berries and roots continued but the dependence on meat, rudimentary among the Anthropoidea, became greatly intensified. The hominid diet was such that it could now be varied within very wide limits and combinations to provide calories and proteins in adequate amount for growth, reproduction and activity. To the adaptability of climatic response was added flexibility of diet.

While the typical hunter's diet is naturally high in animal protein, fairly high in fat and low in carbohydrate, to ensure an adequate energy intake, quite large amounts of food must be eaten. A hunter's daily diet might be composed of protein 400 g, fat 150 g, carbohydrate 60 g, giving 1,600, 1,350 and 240 kCals of energy respectively to make a total of 3,190 kCals. The successful hunter utilising the resources of his environment very fully can secure this nutritional intake. The intensity of exploitation of the game may be judged from the diversity of food eaten by the hominids at Olduvai, *Homo erectus* at Pekin, *Homo sapiens neanderthalensis* and Aurignacian man, as already described.

The fluctuation in the food supply led to the consumption of very large amounts when food was available. This irregular pattern of food consumption (so different from the continuous foraging of the arboreal primates) is characteristic of hunting people of the present day such as the Eskimo, Bushmen and Australian aborigines. It has been suggested that the hunter's "appetite" has persisted into the modern westernised world where it is now entirely inappropriate; people are apt to overeat in a kind of instinctive anticipation of long periods of dearth, which do not in fact occur. This "overeating" is, of course, the major factor responsible for the obesity, overweight and "endomorphic" physiques common in modern western societies but virtually absent in hunting groups.

In its ability to accept strange foods and novel combinations of nutrients, the human organism shows great adaptive capacity. There is also the ability to regulate input to supply. Faced with a sudden reduction or surfeit in the supply of a particular nutrient, the normal level in the body is kept from undue fluctuation by a compensatory decrease or increase in the excretion or degradation of the nutrient concerned; where the intake is very greatly reduced or the output greatly increased there may be some utilisation of whatever stores the body possesses until the balance can be restored. The balance between intake and output of nitrogen, sodium, iron, chloride and calcium, as well as most vitamins, is handled in this way, though not below certain limits of deficiency.

An individual losing a good deal of sodium as salt in the sweat will, during one day, hold back salt from being excreted by the kidney; the urine content of salt will become diluted and over the next few days the sweat will similarly undergo reduction of its salt content. The retention of salt under the stimulation of a negative salt balance is brought about by the action of a hormone (aldosterone) from the adrenal gland. This response is highly appropriate to conditions of desert or tropical activity, since salt loss by sweating would be disastrously high if these regulatory responses could not be made. Thus it turns out that the intake of salt is adjusted at quite different levels for different communities and even in the tropics, a balance can be maintained on an intake of say ten gms a day, no higher than the amount of salt consumed in cool climates.

There is also a large individual variation in the levels at which salt balance can be maintained, even when conditions are standardised, so that genetic differences may well be involved. It might be supposed that genetic selection for efficient salt conservation has favoured some desert peoples, but little study on this point has so far been made.

The rate of growth and the adult body size are influenced by the availability of calories, protein and calcium. Adjustments to shortages are made by reducing the losses of nitrogen and calcium and, in the growing period, by a smaller utilisation leading to a smaller body size.

An individual on a very low calcium intake will be found to have a correspondingly low calcium output; if the individual is a growing child the compensation may not be sufficient to make available enough calcium for bone growth and stunting will be the result. This risk is not run by hunting peoples but it is by many single-crop agriculturalists.

As compared with neolithic agriculturalists and men of temperate and cooler climates, the body size of tropical dwellers now and in the past tends to be relatively small. That hot climates place a limitation on body size seems likely in that, as we shall see, body heat and water regulation may be favoured in lighter rather than heavier individuals.

The possibility that food shortage also might have exerted a strong selective action on body size in some populations cannot be ruled out. In conditions of chronic malnutrition natural selection might have operated in favour of the smaller individuals requiring less food for growth and activity. In this way genotypically smaller women would produce genetically smaller offspring with much less difficulty than women stunted phenotypically by shortage of food during their growing period. The latter would experience difficulty during labour because of skeletal deformity, particularly as a result of pelvic distortion and disproportion. It seems to be the case that ill-nourished women of poor socio-economic class do suffer more difficult and prolonged labour, with greater perinatal mortality, more prenatal and greater maternal mortality, than better-off, better built and taller women; whereas women of comparably small size in communities of genetically lower average bodily stature, produce healthy small babies with little trouble.

Although genetically determined variations in body size in the polymorphic *sapiens* species may well have an explanation in terms of both nutritional and climatic selection, there is little definite evidence of the emergence in particular localities of new genotypes conferring special metabolic or nutritional capabilities or protection against particular vitamin or mineral deficiencies. Two possibilities of this kind should be noted.

Eskimos traditionally had a diet extremely high in fat and low in carbohydrate. European explorers found this diet quite unacceptable; there is, however, some laboratory evidence that fat intake can be gradually raised to high levels. Whether Eskimos do, in fact, digest fat and metabolise fatty acids more efficiently than other populations has not been established.

The second possibility concerns steatopygia, frequently found amongst Hottentot, Bushmen and Andamanese, particularly the women. This character is, in fact, a very variable one in these populations, and by illustrating the most extreme examples racial typologists have given a misleading impression of the prevalence of this character. Moreover its incidence in neighbouring peoples has yet to be investigated. The fat deposits in steatopygia have been shown (Krut and Singer) to be identical in fatty acid composition to that of buttock fat in Europeans and Bantu. There is also evidence that the fat is reduced during lean seasons. Steatopygia may, in fact, be a specialised manifestation of a much more widespread character and confer some selective advantage against food shortage by providing rather more than the usual body fat deposits. Its survival value to the pregnant woman seems very likely.

The high degree of nutritional adaptiveness of modern man is clearly a fundamental and long-standing species property. Any genetic deviations are invariably very serious. Thus, the abnormal genes which lead to disordered amino-acid metabolism (e.g. phenylketonuria, or alkaptonuria) are present in very low frequencies, indicative of the action of strong selective pressure towards their elimination.

BODY AND FACIAL FORM

We do not know in any detail what the

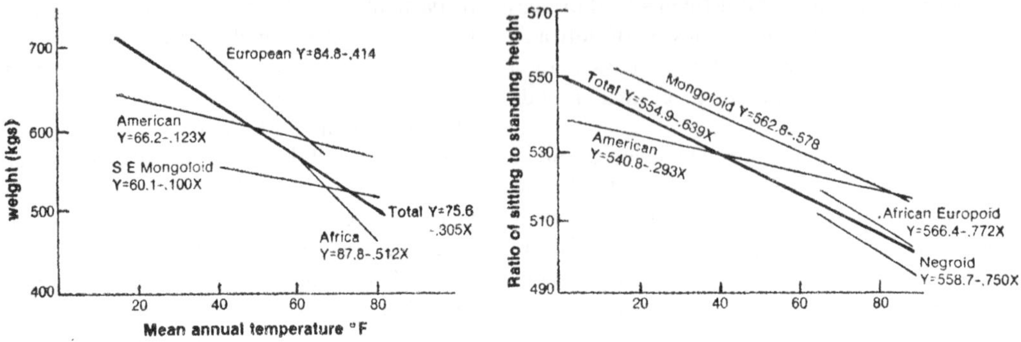

FIGURE 23.1

Relation between mean annual temperature and body weight (right) and body linearity (left).

physique of the hominids in their first tropical phase was really like. All we can say is that these groups, judging from their skeletal remains, comprised quite small-statured individuals. *Australopithecus* is estimated to have been not much more than four feet high. The estimate for the later *Homo erectus* of Java, still a tropical dweller, is a body height of about five feet four inches.

What is certain is that with the widespread dispersion of *Homo* the average value and variation of body size and shape greatly increased. That this represented a developmental adaptability to changes of habitat, particularly climate and nutrition, is a reasonable supposition.

In the relationship of physique to climate man, even today, conforms moderately well to the so-called ecological rules, and this is supported by the statistical evidence. A tendency for tropical populations to be on average slighter and lighter than dwellers in cool climates is in accordance with the so-called zoological rules of Bergmann and Allen; evidence to support this (Fig. 23.1) has been adduced by a statistical comparison of the weights, heights, limb and trunk proportions of present-day populations (Roberts, Newman). The surviving hunting and food-gathering groups conform to these rules. Thus, over the range of climate of Australia, distinct differences can be noted between those living in the uniform hot, humid areas and the desert groups and southern groups. Those in hotter areas are lighter and longer of limb. The African and Asian pygmy populations are all tropical dwellers. *Homo erectus* at Pekin living in a cooler climate was taller than the Javanese representative.

For physical and physiological reasons there would seem to be advantages for people living in predominantly hot climates to remain either of light body weight or linear in build, or both. Where heat is gained from the environment, this gain per unit area of skin is at the same rate for individuals of different physique. The heat produced by metabolism is the same per kg. body weight, but the smaller individual has a greater surface area per kg. body weight available for heat dissipation. The smaller man thus loses heat by sweating more efficiently and his total loss of sweat is less and used more economically. His water and salt requirements are less. The fact that he is physically relatively small is no handicap in enterprises such as hunting which, as we have seen, develop rapidly as a co-operative enterprise.

It is also of advantage for tropical hunters not to accumulate subcutaneous fat and the evidence we have supports this. In "Sheldonian" terms the physique of these peoples is fairly high in both ectomorphy[1] and mesomorphy,[2] low in endomorphy.[3]

Hunting groups, being heavy meat eaters, are in no danger from chronic protein, iron or calcium shortage, as compared to agriculturalists on their high carbohydrate diets. They have no reason to be undersized on these counts. Yet small body size is characteristic of many present day hunting peoples. Pygmies of the Congo, Negritoes of Malaysia, Andamanese Islanders, Philippine Negritoes and Papuans are all meat eaters. They are not phenotypically small and the best hypothesis is that their physique is an adaptation to warm environments.

There is also some suggestion that the configuration of certain facial features, notably nose and eye shape, is in some measure related to the climate. That a correlation exists between breadth of nose and the humidity of the air, as expressed in the mean annual wet bulb temperature or the vapour pressure, seems undeniable. The first hominids show in this, as in other respects, their tropical origin. The nose aperture is notably large in *Australopithecus* and *Homo erectus* in Java. Whether it is any smaller in Pekin man living in cooler climates has not been established. The first Neanderthalers lived during a period of warm climate, but it does not seem that nose shape altered when the climate cooled during the final phase of the last glaciation. This question also awaits a reappraisal. Certainly, in modern man distribution of noseshape is such that very dry areas (desert or cold) are associated with the appearance of narrow nosed peoples. This would seem to carry the functional advantage that the very dry inspired air is easily and quickly moistened (and dust particles trapped), since the lining of the respiratory tract must be kept moist.

PIGMENTATION

We shall probably never know what the skin or hair colour of the first hominids was, nor its range of variability. The possession of a darkish skin colour—and this is not unique to man—seems to be an essential adaptation of tropical and equatorial regions. The later history of the hominids has clearly been attended by great variation in pigmentation, very probably related to climatic factors.

It is necessary to emphasise that the mere visual appraisal of skin colour is not very informative and may often be misleading. An objective study of the colour properties of the skin (by reflectance spectrophotometry), as well as by chemical analysis of the pigments themselves, shows that three main components are always involved—the brown pigment melanin (and this is the most important component, in the lower layers of the epidermis), the red pigment haemoglobin in the skin circulation, and a yellowish pigment called carotene.

The melanin-making system of the human epidermis is a universal property of all varieties of mankind. Genes which suppress this faculty—the genes for albinism—are found in all populations, but albinos were quite rare. Human beings differ genetically in the extent to which melanin is found in the skin, but they are all alike in that melanisation can be intensified by exposure to various agents, of which the most important is the ultraviolet rays of solar radiation. The regional distribution over the skin surface of melanophores (like the distribution of sweat glands), is similar in dark and light skinned people.

The familiar phenomenon of sunburn extends through the stages of reddening of the skin to severe blistering followed by desquamation. The wavelengths provoking these responses are in the ultra-violet range of the spectrum in a region invisible to the human eye. It can be accepted that a certain intensity of melanin pigmentation is universally necessary as a protection against the damaging effects of ultra-violet. An intensification of exposure to ultra-violet brings about further tanning and added protection. This is a true "acclimatising" process. The melanin is produced in the lower layers and the granules

1. Ectomorphy refers to the degree of linearity in body build.
2. Mesomorphy refers to the degree of skeletal and muscular development in body build.
3. Endomorphy refers to the degree of visceral and fat development in body build.

move into the upper layers as tanning proceeds.

There is not only more pigment in dark skin; it is distributed more uniformly throughout the epidermis. It is probable that ultra-violet exposure also increases the thickness of the stratum corneum. Albinos derive their protection against ultra-violet light because of hypertrophy of this skin layer.

Negroes, southern Indians and other dark-skinned people are less sensitive to sunburn. A dose of ultra-violet light which damages light coloured skin to the extent of putting the sweat glands out of action (Thomson), has very little effect on moderately dark or tanned skin.

In sufficient intensity ultra-violet light of the wavelengths that cause sunburn can induce cancers of the skin of albino mice or albino rats. Europeans in northern Australia suffer an incidence of skin carcinoma far in excess of that in Europe. Moreover, those Europeans who tan very poorly are more prone to suffer skin carcinomas early in middle age than those who tan well.

Cancer of the skin is very rare in Negroes in the United States and in South Africa. When cancer of the skin does occur in Negroes it is to be found as often on unexposed as exposed skin, contrary to the finding in Europeans. The incidence of epithelioma in Cape-coloured (White-Hottentot hybrids) groups (Fig. 23.2) is much less than that of the European. In the Argentine nearly all cases of skin cancer occur amongst immigrants, very rarely in Indians or Negroes.

The protective value of melanin against the carcinogenic action of ultra-violet light must have played an important part in the regional differentiation of skin colour.

The selective significance of basal-cell carcinomas is not easy to assess, since they rarely cause severe debility and typically only develop later in life. The melanomas are much more rapidly fatal and a high proportion occur in the reproductive period. The late development of basal-cell carcinomas, however, is no doubt partly due to the protection usually afforded by tanning and clothing and in any case can only be said to characterise the dis-

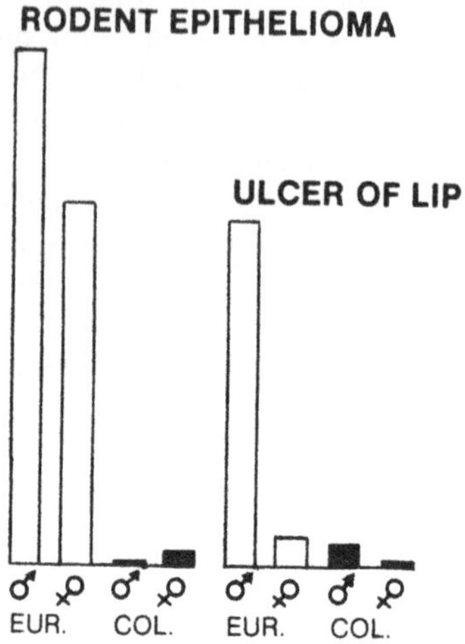

FIGURE 23.2
The extent to which Europeans are more susceptible to carcinogenic action of ultra-violet light compared to Cape-coloured individuals.

ease in temperate and sub-tropical zones. If naked light coloured people were exposed to strong ultra-violet radiation, it seems likely that selection would strongly favour the darker individuals through their lower susceptibility to skin cancer. Nor must the immediate burning effect of solar radiation be neglected. Under natural conditions severe sunburn would be extremely disabling.

Geographical variations in human skin colour seem to be plausibly related to the intensity of ultra-violet. The Sudanese regions of Africa are exposed to the most intense solar radiation and here is found the heaviest pigmentation. Other hot-desert and savannah peoples, such as the aboriginal Australians, who do not possess clothing or have acquired it only recently, tend also to be very dark skinned. Where less heavily pigmented peoples are living under comparable conditions, there is usually evidence that they are relatively recent immigrants, as, for example, in the hot deserts of the New World. The clouded western seaboards of continents in temperate latitudes have the least sunlight. In the Arctic the open dust-free skies of summer and reflection from snow and ice expose the individual to strong ultra-violet radiation. As already mentioned, Arctic peoples tend to be darker than temperate ones.

Thus, the geographical relation between ultra-violet light and skin colour would appear to be determined by the protective effect of dark pigmentation; darker-skinned varieties on the whole are found in regions of higher ultra-violet light intensity. Even in tropical zones there seems to be a difference between jungle-shaded and non-shaded peoples, e.g., African pygmies are lighter coloured on the whole than the surrounding Bantu.

The sun-tanning which is acquired by lighter-skinned individuals as a protective character may be regarded as a "phenocopy" of the genetically determined darker colour of peoples in tropical and equatorial countries. Dark pigmentation must have arisen independently more than once, namely in the dark Europeanoid peoples of south India and Arabia, among Oceanic Negroes, and in Africa, since these peoples do not show close affinity in many of their genetic characters. Relaxation from intense natural selection in the modern era has no doubt permitted the appearance of lighter-skinned populations in parts of southern Africa, in Europe and elsewhere.

While the development of heavy pigmentation in regions of strong insolation can be explained in terms of the protection against the carcinogenic and burning effects of ultra-violet radiation, is there some positive advantage in being light-skinned in regions of low ultra-violet incidence? The nature of this advantage if any is not certainly known.

One theory has it that loss of melanin pigment would have facilitated the manufacture of calciferol (Vitamin D), so important in bone development, which requires ultra-violet light for its synthesis from ergosterol precursors in the skin. Reduction of melanin would facilitate the penetration of ultra-violet radiation into the epidermis in light-skinned people living in regions of low sunlight intensity. This factor might conceivably be critical for growth and survival of those infants where the dietary supply of Vitamin D is very low.

Dark hair colour and the pigmented iris seem plausibly associated with dark skin colour as advantageous characters in regions of strong sunlight, intense ultra-violet radiation and glare. But with the spread of populations to new habitats, diversification in the form and colour of hair and in eye colour has followed. Thus, peoples of both dark and light coloration are now to be found in association with all varieties of body forms, facial features, blood group and other serogenetical characters.

POPULATION CONTROL

Every community suffers a continual struggle for existence—an ecological struggle engendered by demands for food, by climate stress, by pathogens, by danger from predators or by accidents; the struggle is played out in biological terms as adaptive success or its lack, as preferential fitness or elimination, and is finally registered as the demographic experience of the community—of births and deaths, population size and density and composition; in fertility, growth and longevity.

Evidence from many animal communities, including primate bands, shows that natural populations tend to attain a viable density over a recognised territory without actual inter-territorial conflict or warfare. This density is in fact a series of up and down fluctuations around an equilibrium value set by ecological circumstances and resources.

The movement and territorial range achieved by hunting-gathering groups during the terminal phase of the pleistocene period, culminating, as we have seen, in the penetration of vast new areas—North, Central and South America, the southeastern Asian archipelago, Australia and the circumpolar regions—indicate that the expansion of the *sapiens* species went on slowly and inexorably. To be sure, the densities attained remained uniformly low if we may judge from the meagre data on present-day surviving groups. A density higher than one-half square mile per head (two persons per square mile) was probably not often reached. For Britain in the Upper Palaeolithic a value of two hundred square miles per head of population has been estimated—comparable to that of Eskimos and Indians of the northwestern territories. The densities for present-day Australian aborigines is roughly twenty-five square miles per head. In the mesolithic densities were probably higher, about ten square miles per head, and amongst successful hunters such as the Indians of the pampas or British Columbia it reached roughly about one square mile per head.

So slow was the expansion of pre-agricultural populations, and so generally low the density, that it seems reasonable to suppose that the populations must have been subject to control forces. These enabled them to keep in approximate equilibrium with the local 'food-chain' whenever further migration and the splitting off of families or bands become impracticable owing to the occupancy of surrounding territories.

The existence of control forces must mean that pressure on resources, and hence the forces of natural selection (despite the countervailing processes of adaptation), were always at work.

For communities so exposed to the natural environment as primitive hunters and gatherers, multiple natural control processes would operate as they do in animal communities, with the addition of social measures of birth and death control. It is true we have little direct evidence of the way in which these different kinds of control actually operated to determine the population size, composition and longevity, of particular communities. There are, however, a number of pointers, apart from the indirect evidence drawn from animal population ecology. Of particular significance are the reported values for the total fertility, that is, the number of children per completed family. The most frequent figure of four children per mother for hunters and gatherers seems lower than that for advanced agriculturalists, with six or more. Of the four or five actually born in pre-agricultural families, only two or three actually reached maturity. The widespread occurrence of this small family size has been abundantly documented (Carr-Saunders). The family as a whole tended to be small since survival beyond forty years of age was rare. In keeping numbers down, there would appear to have been elimination at all stages in the numbers conceived and born, in the infant and childhood groups and in late maturity.

For the slowing down of the rate of reproduction a combination of biological and sociological factors would operate effectively in primitive society. The intervals between pregnancies are deliberately spread out by prolongation of the breast feeding period aided by restriction on intercourse. As the expectancy of life is short, often as little as twenty-five years, the woman's reproductive period is shortened, so also reducing fertility. A variety of diseases may act as additional causes of infertility, for example, hookworm, anaemia, and malaria. Another factor is genetic, arising from the increased frequency of consanguineous marriages in small inbreeding groups. Due to the increase in homozygosity, the risks of stillbirths and of abortion associated with congenital defect are sensibly increased. Finally, there are accounts of measures to induce abortion deliberately, of which the best known is the report that the Nevada Indians used the herb *Lithospermium ruderale* to prevent conception. How effective the extracts, which can be taken by mouth, really were remains unclear. The fact that claims for many antifertility "bush-tea" preparations have been advanced by primitive peoples indicates a clear realisation of the necessity for population control.

Reduction in the numbers of those born is again the object of both unconscious and deliberate measures. Deliberate infanticide is known to have been practised regularly by communities of hunters in aboriginal America, Australia, Bering Straits and tribal India, and occasionally by many others. Yet for effective reduction of numbers in infancy the usual hazards of disease and malnutrition would ordinarily have been sufficient. Mortality at the weaning period, especially after prolonged and inadequate breast feeding, notoriously makes a major contribution to infantile elimination.

Infectious disease remains the overriding cause of high mortality at all ages; it operated to a much greater extent when man began to crowd into large permanent settlements with the attendant pollution of drinking water, poor sanitation, and easy cross infection between individuals or by insect and other vectors. But even in the pre-agricultural era, men were as susceptible as they are now to infections by viruses responsible for respiratory disorders, influenza, common cold and others; amongst modern hunters and collectors there is a world-wide distribution of identical viral antibodies from Pacific Islanders to Kalahari Bushmen. Because of the multiplicity of viral strains and their repeated modification, mortality would have been high whenever a new pathogen was encountered. The susceptibility of remote communities to diseases like measles, smallpox or tuberculosis, brought in by invaders, migrants or colonisers, has played a large part in the destruction of these societies.

In Africa infection from malaria would seem to have afflicted man from the earliest times. The association of plasmodia with both apes and monkeys makes it likely that man's plasmodia have an anthropoid source, but the exact relationships are still far from clear. This long association led to the establishment through natural selection of a number of haemoglobin variants which in the heterozygote, as we have noted, confer some resistance to malaria. Nevertheless the protection so afforded has been limited and at the present time malaria remains the major killing and debilitating disease.

In primitive society particularly high risks are run by the women during delivery. Under conditions of chronic food shortage, deficiencies of calcium, iron and vitamins will all affect the females especially during the vulnerable periods of puberty and pregnancy. This in turn will adversely affect reproductive performance.

While the composition of the hunters' diet is a very satisfactory one, it is by no means a secure one, whether he be fisherman, hunter, gatherer or all three. Sudden failures through drought, disease, or excessive exploitation of resources must have occurred often as a "control" factor on population numbers, as we may judge from our knowledge of the fluctuation of food supplies of present day hunters.

The net result is that expectation of life throughout the hunter-gathering era was, and still is, very low. Estimates of the age of death have been made for human communities of the Upper Pleistocene and later prehistoric times by means of the stage of tooth eruption, the stage of ossification of the small wrist bones, the degree of union of the epiphyses, and the degree of closure of the cranial sutures. It is possible to group skeletal remains into four or five age-categories: childhood (0-12/13 years), youth or adolescence (12/13-21 years), adults (21-40 years), mature or middle-aged (40-59 years), and aged (60+). Approximate as these estimates are they show that for the most of the period of man's existence the age of death was very much earlier than it has become in recent times in conditions of Western civilisation. Survival beyond the age of forty was a lot of not more than about ten per cent of the population, and about half only lived beyond the age of twenty. According to Vallois, only one of some forty known neanderthal individuals passed the age of fifty; of some seventy-six upper palaeolithic *Homo sapiens* only two did so; of sixty-five mesolithic individuals also only two; amongst ninety-four Silesian neolithic skeletons, only four were over fifty years old. Mortality of this order seems characteristic of food-gathering, hunting and simple agricultural communities of recent and present times. For the Indian Knoll people, a pre-agricultural tribe of the Indian settled community at Pecos, deaths before the age of twenty-one number fifty-seven per cent.

The demographic feature that stands out is the smallness of the family group, particularly the restriction on the number of children coming to maturity. It may be that the incessant need to transport essential household possessions by a man and wife in the nomadic condition of the hunting-gathering life makes it very difficult to carry or tend more than one very small child at any one time. That the groups can persist with reduced fertility may reflect the relatively more favourable living circumstances—less overcrowding, less pollution of living space, more balanced diet —as compared to the denser, more static way of life of later agriculturalists.

With a relatively simple and stable ecological situation, as in the case of hunters and gatherers, it might be expected that quantitative relationships between population density and some function of the environment would be discernible. A rough calculation, based only on energy balance, shows that a food collecting group is bound to occupy its territory at a density of the order of one head per square mile. (Assume that the individual has to move continuously every day in a yearly cycle over a piece of ground in the search for food, obtaining and expending say 1,600 Kcals per day. During active periods he might cover, say, eight miles per day at an average cost of 200 Kcals/mile/hr. If he "explores" a yard width, this represents a coverage of 1/220 square miles per day, i.e., a population density per food collector of 1½ square miles a year.) In fact, food availability for the hunter collector varies very widely from place to place and season to season, and a simple relationship based on continuous slow exploitation of the environment is not likely to hold except in a very general way. Thus, coastal and riverine food collectors tend to be more numerous than inland groups. An overriding population determinant in arid habitats must be the water supply. In Australia, as Birdsell has shown, there is an overall correlation of +0.8 between mean annual rainfall and tribal population density, and he has found a simple equation which gives a relation of density to tribal area and to rainfall.

The rough calculation given above also indicates that if the "extractive" ratio, i.e., ratio of energy needed for food production to energy used per day, is of the order of seventy-five per cent, the density of population will remain at the order of one per square mile. As the ratio falls, so will the density rise. It falls to about twenty-five per cent in simple agricultural communities and today is of the order of three per cent.

24. THE EFFECT OF CHRONIC EXPOSURE TO COLD ON TEMPERATURE AND BLOOD FLOW OF THE HAND

G. MALCOLM BROWN and JOHN PAGE

Reprinted from Journal of Applied Physiology, *5 (1952): 221-27. G. Malcolm Brown is Chairman of the Medical Research Council of Canada. While Professor of Medicine at Queen's University, Kingston, one of his interests was cold acclimatization, particularly as exemplified by the Eskimos of the Canadian Eastern Arctic. He was Director of several Queen's University Arctic expeditions. John Page was a member of the Queen's University Arctic Expedition of 1949-1950 that studied the adaptation of the Eskimos of Southampton Island, North West Territories.*

■ In this selection—the second that deals with physiological adaptations to particular habitats—we momentarily shift our stride. Until now we have been dealing entirely with observations made in natural situations and with inferences made on the basis of these observations. However, there is a serious limitation in confining studies exclusively to natural situations. Nature is very informative, but it does not always communicate well, and often it is difficult to segregate different variables to know precisely which takes precedence in a series of events and which factor is responsible for others. Nature often presents us with "chicken-and-egg" propositions, and it is difficult to determine which came first.

One of the ways by which we can break into these circles is through an experimental approach. It is important to keep in mind that experimentation is not a substitute for observations of natural events; instead, it is a technique—or more accurately a set of techniques—for gaining insight into nature by trying to ravel its tangled skeins, one thread at a time.

Eskimos present some of the most remarkable human adaptations, especially in their material culture. But what of their hands in their terribly low ambient (completely surrounding) temperatures? Is the Eskimo's ability to tolerate extreme discomfort in exposing his hands to cold only the result of training and conditioning, or have physiological changes occurred in this population over time, as a result of natural selection, to make his milieu more habitable for him?

Brown and Page compared a group of Eskimo men with a group of Canadian medical students in a controlled experimental situation and found that the flow of blood in the Eskimo hand was greater, which kept the Eskimo's hand warmer. Under temperature conditions that normally decrease the flow of hand blood, the Eskimo's drop was more gradual than the Canadian's. It can therefore be concluded that significant physiological adaptations, such as increased vasodilation (the expansion of blood vessels), have occurred among the Eskimos to make their milieu habitable for them, although the importance of conditioning and other cultural factors should not be discounted.

The reader who is interested in further exploration of the experimental approach to the study of adaptation can consult several sources. An extensive survey of such work will be found in N. A. Barnicot, "Climatic Factors in the Evolution of Human Populations" *(Cold Spring Harbor Symposia on Quantitative Biology,* 24 [1959]: 115-29). Another study of adaptation to cold was made by B. F. Schonlander, H. T. Hammel, J. S. Hart, D. H. Le Messurier, and J. Steen: "Cold Adaptation in Australian Aborigines" *(Journal of Applied Physiology,* 13 [1958]: 211-18). Other interesting studies are H. T. Hammel's "Response to Cold by the Alacaluf Indians [of Tierra del Fuego]" *(Current Anthropology,* 1 [1960]: 146) and Laurence Irving's "Adaptations to Cold" *(Scientific American* [January, 1966], 214[1]: 94-101). At the other end of the thermometer is Paul T. Baker's "The Biological Adaptation of Man to Hot Deserts" *(American Naturalist,* 92 [1958]: 333-57). ■

VARIOUS AUTHORS HAVE STUDIED the blood flow of the hand at different ambient temperatures and with the hand at different local temperatures. We know of no published data concerning the effect of chronic exposure to low ambient temperatures on hand blood flow. It is the purpose of this paper to present observation on hand temperature as well as hand blood flow in a group accustomed to low ambient temperatures and to compare these with a control group.

METHOD

The group accustomed to low ambient temperatures consisted of 22 healthy male Eskimos on Southampton Island, Northwest Territories; their ages varied from 18 to 40 years. The observations in this group were carried out in July and August of 1949 and 1950 when the outdoor temperatures ranged from 30° to 67° F. The average room temperature of the Eskimo dwellings during July was found to be about 20° C. The subjects chosen were all men who lived the native life of hunting and fishing during the summer and hunting and trapping during the winter. Their dwellings were the usual tent, shack or igloo.

The control data were collected by repeating the experiments in September and October of 1951 on 37 young, healthy men attending Queen's University, Kingston, Ontario. The temperatures in Kingston during those months were roughly similar to those on Southampton Island during July and August.

All experiments were conducted at a mean room temperature of 20° ± 0.5°C. and a relative humidity from 50 to 60 percent. The amount of clothing that was comfortable for the Eskimo was worn by both groups, with the exception that the Eskimo wore seal skin muklucks while the control group wore oxfords. This clothing consisted of a woolen shirt and trousers. The subjects were studied in a fasting state.

The experimental method was patterned after the investigation of temperature and blood flow in the forearm by Barcroft and Edholm. Direct hand blood flow measurements were made with a Lewis-Grant type of venous occlusion plethysmograph. Temperatures were measured by means of thermocouples. Blood pressures were recorded with a sphygmomanometer.

The left wrist and forearm were shaved and a plethysmographic cuff of suitable size was applied. The superficial veins were marked out so that the thermocouple junctions would not be placed near them. A rectal thermocouple was introduced approximately 6 inches into the rectum. The subject then lay on a couch, his trunk and head being elevated to a 30° angle.

At the end of a 30-minute rest period thermojunctions were attached to the thenar eminence, the dorsum of the hand and to the dorsal aspects of the distal phalanges of the first, third and fifth fingers. The plethysmograph was fastened into position and raised to heart level. The appropriate venous occluding pressure was determined for each subject and the experiments were begun 10 to 15 minutes after the rest period. Hand blood flow and temperature measurements were made every 2½ minutes during the first half hour and every 5 minutes thereafter. Pulse and blood pressure were recorded every 5 minutes during the first half-hour and every 10 minutes thereafter. Two groups of experiments were carried out: *a*) Hand blood flow and temperature in room air at 20°C.; the plethysmograph was not ventilated. *b*) Hand blood flow in water-baths at 5° to 45°C. Except for the experiments in the 5°C. water-bath, the observations were made over a 2-hour period.

RESULTS

STUDIES CONDUCTED IN ROOM AIR AT 20°C.

The Eskimos were comfortable throughout the 2-hour experiments while the majority of the control group stated that they felt cool. . . . The hand blood flow and temperature of the Eskimo were greater than those of the control group. In both groups, the blood flow and temperature decreased slightly over the 2 hours. A greater degree of fluctuation occurred in the blood flow in the Eskimo subjects. The rectal temperature of the Eskimo remained constant while that of the control group decreased 0.4°C. on the average during the 2-hour period.

STUDIES ON HAND BLOOD FLOW AT DIFFERENT WATER-BATH TEMPERATURES 5-20°C.

Hand blood flow was observed within 45 to 60 seconds following immersion of the hand and forearm. In the 20°C. water-bath, the blood flow of the control group was quickly and markedly reduced while that of the Eskimo fell more slowly. At 5° and 10°C. marked reduction occurred rapidly in both groups. At all bath temperatures in this range the rate of reduction in average hand blood flow was less in the Eskimo.

At 5°C., the blood flow of the Eskimo achieved a steady state within 15 minutes while 30 to 40 minutes were required at 10° and 20°C. In all baths from 5° to 20°C., the control group attained equilibrium within 10 minutes. When equilibrium had been achieved, there was in only one case a variation of more than 1 cc. in the subsequent determinations of individual blood flows. In the 5°C. water-bath, in one Eskimo a high blood flow occurred 90 minutes after immersion. It was an isolated reading and we are unable to account for it; it may have been an artifact. At the conclusion of observation in the 5°C. water-baths, loud noises and painful stimuli failed to alter blood flow.

At any time during the period of observation at these temperatures, the hand blood flow of the Eskimo was greater than that of the control group. The blood flow at 5°C. was greater than that at 10° and 20°C. in both groups, but this effect was more marked in the control group.

30-35°C.

In both groups studied, no appreciable alteration in blood flow occurred upon immersion in the 30° to 35°C. water-baths. A steady state was achieved within 10 minutes in all individuals. The frequency and magnitude of fluctuations in hand blood flow increased as the temperature of the water increased. The degree of fluctuation was greater in the Eskimo.

The volume of hand blood flow increased as the temperature of the water-bath increased. In this range of bath temperature the blood flow of the Eskimo was greater than that of the control group. At 30° to 33°C., the average blood flows in both groups decreased slightly over the 2-hour period. At 35°C., the Eskimo blood flow remained fairly constant while that of the control group decreased slightly. The volume and pattern of hand blood flow in both groups in water-baths at 32° and 33°C. closely resembled that in room air at 20°C.

38-45°C.

Increased hand blood flow occurred immediately in all control group subjects and the increase was greater at the higher temperatures. A fairly constant pattern was achieved within 30 to 50 minutes. The volume of the blood flow and the frequency and magnitude of fluctuations increased as the temperature increased.

The hand blood flow was studied at 42.5° C. in two Eskimos. The increase in blood flow upon immersion was similar to that seen in the control group. However, the subsequent volume of blood flow was greater than that of the control group and increased gradually over the 2-hour period. The blood flows of the Eskimos fluctuated more at this temperature than did those of the control group.

DISCUSSION

Spealman has demonstrated the influence of the general thermal condition of the body on hand blood flow. He observed that the hand blood flow at any given water temperature is greater the warmer the body. The results of the blood flow measurements which we have carried out on the Eskimo at an ambient temperature of 20°C. are similar to those obtained by other workers who have studied the effect of hand temperature on the hand blood flow in man in temperate climates at an ambient temperature of 24° to 27°C. The blood flow and temperature measurements in our control group are in accordance with those made by other workers at similar ambient temperatures on similar subjects. At their usual summer temperature, the Eskimo has a higher hand blood flow at various water-bath temperatures than does the white man used to a temperate climate; and the results of our studies at one ambient temperature, taken with Spealman's results, suggest that at a given water-bath temperature he has also a higher hand blood flow at various ambient temperatures.

It is to be noted that not only is the hand blood flow greater at low temperatures in the Eskimo than in the white man but also that there is a more precipitous drop in the white man to low levels immediately following exposure to cold. This indicates a greater ability on the part of the Eskimo to maintain hand blood flow for short periods of exposure.

Abramson, Zazeela, and Marrus in comparing hand blood flow measurements at 32° and 45°C. observed that average hand blood flows showed less difference between subjects at 45° than at 32°C. In addition they found that spontaneous changes in the base line usually observed at 32°C. were practically absent at 45°C. They concluded from these observations that minimal fluctuation occurs in hand blood flow at 45°C. We have found, however, in every individual of both groups studied that the frequency and magnitude of fluctuation in successive observations of blood flow was maximal at 42.5° to 45°C. A similar increase in the frequency and magnitude of fluctuations in forearm blood flow at high bath temperatures has been reported by Barcroft and Edholm, and also noted by ourselves.

We have not been able to find any published data concerning the length of time required for the hand blood flow to equilibrate at various water-bath temperatures. The length of time allowed by different workers for the hand blood flow to adjust to a water-bath has varied from 15 minutes to 120 minutes. In all baths between 5° to 35°C., the hand blood flow of the control group subjects reached a steady state within 10 minutes following immersion whereas 30 to 50 minutes were required at the 38° to 45°C. range. The hand blood flow of the Eskimo was slower in equilibrating in the cold and very warm baths.

The vasodilatation which occurs in the skin in response to extreme local cooling was attributed by Lewis to a local axon reflex. Increased blood flow in response to extreme local cooling has been demonstrated plethysmographically in the fingers and in the hand. The failure of additional and distant noxious stimuli to alter the response in our experiments suggests a local mechanism for this vasodilatation, but the observation of Spealman that the volume of hand blood flow was influenced by the general thermal condition of the body even in extremely cold baths, is against this being the only mechanism. Our studies are in accordance with Spealman's observation. We feel that this reaction although it has a local and perhaps chemical basis, as suggested by Lewis, is also influenced by the thermal condition of the body.

It has often been remarked that the Eskimos are better able to work with their bare hands in the cold than are white men. There are certainly many factors concerned in this and perhaps not the least is a remarkable tolerance of discomfort on their part. It seems to us, however, that the increased hand blood flow which we have demonstrated has importance in this regard, and this increased blood flow with the increase in hand temperature may explain the relatively enhanced kinaesthetic sensibility and ability to perform fine movements which have been demonstrated by Mackworth in persons chronically exposed to the cold. Hunter and Whillans have shown that in the cat, exposure of the knee joint to low ambient temperatures is associated with an important increase in the force required to start movement at the joint, and suggested that an increase in the viscosity of synovial fluid and decrease in the flexibility of the joint capsule and tendons are factors which may be responsible. Increased blood flow and temperature would diminish the degree of these changes. There may be other factors such as the difference in the composition of adipose tissue which play a part.

SUMMARY

A comparative study on hand blood flow and temperature has been carried out on Eskimos in the Canadian Eastern Arctic and on medical students living in a temperate climate. One of the effects of chronic exposure of the individual to cold is a reduction in the ambient temperature required for comfort. At this low ambient temperature, the hand blood flow of the Eskimo is twice that of the white man and the skin temperature of his hand is greater. The volume of the hand blood flow of the Eskimo changes more slowly in response to local cold. The degree of spontaneous fluctuation in hand blood flow is greater in the Eskimo and increases in both groups as the local temperature of the hand increases. The alterations which occur in the hand blood flow following chronic exposure to cold would appear to enhance hand function in the cold.

25. HUMAN ADAPTATION TO HIGH ALTITUDE

PAUL T. BAKER

Reprinted from Science, *Vol. 163, pp. 1149-1156. Copyright 1969 by the American Association for the Advancement of Science. Paul T. Baker is Professor of Anthropology at the Pennsylvania State University. He has served as international coordinator for high-altitude research under the International Biological Programme. His principal interests are in the study of human adaptability in desert and alpine habitats. With J. S. Weiner, he is editor of* The Biology of Human Adaptability.

■ As noted in the Introduction to this section, humans are capable of both generalized and specialized adaptations to habitational conditions. The capacity for generalized adaptation has made it possible for the species to exploit a wide range of habitats; the capacity for specialized adaptation underlies the plasticity and flexibility of humans in highly demanding and stressful niches. This selection presents an example of specialized adaptation to high altitude in the Andes Mountains. When people from lower altitudes ascend to the Andean heights, they experience great discomfort. Not only does this represent a significant problem in its own right, it has also had a fortuitous effect that has facilitated this research. As a result of the discomfort they experience at high altitudes, lower-altitude people have left the residents at the higher altitudes relatively "uncontaminated" genetically and culturally, and this has made it possible to study the adaptation achieved by the latter independent of acculturative influences.

The study of specialized adaptation to high altitude raises the broader question, what is sickness? If a native of the Andean alps of Peru was brought into a physician's office in New York, Chicago, or San Francisco and the physician did not know he was an Andean, he would probably be diagnosed as suffering from a disease known as secondary polycythemia. The principal symptoms of this illness are an excess in hemoglobin value (red blood count), a sharp elevation in blood and hemoglobin volume, an abnormal increase in the size of the thoracic (lung) cavity, and an elevation in the production of several important body acids.

Whatever might have brought this patient to a physician's office, his secondary polycythemia is not—for him—a disease. Without the symptoms of secondary polycythemia, he would probably be unable to survive in the alpine habitat of the Andes. What is a disease at low altitudes can be a successful adaptation at high altitudes.

The group studied by Baker are the inhabitants of the Peruvian district of Nuñoa; they represent a biological population within a cultural context. In this sense, the purpose of this research is to examine the biocultural mechanisms of adaptation to the habitational stresses acting on the Nuñoan local culture and population. A major lesson of this investigation is that adaptation among humans is neither biological nor cultural; instead, both operate simultaneously to enable a population to achieve a viable relationship with the habitat. Also noteworthy is the combination of field and laboratory study carried out by Baker and his associates.

Baker concludes this selection with an appropriate note of caution regarding the interpretation of the findings that are presented. The adaptations discussed may be based on differences in genetic structure between the highlanders and lowlanders; they may be based on a lifelong acclimatization to the extraordinary stresses of the habitat; or they may be the products of other unexplored factors. Much work remains to be done in this kind of research.

One of the foremost leaders in the study of adaptation to high altitudes is Carlos Monge, whose work is reported in his book *Acclimatization in the Andes* (Baltimore: Johns Hopkins, 1948) and in his article "Man, Climate, and Changes of Altitude," in *Recent Studies in Bioclimatology: A Group* (Volume 2, number 8 of Metrological Monographs, 1954). The reader should also consult his "Biological Basis of

Human Behavior" in *Anthropology Today: An Encyclopedic Inventory* (edited by A. L. Kroeber [Chicago: University of Chicago Press, 1953]). A very interesting example of physiological adaptations to marine habitats is provided by the diving women of Korea and Japan, who are described in "The Diving Women of Korea and Japan," by Suk ki Hong and Hermann Rahn (*Scientific American* [May, 1967], 216[5]: 34-43). ■

STRETCHING ALONG western South America from Colombia to Chile lies a large section of the Andean plateau, or *Altiplano,* which rises above 2500 meters (about 8250 feet) (Figure 25.1). This area is suitable for human habitation up to the permanent snow line, which is generally above 5300 meters (17,590 feet). There are now more than 10 million people living in this zone, and the historical and archeological records indicate that it has been densely populated for a long time. Indeed, before Europeans arrived, the Inca empire, which had its center in this zone, formed one of the two major civilizations of the Western Hemisphere and, in A.D. 1500, probably contained about 40 percent of the total population of the hemisphere.

FIGURE 25.1

The high-altitude areas of South America and the location of Nuñoa. Shading indicates altitude of 2500 meters (8200 feet) or more.

With such a history, one would assume this to be an ideal environment for man and the development of his culture. Yet, in point of fact, modern man from a sea-level environment ("sea-level man") finds this one of the world's more uncomfortable and difficult environments. The historical records show that such was the case even in the 1500's, when the Spanish complained of the "thinness of the air," moved their capital from the highlands to the coast, and reported that, in the high mining areas, the production of a live child by Spanish parents was a rare, almost unique, phenomenon. Today this environmental zone remains the last major cultural and biological center for the American Indian. The population has an extremely low admixture of genes from European peoples and virtually none from African peoples. The few cities are Hispanicized, but the rural areas retain a culture which, in most aspects, antedates the arrival of the Spaniards.

It would be far too simplistic to suggest that this unique history is explicable entirely on the basis of the effects of altitude on sea-level man. Yet, there is sound scientific evidence that all sea-level men suffer characteristic discomfort at high altitudes, the degree of discomfort depending upon the altitude. There is evidence, also, of long-term or permanent reduction in their maximum work capacity if they remain at these altitudes, and evidence that they undergo a number of physiological changes, such as rises in hemoglobin concentrations and in pulmonary arterial pressure. In a few individuals the initial symptoms develop into acute pulmonary edema, which may be fatal if untreated. On the basis of less complete scientific evidence, other apparent changes are found for sea-level man at high altitudes: temporary reduction in fertility, reduction in the ability of the female to carry a fetus to term, and a high mortality of newborn infants.

With these problems in mind, a group of scientists from Pennsylvania State University, in collaboration with members of the Instituto de Biologia Andina of Peru, decided to investigate the biological and cultural characteristics of an ecologically stable Peruvian Quechua population living in traditional fashion at a high altitude. We chose for study the most stable population known to us at the highest location reasonably accessible. It was hoped that some insight could be gained into the nature of this quite obviously successful and unusual example of human adaptation. In this article I review some of the results available from this continuing study.

The general problem may be defined by three questions: (i) What are the unique environmental stresses to which the population has adapted? (ii) How has the population adapted culturally and biologically to these stresses? (iii) How did the adaptive structures become established in the population?

Our basic method of study, in attempting to answer these questions, was a combination of ecological comparisons and experimental analysis.

THE STUDY POPULATION

The population chosen for study lives in the political district of Nuñoa, in the department of Puno in southern Peru. In 1961 the district had a population of 7750 and an area of about 1600 square kilometers. Geographically, the district is formed of two major diverging river valleys, flat and several kilometers broad in the lower parts but branching and narrow above. These valleys are surrounded by steep-sided mountains. In the lower reaches of the valley the minimum altitude is 4000 meters; the higher parts of some valleys reach above 4800 meters. The intervening mountains rise, in some parts, to slightly above 5500 meters.

The climatic conditions of the district are being studied from weather stations on the valley floors and on the mountain sides at different altitudes. From present records, the pattern seems fairly clear. The lower valley floor appears to have an average annual temperature of about 8°C with a variation of only about 2°C from January, the warmest month, to June, the coldest. This is much less than the diurnal variation, which averages about 17°C. The seasonal variation in temperature is due almost entirely to cloud cover associated with the wet season. Some snow and rain fall in all 12 months, but significant precipitation begins around October, reaches its peak in January, and ends in April. Since the diurnal variation is high, some frost occurs even in the wet months. Mean temperatures fall in proportion to increases in altitude (by about 1°C per 100 meters), but, because of the sink effect in the valleys, minimum temperatures on the valley floors are usually somewhat lower than those on the lower mountain sides.

Except for two small areas of slow-growing conifers, almost all of the district is grassland. Because of the existing climatic and floral conditions, herding has become the dominant economic activity. Alpaca, llama, sheep, and cattle, in that order, are the major domestic animals. Agriculture is limited to the cultivation of frost-resistant subsistence crops, such as "bitter" potatoes, *quinoa,* and *cañihua* (species of genus *Chenopodium*). Even these crops can be grown only on the lower mountain sides and in limited areas on the lower valley floors. In recent years, crop yields have been very low because of drought, but they are low even in good years.

A single town, also called Nuñoa, lies within the district and contains about one-fourth of the district's population; the other three-fourths live in a few native-owned settlements called *allyus,* or on large ranches or *haciendas,* which are frequently owned by absentee landlords. The social structure may be loosely described as being made up of three social classes: a small (less than 1 percent) upper class, whose members are called *mestizos;* a larger intermediate class of individuals called *cholos;* and the Indians, or *indigenas,* who constitute over 95 percent of the population. Membership in a given social class is, of course, based on a number of factors, but the primary ones are degree of westernization and wealth. Race appears to be a rather secondary factor, despite the racial connotations of the class designations: *mestizos*, of mixed race; *cholos*, transitional; *indigenas*, indigenous inhabitants. Biologically the population is almost entirely of Indian derivation.

By Western economic and medical-service standards, this district would be considered very poor. If we exclude the *mestizo*, we find per capita income to be probably below $200 per year. The only medical treatment available in the district is that provided by a first-aid post. The upper class has access to a hospital in a neighboring district. Yet repeated surveys suggest that the diet is adequate, and that death rates are normal, or below normal, for peasant communities lacking modern medicine. Indeed, this population must be viewed as being one nearly in ecological balance with its technology and its physical environment.

Superficial archeological surveys reveal that the central town predated the Spanish conquest, and suggest that the district population has been fairly stable for at least 800 or 900 years.

From this general survey of the Nuñoa population and its environs we have concluded that the unusual environmental stresses experienced by this population are hypoxia and cold; other stresses, more common to peasant groups in general, which the Nuñoa population experiences are specific infectious diseases, the problems of living in an acculturating society, and, possibly, nutritional deficiencies.

HYPOXIA

At elevations such as those of the district of Nuñoa, the partial pressure of atmospheric oxygen is 40 percent or more below the values at sea level. As noted above, such a deficiency of oxygen produces a multitude of physiological changes in sea-level man, at all ages. We therefore attempted to evaluate the native Indian's responses to altitude at all stages of the life cycle.

DEMOGRAPHY

A survey of more than 10 percent of the population revealed an average completed fertility of about 6.7 children for each female. This is quite a high fertility by modern standards, but there was no evidence of voluntary birth control, and cultural practices appeared in many ways designed to provide maximum fertility; under these conditions, 6.7 children is no more than average. This same survey, partially summarized in Table 25.1, did not show an unusually high rate of miscarriage but did reveal two unusual features. (i) The earliest age at which any woman gave birth to a child was 18 in the low valleys and something over 18 at higher elevations. The average age of first pregnancy was also higher for women at the higher altitudes. (ii) The sex ratio was highly unusual in that there was a large number of excess males. Furthermore, there was a higher mortality of females than of males throughout the period of growth. In an associated study of newborns it was found that, for Quechua mothers in Cuzco (altitude, 3300 meters), placenta weights at childbirth were higher and infant birth weights were lower than corresponding weights for comparable mothers near sea level. Finally, an analysis of the Peruvian census showed that, as in the United States, the mortality of newborn infants is higher at higher elevations; this does not appear to be primarily a socioeconomic correlation. From the results so far obtained, we conclude that fecundity and survival through the neonatal period is probably adversely affected by high altitude, even in the native populations of high-altitude regions. However, it is clear that the Nuñoans can still maintain a continuing population increase. Our data do not provide a basis for deciding whether, at high altitudes, fecundity and survival of offspring through the neonatal period are greater for natives than they are for immigrant lowlanders.

GROWTH

Intensive studies on growth were carried

TABLE 25.1

STATISTICS ON REPRODUCTION AND VIABILITY OF OFFSPRING FOR THE DISTRICT OF NUÑOA, BASED ON A SAMPLE OF APPROXIMATELY 14 PERCENT OF THE POPULATION OF THE DISTRICT.

Sample	*Married women*						*Sex ratio (males to females) of offspring*		*Mortality of offspring (%)*†	
	Number in sample	*Mean age (yr.)*	*Mean age at first pregnancy (yr.)*	*Off-spring (No.)*	*Mean number of offspring per woman*	*Mean number of surviving* offspring per woman*	*At birth*	*Surviving**	*Male*	*Female*
Total sample	136	36.2	19.5+	608	4.5	3.2	124	129	30	33
Sample of post-menopausal individuals	31	45+	20.1	207	6.7	4.4	113	146	27	44

*"Surviving" refers to time of the census. †During the period of growth.

TABLE 25.2

STATURE AND WEIGHTS OF NUÑOA INFANTS AND OF INFANTS IN THE UNITED STATES

Age (months)	*Stature, males (cm)*		*Stature, females (cm)*		*Weight, males (kg)*		*Weight, females (kg)*	
	Nuñoa	*U.S.*	*Nuñoa*	*U.S.*	*Nuñoa*	*U.S.*	*Nuñoa*	*U.S.*
6	62	66	61	65	6.9	7.6	6.6	7.3
12	71	75	69	74	7.9	10.1	7.3	9.7
24	76	87	75	87	9.9	12.6	9.0	12.3

out on over 25 percent of the Nuñoa-district children, frow newborn infants to young people up to the age of 21. A number of unusual growth features were apparent shortly after birth. Thus, as shown in Table 25.2, a slower rate of general body growth than is standard in the United States is apparent from a very early age. In addition, developmental events such as the eruption of deciduous teeth and the occurrence of motor behavior sequences occur late relative to U.S. standards. For example, the mean number of teeth erupted at 18 months was 11.5 for Nuñoa infants as compared with 13 for U.S. infants. The median age at which Nuñoa children briefly sat alone was 7 months, and the median age at which they walked alone was 16.2 months. These data were collected by means of the technique developed by Bayley, who reported that the median ages at which U.S. children sat and walked alone were 5.7 months and 13.2 months, respectively.

Some of the growth characteristics in later development are shown in Figures 25.2 and 25.3. In these growth studies it was possible to compare our results with cross-sectional data for groups from lower elevations (Huánuco and Cajamarca, 2500 meters; Lima and Ica, 300 meters). We have also collected some semi-longitudinal data in order to evaluate growth rates. These combined data showed, for Nuñoa children, (i) lack of a well-defined adolescent growth spurt for males, and a late and poorly defined spurt for females; (ii) a very long period of general body growth; and (iii) larger chest sizes, in all dimensions and at all ages, than those of children from lower elevations.

In explanation of the unusual growth aspects of the Nuñoa population, at least three hypotheses may be suggested: (i) all Quechua have an unusual growth pattern, genetically determined; (ii) malnutrition and disease are the prime causes; (iii) hypoxia is the major factor. Our present data are not adequate for testing these hypotheses. However, a number of observations suggest that hypoxia is a major factor. As discussed below, we have been unable to find any evidence of widespread malnutrition or of unusual disease patterns. What data are available on the growth of other Quechua show growth patterns different from those found for the Nuñoans. Finally, hypoxia has been shown to affect growth in a number of animals other than man.

WORK PHYSIOLOGY

The most striking effect of high altitude (4000 meters) on newcomers, apparent after the first few days of their stay, is a reduced capacity for sustained work. This reduction is best measured through measurement of the individual's maximum oxygen consumption. For young men from a sea-level habitat, the reduction is in the range of 20 to 29 percent; the men who had received physical training generally showed a greater reduction than the untrained. Some rise in maximum oxygen consumption occurs during a long stay at high altitudes, but studies extending over periods of as much as a year have failed to show a recovery to even near low-altitude values for adult men.

Maximum oxygen consumption for any individual or group is controlled by a large number of factors, among which the level of continuing exercise is of major importance. Among young men of European descent, mean values for maximum oxygen consumption range from below 40 milliliters of oxygen per kilogram of body weight for sedentary groups, through 45 milliliters per kilogram for laborers, up to more than 55 milliliters per kilogram for highly trained runners. The high degree of variability in this parameter makes it difficult to determine whether the native of a high-altitude region has a work capacity different from that of an individual from a low-altitude habitat, and makes it even more difficult to determine whether the complex physiological differences between groups from high and low altitudes have resulted in a better adaptation, with respect to work capacity, for the high-altitude Quechua.

In order to help clarify these questions, we determined maximum oxygen consumption for a number of carefully selected sam-

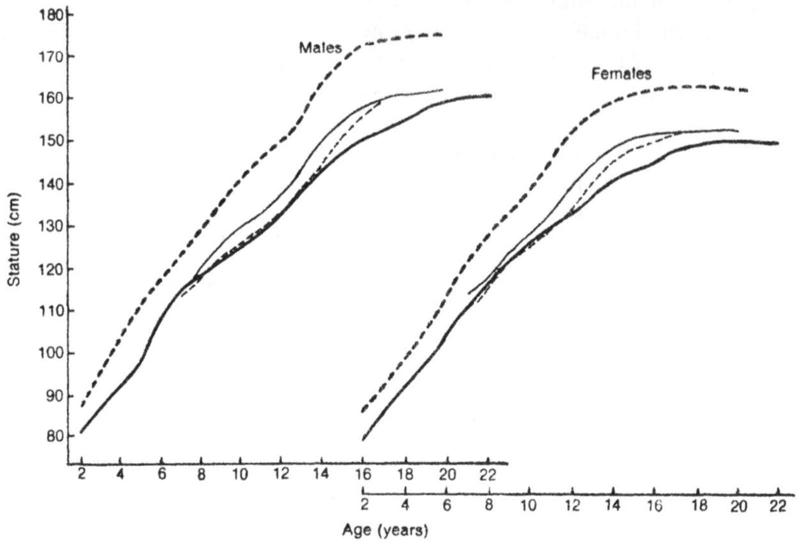

FIGURE 25.2
The growth of Nuñoa children as compared to that of other Peruvian populations and of the U.S. population. (Heavy dashed lines) U.S. population; (light solid lines) Peruvian sea-level population; (light dashed lines) Peruvian moderate-altitude (1990 to 2656 meters) population; (heavy solid lines) Nuñoa population (altitude 4268 meters).

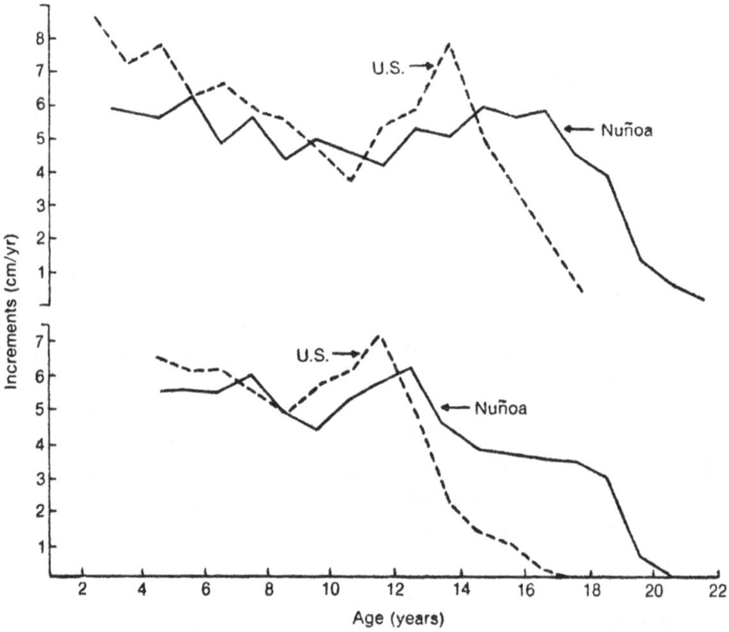

FIGURE 25.3
Rates of general body growth for children in Nuñoa as compared with that for children in the United States. (Top) Males; (bottom) females.

ples of contrasting populations; in all cases the method of determination was the same. Some of the results of these studies are presented in Table 25.3. It should be noted that the individuals referred to as White Peruvian students are so classified on the basis of morphology, and it is quite possible that they contain some admixture of native Quechua genes.

On the basis of results obtained for students alone, one investigator surmised that the differences in maximum oxygen consumption for high-altitude and low-altitude groups might be only a matter of life-long exposure, plus a high state of physical fitness in the highland Quechua. In an article based on partial results, some of the investigators, including myself, pointed out that trained athletes from lower altitudes could, at high altitudes, achieve the same oxygen consumption per unit of body weight as the Nuñoa native. While the data now available support the idea that physical training and life-long exposure to hypoxia act to increase maximum oxygen consumption, these two factors appear insufficient to explain the total results. The data, instead, suggest that a fairly random sample of Nuñoa males between the ages of 18 and 40 have a vastly greater maximum oxygen consumption than a group of reasonably physically fit researchers from the United States. The oxygen consumption of the Nuñoans also significantly exceeded that of young native students from lower altitudes, and equaled that of a group of highly trained U.S. athletes who had spent a month at high altitude. Furthermore, the heart rate and ventilation rate for the Nuñoans remained low. The Nuñoans do walk more than people from the United States, but nothing in the personal history of the Nuñoa subjects suggested that they had experienced physical training or selection comparable to that involved in becoming a college track athlete. To me, the data suggest that a high-altitude Quechua heritage confers a special capacity for oxygen consumption at 4000 meters. In the absence of more precise data, such a conclusion remains tentative. However, the data can certainly be interpreted as showing that the Nuñoa native in his high-altitude habitat has a maximum oxygen consumption equal to, or above, that of sea-level dwellers in their oxygen-rich environments. This conclusion is in agreement with the results obtained by Peruvian researchers.

COLD

THE MICROENVIRONMENT

By the standards of fuel-using societies, the temperatures in the Nuñoa district are not very low, and we would consider the typical daily weather equivalent to that of a pleasant fall day in the northern United States. However, the lack of any significant source of fuel made us suspect that at least some segments of the population might suffer from significant cold stress. As mentioned above, few trees grow in the district, and those that do are slow-growing. At present, these trees are used primarily as rafters for houses; only an occasional member of the upper class uses wood for cooking. The fuel used almost universally for cooking is dried llama dung or alpaca dung. This dung provides a hot but rapid fire, and is burned in a clay stove, which provides little external heat. Since cooking is done in a building separate from the other living quarters, only the women and children benefit from the fire. Bonfires are lit only on ceremonial occasions; the winter solstice is celebrated by a pre-Hispanic ceremony in which many bonfires are lit all over the district. The use of fires during solstice ceremonies throughout the world is often considered an act of sympathetic magic to recall the sun. For Nuñoans it may also recall the warmth.

The houses of the native pastoralists, with walls of stacked, dry stone and roofs of grass, provide no significant insulation. Measurements made within the dwelling units generally showed the temperature to be within 2° or 3°C of the outdoor temperature. This is in sharp contrast to the situation in the adobe houses used by the upper classes, by some agriculturalists in the Nuñoa district, and by

TABLE 25.3

SOME DATA FROM TESTS OF MAXIMUM OXYGEN CONSUMPTION (MAX Vo_2 AT HIGH AND LOW ALTITUDES FOR CONTRASTING POPULATIONS.)

			Subject				*Response**					
Group of subjects	*Altitude at which tested (m)*	*Duration of exposure to high altitude*	*Number*	*Mean age (yr)*	*Mean height (cm)*	*Mean weight (kg)*	*Max Vo_2 (liter/min)*	*Aerobic capacity† (ml/kg/min)*	*Maximum ventilation (liter/min)*	*Ventilation equivalent‡ (V/Vo_2/min)*	*Maximum heart rate (beat/min)*	*Oxygen pulse (ml/beat)*
Nuñoa Quechua	4000	Life	25	25	160	57	2.77	49.1	75	27.3	171	16.0
U.S. white researchers	300		6	30	183	79	3.92	50.4	131	33.7	185	21.2
U.S. white researchers	4000	4 weeks +	12	27	181	75	2.78	38.1	91	32.9	173	16.6
U.S. white athletes	300		6	20	179	71	4.58	64.2	131	28.8	175	26.5
U.S. white athletes	4000	4 weeks	6	20	179	71	3.14	46.6	105	33.7	172	19.4
Quechua from sea level	100		10	22	160	62	3.01	49.3	108	36.2	187	16.7
Quechua from sea level	4000	4 weeks	10	22	160	62	2.67	44.5	87	33.4	190	14.5
Peruvian students												
Quechua	3830	Life	10	23.8	162	60	2.79	46.8	72	25.8	188	15.1
White	3830	Life	13	23.5	169	61	2.62	42.8	74	28.2	186	14.2

*The measurements were made by means of a bicycle ergometer in 10-minute progressive exhaustion tests. †Aerobic capacity is the maximum oxygen consumption per kilogram of body weight and, as such, is the most significant measure available of the success of the individual's (and, by inference, the group's) biological oxygen transport system. It is also assumed to be one of the best measures available for judging an individual's work capacity relative to his body size. ‡Ventilation equivalent is the ventilation volume per unit oxygen uptake per unit time. The lower the value, the greater the relative efficiency in supplying oxygen.

all classes in slightly lower areas of the Altiplano. Adobe houses provide good insulation, and indoor temperatures are frequently 10°C above outdoor temperatures during the cold nights of the dry season.

From this analysis and other observations we concluded that the Nuñoa native depends upon his own calories for heat, and relies, for heat conservation, primarily upon his clothing and upon certain customs, such as spending the early evening in bed and having as many as four or five individuals sleep in the same bed. . . . His clothing is layered and bulky, with windproof materials on the outside. Thus, it provides good insulation for his body. However, the Nuñoa native's wardrobe does not include gloves, and the only foot coverings are sandals occasionally worn by men. The insulating effect of native clothing was tested under laboratory-controlled cold conditions. It was found that at 10°C the clothing increased mean body temperature by about 3°C and raised the temperature of hands and feet despite the lack of gloves and shoes.

From the total assessment we concluded that the Nuñoa native probably experiences two types of exposure to cold: (i) total-body cooling during the hours from sunset to dawn and (ii) severe cooling of the extremities, particularly in the daytime during periods of snow and rain. To assess the degree of stress due to cold we took measurements of rectal and skin temperatures of individuals in samples selected by sex, age, and altitude of habitat. These studies indicated that at night the adult women experience very little stress from cold, whereas adult men showed some evidence of such stress. During the day, women, because they are less active than men, may experience slight stress from cold, whereas men do not, except for their extremities. Active children show no evidence of such stress; however, during periods of inactivity, as at night, their skin temperature and rectal temperature are low. Indeed, at these times, all indices show an inverse relationship between age, size, and body temperature.

PHYSIOLOGICAL RESPONSES

In order to characterize the Nuñoa native's responses to cold, we used three types of laboratory exposures: (i) total-body cooling at 10°C with nude subjects, for 2 hours; (ii) cooling of the subject's hands and feet at 0°C, for 1 hour; (iii) cooling of the feet with cold water. The subjects were Nuñoa males and females, North American white males, and Quechua males from low-altitude habitats.

Some results of the studies of total-body cooling are summarized in Table 25.4. Since the Nuñoa Indians are smaller than either the coastal Indians or U.S. whites, the data are presented in terms of surface area. Viewed in this way, the data show that the native Quechua from a high-altitude habitat produced more body heat during the first hour than individuals from low-altitude habitats, but produced amounts similar to those for such individuals during the second hour. By contrast, heat loss was much greater for the Nuñoans than for members of the other groups during the first hour and similar dur-

TABLE 25.4

EXCHANGE OF BODY HEAT AS EXEMPLIFIED IN HEAT PRODUCTION AND HEAT LOSS. THE VALUES ARE AVERAGES FOR TWO EXPOSURES AT 10°C, EXPRESSED ON THE BASIS OF BODY-SURFACE AREA.

Subjects	*Numbers in sample*	*Heat production (kcal/m²)*			*Heat loss* (kcal/m²)*		
		1st 60 minutes	*2nd 60 minutes*	*Total time*	*1st 56 minutes†*	*2nd 60 minutes*	*Total time*
Whites	19	51.5	62.2	113.7	29.6	11.8	41.4
Nuñoans	26	58.1	65.5	123.6	51.0	12.9	63.9
Lowland Indians	10	54.7	65.5	120.2	30.1	14.7	44.8

*Heat not replaced by metabolic activity. †Because perfect equivalence in body temperature prior to exposure to cold was not achieved, heat loss in the first 4 minutes of exposure has been excluded from this calculation.

ing the second hour. The findings on heat loss are perhaps the more interesting, since they conform with results of two other studies of cooling responses in native Quechua from high-altitude habitats.

When the source of the greater heat loss was closely examined, it proved to be almost entirely the product of high temperatures of the extremities, and these temperatures, in turn, seem to be produced by a high flow of blood to the extremities during exposure to cold. When a comparison was made, as between Nuñoa males and females, of the temperatures of the extremities, the women were found to have warmer hands and somewhat colder feet. In both sexes the temperatures for hands and feet were significantly above corresponding values for white male subjects. The specific studies of hand and foot cooling made with exposures of types ii and iii shed further light on the subject, showing that the maximum differences between populations occurred with moderate cold exposure; that the high average temperatures of hands and feet were the result of a slow decline in temperature, with less temperature cycling than is found in whites; and that the population differences were established by at least the age of 10 (the youngest group we could test).

Since the oxygen exchange between hemoglobin and tissue bears a close positive relationship to temperature, it is clear that, when the temperature of the peripheral tissues remains high, more oxygen is available to these tissues. Therefore, the high temperatures of the extremities of the Nuñoans at low atmospheric temperatures may be considered not only an adaptation to cold but also a possible adaptation to hypoxia.

NUTRITION AND DISEASE

In the complex web of adaptations that are necessary if a peasant society is to survive, adequate responses to nutritional needs and prevalent diseases are always critical. For a people living at high altitude, these responses are important to an interpretation of the population's response to altitude and cold.

NUTRITION

The analysis of nutritional problems in the Nuñoa district has proceeded through a number of discrete studies, including a study of dietary balance in individuals, a similar study for households, an analysis of food intake by individuals, and a study of the metabolic cost, to the community, of food production. Of these studies, the first two have been completed, the third is in the analysis stage, and the fourth is in the data-collection stage. The results to date suggest that the Nuñoa population has a very delicate, but adequate, balance between nutritional resources and needs.

The dietary-balance study was carried out with six native adult males, chosen at random. Food requirements were predicted from U.N. Food and Agriculture Organization standards, on the basis of weight and temperature. The food used in the study consisted wholly of native foodstuffs and was prepared by a native cook. The results showed that, for these individuals, protein, caloric, and fluid balance remained good, and indicated that caloric and protein balance was good prior to the time of the study. The household survey suggested that nutrition for the population as a whole was generally adequate, although the method did not permit conclusions on the adequacy of nutrition for special subgroups, such as children and pregnant women.

The household survey also suggested that the diet might be somewhat deficient in vitamin A and ascorbic acid. Subsequent, more detailed surveys of individuals now cast doubt on the validity of this conclusion and suggest that, if malnutrition exists, it is probably no more common than in U.S. society. Indeed, in the light of the modern concept of "overnutrition," we might even say that the Nuñoans have a better dietary balance than the U.S. population. As noted above, the balance is delicate, and there must be years and times of the year when certain dietary deficiencies exist. Furthermore, the balance is subtle. To cite an example, the basic foods available are very low in calcium, yet adequate calcium is obtained, primarily by use of burned limestone as a spice in one type of porridge.

DISEASE

Our data on infectious disease are particularly inadequate. Health questionnaires are almost useless, since native concepts of health are only partially related to modern medicince. The *indigenas* attribute over 50 percent of all illness and death to *susto,* a word best translated as "fright." As noted above, no regular medical treatment is available, so records are lacking, even for a subsample. In our general survey we encountered the usual variety of infectious diseases and had the impression that respiratory ailments, such as tuberculosis and pneumonia, were common. On the other hand, we did not find evidence of deficiency diseases—not even goiters.

Perhaps the most striking results of the survey were those relating to cardiovascular disease. In the survey, heart murmers were common among children, but no evidence of myocardial infarction or stroke was seen. Casual blood pressures of individuals from age 10 to 70+ were taken. They revealed a complete absence of hypertension, the highest pressure encountered was 150/90 mm-Hg. Other researchers have reported similar results for high-altitude populations, and it has been suggested that hypoxia may directly reduce the incidence of hypertension. In an attempt to trace the etiology of the low blood pressures, we subdivided our sample into a series of paired groups, first into lower- and higher-altitude groups, next into urban and rural groups, finally into more acculturated and less acculturated groups.

Significant differences in the effect of age on blood pressure appear when the sample is divided on the basis of any of these three criteria. However, it is not possible to assess the extent to which the environmental factors are independently related to blood pressure, since the total sample is too small to provide six independent subsamples large enough to give meaningful analytical results. It is our present belief that acculturation is the most significant of the factors, since altitude, urban residence, and acculturation are interrelated within the Nuñoa district population and the group differences are most striking when acculturation is taken as the criterion of subdivision. Children aged 10 to 20 years were classified as "acculturated" if they were in school, whereas children in the same age bracket who were not in school were classified as "unacculturated." Among adults, evidence of schooling, knowledge of Spanish, and use of modern clothing and specific material items, such as radios, were taken as signs of acculturation.

The results of these comparisons are shown in Figures 25.4 and 25.5. Certainly the regressions cannot be taken as evidence that altitude does not affect blood pressure, since none of the Nuñoa males, even those in the acculturated group, have high blood pressures in old age. However, the analysis does show that, within a native population living at high altitude, something associated with the process of acculturation into general Peruvian society leads to significant increases in systemic blood pressure with age. Similar results have been reported for peasant and "primitive" populations at low altitudes, and some researchers have attributed the increase to psychological stress associated with modern culture. Such an explanation might apply to the results of our study, but it does not appear safe to conclude that this is the case before carefully examining nutritional and disease correlates. The available nutritional data are being examined for evidence of possible nutritional differences between the groups.

DISCUSSION AND CONCLUSIONS

From earliest recorded history it has been recognized that men from different populations vary in physical (that is, anatomical) characteristics and in cultures, but it is only recently that the variety of *physiological* differences has been revealed. The physiological differences so far shown are of the same general magnitude as the anatomical variations. That is, the available information suggests a basic commonality with respect to

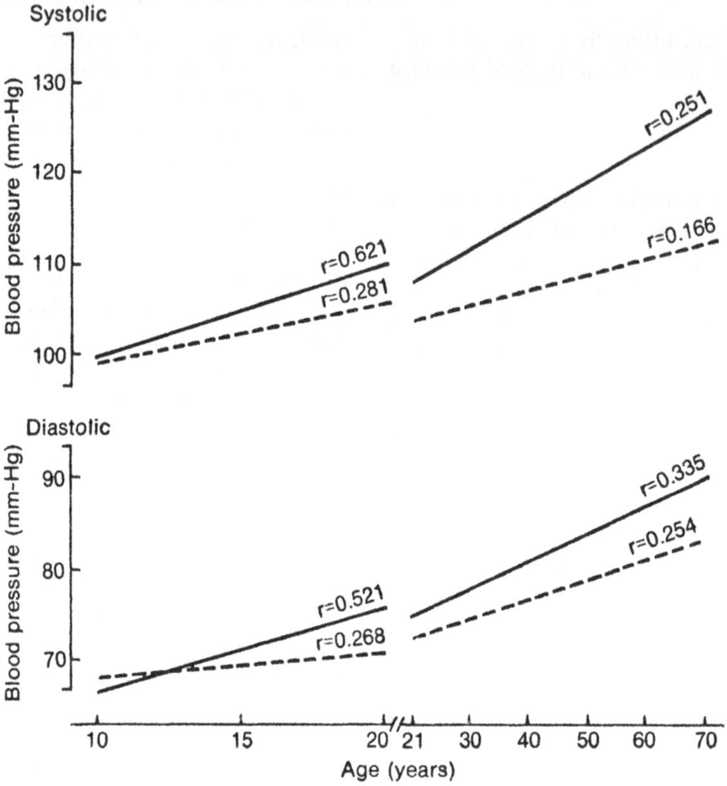

FIGURE 25.4
Changes with age in the blood pressure of male Nuñoa natives, according to level of acculturation. (Solid lines) More acculturated; (dashed lines) less acculturated; *r*, Pearson correlation coefficient.

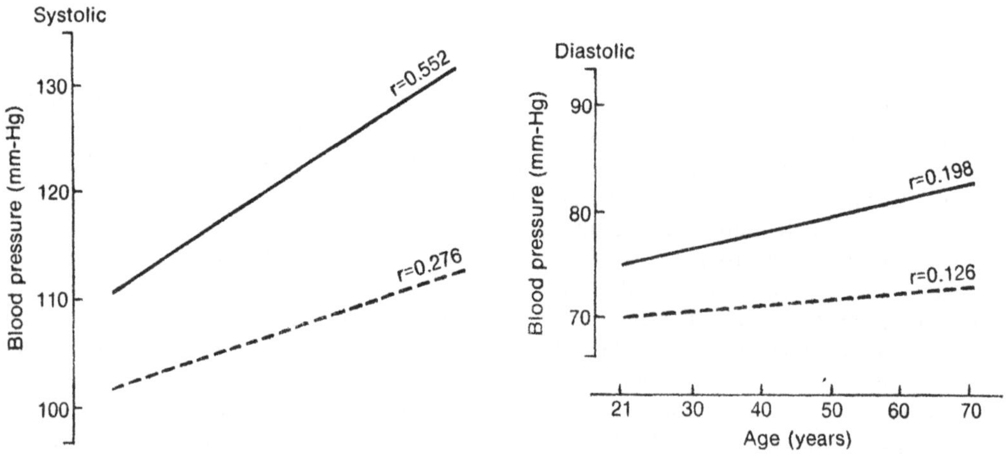

FIGURE 25.5
Changes with age in the blood pressure of female Nuñoa natives, according to level of acculturation. (Solid lines) More acculturated; (dashed lines) less acculturated; *r*, Pearson correlation coefficient.

functions such as temperature regulation, energy exchange, and response to disease, but comparison of different populations has revealed a number of specific variations in response to environmental stress.

Probably the most controversial aspect of these new findings concerns the mechanisms underlying the physiological differences. Because of the time and expense involved in studying the problem, population samples have often been small. Thus, with respect to specific findings, such as the high oxygen-consumption capacity reported by some workers for natives living at high altitudes, it has been suggested that biased sampling explains the difference. Other differences have been explained in terms of short-term acclimatization or of variations in diet or in body composition. Of course, genetic differences and long-term or developmental acclimatization have also been suggested, but the short-term processes have been more commonly accepted as explanations because they are based on known mechanisms.

We believe that a more general application of the extensive and intensive methods used in studying the Nuñoa population is a necessary next step in the search for the sources of differences in functional parameters in different populations. Thus, with respect to the Nuñoa population, a study of growth alone would probably have led to the conclusion that the observed differences in growth were the result of malnutrition.

On the other hand, the results of the growth study considered together with results of detailed studies of nutritional and other responses suggest hypoxia as a better explanation. Similar examples could be cited, from other aspects of the program, to show that a set of integrated studies of a single population can provide insights not obtainable with data pertaining to a single aspect of population biology.

In general, the data are still not adequate to treat the third question originally posed, on the sources of adaptation. Indeed we cannot even clearly differentiate genetic factors from long-term and developmental acclimatization. For this purpose one would require comparable data on several populations that vary in genetic structure, in altitude of habitat, and in other aspects of environmental background. Fortunately, collection of such data is contemplated as part of the Human Adaptability Project of the International Biological Programme. The importance of understanding the sources of population differences seems obvious in a world where the geographical and cultural mobility of peoples is greater than it has ever been throughout history.

It seems clear that the native of the high Andes is biologically different from the lowlander, and that some of the differences are the result of adaptation to the environment. How well and by what mechanisms a lowland population could adapt to this high-altitude environment has not yet been adequately explored. Moreover, almost nothing is known about the biological problems faced by the highlander who migrates to the lowlands.

SUMMARY

The high-altitude areas of South America are in many ways favorable for human habitation, and they have supported a large native population for millennia. Despite these facts, immigrant lowland populations have not become predominant in these areas as in other parts of the New World, and lowlanders experience a number of biological difficulties on going to this region.

In order to learn more about the adaptations which enable the native to survive at high altitudes, an intensive study of a native population is being carried out in the district of Nuñoa in the Peruvian Altiplano. In this area hypoxia and cold appear to be the most unusual environmental stresses. Results to date show a high birth rate and a high death rate, the death rate for females, both postnatal and prenatal (as inferred from the sex ratio at birth), being unusually high. Birth weights are low, while placenta weights are high. Postnatal growth is quite slow relative to the rate for other populations throughout the world, and the adolescent growth spurt is less than that for other groups. The maximum oxygen consumption (and thus the capacity for sustained work) of adult males is high despite the reduced atmospheric pressure at high altitude. All lowland groups brought to this altitude showed significant reductions in maximum oxygen consumption. The Nuñoa native's responses to cold exposure also differ from those of the lowlander, apparently because blood flow to his extremities is high during exposure to cold. The disease patterns are not well known; respiratory diseases appear common, whereas there seems to be almost no cardiovascular disease among adults. Systemic blood pressures are very low, particularly those of individuals living in traditional native fashion. Nutrition appears to be good, but analysis of the nutrition studies is continuing.

The results of these studies are interpreted as showing that some aspects of the natives' adaptation to high altitudes require lifelong exposure to the environmental conditions and may be based on a genetic structure different from that of lowlanders.

26. MINERAL METABOLISM AND BEHAVIOR: ABNORMALITIES OF CALCIUM HOMEOSTASIS

SOLOMON H. KATZ and EDWARD F. FOULKS

Reprinted from American Journal of Physical Anthropology, *Vol. 32, pp. 299-304.* *Solomon H. Katz is Director of the W. M. Krogman Center for Research in Child Growth and Development and Associate Professor of Anthropology and Orthodontics at the University of Pennsylvania; he is also Director of Laboratories of Clinical Research at Eastern Pennsylvania Psychiatric Institute. His principal research interests are in human ecology, child growth and development, and dental and medical anthropology. He has done field work among Alaskan Eskimos, Mexican Indians, and Pennsylvanians. Edward F. Foulks is Associate Professor of Psychiatry at Hahnemann Medical College and Hospital of Philadelphia; he continues to be involved in Arctic research. He was awarded the John Gillin Prize for his Ph.D. dissertation on Alaskan Eskimo Arctic hysteria, which is related to this article.*

■ Adaptations to habitational stress are not exclusively physiological or social. Recently, anthropologists have started to consider the consequences of nutritional factors for behavior, which are especially important as a result of the industrialization of food processing since about the beginning of the 20th century. For example, most minerals and vitamins are lost from breads and cereals in the so-called refinement of flour; the grinding machinery creates such heat from friction that the flour is for all intents and purposes precooked, effectively removing more than two dozen nutrients and about one-third of the original content of iron, vitamin B_1, and niacin.

Vitamin B_1 is necessary for muscular energy, and a deficiency in this nutrient may be responsible for personality changes such as depression, thought confusion, and forgetfulness. Brain cells get their energy from sugar, and glucose cannot be converted into energy without Vitamin B_1. Niacin deficiencies also result in psychological disturbances such as apprehensiveness, suspiciousness, mental confusion, and depression. When such nutritional deficiencies become part of a population's routine diet, these characteristics may become widespread, and adaptations must be made to them.

The problem analyzed in this selection is an Eskimo population's adaptation to calcium deficiency. This deficiency in the diet of the Thule Eskimo of Greenland is due to ecological factors and may be considered a habitational stress like that affecting the Nuñoans discussed in the previous selection. The Thule Eskimo not only lacks sufficient dietary calcium, but sunlight during the winter months is insufficient for adequate synthesis of Vitamin D, a necessary element for the intestinal absorption of what calcium is available.

Lack of calcium affects the crucial and delicate processes of the nervous system, which may result in confused, depressed, and sometimes overacting behavior. The Thule are subject to a mental disorder, *Piblokto,* that is manifest in temporary hysterical behavior and is probably due to calcium deficiency. In most societies, such a psychosis is regarded as highly deviant, and the afflicted are subject to strong social sanctions. The Thule Eskimo, however, have adapted to their habitationally produced condition by sociocultural means. Not only do they attach no social stigma to this form of behavior, but it is sometimes highly valued. In times of stress—such as famine or disease—the rituals performed by shamans strongly resemble the behavior exhibited in *Piblokto.* Thus the behavior precipitated by adverse habitational conditions may be incorporated by the members of a group as an integral part of their culture. When biological adaptation is not possible, sociocultural mechanisms may compensate. *Piblokto* among the Thule is part of their way of life.

One of the first attempts to study *Piblokto* in terms of nutritional and cultural factors is "Mental Illness, Biology and Culture," by Anthony F. C. Wallace, in *Psychological Anthropology*, edited by Francis L. K. Hsu (Cambridge, Mass.: Schenkman, 1972, new edition, pp. 363-402). An important recent study that relates social behavior to nutritional patterns is "Aggression and Hypoglycemia among the Qolla," by Ralph Bolton *(Ethnology*, 12 [1973]: 227-57). Along similar lines are two articles that reflect growing interest in some of these problems: "Evolutionary Implications of Changing Nutritional Patterns in Human Populations," by William A. Stini *(American Anthropologist*, 73 [1971]: 1019-30) and "Infectious Diseases in Ancient Populations," by T. Aidan Cockburn *(Current Anthropology*, 12 [1971]: 45-62). ■

THE BIOLOGICAL BASIS of mental disorder has recently received much attention. Nutritional deficiencies in minerals such as calcium, magnesium, and iodine, in vitamins such as B_1, B_{12}, and D, and essential amino acids have been shown to be associated with a variety of metabolic disorders accompanied by well-documented mental disturbances. Genetic factors also have been implicated recently by Kety and Kallman in the etiologies of schizophrenia and manic depressive psychoses.

However, only cultural anthropologists have made extensive studies of mental disorders in human populations. Naturally, their emphasis has been almost wholly on the cultural factors involved in mental disorder. With the exception of Wallace, little anthropological attention has centered on the biological basis of these disorders.

Traditionally, the physical anthropologist has concerned himself with the problems of human adaptation to environmental variables. It would thus appear appropriate at this time for the physical anthropologist to focus attention on the problem of the biological etiologies of mental disorder and its significance in studies of human adaptation. The study of this problem has two dimensions: The first involves the investigation of the physiological adaptations which also influence the biological functioning of the central nervous system (CNS); the second involves the investigation of the behavioral and cultural adaptations which result from these abnormalities in CNS functioning.

Our group has been using these concepts in the investigation of calcium metabolism and its role in the functioning of the CNS. We have found that abnormalities in serum calcium homeostasis are often associated with a variety of behavioral disorders such as anxiety, depression, and visual hallucinations, as well as overt neuroses and psychoses. These mental disorders have often been successfully treated in cases of hypocalcemia merely by appropriate dietary therapy with supplements of calcium and vitamin D.

THE PHYSIOLOGY OF CALCIUM

Figure 26.1 indicates the many physiological functions of calcium in the body. Calcium is the most prevalent cation in the human body. It is involved in bone formation, muscle contraction and several steps of blood clotting. Two of its other functions have direct relevance to the adequate function of the nervous system. It is involved in membrane permeability to other ions, both during the process of muscle contraction and, as is shown at the right of figure 26.1, in the conduction of a neural impulse. At the synapse or nerve ending, calcium is involved in the regulation of the amount of synaptic transmitter agent, such as acetylcholine, released upon stimulation. The net effect of lowering calcium ion concentration is to increase neural excitability. Clinically, we see this manifested as an increased neuromuscular irritability. This results in an increased reflex activity, and in the more extreme case presents as tetany. A classic sign of tetany is carpopedal muscular spasm.

Figure 26.2 shows how calcium is regulated and compartmentalized in the body. There are three factors important in calcium homeostasis. First is PTH, or parathormone, which is secreted by the parathyroid glands in response to hypocalcemia. Next is TCT, or calcitonin, which is a hormone secreted from the thyroid gland during hypercalcemia. And last is vitamin D, which forms part of the diet and/or is synthesized by the action of ultraviolet light from the sun upon chemi-

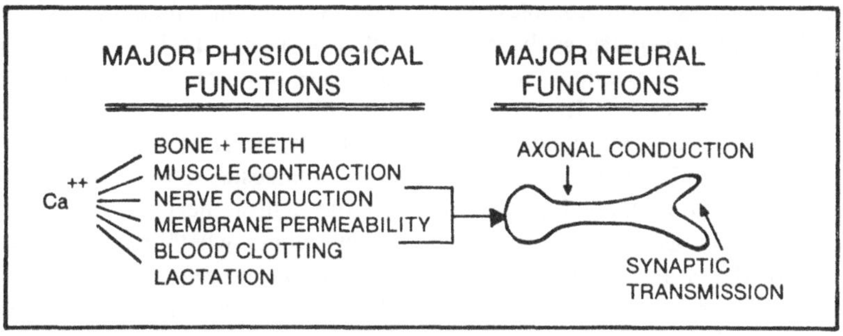

FIGURE 26.1
The major physiological functions of calcium in humans.

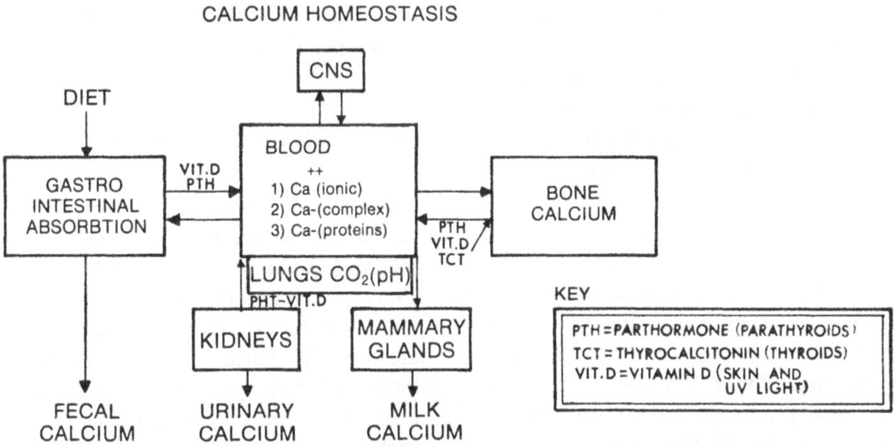

FIGURE 26.2
Calcium ION Homeostasis with all of the major compartments and the known control mechanisms.

cal (sterol) precursors at the level of the skin. This vitamin is fat soluble and may be stored in the liver up to several months.

Calcium is ingested as part of the diet. The richest sources are milk and dairy products. However, in a nutritional analysis of calcium, it is not enough to know the total amount of calcium consumed, other important chemical and physiological factors must be considered. For example, phytate, a chemical contained in a variety of cereal grains, and oxalates found in other plants, may form complexes with calcium in the gastro-intestinal system and prevent its absorption. Furthermore, the amount of phosphates consumed or available for absorption also plays a role in the amount of calcium which can be absorbed. Recently, Phang, Berman, Finerman, Neer, Rosenberele, Fisher and Granger have demonstrated that even the rate of consumption of calcium is also a very important variable in its absorption.

Once in the gastro-intestinal tract (G.I. tract), calcium is absorbed through the action of first vitamin D and then PTH. This order is important, since PTH is ineffective in the absence of vitamin D, whereas vitamin D is effective alone. Calcium, which is not absorbed or is passively secreted in digestive juices, is excreted in the feces.

In the bloodstream, calcium exists in three forms: free ionic calcium which is the physiologically active form, a form complexed to various metabolic intermediates such as lactate and citrate, and a form bound to serum proteins, especially albumin. It has been established that the calcium ion concentration is actively involved in physiological reactions shown in figure 26.1. Unfortunately, there are no methods currently available to measure calcium ion on the scale necessary for a population study. Our laboratory is now on the verge of perfecting such a method using a specially designed calcium ion electrode. Of the protein-bound and complexed forms, the former is the largest and may be the more important since albumin binds calcium in a manner that varys directly with pH. This brings into play the regulatory role of the lungs which help to regulate (blood) pH indirectly by ventillatory loss of carbon dioxide. For example, hyperventilation produces losses of CO_2, a rise in pH and consequently an increased protein binding of "free" ionic calcium. This results in a decrease in "free" ionic calcium. Although hyperventilation is probably a major factor in decreasing ionic calcium, it is important to note that lactate, a calcium complexing metabolite, may also play a role in decreasing calcium ion concentration during anxiety producing circumstances. Pitts' group has postulated that anxiety in certain predisposed individuals triggers a rapid build-up of large circulating concentrations of lactate. Since anxiety reactions are also frequently accompanied by hyperventilation, it is conceivable that if the lactate hypothesis be correct, then anxiety and hyperventilation could act in conjunction with one another to produce an even greater lowering of calcium ion concentration.

The kidneys are also involved in regulating calcium levels. Both PTH and vitamin D act upon the mechanism of renal resorbtion to raise the concentration of calcium in the blood. That calcium which is not resorbed is excreted in the urine. Another source of calcium loss very important in the female is in milk during lactation. Bone, of course, is extremely important as a reservoir of calcium for structural and regulatory purposes. Here vitamin D and PTH act to resorb bone salts and raise serum calcium levels. However, another hormone, TCT, acts to block the action of PTH on bone, but without influencing the kidneys.

Finally, with this general picture of calcium homeostasis, we can examine its effects on the central nervous system. Here, less is known about whether or not any of the hormones serve to regulate calcium secretion into the cerebrospinal fluid or its traversal of the blood-brain barrier. There is indirect evidence that calcification of various regions of the central nervous system takes place as a result of hypoparathyroidism. However, more research at the regulatory level needs to be done in order to clarify further this important aspect of the problem.

ECOLOGICAL CONSIDERATIONS

Given this complex regulatory system, we can begin to question what ecological parameters have direct relevance to the problem. Thus, we would at least expect to find possible disorders of calcium homeostasis in those areas of the world where there is low dietary intake of calcium and/or where sunlight is inadequate for synthesis of vitamin D. The arctic may be such an area.

PILOKTO—A MENTAL DISORDER AMONG THE ESKIMO

Eskimo nutritional surveys conducted by Hoygaard and by Heller and Scott indicate deficiencies in dietary calcium. There are also considerable limitations on natural vitamin D synthesis due to the lack of light during the winter (fig. 26.3). Wallace and Ackerman have suggested that *Piblokto,* a form of mental disorder occurring amongst the Thule Eskimo of Greenland may be based on hypocalcemia that stems from an inability to adapt physiologically to low intake of calcium in the diet. Here it is important to suggest the possibility that additional pathological factors stemming from the ecology of the Arctic such as widespread middle ear infections may exacerbate the symptoms and frequency of occurrence

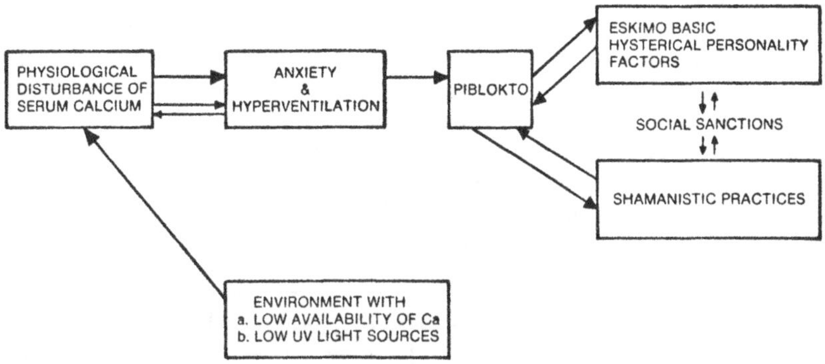

FIGURE 26.3
The relationships among ecology, physiology, behavior and culture using *Piblokto* as an example.

of *Piblokto*. This mental disorder is characterized by a prodromal period of tiredness, uncommunicativeness, and social withdrawal lasting several days. This is followed by an acutely precipitated dissociative state in which the afflicted individual may run about, tear off his clothing, imitate the sounds of animals, perform rituals such as beating sticks or a drum, and make various attempts to defy gravity. This acute dissociative state may last several hours, finally resulting in the individual's falling into epileptiform, generalized tremors with carpopedal spasms. Following this, he either sleeps or arises apparently completely recovered, and resumes his normal everyday duties. There are no social stigmata attached to such attacks. The afflicted person is thought to have lost his soul temporarily to the spirit world. The attacks may occur in either sex at any time of the year, although it is most frequently reported occurring in early spring to women. *Piblokto-Hysteria* is probably precipitated by acute anxiety accompanied by hyperventilation, which results in a decrease in calcium ion concentration climaxed by a classic sign of hypocalcemic tetany, carpopedal spasms. The low nutritional calcium in this case has probably influenced the functioning of the central nervous system and has resulted in behavioral changes.

The Eskimo has made certain adjustments to this phenomenon. Gussow pointed out that the symptomatology of *Piblokto* is mirrored in institutionalized Eskimo group responses in times of stress. This is seen in times of famine or disease when the shaman performs rituals which are reported to parallel closely the behavior seen in *Piblokto*. Gussow further emphasizes the possibility of a circular model of mutual influence for the two behaviors.

Now we can begin to see the full significance of ecology influencing dietary calcium and vitamin D. Under these ecological conditions the normal physiological mechanisms cannot adapt to these deficiencies. The result is a possible abnormal effect of calcium upon the central nervous system. This being the case, a mental disorder results which is expressed in a particular sociocultural context. Here cultural and behavioral adaptations begin to compensate for the inadequate biological adaptation.

BIOCULTURAL ADAPTATION AND HUMAN EVOLUTION

Now we are in a position to ask how natural selection operates in this situation. It is likely, and in fact probably unquestionable, that selection takes place at the biological level, perhaps within the G.I. system, the hormones controlling calcium, the kidneys, or at the level of bone. But it is equally clear in this case that selection also involves the behavioral manifestations, since it is these that interact with the socio-cultural milieu in which adequate mental function is critical. . . .

Figure 26.4 indicates the generalities that can be abstracted from our investigation. There are a large number of ecological variables, such as those influencing nutrition, disease, climate, altitude, diurnal variation and others which influence the function of various physiological systems. When the ecology is so varied that these physiological systems cannot adapt and they begin to influence the biological function of the central nervous system directly, then aberrant behavioral adaptations often take place. Here the socio-cultural system comes into play in human adaptation, and natural selection becomes complicated by processes of cultural adaptation and selection.

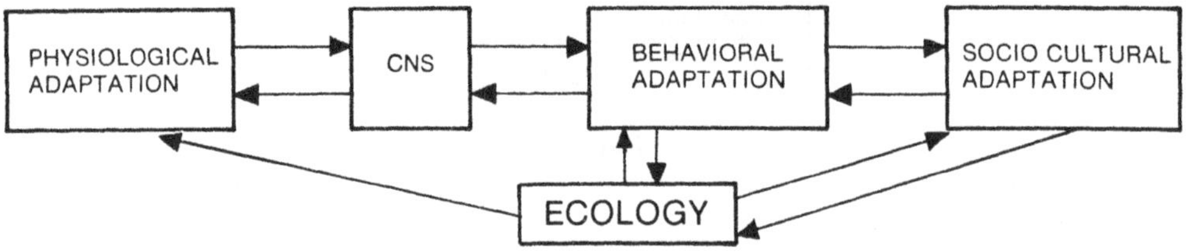

FIGURE 26.4

The general relationships and interactions of ecology, biology, behavior and culture involved in human adaptation.

27. THE SPEARMAN AND THE ARCHER

ALICE M. BRUES

Reprinted from American Anthropologist, *61 (1959): 458-69. Alice M. Brues is Professor of Anthropology at the University of Colorado, after a number of years spent as a teacher of human anatomy at the University of Oklahoma. Her interests range from skeletal variation to genetic polymorphisms. Some of her recent publications present computer-derived models of selection and evolution in man.*

■ The natural habitat in which people live constitutes only one—albeit an important—aspect of their environment. When a group introduces a new source of energy or a new mode of gaining a livelihood into the habitat that its members seek to exploit, they thereby alter their environment. By altering their relationship with their habitat they change the total balance of forces to which they must adapt. Just as people must adapt to the pressures of their habitats, they must adapt also to the implements they use for gaining a livelihood from the habitat. As will be seen below (Part VI) and in the accompanying volumes, one of the most important adaptations that people have to make to the introduction of new sources of energy for gaining a livelihood is in their organization of social relations. In some cases, however, physiological adaptations are important accompaniments of technological factors.

The evolution of locomotion, of the hand, and of perception were indispensable to the development of weapons such as the spear and the bow and arrow, which are more efficient than the bludgeon or the hand-axe. The spear and the bow are not only efficient as destructive weapons, they also are important for the survival of the hunter. If you wish to kill a lion or a bear and attack him with only a bludgeon or some other implement that requires you to get very close to the beast, your proximity obviously endangers you. A spear or bow makes it possible to maintain a healthy distance from the prey until he is dead or sufficiently weakened from loss of blood to make it safe to kill him.

Can the same person be equally good as a spearman and an archer? Is it merely accidental that the spear is found among people in some parts of the world and the bow and arrow among others? The bow seems clearly more advantageous than the spear; why did spearmen who knew about the bow and arrow retain their spears instead of adopting the more efficient weapon?

In this selection Brues shows that there is an intimate relationship between body build and weaponry. Before early man developed spears and bows—when his armament was confined to bludgeons—a large body size was the most effective adaptation in stature, but other body builds are more effective adaptations for the use of projectile weapons. For example, a linear (ectomorphic) body build is most adaptive for a spearman because it gives him the physical proportions that enhance the speed leverages in the body. This build, more than any other, meets the velocity requirements that are necessary for spearing. Hence, after the spear was introduced into the technology of early man, a linear body build had a decided selective advantage: Such people would survive longest and produce the most offspring. Brues also surmises that one of the reasons Neanderthal man lost out in conflict with *Homo sapiens* was that Neanderthal was heavily built and geared more to power than to speed. Even if he learned about spears, he could not have used them so effectively as his ectomorphic competitors.

The next major invention was the bow, but the velocity of the arrow does not depend on the hand's speed of motion. The amount of energy stored in a drawn bow depends on the length of the draw, which is favored by short limb segments and relatively short and thick muscles—the very opposite of the most favorable body build for throwing a spear. Hence the bow spread most rapidly among extremely muscular people of short stature with well-developed shoulders and upper extremities (a mesomorphic physique). A fascinating test of this hypothesis, in respect to the Bushmen of South Africa, is offered by Tobias in Selection 28.

Brues' paper again raises Medawar's question in Selection 4: What is fitness? She suggests that not all superiority is superior: The bow and arrow theoretically is a superior weapon to the spear, but whether it is superior in any given case depends on the hunter's body build.

Some of the more general problems that underlie this inquiry are discussed extensively in "Ecology and the Proto-Hominids" by George A. Bartholomew and Joseph B. Birdsell (*American Anthropologist*, 55 [1953]: 481-98). Specific case studies in stature and adaptation are to be found in three articles by Marshall T. Newman: "The Application of Ecological Rules to Racial Anthropology of the Aboriginal New World" (*American Anthropologist*, 55 [1953]: 311-27), "Adaptation in the Physique of American Aborigines to Nutritional Factors" (*Human Biology*, 32 [1960]: 288-313), and "Evolutionary Changes in Body Size and Head Form in American Indians" *American Anthropologist*, 64 [1962]: 237-57). ■

INTRODUCTION

BODY BUILD PRESENTS many interesting aspects to the student of selection in man. Various local and minor groups have become specialized in body size or body shape, though most tend to cluster about a general average for the species, with a considerable degree of variation within each group. As between major races, or populations on a continental scale, the differences are unclear; any attempt to characterize a major race by a typical physique can be countered by the citing of exceptions. In addition, body build has a fairly apparent relation to the environment and man's energy exchanges with it, though the very degree of polymorphism which it exhibits leads us to suspect that its survival value involves numerous elements, some of which are rather obscure. Recent clinically slanted studies of body build have in fact emphasized the durable quality of physiques which are not considered strong in the muscular sense. These findings can explain in part the retention by most populations of considerable genetic variability with respect to body build. If, as is probable, the implications of body build for health and longevity are more complex than has yet been demonstrated, the polymorphism of body build may prove to be influenced by a balance between physiological factors which work at cross purposes. Such a balance could be unrelated to specific environment; or it could be profoundly affected by it, as in relation to temperature, nature of available foodstuffs, or the presence of specific transmissible diseases or occupational hazards.

A further possible element in selection, with which this paper proposes to deal specifically, is the relation of body build to the efficiency with which different activities can be carried out, particularly those activities which vary with culture and means of subsistence. In our recent civilization we may justify the maintenance of polymorphism in body build as suitable to the carrying out of a variety of occupations; but this explanation cannot be projected to ancient and primitive peoples who lack such a degree of occupational specialization. In such groups we would expect that the characteristic tools and techniques of each economy and cultural stage would establish an optimum type of body build toward which selective trends would aim during that period. The present essay will attempt to explore the implications of this thesis with respect to the evolution and distribution of the varieties of body build.

If any doubt remains in regard to the potentialities of selection in body build, it may be resolved by consideration of changes wrought in domestic animals. Smaller species such as the chicken and rabbit have been brought to body weights commonly three or more times that of the wild progenitors. Dwarf or pygmy strains have also been bred out from time to time, the most recent being a minuscule pig intended as a laboratory animal. Body build selection in the Sheldonian sense is also apparent—as the development of ectomorphy[1] in the race horse as opposed to mesomorphy[2] in the draft horse. Among cattle the beef breeds have been selected for mesomorphy in order to increase

1. Ectomorphy is characterized by linearity and fragility; small delicate bones; slight, "thready" muscles; relatively long thorax; drooping, narrow shoulders; relatively long end parts of extremities.

2. Mesomorphy is characterized by squareness and hardness; heavy trunk, massive limbs and muscles; broad shoulders; relatively large and heavy end parts of extremities, as in thick heavy wrists, hands, fingers.

the bulk of edible muscle, and a generation ago the domestic pig could have been put forth as the epitome of endomorphy.[3] However, with the increase in use of vegetable oils and the declining price of lard, hog breeders have rapidly faced about, and the endomorphic lard hog has been replaced by a more mesomorphic type which produces better chops. Here body build has been shown susceptible to selection even to the extent that the trend is readily reversible. There is no reason for supposing that the potentialities of selection for body build in man have been any less, though natural selection even under cultural conditions will probably not be as rigorous or rapid as artificial selection, particularly as practiced by the genetically sophisticated breeders of the present time.

Various possible selective influences on body build have been studied in many animals, some in fact being suggested and demonstrated (as well as one can demonstrate the progress of evolution) by comparison of the build of various species with their capabilities and habits. One of the most obvious and first to be noted is the advantage of slender build with long limb segments in the attainment of speed of movement. Another factor is "strength," usually a rather poorly defined term, with respect to the self-defense of an individual who stands his ground instead of fleeing. It is notable that carnivores, which must generally reach their prey by speed and dispatch it by direct force, tend to remain average in build. The predatees, however, are prone to reject this compromise and choose a policy either of flight, with specialization for speed, or of static defense, with a tendency to increased body size. In fact, one of the most insidious of evolutionary traps is the sacrifice of mobility for indestructible size, as seen in our closest kin the gorilla. As we shall see later in the analysis of factors of strength, lateral build also is favorable to static defense, and is developed concomitantly with increasing size in many cases. Large size and comparatively lateral build are also favored by cold environment, since they tend to retard heat loss by reduction of relative surface area. Small body size may be an advantage if an animal can slip into crevices not large enough for others to pass; it has been suggested that the development and survival of pygmy types of man in dense jungle environments has been favored by selection on this basis. Endomorphy presents advantages in a cold climate because of the favorable insulating effects of the subcutaneous fat layer. Endomorphy has also been suggested as an asset where food supplies vary greatly from season to season, since the endomorph presumably stores excess aliment more readily.

3. Endomorphy is characterized by softness and roundness; small bones; relatively long trunk; chest wide at the base with a high and faint waistline; short, tapering limbs; weak, small hands and feet.

In evaluating physical efficiency from the complete biological standpoint, we must also take into consideration the dietary requirements of different sized individuals. In the rigors of primitive culture a large body may be very costly to maintain. Body bulk must be used efficiently in terms of its food-getting capacity; for it will little profit a primitive hunter if his hunting proceeds increase 10 percent while his appetite goes up 11 percent. For this reason the nicest adjustment of body to activity will be by change of shape rather than size. In the course of this study the author had occasion to examine anthropometric data on modern athletes distinguished in various sports. It revealed little except that they were all fine, large individuals compared with the general population—in fact, veritable giants compared with most peoples in a primitive population. One suspects that if their athletic accomplishments were strictly weighted according to the amount of food it took to keep them going, they might yield their world's records to their smaller competitors. Most of them would undoubtedly be expensive ornaments in a primitive society. Certain data on stature in relatively modern populations, in fact, suggest that there is a tendency for the larger individual (and race) not to work, but instead to bully others into working for him.

WORK, FORCE AND ENERGY

In order to understand the potentialities of

various body types for various cultural techniques, we need to analyze the concept of "strength" in physical activities. The word "strength" is not found in the physicist's vocabulary; it is a lay term which inextricably confuses the physical concepts of work, force, and energy. Force is the simplest of this constellation of concepts; it is measured in pounds, and in its simplest form is that influence which a heavy object exerts by resting on something; force can also be exerted in various directions and by various means other than the collaboration of mass and gravity. The concept of force leads to that of pressure, measured in pounds per square inch, which depends on the area through which a force is exerted on a surface; pressure, so defined, is important in overcoming the resistance of an object which is to be broken or altered in shape. Force and pressure are measures of a momentary state only. When force acts so as to move something for a distance, we say that work has been performed. That which can produce work, when properly directed or released, is known as energy. Energy may be electrical, chemical, or of other forms; at present we are mainly concerned with mechanical energy, which may be potential, if embodied in a configuration of matter under tension, as in a drawn bow; or kinetic, if embodied in motion, as in a flying spear. Kinetic energy is the most readily measurable of energy forms, being expressed by the product of the mass and the velocity of a moving object, and from it we infer the amount of energy that was necessary to set the object in motion. Energy is conserved through various transmutations between potential and kinetic energy and work performed; in the case of energy manifested by the animal body, it is ultimately derived from the chemical energy present in foodstuffs, and mediated by the action of muscle.

The peculiar role of muscle tissue in energy exchange is its capacity to shorten against resistance. This may produce free motion in a part of the body or something grasped by it; more exactly, it accelerates the mass of a portion of the body and attached object and causes it to attain a velocity, thus transforming chemical into kinetic energy. If the motion produced is resisted by some elastic structure, or in other analogous ways, much of the energy is stored as potential energy. A muscle, contracting, exerts a certain amount of force; and the force which it can exert under maximum effort depends on what is called its physiological cross-section. The latter, except where certain special arrangements of muscle fibers exist, is equivalent to the cross-section area of the muscle itself. This force of contraction, it should be noted, is the pull in pounds as exerted at the point on the skeleton where the muscle inserts. The length of a muscle does not affect the force of the pull which it exerts, but, since muscle fibers can only contract a certain fraction of their length, does determine the distance over which motion at the insertion of the muscle can be produced. A short stout muscle will exert greater force over a less distance, and a long thin muscle less force over a greater distance, if their total bulk (cross-section × length) is the same, their capacity for work (force × distance) is the same, although their exact properties and capacities will differ because of their shapes.

In nearly all cases the force exerted by the contracting muscle is not applied directly to the outside environment, but is mediated by certain lever arrangements within the skeleton. One of the simplest examples which can be given—as well as one applicable to many work operations—is the contraction of the biceps muscle producing bending of the elbow. The muscle inserts into the radius below the elbow joint, and the ultimate force against the environment is generally exerted by the hand. In this arrangement the hand always moves through a much larger arc than does the point of muscle insertion, and by simple mechanical law the movement of the hand while greater in distance is less in force. The exact ratio of reduction of force in this arrangement depends on the relative lengths of the segment of the skeleton lying between the joint axis and the muscle attachment, and the length of the segment between the muscle attachment and the part of the hand with which the force is finally applied. If the total length of the forearm is reduced, with the muscle attachment remaining in the same place, the force of the hand movement is increased and the distance of movement decreased. Given the same force, as measured at the muscle attachment, and the same rate of contraction, there is an inverse ratio between the force of pull of the hand, as it would be measured by a tension dynamometer, and the speed with which the hand can be moved in a flailing or throwing action. The power leverage is accomplished by a short forearm, the speed leverage by a long forearm. Since a long thin muscle, as we have seen above, exerts less force, but over a greater distance, than a short thick one, increasing length of the proximal as well as the distal segment of the limb accentuates speed and lessens force. Conversely, the shorter proximal muscle as well as the shorter distal limb segment increases force. It is easy to see then why the stocky individual is thought of as "strong" in terms of the force with which he can exert momentarily in lifting a weight or crushing a resistant object, a common lay concept of strength. (Tappen even found an achondroplastic dwarf among his champion weight lifters.) This is in spite of the fact that a tall slender individual of the same weight will have approximately the same work capacity (dependent on total muscle bulk) though his muscular system, due to shape of muscles and skeletal leverages, cannot at any moment exert as many pounds of force on a resistant object. The apparent paradox that the physically "strong" individual is not always constitutionally strong appears in this light to be based on an oddity of definition. This individual is "strong" not because he is made of different material but simply because he is of a different shape.

MECHANICS OF ESSENTIAL ACTIVITIES

The most primitive of the mechanical actions in which body form is selective is locomotion. The striking variations in limb size associated with brachiation versus terrestrial progression have been exhaustively discussed in relation to the evolution of the higher primates. It should be pointed out, however, that terrestrial progression is not precisely the same thing in all environments. Running is the most rigorous of bipedal acts and the one requiring the highest specialization for bipedal progression; yet there are many environments in which running can be performed only intermittently, and only in fairly open country can running speed alone be relied on for either pursuit or flight. Elsewhere one must have the ability to pass obstacles by leaping, climbing, or in the case of light brush, crashing through. Of these activities, only leaping is consistant with the type of build which is optimum for running. It appears likely that the original specialization of the human leg took place with *Australopithecus* and related forms, in an open prairie country where continuous running and leaping were possible. Such an environment is conducive to running specializations in all animals, and generally results not only in leg lengthening but in overall slender build. This effect would hardly be expected in a terrestrial animal dwelling in a tropical forest where progression among underbrush and vines is slow and crooked, except for the animal which has sufficient bulk to tear through obstacles by sheer weight. This environmental difference probably determined the divergence in build between the gorilla and the progressive terrestrial hominids.

An environment which offers special problems for terrestrial locomotion is that of the northern forest. Here large tree-trunks, not subject to decay as in warm climates, lie fallen and sometimes piled up two and three deep, so that walking, let alone running, cannot be performed for more than a few yards continuously; in fact, the technique of choice often is to traverse the tops of the barricades and avoid descent to the ground. Travel in such an environment involves as much or more climbing than walking, so that although terrestrial in the technical sense, it requires some of the talents of an arboreal animal. Probably the evolutionary lag in the limb skeleton of Neanderthal man is related to the fact that he remained close to the edge of the continental glacier in a zone which produced this type of forest.

Under primitive conditions a critical operation for survival is the use of weapons, particularly against larger animals and man. Their use against a human enemy is relatively rare, but particularly critical when conflict occurs. In killing animals for food, especially the larger ones, much is at stake in a single motion both as regards the safety of the hunter and the amount of food represented by the victim. It should be remembered also that earlier hominids and men, with their less developed armamentarium, were not dealing with animals which had acquired a "fear of man," and were in real danger of attack upon themselves. Hence efficiency of weapon use, if related to body build, makes the latter a selective factor of considerable importance.

The aggressive and defensive techniques of a terrestrial hominid in the pre-weapon stage would probably have to follow the pattern of the gorilla, whose destructiveness is in proportion to the amount of squeezing or crushing force exerted momentarily on the fragile parts of the victim. Such a technique places a premium on large muscle bulk combined with power leverages in the skeletal and muscular arrangements. The end result of selection under such circumstances is large size with lateral build and limb segments short relative to muscle bulk. This physique in a pronounced form drastically reduces speed of locomotion, and such an animal finally becomes a herbivore, since, though he could kill anything he could catch, he cannot catch anything.

The first presumed hominid weapons are the ungulate femora described by Dart as the weapons of *Australopithecus.* These fall into the general class of "blunt instruments" and are associated with a besticidal (and homicidal) technique which we shall refer to as "bludgeoning." In this technique the typical weapon is designed to crush some part, most effectively the skull, rather than to penetrate deeply, and hence does not have a sharp point. Such a weapon is not generally thrown, since its crudity makes it advisable to retain it in the hand for repeated blows as needed. It seems reasonable to assume that as long as artifacts are blunt or axe-like and not apparently adapted to tipping a projectile, some form of the bludgeoning technique was in use.

Bludgeoning does not require any very specialized application of energy. Within reasonable limits, the destruction wrought on the victim will depend on the total amount of energy absorbed; that is to say, a four-pound club moving at the rate of ten feet per second will probably smash as much skull as a two-pound club moving at the rate of twenty feet per second, and so on, unless the velocity is so low that the blow simply pushes. Thus the effectiveness of the bludgeon can remain the same with two factors reciprocally varying, namely, the weight of the object accelerated and the velocity which it attains. In this case neither momentary force nor speed of action need be at a maximum; rather, the total amount of energy embodied in the moving weapon, and therefore the total amount of muscular work performed, determines the effectiveness. Thus the determining factor in terms of body structure will be the aggressor's total bulk of muscle, rather than specific proportions of leverages. Equal matching of opponents under these circumstances could be adequately brought about by weight classes as in modern boxing. A reasonable amount of variation as between linear build with speed leverages and lateral build with power leverages, would not affect efficiency. Therefore, during the stage when bludgeoning was the typical weapon technique, large body size would be a favorable selective factor without particular favor to any extreme of build, though probably at first laterality of build would be carried over as a concomitant of some continued use of wrestling or crushing behavior.

The invention of projectile weapons introduces selective factors other than total muscle bulk. It has been a general trend in the development of projectile weapons, from spear to bullet, that the size of the projectile has decreased and the velocity increased. There are sound reasons for this. The amount of kinetic energy embodied in the projectile, and consequently the amount of destruction that it can produce in the object which brings it to a stop, is, as in the case of the bludgeon, a product of the mass of the projectile and its velocity. Hence the size of the weapon can profitably be decreased if its velocity increases. There are definite advantages to this change. The moving projectile is always subject to the force of gravity, which draws it off course and eventually brings it to the ground. In the case of the light but rapidly moving projectile, the influence of gravity is far less in proportion to the momentum carrying it in the direction of its aim. Hence the range and accuracy of a projectile can be greatly increased by trading weight for speed, while its destructive power remains the same. As a result, the history of projectile weapons is a succession of inventions for increasing the speed at which the projectile flies. (A secondary effect, due to aerodynamic considerations, is increase in the specific gravity of the projectile.)

In the case of the spear, the first of man's projectile weapons, maximum efficiency will be attained by increasing to a maximum the velocity of the spear as it leaves the hand. There were probably intermediate stages from the bludgeon, through some instrument used as a pike, to the spear thrown first crudely over a short distance (as the bludgeon or club might be sometimes thrown) to the true spear designed to be propelled at high speed from a considerable distance. There is a critical point in this process which is related to the form of the tip of the weapon. In bludgeoning, where more than one blow is generally struck, it is desirable that the weapon shall not penetrate; if it does, it is difficult to withdraw for successive blows. In the fully developed spear, the point is made sharp and small, so as to give a maximum pressure at the point of contact and penetrate deeply, preferably to some vital spot. It cannot then be withdrawn readily, and if a second try is made a spare weapon may be needed. As soon as a sharp point is placed on a weapon, the user is committed to a technique of maximum penetration with accurate aim. He not only can operate at a distance but operates better so, since if actual bodily contact with the victim occurs, aim is difficult and the inability to withdraw the weapon becomes dangerous. This is probably why the pike has never been popular as a weapon except for purposes such as boar-hunting, where danger is construed as sport—a concept foreign to the primitive, in most cases.

We may assume, then, that as soon as we find weapon points designed for penetrating ability, a course has been set for the development of maximum projectile velocity in the hunting technique. (How much earlier this may have taken place we cannot know, since it is likely that the first weapons designed for throwing were merely sharpened and that the adaptation of stone work to a narrow spear point was rather later than the first use of the sharp-pointed weapon.) The importance of velocity in the use of the projectile immediately places a premium on speed leverages in the body. We should expect, then, that concomitant with refinement in shape and increased penetrating power in stone artifacts, a situation is arising in which linear body build is becoming the most efficient type for weapon manipulation. This raises several interesting questions with regard to the spear as a typical weapon. The physique of a group might influence its likelihood of adopting the spear as a standard weapon after it had become known to them, even though it was potentially a superior weapon to the bludgeon. This would produce a kind of group selection in which more linear races would have an advantage because of their ability to exploit the newer weapon; we might imagine, for instance, that one of the weaknesses of the Neanderthals in their final conflict with *Homo sapiens* was not simply that the latter had superior weapons of projectile nature, but that the Neanderthals could not have used these weapons to advantage even after attaining knowledge of them, because of their heavy build and adaptation to power rather than speed. (We might call them "muscle-bound"; this term is the derogatory synonym of "strong" and denotes the loss of speed and agility which accompanies specialization in the direction of force.) Later, within a group in which the spear had become a standard weapon, there would be selection in favor of the individuals who, because of linearity of build, were able to attain the maximum in range and accuracy with it, and thereby enjoy a better food supply. (It should be noted that any improvement in hunting technique thins out the supply of game and increases its wariness, so that the less well equipped group or individual can no longer survive with a technique that was formerly adequate.) In either case, whether competition was within the group or between groups, the change-over from bludgeon to spear would eventually cause the proportion of linear individuals within the species to increase, by creating a new standard for optimum body build. The heavy physiques which had been most efficient in the use of the bludgeon would now be selectively discriminated against, while individuals of linear build would multiply. An actual decrease of body weight would be an asset since it would decrease food requirements.

A second stage in the use of the spear involves development of a throwing stick which artificially extends the length of the limb and so increases the velocity of the spear. Within limits this extension of the throwing arm appears not to decrease accuracy appreciably. This affords a means of compensating for the disadvantage of a short arm in the use of the spear. Possibly it was devised as a means of adapting the spear to the use of peoples of more lateral build, who would not have been apt to have developed the spear themselves but might have received it from others. The possibility that the throwing stick represents a compromise with body build finds confirmation in the fact that the very linear spear-users of Africa generally throw the spear with the bare hand; apparently the throwing-stick has little to offer to a physique with maximum built-in speed leverages.

The next major invention in weapons was the bow, which offered a means of further increasing velocity and allowing reduction in mass of the projectile. The mechanics of the bow are totally different from those of the spear and must be carefully considered in relation to body build. As we have seen, the determining factor in the efficiency of the spear is the velocity with which the weapon leaves the hand, and it is favored by linear build and bodily leverages conducive to speed. High velocity is the desideratum of the arrow also; but the velocity of the arrow is not dependent on speed of motion of the hand. The energy imparted to the arrow, which, making due allowance for the mass of the arrow, determines its velocity, is stored as potential energy in the drawn bow and is in no way affected by the speed with which the bow was drawn. (In fact, some late and powerful types of crossbow were wound up with cranks.) The amount of energy stored in a given bow varies with the length of the draw; and since the pull becomes harder the further the bow is drawn, the critical factor is the maximum absolute pulling force which the drawing arm can exert just before the arrow is released. (More exactly, perhaps, the force which can be exerted with sufficient ease that aim is not impaired.) Since the possible length of draw is limited by the individual's arm length, best results are obtained if the stiffness of the bow is adapted to the individual archer's pulling ability so that he is exerting his maximum pull at the optimum length of draw. The importance of momentary force in efficiency of use of the bow entirely alters the selective advantage of body build. The archer requires a power leverage in the arm, which is favored by short limb segments and relatively short and thick muscles; the exact opposite of the most favorable structure for throwing a spear. We must imagine, then, that any selective effects of weapon use on body build were reversed when the bow supplanted the spear. The incomplete distribution of the bow and its failure to become the dominant weapon among some peoples who knew it shows that for some groups it was not worth while to make the transition, in spite of the theoretical superiority of the bow. The bow probably developed and spread most rapidly among peoples who were of short stature and relatively mesomorphic, and by the process of selection made them more consistently so over the course of time. Insofar as selection could be even more specific, we would expect the use of the bow to favor the increase of individuals whose laterality of build was particularly well expressed in the shoulder and upper extremity ("omomorphy"). Our present esthetic standards for the male physique seem still to reflect the influence of an age of archery.

The ectomorph's adaptation of the bow is to increase its length, as in the famous English long bow. Here the distance between the position of the string at rest and when drawn is increased. In this arrangement less force, multiplied by greater distance of draw, can store the same amount of energy. This, however, involves very considerable loss of efficiency. When the length of the bow is increased, the thickness must be increased also, to prevent the draw becoming too easy. This markedly increases the total mass of the bow, and as a result much of the energy stored in the bow is expended in accelerating the free ends of the bow itself. In order to transmit a greater proportion of the stored energy to the arrow, the mass of the arrow must be increased. This change—greater mass and less velocity—is a backward step as far as efficiency of a projectile weapon is concerned. It is for this reason that the short Turkish bows attain a greater maximum performance than heavier bows.

Influences of physique may perhaps be seen in other aspects of the hunting technique associated with various weapons. Neanderthal man is often and probably correctly pictured as capturing game by traps or surrounds in which the animals were dispatched by bludgeons. This is a plausible picture, since Neanderthal man's skeleton is not that of a swift runner. Inability to outrun the game, and a need to approach close to kill it, would require some special devices. (In justice to the species, it should be noted that such a type of hunting would require a greater degree of intelligent planning and cooperation than was needed by their fleeter contemporaries.) If Dart's suppositions are correct, the picture we have of the hunting Australopithecines is anomalous. These hominids apparently ran down game with their newfound bipedal celerity and then clubbed it to death. A light running and leaping animal should not have to carry a heavy club in his hand, and furthermore he cannot use it as efficiently as a larger and heavier individual. Perhaps we may see two ways of solving this dilemma, as shown by two types: the Neanderthal type, who kept the bludgeon and developed a heavier physique to go with it; and the sapiens type, who invented the spear and then were directed by selection toward a linear build, with further improvement of running ability. It is interesting to note that recent recrudescences of the bludgeon principle, the clubs and maces of medieval Europe and Polynesia, have appeared among peoples of comparatively large stature and balanced physique. The typical association of the spear with approach to game by running, and the bow with approach by stalking, is interesting. This is undoubtedly due in part to the fact that throwing of the spear can more readily be combined with running in one continuous movement, while aiming of a bow requires a stationary moment during which already alerted game can increase its distance. However, it would appear that the habitual spear-thrower, because of his linear build, would as a rule be a swifter runner than the typical archer.

With the coming of the Neolithic economy there are new instruments of culture and changed emphases on old ones. The hunting technique becomes less important as hunting itself becomes a sideline; and Neolithic man becomes more and more a wielder of the hoe or other soil-stirring apparatus. The mechanical principles involved in the use of projectile weapons are no longer pertinent. Destroying the cohesion of the soil is work, in the physicist's as well as the layman's sense, and the amount accomplished is in proportion to the amount of muscular energy applied. The hoe is in effect a bludgeon, which requires neither speed of action nor necessarily great momentary force but rather an uninspiring, back-breaking combination of both. The wood-cutting axe of the Neolithic is also a bludgeon with respect to its manner of use and demands on physique. The optimum food-getter of the Neolithic economy is not the long-limbed spearman or the broad-shouldered archer, but a sturdy peasant of medium build. Of course, beginning in the Neolithic and continuing into modern times, there has been an increasing development of means for the individual or the group to escape work in the sense that is represented by the hoe and the axe. Herding, cultivation of vine and tree crops, and finally special trades, have been outlets for groups or individuals of less work capacity. However, well into the 16th century, human power harnessed by the treadmill or galley was a simple commodity of industry in which each individual was roughly worth his weight in muscle. It is interesting to speculate whether the lead taken by North European groups in the development of an agricultural-mechanical civilization may have resulted from their having been retarded by their marginal position in receiving specialized weapons, and consequently having passed, with a minimum interval, from the bludgeon stage of hunting to the threshold of the Iron Age. Thus, bypassing the stage of linearity which would have resulted from a long spear-using period, or the wide-shouldered specialization of a period of archery, they approached the heavy toil of an early civilization with a physique preadapted to it. If Coon is correct, there may have been appreciable continuity of Neanderthal blood as well.

CONCLUSIONS

Since this paper is presented as an essay and not as a finished study, the term "conclusions" is used with some hesitancy. Perhaps only one conclusion should be drawn: that the customary concept of man as physically unspecialized should be regarded with doubt. In contrast to animals whose way of life is rigidly determined by their physical bodies, we are impressed with the physical versatility of man. However, it is interesting to speculate that even within our own species there has been some correlation between physique and habitual activity, resulting in a reciprocal influence between culture and body build. This influence may take several forms: a dominant weapon or tool may alter the average physique of a race using it over the course of time by giving a selective advantage to individuals of a body build best adapted to its use. It may also alter the numerical proportions of conflicting races of which one is physically better adapted to the use of a valuable implement—this better adaptation perhaps being itself the result of a long intra-group selection. And differences in physical type between races may retard the transmission from one group to another of a new tool or weapon, and with it a whole new way of life.

A series of slightly disconnected suggestions is made on the basis of this hypothesis. *Australopithecus* is pictured as imperfectly adapted physically to his way of life, having a light linear build appropriate to his running habits and open plains habitat, but resorting to a bludgeon-like weapon which would be more effectively used by a heavily built individual. Neanderthal man was physically well adapted to his weapons, having the heavy muscle-bound physique best adapted to the use of blunt crushing implements. In contrast technically and physically were the precursors of *Homo sapiens*, who developed the spear as the first projectile weapon and were then selectively directed toward greater linearity of build, at the same time developing and enhancing the bipedal skill of *Australopithecus* more rapidly than did *Homo neanderthalensis*. This "linear" stage in the evolution of physique has been preserved, possibly even exaggerated, in the contemporary Australian, who is still a spearman. Most of the peoples of Africa reflect this stage to a greater or lesser degree. It is suggested that the submergence of *Homo neanderthalensis* resulted from the unequal contest between the newer projectile weapon and the older bludgeon, to which Neanderthal man was bound because of his physical adaptation to it perhaps as much as by ignorance or conservatism. Soon the more refined projectile weapon, the bow, appeared, and though its exact origin may be doubtful, there is no doubt that it reached its highest development among the central Mongoloids, who show in lateral build and strong shoulders the highest adaptation to its use. The bow became widely known but in spite of its technical advantages was not everywhere adopted as the principal weapon; and in the case at least of the extremely linear peoples of East Africa, the reason for its rejection appears to be that the prevailing physique was so preferentially adapted to the use of the spear.

At the time of the beginning of agriculture a new selective trend appeared, in the direction of a generally heavy build, capable of sustained labor. The physical types most highly specialized for the bow and the spear were poorly adapted for agricultural pursuits, the very linear spearman being the least effective, though the idealized archer, with light hips appended to his broad shoulders, will also be somewhat inadequate for heavy labor. Where ecological conditions were favorable, herding became the high culture of the most linear Africans and the most lateral Mongoloids. Other groups of these same races compromised with agriculture, probably with a slow subsequent modification of body build. Since body build is difficult to judge from the skeleton, it is interesting to note that body proportions are to some extent reflected in the cranium, so that in the marginal position and decline of dolichocephaly, both in the Old and New Worlds, we may see reflected the overwhelming of the spearman by the archer and finally by the agriculturist.

All of these suggestions are speculative and should be critically questioned in principle, as well as with respect to those details which are found to have been misrepresented as a result of overgeneralization. It is hoped, however, that these ideas will illuminate the complex problem of selection in man. Any selective effects suggested here must be considered jointly with other types of selection; to mention only one of particular prominence, selection in relation to climate. And finally, adequate evaluation of all hypotheses concerning selection in man will require the cooperation and interest of ethnologic field workers who are able to observe at first hand the interactions of man with his natural and cultural environment.

28. BUSHMAN HUNTER-GATHERERS: A STUDY IN HUMAN ECOLOGY

PHILLIP V. TOBIAS

Reprinted from D. H. S. Davis (Ed.), Ecological Studies in Southern Africa (Monographiae Biologicae, *Vol. XIV) (The Hague: W. Junk, 1964). Phillip V. Tobias is Professor and Head of the Department of Anatomy at the University of the Witwatersrand, Johannesburg, and President of the Institute for the Study of Man in Africa. He has been engaged in researches on the Kalahari Bushmen and other peoples of Subsaharan Africa since 1951, and the study of fossil man has engaged his attention in recent years. He has published many works on the past and present inhabitants of Africa and is the author of* Chromosomes, Sex-Cells, and Evolution; Man's Anatomy *(with M. Arnold); and* Olduvai Gorge, 1951-1961. *He is editing a large work that will be entitled* Studies on the Biology of the Bushmen.

■ We have seen that man's biological constitution is malleable, within limits, but Baker's findings in the Andean alpine population, Brown's and Page's among the Eskimos, and Brues' work should be used cautiously in attempts to generalize about the processes of human adaptation. *Homo's* biological plasticity is not necessarily the principal source of his adaptability, and not every unusual physiological feature is necessarily evidence of biological adaptation.

One of the most intriguing phenomena that confronted anthropologists and others who came into contact with the "Bushmen" of South Africa was their physical structure, and especially their steatopygia—large deposits of fat over their buttocks. It was frequently assumed that steatopygia had some kind of adaptive value, especially perhaps related to shortages of water. For what other reason would such a phenomenon occur in a population? The habitat of the Bushmen is so harsh, the reasoning seemed to go, and their physiques so unusual (by European standards) that there must be some relationship between the two.

This selection demolishes this mode of reasoning, or at least sets important limits on it. Basing his analysis on an examination of the geographical and habitational distribution of the Bushmen, Tobias shows that none of the pressures to which steatopygia supposedly is an adaptation is consistently found in all places in which Bushmen are found. For example, Bushmen had lived in parts of Africa in which water was abundant; in fact, it is only in recent times that they became almost exclusively desert-dwellers. Hence steatopygia could not have arisen as a form of portable water-storage. As Tobias observes, however, it might be an adaptive storage of fat and protein.

To understand how the Bushmen survived and flourished within the limitations of their habitat, Tobias asserts that we must understand their cultural adaptations, which predominate in making their survival possible. An important lesson that is underscored by this selection, as well as by many of the earlier selections, is that the study of human adaptation must always consider cultural as well as biological factors. All group adaptations must be examined in terms of the hypothesis that an aspect of man's uniqueness is his attempt to free himself from the restricting limitations of his habitat by means of his culture.

Another lesson that can be drawn from the Tobias selection is that our ethnocentric and racial biases may influence our understanding of such people as the Bushmen. Until recently, anthropologists have been almost exclusively white. Did this have anything to do with the tendency to think of Bushmen's adaptations in biological terms rather than in terms of their inventive and ingenious intelligence?

In the second part of his essay, Tobias applies some of Brues' formulations pertaining to the relationship between body build and weaponry (Selection 27). It will be recalled that Brues suggested that the most adaptive body build for an archer is mesomorphic, but the Bushman, although an archer, is of ectomorphic (linear) build. This seems either to contradict or at least to provide an exception to Brues' hypothesis. After closer examination, however, Tobias notes that the contradiction is more apparent than real. The Bushman's bow is small and slight and his arrows are fragile;

they cannot seriously wound an animal. Instead, the Bushman relies on poisoned arrows and his remarkable tracking ability; after his arrows scratch the prey, he may track it for many days, while the poison slowly takes effect. Tobias concludes that the Bushmen are not an exception to Brues' hypothesis.

Clearly, then, it is the Bushman's ingenuity—his cultural adaptation—that has made life in the Kalahari Desert possible for him (although there have been some significant physical adaptations, such as the thin layer of subcutaneous fat). Tobias reports on the Bushman's cultural adaptations in the last part of this selection.

The student who wishes to read more about adaptation in South Africa should consult the volume in which this selection originally appeared. Other sources are L. Oschinsky's *The Racial Affinities of the Baganda,* (Cambridge W. Heffer, 1958) and Gabriel W. Lasker's "Migration and Physical Differentiation" *(American Journal of Physical Anthropology,* 4 [1946]: 273-300). ■

THE BUSHMEN OF southern Africa are a hunting and food-gathering people. As such, they provide ecologists with an opportunity for studying a present-day society in a pre-Neolithic cultural context. At the same time, large sections of the surviving Bushmen are absorbing elements of Neolithic culture, by the adoption of some settled, pastoral habits from European, Colored and African pastoralists: this other, changing face of Bushman economy provides a rare chance of studying the dynamics of ecological change-over from food-gathering to food-producing. Both of these aspects of the ecology of the Bushmen will be explored in the present essay.

At the outset, it should be stressed that, in their application to Man, the biological elements of general ecology are overlaid and tempered by cultural factors, the effects of which may be most difficult to dissect away from the biological. Marston Bates has lucidly analyzed the fields of human ecology: he shows that ecology is being applied as a label in at least five different sorts of human studies. These may be characterized as medical epidemological, geographical, sociological, behavioural and anthropological. The last would seem to come closest to the concept of ecology as originally defined by Ernst Haeckel and as understood by most botanists and zoologists. It is mainly this interpretation of human ecology which will be followed here. The human organism will be "regarded as a whole unit functioning in its environmental context" and in this survey, "biological propositions . . . (will) be examined anew in the cultural context."

Let me first summarize the features of the Bushman's morphology, some or all of which characters have at one time or another been considered to be of adaptive value. The Bushman is small of stature, the height in various groups ranging from 1,489 mm (58.7 inches) to 1,611 mm (63.5 inches). Accompanying this small stature are postural peculiarities, most notably a marked lumbar lordosis (hollowing of the lumbar spine) which throws the sacrum back. Although slender of build, Bushmen from puberty onwards develop isolated fat deposits over the buttocks (steatopygia) and over the thighs (steatomeria); in females these fat lumps may be appreciable, varying from group to group and, on the average, not as bulky as in Hottentot women. The skin-color is yellowish; the hair tightly spiralled and arranged in isolated tufts or "peppercorns;" the eyes often slanting and possessing epicanthic and other folds. The labia minora of the female are grossly elongated and may protrude as much as 80—90 mm beyond the lower limit of the labia majora. In the male the penis stands out almost at right angles to the vertical. In addition, especially in the head and skull, the adult Bushmen show a great number of infantile morphological features. These have been enumerated elsewhere.

THE NUMBERS OF SURVIVING BUSHMEN

It is most difficult to obtain an accurate estimate of the numbers of surviving Bushmen. This is partly due to their nomadic life in a relatively inaccessible part of southern Africa; partly, too, to the fluctuating size of bands from one season to another and partly, indeed, to the difficulty of deciding who is a Bushman. Before any attempt at enumerating Bushmen could be made therefore, it was necessary to define the Bushman. Three different canons have been used by students of the Bushmen: firstly, there is the Bushman way of life, the hunting and food-gathering economy; secondly, there are the very characteristic Bushman click-languages; and thirdly, there is the concept of the "Bush race" or "Bush physical type" which is used by physical anthropologists. Ever since Dart demonstrated that /? Aunia and ≠ Khomani Bushmen of the south show the influence of several physical types, it has been clear that the Bush type is only one of the racial elements which has gone to mould the surviving Bushmen. The variety of physical types in the structure of different groups of Bushmen is probably greater even than Dart estimated, and recent analyses have suggested no fewer than five or six strains.

This dilemma of definition has been masterfully discussed by Schapera. He has shown that, although no single feature, whether cultural, linguistic or physical, is adequate to characterize all those commonly included as Bushmen, language is perhaps the most reliable single criterion. In my estimates of the Bushman population, I have added to the language criterion an additional yardstick, namely the common recognition of individuals and tribes as Bushmen or Sarwa, both by themselves and by their neighbours. This second rule of thumb is not perhaps as misleading as it might seem at first sight, for where they live in the midst of other people, settled Bushmen who have adopted pastoral habits commonly deny that they are Bushmen. One is therefore somewhat less likely to overestimate the number of Bushmen. In practice, as Joyce found, the double rule serves to determine most Bushmen or Sarwa.

It may be added that with the increasing adoption by Bushmen of more settled pursuits, the cultural criterion is becoming more and more difficult to apply: this development, though inevitable, is the more interesting when one recalls that, historically, the recognition of Bushmen as distinct from Hottentots rested solely on the different ways of life of the two groups!

It has long been customary to regard the Bushmen as a vanishing race. In 1924, Sollas and, in 1930, Elliot Smith were "mourning the disappearance of this remarkable race," which was reduced to "a few scattered groups still lingering in the Kalahari Desert." Although Dornan estimated that as many as 10,000 Bushmen might be alive in 1925, he nevertheless stated, "They are a dwindling race and in a comparatively short space of time, they will have ceased to exist as a separate people."

In 1930, Schapera placed the total number of Bushmen still in existence "at a conservative minimum of about 7,000 to 7,500." This figure is still quoted. Nevertheless, in 1939, Schapera published a greatly increased estimate of 30,000 Bushmen.

In 1951, when I was seconded as physical anthropologist to the French Panhard-Capricorn Expedition, I began to collect data from all those territories in which Bushmen were known to occur. As a result of direct counts, official and unofficial, in some districts, estimates in others, and records of numbers of Bushmen inoculated in a diphtheria campaign, it was possible to piece together district returns and estimates, and territorial returns, to a grand total of 55,000 souls. These were distributed as follows:

Bechuanaland Protectorate	31,000
South West Africa	20,311
Angola	4,000
Northern Rhodesia	200
Republic of South Africa	20
Grand Total	55,531

These surprisingly high figures do not necessarily indicate an absolute increase in the number of Bushmen over the earlier smaller estimates. Although an increase may have occurred, it cannot be stated with assurance that the contrast reflects anything but the extreme difficulty of collecting reliable demographic statistics on the Bushmen. The new figure is the result of a more precise and regionally localized series of computations than those made hitherto. Further refinements and minor modifications of the estimate remain to be made, but it is unlikely that these will change the total significantly.

We have at present no means of knowing whether the Kalahari Bushman population is growing rapidly or slowly, remaining static or falling. It is commonly believed that the numbers of Bushmen have been falling rapidly; this was undoubtedly true for those segments of the Bushman population which were subjected to almost systematic extermination, in the Cape Province, Natal, Orange Free State and Basutoland. These massacres of Bushmen took place at the hands of Europeans, Hottentots and Bantu-speaking Africans and there is little doubt that tens of thousands of Bushmen were killed in little over 200 years. This crisis in the life of the Bushmen as a group makes it all the more difficult to assess population trends among them. A more recent crisis was a smallpox epidemic about 1951, which took toll of large numbers of "wild" Bushmen. It would seem, however, that where Bushmen have taken to living on farms, under more settled conditions, with an assured supply of food and water and some medical protection, numbers are increasing rapidly. This is the impression of administrative officers who have Bushmen in their districts; numerical proof of this tendency is badly needed.

It is to be expected that the changeover from hunting to pastoral life will lead to a substantial increase in the Bushman population in the next decade or two. Cappier has shown how rapid is the growth of the Bantu-speaking agriculturalists and pastoralists of southern Africa at present. The average yearly increase among southern and eastern Bantu is many times higher than that of the more slowly increasing Australian aborigines, Lapps and American Indians. Of the Bantu-speaking Africans, those in Bechuanaland have one of the highest average annual increases, being surpassed only by those of southern Rhodesia. These Bechuana are the closest African neighbours of the majority of Kalahari Bushmen; and their figures for population growth suggest that a dramatic increase in the number of Bushmen is to be expected after most Bushmen have passed through the agrarian revolution.

IS THE BUSHMAN AN ECOTYPE?

The two most striking points about the Bushman are that he possesses a peculiar and, in some respects, extreme morphology and that he inhabits a restricted and, in some respects, extreme environment. At a time of somewhat facile attribution of physical traits to geographical and ecological determinism, it is but a short step from the recognition that the stunted, steatopygous Bushman lives in the desert, to the claim that his peculiarities are the specialized or degenerate products of desert conditions.

The hypothesis that these curious morphological features are desert-determined has bedeviled the literature for decades and largely distracted attention from other aspects of the Bushman's ecology. Thus Hooton suggested that the steatopygia and its accompanying marked lumbar lordosis are an evolved means of overcoming drought. Marett proclaimed the Bushman as "the one form of man specialized for desert conditions" and, among many other reasons, held that steatopygia represented a peculiar capacity to economise water. He also tried to relate the Bushman's small stature to a postulated low activity of the anterior lobe of the pituitary, which, in turn, he related to the need to check diuresis. The yellow skin and epicanthic fold of the Bushman, Marett regarded not as a sign of Mongoloid admixture, but "as a primary character evolved in the desert cradle-land of this race," and the "peppercorn hair" as "an adaptation to withstand heat." Coon cited the Bushman as a human illustration of "Rensch's desert-fat rule," namely that fat in hot desert-dwellers "is deposited in lumps, where it will not interfere with body-heat loss or locomotion." Broom regarded the Bushmen as the degenerate descendants of an earlier prehistoric African race, though he did not directly attribute their degeneration to desert conditions.

It is implicit in such views that the Bushman is not only a present-day desert-dweller, but that he is an ancient son of the desert. It is tacitly assumed that he has developed his morphological attributes as adaptations to desert conditions. It will be my object in the first part of this paper to show that this notion is widely at variance with the facts. A glance at the present and past distribution of the Bushmen will suffice to show that, formerly, the Bushmen inhabited a wider range of ecological regions than those to which they have today been confined and that they cannot therefore be regarded as a desert ecotype.

THE ENVIRONMENT OF THE BUSHMEN

PRESENT-DAY DISTRIBUTION

The bulk of the surviving Bushmen are in Bechuanaland Protectorate and in South West Africa, with an important subsidiary group in Angola. Small numbers of Bushmen may still be encountered in the Rhodesias and in the Republic of South Africa.

Schapera's definition of the limit of their distribution is still largely valid:

> On the north they extend into South-East Angola up to about 15 degrees S. latitude, on the east into the Tuli and adjacent districts of Southern Rhodesia, on the south into the region of the Nossob and Molopo Rivers, and on the west to the Etosha Pan in South-West Africa. Small isolated remnants are also met with in the vicinity of Lake Chrissie in the Eastern Transvaal, in the coastal province of Mossamedes in South-West Angola, and in the Namib desert strip along the coast of South-West Africa.

Most of the surviving Bushman population lives in the Kalahari basin, described by King as "perhaps the greatest expanse of sand-veld in the world." This semi-desert zone has a low average annual rainfall of 5 to 15 inches, but years of drought are frequent. The sand of red, yellow and white hues, is covered by a mantle of grass and xerophytic scrub, bare stretches of sand being rare. In the northern Kalahari, the rainfall rises to nearly 30 inches; the vegetation is dense, gradually increasing until it merges with the woodland-savanna to the north.

Thus, most Bushmen live under arid, semi-desert conditions, where water is scarce, droughts are frequent and the vegetation is drought-conditioned though surprisingly extensive and varied for regions of such low rainfall. Much game is encountered including springbok, gemsbok, blue wildebeest, eland, giraffe, kudu, and other antelopes; a variety of carnivores; and water-loving species in the swamp area to the north.

Natural water is scarce, there being not many permanent water-sources. River-beds are nearly always dry, and as King has pointed out, even at times of flood, virtually no surface outflow escapes from the Kalahari.

Such is the physiographic background of the modern Bushman. How he copes with life in these circumstances will be considered below.

EARLIER DISTRIBUTION

As late as the middle of the 19th century, the Bushmen were dispersed over a much greater area of southern Africa than at present. From early official records, the Dutch settlers first encountered Bushmen 50 miles to the north of the Cape settlement in April 1655. At first known as Sonquas, they came to be called Bosjesman some time later. According to Maingard, the earliest recorded reference to the name is an entry in the account of the first journey of Olof Bergh, under the date, 4th November 1682: "some Hottentots, being Sonquas alias bosjesmans." Both from their physical appearance and from their mode of existence, there is no doubt that these were members of the Bush race living at a hunting and gathering level.

Subsequent official records, as well as the accounts of such observant travellers as Kolben, Sparrman, Le Vaillant, Barrow, Campbell, Livingstone, Chapman and Baines, have made it possible to build up a more complete picture. At the time of the first white settlement (1652), Bushmen were living in many districts, especially mountainous areas, of the Cape Province (both western and eastern Province), the Orange Free State, Natal and Basutoland, from some of which areas they did not entirely disappear until about the beginning of the present century.

The three centuries of permanent white settlement have brought a toll of hostility, extermination and hybridization to the Bushmen, at the hands of Bantu-speaking tribes advancing down the east coast, European settlers expanding northwards and eastwards from the Cape of Good Hope and Hottentots harrying the Bushmen in the hinterland. The virtual extinction of Bushmen in the more genial areas has left only those who inhabit the most arid parts of the sub-continent. So, it is only since recent historical times that the Bushman has been almost exclusively a desert-dweller.

Before these events wiped the Bushman off most of the face of southern Africa, he inhabited a wide range of ecological regions. Far from being confined to the semi-desert of the Kalahari, we find Bushmen inhabiting the low scrub country of the Karoo, Bushmanland and Namaqualand; the high central plateau with its gradient of Lowveld, Bushveld, Highveld and the less elevated Middleveld; the southern winter rainfall zone with its shrub flora and belt of heavy forest; the tropical and subtropical belt extending southwards from Mozambique down to Natal; and the high, cold, montane zone of the Basuto highlands. All evidence—from published accounts, exhumed skeletons and rock paintings—points to the Bushmen of these other areas as essentially the same anatomical entity as the present Kalahari dwellers.

The evidence of exhumed remains suggests that, in proto-historic and prehistoric times, the distribution of people of Bushman morphology was even more widespread and may well have included central and East Africa. The scanty records from this earlier period include that of the Moslem chronicler, Massoudi of Bagdad: in The Golden Meadows (A.D. 915), he spoke of the little Wak-wak people, identified as Bushmen, in the parts around Sofala, while Idrisi, in his map (A.D. 1154) of south-east Africa, showed the Wak-wak inhabiting the Sofala coast.

THE BUSHMAN NOT A DESERT ECOTYPE

Formerly inhabiting this wide variety of ecological zones, the Bushman can clearly not be regarded as an ecotype; he is a geotype as, indeed, are most other primary and secondary races of mankind.

The conclusion that the Bushman developed his infantile (or neotenous) features before he became exclusively a desert-dweller rules out the possibility that neoteny adapted the Bushmen specifically to desert life. We cannot even say that neoteny is a necessary adaptation to the hunting and food-gathering way of life, for there are hunters and gatherers like the Australian aborigines who do not show these infantile features. Even the yellow skin color, it has been suggested, so far from adapting the Bushmen to excessive solar irradiation, is a positive handicap, for the Bushman is prone to sunburn.

Dwarfing, too, cannot be regarded as a consequence of life under desert conditions, because of the former distribution of Bushmen in some of the most fertile and well-watered parts of southern Africa. Bushmen and Pygmies share this dwarfing with a number of other mammals and it is possible that Late Pleistocene conditions in parts of Africa may have favored and strongly selected dwarf forms. Mr. G. Silberbauer, officer-in-charge of the Bushman Survey of Bechuanaland Protectorate, informs me it is his impression that the taller members of Bushman groups are almost invariably poor hunters, are clumsy, and enjoy little prestige. This raises the thought that cultural selection may have favored shortness of stature, apart from natural selective factors.

As far as concerns steatopygia, its occurrence in other ecological regions—as testified, for example, in many of the human figures depicted in the rock paintings of the Basuto Highlands—rules it out as an adaptation peculiar to conditions of life in the desert. It is interesting that Upper Palaeolithic cave paintings of North Africa and of Europe also show steatopygous human figures. An explanation couched in terms of sexual selection has been proposed, for large buttocks are highly prized sexual goods among Bushmen At the same time, it seems as though steatopygia does diminish under conditions of impaired nutrition and water shortage, such as obtain in old age or severe drought, or even during the annual dry season. Perhaps steatopygia is an adaptive trait associated with a hunter gatherer economy, with its attendant periodic shortages of food, especially of fat and first-class protein. Such a view could be sustained even without assuming the Bushman to be a desert-dwelling ecotype; and without any need for considering this particular feature as the only possible adaptation to a hunting and food-gathering economy. Sexual selection could have been added to natural selection in heightening the selective advantage of steatopygia, so accelerating the establishment of the feature and entrenching it in the population gene-pool. It is a thought perhaps worth stressing that, although natural selection still operates in modern mankind—as Dobzhansky and Allen have shown—selection pressure is likely to be greatly enhanced if sexual, domestic or cultural selection acts in the same direction as natural selection. If Man, as it were, makes a cultural or sexual virtue of biological necessity, the whole process of selective evolution will be accelerated.

This notion is akin to the idea of Man being considered as a domestic animal, in whom evolutionary change and race formation may proceed as rapidly as in other domestic animals.

Steatopygia would seem to provide an excellent example of a selectively advantageous feature, the intensity of selection of which has been heightened by sexual or cultural selective pressures. Perhaps shortness, too, has developed fairly rapidly under the combined impact of these two types of selective agency. On this view, it becomes at least possible to see how a Bush type could have developed from his morphological antipode, Rhodesioid man, in perhaps as short a space of time as 50,000 years.

GENETIC ADAPTATION

We have outlined above some of the main features of the physical environment of the Bushman and we have also been led to conclude that none of the morphological features typifying Bushmen, whether pygmoid stature, or the cluster of infantile or retarded anatomical features, or steatopygia, seems to be an adaptation to the desert. This is not to say that the Bushman shows no anatomical adaptation whatever.

The Bushman's body build, as reflected in his height-weight ratio, seemingly differs but little from that of other African peoples living in hot, though not desertic conditions. People in hot climates tend to have a lower body weight than those in temperate and colder climates. Coupled with a relative increase of body surface area, lower weight makes for a much greater cooling surface relative to body weight. This feature the Bushman shares with other Africans. The fact that the present-day desert-dwelling Bushmen do not show it in more marked degree than other Africans who do not live under desert conditions confirms again the absence of anatomical features specifically related to life in the desert.

The Bushman likewise shares a low, broad nose with other Africans. This feature is correlated with external climatic conditions. The observation that many Bushmen possess even lower and broader noses than do most other groups in Africa has been made in repeated craniometric and cephalometric studies. Whether this difference is due to a distinct desert adaptation by the Bushman, or whether it is to be seen as part of the greater infantilism of the Bush head and skull, remains to be determined. At the moment, it would seem to be the only morphological difference between Bushmen and other Africans which may be of specific adaptive value under hot, desertic conditions.

Alice Brues, in a stimulating essay [Selection 27], has suggested that, within the human species and its immediate precursors, there has been some correlation between physique and habitual activity, resulting in a reciprocal influence between culture and body build. In other words, she is suggesting that we should seek evidence for bodily adaptation not only to the physical environment but to the "cultural environment."

> This influence may take several forms: a dominant weapon or tool may alter the average physique of a race using it over the course of time by giving a selective advantage to individuals of a body build best adapted to its use.

It may also alter the numerical proportions of conflicting races of which one is physically better adapted to the use of a valuable implement—this better adaptation perhaps being itself the result of a long intra-group selection. And differences in physical type between races may retard the transmission from one group to another of a new tool or weapon, and with it a whole new way of life.

As examples, she suggests that the heavy muscle-bound physique of Neanderthal man was physically well adapted to the use of bludgeon-like weapons; the linearity of the precursors of *Homo sapiens* to the use of spears as the first projectile weapons; the lateral build and strong shoulders of the central Mongoloids to the use of the bow; and a generally heavy build to the sustained labour required by an emergent agriculture. Most of the peoples of Africa, on this analysis, have retained to a greater or lesser extent the linear build of the idealised spearman.

Where does the Bushman stand? He is primarily an archer; yet his body build is slight and linear with stringy musculature. At first sight, this combination would seem to contradict Brues's suggested correlation. However, on closer examination, an explanation suggests itself: the Bushman bow is small and light, his arrows rather fragile. They can never seriously wound any animal which they strike; reliance is placed rather on the poisons smeared over the arrow-points. Instead of requiring heavy shoulders to wield large, strong arrows from big bows, the Bushman relies rather on his fleet-footedness and extraordinary tracking ability. These qualities enable him to come very close to a herd before letting fly his arrows. A mere scratch is generally sufficient for the poison to penetrate the animal but its effect is slow and the animal is able to run away with the herd. It may be two or three days before the Bushmen, who have been following all the while on foot, are able to close in to the now dying animal to deliver the coup-de-grace. So his very hunting methods have placed a premium upon other qualities than broad shoulders—namely smallness and lightness of build, staying power (which is so commonly associated with the ectomorphic somatotype) and an acute reliance on veld-craft in the tracking of wounded animals.

Perhaps, then, it is to his own culture that the Bushman's body-build has become adapted, rather than to the immediate challenge of the desertic environment. This essentially new idea of Brues requires to be worked out over many more groups before it will be possible to say how much validity such an interpretation may have in the Bushmen.

In brief, while the Bushman shows some general African adaptations to hot climates, the very structural features which make him racially distinctive from other Africans are not attributable to desert adaptation. In this respect, he differs from, say, the high-altitude dwellers of the Andes.

ACCLIMATIZATION

Whereas genetic adaptation is probably a long-term process, acclimatization is a shorter-term adaptation of the physiological processes within each individual. Is it possible that the Bushmen are well-acclimatized to conditions of the desert, even though their morphology may not be genetically adapted thereto? This question has been one of those which the Kalahari Research Committee of the University of the Witwatersrand has sought to answer in recent studies on the Kalahari Bushmen. One of the collaborators of this Committee, the Applied Physiology Laboratory of the Transvaal and Orange Free State Chamber of Mines, has devised a special portable climate-chamber in a tent, in order to assess the response of Bushmen to a variety of conditions.

As a result of these studies, Wyndham and Morrison have pointed out that as far as heat regulation is concerned, Bushmen in their natural state fulfil two important criteria for acclimatization to heat. In the first place, they are exposed to high temperatures and fairly intense radiation. In the second, they are nomadic hunters and, in tracking down and pursuing wild game, they exercise actively in these temperatures. Man adapts to heat by an increased volume of sweat, by greater dilution of sweat to conserve salts in the body, by a decrease in the heart-rate during activity and by dilution of the blood. Such internal adjustments affecting the individual's body during his life-time undoubtedly help the Bushmen to cope with hot conditions in the Kalahari. Comparison of Bushman responses to a standard heat-stress condition, with those of unacclimatized and highly acclimatized Europeans, suggests that the Bushman is partially acclimatized to heat. A similar state of acclimatization was found in a study of Arabs of the Chaambra tribe and of Frenchmen in the Sahara desert, and in Aborigines and Europeans in the hot, humid tropics of northern Australia. The small stature and light weight of the Bushman does not confer any advantage on him in his heat responses compared with the much heavier European.

The extreme cold of the Kalahari nights in winter prompted a further investigation to determine whether the Bushmen show a qualitative difference in response from that of modern Europeans. Like their Australian aboriginal counterparts, the Bushmen make scant use of clothing and shelter, yet are subject to severe cold. Scholander *et al.* had, indeed, found a significantly different response to cold in aborigines as compared with Europeans. In its work on the Bushmen, the Kalahari Research Committee exposed Bushmen and Europeans to winter night temperatures. It was found that the metabolic rate of Bushmen increases with decreasing air-temperature as it does in Europeans. The rate of decrease in rectal temperature was the same in Bushmen and Europeans. On the other hand, skin temperatures tended to be lower in the Bushmen (with little subcutaneous fat to insulate against heat flow from the deep tissues to the skin) than in the Europeans (who had a thicker insulating layer of subcutaneous fat). In fact, the main conclusion of the studies to date is that the only essential difference between Bushmen and Europeans in their physical and physiological equipment against cold seems to be the lesser subcutaneous fat deposit in the Bushmen. These general conclusions have since been validated in a second series of studies under even more rigidly controlled conditions, with respect to diet and rest before the metabolic studies were made. A comparison has also been made with Bantu-speaking Negroids and no qualitative or quantitative differences from the Bushman responses have been shown.

CULTURAL ADAPTATION

The Bushman has undergone another type of adaptation: he has adjusted his pattern of culture to his harsh, desertic terrain. Cultural acclimatization is swifter and simpler for resourceful man than genetic or physiological adaptation. As Marston Bates has pointed out, ecology in man includes the study of climate, soil, vegetation and all the other things which it connotes in animals, together with the additional cultural factors which modify physical and biological factors. Culture modifies the man, but also modifies the environment. Hence, we view culture "at one moment as a part of the man and, at another moment, as a part of the environment." Thus, an instant ago, we were considering the weapons of the Bushman as a part of his environment to which, perhaps, he has had to make bodily adjustments. Now, let us consider the reverse of the coin: the culture in which the Bushman clothes himself as a protection against the physical environment.

It is not, we have seen, any greater proclivity for metabolic adjustment which protects the Bushmen against the cold; instead, his culture helps him to build a microclimate around him on cold winter nights. There are three distinct elements in this process: he makes skillful use of his "skerm" or hut as a windbreak; of his skin-cloak within which he curls himself up into a very small compass, tucking the cloak in around the head and trunk; and of his fire, which he keeps going all night while he sleeps either with his feet, his face or his back towards it. Sometimes, a fourth element is added, in that he scoops out a shallow bed within which he is protected from cold air-movements. Wyndham and Morrison have shown that the microclimate under the cloak over the trunk is approximately 25°C-26°C and therefore very close to the thermoneutral zone of 28°C-30°C for naked man at rest.

Interesting is the fact that, of the three main devices used by the Bushmen—the windbreak, skin-cloak and fire—only windbreak and fire are used by the aborigine of Australia. The aborigine does not use the cloak, though, according to information kindly supplied by Professor A. A. Abbie, he frequently sleeps with his dogs for extra warmth, sometimes speaking of a "two-dog or four-dog night," according to the temperature.

There are many other examples of cultural adaptation among the Bushmen but one must suffice for our present purpose. He coats his skin with plant juices and fats and, when obtainable, animal fat or blood and accumulates a fine layer of Kalahari sand on the surface: this protective mail assists his physiological adjustments to hot conditions by protecting him from undue absorption of solar radiation and from the desiccating effects of hot, dry winds. For instance, an ointment made from the tsama melons is smeared over the body and vigorously rubbed into the skin. "In the very dry atmosphere of the Kalahari in winter time, this treatment has about the same beneficial effect as the application of cold cream to the exposed skin surfaces of the European." Other cultural habits affect his diet and water supply and will be considered below.

FOOD AND WATER

There are three cardinal aspects of the Bushman's quest for food and water: hunting, gathering and water-storage.

(*a*) *Hunting*

(1) Size of hunting bands

The size of Bushman bands varies considerably, not only from tribe to tribe but from time to time. The reported groupings range from three to four families up to about one hundred individuals. Summarizing the evidence, Schapera concludes that among the north-western tribes the bands are on the whole larger than any of the other tribes; this fact is probably correlated with the better food resources of the area, as compared with those of the central Kalahari. When one considers the question of population pressure, one is struck by a number of social and cultural controls which, directly or indirectly, influence the size of groups in a Bushman population.

Firstly, observers are agreed that the band structure is extremely flexible. It is at its maximum in times of plenty. During scarcity, a single band may break up into several hunting and gathering bands; these re-unite when food is once more in good supply. From this it follows that Bushmen living in drier circumstances may spend a greater part of their time in small groups; whereas those in more congenial areas may be grouped in larger bands for much of the time. This fluidity would account for the very variable band-sizes reported by different travelers and observers.

Secondly, it is said that permanent splitting of bands into two or more groups is a feature among the Kung. Such a division is seemingly accompanied by a degree of ceremonial and a wise elder from outside the initial band concerned carries out the division.

Thirdly, the custom of killing one or both members of a pair of twins is another Bushman practice which restricts population growth to a slight degree. Again if a mother died in childbirth, the baby was often buried alive with the mother. Seiner, quoted by Schapera attributed the high infant mortality to another cultural device reported by him alone, among the Kung: after birth a child is handed to another woman for suckling, as it is believed to cause the death of both mother and child if it is put on its mother's breast too soon after birth. If no other breasts are in milk, the child may die of hunger before it is returned to its mother's nipple.

Fourthly, the abandonment of the aged and infirm may have some survival value, if a band is on a forced march under conditions of hunger and scarcity. An old person unable to keep pace may be placed in a screen of bushes, provided with firewood and food and water, if available, and deliberately abandoned. If food is found soon, the old one may be rescued; if not, death follows and the hyaenas complete the next phase in the cycle of nature. Such a practice is known, too, among the Hottentots.

Cannibalism, on the other hand, is not practised among the Bushmen, nor are there indications or records that it was formerly practised.

(2) Nomadism

As with hunting people generally, the Bushmen are nomads. They follow the game if it moves in search of water and better grazing in the dry season. A corollary of this nomadism is that the Bushman owns few possessions: he travels light. Apart from his skin-apron and cloak, a small skin bag, tortoise-shell, ostrich egg-shell beadwork, bow and arrows, knife, flint and water-containers (e.g., ostrich egg-shells and calabashes), he generally has no other material goods. His shelters are correspondingly rude and temporary, as they are frequently abandoned.

(3) Territoriality

The nomadic habit is pursued with strict territorial limits. Territoriality applies among bands of the same tribe and between different tribes. Intertribal territorial bounds are sometimes reinforced by social attitudes, such as the traditional enmity between the Auen and Naron. Under special conditions—such as an abundance of food—these bounds and the accompanying enmity are forgotten. Thus, in an analysis of Bushman marriages on the line of European-owned farms at the junction between the Auen and Naron territories, I found that almost two-thirds of marriages were intertribal and interracial, although a few miles back from the farms on either side, the old enmity of the other tribe persisted.

A second example of the breakdown of territoriality under conditions of abundance was quoted by Passarge. At the time of the great epidemic of rinderpest in 1897, thousands of animals were dying in the zone between the two tribal areas, numbers of Naron and Auen came together peaceably to collect food.

(4) Weapons

We have already referred to the bow and poisoned arrows. Many Bushmen have today adopted the spear and, especially in the north, the throwing-stick (knobkierie). The Water-Bushmen of the Okovango swamp have fishing-traps and nets of a pattern borrowed from a local Bantu tribe, the Mambukushu.

(5) Dogs

The only domestic animal occasionally encountered among "wild" Bushmen is the dog. Dogs are barely tolerated in the encampment, being half-starved and cowed, but on the hunt they provide excellent assistance to the Bushmen.

(6) Other sources of animal protein

A great many sources of animal protein are ingested apart from small and large game. Thus, Bushmen will catch and eat tortoises, frogs, snakes, birds and birds' eggs, hare and springhare, porcupines, locusts, beetles, termites, flying ants and ants' eggs. In an analysis of such supplementary foods among the Bambuti Pygmies, Fischer was able to demonstrate that their diet of similar small animals, to the extent of about one-third, supplemented by two-thirds of vegetable matter, was adequate in calorific and protein content.

(b) Gathering

"Veldkos" or gathered foods provide, on the one hand, calories, vitamins, plant proteins and other forms of nutriment and, on the other hand, a source of water. A very detailed study of the food-plants of the Kung Bushmen of South West Africa has been made by Maguire while the salient features of the botanical origin of the Bushman diet are summarized elsewhere by Story.

(c) Water

The Kalahari rainfall is sudden and sporadic. The Bushman stores all the water he can in empty ostrich egg-shells and calabashes, carefully burying them in the cool earth or concealing them in a tree against a later day of thirst. Such supplies are rapidly exhausted during the dry season, and he resorts to eating quantities of water-rich plants, like tsama melon (*Colocynthis citrullus*), the Gemsbok cucumber (*Colocynthis naudinianus*) and other species with juicy storage organs, as well as the rumen-contents of buck. In a very few areas, ground-water lies just below the surface: the Bushmen dig a hole at such places and suck out the water with a hollow reed. The lower end of the reed is protected by a filter of grass. The water is then carefully transferred to egg-shell containers.

It has always been questionable whether such precarious sources of water could tide the Bushmen over an exceptionally dry season. A recent expedition of the Kalahari Research Committee (July 1959) gave the opportunity to see an alternative source of water ensuring survival. Dotted about in the Kalahari are several permanent springs, which are undoubtedly ancestral watering-places of the Bushmen. If the tsama and other water-rich plants fail and all other sources of water are spent, the Bushmen make forced marches to the nearest such natural source and encamp near it until the drought breaks. The very severe drought through which the Kalahari passed in the five years from 1955 to 1959 led hundreds of Bushmen to flock in to such watering-places. It is not surprising that on these forced migrations towards water and survival, the aged and the infirm are abandoned on the way.

An interesting sidelight is that European and Bastard farmers today have their farms on many of the natural springs. One consequence is that some of the springs are drying up, but this poses no serious problem for the farmers, since water is readily available on boring. A second, more important consequence is that, when the thirsting Bushmen come in today to seek water at their old sources, they find not only water but all the concomitants of a pastoral farming community—goats, cattle, chickens, a regular source of food and water, and settled life.

THE BUSHMEN IN TRANSITION

After the change-over from food-gathering to food-production in the eastern Mediterranean some 9,000 to 10,000 years ago, the new mode of life slowly spread among the neighbouring hunter-gatherers, until it had been adopted all over Europe, North Africa and south western Asia. A similar transition is now going on among the surviving Kalahari Bushmen. More and more of them are abandoning their pristine ways in favour of the assured food- and water-supply of the farms. "Wild" Bushmen who have had little or no contact with European and other farmers are becoming fewer and further between. Even between 1951, when I had the privilege of participating in the French Panhard-Capricorn Expedition, and 1958-59, when a series of expeditions was mounted by the Kalahari Research Committee, there was a noticeable decline in the number of "wild" Bushmen.

In 1959, we were obliged to classify the Kalahari Bushmen into at least five main categories, according to the amount of contact experienced:

(1) Those born in the veld, who have remained "wild," having little or no contact with settled and pastoral ways. The numbers in this category are declining seriously: even an apparently very isolated group which we encountered north of the Sonop Koppies possessed a few goats!

(2) Some born on the veld come in periodically to the farms or boreholes for food and water in dry seasons, but return always to the veld.

(3) Of the Bushmen born in the veld, some have taken up residence on farms, but journey forth into the veld to hunt and collect wild foods every now and again.

(4) Some of the veld-born Bushmen now live on farms, and never wander away to the wilds.

(5) Finally, there is a new generation of Bushmen born on the farms, who know no other life than farm life, do not hunt or collect wild foods, and are known in the Kalahari as "tame" Bushmen.

As a result of five successive years of drought, the movement of "wild" Bushmen in to the water-holes has been greatly accelerated. Because the farms are on their old permanent water-places, the Bushmen are acquiring new ideas, new habits of life, along with victuals and water. They are passing through another Neolithic revolution, which is transforming their way of life from a nomadic, hunting and food-gathering, subsistence economy, to a participation in a more settled, pastoral, food-storing existence.

Already, the changes in diet consequent

upon the adoption of some pastoral habits over the past 50 years have made themselves felt in an increase in the height of the Bushmen. The Bushmen who have been measured since 1950 are taller on the average by over an inch, than those who were measured before 1915. This secular trend applies to both males and females and is the first demonstration of such a trend in any Negroid or Khoisanoid people.

CONCLUSION

Our final picture is of a people whose curious, genetically-controlled anatomy plays little part in adapting them to desert life, whose physiological responses acclimatize them to hot, dry conditions and whose cultural pattern, primitive though it may seem, finally makes it possible for them to survive in the desert. It is not the slow, structural, genetic adjustment nor the smooth and reversible functional accommodation, but the swift, intelligent, cultural adaptation which has permitted the Bushman to cope with the rigours of his environment. It is remarkable that, even in a pre-Neolithic economy like that of the Bushman, culture predominates over biological considerations in ensuring survival. Professor R. A. Dart in his contribution on "The Ecology of the South African Man-Apes" has limned the beginning of that critical change-over from biological to cultural evolution; in this essay I have attempted to show the latter end of that Pleistocene-long development—the emergence of a man whose inventive genius and flexibility have largely ruled out the possibility of extinction through too narrow an adaptedness to a circumscribed biological niche.

It seems that it can be only a matter of time before the old traditional life will have disappeared altogether. Meantime, anthropologists and human ecologists are provided with a unique opportunity of studying the dynamics of the transition from the palaeolithic existence of pastoral life, such as was taking place in the river basins of the eastern Mediterranean thousands of years ago.

V. SPEECH AS A HUMAN ADAPTATION

FEW EVOLUTIONARY DEVELOPMENTS in the emergence of *Homo sapiens* have done more to free him from the limitations of his biological constitution and the confines of his habitats than language. Few other events, or series of events, have had a greater effect upon the species' relationships to its habitats. The emergence of speech was an adaptive watershed in evolution because it enabled man to create systems of communication in which he is able to combine and recombine elements of language to an extent that is unknown elswhere among living forms.

Nonhuman animals also have language, but there are important differences between nonhuman and human languages. One of the most important of these differences is in the use that humans and nonhumans can make of the signals that compose their languages. Nonhuman language signals almost always are genetically fixed and usually have a one-to-one correspondence between a signal and the message conveyed. When a bee searching out a source of food dances in a particular way and emits a particular sound in combination with the movement, the signal can have only one meaning for the other bees in the group. Words, nonword sounds, and movements among humans can have similarly arbitrary meanings, but a word (or other symbol) in a human language can also have many meanings simultaneously. We can, for example, use the word "snow" to refer to a particular kind of preciptation or to making a particular kind of impression on an instructor in an essay examination. The sound of "b-a-r-e" can refer to a lack of cover or to an animal. Similarly, a single word can have positive or negative connotations, depending on the social context in which it is used. For example, "s.o.b." can be a term of endearment between friends or an expression of hostility. "Aggressive" can have a positive meaning in American culture when it is used to describe a ballplayer or a business man, or a negative implication when it is used to refer to a person's informal social relations. Human language is unique in its flexibility.

Another important aspect of the adaptive importance of man's languages is his ability to speak intelligently about things and events that are outside his immediate experience; he can even speak about them foolishly and still make himself understood. We can speak about the United Nations, feudalism, or extinct civilizations although we have not experienced them directly. Furthermore, similar ideas and categories—within limits that are discussed below—can be put into any language. This is another unique feature of the human capacity for speech.

As with everything else that is uniquely human, the question of origin—when and how the capacity for language emerged in the course evolution—has long been intriguing. Because of the absence of any physical remains of speech, however, we have to content ourselves with the following assumption. At some point in the course of evolution—probably about the time of the emergence of erect bipedal locomotion, the use and modification of tools, and an increase in the capacity of the braincase—the signaling and communicative patterns of one or a few protohuman forms improved in the direction of human language as it is known today. Implicit in this assumption is the further assumption that natural selection favored the survival of populations that were able to improve on their ancestors' modes of communication. One of the advantages of such an improvement must have been an ability to organize groups effectively that was greater than could have been possible with nonhuman signaling systems.

Closely related to man's capacity for speech is his unique capacity for self-consciousness. The two are inseparable, and the study of their relationships to each other as aspects of adaptation is a newly opened frontier in anthropological inquiry. One way to conceptualizing man's capacity for self-consciousness is to say that animals know, but only man knows that he knows. Animals are aware of themselves, but only man tries to express this self-awareness symbolically. Would man's concern with ethical and moral problems, which is only one aspect of his ability to think of himself in third-person terms, be very different if the evolution of his capacity for speech had taken a different course? The answer to this must be yes, but it is entirely speculative at the present stage of theory.

Because of recent developments, however, there is every reason to expect that future research in this area of study will have a better empirical grounding. The problem is important for several reasons and is of central concern for the authors of many of the selections in this section. Perhaps the most important reason for the significance of the question is that it reflects one of the outstanding features of being human: in pursuing this line of inquiry we are being self-conscious about our self-consciousness. On the practical side (who except man can distinguish between theoretical and practical problems?), the question preoccupies many thinkers who are concerned with the development of a worldwide moral order: What is the largest group with which the individual can concern himself in adopting ethical and moral standards?

In many of the selections that dealt with nonhuman primates (Part II) we saw that even nonhuman group life is governed by rules, prescriptions, and proscriptions; the bees and ants teach us that this is not confined to mammals. We can therefore conclude that man is similarly genetically programmed for patterns of social living that are governed by rules and regulations, although no other species exhibits man's flexibility and adaptability in this regard. This continuity from our nonhuman past raises some important questions. A member of an ant colony or a baboon troop probably cares not a whit about the fate of ants in other colonies or baboons in other troops, and in many of the world's human societies it appears that the individual is not particularly concerned with the fate of *sapiens* who are not members of his group, unless they are closely interdependent. But we know that as man's mastery over his habitat increases he is able to expand his horizons and his awareness of others. Is there a limit to this expansion? Are there biologically imposed restraints on man's capacity for thinking, speaking, and worrying about the fates of people who are distantly removed from him and on whom he is not materially dependent? In terms of his constitutional endowment, is man capable of converting the tenet that all men are brothers into a standard of action, or must it be confined to a weekly ritualistic incantation?

Of course, these are not the only reasons for the study of the emergence and evolution of the human capacity for speech. Because social institutions change at an increasingly rapid rate and as cultural horizons expand along with advances in the technologies of travel and communication (to say nothing of political stimuli to intersocietal involvements), it is necessary to focus on the capacity for speech as an important mechanism in man's maintenance of a viable relationship with his milieu. This milieu is not only physical, it is also social; it entails not only the acquisition of a livelihood but the maintenance of poltical relations.

The evolution of speech is important in this context not only because it involves making communicative sounds and receiving socially relevant signals but also because it is inseparable from the capacity to think, to abstract from concrete experience, to formulate new ideas, to know, and to feel. Language is the symbolic system by which man expresses his capacity to see himself as an object in a world of objects other than himself. Furthermore, it is now generally accepted by anthropologists that man's speech is affected by what he thinks and feels and that a language limits what its speakers can know and express. Speakers of a language can think and feel only what they can say. This applies as much to speakers of modern English as to speakers of Bantu languages.

If (as seen in Selection 33) a language does not have a term for a particular color, the speakers of this language will be unable to make the same color distinctions as people who have such a term. If language does not have a set of terms for different kinds of monetary investment, will the ordinary member of the group that speaks this language be able to think in terms of long-range economic planning? In all probability, he will not; and this is one of the many problems that confront new nations that are trying to move from a pre-industrial level into an industrial stage of development.

Many societies have elaborate terminologies for the sense of guilt a person is supposed to experience as a result of failure to perform family and other kinship obligations. At the same time, many of these societies have no linguistic idiom that refers to guilt in connection with nonfamilial activities, such as extramarital sexual relations. Many missionaries and other religiously oriented travelers have felt that the members of such societies lacked moral feelings, especially in nonfamilial situations. This is especially noticeable in their writings about sexual mores in these societies, in which they often asserted that there was an apparent absence of moral standards—by which they meant the Western notion of "sin"—in sexual relations. Nothing, of course, could be further from the truth. Correlatively, although many people in non-Western societies found some of the ethical standards of the missionaries to be attractive, they were perplexed (if not bored) by the sexual ethics of the Westerners.

These illustrations raise an important but difficult problem. Clearly, man's capacity for language is one of his most important instruments of adaptation. But does this mean that all languages are equally adaptive? Do all languages serve equally well in preserving and perpetuating life? Do all languages make man's milieus equally habitable? Are all languages equally effective in making man fit to live in the habitats in which he tries to live, including changing habitats? A fundamental premise in most anthropological studies of speech is that all languages are equally adaptive. However, the study of adaptation in all aspects of life suggests that this premise may be untenable.

It may seem unimportant that some languages do not allow their speakers to distinguish particular colors, to engage in long-term economic investment plans, or to experience guilt in connection with extramarital sexual relations. (Some readers may even feel that such inabilities are more adaptive than otherwise.) Are there maladaptive features in the English language as it is spoken currently? This question, however, involves a special difficulty because our ability to point to a problem means we have the terminology with which to identify it, which suggests that the language is adaptive because it meets all of our immediate needs. Conversely, if we do not have the words with which to identify a problem, how can we discuss it?

Fortunately, at least for our present purposes, there is at least one phenomenon in social life for which we have no adequate terminology. (The fact that the problem can be discussed means that an adaptive linguistic response will be made to it at some future date, but we have not yet done so.) This problem can be conceptualized in the following terms. Ever since the foundation was laid, about a century ago, for modern existentialist philosophies, an abiding concern of many thinkers has been the phenomenon of individual self-awareness and the sense of being. Especially among members of the younger generations since the 1960s in Western societies, there have been many attempts to find means of expressing the experience of personal being—the sense of individuality, aspects of self-objectification, awareness of self. Our language has a fairly complete vocabulary with which the individual can express different emotional states: I am happy, I hurt, I'm cooling it, I am enraged, I am embarrassed, I'm in love (usually said as one word). But how does one express the experience of the full sense of being, of personal self-realization, or of personal self-fulfillment? It would be almost meaningless to say "I be" or "I am being;" if one said "I am" or "I am being" the listener would automatically ask "I am what?" or "I am being what?" In a social system in which self-realization and self-objectification are becoming increasingly important standards for the individual's evalution of his relationship to the world around him, there is no terminology by which this can be conveyed to others or formulated for oneself. The new jargon of this standard of experience—for example, to say "I am turned on"—is insufficient for meeting the new sociocultural pressures because it continues the linguistic procedure suggesting an outside (and impersonal) agency that is responsible for the sensation. There is no terminology that can place the beginning and end of an experience entirely within the individual. This is an example of a language that is not yet adapted to a particular pressure of the sociocultural milieu.

Linguistic adaptation illustrates—perhaps more than any other sphere of social behavior—the proposition that the milieu within which man adapts is as much sociocultural as physical. It is misleading, in the study of adaptation, to speak of a genetic, structural, or behavioral pattern as adaptive and let the matter go at that; we must carefully examine the total context in which the pattern appears and for which it is adaptive. The speakers of any language can easily invent new words to refer to different types of yams, snow, illnesses, and means of transportation. It is when languages must cope with pressures that emanate from the sociocultural aspect of the environment that they meet the severest test of adaptation. There is a vast difference between finding a word to denote a physical object or experience and developing a conceptual terminology that will enable people to communicate experiences that have no apparent physical referents.

The principal reason for this difficulty is the indivisible feedback between language and thought: People can say only what they think, and they can think only what they say. In most societies people are able to think and say whatever is needed because the physical and sociocultural aspects of their environments are more or less constant; but when these factors change, new pressures are exerted on the language to make it more "fit" for the new environment.

Thus, although all languages are comparable (in large part for reasons that are discussed by Lenneberg in Selection 30), they are not necessarily equally adaptive. This is the point of view taken by Hymes in Selection 31, and it raises important questions about the adoption of new languages in nations that are moving rapidly from a horticultural level of cultural development into the modern world of guided missiles, satellites, computer technology, and complex political systems—developments that are discussed by Greenberg in Selection 32.

Hockett and Ascher (Selection 29) discuss the emergence of speech in the course of evolution in terms of "the human revolution." This imagery is well drawn. It must be borne in mind, however, that although every revolution by definition changes man's relationship to his habitat and sometimes enables him to alter the habitat, these changes are never permanent or static. Instead, they lead to new conditions to which subsequent adaptations must be made. The evolutionary emergence of speech was a revolutionary adaptation because it enabled man to change his relationship to the habitat and, in many instances, to change the habitat. These modifications had further repercussions, ranging from the formation of kinship groups to the development of civilizations and imperialistic policies. Not all languages have been equally adaptive in meeting the challenges of all these repercussions.

As we have seen, the capacity for communication was well established before man made his appearance. The evolution of this feature of man's constitution, perhaps more than anything else, has made human sociocultural life possible. It must be viewed, however, as one feature of human behavior. It is closely related to the evolution and adaptation of other features, but it has its own history and components. Because of its indispensability in human life, we devote considerable attention to speech.

Two excellent introductions to some of the issues that were raised above can be found in *Horizons of Anthropology* (edited by Sol Tax [Chicago: Aldine, 1964]): "Language and Thought," by Susan M. Ervin, and "A Perspective for Linguistic Anthropology," by Dell H. Hymes. Basic to all such discussions are two keystone works: *Language,* by Edward Sapir (New York: Harcourt, Brace, 1949), and *Language, Thought and Reality,* by Benjamin Lee Whorf (edited by J. B. Carroll [Cambridge, Mass.: Technology Press of M.I.T., 1956]). The reader who wishes to explore further can consult a collection of highly technical writings: *Communication and Culture* (edited by Alfred G. Smith [New York: Holt, Rienhart and Winston, 1966]). A much more extensive collection, which covers almost the entire range of the relationship between anthropology and linguistic studies, is *Language in Culture and Society: A Reader in Linguistics and Anthropology* (edited by Dell H. Hymes [New York: Harper & Row, 1964]). Recent advances in the study of communications systems among bees are described in "The Evolution of Bee Language," by Harold Esch (*Scientific American* [April, 1967], 216[4]: 96-104).

Several excellent works can serve as introductions to the basic work of linguistics: *Language,* by Leonard Bloomfield (New York: Holt, Rinehart, 1954); *An Introduction to Descriptive Linguistics,* by H. A. Gleason (New York: Holt Rinehart and Winston, 1961); *A Course in Modern Linguistics,* by C. F. Hockett (New York: Macmillan, 1958); and *Historical Linguistics: An Introduction,* by Winfred P. Lehmann (New York: Holt, Rinehart and Winston, 1962). Ignace J. Gelb's *A Study of Writing* (Chicago: University of Chicago Press, 1952) is useful to those who are interested in the development of writing. An indispensable work for the reader who is concerned with the issues involved in symbolic communication is Suzanne K. Langer's *Philosophy in a New Key* (Cambridge, Mass.: Harvard University Press, 1957).

29. THE HUMAN REVOLUTION

CHARLES F. HOCKETT and ROBERT ASCHER

Reprinted from Current Anthropology, *5 (1964): 135-47. Charles F. Hockett, Professor of Linguistics and Anthropology at Cornell University, has worked in almost every phase of pure and applied linguistics. The author of several books, his central concern is set forth in the present paper and in other papers on closely related topics. Robert Ascher is Associate Professor of Anthropology at Cornell University. His interests are expressed in publications on size estimation of prehistoric populations, human evolution and interstellar communication, ethnographic analogy in archeology, computer-aided chronological ordering, and the logic of the experimental method in archeology. He is currently investigating biocultural evolution and problems of relative time.*

■Although we have been reading about whole groups or populations—sometimes whole species—and their processes of adaptation, we have seen that such an exclusive focus can create an erroneous impression. It is not an organism or a form as a whole that evolves and adapts, and man did not appear suddenly on the scene as a completely new and unique adaptation in organic nature. Instead, he inherited most of his capacities or potentials from his primate past, and each capacity had its uniqueness resides. in only one small feacareful examination of the human hand reveals that it is unique in the world of nature, but its uniqueness resides in only one small feature; all others had been well established in our prehuman ancestry (Selection 16). Similarly, not all of the human organism adapts to the deprivation of oxygen at alpine heights; only a few features adapt. The emergence and recombination of specific features denote the appearance of new total forms on the evolutionary scene.

In this selection we turn to an attempt by Hockett and Ascher to reconstruct the course of hominoid evolution to the development of human culture in the context of language. Their central question is: How did the emergence and evolution of language contribute to turning nonhumans into humans? After a brief discussion of the nature of the evidence and the methodology employed in exploring this problem, the authors discuss some of the important physiological capacities that our terrestrial hominoid ancestors inherited from their own tree-dwelling ancestors. Then they turn to the emergence and evolution of speech.

Hockett and Ascher suggest that the protohominoids did not have the power of speech that is characteristic of hominoids. Speech, as we know it, evolves from a "call system" that has a repertory of a half dozen or so recognizable signals, each of which is appropriate to a different situation. They then distinguish between a call system and language. The central point of their paper, to which they lead up carefully and systematically, is that this feature of evolution could not have occurred if grasping, manipulating, and carrying had not been transferred from mouth to hand, leaving the mouth free for chattering. On the basis of these inferences, Hockett and Ascher attempt to portray what must have occurred in the evolution of language.

Accompanied by one of the most memorable displays of humor in anthropological writing (in which they speculate briefly on what human speech would be like if physical evolution had taken a slightly different course), Hockett and Ascher give us an important model for the study and understanding of adaptation. Their analysis reminds us that, although the study of evolution and adaptation must focus on one feature at a time (in this case speech), we also must keep in mind the total set of adaptive changes that made the emergence of the feature possible. They stress the importance of the total context of the shift from arboreal to terrestrial living, among other factors that enabled speech to emerge. If human culture as we know it would be impossible without language, it follows that human culture as an adaptive system would be vastly different if other, more fundamental changes had not first taken place.

In a very important article, "The Origin of Speech" *(Scientific American* [September, 1960], 203[3]), Hockett explores in detail the differences between the "signaling" communications of nonhuman animals and the abstract symbols of man's language. Along similar lines is Leslie A. White's "The Symbol: The Origin and Basis of Human Behavior" *(Etc.: A Review of General Semantics,* 1 [1944]: 229-37). The reader also might find it worthwhile to consult Karl von Frisch's *Bees: Their Version, Chemical Senses, and Language* (Ithaca, N.Y.: Cornell University Press, 1950), and Chapters 16 and 17 of *Primate Behavior* (edited by Irven DeVore [New York: Holt, Rinehart and Winston, 1965]): "Communication in Monkeys and Apes," by Peter Marler, and (somewhat more difficult) "Primate Signaling Systems and Human Languages," by Jarvis Bastian. ■

THIS ESSAY ATTEMPTS to set forth the story of the emergence of the first humans from their prehuman ancestors. A special feature is that we have tried to incorporate the various steps and stages of the evolution of language into the total picture.

We dedicate this essay to the memory of Paul Fejos, whose encouragement, over a number of years, played an important part in bringing the work to fruition.

The inquiry into human origins is a collective task to which hundreds of investigators have contributed. Virtually none of the proposals in the present paper are our own. Even for the ways of thinking about the evidence that seem to be fruitful, we are completely indebted to our predecessors. We do accept responsibility for the particular way in which we have chosen among alternative theories, and for the way in which we have tied them together. We believe that the time is ripe for a synthesis of this sort, if only as a clear point of departure for the further investigation of both method and detail.

The term "revolution" in our title is not intended to be flamboyant. A revolution is a relatively sudden set of changes that yields a state of affairs from which a return to the situation just before the revolution is virtually impossible. This seems to be the sense of the word intended by V. Gordon Childe when he speaks of the "Neolithic Revolution" and of the "Urban Revolution." But these two revolutions were experienced by our fully human ancestors. The second could not have occurred had it not been for the first. The first could not have taken place had it not been for an even earlier extremely drastic set of changes that turned nonhumans into humans. These drastic changes, as we shall see, may have required a good many millions of years; yet they can validly be regarded as "sudden" in view of the tens of millions of years of mammalian history that preceded them.

For the reconstruction of human evolution we have evidence of two sorts, plus certain firm and many tentative principles of interpretation.

One kind of evidence is the archeological, fossil, and geological record. The fossil record of our own ancestry is still disappointingly sparse for the bulk of the Miocene and Pliocene. It seems unlikely that such records can ever be as complete as we might wish. But techniques of interpretation improve, and we suspect that the archeological record, in particular, holds an as yet unrealized potential.

The second kind of evidence is the directly observable physical structure and ways of life of ourselves and of our nearest nonhuman cousins, the other hominoids of today. Chimpanzees, gorillas, orangutans, gibbons, siamangs, and humans have ultimately a common ancestry not shared with any other living species. We shall refer to their most recent common ancestors as the *proto-hominoids.* Since all the hominoids of today constitute continuations of the proto-hominoids, we can attempt to reconstruct something of the physical structure and of the lifeways of the common ancestors by comparing those of the descendants. Such an effort at reconstruction must at the same time propose realistic courses of development from the ancestral group down to each of the directly observable descendant groups, and must make proper provision for those strains known only through fossils or archeological remains.

The method is very much like the comparative method in historical linguistics—and, as a matter of fact, it was first devised in the latter context, only subsequently transferred to the domain of biological evolution. The term "comparative" appears also in "comparative morphology" (or "comparative anatomy");

we must therefore emphasize that the method of which we are speaking applies not only to gross anatomy but also to the fine-scale phenomena dealt with in biochemistry, and not only to structure but also to behavior.

In any domain of application, a comparative method shares with all other historical methods the fact that it can yield reliable results only insofar as one can be sure of certain key *irreversible* processes. Given information about stages *A* and *B* in the history of a single system, we can posit that stage *A* preceded stage *B* if and only if the change from *A* to *B* is the sort that happens, while a change from *B* to *A* is impossible or highly improbable. In historical linguistics, the requisite irreversibility is afforded by sound change. The philologists of the late 19th century were correct when they characterized sound change as slow, constant, inexorable, and beyond conscious control; for, as we shall see later, it is a necessary by-product of a crucial design feature of all human language, and could not be eliminated save by altering language into something unrecognizable. Whenever sound change leads to the repatterning of the phonological system of a language—and this has happened about 100 times in English between King Alfred's day and our own—the consequences ramify through every part of the language; soon the results are so scattered, so subtle, and from the point of view of effectiveness of communication so *trivial*, that a return to the state of affairs before the repatterning has, in effect, probability zero.

The situation in biological evolution is much more complicated, with no simple analogue for sound change. Is a particular organ in a particular species (living or fossil) vestigial or incipient? Is the swimming bladder of current teleosts a former lung, or is the lung of lungfishes a one-time swimming bladder? Evolutionists are plagued by such questions. The answers are often obtainable, but not through any simple formula. A new fossil does not automatically resolve the dispute, since one's opinions as to lines and directions of development will affect one's notions as to how the new fossil is to be fitted into the picture.

For the *mechanisms* of change we are in less trouble. We have now a good understanding of genetics, and also of the traditional transmission of lifeways. The latter was once believed to be exclusively human, but this is not so. At least for land mammals and for birds, genetics and tradition work in a constant dialectic complementation, neither being wholly responsible for anything. We are also clearer about a point that used to be quite obscure: the domain (so to speak) within which these two mechanisms operate is not the individual but the community, which has a gene pool, a distribution of phenotypes, and a repository of lifeways, and which, as a functioning unit, faces the problems of survival.

The greatest pitfall in evolutionary thinking stems from the keenness of hindsight. For example, we know that long ago, over a long period of time, our own ancestors abandoned the trees for the ground and developed effective machinery for bipedal locomotion. This seems beyond dispute, because the prehominoid primates were arboreal and we ourselves are bipedal ground walkers. But when we ask *why* this change, we must remember that our ancestors of the time were not striving to become human. They were doing what all animals do: trying to stay alive.

Thus, in searching for causes of the change we must look to conditions pertaining at the time. There are only two possibilities. The conditions at that time may have been such that minor variations in gait and posture had no bearing on survival. We should then class the change that actually did take place as fortuitous. Or, the conditions of life at the time may have positively favored selection for bipedal locomotion and upright posture. If this is what happened, then the change was adaptive. By definition, a change that was neither adaptive nor fortuitous would lead to the extinction of the strain that underwent it, and in the present instance we know that that did not happen.

The most powerful antidote for the improper use of keen hindsight is a principle that we shall call "Romer's Rule," after the paleontologist A. S. Romer who has applied it so effectively—without giving it any name —in his own work. We phrase this rule as follows:

> The initial survival value of a favorable innovation is conservative, in that it renders possible the maintenance of a traditional way of life in the face of changed circumstances.

Later on, of course, the innovation may allow the exploration of some ecological niche not available to the species before the change; but this is a consequence, not a cause.

One of Romer's examples concerns the evolution of Devonian lungfishes into the earliest amphibians. The invasion of the land was feasible only with strong fins (which in due time became legs). But strong fins were not developed "in order to" invade the land. The climate of the epoch was tempestuous; the water level of the pools in which the lungfishes lived was subject to sudden recessions. There was thus selection for those strains of lungfishes which, when stranded by such a recession, had strong enough fins to *get back to the water.* Only much later did some of their descendants come to stay ashore most of the time.

It is worthy of note that Romer's Rule is not antiteleological. We are permitted to speak in terms of purposeful behavior whenever we are dealing with a system that incorporates negative feedback. Individual organisms, and certain groupings of organisms (the kinds we call "communities"), are such systems. There is nothing wrong in asserting that a stranded Devonian lungfish tried his best to get back to the water. We are forced, however, to distinguish carefully between purposes and *consequences,* and we are not allowed to ascribe "purposefulness" to any such vague and long-continuing process as "evolution."

No principle, no matter how universal, answers all questions. Romer's Rule cuts as keenly as any razor ever devised by Occam to expose, excise, and discard unworkable pseudo-explanations. Yet it is applicable, in a sense, only after the fact. For example, in this paper we follow majority opinion and trace man's ancestry back to a point of separation from the ancestors of the great apes, the gibbons, and the siamangs. Having assumed this, we elaborate one of Romer's own suggestions as to how some of the early developments may have come about. Suppose, however, that new fossil finds should convince us that man is actually more closely related to some other group of surviving primates. We should then be confronted by a different set of putative historical facts requiring explanation; but we should evoke the same Rule as we sought that explanation. The Rule does not tell us which line of descent to postulate.

THE PROTO-HOMINOIDS

From the location, date, and morphology of the fossil dryopithecine *Proconsul* we infer that the proto-hominoids lived in East Africa in the Middle or Lower Miocene, or, at the earliest, in the Upper Oligocene. This does not mean that *Proconsul* himself—in any of the strains or species so far identified—was a proto-hominoid; indeed, he is not a good candidate as an ancestor of the gibbons and siamangs, to whom, by definition, the proto-hominoids were ancestral. But *Proconsul* was clearly an *early* hominoid, and at the moment he is the best fossil evidence available for the date and provenience we seek.

The proto-hominoids inherited certain crucial capacities from their totally tree-dwelling ancestors. It is the arboreal pattern that developed the keen accommodative vision characteristic of the higher primates, de-emphasized the sense of smell, turned forelimbs into freely movable arms with manipulative hands, and built brains somewhat larger than the average for land mammals.

The balance of the characterization we are about to give—what Count would call a "biogram" of the proto-hominoids—derives mainly from the comparative method applied to what we know of the hominoids of today. We shall not give all the evidence in detail. Furthermore, for the sake of vividness we shall allow some interpolations of a degree of precision that may be unwarranted. The proportion of guesswork in each statement will, we think, be fairly obvious.

Like most of their descendants, the proto-hominoids were hairy. Like all of them, they were tailless. They were smaller than we are, though not so small as present-day gibbons, whose size has decreased as an adaptation to brachiation. They had mobile facial muscles; they had neither mental eminence nor simian shelf (nor mastoid processes); they had large interlocking canines, and could chew only up and down; their tooth pattern was: $\frac{2:1:2:3}{2:1:2:3}$ It seems likely that there was little sexual dimorphism, although on this the comparative evidence is conflicting. The chromosome count was somewhere in the forties.

They lived in bands of from ten to thirty, consisting typically of one or a very few adult males plus females and offspring. They had a roughly defined nucleated territoriality: that is, the territory within which the members of a band moved about had only roughly demarcated boundaries, but centered on the specific arboreal sites in which they built their nests. The total population was probably never very great, nor very dense, from the proto-hominoids all the way down to the first true humans.

They were expert climbers and spent much of their lives in the trees of the tropical or subtropical forests which were their habitat, certainly building their nests in the trees and sleeping there. Like rodents, they climbed up a tree head first; unlike rodents, they climbed down stern first. They slept at night, from dusk to dawn, which in the tropics means nearer to one-half of each twenty-four-hour period than to the one-third characteristic of ourselves in recent times. They were active during the day. Some activities, particularly the constant search for food, led them not only among the trees—in which they may have brachiated, but with no great expertness—but also quite regularly to the ground below. On the ground, they could stand with a semi-upright posture (erect enough to raise their heads above shoulder-high grass to look about), and they could sit with arms free for manipulative motions; they could walk on all fours and could run on their feet, but bipedal walking was infrequent and awkward.

Occasionally they would pick up a stick or stone and use it as a tool. Judging from modern chimpanzees, they may have reshaped such tools slightly, using nothing but their hands and teeth to do so, and may have carried a tool for a short distance for immediate use, thereafter discarding it. They carried other things too, in mouth or hands or both, in connection with nest-building; and at least the females, perhaps on occasion the males, carried infants.

Their diet was largely vegetarian, supplemented by worms and grubs, and sometimes by small mammals or birds that were injured or sick and thus unable to escape. (We might call this "*very* slow game.") They scavenged the remains of the kills of carnivores whenever they could. Unlike all other mammals except the Dalmatian coach hound, their bodies produced no uricase; hence uric acid was not converted into allantoin before secretion in the urine, and had a chance to accumulate in the bloodstream. The structural formula of uric acid is something like that of caffein and, like the latter, it seems to be a mild brain stimulant. Since this type of purine metabolism is shared by all the hominoids, it can hardly explain our own unusual brilliance; but it may help to account for the generally high level of hominoid intelligence as compared with other primates and other mammals.

The males had the pendulous penis typical of the primates. Copulation was effected exclusively with the dorsal approach common to land mammals in general. Gestation required about thirty weeks. The uterus was single-chambered, and twinning was as rare as it is for us today. The placenta was of the single-disc type. The young required and received maternal care for many months. Mammary glands were pectoral; nursing females held infants to the breast in their arms, though doubtless the infant clung to the mother's fur also. The eruption of permanent teeth began perhaps at two and one-half or three. Menarche was at eight or nine years; general growth stopped for both sexes at nine or ten. The females showed a year-round menstrual cycle rather than a rutting season. Inbreeding within the band was the rule. The life-span was potentially about thirty years, but death was largely from accident, disease, or predation, or a combination of these, rather than old age. Corpses were abandoned, as were members of the band too sick, injured, or feeble to keep up with the rest, and were disposed of by predators or scavengers. Adult males were sexually interested in females and "paternally" interested in infants, but without any permanent family bond, and without any jealousy when they were themselves sexually satisfied.

Relations with adjacent bands were normally hostile to neutral, rarely if ever friendly; yet there was surely enough contact to provide for some exchange of genes. Social differentiation within the band turned largely on age and sex, secondarily on physical strength. In case of conflict of interest within the band, the huskiest adult males normally got their way. Collective activities required intragroup coordination, effected by various forms of communication—patterns of body motion, pushing and prodding, changes of body odor, and vocal signals. The conventions of these forms of communication were transmitted in part genetically, but in some part by tradition, acquired by the young through guided participation in the ways of the group. This implies also a certain capacity to learn from experience, and to pass on any new skills thus acquired to other members of the band by teaching and learning, rather than merely by slow genetic selection. But we may assume that usually there was very little new in any one lifetime thus to be learned or passed on.

A kind of activity called *play* is widespread among land mammals, and obviously intensified among primates; we can be sure that the proto-hominoids indulged in it, at least before maturity. It is very hard to characterize play precisely, beyond saying that it resembles one or another serious activity without being serious. Play at fighting, observable for example among dogs, goes through much the same gross motions as true fighting but the participants receive no injury. Sexual play has the general contours of courtship, but ends short of coitus or with mock coitus. We suspect that play is *fun,* for any species that manifests it, and that that is the immediate motive for indulging in it. But play is also genuinely pedagogical, in that the young thereby get needed practice in certain patterns of behavior that are biologically important for adult life.

The proto-hominoids did not have the power of speech. The most that we can validly ascribe to them in this respect is a call system similar to that of modern gibbons. Even this ascription may be stretching the comparative evidence somewhat. It is not hard to assume that a line of continuity from the proto-hominoids to the gibbons should have maintained such a call system essentially unchanged. It is also quite reasonable, as we shall see, to explain the evolution of a call system into language among our ancestors. The difficulty is to account for the apparently less highly developed vocal-auditory signaling of the great apes. Our hypothesis for the proto-hominoids suggests that the communicative behavior of the great apes may be somewhat more subtle and complex than has yet been realized. Be this as it may, we posit a call system for the proto-hominoids because we know no other way to proceed.

The essential design features of a call system are simple. There is a repertory of a half-dozen or so distinct signals, each the appropriate vocal response—or the vocal segment of a more inclusive response—to a recurrent and biologically important type of situation. Among gibbons, one such situation is the discovery of food; another is the detection of danger; a third is friendly interest and the desire for company. A fourth gibbon call apparently does nothing but indicate the whereabouts of the gibbon that emits it: this call keeps the band from spreading out too thin as it moves through the trees. One can guess at other possible situations appropriate for a special call: sexual interest; need for maternal care; pain. Band-to-band differences in calls may help to distinguish friend from alien.

A single call may be varied in intensity, duration, or number of repetitions, to correlate with and give information about the strength of the stimulus which is eliciting it. However, the signals of a call system are *mutually exclusive* in the following sense: the animal, finding himself in a situation, can only respond by one or another of the calls or by silence. He cannot, in principle, emit a signal that has some of the features of one call and some of another. If, for example, he encounters food and danger at the same time, one of these will take precedence: he is constrained to emit either the food call or the danger call, not some mixture of the two.

The technical description of this mutual exclusiveness is to say that the system is *closed*. Language, in sharp contrast, is *open* or *productive*: we freely emit utterances that we have never said nor heard before, and are usually understood, neither speaker nor hearer being aware of the novelty.

A call system differs from language in two other ways, and perhaps in a third. (1) Gibbons do not emit, say, the food call unless they have found food (or, perhaps, are responding to the food call from another gibbon, as they approach for their share of it). Furthermore, the gibbon that finds food does not go back to headquarters and report; he stays by the food as he emits the call. A call system does not have *displacement*. Language does: we speak freely of things that are out of sight or are in the past or future—or even nonexistent. (2) The utterances of a language consist wholly of arrangements of elementary signaling units called *phonemes* (or *phonological components,* to be exact), which in themselves have no meanings but merely serve to keep meaningful utterances apart. Thus, an utterance has both a structure in terms of these meaningless but differentiating elements, and also a structure in terms of the minimum meaningful elements. This design feature is *duality of patterning*. A call system lacks it, the differences between any two calls being global. (3) Finally, the detailed conventions of any one language are transmitted wholly by the traditional mechanism, though, of course, the capacity to learn a language, and probably the drive to do so, are genetic. On this score we are still in ignorance about the gibbons. Regional differences in gibbon calls have been noted, but various balances between tradition and genetics can yield that. We believe it safer to assume that proto-hominoid call systems were passed down from generation to generation largely through the genes, tradition playing a minor role. This assumption is the conservative one—it gives us more to try to explain in later developments than would any alternative.

This completes our characterization of the proto-hominoids, which can now serve as point of departure for the story of our own evolution.

OUT OF THE TREES

Some of the descendants of the proto-hominoids moved out of the trees and became erect bipeds. Romer's description of how this may have begun affords another example of the application of the Rule we ascribe to him.

Geological evidence suggests that at one or more times during the East African Miocene a climatic change gradually thinned out the vegetation, converting continuous tropical forest into open savannah with scattered clumps of trees. As the trees retreated, some bands of hominoids retreated with them, never abandoning their classical arboreal existence; their descendants of today are the gibbons and siamangs. Other bands were caught in isolated groves of slowly diminishing extent. In due time, those bands whose physique made it possible for their members to traverse open country to another grove survived; those that could not do this became extinct. Thus, for those bands, the survival value of the perquisites for safe ground travel was not at all that they could therefore begin a new way of life out of the trees, but that, when necessary, they could make their way to a place where the traditional arboreal way of life could be continued. The hominoids that were successful at this included those ancestral to the great apes and to ourselves.

Sometimes the band forced to try to emigrate from a grove would be the total population of that grove. More typically, we suspect, population pressure within a diminishing grove would force bands into competition over its resources, and the less powerful bands would be displaced. Also, when a migrating band managed to reach another grove, it would often happen that the new grove was already occupied, and once again there would be competition. Thus, in the long run, the trees would be held by the more powerful, while the less powerful would repeatedly have to get along as best they could in the fringes of the forest or in open country. Here is a double selective process. The trees went to the more powerful, provided only that they maintained a minimum ability to traverse open country when necessary: some of these successful ones were ancestral to the great apes of today. Our own ancestors were the failures. We did not abandon the trees because we wanted to, but because we were pushed out.

We are speaking here of displacements and movements of whole bands, not of individual animals. There is one thing that surely accompanied any band whenever it moved: the essential geometry of its territoriality. At any halt, no matter how temporary, whether in the trees, under the trees, or in open country, some specific site became, for the nonce, "home base"—a GHQ, a center, a focus, relative to which each member of the band oriented himself as he moved about. Headquarters was the safest place to be, if for no other reason than the safety of numbers. In a later epoch—though doubtless earlier than will ever be directly attested by archeology—headquarters among our own ancestors came to be crudely fortified, as by a piled ring of stones; it became the place where things were kept or stored; in due time it became house, village, fort, city. But earliest of all it was *home*. The tradition for this sort of territoriality is much older than the proto-hominoids, and has continued unbroken to the present day.

It is at this point in our story that we must stop referring to our ancestors as "hominoids" and start calling them "hominids." Of course, all hominids are hominoids; but we have now seen the sorting-out of the pre-apes from the pre-humans, and when we wish to speak exclusively of the latter the appropriate term is "hominid."

CARRYING

It is no joke to be thrown out of one's ancestral home. If the next grove is only a few miles away, in sight, then one has something to aim for; but sooner or later movements must have taken place without any such visible target. Treeless country holds discomforts and dangers. There may not be much food, at least not of a familiar sort. There may be little available water, for the trees tend to cluster where the water is more abundant. And there are fleet four-footed predators, as well as herbivorous quadrupeds big and strong enough to be dangerous at close quarters. One cannot avoid these other animals altogether, since their presence often signals the location of water, or of food fit also for hominid consumption. The quest for food must be carried on constantly, no matter how pressing may be the drive to find a new grove of trees in which to settle. It is a wonder that any of the waifs of the Miocene savannah survived at all. Enormous numbers of them must have died out.

The trick that made survival possible for some of them was the trick of *carrying*. The proto-hominoids, as we have seen, probably carried twigs and brush to make nests, and certainly carried infants. Also, they had fine arms and hands usable for carrying as well as for climbing, grasping, and manipulating; and the comparative evidence suggests that they occasionally picked up sticks or stones to use as tools. These are the raw-materials for the kind of carrying to which we now refer. But it takes something else to blend them into the new pattern. In the trees, hands are largely occupied with climbing. The infant-in-arms grabs onto the mother when the latter needs her hands for locomotion. The twig being taken to the nest is transferred to the mouth when the hand cannot at the same time hold it and grasp a tree branch. One puts down one's ad-hoc tool when one has to move.

The conditions for carrying are no better on the ground than in the trees if the hand must revert to the status of a foot. But if bipedal locomotion is at all possible, then the hand is freed for carrying; and the survival value of carrying certain things in turn serves to promote a physical structure adapted to bipedal locomotion.

Two sorts of ground carrying in the hands may have been extremely early; there seems to be no way of determining which came first. One is the carrying of crude weapons; the other is the transportation of scavenged food.

The earliest ground-carrying of weapons may well have been a sort of accident. Imagine an early hominid—perhaps even a prehominid hominoid—sitting on the ground and pounding something (a nut, say) with a handy stone. A predator approaches. Our hero jumps up and runs away as best he can on two legs—there are no trees nearby to escape into—but keeps his grasp on the stone for no better reason than that he does not need his hand for anything else. Cornered, he turns, and either strikes out at the predator with the hand that holds the stone, or else throws it. The predator falls or runs off, and whatever in our hero's genes or life experience, or both, has contributed to his behavior stands a chance of being passed on to others.

The first carrying of scavenged food back to headquarters (instead of consuming it on the spot) may also have been a sort of accident. A scavenging hominoid is eating the remains of a predator's kill where he has found it, and is surprised by the predator who is coming back to make another meal from the same kill. The hominoid runs off towards headquarters, still holding a piece of meat in his hand. In due time, he or his successors develop the habit of carrying the spoils off without waiting for the predator to turn up.

As described, these two early kinds of hand-carrying involve movements of a single animal *within* the band's territory. The carrying-along of things as the whole band moves is another matter, and probably a later development. Surely the earliest carrying of this latter sort was of unshaped weapons of defense. Yet other things might have been taken along. Extra food would be a great rarity, but if some were taken along because no one happened to be hungry as a movement began, it would be important if the band reached a particularly barren region. Water-carrying would have been extremely valuable—primates in general have to drink at least once a day, in contrast to some mammalian species which can store up several days' supply. Short hauls of small quantities of water cupped in the large leaves of tropical plants may have been quite early; large-scale water transport as a whole band moves must have been a great deal later, since it requires technologically advanced containers.

The side-effects of carrying things in the hands are of incalculable importance. We have already seen that its immediate practical value helped to promote bipedal walking, which in turn selected both for carrying and for an upright posture that rendered bipedal walking mechanically more efficient. A less obvious consequence is that carrying made for a kind of behavior that has all the outward earmarks of what we call "memory" and "foresight": one lugs around a heavy stick or stone despite the absence of any immediate need for it, as though one were remembering past experiences in which having it available was important and were planning for possible future encounters of the same kind. Taking scavenged meat back to headquarters without waiting for the predator to return to his kill also looks like foresight. We do not mean to deny the validity of the terms "memory" and "foresight." The point is that the outward earmarks surely came first, and only over a long period of time *produced* the psychological characteristics to which these terms refer.

A third consequence of carrying and of wandering was a change in dietary balance.

The first tools to be carried were defensive weapons. Often enough, no doubt, the use of these weapons against a predator, even if successful, would only scare him off. But sometimes the predator would be killed. Why waste the meat? We can also suppose that the wandering Miocene or Pliocene hominids occasionally found themselves in open country where no suitable plant food was available. Herbivorous animals could eat the grass; quadruped predators could eat the grazers; and the hominids, if they were lucky, could eat the grazers or the predators, or else starve. Thus the hunted became the hunters, and weapons of defense became weapons of offense.

The gradual increase of meat in the diet had important consequences of its own, to which we will turn after noting one further direct consequence of hand-carrying.

The use of the hands for carrying implied that the mouth and teeth, classically used for this by land mammals, birds, and even reptiles, were freed for other activities. It can quite safely be asserted that if primate and hominid evolution had not transferred from mouth to hand first the grasping and manipulating function and then the carrying function, human language as we know it would never have evolved. What were the hominids to do with their mouths, rendered thus relatively idle except when they were eating? The answer is: they chattered.

Remember that the proto-hominoids are assumed in this account to have had a call system, and that that system would not have been lost by the stage we have now reached. The hunting of dangerous animals is a challenge even with advanced weapons. With primitive weapons there is a great advantage if it can be done collaboratively. But this calls for coordination of the acts of the participants. Their hands hold weapons and are thus unavailable for any complicated semaphor. Their visual attention must be divided between the motions of the quarry and those of the other participants. All this favors an increase in flexibility of vocal-auditory communication.

Other factors also favor such an increase. Meat is a highly efficient and compactly packaged food, as compared with uncultivated plants. A small kill may not go very far, but with collective hunting larger quarry were caught. After such a large kill, there is often more food than can be consumed even by all the direct participants in the hunt. Sharing the food among all the members of the band comes about almost automatically, in that when the hunters themselves are sated they no longer care if the rest take the leavings. Thus the sharing of meat makes for the survival of the whole band. Collective hunting, general food-sharing, and the carrying of an increasing variety of things all press towards a more complex social organization, which is only possible with more flexible communication. These same factors also promote what we vaguely call the "socialization" of the members of the band.

Another development bearing on the quality, if not the degree, of hominid socialization must have taken place during this same period. At some point during the slow morphological shift to efficient upright posture, the frontal approach for copulation must have first become anatomically possible, and it was doubtless immediately exploited. It may even be imagined that, for certain strains of the hominids at certain times, the expansion of the gluteus maximus rendered the dorsal approach so awkward that the invention of the frontal approach had the conservative value required by Romer's Rule. Humans have never shown much tendency to confine themselves to this position for intercourse, but it does seem to be universally known, and is almost exclusively human. Just how this change may have affected hominid lifeways is not clear. Our guess is that it changed, for the adult female, the relative roles of the adult male and of the infant, since after the innovation there is a much closer similarity for her between her reception of an infant and of a lover. This may have helped to spread the "tender emotions" of mammalian mother-infant relations to other interpersonal relationships within the band, ultimately with such further consequences as the Oedipus complex.

OPENING OF THE CALL SYSTEM

We have seen a changing pattern of life that would be well served by a vocal-auditory communicative system of greater complexity and subtlety. Now a call system can become more flexible, within limits, through the development of totally new calls to fit additional types of recurrent situations. But it cannot take the first step towards language as we know it unless something else happens: through a process about to be described, the closed system becomes open.

Let us illustrate the way in which this can come about by describing what may occasionally happen among the gibbons of today—although, to be sure, such an occurrence has never been observed. Suppose a gibbon finds himself in a situation characterized by both the presence of food and the imminence of danger. The factors are closely balanced. Instead of emitting either the clear food call or the unmistakable danger call, he utters a cry that has some of the characteristics of each. Among gibbons such an event is doubtless so rare and unusual that the other members of the band have no way of interpreting it; thus, the consequences are negligible. But if we suppose that the early weapon-carrying hominids had a somewhat richer call system (though still closed), functioning in a somewhat more complex social order, then we may also assume that this type of event happened occasionally, and that sooner or later the other members of a band responded appropriately, therefore handling an unusually complex situation more efficiently than otherwise. Thus reinforced, the habit of *blending* two old calls to produce a new one would gain ground.

Indeed, we really have to believe that this is what happened, because the phenomenon of blending is the only logically possible way in which a closed system can develop towards an open one. Let us represent the acoustic contours of one inherited call arbitrarily with the sequence of letters *ABCD* and those of another with *EFGH*. All we mean by either of these representations is that each call possesses two or more acoustic properties on which primate ears could focus attention; it does not matter just how many such acoustic properties are involved nor just what they are. Suppose that ABCD means "food here," while *EFGH* means "danger coming." Finding both food and danger, the hominid comes out with *ABGH*. If this new call becomes established, then the 2 old calls and the new one are all henceforth *composite,* instead of unanalyzable unitary signals. For, in *ABCD,* the part *AB* now means "food" and the part *CD* means "no danger"; in *EFGH, EF* now means "no food" and *GH* means "danger"; while *ABGH* means "food and danger" because *AB* and *GH* have acquired the meanings just mentioned. One might eventually even get *EFCD,* obviously meaning "no food and no danger."

It must be asked whether this mechanism of blending can really turn a closed system into an open one. The answer is that it can start the transformation (while no other known mechanism can), but that further developments must follow. Consider the matter for a moment in a purely abstract way. Suppose the initial closed system has exactly ten calls, and that each is blended with each of the others. After the blending, there are exactly 100 calls. From one point of view, a repertory of 100 calls—or of 1,000, or of ten million—is just as closed as is a system of 10 calls. A second point of view is more important. Each of the hundred possible calls now consists of 2 parts, and each part recurs in other whole calls. One has the basis for the habit of *building* composite signals out of meaningful parts, whether or not those parts occur alone as whole signals. It is this habit that lies at the center of the openness of human languages. English allows only a finite (though quite large) number of sentences only two words long. But it allows an unlimited number of different sentences because there is no fixed limit on how long a sentence may be.

Surely the opening-up of the closed call system of our ancestors required literally thousands of years, just as all the other developments on which we have touched came about at an extremely leisurely pace. It is irrelevant that the production of a single blend, or the momentary accidental carrying of a stick or stone in the hand, is a brief episode. A potentially crucial type of event can recur numberless times with no visible effect, or with effect on a band that later becomes extinct for unrelated reasons, for every one occurrence that has minuscule but viable consequences. When the opening-up of the formerly closed call system was finally achieved, the revolutionary impact on subsequent developments was as great as that of hand-carrying.

For one thing, the detailed conventions of an open system cannot be transmitted wholly through genes. The young may emit some of the calls instinctively. But they are also exposed to various more or less complex composite calls from their elders, and are obliged to infer the meanings of the parts, and the patterns by which the parts are put together to form the whole signals, from the acoustic resemblances among the calls they hear and from the behavioral contexts in which they are uttered. (To this day, that is how human infants learn their native language.) Thus, the development of an open system puts a premium on any capacity for learning and teaching that a species may have, and selects for an increase in the genetic basis for that capacity.

If the conventions of a system have largely to be learned before the system can be efficiently used, then much of that learning will eventually be carried on away from the contexts in which the utterances being practiced would be immediately relevant. We recall the general mammalian phenomenon of play. The development of an open, largely traditionally transmitted, vocal-auditory communicative system means that *verbal play* is added to play at fighting, sexual play, and any other older categories. But this, in turn, means that situations are being talked about when they do not exist—that is, it means the addition of displacement to the design features already at hand. Speaking of things which are out of sight or in the past or future is very much like carrying a weapon when there is no immediate need for it. Each of these habits thus reinforces the other.

What was formerly a closed call system has now evolved into an open system, with details transmitted largely by tradition rather than through the genes, and with the property of displacement. Let us call such a system *pre-language*. It was still not true language, because it lacked the duality of patterning of true language. Nothing like pre-language is known for sure in the world today. Any hominid strain that developed its vocal-auditory communication only to this stage has become extinct. If we could hear the pre-language of our forerunners, it would probably not sound like human speech. It would sound much more like animal calls, and only very careful analysis would reveal its language-like properties.

The development of openness, with the various consequences already mentioned, either accompanied or paved the way for some radical developments in tool habits. We imagine that tool *manufacture*—as over against the using and carrying of tools—received its single greatest impetus from this source. If carrying a weapon selects for foresight, shaping a rough weapon into a better one indicates even greater foresight. The manufacturing of a generalized tool—one designed to be carried around for a variety of possible uses—and the development of tools specialized for use in the making of other tools, certainly followed the inception of pre-language. Weapon-making and tool-shaping are further activities at which the young can play, as they learn their communicative system and other adult ways by playing with them.

We must suppose that the detailed conventions of pre-language underwent changes, and became differentiated from one band to another, much more rapidly than had the earlier call system from which it sprang (though perhaps much more slowly than languages change today). Both of these points are implied by the increased relative role of tradition as over against genetics. New blends were not uncommon. They introduced new patterns for combining elements into whole signals, and old patterns became obsolete. Any such innovation of detail spread naturally to all members of the band in which it occurred, but not readily, if at all, from one band to another. If a band fissioned into two bands—this must have happened repeatedly throughout hominoid and hominid history—the "daughter" bands started their independent existence with a single inherited pre-language, but innovations thereafter were independent, so that in course of time the two daughter bands came to have two "mutually unintelligible" pre-languages. This is exactly—except for rate of change—what has happened to true human languages in recent millennia; we must assume that the phenomena of change and of divergence are as old as the emergence of pre-language.

THE INCEPTION OF DUALITY

Something else had been happening during prehominid and hominid evolution up to this point. In apes, the glottis lies very close to the velum, and articulatory motions anything like those involved in human language are structurally awkward. The development of upright posture, with the completion of the migration of the face from the end to the ventral side of the head, turns the axis of the oral cavity to a position approximately at right angles to the pharynx, and introduces a marked separation of glottis from velum. Hundreds of generations of chattering, first in a call system and then in pre-language, increases the innervation of the vocal tract and enriches the cortical representation of that region. The stage is set for the development of the kinds of articulatory motions familiar today.

Now, neither of these changes leads directly and inevitably to duality of patterning. Indeed, the first change is in no sense logically required if duality is to develop; in a way, it was fortuitous, since it was a by-product of changes taking place for a totally different set of selective reasons. In another species with a different earlier history, duality might use some other apparatus. If early primate history had for some reason promoted precision of control of the sphincter, and of the accumulation and discharge of intestinal gas, speech sounds today might be anal spirants. Everything else about the logical design of human language could be exactly as it actually is. The failure to distinguish in this way between the logically possible and the historically actual has led many investigators astray: they infer, for example, that our ancestors could not have had language until the articulatory apparatus had evolved to what it is now. They then interpret fossil jaws in invalid ways—and offer inadequate explanations of why the speech parts should have changed their morphology as they actually have during the Pleistocene.

However, the two changes described above did set the stage in a certain way. The hominids were in a state in which, if duality did develop, the machinery used for it was in all probability going to be the kind of articulatory motions we still use.

We can envisage the development of duality as follows. Pre-language became increasingly complex and flexible, among the successful strains of hominids, because of its many advantages for survival. The constant rubbing-together of whole utterances (by the blending mechanism described earlier) generated an increasingly large stock of minimum meaningful signal elements—the "pre-morphemes" of pre-language. Lacking duality, however, these pre-morphemes had to be holistically different from one another in their acoustic contours. But the available articulatory-acoustic space became more and more densely packed; some pre-morphemes became so similar to others that keeping them apart, either in production or in detection, was too great a challenge for hominid mouths, ears, and brains. Something had to happen, or the system would collapse of its own weight. Doubtless many overloaded systems did collapse, their users thereafter becoming extinct. In at least one case, there was a brilliantly successful "mutation"; pre-morphemes began to be listened to and identified not in terms of their acoustic gestalts but in terms of smaller features of sound that occurred in them in varying arrangements. In pace with this shift in the technique of detection, articulatory motions came to be directed not towards the generation of a suitable acoustic gestalt but towards the sufficiently precise production of the relevant smaller features of sound that identified one pre-morpheme as over against others.

With this change, pre-morphemes became true morphemes, the features of sound involved became phonological components, and pre-language had become true language.

Although brilliant and crucial, this innovation need not have been either as sudden or as difficult as our description may seem to imply. With openness, but as yet without duality, the hearer is already required to pay attention to acoustic detail, rather than merely to one or another convenient symptom of a whole acoustic gestalt, if he is to recognize the constituent pre-morphemes of a composite call and thus react appropriately to the whole call. In a pure call system, the beginning of a call may be distinctive enough to identify the whole call; the rest does not have to be heard. In pre-language, one cannot predict from the beginning of a call how it will continue and end. This clearly paves the way for duality. It is then, in one sense, but a small step to stop regarding acoustic details as *constituting* morphemes and start interpreting them as *identifying* or *representing* morphemes.

Here, as for all the other developments we have mentioned, we must remember Romer's Rule. The ultimate consequences of the inception of duality have been enormous. But the immediate value of the innovation was conservative. It rendered possible the continued use of a thoroughly familiar type of communicative system in a thoroughly familiar way, in the face of a gradual but potentially embarrassing increase in the complexity of the system.

The emergence of true language from a closed call system, by the steps and stages we have described, should properly be thought of not as a replacement of one sort of communicative system by another, but rather as the growth of a new system within the matrix of the old one. Certain features of the proto-hominoid call system are still found in human vocal-auditory behavior, but as accompaniments to the use of language rather than as part of language. The proto-hominoids could vary the intensity, the pitch, and the duration of a single call. We still do this as we speak sentences in a language: we speak sometimes more loudly, sometimes more softly, sometimes in a higher register and sometimes in a lower, and so on. Also, we use certain grunts and cries (*uh-huh, huh-uh, ow!*) that are not words or morphemes and not part of language. These various *paralinguistic* phenomena, as they are called, have been reworked and modified in many ways by the conditions of life of speaking humans, but their pedigree, like that of communicative body motion, is older than that of language itself.

The phenomenon of sound change, mentioned briefly at the outset of this paper, began immediately upon the transition from pre-language to true language, continues now, and will continue in the future unless our vocal-auditory communication crosses some currently unforeseeable Rubicon. The phonological system of a language has almost as its sole function that of keeping meaningful utterances apart. But a phonological system is a delicately balanced affair, constantly being thrown into slight disbalance by careless articulation or channel noise and constantly repatterning itself in a slightly altered way. It is perfectly possible, in the course of time, for two phonemes to fall together—that is, for the articulatory-acoustic difference between them to disappear. Obviously, this changes the machinery with which morphemes and utterances are distinguished. The interest this holds for us is that it affords an example of the workings of Romer's Rule in a purely cultural context instead of a largely genetic one.

What happens seems to be about as follows. A particular phonemic difference is slowly eaten away by sound change, to the point that it is no longer reliable as a way of keeping utterances apart. This is the "changed circumstances" of Romer's Rule. The speakers of the language develop, by analogy, a way of paraphrasing any utterance that would be potentially ambiguous if uttered in the traditional way. The paraphrase is the "innovation" of the Rule. The value of the paraphrase is that the speakers can thereby continue to speak in largely the same way they learned from their predecessors. The innovation is minor and trivial, but effective in that if the phonemic contrast disappears entirely, ease of communication is in no way impaired. The inevitable and continuous process of sound change never reduces the machinery of a language to zero. A compensation of some sort is developed for every loss of contrast.

CHRONOLOGY

We have now outlined a plausible evolutionary sequence leading from the proto-hominoids to our earliest truly human ancestors. For we assert that as soon as the hominids had achieved upright posture, bipedal gait, the use of hands for manipulating, for carrying, and for manufacturing generalized tools, and language, they had become men. The human revolution was over. Two important questions remain. How long did the changes take? How long ago were they completed?

It is certain that the changes we have talked about did not begin before the time of the proto-hominoids. But at present we have no way of knowing how much later than that was their inception. Conceivably the hominids of the Middle or Upper Pliocene, though already separated from the pongids, were very little more like modern man than were the proto-hominoids.

On the other hand, we are convinced that all the crucial developments of which we have spoken had been achieved by about one million years ago—that is, by the beginning of the Pleistocene.

The most important evidence for the date just presented is the *subsequent* growth of the brain, attested by the fossil record. The brain of *Australopithecus* is scarcely larger than that of a gorilla. But from about three-quarters of a million years ago to about forty thousand years ago, the brain grew steadily. Part of this increase reflects an overall increase in body size. Allowing for this, there is still something to be explained. Was the increase in relative size fortuitous or adaptive?

It is utterly out of the question that the growth was fortuitous. A large brain is biologically too expensive. It demands a high percentage of the blood supply—12% in modern man, though the brain accounts for only about 2% of the body's volume—and all that blood, in an upright biped, must be pumped uphill. It requires an enlarged skull, which makes for difficulty during parturition, particularly since the development of upright posture resculptures the pelvis very badly for childbirth. This cost cannot be borne unless there are compensations.

We must therefore assume that if a species has actually developed a bigger and more convoluted brain, with a particularly sharp increase in the forebrain, there was survival value in the change. For our ancestors of a million years ago the survival value of bigger brains is obvious if and only if they had *already* achieved the essence of language and culture. Continued growth would then be advantageous up to a certain maximum, but thereafter unprofitable because it made for excessive difficulties in other respects but yielded no further usable gain in brainpower.

The archeological and fossil record supports our date, or even suggests that we have been too conservative. Until recently, the earliest obviously shaped tools that had been dug up were not quite so ancient, but they implied an earlier period of development that was not directly attested. Now, however, we have the direct evidence of at least crudely shaped stone tools in association with hominid fossils from Bed I at Olduvai, for which a maximum date of one and three-quarters million years ago is seriously proposed. What is more, the Australopithecines show the typically human reduction in the size of the canine teeth, formerly used for cutting and tearing; and this reduction could not have been tolerated had the hominids not developed tools with which to perform such operations.

It might be suggested that, although all other crucial innovations of the human revolution were as early as we have proposed, the inception of duality may have been later.

There are two reasons why we think that duality is just as old as the rest.

One side-effect of brain growth is that the top of the head is pushed forward to form a forehead. We do not see why this should in itself entail a recession of the lower part of the face, to yield the essentially flat perpendicular human physiognomy which, with minor variations, now prevails. In terms of the balancing of the head above an upright body, perhaps the recession of the snout and the decrease in its massiveness are useful. If cooking is a sufficiently old art, then perhaps this external predigestion of food at least rendered possible the reduction in size of teeth and jaws. But it seems to us that these factors still leave room for a further influence: that of the habit of talking, in a true language that uses the kinds of articulatory motions that are now universal, requiring precise motions of lips, jaw, tongue, velum, glottis, and pulmonary musculature. If true language can be assumed for our ancestors of a million years ago, then it is old enough to have played a role in the genetically monitored evolutionary changes in what we now call the "organs of speech." And if this is correct, then "organs of speech" is no metaphor but a biologically correct description.

Our other reason for believing that duality of patterning, and the modern type of sound-producing articulatory motions, are very old, turns on time, space, and degrees of uniformity and diversity. The fossil record shows that the human diaspora from East Africa cannot be much more recent than the Middle Pleistocene. This means that several hundred thousand years have been available for a genetic adaptation to a wide variety of climates and topographies. Yet man shows an amazingly small amount of racial diversity—far less, for example, than that of dogs, which has come about in a much shorter span of time. (Of course, the difference in generation span between men and dogs must be taken into account; but when one allows liberally for this the comparison, though less striking, still seems valid.)

There is this same striking lack of diversity in certain features of language. Though we have no fossils, our observations of the languages of today, and of those few attested by written records during the past few millenia, have some relevance. Almost every type of articulation known to function in any language anywhere recurs in various other languages, with no significant pattern of geographical distribution. Phonological systems—as over against individual speech sounds—show much less variety than could easily be invented by any linguist working with pencil and paper. This uniformity precludes the independent invention of duality of patterning, and of modern articulatory motions, in two or more parts of the world. The crucial developments must have taken place once, and then spread. The innovations could have been either recent or ancient, except for an additional fact: in every language, the phonological raw materials are used with remarkable efficiency. This speaks for great antiquity, since we cannot imagine that such efficiency was an instant result of the appearance of the first trace of duality.

True diversity is found in more superficial aspects of language, and in all those other phases of human life where tradition, rather than genetics, is clearly the major mechanism of change and of adaptation. We are thus led to a familiar conclusion. The human revolution, completed before the diaspora, established a state of affairs in which further change and adaptation could be effected, within broad limits, by tradition rather than genetics. That is why human racial diversity is so slight, and it is why the languages and cultures of all communities, no matter how diverse, are elaborations of a single inherited "common denominator."

ADDITIONAL PLEISTOCENE CHANGES

The further consequences of the human revolution include, in the end, everything that we have done since. Only a few of the more striking (and earlier) of these subsequent developments need to be mentioned here.

Language and culture, as we have seen, selected for bigger brains. Bigger brains mean bigger heads. Bigger heads mean greater difficulty in parturition. Even today, the head is the chief troublemaker in childbirth. This difficulty can be combatted to some extent by expelling the fetus relatively earlier in its development. There was therefore a selection for such earlier expulsion. But this, in turn, makes for a longer period of helpless infancy —which is, at the same time, a period of maximum plasticity, during which the child can acquire the complex extra-genetic heritage of its community. The helplessness of infants demands longer and more elaborate child care, and it becomes highly convenient for the adult males to help the mothers. Some of the skills that the young males must learn can only be learned from the adult males. All this makes for the domestication of fathers. This, together with the habit of paying attention to past experiences and future contingencies (which we have seen arising in the context of play, of tool-carrying, of the displacement of pre-language, and of tool-making), promotes male jealousy. The seeds of this may have been earlier, but it now becomes eminently reasonable for a male to reserve a female, even when he is not sexually hungry, that she may be available when the need arises.

In the developments just outlined we can also see contributing sources for the complex restrictions and rituals with which human sexual relations are hedged about. These include not only all the rules of exogamy and endogamy and the varying principles controlling premarital and extramarital relations, but also the whole matter of taste—some individuals of the opposite sex are attractive, others unattractive, according to criteria learned from one's community. Any male past puberty, and any female between menarche and menopause, can, in a matter of seconds, stand a good chance of launching a new human. But child care requires time and energy thereby unavailable for other important activities. From this stem such varied modern institutions as celibate orders and beauty contests.

Among the proto-hominoids the band leaders were the strongest adult males. Language, in particular, changes this. The oldest members of the band, strong or feeble, are valued because they have had time to learn more. They are repositories of information on which the community can call as it is needed. This use of the elderly as encyclopedias perhaps helps to select for a greater life span, though the pedomorphism discussed earlier may also have played a part in bringing about this result. Certainly the increased social utility of the elderly promotes a protection of the old and feeble by the young and strong; it may contribute to doing something positive about the disposal of the dead.

As soon as the hominids had achieved a reasonably effective bipedal *walking* gait—not running, which is useful only for fast coverage of short distances—they had the basic wherewithal for migrating slowly throughout all the continental territory to which they could adapt their lifeways. For the invasion of some climatic zones, protection against the cold is necessary. There are various physiological ways of doing this, but the hominids developed an additional device: clothing.

The Chinese variety of *Pithecanthropus* used fire for warmth. By his epoch, then, the hominid invasion of cold climates had begun. But we suspect that clothing was a much earlier invention, already available when it was first needed for warmth.

Clothing serves roughly three functions: protection, as against the cold; modesty and vanity; and *carrying*. The last of these functions was, we suggest, the one of earliest relevance. If one's way of life rests on hand-carrying, and if the number and variety of things to be carried is increasing to the point of awkwardness, then the invention of a device that helps one carry things has the conservative survival value required by Romer's Rule. The first clothing-as-harness may have been nothing more than a piece of vine pulled from the trees and draped over the shoulder or around the waist. Later, when the hominids were regularly killing small animals, the hides—useless as food—might have been put to this use. A hide cannot be eaten, but if one is hungry enough there is some nourishment to be obtained by chewing at it. Almost as early as the first use of hides as harness, it may have been discovered that a hide that has been chewed is more flexible and comfortable to wear than one that has not. This way of processing hides was still widespread only yesterday.

It is unlikely that any direct archeological evidence of these posited early clothing developments will ever turn up. But if clothing of sorts is actually that ancient, then it was already available, with only minor modifications, when it was first needed to help explore ecological niches characterized by cold. It may even be old enough to have played a part in permitting the development of the relative hairlessness characteristic of all strains of *Homo sapiens* today.

30. THE CAPACITY FOR LANGUAGE ACQUISITION

ERIC H. LENNEBERG

Reprinted from Jerry A. Fodor and Jerrold J. Katz (Eds.), The Structure of Language: Readings in the Philosophy of Language *(Englewood Cliffs, N. J.: Prentice-Hall, 1964). Eric H. Lenneberg is Professor of Psychology and Neurobiology at Cornell University. He has made extensive investigations into the relationship between language and brain mechanisms, which have been summarized in his book,* Biological Foundations of Language. *His research into language mechanisms has led him into more general studies of brain maturation and behavior.*

■ There are many sources of excitement in science, one of which is the discovery of new facts, especially when they expand scientific horizons and lead to further discoveries. An example of this is the research that led to deciphering the genetic code or learning the relationship between rules of residence at marriage and the organization of groups in society. Another source of excitement is fresh and speculative insight (also known as "sticking one's neck out" or "going out on a limb"). Indeed, speculative risk-taking is usually the foundation for empirical discoveries of the first type, and this paper by Lenneberg is an example of such a groundbreaking "hunch."

Many students of evolution and language are skeptical of Lenneberg's thesis, but few are ready to reject it out of hand (or to adopt it completely) until a way is found to evaluate his notions empirically. Lenneberg contends there is a biological foundation in the individual that shapes the development of speech in the child. This hypothesis has profound implications for the study of the evolution of language because it suggests not only that the emergence of man brought various capacities for the development of language but also that man's biological constitution has set limits on the nature of language and the direction of the development. Lenneberg, for example, observes that one of the similarities among all human languages is that none of them incorporates the calls of birds, the sounds of animals, or the crying noises of babies, even though men are capable of imitating these noises. All languages have rules of syntax; all join words together instead of using single-word utterances ("I like you" instead of "like"). Are these similarities accidental or are they due to some aspect of man's biological constitution? Lenneberg feels that the available evidence supports the latter hypothesis and points to an innate drive for symbolic communication. This is not to say, as he takes care to point out, that language is an inherited phenomenon, but only that a biologically based capacity or predisposition for language seems to account for its universality. In other words, Lenneberg suggests, language is not wholly cultural; it has an important biological component. It is an example of genetic programming.

The reader can follow this line of analysis further in Basil Bernstein's "Aspects of Language and Learning in the Genesis of Social Process" *(Journal of Child Psychology and Psychiatry*, 1 [1961]: 313-24). This paper is reprinted with excellent suggestions and bibliographical references in *Language in Culture and Society* (edited by Dell H. Hymes [New York: Harper & Row, 1964]). The reader also should look into the writings of Jean Piaget, especially his *Language and Thought of the Child* (London: Routledge and Kegan Paul, 1962), *Judgment and Reasoning in the Child* (Cleveland: Meridian Books, 1955), and *The Moral Judgment of the Child* (New York: Collier Books, 1962). ■

THERE IS A TENDENCY among social scientists to regard language as a wholly learned and cultural phenomenon, an ingeniously devised instrument, purposefully introduced to subserve social functions, the artificial shaping of an amorphous, general capacity called *intelligence.* We scarcely entertain the notion that man may be equipped with highly specialized, biological propensities that favor and, indeed, shape the development of speech in the child and that roots of language may be as deeply grounded in our natural constitution as, for instance, our predisposition to use our hands. To demonstrate the logical possibility—if not probability—of such a situation is the purpose of this paper. It is maintained that clarity on the problem of the biological foundation of language is of utmost importance in formulating both questions and hypotheses regarding the function, mechanism, and history of language.

The heuristic method to be employed here will be analogous to procedures employed in studying processes too slow and inert to be amenable to laboratory experimentation, notably biological evolution. The reasoning of our argument may gain by a few general statements on this type of theory construction and by a review of the basic, modern principles evoked in current discussions of evolution.

In many scientific endeavors we are faced with the problem of reconstructing a sequence of events from scattered, static evidence. The writing of geological, phylogenetic, and cultural histories is alike in this respect. But our treatment of geological and phylogenetic history differs from cultural history when it comes to "explaining" the causal relationships that hold between the events.

In geology we may trace cycles of elevation of the continent: subsequent leveling by erosion, followed by sedimentation at the bottom of the sea, and then recurrent elevation of the once submerged land, far above the level of the sea, resulting again in erosion and so forth. We cannot *explain* these sequences in terms of purpose, for purpose assumes a planned action, a pre-established end. Erosion, for instance, serves no more the *purpose* of establishing a balance than the eruption of a volcano serves the purpose of making erosion possible. It is appropriate to speak about disturbed and re-established equilibria; but the use of the word *purpose* has the common connotation of striving toward a goal, and, therefore, ought to be reserved for pieces of behavior that do indeed aim at a pre-established end without, however, being bound by nature to reach such ends by pre-established means.

In our discussions of phylogeny we must be as careful to avoid teleological explanations as in the case of geological history. Yet, many a time we seem to have no small difficulty in living up to this idea. It seems so reasonable to say that the *purpose* of man's increased cranial vault is to house a large brain; and that the *purpose* of a large brain is the perfection of intelligence. We must take exception to this formulation because it implies finality in evolution or, at least, the assumption of a pre-established direction and end. The geneticist looks at evolution as the interplay between a *random* process and certain constraining factors. The random process is the blind generation of inheritable characteristics, i.e., mutations, while all the constraining factors have to do with viability of the individual or the species as a whole. Of the many new traits that may chance to appear, the great majority will have a lethal effect under given environmental conditions and are thus of no consequence for evolution. But occasionally there is one that *is* compatible with life and will thus result in perpetuation, at least over a limited period of time.

Attempts have been made to discover whether specific types of mutation could be regarded as adaptive responses of the germ plasm to environmental necessities, but I believe it is fair to say that so far results are not sufficient to conclude that there is a generally adaptive directionality in mutations. Dobzhansky states: "Genetics . . . asserts that the organism is not endowed with providential ability to respond to the requirements of the environment by producing mutations adapted to these requirements."

If it is conceded that variability of inheritable traits due to mutation does not reflect direct responses to *needs,* it is quite conceivable that we may find characteristics that are compatible with life under prevailing conditions but that have no heightened adaptive value and can therefore not be explained in terms of utility to the organism. The differentiating characteristics of human races may be cases in point. The shape of skulls or the textures of hair cannot be rated by usefulness; nor can those mutations that have resulted in new species without extermination or limitation of the older forms.

The problem is more complicated when we observe a long and linear evolutionary trend, for instance the more or less steady increase in the body size of a species. When such a linear development occurs, we say that the evolved trait is *useful* to the species. I would like to stress, however, that the word *useful* (or reference to utility) must be employed with great care in this context and not without careful definition lest it be confused with purposiveness. In case of gradually increasing body size (take for instance the history of the horse), an individual animal stays alive if it is of a certain size, whereas an individual may perish if it falls short of the size that is critical at the time it is born. Since the individual cannot alter its inherited size, it also cannot change its fate of starving or being killed before maturation. Thus, no matter how *useful* it may be to be large, this state of affairs cannot be reached by purposeful striving of individual animals. Much less can we conceive of a super-individual entity (such as the species as a whole), making *use* of this or that trait in order to "insure the continuation" of the species. Something can become useful after it has come into being by a random process; but to make systematic use of a trait, such as size, seems to imply foresight and providence not usually accorded to the driving forces of genetics.

The situation is quite different when we come to a discussion of cultural history. Here, explanations in terms of long-range purpose and utility often are in order because man, indeed, does have final ends in view which he strives to achieve by this or that means. Frequently there are even explicit criteria for usefulness in reaching a goal such as reduction of physical effort, maximizing gratification, or introducing order and manageability into a certain situation. In the development of coin money, for instance, there may have been some trial and error in the course of history, but many changes were introduced by fiat with the explicit purpose of facilitating economic intercourse. In other words, the development of coin money is the direct result of a certain property of human behavior, namely purposiveness. Or, more generally, it may be said that the phenomenon of *culture per se* is the outgrowth of this characteristic trait. But this should not obscure the fact that man and his abilities are also the product of biological evolution and that many of his traits are genetically determined and as such their existence must not be explained in terms of purpose. For instance the alternations between sleep and wakefulness, the shedding of tears, the closure of the epiglottis in swallowing, or any other unconditioned responses cannot be considered as the outcome of rational invention, as the end product of a purposeful striving just as any other genetic phenomena must not be accounted for in this way.

It is well to remember that purposiveness is a trait that is itself the result of evolutionary history, of phylogenetic development. Rudimentary forms of short-term purpose are observable as far back as the invertebrates. It is the ability to strive toward a goal (say nest building) by more than a single rigid action pattern. It is an ability to take advantage of specific environmental conditions in the accomplishment of certain tasks. For instance, birds are not confined to one specific type of material in the construction of their nests; the use of tiny shreds of newspaper incorporated in these structures is not an uncommon finding. Purposiveness requires anticipation or expectancy together with a flexibility in the choice of routes that lead to the goal.

The purposiveness displayed in man's activities differs from that seen in lower animals not so much in quality as in degree. No other animal seems capable of performing actions with such long-range purpose as is seen in our sociocultural activities. Not even such activities as nest building in birds which may last for days and weeks are the result of long-range purpose. This has been described by Tinbergen and commented on by Thorpe. The nest is merely the end result of a very long series of individual tasks where each accomplishment seems to trigger off a striving for the fulfillment of the next task, and there is evidence that purposiveness, as defined above, does not actually extend over the entire plan; each task has its own characteristic, short-term purposiveness.

Our objective now is to examine language and to decide in which of its aspects we must assume it to be a genetically determined trait and in which of its aspects it might be the result of cultural activity. Insofar as it is revealed to be a biologically determined affair, we cannot explain it as the result of a purposefully devised system; we may not claim that *the reason* a child learns it is the inherent possibility of providing pleasure, security, or usefulness; or that language has this or that property because this was found in pre-historic times to serve best the purpose of communication. Any hedonistic or utilitarian explanation of language is tantamount to claiming that speech as such is a cultural phenomenon or, at least, that it is the product of purposive behavior. Whereas, a demonstration that language is at least partly determined by innate predispositions would put serious constraints on utilitarian explanations of language and would instead focus our attention on physiological, anatomical, and genetic factors underlying verbal behavior.

Before embarking on the actual argument, a brief warning may be in place. The distinction between genetically determined and purposive behavior is *not* the same as the distinction between behavior that does or does not depend upon environmental conditions. The following example based on work by B. F. Riess will illustrate this point; the quotation is due to Beach.

> The maternal behavior of primiparous female rats reared in isolation is indistinguishable from that of multiparous individuals. Animals with no maternal experience build nests before the first litter is born. However, pregnant rats that have been reared in cages containing nothing that can be picked up and transported do not build nests when material is made available. They simply heap their young in a pile in a corner of the cage. Other females that have been reared under conditions preventing them from licking and grooming their own bodies fail to clean their young at the time of parturition.

From this example it is obvious that innate behavior may be intimately related to or dependent upon the organism's interaction with its environment, yet the action *sequence* as a whole (first carrying things in an unorganized way; then, when pregnant, carrying things in an organized way so that the end product is a nest) is innately given. In other words, it would not be reasonable to claim that the young female rat carries things around because she is planning to build a nest if she should be pregnant and that she is purposefully training herself in carrying around material to be better prepared for the eventualities in store for her.

On the other hand, *purposive* behavior may only be very indirectly related to environmental conditions and thus give the impression of completely spontaneous creation; the composition of the Jupiter Symphony is an example.

In our discussion of language, we shall proceed in the following way. We shall juxtapose two types of human activities, one of which we have good reasons to believe to be biologically given, i.e., walking, while the other one we can safely assume to be the result of cultural achievement and thus a product of purposiveness, namely, writing. By comparing these two types of activities, it will be shown that there are at least four good criteria which distinguish in man biologically determined from culturally determined behavior. When these criteria are applied to language, it will be seen that verbal behavior in many important respects resembles the biological type, while in other respects it bears the sign of cultural and purposive activity. Since the culturally determined features in language are widely noted, the discussion will emphasize innate factors more than cultural ones.

Bipedal Gait	*Writing*
CRITERION 1	
No intraspecies variations: The species has only one type of locomotion; it is universal to all men. (This is a special case of the more general point that inherited traits have poor correlations—if any—with social groupings: cf. black hair or protruding zygoma.)	*Intraspecies variations correlated with social organizations*: A number of very different successful writing systems have co-existed. The geographical distribution of writing systems follows cultural and social lines of demarcation.

CRITERION 2

No history within species: We cannot trace the development of bipedal gait from a primitive to a complex stage throughout the history of human *cultures*. There are no geographical foci from which cultural diffusion of the trait seems to have emanated at earlier times. All human races have the same basic skeletal foot pattern. For significant variations in gait, we have to go back to fossil forms that represent a predecessor of modern man.	*Only history within species*: There are cultures where even the most primitive writing system is completely absent. We can follow the development of writing historically just as we can study the distribution of writing geographically. We can make good guesses as to the area of invention and development and trace the cultural diffusion over the surface of the globe and throughout the last few millennia of history. The emergence of writing is a relatively recent event.

CRITERION 3

Evidence for inherited predisposition: Permanent and customary gait cannot be taught or learned by practice if the animal is not biologically constituted for this type of locomotion.	*No evidence for inherited predisposition*: Illiteracy in nonWestern societies is not ordinarily a sign of mental deficiency but of deficiency in training. The condition can be quickly corrected by appropriate practice.

CRITERION 4

Presumption of specific organic correlates: In the case of gait, we do not have to *presume* organic correlates; we *know* them. However, behavioral traits that are regarded as the product of evolution (insects) are also thought to be based on organic predispositions, in this case, on the grounds of circumstantial evidence and often in the absence of anatomical and physiological knowledge.	*No assumption of specific organic correlates*: We do, of course, assume a biological capacity for writing, but there is no evidence for innate predisposition for this activity. A child's contact with written documents or with pencil and paper does not ordinarily result in automatic acquisition of the trait. Nor do we suppose that the people in a society that has evolved no writing system to be genetically different from those of a writing society. It is axiomatic in anthropology that any normal infant can acquire all cultural traits of any society given the specific cultural upbringing.

BIOLOGICAL AND SOCIOCULTURAL FACTORS IN LANGUAGE

Let us now view language in the light of the four criteria discussed above in order to see to what extent language is part of our biological heritage.

1 FIRST CRITERION: VARIATION WITHIN SPECIES.

One of the major contributions of modern linguistics was the dispelling of the eighteenth-century notion of a *universal grammar* which, at the time, was based on the assumption of a universal logic. In America it was particularly the descriptivist school initiated by Franz Boas that has been most active during the last thirty years in demonstrating the truly amazing variety of phonological, grammatical, and semantic systems in the languages of the world. These workers have shown how the traditional method of describing languages in terms of logic must be abandoned for more objective, formal, and unprejudiced analyses; they have shown that lexicons of different languages are never strictly comparable; in fact, they have made us aware of the difficulty inherent in such notions as *word, tense,* or *parts of speech.* Thus, today anyone interested in language and speech is keenly aware of the great diversification of linguistic form in the languages of the world, and it is commonly acknowledged that their histories cannot be traced back to a common "*ur-language.*" In the light of this realization it is very remarkable to note in some respects all languages are alike and that this similarity is by no means a *logical* necessity. Following are three points in which languages are identical; they are, however, not the only similarities.

1.1 *Phonology.*

Speech is without exception a vocal affair, and, more important, the vocalizations heard in the languages of the world are always within fairly narrow limits of the total range of sounds that man can produce. For instance, we can faithfully imitate the noises of many mammals, the songs of a number of birds, the crying noises of an infant; yet, these direct imitations never seem to be incorporated in vocabularies. There is onomatopoeia, to be sure; but onomatopoetic words are never faithful imitations but phonemicized expressions. This is precisely the point: all languages have phonemic systems; that is, the morphemes of all languages can be further segmented into smaller, *meaningless,* components of functionally similar sounds. Words and morphemes are constituted in all languages by a sequence of phonemes. This is not a matter of definition or a methodological artifact. One can visualize a very complex language in which the symbol for *cat* is a perfect imitation of that animal's noise (and so on for other mammals), for *baby* the infant's characteristic cries, for a *shrew* scolding yells; the *size* of objects could be represented by sound intensity, *vertical direction* by pitch, *color* by vowel quality, *hunger* by roaring, *sex* by caressing whimpers, and so on. In such a language we would have morphemes or words that could not be segmented into common, concatenated sound elements. Most words and perhaps all morphemes would constitute a sound-Gestalt *sui generis* much the way pictograms and idiograms cannot be analyzed into a small set of letters.

It would be interesting to see whether parrots speak in phonemes or not; if they do not speak in phonemes (as I would assume), we would have an empirical demonstration that the phonemic phenomenon is neither a methodological artifact nor a logical necessity. One could, for instance, take a parrot who was raised in Brazil and who has acquired a good repertoire of Portuguese phrases and words, and suddenly transplant him to an English-speaking environment where he would add English bits to his stock of exclamations. If the first few words are pronounced with a heavy Portuguese accent, we would have evidence that the bird generalizes his Portuguese habits, that is, that he has actually learned Portuguese phonemes which he now uses in the production of English words. However, if his English acquisitions sound at once *native,* it would appear that the parrot merely has an ability for imitating sounds without deriving from it a generalized habit for the production of speech.

Whether this experiment is practically possible, I do not know. It is related here rather to highlight the problem at stake. It also suggests some empirical research on human subjects. If foreign accents are a proof for the existence of phonemes and if the child at three is said to speak phonemically (which every linguist would have to affirm), we would expect him to have an English accent if he is suddenly asked to pronounce say a simple German word—provided he has never heard German before. This is a project that could be done quite easily and objectively and which would be very revealing. (We are not speaking here of the young child's ability quickly to learn foreign languages, i.e., learn more than a single phonemic system. This is a different problem that will be discussed in greater detail.)

1.2 *Concatenation.*

This term denotes the phenomenon of stringing up morphemes or words into a complex sequence called *phrases, sentences,* or *discourse.* No speech community has ever been described where communication is restricted to single-word discourse, where the customary utterance would be something like "water!" or "go," or "birds"; where it would be impossible, for example, to give geographical directions by means of concatenated, independent forms. Man everywhere talks in what appears to be a "blue streak."

1.3 *Syntactic structure.*

We know of no language that concatenates randomly, that is, where any word may be followed by any other. There are contingencies between words (or, languages have typical statistical structures) but this in itself does not constitute grammars. We can program stochastic processes into machines such that they generate symbols (e.g., words) with the same statistical properties as that noted for languages; yet these machines will not "speak grammatically," at least not insofar as they generate new sentences. It is generally assumed by linguists—and there are compelling reasons for this—that there must be a finite set of rules that defines all grammatical operations for any given language. Any native speaker will generate sentences that conform to these grammatical rules, and any speaker of the speech community will recognize such sentences as grammatical. We are dealing here with an extremely complex mechanism and one that has never been fully described in purely formal terms for any language (if it had, we could program computers that can "speak" grammatically); and yet, we know that the mechanism must exist for the simple reason that every speaker knows and generally agrees with fellow speakers whether a sentence is grammatical or not. (This has nothing to do with familiarity or meaning of an utterance. One may easily demonstrate this by comparing Chomsky's two sentences, "colorless green ideas sleep furiously" and "furiously sleep ideas green colorless," where neither of the sentences are likely to have occurred prior to Chomsky's illustration, yet one is recognized as grammatical and the other not.) Note that types of sentence structures are as variable as speech sounds among languages of the world, but the phenomenon of grammar as such is absolutely universal.

1.4 *Conclusion.*

The importance of the universality of phonematization (evidenced by the universality of small and finite phoneme stocks), of the universality of concatenation, and of the ubiquitous presence of grammar cannot be overestimated. Consider the vast differences in the forms and semantics of languages (making a common and focal origin of language most unlikely); consider the geographical separation of some human societies that must have persisted for thousands of years; consider the physical differentiation into a number of different stocks and races. Yet, everywhere man communicates in a strikingly similar pattern. There are only two kinds of conclusion that can be drawn from this situation. Either the similarities are due to the fact that, by happenstance, identical principles of communication have developed completely independently over and over, hundreds of times—an extremely improbable supposition, or the universal phenomena reflect some trait that is related to the genetic mutation that has constituted the speciation of *Homo sapiens* and are, therefore, of a venerable age. I should like to take the latter view, and I feel strengthened in this position by the evidence that follows.

Perhaps someone would like to argue that a third explanation *is* possible, namely, that languages are alike because everywhere it was discovered that there is an "optimal way of oral communication" and that languages as we find them simply reflect optimization of conditions. This statement is either false or else it turns out to be simply a different formulation of my second alternative. It is objectively not true that languages are the most efficient communication systems possible. From an information-theoretical point of view, they are very redundant; as far as their grammars are concerned, they seem to be "unnecessarily complicated" (the simplicity of English grammar as against, say, Navaho is certain to be an illusion); in semantic efficiency they leave much to be desired. They can only be said to be ideally efficient if we add "given man's articulatory, perceptual, and intellectual capacity." But with this concession we have admitted that man's pattern of speech is determined by his biological equipment, a point that will be further expanded in connection with the fourth criterion.

2 SECOND CRITERION: HISTORY WITHIN SPECIES.

Languages, like fashions, have histories, but nowhere does the historical evidence take us back to a stage where the phonemic mode of vocalization was in its infancy: we have no records testifying to an absence of grammar; we have no reason to believe that there are places or times where or when concatenation had not been developed. Perhaps this ought to be attributed to the rather recent development of written records. Yet, a lingering doubt remains: writing can be traced back some five thousand years, and, while the earliest written records give us few clues about the language they represent, some of our linguistic reconstructions reach back to about the same era. This is a time span that comprises about one tenth of the age of the earliest evidence of Levalloiso-Mousterian culture (some 50,000 years ago) and the appearance of fossil forms that may be considered to be the direct ancestors of modern man. Thus, the oldest documented history of languages may be short when compared with palaeontological history; but it would not be too short to demonstrate trends in the development of, for instance, phonematization if this phenomenon *did have a cultural history*. We might expect that historical phonemic changes follow a general pattern, namely, from a supposedly *primitive* stage to one that could be called *advanced.* But the phonemic changes that we actually find—and they occur rapidly (within periods of 10 to 15 generations), frequently, and continuously—seem to follow no universal line and have, by and large, a random directionality; we cannot make predictions as to the qualitative changes that will occur in English 300 years hence.

The concatenating phenomenon is, historically, completely static. Throughout the documented history there is evidence that concatenation must have existed in its present complex and universal form for at least some 5,000 years and most likely considerably longer.

The history of syntax is the same as that of phonemes. Our oldest linguistic reconstructions are based on reliable evidence that there was *order* in the concatenation of forms, that there were rules and regularities governing the sequences of morphemes which from a formal point of view cannot have been much different from grammatical processes of modern languages. We are not speaking here of specific grammars, but merely of the grammatical phenomenon as such. Syntax changes as rapidly and widely as phonemic structures, but, again, we cannot discern any constant and linear direction. At the most, there is a certain cyclicity, one grammatical type perhaps alternating with another. The so-called "analytical languages," such as Chinese and English, were preceded by synthetic types; and there is reasonable evidence, at least for Indo-European, that the grammatical *synthesis* as seen in ancient Greek was preceded by a more analytic stage (inflectional endings having been derived from once independent words). We cannot be sure, however, whether synthesis *generally* alternates with analysis; indeed, the very polarity expressed by these two terms is not very well defined in grammatical theory. It is widely agreed today that no typology of modern grammars reflects stages of absolute, nonrecurring grammatical development. Nor do we have any means for judging one grammatical system as more primitive than another.

Contrast to this situation the forms found in the animal kingdom. Species *can* be ordered in terms of anatomical simplicity (which we equate with primitivity) so that an arrangement from low to high forms results; and since phylogenetic stages are assumed to be unique and nonrecurring, we can construct phylogenetic history merely from taxonomy. But this reasoning may not be extended to linguistics. No classification of languages in terms of structural type (such as *synthetic, analytic* or *agglutinative*) provides us with a theory for a universal development of language.

There can be no question today that we are unable to trace languages back to an ungrammatical, aphonemic, or simple imitative stage; and there is, indeed, no cogent reason to believe that such a stage has ever existed. This does not imply a nineteenth-century assumption of an instinct, particularly not an instinct for specific languages. Obviously, the child's acquisition of Chinese consists in the acquisition of certain culturally evolved traits. But a phenomenon such as phonematization *per se* need not be thought of as a cultural achievement, need not constitute the summation of inventions, need not have resulted from a long series of trial and error learning in communication.

To put my point more bluntly: the absolutely unexceptional universality of phonemes, concatenation, and syntax and the absence of historical evidence for the slow cultural evolvement of these phenomena lead me to suppose that we have here the reflection of a biological matrix or Anlage which forces speech to be of one and no other basic type. From the point of view of genetic theory we would not have to expect a *gradual* and *selective* process culminating in present-day languages. Mutations are thought of today as sudden intracellular reorganizations in germ plasm, resulting in changes of the gross anatomical structures and also in radical, innate, neuronal re-organization manifested by highly specific behavioral patterns.

3 THIRD CRITERION: EVIDENCE FOR INHERITED PREDISPOSITION.

The obvious experiments for testing the question, to what degree language is inherited, cannot be performed: we may not control the verbal stimulus input of the young child. However, pathology occasionally performs some quasi-experiments, and, while anomaly frequently introduces untoward nuisance variables, it gives us, nevertheless, some glimpses into the immensely intricate relation between man's nature and his verbal behavior.

Just as we can say with assurance that no man inherits a propensity for French, we can also and with equal confidence say that all men are endowed with an innate propensity for a type of behavior that develops automatically into language and that this propensity is so deeply ingrained that language-like behavior develops even under the most unfavorable conditions of peripheral and even central nervous system impairment.

Language development, or its substitute, is relatively independent of the infant's babbling, or of his ability to hear. The congenitally deaf who will usually fail to develop an intelligible vocal communication system, who either do not babble or to whom babbling is of no avail (the facts have not been reliably reported), will nevertheless learn the intricacies of language and learn to communicate efficiently through writing. Apparently, even under these reduced circumstances of stimulation the miracle of the development of a feeling for grammar takes place.

There is another important observation to be mentioned in connection with the deaf. Recently I had occasion to visit for half a year a public school for the congenitally deaf. At this school the children were not taught sign language on the theory that they must learn to make an adjustment to a speaking world and that absence of sign language would encourage the practice of lip-reading and attempts at vocalization. It was interesting to see that all children, without exception, communicated behind the teacher's back by means of "self-made" signs. I had the privilege of witnessing the admission of a new student, eight years old, who had recently been *discovered* by a social worker who was doing relief work in a slum area. This boy had never had any training and had, as far as I know, never met with other deaf children. This newcomer began to "talk" sign language with his contemporaries almost immediately upon arrival. The existence of an innate impulse for symbolic communication can hardly be questioned.

The case history of another handicapped child gives an illustration that true organic muteness in the presence of good hearing is no hindrance for the development of a speech comprehension that is ever so much more detailed than, for instance, a dog's capacity to "understand" his master. This was a five-year-old boy who, as a consequence of fetal anoxia, had sustained moderate injury to the brain pre-natally, resulting in an inability to vocalize upon command. When completely relaxed and absorbed in play he was heard to make inarticulate sounds which at times appeared to express satisfaction, joy, or disappointment (when a tall tower of blocks would tumble to the floor). But the boy has never said a single word, nor has he ever used his voice to call someone's attention. I was once able, after considerable coaxing and promises of candy, to make him say "ah" into a microphone of a tape recorder. The tape recorder had a voltmeter with a large pointer that would make excursions with each sound picked up by the microphone. The child had been fascinated by this and had learned to make the pointer go through an excursion by clapping his hands. After his first production of the sound "ah" he was able to repeat the sound immediately afterwards, but when he came back the next day, he tried in vain to say "ah," despite the fact that he seemed to be giving himself all the prompting that he could think of, like holding the microphone in both hands and approaching it with his mouth as if to say "ah." A series of examinations revealed that this boy had a remarkable understanding of spoken English; he could execute such complex commands as "take a pencil and cross out all A's in this book," "look behind the tape-recorder and find a surprise" (this was a tape-recorder instruction delivered in the absence of the experimenter), "point at all pictures of things to eat." He was able to distinguish pronouns ("touch my nose; touch your nose"), to show one, two, three, four, or five fingers; he could distinguish between a question and a declarative statement by nodding a yes-or-no answer to the question but not to the declarative sentence. He would even nod yes or no correctly when asked about situations that were spatially and temporally removed. This is discrimination learning but on a plane that requires a much more intricate understanding and sensory organization than the simple association of an object and a sign.

These examples do not *prove* that language is an inherited phenomenon. But they do point to the degree of man's preparedness for speech, a preparedness which seems to be responsible for the universality of the speech phenomenon.

4 FOURTH CRITERION: PRESUMPTION OF SPECIFIC ORGANIC CORRELATES.

From the title of this section it should not be inferred that we wish to draw a sharp line between behavior with and without organic basis. Thought and emotion have no less an organic basis than breathing or the tonic neck reflex. Yet there is a difference between the former and the latter types which can be described in empirical terms. In drawing the distinction we must not forget that we are dealing with a difference of degree, not quality.

4.1 *Onset and fixed developmental history.* Any innate reflex activity and sensory irritability appears at a characteristic moment in an individual's pre- or post-natal maturational process and follows a typical natural history throughout life. For instance, rudiments of the tonic neck reflex have been observed in a 20-week-old embryo; during the second half of fetal life this reflex seems to be well-established, and it is strongest during the first eight post-natal weeks, with a peak of activity during the fourth week. At 12 weeks the reflex is less conspicuous and it is normally absent by the twentieth week. If the tonic neck reflex is observed at a later period, it is usually a sign of neurological disorder or pathognomonic retardation. Another example is manual dexterity: our hands become increasingly skillful throughout infancy, greatest control being achieved during young adulthood after which time there is a steady decrease which is accelerated about the fifth decade. Also the acuity of sensory perception follows characteristic age curves. Sensitivity to a number of acoustic stimuli is very low at birth, rapidly reaches a peak during the second decade and then steadily declines throughout the rest of life.

In the case of human behavior it is not always easy to rid ourselves of our pervasive and often quite irrational belief that all of our activities are the result of training. For instance, in the case of walking on two feet it is popularly believed that this is the result of the social environment. People who hold this view earnestly propose that the healthy child learns to walk between its 12th and 18th month because this is the time during which the mother is expected to teach her child this accomplishment. Speculation is often carried to the extreme where it is assumed that children brought up in social isolation would probably be seen with different modes of locomotion than is actually observed. That this need not even be regarded seriously as a possible hypothesis may be seen from the developmental events alone. Gesell and associates write:

> Although incipient stepping movements occur during the first week [after delivery!], they are more marked and appear with greater frequency at about 16 weeks. At this time also the infant pushes against pressure applied to the soles of the feet. At 28 weeks he makes dancing and bouncing reactions when held in the upright [N.B.] position. Flexion and extension of the legs are accompanied by raising the arms. At 48 weeks the infant cruises or walks, using support.

We get a flavor here of how deeply walking is based on reflexes that must, under all circumstances, be called *innate*. Also, walking is not an isolated event in the child's developmental history. It is merely one aspect of his total development of motor activity and posture. Compare the same authors' description of the development of the upright position:

> [Stiffening of] the knees occurs before full extension of the legs at the hips. At 40 weeks the infant can pull himself to his knees. He can also stand, holding onto support. At 48 weeks he can lift one foot while he supports his weight on the other, an immature anticipation of a three-year-old ability to stand on one foot with momentary balance. At this age he can also pull himself to standing by holding onto the side rails of the crib. In standing he supports his weight on the entire sole surface.

Anyone who has observed a child during the second half of his first year knows that there is continuous activity and exercise, so to speak, and that most accomplishments occur spontaneously and not as a response to specific training (for instance, climbing out of the playpen).

The most suggestive (even though not conclusive) evidence for this point comes from animal experiments. Thorpe reports on an experiment by Grohmann in which young pigeons were reared in narrow tubes which prevented them from moving their wings. He writes:

> Thus they could not carry out the incipient flights which would naturally be regarded as in the nature of practice. When Grohmann's control birds, which were allowed free practice flights every day, had progressed to a certain point, both groups were tested for flying ability, but no difference was found between them. In other words, the instinctive behavior pattern of flight had been maturing at a steady rate, quite irrespective of the birds' opportunity of exercising it. Those that had been kept in tubes had reached just the same stage of development as those that had what appeared to be the advantage of practice. There is little reasonable doubt that at a later stage further skill in the fine adjustment of flight is acquired as a result of practice . . . ; [Grohmann's] work . . . suffices to show how cautious one must be in interpreting what appears to be the learning behavior of young birds.

Coghill has shown how the primary neural mechanism of swimming and walking in Amblystoma is laid down before the animal can at all respond to its environment. Also, it is common knowledge that neonate colts or calves can stand immediately after birth and that most quadrupeds can either take a few steps or at least go through walking motions within the first few hours of life. If locomotion is innate in such a great variety of vertebrates, why should man be an exception?

The developmental history is not always perfectly synchronized with the advance of chronological age so that we often have the impression that individual maturational phenomena, such as control over equilibrium in stance or the onset of menstruation, occur more or less randomly within a given period. This is probably an erroneous notion arising from our lack of information on other concomitant developmental aspects. If we had complete and accurate longitudinal case histories (instead of dealing with data gathered in cross-sectional surveys), developmental histories would probably reveal fairly constant sequences of events.

Contrast now the appearance and history of acquired behavior. A child waves goodby when he is taught to do so. Some children may learn it before they can speak; some may learn it only in school; and in some cultures, it may never be practiced at all. Another characteristic of acquired habits or skills is that they may be lost at any time during the individual's life so that neither onset nor disappearance of the phenomenon fits into an established place of the life cycle.

When the development of speech is considered in this light, it appears to follow *maturational* development. Cultural differences seem to have no effect on the age of onset and mastery of speech. Unfortunately, completely reliable data on cross-cultural comparison of language development are still a desideratum, but a check through pertinent literature in anthropology and child development have revealed no contrary evidence. Nor have the author's personal experience with two North American Indian tribes (Zuni and Navaho) or his inquiries from natives of non-English speaking countries cast the slightest doubt on perfect chronological commensurability of language development throughout the world. This is also congruent with our present belief that a normal child will learn any language with the same degree of ease, whereas a child who has failed to learn the language of his native land by the time he is six, also could not learn a foreign *simpler* language without trouble. We have to conclude from this that natural languages differ little in terms of complexity when regarded from a developmental point of view.

Compare this situation with writing. Writing does not develop automatically at a specific age and it also seems that various cultures have developed writing systems of varying degrees of difficulty. For instance, the *petroglyphs* left behind by the North American Indians of the Southwest can be roughly interpreted even today by the naive observer. The picture of a woman, an infant, and two feet in a certain direction is most likely a message involving a mother, child, and walking. Narrowing down the meaning of this inscription is easier than one written in Runes. Knowledge of Chinese characters requires greater study than that of the Roman alphabet. I have also made some clinical observations that deserve mention in this connection.

Neuro-psychiatrists are familiar with a condition that is referred to in the American medical literature as *specific reading disability*. It consists of a marked congenital difficulty in learning to write. Intensive drill will sometimes correct this deficit but cases have been reported where writing was never acquired despite a normal IQ as measured by the usual tests. I have examined eight such cases (who were seen in a neuro-medical outpatient department) in order to find out whether these patients had learned some more primitive type of graphic representation. It appeared that none of them had the slightest difficulty in understanding such symbols as arrows pointing in certain directions, simple representations of stars, hearts, or crosses; nor was there any difficulty in interpreting simple action sequences represented by three very schematic stick-men designs.

Presumably, these subjects have difficulty with some aspects of English orthography but not with visual pattern recognition or the interpretation of graphic symbols. The condition, therefore, is not actually a general *reading difficulty* but merely a difficulty with certain, at present unidentified, associative processes involved in *our* type of writing system. It would be interesting to know whether other countries have the same incidence and types of "specific reading disability" as encountered in England and the United States.

Let us now take a closer look at the longitudinal development of language acquisition. Unfortunately, we only have data gathered within our own culture; but even this much will be instructive, and there is, indeed, very little reason to believe that the main phenomena should differ significantly in non-English-speaking communities.

All children go through identical phases in the process of acquiring speech. First, they have a few words or phrases, never longer than three syllables, that refer to objects, persons, or complex situations. At this stage they may have a repertoire of fifty short utterances that are somewhat stereotyped and are never combined one with the other. All attempts to make the child string up the words that he is known to use singly will fail until he reaches a certain stage of maturation. When this is attained, the combining of words seems to be quite automatic, that is, he will surprise the parents by suddenly putting two words together that may not have been given him for repetition, in fact, that may often sound queer enough to make it quite unlikely that anyone in the child's environment has ever spoken these words in just that sequence. "Eat cup" may mean "the dog is eating out of the cup" or "is the dog eating the cup?" and so on. Whatever was meant by this utterance (which was actually heard), it is a sequence of words that nobody had used in the particular situation in which the words were spoken. As the child grows older, longer phrases are composed of individual vocabulary items which had been in the child's repertoire for many months, sometimes years.

Other aspects of language exhibit a similar developmental constancy. There are certain sentence structures that are virtually never heard during the first three years of life (for instance, conditionals or subjunctives). The frequency of occurrence of words shows certain characteristic constancies for child language, which, interestingly enough, are somewhat different from the frequency of occurrences of adult speech. In English, the most frequently occurring words are the articles *a* and *the*; yet, the child's first words never include these. (There is an active process of selection going on that must not be confused with mechanical parroting.) There is also a fairly constant semantic development. Children seem to begin speech with very characteristic semantic generalizations. The word *car* may be extended at first to all vehicles (a child of my acquaintance once pointed to a plane and said car); *dog* to all animals; *daddy* to all people or all men. But there is already an ordering activity apparent that is characteristic of speech as a whole.

Also, the usage of words has a characteristic history. All observers of longitudinal child-language development have reported a difficulty in naming colors correctly at an early stage. The curriculum of many public kindergartens includes special training in color naming. This characteristic difficulty for the child at 2½ to 3½ years of age is the more interesting as color words are among the most frequently occurring words in English, and it is hard to see that their correct use should have smaller reinforcement value than words such as *big* and *small, hot* and *cold,* or *heavy* and *light, wet* and *dry,* all of which are words used correctly before color words. Of course, we do not take this observation to mean that something like a special structure has to mature which is particularly involved in color naming. The point here is that naming is a complex process which presents varying degrees of difficulty or, in other words, which depends upon a number of skills that develop at a slow rate. All we can say at our present state of knowledge is that on his second birthday the child does not ordinarily have the capacity to learn to name four basic colors consistently and correctly, whereas he develops this capacity within the next two or three years. (To make a distinction between *concrete* and *abstract* names is of little help since we only have *post hoc* definitions of these terms.)

Another line of evidence that would support the thesis that language-learning follows a maturational course is the phenomenon of foreign accents; it seems as if the degree of accent correlates fairly well with the age during which a second language is acquired. The following case will illustrate the point: Mr. R. W., whose major interest is the study of language, is a middle-aged graduate of one of this country's universities. He was born and lived in Germany until he was twelve years old when his family emigrated from Germany to a Portuguese-speaking country where he spent the next ten years. Within two years after his arrival in the new country he had such a perfect command over the second language that his foreign background was never suspected when he spoke to natives. At the age of 22 he came to the United States where he was at once obliged to speak English exclusively. From then on he had no further opportunity to speak Portuguese, and only occasionally (never more than a few hours at a time) has he spoke German since his arrival in this country. The result is interesting. His ability to speak English has completely displaced his facility in Portuguese and even the availability of his German vocabulary seems to have suffered in the course of the years. Yet, his pronunciation of English is marked by a gross and virtually insuperable foreign accent while his German continues to sound like that of a native and his Portuguese, as evidenced in the pronunciation of isolated words, continues to have the phonological characteristics of perfect Portuguese. (Yet, this person has heard and spoken more English during his life than either German or Portuguese.)

Here again it would be important to verify empirically the plasticity for the acquisition of languages throughout an individual's life history. Systematic research on immigrant families and their progress in learning English as a function of the age of the learner would seem to be a quite feasible and interesting study.

Before leaving the subject of fixed developmental histories in language learning, we must briefly consider those cases where language does not develop normally. Speech disturbances are among the most common complaints of the pediatric patient with neurologic disorders. It is precisely the area of speech disorders in childhood which can shed the most light on the nature of language development; yet, despite a very prolific literature, the most elementary observations have either not yet been made or the reports cannot be used reliably. This is primarily due to the imprecise terminology common in these studies, to a predilection for subjective interpretation, to the complete absence of complete and accurate case reports of longitudinal descriptions (instead of the now fashionable cross-sectional studies of many hundreds of subjects), poor categories for classification, and similar other shortcomings. The only aspect in which little or no further spadework needs to be done in this respect is the establishment of norms for speech development. We have little trouble today in deciding whether or not a patient's speech is normal for his age.

If speech disturbances were viewed as nature's own experiments on the development of speech, a wide variety of observational research projects could be formulated, details of which need not be gone into here. Suffice it to point out that research could easily be conducted that would constitute direct verification of (or means for refining) the view that language development follows a characteristic, natural history. It would be very revealing, for instance, to know exactly under what circumstances the present practice of speech "therapy" (which is strictly speaking a training procedure) is successful. This would include detailed description of the patient's condition before and after treatment, perfectly objective evaluation of his improvement, and an accurate assessment of the role that specific speech therapy played in the course of the condition. In reporting on the patient's condition, it is not enough to mention one or another type of speech defect, but a complete inventory of the subject's speech facility ought to be given in addition to a complete and accurate report of his clinical and developmental status. Speech is so complicated a matter that we must not be surprised that a full case report is meaningful only after collecting most meticulous data on hours of patient testing and observations. A few scattered clinical notes, a random collection of psychological test results and a global statement of "improvement" is meaningless in this field. Among the most important objectives of a "log" of a long course of speech therapy is the determination whether language must be taught and learned in terms of a hierarchy of levels of complexity, whether it is essential that one set of skills precedes another, or whether almost anything can be taught and learned in a wide variety of orders. If there *is* a hierarchy of complexity, what are the linguistic correlates, what are the factors that make some linguistic aspects "easy," and what make them "difficult"? Answers to these questions would be major contributions to our present state of knowledge.

In conclusion, I would like to suggest (subject to further verification) that the development of speech does not proceed randomly; there are certain regularities that characterize speech at certain stages of development, but empirical work still needs to be done on the individual differences that may also be observed. Moreover, we know that language development, viewed cross-culturally, has never been said to deviate essentially from development in Western cultures, and if we accept temporarily and subject to further work this indirect evidence, it would be more reasonable to assume that the acquisition of language is controlled by a biologically determined set of factors and not by intentional training, considering that cultures differ radically in their educational procedures.

4.2 *Dependence upon environment.*

Thorpe describes the behavior of a hand-reared Tawny Owl "which, after being fed, would act as if pouncing upon living prey although it had never had the experience of dealing with a living mouse." This is not an isolated instance. Ethologists are familiar with this and similar types of action patterns that are usually triggered by so-called "innate releasing mechanisms." Thorpe notes that for every action pattern there is an ideal training stimulus such that every time it acts upon the animal, the latter will go through the entire action pattern. However, it is said that if the animal has not encountered the ideal stimulus in his environment for some time, the threshold for the release of the action pattern is lowered so that a stimulus that ordinarily does not evoke the patterned response is now capable of so doing. In the complete and continuing absence of suitable stimuli for the release of the mechanical action pattern, the threshold is lowered to a zero point, that is, the action pattern will go off in the complete absence of any environmental stimulus. This is the significance of the behavior of Thorpe's hand-reared Tawny Owl. But the absence of environmental stimuli does not imply absence of stimulation. Just as there is no effect without a cause, so there is no biological activity without a stimulus. In the owl's case, the stimuli must be assumed to be within the organism, i.e., be reducible to chemico-physical events. Again, this is nothing that is peculiar to innate action patterns because the behavior of pigeons that learn to pick at certain spots is also the direct result of chemico-physical reactions that take place within the bird's body. But differences there are. Compare the owl's pouncing behavior with Skinner's rat that learned to "purchase" a token which it would drop into a food-dispensing machine. In the case of the rat, various bits of spontaneous rat behavior have been artificially (from the rat's point of view, *randomly*) chained so that the *sequence* of rat-behavior-bits has a *perfect* correspondence to a sequence of environmental events. Or, in other words, every individual bit of behavior making up the food-purchasing sequence was at one time preceded by a distinct environmental stimulus or at least linked to a reinforcing event. The only reason the total food-purchasing behavior appears in the sequence that it does is that environmental stimuli and reinforcements have been arbitrarily arranged in a particular order. But the owl's pouncing behavior, which may not be as complex an affair as the purchase of food but still is elaborate enough and may last for as long a time as it takes the rat to purchase food, cannot be decomposed into bits of individual behavior components that can be rearranged into any combination and sequence. The sequence is completely fixed. This is a very important point. These days of electronic computers have made it fashionable to use electronic metaphors. We may say that innate behavior, such as the owl's pouncing is *programmed* into the organism. Environmental conditions may trigger the sequence (or perhaps forcefully prevent it), but once it goes off it follows a prescribed course. It hardly needs to be pointed out that food purchasing is different.

If we were asked whether instinctive behavior, such as the predisposition for nest-building, or the pouncing of the owl, is primarily based on organic factors, we could hardly fail to answer in the affirmative. Since the environment alone is either insufficient for producing the behavior (dogs are not stimulated to build nests), or in some cases quite unnecessary, the action pattern must have an internal cause. This statement must be true even if we shall never discover the neuro-anatomical basis of the behavior.

Let us now consider how language development fits into the scheme. The purpose of the following discussion will be to show that the rat who learns to press a lever or to purchase food gives us no more insight into the process of language acquisition than, for instance, thorough observations on the nest building habit of the rat or the acquisition of flight in the bird.

The constancy in language developmental histories is merely an indirect cue for the deep-seated nature of language predispositions in the child. Much stronger arguments can be marshaled.

First of all, in the case of the food-purchasing rat, the sequence of behavior is preplanned by the trainer and in that sense it has a rational aim. But language "training" and acquisition cannot possibly be the result of rational pre-planning because no adult "knows" how he generates new grammatical sentences. This fact cannot be appreciated except by sophisticated analysis of the principles of language. In the current explanations of language-learning we hear a good deal of how the supposedly random babbling of the infant is gradually shaped into words by the trainer's waiting for the accidental appearance of certain sounds which can then be reinforced, and thereby elicited with greater frequency, and how from this procedure the infant learns to imitate in general. This conception of speech acquisition is unsatisfactory from many viewpoints; for the time being, we merely point out that *imitation* (whatever psychological processes this term might cover) may be a part of language-learning but by no means its most important aspect. Speech activity is virtually never a mechanical play-back device. This is most readily seen on the morphological level, where children will automatically extend inflexional suffixes both to nonsense words and to words that have irregular forms such as *good-gooder, go-goed, foot-foots.* Not quite so obvious, but in a sense much more striking, is the generalization that takes place in syntactic matters. Here it becomes quite clear that there must be a second process in addition to imitation, for the language of children is not confined to stereotyped sentences. Children ask questions that have never been asked before ("What does blue look like from in back?"), make statements that have never been stated before ("I buyed a fire dog for a grillion dollars!"), and in general apply grammatical rules that only few adults could make explicit ("I didn't hit Billy; Billy hit me!").

The phenomenon of morphological generalization puts great strains on a simple referent-symbol association theory of language. The *-s* suffix of the third person singular ("he go[e]*s*") has no demonstrable referent taking this word in its literal meaning; nor the *s* of plurality, the *-ed* of the past tense, the *-er* of the comparative. The referent of the "small" words such as *the, is, will* is completely nebulous, and neither training nor learning can possibly be the result of any kind of referent-symbol contiguity, that is, the proximity of the words *the* and *man* welds them into a unit. As long as ten years ago, Lashley thoroughly demonstrated the impossibility of explaining syntax on the grounds of temporal contiguity association, and he has pointed to the generality of his observations on language with respect to other motor behavior. Lashley's argument is so compelling that little can be added to it. More recently, Chomsky has demonstrated from a purely formal approach that grammatical sentences cannot be the product of stochastic processes in which the probability of occurrence of an element (morpheme or word) is entirely determined by preceding elements, and Miller and Chomsky have discussed the psychological implications of this observation. We have neither a good theoretical model nor any practical insights into how we could teach an organism to respond to plurality, third-personness, past-ness, let alone how we could train him to use these responses in the correct order and verbal contexts within original sentence constructions. Consequently, both the teaching and learning of language cannot simply be explained by extrapolating from rat and pigeon experiments where all earning follows an explicit program.

All that we have said about production of speech is equally valid for the understanding of speech. The baby can repeat new words with great ease and be satisfied with his own baby-talk replica of the adult prototype, because he seems to perceive adult words not like a tape recorder but like a "phoneme-analyzer." He recognizes the functional similarity between phones and between his own reproduction of the adult speech sounds, and this enables him to disregard the very marked, objective, physical differences between a baby's voice and a middle-aged man's voice. Chomsky and Miller regard the child at three as a machine that can make syntactic analysis of the input speech. Obviously, children are not given rules which they can apply. They are merely exposed to a great number of examples of how the syntax works, and from these examples they completely automatically acquire principles with which new sentences can be formed that will conform to the universally recognized rules of the game. (We must not be disturbed by the fact that a transcription of a child's speech—or adult's speech for that matter—would be quite unpolished stylistically. There might be incomplete sentences and every now and then ungrammatical constructions resulting primarily from beginning a sentence one way and finishing it another. The important point here is that words are neither randomly arranged nor confined to unchangeable, stereotyped sequences. At every stage there is a characteristic structure.)

A word on the problem of motivation is in place. Animals are not passive objects upon which the environment acts. Their peripheral sensitivities are centrally controlled to the extent that, for instance, a certain odor may at one time have an arousing effect upon an individual animal but at another time (say after consuming a satiating meal) leave it inert. Moreover, *ability to stimulate* is not an objective physical property such as weight or temperature. It can only be defined with reference to a given animal species. A tree might stimulate a monkey to do some acrobatics, a beaver to start gnawing, and a grandmother to rest in its shade (where the latter is merely a subspecies). Motivation for action resides in the physiological state of the organism and in some instances can be immediately correlated with clear-cut states of deprivation, say of food or sex. Ordinarily it is false to assume that the environment *produces* a given type of behavior; it merely triggers it. There are many ways of chasing and eating a rabbit, and even though all of its predators may be motivated by the same physiological drive, hunger, the mode of catching and consuming the rabbit will bear the characteristic stamp of the predator's species.

In view of this, it seems reasonable to assert that there are certain propensities built into animals and man to utilize the environment in a fairly species-specific way. Sometimes this is obscured (a) because of individual differences in behavior traits and (b) because behavior is also affected, within limits, by environmental variations (such as availability *either* of little sticks, *or* of leaves, *or* of rags for the building of nests; analogously, because a child may grow up *either* in a Chinese-, *or* in a German-, *or* in a Navaho-speaking environment).

The appearance of language may be thought to be due to an innately mapped-in *program* for behavior, the exact realization of the program being dependent upon the peculiarities of the (speech) environment. As long as the child is surrounded at all by a speaking environment, speech will develop in an automatic way, with a rigid developmental history, a highly specific mode for generalization behavior, and a relative dependence upon the maturational history of the child.

It may seem as if we were begging the question here: If speech develops automatically provided a speech environment is given, how did the speech environment come about originally? Actually, we are in no greater logical trouble than is encountered by explanations of any social phenomenon in biology, for instance, communal life as the evolution of herds, flocks, or schools. Compare also the colonial life of ants, the family formation of badgers, the social stratification of chickens. Nor is human language the only form of communication that has evolved in the animal kingdom. Bees and many species of birds have communication systems, and in none of these cases do we find ourselves forced to argue either that these communication systems (or the social phenomena) are the result of purposeful invention or that an individual of the species undergoes a purposeful training program to acquire the trait. If in the case of lower animals we assume without compunction that the communicating trait is the result of an *innate predisposition elicited by environmental circumstances,* we have no reason to assume *a priori* that the language trait of man is purely acquired behavior (not predetermined by innate predispositions). We are making no stronger a claim here than what is expressed by Dobzhansky's words:

> The genetic equipment of our species was

molded by natural selection; it conferred upon our ancestors the capacity to develop language and culture. This capacity was decisive in the biological success of mans as a species; . . . man . . . has become specialized to live in a man-made environment.

CONSEQUENCES FOR THEORY AND RESEARCH

The great achievement of contemporary psychology was the replacement of mentalistic explanations by mechanistic ones and the simultaneous insistence upon empirical testability of hypothesized laws. In the search for laws of behavior it seemed at once desirable to discover the most universal laws since this alone, it was thought, could give our theoretical edifice insurance against *ad hoc* explanations. Many behaviorists have explicitly renounced interest in those aspects of behavior that are specific to one species and consequently confine themselves, by program, to what is universal to the behavior of *all* organisms. This attitude has cost the science of behavior a price; it has made it difficult to recognize the very intimate connection between the behavior repertoire of a species and its biologically defined constitution, that is, its anatomy and physiology.

The treatment of language by behaviorists is an excellent example of this situation. The literature, including experimental reports, in the area of verbal behavior is very voluminous and cannot be reviewed here. In general it may be characterized as a gigantic attempt to prove that general principles of association, reinforcement, and generalization are at work also in this type of behavior. The basic process of language acquisition is roughly pictured as follows: The child associates the sounds of the human voice with need-satisfying circumstances; when he hears his own random babbling, these sounds are recognized to be similar to those uttered by the adults so that the pleasure or anticipation of pleasure associated with mother's voice is now transferred to his own vocalizations. Thus, hearing his own sounds becomes a pleasurable experience in itself, the more so as mother tends to reinforce these sounds, particularly if they by chance resemble a word such as *Dada*. This induces a quantitative increase in the infant's vocal output. Soon he will learn that approximating adult speech patterns, i.e., imitating, is generally reinforced, and this is thought to put him on his way toward adult forms of language. Admittedly, this account is a gross simplification of what has been published on the subject, but the basic mechanisms postulated are not violated. Many psychologists have noted that the concept of imitation is not satisfactory in an explanation because it is precisely the process that needs to be counted for. I am in agreement with this objection but would add to the current views on the problem of language acquisition that there is a host of other questions that have not even been recognized, let alone *answered*. A few illustrations follow.

(1) The perception of similarities is a general psychological problem closely related to the problem of generalization which, however, in the perception of speech sounds plays a particularly prominent role. Acoustically, the sounds of a two-month-old infant are totally different from those of the mother; how then can it become aware of similarities between his and his mother's voices? There is also great random variation in the acoustic nature of phonemes. The identical physical sound is in one context assignable to one phoneme and in another context to another phoneme. This is even true for the speech of one individual. Thus, phoneme identification is dependent upon analysis of larger language units thus calling for a sound-Gestalt perception which may well be based on highly specialized sensory skills. We cannot be sure, for instance, whether a dog that has learned to respond to some twenty spoken commands responds to these words phonemically or whether he responds to secondary extralinguistic cues such as its master's movements. (This is an empirical question and the evidence so far is in favor of the latter.)

(2) Even if we agree that we do not know how the process of imitation works, everyone has to admit that in some way the child learns to behave like those around him, that is, to

imitate. Bracketing the problem of imitation *per se,* there is a still more primitive problem: why does the child begin to "imitate" in as highly characteristic a way as he does? His first goal does not appear to be a replication of the motor skill—he does not at first simply parrot—but his first accomplishment is to *name* objects; in fact, the motor skill lags significantly behind the naming. There is nothing necessary or obvious about this. Talking birds do the exact opposite—if they learn to name at all, for which there is, again, no good evidence. Reinforcement theory does not explain this; to the contrary, from the common psychological accounts of the beginning of verbal behavior the perfection of the motor skill intuitively ought to have preference over the more abstract naming skill in the infant's learning agenda. The naming of objects, that is, to learn that there is a general class of objects called *cup* is notoriously difficult for animals.

(3) Most terrestrial vertebrates make noises, and in mammals these are produced through the larynx and oro-pharyngeal cavity. Without exception these acoustic signals serve some biological function which, in their homologous form in man, would relate to emotions. Examples are courting, territoriality, warnings and danger signals, anger, care for the young. It is extremely difficult and for many species reportedly impossible to train animals to use these vocal signals for instrumental conditioning. It is not possible, to my knowledge, to teach a dog to howl in order to obtain a morsel of food; a tomcat to make courting noises to avoid shock; a rat to squeal in order to have doors opened in a maze. There are many indications that human vocalization is phylogenetically also related to the expression of emotions; yet, in the course of normal development a child begins to make use of his vocal apparatus independently from his emotions. Why is this so?

(4) The general problem of attention has haunted practically every research in psychology, and so we are not surprised to encounter it also in connection with language acquisition. The apes that were raised in human homes failed to develop speech, partly, it was thought, because they could not be induced to pay attention to the relevant cues in their environment. But why do all children without any special training automatically attend to these cues?

(5) It is well known that there is a nearly perfect homology of muscles and bones in the head and neck of mammals and the geometry of the oral cavity of the great apes is sufficiently similar to that of man to make it potentially and physically possible to produce speech sounds. Except for a report on a single chimpanzee, who could whisper a few "words" in heavy and, to the outside, incomprehensible chimpanzee accent, no chimpanzee or other primate has been able to learn to coordinate respiration, laryngeal, and oral mechanism with the speed, precision, and endurance that every child displays. What is the extraordinary skill due to? Does it merely depend on practice, or are there physiological predispositions?

Many more questions of this kind could be asked. They are all essential to our understanding of language and speech yet we have no answers to any one of them. Present-day psychology tends to brush these problems aside by simply admitting that it is in the biological nature of man to behave in this way and not in that, and that biological aspects of behavior may be disregarded in the psychological treatment of it. But such a position endangers the discovery value which a psychological description of behavior may have. It threatens many a conclusion to boil down to the triviality that children learn to speak because they are children and that all children learn to speak provided they are healthy and live in a normal environment.

If, on the other hand, the study of speech and language is from the outset seen as a study in biology (including the study of the interaction between heredity and environment), we can hope to combine research on questions such as those posed above with those that are customarily asked in psychology and thus to obtain new insights into the nature of man. It is true that this approach will not allow us to generalize our findings to all species or to speak about "the organism" in general. But I see no reason why the difference between species and their behavior should be less interesting or pertinent to a general science than the similarities.

SUMMARY AND CONCLUSION

The behavior repertoire of many animals depends upon certain biological predispositions. On the one hand, the animal may be constitutionally pre-destined or have an Anlage for the exercise of given behavior patterns, or, on the other hand, it is innately turned to react to specific environmental stimuli in a species-characteristic fashion. In a sense, all of man's activities are a consequence of his inherited endowments including his capacity for culture and social structure. But some of his behavior patterns, for instance, bipedal gait, are based upon very specific anatomical and physiological predispositions, whereas other patterns, such as writing, are based on more general capacities of motor coordination, perception, and cognitive processes. In the present article, criteria were developed to distinguish behavior patterns based on specific predispositions from those based on general ones. When these criteria are applied to language, one discovers that it falls between these two poles, though considerably closer to the side of special predispositions than to its opposite.

Since it is proper to speak of language as species-specific behavior, we are implicitly postulating a biological matrix for the development of speech and language. This is tantamount to an assumption that the general morphology characteristic of the order *primates* and/or universal physiological processes such as *respiration* and *motor-coordination* have undergone specialized adaptations, making the exercise of this behavior possible. At present, there is scanty evidence for this because proper questions that might lead to decisive answers—either for or against the hypothesis—have not been asked. Let us hope that the present formulations help us to ask such novel questions.

31. FUNCTIONS OF SPEECH: AN EVOLUTIONARY APPROACH

DELL H. HYMES

Reprinted from Frederick C. Gruber (Ed.), Anthropology and Education *(Philadelphia: University of Pennsylvania Press, 1961). Dell H. Hymes is Professor of Anthropology and Curator of Linguistic Anthropology of the University Museum at the University of Pennsylvania. His chief theoretical concern is the interpretation of language from an ethnographic and evolutionary standpoint. Much of his work has been done with American Indian languages and folklore, especially of the Pacific Northwest. He is the editor of* Language in Culture and Society, The Use of Computers in Anthropology *and* Re-Inventing Anthropology. *His most recent book is* Traditions and Paradigms: Studies in the History of Linguistics.

■ It is frequently asserted that every language possesses all the vocabulary and concepts that are needed by the people who speak it, and linguists often assume that all languages are equally well adapted to the needs of their speakers. These assumptions are challenged in the following paper.

Human culture as we know it would be impossible without language, which is one of the most important instruments of adaptation. Because every adaptation must successfully meet the specific challenges of a specific habitat, a language is adaptive to the extent that it enables people to cope with particular pressures. Hence we must conclude that if the demands of the habitat change, a group's language, among other aspects of its culture, must change accordingly. For example, if a society acquires horses it must develop a vocabulary to denote different kinds of horses and equine equipment and conditions.

A language, however, is more than its vocabulary, grammar, sound structures, and so forth; it is also a repository of the modes of thought, the notions of causation, and the conceptual and cognitive categories of the culture. These too must change if the pressures upon the culture undergo change. If linguistic change enables people to meet new habitational demands successfully, we can say that the language is adaptive. But if the habitational challenges are altered and the language does not change to enable people to meet them—and especially if the language impedes people in trying to meet these pressures—we can say that the language is maladaptive. The notion that all languages are inherently equally adaptive is rooted, in part, in the assumption that habitats are unchanging (Hymes notes other sources of this assumption in his paper). But it is not only the material and the physical habitat that changes; social and cultural milieus also change frequently.

Almost all societies are and always have been in contact with others. Until relatively recently in history, societies in contact have generally been at approximately the same level of technological adaptation. If they did not speak languages that were mutually intelligible in vocabulary and grammar, they at least "spoke each other's language" on the conceptual level. But with the advent of imperialism among industrial nations and their conquest of much less advanced societies, and with the creation of dominant groups in the latter that were based on the conquerors' cultures, the subject people were faced with entirely new and different cultural realities from those they had known theretofore.

The ethics and morality of imperialism aside, the challenges and pressures of conquerors and of industrial development constitute a new reality for peasants. If they are to survive they must adapt to the new sociocultural reality, and this demands linguistic adaptation in addition to adaptation in other spheres. In many instances their languages have proved to be maladaptive. This is the substance of Hymes' paper, which discusses an example, the maladaptive language of a group in Mexico.

The position set forth by Hymes in the following paper is not commonly held among anthropologists. There are, however, several sources that, although not directly addressed to this thesis, lend considerable support to it. An excellent starting point is Dorothy Lee's "Being and Value in a Primitive Culture" (*Journal of Philosophy,* 46 [1949]: 401-15). In a similar vein are Chapters 8 and 9 ("The Tongue of the People" and "The Navaho View of Life") in *The Navaho,* by Clyde Kluckhohn and Dorothea Leighton (Cambridge, Mass.: Harvard University Press, 1946). The point of view set forth by Hymes in the following paper calls to mind the Sapir-Whorf hypothesis about the relationship between language and culture. In this connection, a worthwhile collection of papers can be found in *Language in Culture* (edited by Harry Hoijer [*American Anthropologist* (N.S.), 56(6), Part 2, December, 1954]). Of direct relevance to many of Hymes' concepts is Joseph H. Greenberg's *Essays in Linguistics* (Chicago: University of Chicago Press, 1957), especially Chapter 5: "Language and Evolutionary Theory". ■

LET ME BEGIN BY stating the thesis that lies behind my title. I want to controvert two widely accepted views; first, that all languages are functionally equivalent, and second, that all languages are evolutionarily on a par. I want to maintain that the role of speech is not the same in every society, and that the differences can best be understood from an evolutionary point of view; that we must understand speech habits as functionally varying in their adaptation to particular social and natural environments, and recognize that there are ways in which some languages are evolutionarily more advanced than others. Letting "speech habits" stand for the gamut of linguistic phenomena and "functions" for the varied roles these play, I am arguing for an evolutionary, comparative approach to functions of speech. Such an approach does not now exist in anthropology. I shall indicate reasons for the present neglect, and try to show that by overcoming it, anthropology will contribute to both its own theory and the foundations of education.

There can be, particularly, a contribution to some problems of education in the rapidly changing modern world, especially in underdeveloped and linguistically complex areas. We are all aware that, given the great surge throughout the world toward social and economic progress, the only feasible goal is for all to share as equitably and peacefully as possible in the fruits of industrialized civilization. To attain this goal in many areas requires the introduction of new educational forms and content, and we must help in this introduction while maintaining and enhancing the quality of education as part of a democratic way of life in our own country. And while success depends much upon problems which are political and economic, it also involves problems which have to do with the functions of speech.

As an instance of a problem encountered widely, we can cite the Mezquital Otomi of Mexico. The need here is through education to enable a group to overcome its poverty and isolation from the national society. Dr. Manuel Gamio, father of applied anthropology in Mexico, once commented that Otomi cultural character had changed very little during the twenty-five years of his active work among them. Part of the problem is an arid environment, but a missionary linguist writes that "the comparatively high degree of monolingualism in the tribe, forming an immediate barrier to fusion with the official system of education executed in Spanish, a language foreign to most of the members of the tribe, is an obstacle to progress tantamount to the imposing economic one." And the obstacle of monolingualism in such a case can best be overcome with the help of an adequate analysis of the functions of speech in the situation. How this is so will be easier to see after we have first examined the functions of speech from a general point of view. Let us do this by considering what it means for a child (or an adult) to master the speech habits of a group, to function as a linguistically normal member of it.

FUNCTIONS OF SPEECH

When we think of learning a language, we may think first of rules of pronunciation, grammar and vocabulary; but there is clearly more than this to the acquisition of a form of speech. A person could master these rules but still be unable to use them. He could produce any possible utterance but not know which possible utterance to produce in a given situation, or whether to produce any. If he spoke, he might say something phonologically, grammatically, and semantically correct, but wrong, because inappropriate. He might find hearers (or correspondents) "taking him the wrong way" or responding in ways that indicated that, although understood, he was not a normal member of the speech community.

In a society, speech as an activity is not a simple function of the structure and meanings of the language or languages involved. Nor is the speech activity random. Like the languages, it is patterned, governed by rules; and this patterning also must be learned by linguistically normal participants in the society. Moreover, the patterning of speech activity is not the same from society to society, or from group to group within societies such as our own.

The nature of such patterning, as well as its cross-cultural variation, can be brought out by considering four aspects of it: (1) in terms of the materials of speech, there is the patterning of utterances in discourse; (2) in terms of the individual participants, there is the patterning of expression and interpretation of personality; (3) in terms of the social system, there is the patterning of speech situations; and (4) in terms of cultural values and outlook, there is the patterning of attitudes and conceptions about speech. Let us briefly take up each in turn.

1. Beyond the syntactic structure of sentences (with which grammars usually deal), utterances have an organization into what we may call *routines*. By *linguistic routine*, I refer to sequential organization, what follows what, either on the part of a single individual or in interchange between more than one. Routines range from reciting the alphabet, counting, and greeting, to the sonnet form, the marriage ceremony, and the direction of a buffalo hunt. Obviously, societies and groups differ both in the content of equivalent routines, such as those for greeting, and in the kinds and numbers of their routines. The more complex the society, the greater the number and variety of routines, and the greater the variation in control of routines by individuals.

2. Persons who participate in speech activity learn the patterning of its use as medium for personality and role-playing. Cues expressed and perceived in speech may enable individuals to place, and to adjust quite subtly to, each other. This complex process ranges from tempo and general handling of voice dynamics to choice of expressions and overall style. The individual learns both signaling patterns outside language proper and the integration of these in speaking (and, correspondingly, in writing). Such signals differ from group to group, of course, and can be misinterpreted, either in themselves or as part of other behavior. Thus, as James Sledd observes:

> British speakers have far more final rising pitches in statements than do Americans, whose favorite intonation pattern /231#/ sounds brusque to British ears. Britishers are also likely to use a greater range of pitches than Americans, more frequent and extreme pitch changes, and more numerous expressive devices. . . . In some parts of the United States, an adult male who talked so would be suspect.

In a school for Mesquaki Fox Indian children near Tama, Iowa, many white teachers who probably regard their classroom behavior as normal, have had loudness of voice, together with verbal directness, interpreted by Indian pupils as "mean" -ness and a tendency to "get mad."

The relative importance of speech to personality, vis-a-vis other modes of activity and communication, varies from person to person and group to group; and so do the range of expression and interpretation of personality possible in speech, the extent to which speech is a form of gratification (oral or other), and the importance of speech for role performance and attaining rewards, especially those depending on personal interaction. Among the Ngoni of Africa, rules of speaking etiquette are strict, and skill in speech is greatly encouraged, for such skill is considered part of what it means to be a true Ngoni. Contrast this to conceptions of the "strong, silent type," the "man of few words," etc., in sectors of American society.

3. Social systems are often regarded as patterned relationships among roles and among groups such as families, lineages, and

corporations; and there are speech patterns diagnostic or characteristic of particular roles and groups, just as there are speech patterns diagnostic or characteristic of particular personalities. These of course differ cross-culturally in content and relative importance. If we also look at a social system in terms of the behavioral activity involved, we can see it as a network of interaction in situations or behavior settings, and can discover related patterns of speech. For example, societies differ in the settings in which speech is prescribed, proscribed, or simply optional. We so commonly think a social situation requires something to be said that writers have described this as a universal need. Certainly, were someone to come to your or my house, sit silently for half an hour, and leave still silent, we should not consider it normal. In some American Indian groups this would constitute an acceptable social visit. For them, physical presence is enough; the situation is defined as one in which speech is not necessary when one has nothing to say.

If we look at a social network in terms of speech settings, we can discriminate a set of factors whose interrelations may serve to describe its patterns of speech activity, and so provide a basis for comparing the functions of speech in different social systems. These factors can be termed: a *sender* (or *source*); a *receiver* (or *destination*); a *message* (viewed in terms of its form or shape); a *channel*; a *code*; a *topic*; a *context* (setting, situation, scene). All are compresent in speech activity. Societies differ in what can function as an instance of each factor, and in the relations of appropriateness which obtain among the factors in given cases. There is a system of speech activity in a society, then, because not all possible combinations of particular senders, receivers, message forms, channels, codes, topics, and contexts can occur.

A teacher in a school for Navaho children may discover that one boy cannot speak to a girl classmate because she stands in a certain kinship relation to him. A society may traditionally permit only certain individuals to use the channel of writing; and among the otherwise nonliterate Hanunoo of the Philippines, writing is used only among young people in courtship and love affairs. Education for birth control may encounter the barrier that such a topic cannot be discussed among or in the presence of both sexes, including husbands and wives. A teacher may misinterpret an ornate and allusive style in an examination as an attempt to conceal ignorance of the answers, not realizing that Puerto Rican students may deem it the only style appropriate to such an occasion. One teacher in a project of fundamental education may find it hard to teach children to define the classroom situation as one in which they do not talk to each other, and in which they speak to her only when asked or acknowledged. In another society a teacher may find it equally difficult to bring children to define the classroom situation as one in which they can speak at all, they having learned to regard instruction as a situation in which they function as receivers only.

These scattered examples must suffice here, except to note, under the *code* factor, the importance of levels, styles, and functional varieties of a language, and in some societies, of entirely different languages, in relation to particular settings, channels, senders, and receivers. Here rules of appropriateness may make a great difference, especially if they differ for teacher and students because of differences in class or cultural background.

4. Cultural attitudes and conceptions regarding speech differ notably from society to society and also from class to class. The pattern of such attitudes and conceptions permeates the role of speech in personality and social structure. Reciprocally, differences with regard to the interest in and valuation of speech (or of a particular linguistic code) may have correlates in differences with regard to how speech enters into the socialization and early education of children. It is clear from ethnographic sources that societies differ as to their conceptions of children as users of language, and of the process of language learning; as to the stage in children's speech development at which major socialization pressure is exerted; as to the extent to which interest in speech and speech play is

encouraged or discouraged; as to the extent to which speech is a mode of reward and punishment for children; and as to the portions of culture which are linguistically communicated. Some societies are permissive about eating and toilet training until the child can understand verbal explanations, whereas others conceive a newborn child as capable of understanding speech, and lecture it from the cradle. Adult skills are transmitted verbally for the most part in many societies, but among societies such as the Kaska of northwestern Canada, children learn them almost wholly through observation and imitation. Differences as to the functions of speech in adult life probably are related to such differences as these in the functions of speech in childhood, but there has not been the systematic comparative study which would permit us to be sure.

In any event, although we tend to think first of cases in which language has been integral to a group's sense of identity and unity, and in which it is thus a focus of pride, it is clear that here also the function of speech may vary. Four distinct functions and three correlated attitudes have been differentiated by students of the development of standard languages, and these can be applied generally to all languages. The first two functions are *separatist* and *unifying*, jointly associated with an attitude of *language loyalty*. There is a *prestige* function associated with *language pride*; and a *frame of reference* function associated with *awareness of a norm*.

Two South American peoples contrast sharply with regard to the separatist and unifying functions. The Fulnio of Brazil have abandoned their homes several times in the past three centuries to avoid assimilation by Brazilian national society. The preservation of their language and an annual religious ceremony have been the basis as well as the symbol of their distinct identity. The Guaqueries, a Venezuelan group, seem to have abandoned their langauge and native religion perhaps as early as the eighteenth century, but the society thrives as a distinct identity within the Venezuelan nation, through maintenance of a special socioeconomic base.

Two North American groups contrast sharply with regard to the prestige function. Three centuries ago a Tewa-speaking group fled from the Spanish to find refuge on one of the Hopi mesas in Arizona. There, as the Hopi-Tewa, they have maintained a position as a specially regarded and privileged minority—a situation in which attitudes toward language have been a major factor. Loyalty to their Tewa dialect has had a separatist and unifying role, as has their persuasion of the Hopi, through the reiteration of a myth and constant ridicule, that no Hopi can learn their language. They in turn have a reputation as polyglots for their own knowledge of Hopi, and often of Navaho and English; they have maintained pride in their language, and have won linguistic prestige for it and themselves. In contrast, the Eastern Cherokees of North Carolina, a remnant group, retain their language in large part, but without pride. It is a source neither of prestige nor of unity, persisting only in a separatist function with a negative, anti-White language loyalty.

In Mexico the Zapotecs of the Isthmus of Tehuantepec resemble the Hopi-Tewa in the fact that, as a group bilingual in Spanish and Zapotec, they retain pride in their first language and national identity; and these are accorded prestige by those around them. In contrast, language has been salient in the cultural persistence of the Otomi against Spanish pressure, but the Otomi have accepted an outside valuation of their language as inferior to Spanish, and feel no prestige in its use. Language loyalty to Otomi makes imposition of education in Spanish alone impossible, but acceptance of prestige for Spanish alone makes education in Otomi alone unacceptable; it suggests an attempt to keep them in an inferior status. Bilingual education, using diglot texts, wins acceptance, by reassurance that knowledge of Spanish is the end in view. (Such bilingual education may, of course, come to enhance the prestige of Otomi.)

How groups differ in the degree to which a language serves as a frame of reference in the sense of awareness of a norm is noticeable in attitudes among themselves toward incorrectness or slovenliness of speech. Among some American Indian groups such as the Washo and Paiute, a child might receive as a nickname a word it frequently mispronounced. Attitudes towards correctness among foreigners may depend upon the identity of the speaker. Many Frenchmen find a Spanish or Italian accent charming but a Germanic accent unbearable. The choice of teaching personnel and procedures obviously would pose a different problem among linguistic sticklers, such as the Ngoni of Africa, from that posed among linguistically more laissez-faire peoples.

The functions of writing systems in these respects are often significant for attempts to introduce literacy and new education. Often, as among the Otomi, a successful orthography for the native language, if it is to be easily accepted, must depart from scientific accuracy to resemble a prestigeful other written language. Native conceptions (folk-linguistics) enter too, as when the writing of tones with accent marks was found impractical among the Soytaltepec Mazatec of Mexico. They conceive of tones, not as high and low, but as thick and thin, and it makes no sense to try and teach them the rule that the mark for the "high" tone slants up and that for the "low" tone slants down. Printing expense and legibility make the use of their own metaphor of thick and thin impractical also, but superscript numbers for tones have proved successful.

Even in this cursory survey, we see that the role of language and linguistic activity can vary greatly from group to group, and we can begin to see more clearly how this variation matters for practical problems of raising the educational level of the world. In programs of fundamental education such as UNESCO has sponsored, for example, literacy must often be introduced. To this end, one among several dialects or languages often must be chosen as the medium of education, and often an orthography must be selected or constructed. Whether literacy is already present or not, new speech habits and verbal training must be introduced, necessarily by particular sources to particular receivers, using a particular code with messages of particular topics and in particular settings—and all this from and to people for whom there already exist definite patternings of linguistic routines, of personality expression via speech, of uses of speech in social situations, of attitudes and conceptions toward speech. It seems reasonable that success in such an educational venture will be enhanced by an understanding of this existing structure, because the innovators' efforts will be perceived and judged in terms of it, and innovations which mesh with it will have greater success than those which cross its grain.

There is direct analogy with the fact that one perceives the sounds of another language in terms of the structure of sounds in one's own. This phenomenon—perception of another system in terms of one's own—has been studied by linguists as *interference* between two systems, most notably by Uriel Weinreich of Columbia University in his book *Languages in Contact*. When both systems in question are known, it is possible to predict quite accurately where and what kind of interference will occur, and what kinds of substitutions and interpretations will be made, as speakers of one language learn the other. In consequence, it is possible to design materials for the teaching of one language specifically for speakers of another, and to anticipate the particular advantages and disadvantages their own system will confer in the task.

This suggests the nature of the contribution that anthropology can make to such problems in education. It would be a matter of applied anthropology, defined as "the formal utilization of social science knowledge . . . to understand regularities in cultural processes and to achieve directed culture change." An adequate comparative study of the functions of speech, would imply a descriptive science of the functions of speech just as there is a descriptive science which deals with language structures. Such a science would provide a basis for detailed analysis of the differing systems of speech activity which meet in an educational situation, and such analysis would make it possible to predict or at least to anticipate more effectively the interference which a program of literacy, bilingual education, and so forth, would encounter. Even the broad conceptual analysis outlined above can help by calling attention to aspects of the problem, such as the Mesquaki children's perception of teachers as "mean," or the need for a successful written form of Otomi to resemble that of Spanish. But there must be detailed empirical studies, from which can emerge a more refined theory and the descriptive science I have advocated.

It is remarkable that no such comparative study of speech functions exists. Anthropology is noted for just this sort of cross-cultural perspective when it is a matter of religion, of kinship, of sexual behavior, of adolescent crisis. Why not when it is a matter of the functioning of speech in society?

The answer lies in the theoretical perspective on the functioning of speech now usual in anthropology, a perspective which is non-evolutionary and minimizes cross-cultural variation. So I must now sketch an evolutionary perspective, as framework for a short critique of current anthropological views and a basis for a broader concluding interpretation of the educational aspects of functions of speech.

AN EVOLUTIONARY PERSPECTIVE

There is no single or monolithic body of evolutionary theory in anthropology and biology, but there is a body of recent literature from which we can single out some essential features for application to speech.

First, there is an essential distinction between two kinds of evolutionary study, namely of *specific* evolution and of *general* evolution. Specific evolution is concerned with individual lines of evolution, the development and adaptation of particular groups in particular environments. General evolution is concerned with the course of evolution as a whole. It abstracts from and often cuts across individual lines of evolution to consider types, as these have emerged and as these represent broad levels of evolutionary advance. General evolution might consider the relation between the mammal and marsupial types, and the advance which led to the dominance of the former. Specific evolution would examine such questions as the adaptation of the whale to marine existence, the bat to flight, and the radiation of the kangaroo line of marsupials into various ecological niches in Australia. In terms of specific evolution, advance means improved adaptation to the particular environment relevant to a group and in relation to those with whom it is in direct competition in that environment; outside this context it is relativistic. In terms of general evolution, advance means progress which emerges in the course of specific evolution and has consequences for it, but considered in a broader spatial, temporal, and environmental context. Thus a familiar case is the successful adaptation to a specific environment which proves fatal in the long run. Criteria for general evolutionary advance include "change in the direction of increase in range and variety of adjustments to its environment" and succession of dominant types.

When we study evolution specifically, we find that its focus is upon a population, and a set of traits associated therewith; that it analyzes the variation in traits within a population and the differential retention of traits within the population over a period of time; that it interprets this process through the adaptation of the population within its environment (and of the traits to one another), in connection with the pressures which selectively affect the retention of traits and hence this adaptation. It sees a population and its characteristics as participating in a continuous process of change, and it interprets the change and the characteristics of the population in broadly contextual and functional terms.

It is easy enough to see linguistic change in these terms: a speech community has a certain set of speech habits, whose incidence varies within the population and which are differentially retained, as a result of selective pressures (such as the social and natural environment, prestige of speakers, customs such as tabu and word-play, and internal requirements for maintenance of the linguistic code), the whole being adaptive both to the environment of the speech habits and to the maintenance of the code.

If we carry through such a view, however, we can find ourselves in a stance quite different from that typical of the attitude toward language today. Our broad category of *speech habits* in relation to a *population* as the unit of primary focus, does not in the first instance isolate the formal structure of the linguistic code, the usual object of linguistic attention, from the patterning of the uses of speech. Both would equally be analytical abstractions from the same phenomenal reality, the speech activity of the population. We begin by considering the totality of the speech habits of the population, and so subsume at first the presence not only of different types, varieties, and dialects, but even of different languages as parts of the whole.

Since one readily takes a functional view of traits involved with selective pressures, it becomes quite natural to analyze individual speech habits or sets of habits, including those of separate languages, in terms of competition within the environment of a population, and to see this competition as turning partly on the merits of the habits themselves. Such a view requires one to consider what the relevant environment for the adaptation of a population's speech habits really is at present, and is necessarily concerned with the real locus of change of such habits in the speech activity of definite individuals living in a definite society. Such an evolutionary view will direct attention toward the variation within and between the speech habits of populations, and will give due importance to the differences in the functions of speech associated with a set of speech habits of a population, to the consequences of this variation, and to the evolutionary survival, development, or disappearance of traits or sets of traits.

In short, the evolving units are sets of speech habits as characteristics of populations—units which sometimes will, and sometimes will not, coincide with the historical units known to us as languages. We begin by examining natural totalities of speech habits firmly embedded in environmental context, cultural and physical, as adaptive to that context, as an integral part of the whole sociocultural adaptation of the population.

If we examine our recurrent example, the Mezquital Otomi, from this perspective, we find that an Otomi child grows into, and acquires its education as a member of, a population whose speech habits comprise a majority of Otomi provenience, a minority of Spanish origin. The child becomes well aware of the competition, selection, and specialization among these two sets of habits. The Otomi habits are dominant in the sphere of subsistence and in most of social life, occupying a privileged position in the early socialization of young and the loyalties of all, and generally in the tribal environment within which its adaptation has almost wholly taken place. But Spanish habits are dominant in certain situations such as the market and the classroom, and as the relevant environment of Otomi life and speech habits shifts and enlarges, the position of Spanish speech habits is enhanced, and more situations are encountered in which Spanish has selective advantage. The prestige which all accord to Spanish speech habits, and the experienced relatively greater utility of these situations in the expanding sphere of the environment, underlie the general expectation that the relative function of Spanish will continually expand—an expectation which in turn helps bring about the expected state of affairs. As for development toward filling the enlarged environment on the part of Otomi speech habits, this occurs only indirectly as a by-product of bilingual education and the inculcation of Spanish.

Although the published analysis of the Otomi case is one of the few such, it is brief, and its focus is not upon the kinds of question and of data which an evolutionary perspective requires. The discussion is cast in terms of the proposition that all languages are functionally equivalent and equal, neutral in their own cultural settings; and Otomi is seen as having been forced into a status of ascribed inferiority, because of inferiority ascribed to its associated culture. This is considered to be entirely a social-psychological matter of prestige, having no basis in anything connected with Otomi as a language. Now this assumption is remarkable, since the discussion does mention concrete *linguistic* differences in the functional value of Otomi and Spanish in certain situations; and despite the declared equality of Otomi, the possibility that Otomi can be developed to meet the modern educational needs of its speakers is not considered.

While the superiority of Spanish in the situation is partly a matter of attitude and prestige, it is at best ingenuous not to see it as also partly a matter of the actual linguistic superiority of Spanish. The failure to see this is in part due to heavy reliance on the view that there is no evolutionary superiority among languages, while in fact the superiority of Spanish in the situation is in part a consequence of its being one of a type of evolutionarily more advanced language. Let us now turn to the study of general evolution, which deals with this question.

We mentioned "increase in range and variety of adjustments to environment" and succession of dominant types as two criteria of evolutionary advance as between general types. When such criteria are applied to culture, it is generally agreed that some cultures are technologically more advanced than others. Vocabulary is the linguistic analogue of technology. Clearly the lexical content of standard languages shows increase in range and variety of adjustments to environment in comparison to dialects or regional or minority languages, where these are not supported by a standard language outside the situation. World languages such as English, French, Spanish, Russian, and Chinese show such increase, and have spread as representatives of a dominant type, quite apart from military conquests. The existence of linguistic science itself, and of the self-consciousness and awareness and control which go into the construction of logics and systems of mathematical manipulation, argue for the advanced status, as a type, of languages which participate in, and indeed make possible, such activities. The same holds for linguistic routines in philosophy, literature, religion, and science. It has been argued that mathematics, logic, *et al.*, are not language itself, but "post-language." Even if this view is taken, it remains that ordinary language is the medium in which "post-language" systems must ultimately be interpreted, and not all natural languages can perform this function. Indeed, differentiation and specialization of function is an important aspect of evolutionary change in languages, as has long been recognized by students of the development of standardized languages. Of course, increase in number and diversity of functions of a language is a response to change in other aspects of a culture. This is true also of increase in the content and complexity of the vocabulary of a language. Some scholars may point to this fact as a reason for disregarding such changes as not properly a linguistic problem, or as not part of language. The argument does not hold. Many, if not all, linguistic changes have sociocultural roots. Lexical borrowing, a standard topic in linguistics, is a case in point. As part of general evolution in language, then, increased complexity in the lexical content and functions of a language cannot be disregarded. The response is linguistic, even if the stimulus is not.

We may bypass the question of evolutionary advance in grammatical features, a question raised especially in this century by leading French linguists, and also the question of increased efficiency and economy in language evolution. Both possibilities may be regarded, not as disproved, but as unproved. The reality of general evolutionary advance in the sphere of language seems clear.

It may be pointed out that any set of speech habits is capable of expanding in content and functions sufficiently to serve a complex civilization and its associated systems of thought. Yes, of course, *potentially* it can so serve; but we must distinguish between potential and actual development, recognizing that some languages are actually of the more advanced type while others are not.

If this distinction is valid, why is it not part of the common perspective of linguistics and anthropology today or at least a subject of discussion? Particularly now that an evolutionary perspective toward culture is being renewed in anthropology, how can a part of culture, language, be omitted? The answer lies in a dominant outlook, whose focus is upon the single language and its most highly formal, structured aspect, its grammar and phonology considered in abstraction in the first instance apart from cultural and natural context. Vocabulary, the aspect most closely tied to this context, is likely to be treated as residual. The thrust has been to exclude from central concern those realms of phenomena and bits of data which do not seem to fit into formal structures. Such structures are abstracted from variation, the occurrence of which, though not denied, has been submerged under the dominant presumption of regularity and homogeneity throughout a speech community. As a background against which to set off the structure of the formal code, speech activity commonly has been considered random, perhaps a matter of individual and unpredictable choice. As for the functions of speech, these have been seen as universally equivalent; while competition between sets of speech habits, languages, or parts thereof, has frequently been taken to be a purely social matter, not a matter involving the adaptive merits of the habits involved, and often interpreted under the blanket term of an unanalyzed differential "prestige." Thus important questions about linguistic change have gone unanswered because the focus is not upon the actual locus speech change, and the relation of language to culture becomes a theoretical problem.

Now there are exceptions to each part of the picture I have sketched. It is not a question here of a monolithic ideology. It is a question of emphases, a dominant outlook and direction, and the terms in which matters tend to be couched. Moreover, one must understand that these arose in answer to definite needs, and that with them great advances in knowledge of language have been made.

I see the dominance of the view just outlined as arising out of a battle that had to be won early in this century for the autonomy and legitimacy of formal linguistic structure as an object of study in its own right, as distinct from historical and psychological problems and explanations. And this carried with it, especially in anthropology, the implication of equality of all languages for such study. This autonomy of structural linguistics is a theme in two classics, the *Cours de linguistique générale* of Ferdinand de Saussure and the *Language* of Edward Sapir. The need for this focus carried with it the deemphasis upon cultural entanglements that we have noted, and, especially in anthropology, an emphatic non-evolutionary view. There still linger misconceptions about the existence of so-called "primitive" languages, whose meager vocabularies must be eked out by gesture, which lack grammars, definite systems of sounds, and abstract terms, and which are more variable and change more rapidly because of being unwritten. All this was demonstrably untrue, and stood in the way of a general science of linguistics, whose material must be the rich diversity actually at hand in the world's languages. "Equality, diversity, relativity" became a linguistic theme.

The rejected notions about "primitive" languages (along with equally mistaken notions about the superiority of an Indo-European type, or of one of its exemplars, Latin, and the use of such as ideal models for description) were of course evolutionary in one sense. But now the notion of evolution was rejected *in toto*. Whatever differences might obtain between simpler and more advanced cultures, no correlative difference was found to hold between the structures of their languages. "When it comes to linguistic form, Plato walks with the Macedonian swineherd, Confucius with the head-hunting savage of Assam." No measures which would satisfactorily rate language structures as more or less advanced appeared that were free from cultural bias. One evolutionary typology would have put Chinese at the bottom of the scale—a patent absurdity. Also, efforts by distinguished scholars, such as Jespersen, to show trends toward progress in efficiency, suffered from limitation in data to two language families, Indo-European and Semitic, and from inadequacies of conceptual analysis. In view of these inadequacies and errors, it is not surprising that the evolutionary notions of the day dropped out. And, despite present-day attacks on that generation of cultural anthropologists for being anti-evolutionary, we must remember that evolutionary theory in biology then was rather disunited and uncertain, not at all the vital force and stimulus it is today; and that to be against certain evolutionary stereotypes was to adopt a democratic and progressive stance.

Now that the battle against the mistaken evolutionary ideas has been won, and the study of formal structure well established, it is time to take up the evolutionary question again. And we can see that the fight against notions of "primitive" languages, whose echoes still reverberate today, confused three levels of evolutionary advance, and so jumped too quickly to the conclusion that all languages are evolutionarily equal. There is the level of "primitive" languages, proto-forms below the status of full languages; then the status of full languages; and finally, the advanced status we have indicated as occupied by world languages and some others. The fight against misconceptions about "primitive" languages did not distinguish the two latter stages, so that to deny the equality of all languages was taken to imply that the less advanced were "primitive." No known languages are. All known languages have achieved the middle status. All languages have achieved the level of basic or primary efficiency, such that they can fully adapt, in time, to the needs of any population. In this sense all languages are potentially equal, as we observed above, and hence capable of adaptation to the needs of a complex industrial civilization. This is just what has happened historically in the case of English, which in its Old English period would certainly not have been adequate to modern technology and science. But not all languages are equally efficient compared with one another, either in terms of specific evolution in meeting particular needs, or in terms of general evolution in meeting the needs of modern complex civilization. Already for many local languages, and ultimately for all, the direction of change in the world is one which is making modern complex civilization part of their relevant environment, within whose context they must compete. The ideal image of a single "neutral" language in a single, homogeneous cultural context hardly holds any longer. For populations such as the Otomi the relevant environment is one in which sets of habits of differing origins compete as means of developing the new forms and content of speech activity—among which education may be included—that successful adaptation requires.

In all this, equality in primary or basic efficiency is not enough, nor is the difference purely a matter of social, nonlinguistic factors. In the particular time and place of competition, one set of speech habits is as such functionally superior. Partly the superiority is mutual and relative, specific to particular niches, e.g., that of Spanish routines for the market place, and of Otomi for the usual subsistence activities of tribal life. But partly it is a matter of general superiority, and here the perspective of general evolution provides a sober and realistic attitude. Otomi, like Anglo-Saxon and many languages around the world today, could become a medium of technology, science, and philosophy. Just so, any normal infant, wherever born, could participate in any culture, however complex. But the human infant need only be raised in the cultural environment for potentiality to be realized in one lifetime, while the realization of the potential of languages often takes much longer. Even granted the will, the cost in money, personnel, and time may be prohibitive for a poor nation or a newly struggling one. This is a poignant fact, for the decisions that must be made in view of it are often hard.

There is a widespread respect for cultural autonomy and integrity. And, as the

UNESCO report *The Use of Vernacular Languages in Education* shows, a child learns to read most efficiently if taught in its native language first, even when it is then to learn to read in a second language. Yet what if, once literate, the child finds that there is nothing to read in this first language, because nothing has been written and the country cannot afford to duplicate the needed educational materials many times over in different languages? (The absence of the needed written materials is a problem even in such places as Egypt and Puerto Rico, where certain aspects of advanced education have had to be conducted in English.) And there are many people of great talent whose efforts to develop a literature in a local language, as for example in Ghana, must, for similar reasons, come to naught. The languages could so have developed, but they have not done so in their existing adaptation, and now it is too late.

So the selective pressure among the languages of the world continues in environments rapidly changing with technological and social revolutions. From a scientific and humanistic point of view, it is a hard loss to see much of this diversity disappear or become constricted in local uses, not because of inadequacy in its own terms, but because the terms have changed, and the chance for development through creative nationalism is lost to all but a handful. The scientific value of a language is independent of its political importance, just as the scientific import of a plant or animal does not depend upon its utility as food; and accordingly, some linguists are devoting their energies to recording and analyzing the languages about to disappear, so that future theories about language can have the broadest possible base, and so that we may come as close as possible to enjoying the full light that language can shed on the range of human nature and creativity. With plants and animals, discovery of a new process or type or rare form is worth more to science than millions of additional pigs and potatoes; and so it is with languages.

The reduction of linguistic diversity is a loss for humanistic educational values too, and perhaps a matter of concern for our own future adaptation. Any one form of language of necessity selects a small portion of the total range of ways of categorizing and analyzing experience that language can embody. To a large extent, the growth of science transcends the framework of any one language, but insofar as the particulars of our first language shape our later thought and use of language, the existence of diverse languages is of value as a means of transcending the perspective of any one, valuable perhaps even to mankind as a reservoir of potential change. And for such transcendence, records of past languages are never so generally effective as living examples. Let us hope, then, that the attrition of the world's languages will leave us not entirely impoverished, but still with some store of diversity.

CONCLUSION

To sum up: I believe that an evolutionary approach to speech can be unifying and vivifying. In linguistics itself it can, by its generality and functional perspective, integrate many separate concerns—genetic classification, linguistic areas, dialectology, bilingualism, standard language studies, linguistic acculturation, and the like—that deal with language change. In anthropology it can remove the embarrassing contradiction between an evolutionary view of culture and a nonevolutionary view of culture's part, language, and point toward integration of linguistic and other anthropological studies. In education it can, for instance, provide perspective on questions of correctness in speech. But chiefly, for the problems of education in large parts of the world, it can contribute to linguistics and anthropology by focusing on speech habits in relation to populations; by emphasizing a process of change through variation, adaptation, and selection; and by providing a framework and incentive for a descriptive science of the functions of speech. From this, I hope, will come the comparative perspective which anthropology should provide on the ways in which speech activity enters into the process of education.

32. URBANISM, MIGRATION, AND LANGUAGE

JOSEPH H. GREENBERG

Reprinted from Hilda Kuper (Ed.), Urbanization and Migration in Africa *(Berkeley: University of California Press, 1965). Joseph H. Greenberg is Professor of Anthropology, Chairman of the Committee on African Studies, and Acting Chairman of the Committee on Linguistics at Stanford University. One of his major interests has been the classification of languages, such as those of Africa and, more recently, the non-Austronesian languages of the Pacific. He has worked extensively in theoretical linguistics on the universals of language. He is the author of* Languages of Africa *and* Essays in Linguistics *and the editor of* Universals of Language.

■ Our focus in the study of adaptation is on the responses of groups to changing milieus. Sometimes we conceptualize these as variations in habitat from one region to another, as differences in habitational pressures at approximately the same time for two or more populations in the same species. At other times we study a habitat that is undergoing drastic modification within the experience of a single group. This can occur, for example, if there is an abrupt change in the natural habitat, like a sudden but prolonged drought. It can also occur, as in the case of the Otomi described by Hymes, if a group's political-economic milieu has been so radically altered through conquest that new adaptations in language, among others, are required.

In this selection Greenberg describes another type of change in the sociopolitical component of the environment that necessitates linguistic adaptation. A frequent accompaniment of increasing socioeconomic and political complexity is the growth of urban centers with a continuous movement of people between rural and urban places. Such developments antedate colonialization in Africa—from which Greenberg draws his data in this selection—and other parts of the world. One of the consequences of urbanization is that people from different ethnic and linguistic backgrounds come together in their daily activities. Their ability to communicate with each other demands that they adopt a common language in place of the multitude of languages that characterized their diverse ethnic roots. (There is a passing reference to this in Genesis 11:8-9, in connection with the "confusion" of languages in the urban center at "Babel.") Also, the integration of tribal or other groups into a nation-state requires the development of a common language. This is necessary not only to enable the rulers to communicate effectively with the ruled but also, and equally important, to help create a sense of national unity.

Greenberg makes it abundantly clear that this is no easy matter; it goes much deeper than merely learning a new vocabulary and grammar. Each group, understandably, wants its language to be adopted as the standard and each resents the possibility of the dominance of a neighboring group's language. Some languages, like the "pidgin" of the colonial situation, carry painful scars of the whiplash of the master-servant relationship, Paralleling this is the symbolization of elite status through a command of the languages of the European colonialists. Often, to reduce the eruption of hostilities between diverse linguistic groups (as occasionally happens in India, Belgium, and other countries), a neutral linguistic ground must be found. Such developments are adaptations to pressures created by changing socioeconomic and political aspects of the environment. Thus what began as a purely physiological adaptation to terrestrial living, as described by Hockett and Ascher, is now governed almost exclusively by sociocultural factors.

In the concluding portion of this selection Greenberg notes several major parameters of sociolinguistics, a relatively recent development in the study of language and one in which Greenberg himself has played an important role. The reader can consult the following works: "Dialect, Language, Nation," by Einar Haugen (*American Anthropologist,* 68 [1966]: 922-35); *The National Language Question:*

Linguistic Problems of Newly Independent States, by R. B. LePage (London/New York: Oxford University Press, 1964); *India: The Most Dangerous Decades,* by Selig Harrison (Princeton: Princeton University Press, 1960); *Linguistic Diversity in South Asia* (edited by Charles A. Ferguson and John Gumperz [Publication 13, Indiana University Research Center]); *Language in Africa* (edited by John Spencer [Cambridge: Cambridge University Press, 1963]); and *Languages in Contact,* by Uriel Weinreich (The Hague: Mouton, 1963). ■

INTRODUCTION

AMONG THE MOST conspicuous phenomena noted by observers of the present-day African scene are the large-scale seasonal migrations and the related growth of urban centers. For, although most of the workers return to the rural tribal areas and not all migrate to urban centers (many work on farms), it is nevertheless true that the urban centers of Africa are growing at a significantly faster rate than rural areas. This expansion of population is a result not only of natural increase, but to a considerable extent of recruitment of immigration from the countryside. In other words, permanent urbanization of a substantial and increasing part of the population is now a continent-wide phenomenon.

Urbanism and migration are, of course, not exclusively colonial and postcolonial events in Africa. In particular, the historical record discloses the existence of large-scale agglomerations, especially in the western Sudan, the present Western Region of Nigeria, the east coast Swahili-speaking ports, Ethiopia, and perhaps even in Bantu Monomotapa in present Southern Rhodesia. Moreover, in some instances at least, the ethnic heterogeneity bears witness to the role of migration in the formation of such cities as Timbuktu, which even in the precolonial period contained Arab, Tuareg, and Songhai elements, to mention the most important.

Still the present movement toward urbanization differs in both quantity and quality from that of the precolonial period. In sheer size and number the cities are usually of a different order of magnitude. Moreover the functions they serve, directly or indirectly, result from the impact of the West. Formerly largely commercial, or even raising their own food, as did the pre-British Yoruba cities, they now perform industrial or service and administrative functions that are of relatively recent and exotic origin. The link with the past can be seen, however, for these cities, particularly in West Africa, are often older communities, not new foundations, and inevitably the preexisting conditions continue to exert their influence. It has been pointed out, as a further historic connection, that the routes and the destination of migrants display, in many instances, the persistence of pre-European patterns.

LINGUA FRANCAS

It is obvious that such large-scale movements of people as are now taking place must inevitably have repercussions on the most basic aspects of the societies involved; among these is the communication system, or language. In the present paper I attempt to give some notion of the nature and scope of these changes. I believe it will also appear that communication is an area in which we are short of basic information, no doubt because the problems are by their very nature interdisciplinary. It is only recently that linguists have begun to talk seriously about an area of research which they call sociolinguistics, and its very existence is probably unknown to many workers in other social sciences. I have attempted to discover the pertinent data from a fairly full, though not exhaustive, study of the literature on urbanism and migration, and to supplement these with my own personal observations. The inadequacy of the data will be sufficiently obvious. It is to be hoped, then, that by pointing to questions that cannot be fully answered on present evidence, this study may make a modest contribution toward outlining the problems of sociolinguistics in an African setting and toward sharpening the awareness of practitioners of social sciences, other than linguistics, to a set of problems that I believe to be relevant to their own research interests.

The most obvious, one might say elemental, problem of communication which arises for the migrant is that of a common language with his employer and others with whom he comes into contact, whether in an urban or a rural situation. The overriding sociolinguistic fact about Africa is simply its vast language diversity. The conventional number of 800 separate languages for the continent is certainly an underestimate. These languages are for the most part the primary spoken languages in relatively restricted areas. Hence migrations of more than local scope are bound to bring populations with divergent native languages into the same urban or rural areas. With the addition of European employers and administrators, as well as clerical personnel who, if not Europeans, are likely not to be indigenous to the area in which they carry on their work, the existence of linguistic heterogeneity tends to be the rule. Among the few exceptions are the Yoruba-speaking cities of the Western Region of Nigeria which, as we have seen, existed in precolonial times. The Yoruba-speaking area is sufficiently extensive for these cities to have retained in large measure their linguistic and cultural homogeneity. Yet even here the necessity of dealing with Europeans and, in certain cities, with a significant influx from the Eastern Region, mainly of Ibo-speaking people, creates a communication problem of the usual type, though of less significant proportions.

The usual solution to this problem is the so-called lingua franca. Terms in this area are not very well standardized. In the present context nothing is being asserted about the linguistic nature of the lingua franca. It may be a standard form of some existing language, or a "pidginized" form, or even, conceivably, a new creation. All that is meant here by a lingua franca is a language used for purposes of communication between people; it is not the first language of both communicating parties, but on occasion may be the first language of one party or of neither party. A given area might thus have more than one lingua franca, as so defined, and this situation occasionally exists.

Most frequently, however, there is a single lingua franca which tends to be dominant over a substantial area. This solution is rational, in a sense. Particularly if the lingua franca is the first language of the numerically largest group, it is the solution that requires the least amount of second-language learning. There is a further psychological advantage, for a single lingua franca is likely to be, in many instances, a language foreign to both speakers. Thus neither has to make the compromise of speaking the other's language. That this element is important in interpersonal and intergroup relations may be seen from the fact that in most instances Europeans, who were in the dominant position, did not learn the African language. It was the African who had to make the linguistic adjustment, a situation that sometimes produced resentment on the African side. In other instances, the use of an African language or, even more, of a pidgin is viewed as "talking down" to the African, and likewise causes resentment.

The single lingua franca tends to become the dominant solution not because anyone plans it that way, but because, once a language has a head start by being the language of a numerically important group, particularly the locally dominant one, others discover the advantage or even the necessity of learning it. Once it becomes at all widespread, it has an advantage over other possible lingua francas so that its expansion continues. There is thus a dynamic quality to the spread of a lingua franca. It tends to accelerate after the initial stage is passed. Once well established, it is likely to be the subject of certain policy decisions making it an official regional language, for example, or a language of school instruction. Such adoption gives the lingua franca a further impetus. The only thing that is likely to arrest its spread is a rival lingua franca. A lingua franca, however, may spread very slowly when it encounters a language with a large number of speakers in a compact area, particularly if there is relative isolation from developing urban-industrializing trends.

An example is Kanuri, dominant in Bornu Province in the Northern Region of Nigeria. Itself not expansive—it is rarely spoken by non-Kanuri—it has offered solid resistance to Hausa, the dominant lingua franca of the Northern Region. The demographic and economic factors are obvious. The Kanuri are a substantial population occupying a fairly large area, relatively undeveloped in the modern sense and hence with relatively restricted contacts with non-Kanuri speakers. A historical-psychological factor, however, also plays a definite role: Hausa was the language of the Fulani-dominated Muslim empire of Sokoto, which fought for supremacy with the Kanuri empire of Bornu in the pre-European period. This traditional attitude of hostility still finds expression in an unwillingness to recognize the dominant position of Hausa and to accommodate to it.

Along with urbanism, lingua francas existed in the precolonial period. In the well-attested instance of Timbuktu, for example, Songhai functioned as the lingua franca and was known as a second language by the resident Bella Tuareg and Arabs. Indeed, it was largely preexisting lingua francas that spread as the result of repression of internal conflict and of the expanding trade and industrialization of the colonial period. In the pre-European period, languages spread in response to political and commercial needs. The large Muslim empires were always ruled by an elite from a particular ethnic-linguistic group. Other groups learned the language of the dominant group, just as more recently they have learned European languages. The expansion of internal and external trade tended to favor traders of the dominant group. It thus became expedient to learn their language in order to carry on trade. The empire of Mali, for example, which flourished in the Middle Ages in the western Sudan, had a dominant core of Malinke speakers. It seems reasonable to attribute the wide spread of the very closely related Malinke-Bambara-Dyula complex to its dominant position in this empire. Malinke was widely employed by traders, and is today the dominant lingua franca over extensive areas of West Africa, particularly of Mali, which takes its name from the old empire. Again, Hausa spread over an extensive area as a lingua franca in the pre-European period largely through its linguistic dominance of the vast Fulani empires of Sokoto and Gwandu. Together with trading, the mechanism of large-scale slave raiding, followed by linguistic and cultural assimilation of the diverse pagan groups, was a dominant factor in the precolonial spread of Hausa.

While several African languages thus became established as lingua francas through political and commercial factors in the interior of West Africa, in the centuries preceding the explorations and the colonial expansion of the nineteenth century quite different lingua francas developed on the coast. Here the contact of Europeans and Africans led to the development of "pidginized" forms of European languages, mainly English, French, and Portuguese. The manner in which these languages were formed still presents important, and as yet unsolved, historical problems. It does seem safe to assume that these pidgins were widely disseminated in Africa before being brought to the New World.

A pidgin that becomes the first language of a population is called a Creole. Such Creoles became the dominant local language in a number of Caribbean islands—Haiti, Curacao, and Trinidad—and on the South American mainland. Taki-Taki is the English-based Creole of Surinam. The repatriated slaves who formed the original population of Freetown at the end of the eighteenth century likewise adopted their own common tongue, pidgin English, which in Africa itself thus became the first language of a population. Events of the colonial and postcolonial periods have fostered two major developments in the communication situation as thus outlined. The suppression of intertribal warfare and the establishment of rail and road communications have led everywhere to the expansion of indigenous lingua francas. These have usually been languages that had performed these functions in earlier times, such as Hausa, Malinke, and Swahili. Some African languages, however, usually in modified, pidginized form, have arisen in the European contact situation. Examples are Sango, the lingua franca of the Central African Republic; a pidginized form of Ngbandi; and Lingala, a similarly modified form of Bangala, spoken in the Middle Congo. Such new lingua francas do not seem to have originated in West Africa. I believe that the reason is the preexistence of urban centers whose heterogeneous populations already had well-established lingua francas in contrast with the situation in most of central and eastern Africa.

The spread of lingua francas has been so extensive under the impact of Westernization that there is now hardly an area in Africa which does not have a dominant lingua franca. All our existing maps are first-language maps. There is a great need for precise information on the areas of dominance of lingua francas. We need to know not only the geographical area embraced, but the extent to which the lingua franca is known, the degree of its command, and its distribution in relation to social stratification.

A second major result of European contact has been the introduction of European languages themselves as lingua francas. For example, the dominant lingua franca of both the Western and Eastern regions of Nigeria is English. The introduction of European languages was partly on the level of the "picking up" of more or less pidginized versions of European languages by the illiterate, who in the beginning, at least, used the traditional pidgin of the slave-trade period. At the same time languages such as English and French were being taught in mission and governmental schools, and have been in undisputed use for university-level education. These dual forms of European languages led to the denigration of pidgins, which became associated with illiteracy and, above all, with the colonial master-servant relationship. On the other hand, a full spoken command of standard English, French, or Portuguese is the sign of the African elite. Though nowhere, except in Liberia, is any European speech the first language of Africans, the ability to speak a European tongue becomes a supraregional, supratribal, even supranational mark of a new elite whose badge of membership is education, as proven by fluency in the European language employed in higher education.

Statements concerning the language situation of migrants are not very common in the literature. In general, it would seem, the migrant acquires a reasonably practical command of the dominant language of the area to which he comes. Thus, Abdoulaye Diop informs us that of ninety-six Toucouleur (linguistically Fulani) in Dakar, ninety-one had learned Wolof, the language of the African inhabitants of the city. Again, Audrey Richards found that a knowledge of Luganda was sufficient in interviewing most of the migrants in Uganda. There are, however, exceptions. Skinner and others report that the Mossi migrants in Ghana normally do not learn the local language (usually one of the Akan group). The migrants from the north have a common linguistic bond in their own regional lingua franca, Hausa, and therefore tend not to associate much with the Ghanaians. This linguistic insulation, so to speak, considerably restricts the effect of migration on traditional tribal life. According to Skinner, "Even when the loud-speaker trucks of the various Ghanaian political parties visit the farm areas where most of the migrants work, the Mossi seldom understand the language being used."

LANGUAGE AND SOCIOPOLITICAL GROUPINGS

Thus far, language has been considered mainly from the utilitarian point of view, as a means of communication. It is, however, more than that. Language is perhaps the most important single criterion of group identification, at least among groups sufficiently large to play a political role. For example, in Africa "tribe" is defined, with very few exceptions, in terms of first language. If a common language did not have this important function, we might expect that people who had learned a foreign lingua franca of wider usefulness than their tribal language would forthwith abandon their first language, or at least not bother to see that their children learned it. Of course, there are sources of tribal cohesion other than language, but Africans themselves are aware that the loss of their linguistic heritage would almost inevitably follow their loss of tribal identity. In interviews I conducted in the Plateau Province of the Northern Region of Nigeria, a highly multilingual area in which Hausa is the undisputed lingua franca, practically all informants with children said that they taught them their tribal language as their first language, and Hausa somewhat later. In the words of one informant, "If we abandoned our own language, we would become Hausa just like the rest." Pagans said that their ancestors would be greatly angered by the abandonment of their language. Several illiterate informants ventured the opinion that the language itself had a positive aesthetic aspect. There was, however, no hostility to the learning of Hausa. In fact, my informants expressed a unanimous desire for their children to learn Hausa because it was the medium of instruction in the lower grades and because ignorance of Hausa condemned a man to a restricted and economically marginal traditional agricultural existence.

We can think of linguistic continuity of a group in the urban environment as most perfectly maintained by group endogamy and the teaching of the tribal language to the children. Concomitantly, a lingua franca may be current in the group without producing any evident movement toward assimilation. Such a situation, called by linguists "stable bilingualism," has apparently existed in Timbuktu, as previously noted. The Tuareg and the Arabs have been bilingual for centuries, employing Songhai and their own language without loss of ethnic identity or serious impairment of group membership. Tribal intermarriage on a wide scale, of course, tends to undermine this continuity. The evidence seems to show that up to now tribal endogamy has prevailed markedly over tribal exogamy in the urban centers of Africa. Systematic data seem to be lacking in regard to the language or languages spoken by the offspring of intertribal marriages, and in regard to the relationship of this question to tribal identification. As most people in West Africa are patrilineal, and as the bride price is interpreted as implying possession of the children by the father's group, we may conjecture that in most instances the child learns the language of the father, along with the language of the mother and sometimes an external lingua franca. Skinner describes the reluctance of the patrilineal Mossi to marry women of the matrilineal Ashanti, presumably because both groups would claim the children under native law.

The phenomenon of "passing" is a more direct mechanism for changing group affiliation than is the use of children. The situation in Africa is quite the opposite from the passing of Negroes in the United States, where physical type, not language, is the problem. There is occasional evidence in Africa of such passing by the adoption of language, dress, and other distinguishing marks of tribal membership. Audrey Richards describes migrants in Uganda whose Luganda is good enough for them to pass by claiming ancestry in some other district and thus becoming members of a group with superior local prestige and economic possibilities. Banton describes how in Freetown, where non-Creole Africans are ranked in prestige in order of degree of adherence to Islam, members of the still largely pagan Temne tribe joined Mandinka and Aku (Yoruba) associations, and acquired the appropriate languages competently enough to pass into these groups. As Banton points out, however, an "African does not lightly renounce his tribe for another," and a countermovement led to the formation of a Temne young men's association to avert further depletion of Temne tribal strength in Freetown. Although instances of this kind have no doubt occurred more widely in Africa than has been reported in the literature, in neither of these two, nor presumably in others, was the phenomenon of sufficient scope to result in the complete absorption of a tribal group.

The attitude toward the tribal language as against the lingua franca is very different. The former is connected with a sense of group identity, of loyalty to traditional ways and to ancestors; the latter is a utilitarian instrument important—indeed, often absolutely essential—to getting ahead in the world, but not as yet setting up a real bond of solidarity.

It does not follow that the indefinite survival of tribal languages is thereby assured. Innumerable languages have become extinct in the course of the world's history. The preponderance of existing evidence from Africa, however, is that exceedingly few languages are in danger of immediate extinction, and any political or social planning that would count on the loss of tribal identity through the universal use of a lingua franca in the next generation, at least, is not realistic.

It is useful to draw a distinction here between tribalism as a political form based on a territory and marked by traditional customs and by political organization in terms of chiefly office, and the wider notion of group identity which, for want of a better term, I will call "ethnicity." Thus the Welsh are an ethnic group but not a tribe. It is entirely possible that the traditional tribal groupings of Africa are evolving toward ethnic groups of this kind, although it is clear that traditional tribal organization has, on the whole, shown remarkable resilience and adaptability. The tribe, as has been pointed out, may, through the medium of associations or tribal unions, re-create itself in urban centers and strengthen and maintain its ties with the rural hinterland.

Africans seem quite conscious of the mechanisms necessary for the maintenance of group identity. Rouch aptly compares the behavior of some tribal groups, with migrant members in cities or on farms elsewhere, with that of expatriate Europeans. If indeed there is group endogamy, if the children are sent back to the villages after a certain age to learn the traditional customs and if the language is maintained, then, in effect, the emigrant groups become "colonies" in the original Greek sense. As long as distance is not too great for fairly continuous visiting and intermarriage, the prediction is that linguistic unity can be maintained indefinitely. Once the ties of fairly continuous communication weaken, the emigrant's language will diverge and ultimately become unintelligible to the home group. This process has often taken place in the past in Africa, as elsewhere, but it requires centuries, not decades.

Finally, it should be noted that language may be the subject of conscious planning and policy. Until now, decisions on matters of language in Africa have been made either by missions or by colonial governments. The former, by choice of specific dialects as standard forms of a language, and of certain languages rather than others for use in instruction in mission schools, as well as by the orthographies they introduced, exercised a very real, if haphazard, influence. The very act of creating an orthography and using a language for literary purposes gives it a certain prestige and an attendant advantage in other situations. The influence was haphazard because decisions were habitually made without sufficient knowledge of the degrees and the kinds of dialectal variations, or of the attitude held toward the various dialects by those who spoke the language. The form of speech current near the mission station was normally chosen without any understanding of the long-term consequences. Mission influence on language may be called haphazard also because of the lack of overall interdenominational planning. Consequently we even have Protestant and Catholic forms of orthography for certain languages (e.g., Ibo).

The other important policy agent is government. The French policy of using French exclusively in education and administration has tended both to facilitate the spread of French as a lingua franca and to bring loss of prestige to African languages; the latter is a factor still to be reckoned with in areas of former French rule. On the other hand, the official recognition and use of Swahili in East Africa, particularly in Tanganyika, and of Hausa in the Northern Region of Nigeria, under British administration, have strengthened processes already at work tending to the spread of these languages.

The question of language, then, has more

than local import. Because tribalism, as a basic political factor in Africa, is tied to the question of the survival of communities, each with its own peculiar linguistic heritage, the question of language becomes a fundamental one for the newly independent African states. A degree of linguistic unity is a presupposition of European nationalism, occasionally violated but then always with some derogation of the feeling of national unity, as in Belgium, for example. In Africa, outside Somalia and the Malagasy Republic, this linguistic unity is lacking and, if the present analysis is correct, unattainable in the reasonably near future. Here I propose merely to point out that there is a problem of language in relation to nationalism, for it has ramifications that would take us far beyond the present topic.

SOCIOLINGUISTICS

This brief review of what I believe to be the salient problems regarding the role of language in the urban and migratory situations of contemporary Africa should serve to emphasize the point made earlier, that relevant data on a good many essential topics are scarce. It may prove useful, in summary, to indicate the main aspects of the "language situation," the basic topic of investigation of sociolinguistics.

These aspects may be enumerated by means of a rough division on the basis of the disciplines that seem best equipped to handle particular problems. The purely linguistic factors include the classification of languages and dialects, judgment as to whether a language is to be considered a pidgin or a Creole, extent of vocabulary development, linguistic complexity as a factor in ease of learning for non-native speakers, and information regarding the existence of standardized literary forms and their relation to existing spoken dialects. The purely demographic aspect concerns the distribution of speech forms in the area under study, and the extent of multilingualism and of literacy in one or more languages. Ideally, such information should be included in censuses. Until now, only first-language information has been available, but, in my experience, it has seldom been even reasonably accurate. Sociocultural facts concern the distribution of first and other languages in relation to occupation, social stratification, ethnic origin, and religious affiliation, as well as types and frequency of language choices for offspring from mixed marriages. The social-psychological aspect includes attitudes toward the prestige ratings of languages as might be indicated by attitude tests, semantic differential, and so forth. Of more individual psychological import are the motives for learning new languages, and the manner in which they are learned. Africans pick up new languages without formal instruction, but we know almost nothing of the processes involved. Finally, a political aspect may be recognized. We may include here the policies of governments and private agencies in language matters, and the relation of language questions to political parties, national aspirations, and so on.

This essay is intended, of course, merely as a rough outline. It should be evident, however, that the language situation, taken in a broad sense, is a substantial part of what economists call the "infrastructure" of development. As such it is, I believe, at once one of the most important and the least studied factors in the contemporary African situation.

33. LINGUISTIC RELATIVITY AND DETERMINISM

ROGER BROWN

Reprinted from Words and Things *(New York: The Free Press, 1958). Roger Brown is Professor of Social Psychology and a member of the Center for Cognitive Studies at Harvard University. His research has been concentrated on problems of the psychology of language, but his writing and teaching range more widely in social psychology. He is also the author of* Social Psychology.

■ One of the most important discoveries of anthropological linguists was that although language makes it possible for people to communicate their perceptions, desires, feelings, and knowledge to each other, every language also places limitations on what can be communicated. In Welsh, for example, the word "glas" covers everything that in English is called blue, some of the colors that are called green, and some that are called gray; "illwyd" covers the rest of what in English is called gray, all brown, and some of what is called red. This does not mean that Welsh eyes cannot distinguish the same colors as English eyes; they can, but Welsh culture, through its language, does not require—or allow—them to make the same color distinctions as English-speaking people. Thus the subjective experience of color is different among speakers of Welsh and English. The notion that language contributes to the definition or quality of experience, in addition to reflecting the basic assumption and orientations of a culture, is known as the Whorf Hypothesis, after Benjamin Lee Whorf, whose major writings were brought together in a volume entitled *Language, Thought, and Reality* (Cambridge, Mass.: Technology Press of Massachusetts Institute of Technology, 1956).

Another conclusion to be drawn from the Whorf Hypothesis, as Brown observes in this selection, is that not only are people's basic cultural assumptions and orientations locked into their languages, but a language is an important contributor to the perpetuation of these assumptions and orientations. This is one of the reasons that it is so difficult for people like the Otomi of Mexico, described by Hymes, to make the necessary linguistic and intellectual adaptations to a modern sociocultural environment. Hence a group's language is an important part of its adaptation.

Brown carries the Whorf Hypothesis a step further here: He hypothesizes that a category such as snow, which is used very often, is more "available" to a speaker than one that is used less often. When a speaker uses a category frequently, the words he uses to denote the category are highly "codable": They are short, people respond to them consistently and quickly, and they refer to particular stimuli (for example, the texture and consistency of different types of snow) as well as to the category as a whole. Thus an Eskimo would be expected to have a wide repertory of terms for referring to varieties of the category "snow." He needs this vocabulary not only to classify different snows for himself but also to communicate his perceptions to other people. These terms, however, not only enable him to communicate effectively with other Eskimos about snow; they also make it possible for him to perceive many more kinds of snow than a person who has only one word for snow.

What would our reactions be to people of different skin color and hair texture if we had absolutely no vocabulary to denote racial differences? Would this not place a strong limitation on our personal reactions to these differences, and would this not make a difference in our social policies?

In addition to many of the works cited earlier, such as those by Sapir and Whorf, two articles have gained great popularity in recent years among anthropologists who have been concerned with these phenomena: "Hanunoo Color Categories," by Harold Conklin (*Southwestern Journal of Anthropology,* 11 [1955]: 339-44), and "The Diagnosis of Disease among the Subanun of Mindanao," by Charles O. Frake (*American Anthropologist,* 63 [1961]: 113-32). Both are reprinted in *Language in Culture and Society* (edited by Dell H. Hymes), which has an extensive and helpful bibliography (New York: Harper & Row, 1964). ■

IT IS POPULARLY believed that reality is present in much the same form to all men of sound mind. There are objects like a house or a cat and qualities like red or wet and actions like eating or singing and relationships like near to or between. Languages are itemized inventories of this reality. They differ, of course, in the sounds they employ but the inventory is always the same. The esthetic Italian deals in euphonious vowels while the German is addicted to harsh consonant groupings but the things and actions and qualities named are the same in both tongues. We are confirmed in this view by our first foreign language textbooks which present us with lists of French, German, or Latin words standing opposite their exact English equivalents. We are encouraged by this view to believe that the world will soon recognize the desirability and practicability of a universal auxiliary language. Since the only barrier to international communication is a disagreement on code, this disagreement ought speedily to be resolved.

There are, of course, poetic persons who claim to find in each language some special genius that peculiarly fits it for the expression of certain ideas. But the majority of us are at a loss to understand how this can be, since there is apparently a relationship of mutual translatability among the languages we learn. To be sure, we can see that one lexicon might contain a few more items than another. If the Germans were to invent a new kind of machine, and we had not yet thought of such a machine, their dictionary would contain one more entry than ours until we borrowed the word or named the machine for ourselves. But these inequalities are in the lexical fringe. They do not disturb the great core of common inventory.

This popular view of the relation between language and thought successfully separates message and code. It does not fall into the error of believing that the names we learn as children are natural and inevitable attributes of the entities named. That kind of extreme glotto-centrism is only possible to a linguistically isolated society. The popular view has thoroughly grasped the conventionality—the cultural status—of linguistic codes. That the kinds of messages to be coded might also be culturally determined is a less familiar idea. It is downright exotic if meant to imply the notions about the physical world itself are no more than conventions.

A thoroughgoing linguistic relativity has, in recent years, been proposed by Benjamin Lee Whorf. It is his belief that each language embodies and perpetuates a particular world view. The speakers of a language are partners to an agreement to perceive and think of the world in a certain way—not the only possible way. The same reality—both physical and social—can be variously structured and different languages operate with different structures. If there is anything in such a view, the establishment of a universal auxiliary language must seem a remote goal. For, if Whorf be correct, the peoples of the world do not disagree on words alone, but also on what they have to say. However, Whorf's documentation of his view comes from American Indian tongues, chiefly Hopi. Perhaps linguistic relativity is true for people speaking totally unrelated languages, but it may have no application within the Indo-European orbit to which we and the other western nations belong.

Whorf champions a second proposition concerning language and thought which makes an even more dramatic break with popular belief than does the relativity thesis. A man's language, in Whorf's opinion, is a principal determinant of his mode of thought. The language of a people not only embodies their world-view but also perpetuates that view. The languages of the world are so many molds of varying shape into which infant minds are poured. The mold determines the cognitive cast of the adult.

The evidence for relativity and determinism comes largely from anthropological linguistics and is extraordinarily difficult to interpret. We will examine the kinds of data presented and judge what sorts of conclusions they justify.

THE LITERAL TRANSLATION

The Nootka language of Vancouver Island is polysynthetic. A single word will often express something for which a sentence is required in English. Where English arranges a sequence of free forms to say "He invites people to a feast," Nootka makes affixations to a basic verb stem to yield something that may be literally translated: "Boiled eaters go for [he] does." How very unlike our way of thinking! It is as strange as the Apache way of saying it is a "dripping spring," which Whorf finds to be "as water or springs whiteness moves downward." Evidently these people do not live in the same world we do.

In elementary language classes it is well known that uncomprehending literal translations can yield quaint and comical results. To go through a German or French sentence morpheme by morpheme will often result in a rather odd English sentence. The French say *"Comment allez-vous?"* which is literally "How go you?" where we say "How are you?" They use a verb of action where we use a verb of being. Does this mean that the French are a kinetic and we a sedentary people? French teachers do not draw that conclusion but rather instruct their pupils to translate more freely. *"Comment allez-vous?"* is said to be best rendered as "How are you?" Retaining the integrity of each morpheme is to lose the sense of the whole. In the classroom where European languages are taught it is customary to assume that psychological processes are fundamentally the same and then to liberalize translation procedures so as to guarantee equivalent meanings. With Shawnee, Nootka, Apache, and Hopi (though never with the European languages), Whorf changed this usual procedure, insisted on a literal translation, and concluded that world views differ.

The most entertaining literal translation I know is that provided by Mark Twain for his address to the Vienna Press Club of November 21, 1897. In German the speech is called *"Die Schrecken der Deutschen Sprache,"* and in English "The Horrors of the German Language." The following excerpts from the literal English translation indicate how strange the German mind must be.

> I am indeed the truest friend of the German language—not only now, but from long since—yes, before twenty years already. . . . I would only some changes effect. I would only the language method—the luxurious, elaborate construction compress, the eternal parenthesis suppress, do away with, annihilate; the introduction of more than thirteen subjects in one sentence forbid; the verb so far to the front pull that one it without a telescope discover can. With one word, my gentlemen, I would your beloved language simplify so that, my gentlemen, when you her for prayer need, One her yonder-up understands.
>
> . . . I might gladly the separable verb also a little bit reform. I might none let do what Schiller did: he has the whole history of the Thirty Years' War between the two members of a separate verb in-pushed. That has even Germany itself aroused, and one has Schiller the permission refused the History of the Hundred Years' War to compose—God be thanked! After all these reforms established be will, will the German language the noblest and the prettiest on the world be.

On the basis of the premises that Whorf applies to American Indian languages, we should be compelled to conclude that the German has a cognitive psychology very unlike our own. Of course we do not operate with these premises, but make liberal translations which represent the German mind to be very like our own. What justification can there be for operating with different assumptions in the two cases?

To begin with, American and German cultures have an obvious close resemblance and known historical ties. There are numerous bilingual persons and countless translated documents. The cultures of the modern western nations differ greatly and obviously from the cultures of the Shawnee, Hopi, and Apache. Persons truly bilingual with one of these Indian languages are uncommon, and those who exist have often said that thinking is different in the Indian language. It is not surprising, therefore, that Whorf works with different premises in approaching what he calls Standard Average European languages and the strange tongues of America and the Far East.

However, Whorf could very easily have done the wrong thing in using special premises with Indian languages. Differences of material culture and social custom do not guarantee distinct cognitive psychologies. There are few bilinguals, after all, and the testimony of those few cannot be uncritically accepted. There is a familiar inclination on the part of those who possess unusual and arduously obtained experience to exaggerate its remoteness from anything the rest of us know. This must be taken into account when evaluating the impressions of students of Indian languages. In fact, it might be best to translate freely with the Indian languages, assimilating their minds to our own. On the other hand, the error may be in our usual approach to the European languages. Perhaps Whorf's Indian premises ought to be applied to French and German and Latin, and we should speak of the psychological differences between peoples who put adjectives before nouns and those who put them after. Our impression that the Europeans share the same cognitive psychology could be the sum of many clumsy tourist observations. We don't have the basic knowledge of psychology and language that would enable us to decide intelligently on the premises that ought to underlie translation.

The evidence of the literal translation does not establish linguistic relativity. The relativity is assumed in the premises underlying Whorf's "unsympathetic" translations. A more familiar set of premises—those used with European languages—would transform all of the translation data into evidence for an opposite thesis of linguistic absolutism. We turn, therefore, to some less equivocal data, not involving philosophies of translation.

CASES OF A NAME AND THE LACK OF A NAME

The discrepancy most interesting to Whorf is the case in which one language has a single category and a single name where another language has more than one category and more than one name. In Hopi there is a single word for all flying things except birds. Where we say *aviator* and *butterfly* and *airplane* the Hopi can use a single word. Here is a region of experience more differentiated by names in English than in Hopi. It is not difficult to think of other areas in which American English is category-rich. Perhaps the world of the automobile is the supreme example. We have the convertible, the club coupe, the hardtop, the sedan, the four-door, the two-door, the station wagon, the runabout, the sport car, the limousine, and many others all exclusive of the familiar trade names. It is not, however, a general property of English or Indo-European languages to be more differentiated than other languages. There have been those who held conceptual differentiation to be a sign of linguistic maturity. However, the knowledge we have of this kind of comparative linguistics does not suggest an evolutionary scale from low to high differentiation. In any comparison of two languages (A and B) there seem always to be ranges of experience more differentiated in A than in B and other ranges more differentiated in B than in A. The Laplander names far more kinds of snow; the Wintu names more varieties of cattle; the Indians of Brazil more kinds of parrots and palm trees than does the speaker of English.

Let us take the most familiar example of this kind of semantic difference and try to determine what psychological conclusions are justified. Whorf notes that the Eskimo lexicon uses three words to distinguish three varieties of snow for which English does not have three single-word equivalents. We should use *snow* for all three. Does this mean that the Eskimo sees differences and similarities among snows that we are unable to see?

There is evidence to indicate that the speaker of English *can* classify snows as the Eskimo does. If we listen to the talk of small boys it is clear that they pay attention to at least two kinds of snow—the *good packing* and the *bad packing*. Whorf himself must have been able to see snow as the Eskimos do since his article describes and pictures the referents for the words.

It seems always to be possible to "name" a category that falls within our experience though we may not have a word for the category. Whorf calls the Hopi category that includes aviator, butterfly, and airplane the "flying class minus birds." Dorothy Lee speaks of one of the Wintu cattle-classes as the "speckled white and gray class." Hockett translates the northern Mandarin Chinese *"gwo"* as "fruits-and-nuts." Murdock has studied kinship terminology in 250 societies; he notes that the English word "aunt" applies to four distinct biological relationships. We don't have separate words for these while some other languages do. The absence of words is not the same as the absence of names. Murdock calls the four relationships "father's sister," "mother's sister," "father's brother's wife," and "mother's brother's wife." In all our examples of denotational discrepancy, it is not correct to say that one language has names for distinctions which another language cannot or does not name. It is always possible to name the categories in both languages so long as the nonlinguistic experiences are familiar. Since members of both linguistic communities are able to make differential response at the same points, we must conclude that both are able to see the differences in question. This seems to leave us with the conclusion that the world views of the two linguistic communities do not differ in this regard.

Although three kinds of snow, four kinds of aunt, and unlimited varieties of cattle can be named in English as in Eskimo or Wintu, the English names are phrases rather than single words. Zipf has shown that there exists a tendency in Peiping Chinese, Plautine Latin, and American and British English for the length of a word to be negatively correlated with its frequency of usage. This is true whether word length is measured in phonemes or syllables. It is not difficult to find examples of this relationship in English. New inventions are usually given long names of Greek or Latin derivation, but as the products become widely known and frequently mentioned in conversation the linguistic community finds short tags for them. The *automobile* becomes the *car* and *television* shrinks first to *video* and eventually to *T.V.* Three-dimensional movies are predictably called *3-D.* In France *cinématograph* has dwindled to *cinéma* and, at last, to *ciné.* Within a linguistic subculture, words having a high frequency in local usage may abbreviate though in the larger community they remain unaltered. At Harvard, *psychology* is *psych.*, *social relations* is *soc. rel.*, and *humanities* 2 is *hum.* 2. For the readers of American psychological journals *subjects* are *S's* and *experimenters* are *X's.* We shall speak of categories having single word names as more *codable* than categories named with a phrase.

Doob has suggested that Zipf's Law bears on Whorf's thesis. Suppose we generalize the finding beyond Zipf's formulation and propose that the length of a verbal expression (*codability*) provides an index of its frequency in speech, and that this, in turn, is an index of the frequency with which the relevant judgments of difference and equivalence are made. If this is true, it would follow that the Eskimo distinguishes his three kinds of snow more often than Americans do. Such conclusions are, of course, supported by extralinguistic cultural analysis, which reveals the importance of snow in the Eskimo's life, of palm trees and parrots to Brazilian Indians, cattle to the Wintu, and automobiles to the American.

I will go further and propose that a perceptual category that is frequently utilized is more *available* than one less frequently utilized. When the Eskimo steps from his igloo in the morning I expect him to see the snow as falling into one or another of his single-word-named categories. For the American who is only able to name these categories with a phrase (low codability) I do not expect such ready categorization of snows. If, however, the American were subjected to a discrimination learning experiment, if he were studying the Eskimo language, or if the perceptual structure were otherwise made worth his while, he could see snow as the Eskimo does. It is proposed, really, that categories with shorter names (higher codability) are nearer the top of the cognitive deck—more likely to be used in ordinary perception, more available for expectancies and inventions.

I expect to find the same relationship between category codability and availability within one language as in comparisons of two languages. Within one language one can compare the codability of a single category for different speakers. Suppose I point out a number of different dogs at the "Dog Show" and ask someone to tell me what he would call this class of dogs. One person may say "a breed with reddish fuzzy hair" and another may say "chows." This category is more codable for the second person and I should suppose that the classifying principle involved is more available to him. If both individuals had seen the same dog bite someone, I think that the man who says *chows* would be more likely to store the experience as something to be expected of chow dogs than would the man who must name this category with a phrase like *reddish, fuzzy haired dogs.* Many other things equal, the presence in someone's vocabulary of a one-word name for a category instead of a phrase name should indicate a superior cognitive availability of the classifying principle involved. The man who identified certain clouds as *cirrus* should be more likely to form expectancies involving this type of cloud than the man who calls them *wispy, horse-tail clouds.* The man who readily identifies a set of faces as *Jews* should be more prone to form expectancies about Jews than the man who names the same array *a lot of people, most of them are rather dark, quite a few are wearing button-down shirt collars.* One way to operationalize availability is in terms of the readiness to use a principle of categorization in forming an expectancy. This readiness can be discovered without recourse to linguistic behavior by observing the generalization of an expectancy to new instances. The man who has the word *chow* when he has seen a chow bite someone avoids members of this breed. The man who does not have the word *chow* but has seen the same dog bite someone steers clear of unlicensed dogs, or sleeping dogs, or all dogs.

The codability variable needs to be expanded in two directions. In the first place, the length of a name need not be the only index of codability. When the name is a phrase I expect subjects to hesitate before naming, to disagree among themselves on the name, and to be inconsistent from one occasion to another. When the name is a single word I expect subjects to respond quickly, in perfect accord with one another, and consistently from time to time. These expectations have been checked and the result will shortly be described. Secondly, the notion of codability can be applied to particular stimuli as well as to categories. Category codability involves the naming of a collection or class of particular stimuli. When category codability is low it has been suggested that the principle of classification is relatively unfamiliar or unavailable. Particular stimuli might also be named and a codability score assigned to each stimulus. How should the two codabilities be related to one another? We will explore these extensions of codability with reference to color perception and color terminology.

Sensory psychologists have described the world of color with a solid, using three psychological dimensions: hue, brightness, and saturation. The color solid is divisible into millions of just noticeable differences. The largest collection of English color names runs to less than 4,000 entries, and of these only eight occur very commonly. Evidently there is categorization of colors among speakers of English. It seems likely that all human beings with healthy visual apparatus will be able to make much the same set of discriminations. This ability is probably standard equipment for the species. Whatever individual differences do exist I do not expect to be related to culture, linguistic or nonlinguistic. This is not to say that people everywhere either see or think of the color world in the same way. Cultural differences probably operate on the level of categorization rather than controlled laboratory discrimination.

Explorations in the "Human Relations Area Files" turn up many reports of differences on this level. Seroshevskii, for instance, has reported that in the Iakuti language there is a single word for both green and blue. This makes the kind of denotational discrepancy with English that we have been describing. A region of experience is lexically differentiated in one culture but undifferentiated in another. If Iakuti informants were asked to name the range of colors that we call *blue* they could only do so by circumlocution, having no single word for that particular array of colors. When we are asked to name the large group of colors called by a single word "X" in Iakuti we must resort to *the blues and the greens*—a phrase rather than a word. It is presumed that these differences of codability locate differences of cognitive availability.

Suppose, now, that the task is changed and subjects speaking English and Iakuti, respectively, are asked to name individual color chips rather than classes. Let them be asked to give the usual name for each color—not feeling it necessary to give a distinctive name to each. English speaking subjects ought to find some hues that are unequivocally *blue* and others that are unequivocally *green*. These should be named promptly, with a single word and with perfect accord among subjects. They would be the stimuli having highest codability. Such stimuli will be centrally located in the categories called by the names *blue* and *green*. There would be other colors which speakers of English would name hesitantly and with disagreement from one another—some calling them *green* and some *blue*, while many would create such phrases as *greenish-blue* or *bluish-green* or a *mixture of green and blue*, or *half green and half blue*. Such low codability stimuli will be peripheral to the categories named by the single-word names which are included among the names accorded the stimuli. Concretely, for the present case, the stimuli will be peripheral to the *blue* and *green* categories; *blue* and *green* being the only single-word names assigned the stimuli. The stimuli of low codability may be central to certain other categories. They may be good instances of the category *bluish-green*. However, this category has a lower availability for the society in question than the more codable categories *blue* and *green*.

Iakuti speakers asked to name the same stimuli ought to behave somewhat differently. Their region of highest codability ought to correspond to a region of low codability for English speakers. The blue greens could be central to a highly available category for the Iakuti. They would be peripheral to available categories for English speakers and central to unavailable categories.

Some of this guesswork was put to test by Brown and Lenneberg in their "A Study in Language and Cognition." The entire series of Munsell colors for the highest level of saturation ("chroma" as Munsell calls it) was shown to ten judges who were asked to map out the respective regions they would call by the names *red, orange, yellow, green, blue, purple, pink,* and *brown*. These are the color names that occur most frequently in English. The judges were also asked to select within each region a single color chip constituting the best instance of the color in question. There was very high agreement among the judges on the mapping of the color areas and also in the eight ideal instances. The eight chips most commonly selected constituted the core of a list of twenty-four test colors. The remaining sixteen chips were selected so that the total twenty-four might provide as even a coverage as possible of the color space.

The twenty-four test colors were shown to a subject one at a time. He was instructed to name the color as he would if describing it to a friend. The experimenters recorded the subject's reaction time (from exposure of the color to pronunciation of a name) and also his response. There were twenty-four subjects—all Harvard and Radcliffe students, all with normal color vision, and all speaking English as their native language.

Five measures were taken: (a) The average length of naming response to each color was obtained by counting syllables. (b) The average length was also obtained by counting words. (c) The average reaction time for each color. (d) An index of the degree to which subjects agreed with one another in naming each color. (e) An index of the degree to which subjects agreed with themselves in naming each color on two occasions.

The intercorrelations among these meas

ures were all in the direction we anticipated when considering the possibility that codability might be measured by a variety of indices. Most of the correlations were significant, with .355 being the lowest. Colors that evoked long names (whether measured in phonemes or syllables) were named with hesitation, with disagreement from one subject to another, and with inconsistency on the part of a single subject from one occasion to another. The social norm with regard to a particular color is revealed by the data on agreement among subjects. Where there is no clear norm for naming, the disagreement within the community is matched in the behavior of a single subject by inconsistent naming from one time to another. The conflicting habits in the community may be presumed to exist within the subject and it is the internal conflict that causes a delay in reaction. These data provide excellent illustration of the interiorization of social norms.

The correlation matrix yielded a single general factor which Brown and Lenneberg called *codability*. The fourth index, the degree of agreement between subjects, had by far the largest factor loading and is therefore suggested as the best measure of codability. For interlinguistic comparisons the name-length index will serve so long as it is a matter of contrasting a phrase with a single word. When a better measure is practicable, and where the contrast involves single words on both sides or phrases on both sides, the agreement index is to be preferred. When it is a question of comparing the codability of the same category for different individuals in one linguistic community the index must be either length of name or reaction time; it cannot be agreement.

Examining the color maps of the original ten judges, Brown and Lenneberg found that the particular chips of highest codability occupied spatially central positions within the regions named by the eight color words. The chips of highest codability were in fact the ideal instances of these eight categories. Colors of lowest codability always fell within an area transitional between two regions. They occupied the anticipated peripheral loci. *Centrality* and *peripherality* of position within a category can be given a neat visual representation in the case of color since the dimensions of this region of experience are known. It is not so clear how to plot a category like chair because we are not sure of the defining attributes of this category. In one direction chairs probably shade off into couches and, in another direction, into tables. Centrality will always mean a position defined by the optimal values of the attributes determining category membership, while peripherality will mean any falling away from these values. It is proposed (with support from color categories) that central stimuli will be highly codable and peripheral stimuli less codable.

These thoughts are offered as blueprints for a set of psychological laws relating category codability to category availability and stimulus codability to category centrality. The laws are expected to hold for comparisons of different regions of experience within one linguistic community, for comparisons of different individuals from the same community for the same region of experience, and also for comparison of the same region of experience in different linguistic communities. The proposals are couched in a form that makes intercultural validity possible. They predict a relationship between variables that are defined so as to apply to any society.

A summary, then, of the so-called case of a name and the lack of a name. To begin with, this case proves to be better described as that of a short name and a long name. Length of name, in turn, yields to a dimension of codability defined by length of name, to be sure, but also by agreement and quickness in naming. Finally, it is proposed that more codable categories of experience are also more available and that more codable stimuli are centrally located in available categories.

34. SIGNIFYING AND MARKING: TWO AFRO-AMERICAN SPEECH ACTS

CLAUDIA MITCHELL-KERNAN

Reprinted from Directions in Sociologuistics: The Ethnography of Communication *edited by John J. Gumperz and Dell Hymes. Copyright © 1972 by Holt, Rinehart and Winston, Inc. Reprinted by permission of Holt, Rinehart and Winston, Inc. Claudia Mitchell-Kernan is an Assistant Professor of Anthropology at the University of California, Los Angeles. Her research interests are in Sociolinguistics and Urban Anthropology. She has done field work in Manu'a, American Samoa, Belize, Central America and in urban United States. She is the author of many articles on speech usage among Black Americans and of a monograph,* Language Behavior in a Black Urban Community.

■ Speech is social behavior, like marriage, economic cooperation, and the control of deviant behavior. In a society like the United States, in which there are many social groups that for one reason or another are socially, economically, and politically separate, the members of each of these groups differ from each other in their marriage and household relationships, economic values, modes of friendship—and in their speech.

Because of economic and political policies in the United States, black people remain socially distinct in our society. Thus it is to be expected that they are also characterized by distinctive linguistic behavior, as are Amish, Indians, many Jews, elite whites, and others. Linguists now generally agree that the language of urban blacks is an independent dialect of English that, like other dialects of the language, has its own rules of grammar and pronunciation. Again, as in the case of other dialects, these rules must be understood in the context of the speakers' history and their current status within the society as a whole.

A commonly held misunderstanding of this dialect is that it is a corruption of "proper" (or dictionary) English, but Mitchell-Kernan shows that this is not so. She illustrates how the dialect includes an elaborate repertory of skills of which signifying (the use of indirection) and marking (characterization) are examples.

The significance of dialects is that they represent particular adjustments within the overall adaptation of human speech. Linguistically, the dominant American culture is represented by "standard" English. Urban blacks, forced to maintain a separate and distinctive identity, impose their own values on the society's dominant speech patterns. The result is not a simple act; it is an elaborate and complex system of communication in which the identity and social status of the individuals engaged in an interaction give words meanings that may not be found in dictionaries.

In "Social Influences on the Choice of a Linguistic Variant" (*Word,* 14 [1948]: 47-56), John L. Fischer examines some of the factors that give rise to linguistic variation by analyzing the pressures in a modern New England community to adopt different speech forms. But these phenomena are not confined to particular age or ethnic groups; many persons will recall the storm that was raised after the publication of Webster's *Third International Dictionary;* see, for example, *Dictionaries and That Dictionary,* by James Siedd and Wilma R. Ebbit (Chicago: Scott, Foresman, 1962), and Karl W. Dykema's "Cultural Lag and the Reviewers of Webster III" (*AAUP Bulletin,* 49 [1963]: 364-69). It must be borne in mind, however, that language is not always verbal. Jurgen Ruesch and Weldon Kees, in *Nonverbal Communication: Notes on the Visual Perception of Human Relations* (Berkeley: University of California Press, 1956), photographically demonstrate how symbols convey socially relevant messages in different situations. ■

IN A LINGUISTIC COMMUNITY which is bilingual or bidialectal, the code in which messages are conveyed is likely to be highly salient both to members of the community and to the ethnographer. The languages spoken tend to be named, and individual speakers, who speak one or the other dialect in particular settings, identified as belonging to one or more groups. The fact that more than one language is spoken, that various social categories of people use specific languages in certain settings when discussing particular topics with members of other social categories, is a significant point of departure.

Aside from language or grammar *per se*, there are, however, other aspects of the communicative competence of such a group which require analysis. The appropriate beginning point for an investigation may be the analysis of the components which are emphasized by elaboration in a variety of speech forms. Well-elaborated components comprise a basis for selection among alternates. The pattern of such selection reveals crucial social information.

Hymes notes that precedence of components may differ from case to case, and such differences may be a basis for the classification of sociolinguistic systems. Such hierarchies of precedence may depend not simply on apparent casual direction in the interrelationships between components but also on the cultural focus (salience-emphasis) upon one or more of the components.

The artistic component is significant in black English. The salience of consideration of the artistic characteristics of speech acts in black English is evidenced by both the proliferation of terms which deal with aspects of verbal style and the common occurrence of speech routines which may be labeled by these terms. The artistic characteristics of a speech act are the characteristics that have to do with the *style* of the speech act, i.e., with the way in which something is said rather than with such components as the topic or the interlocutors. Moreover, the very term art carries connotations of value or judgment of appreciation (or nonappreciation).

The speech acts which will be described here are among the many which are given labels in black English. The terms themselves are sometimes descriptive of the style of the speech act. A partial list of such terms is: *signifying, rapping, sounding, playing the dozens, woofing, marking, loud-talking, shucking,* and *jiving*. Some of these terms are variants used in particular geographic areas. Undoubtedly, other variants exist.

I shall deal in detail with two of these speech acts, treat their stylistic aspects, and attempt to relate the artistic characteristics to the other components which together comprise the speech act. I will describe how these speech acts are used and demonstrate that concern with style and value of artistic merit on the part of speakers of black English influences the other components. Specifically, I will show that this concern has a direct effect upon the choice of the linguistic code in certain conversational settings and frequently explains the use of black dialect forms.

Value regarding verbal art in black English is evident not only from the high frequency of occurrences of nameable artistic variants but also from the comments on such variants in ongoing conversations, including stated values regarding speech use and judgments of the ability of particular speakers that are based upon considerations of artistic merit and style. Concern with verbal art is a dominant theme in black culture, and while these speech acts do not have style as their sole component, style is nevertheless the criterion which determines their effective use.

SIGNIFYING

A number of individuals interested in black verbal behavior have devoted attention to the "way of talking" which is known in many black communities as *signifying*. Signifying can be a tactic employed in game activity—verbal dueling—which is engaged in as an end in itself, and it is signifying in this context which has been the subject of most previous analyses. Signifying, however, also refers to a way of encoding messages or meanings in natural conversations which involves, in most cases, an element of indirection. This kind of signifying might be best viewed as an alternative message form, selected for its artistic merit, and may occur embedded in a variety of discourse. Such signifying is not focal to the linguistic interaction in the sense that it does not define the entire speech event. While the primacy of either of these uses of the term *signifying* is difficult to establish, the latter deserves attention due to its neglect in the literature.

The standard English concept of signifying seems etymologically related to the use of this term within the black community. An audience, e.g., may be advised to signify "yes" by standing or to signify its disapproval of permissive education by saying "aye." It is also possible to say that an individual signifies his poverty by wearing rags. In the first instance we explicitly state the relationship between the meaning and the act, informing the audience that in this context the action or word will be an adequate and acceptable means of expressing approval. In the second instance, the relationship between rags and poverty is *implicit* and stems from conventional associations. It is in this latter sense that standard English and black usage have the most in common.

In the context of news analyses and interpretation we hear the rhetorical question, "What does all of this signify?" Individuals posing this question proceed to tell us what some words or events mean by placing major emphasis on the implications of the thing which is the subject of interpretation and, more often than not, posing inferences which are felt to logically follow. Such interpretations rely on the establishment of context, which may include antecedent conditions and background knowledge as well as the context in which the event occurred.

The black concept of *signifying* incorporates essentially a folk notion that dictionary entries for words are not always sufficient for interpreting meanings or messages, or that meaning goes beyond such interpretations. Complimentary remarks may be delivered in a left-handed fashion. A particular utterance may be an insult in one context and not in another. What pretends to be informative may intend to be persuasive. Superficially, self-abasing remarks are frequently self-praise. The hearer is thus constrained to attend to all potential meaning carrying symbolic systems in speech events—the total universe of discourse. The context embeddedness of meaning is attested to by both our reliance on the given context and, most importantly, our inclination to construct additional context from our background knowledge of the world. Facial expression and tone of voice serve to orient us to one kind of interpretation rather than another. Situational context helps us to narrow meaning. Personal background knowledge about the speaker points us in different directions. Expectations based on role or status criteria enter into the sorting process. In fact, we seem to process all manner of information against a background of assumptions and expectations. Thus, no matter how sincere the tone of voice affected by the used car salesman, he is always suspect.

Labeling a particular utterance as signifying thus involves the recognition and attribution of some implicit content or function, which is potentially obscured by the surface content or function. The obscurity may lie in the relative difficulty it poses for interpreting (1) the meaning or message the speaker is adjudged as intending to convey; (2) the addressee—the person or persons to whom the message is directed; (3) the goal orientation or intent of the speaker. A precondition for the application of the term *signifying* to some speech act is the assumption that the meaning decoded was consciously and purposely formulated at the encoding stage. In reference to function the same condition must hold.

The following examples of signifying are taken from natural conversations recorded in Oakland, California. Each example will be followed by interpretations, intended to clarify the messages and meanings being conveyed in each case.

1. The interlocutors here are Barbara, an informant; Mary, one of her friends; and the researcher. The conversation takes place in Barbara's home and the episode begins as I am about to leave.

BARBARA: What are you going to do Saturday? Will you be over here?
R: I don't know.
BARBARA: Well, if you're not going to be doing anything, come by. I'm going to cook some chit'lins. [Rather jokingly] Or are you one of those Negroes who don't eat chit'lins?

MARY: [Interjecting indignantly] That's all I hear lately—soul food, soul food. If you say you don't eat it you get accused of being saditty [affected, considering oneself superior].
[Matter of factly] Well, I ate enough black-eyed peas and neck-bones during the depression that I can't get too excited over it. I eat prime rib and T-bone because I like to, not because I'm trying to be white. [Sincerely] Negroes are constantly trying to find some way to discriminate against each other. If they could once get it in their heads that we are all in this together maybe we could get somewhere in this battle against the man.
[Mary leaves.]

BARBARA: Well, I wasn't signifying at her, but like I always say, if the shoe fits, wear it.

While the manifest topic of Barbara's question was food, Mary's response indicates that this is not a conversation about the relative merits of having one thing or another for dinner. Briefly, Barbara was, in the metaphors of the culture, implying that Mary (and/or I) is an assimilationist.

Let us first deal with the message itself, which is somewhat analogous to an allegory in that the significance or meaning of the words must be derived from known symbolic values. An outsider or nonmember (perhaps not at this date) might find it difficult to grasp the significance of eating chit'lins or not eating chit'lins. Barbara's "one of those Negroes that" places the hearer in a category of persons which, in turn, suggests that the members of that category may share other features, in this case, negatively evaluated ones, and indicates that there is something here of greater significance than mere dietary preference.

Chit'lins are considered a delicacy by many black people, and eating chit'lins is often viewed as a traditional dietary habit of black people. Changes in such habits are viewed as gratuitous aping of whites and are considered to imply derogation of these customs. The same sort of sentiment often attaches to other behaviors such as changes in church affiliation of upwardly mobile blacks. Thus, not eating or liking chit'lins may be indicative of assimilationist attitudes, which in turn imply a rejection of one's black brothers and sisters. It is perhaps no longer necessary to mention that assimilation is far from a neutral term intraculturally. Blacks have traditionally shown ambivalence toward the abandonment of ethnic heritage. Many strong attitudes attached to certain kinds of cultural behavior seem to reflect a fear of cultural extermination.

It is not clear at the outset to whom the accusation of being an assimilationist was aimed. Ostensibly, Barbara addressed her remarks to me. Yet Mary's response seems to indicate that she felt herself to be the real addressee in this instance. The signifier may employ the tactic of obscuring his addressee as part of his strategy. In the following case the remark is, on the surface, directed toward no one in particular.

2. I saw a woman the other day in a pair of stretch pants, she must have weighed 300 pounds. If she knew how she looked she would burn those things.

Such a remark may have particular significance to the 235-pound member of the audience who is frequently seen about town in stretch pants. She is likely to interpret this remark as directed at her, with the intent of providing her with the information that she looks singularly unattractive so attired.

The technique is fairly straightforward. The speaker simply chooses a topic which is selectively relevant to his audience. A speaker who has a captive audience, such as a minister, may be accused of signifying by virtue of his text being too timely and selectively apropos to segments of his audience.

It might be proposed that Mary intervened in the hope of rescuing me from a dilemma by asserting the absence of any necessary relationships between dietary habits and assimilationist attitudes. However, Barbara's further remarks lend credence to the original hypothesis and suggest that Mary was correct in her interpretation, that she *was* the target of the insinuation.

BARBARA: I guess she was saying all that for your benefit. At least, I hope she

> wasn't trying to fool me. If she weren't so worried about keeping up with her saditty friends, she would eat less T-bone steak and buy some shoes for her kids once in a while.

Although Mary never explicitly accuses Barbara of signifying, her response seems tantamount to such an accusation, as is evidenced by Barbara's denial. Mary's indignation registers quite accurately the spirit in which some signifying is taken.

This brings us to another feature of signifying: The message often carries some negative import for the addressee. Mary's response deserves note. Her retaliation also involves signifying. While talking about obstacles to brotherhood, she intimates that behavior such as that engaged in by Barbara is typical of artificially induced sources of schism which are in essence superficial in their focus, and which, in turn, might be viewed as a comment on the character of the individual who introduces divisiveness on such trivial grounds.

Barbara insulted Mary, her motive perhaps being to injure her feelings or lower her self-esteem. An informant asked to interpret this interchange went further in imputing motives by suggesting possible reasons for Barbara's behavior. He said that the answer was buried in the past. Perhaps Barbara was repaying Mary for some insult of the past, settling a score, as it were. He suggested that Barbara's goal was to raise her own self-esteem by asserting superiority of a sort over Mary. Moreover, he said that this kind of interchange was probably symptomatic of the relationship between the two women and that one could expect to find them jockeying for position on any number of issues. "Barbara was trying to *rank* Mary," to put her down by typing her. This individual seemed to be defining the function of signifying as the establishment of dominance in this case.

Messages like the preceding are indirect not because they are cryptic (i.e., difficult to decode) but because they somehow force the hearer to take additional steps. To understand the significance of not eating chit'lins, one must voyage to the black social world and discover the characteristics of social types referred to and the cultural values and attitudes toward them.

The indirect message may take any number of forms, however, as in the following example:

3. The relevant background information lacking in this interchange is that the husband is a member of the class of individuals who do not wear suits to work.

WIFE: Where are you going?
HUSBAND: I'm going to work.
WIFE: (You're wearing) a suit, tie, and white shirt? You didn't tell me you got a promotion.

The wife, in this case, is examining the truth value of her husband's assertion (A) "I'm going to work" by stating the obvious truth that (B) he is wearing a suit. Implicit is the inappropriateness of this dress as measured against shared background knowledge. In order to account for this discrepancy, she advances the hypothesis (C) that he has received a promotion and is now a member of the class of people who wear suits to work. B is obviously true, and if C is not true, then A must also be false. Having no reason to suspect that C is true, she is signifying that he is not going to work and moreover, that he is lying about his destination.

Now the wife could have chosen a more straightforward way of finding an acceptable reason for her husband's unusual attire. She might have asked, e.g., "Why are you wearing a suit?" And he could have pleaded some unusual circumstances. Her choice to entrap him suggests that she was not really seeking information but more than likely already had some answers in mind. While it seems reasonable to conclude that an accusation of lying is implicit in the interchange, and one would guess that the wife's intent is equally apparent to the husband, this accusation is never made explicit.

This brings us to some latent advantages of indirect messages, especially those with negative import for the receiver. Such messages, because of their form—they contain both explicit and implicit content—structure interpretation in such a way that the parties have the option of avoiding a real confrontation

(Brown provides a similar discussion). Alternately, they provoke confrontations without at the same time exposing unequivocally the speaker's intent. The advantage in either case is for the speaker because it gives him control of the situation at the receiver's expense. The speaker, because of the purposeful ambiguity of his original remark, reserves the right to subsequently insist on the harmless interpretation rather than the provocative one. When the situation is such that there is no ambiguity in determining the addressee, the addressee faces the possibility that if he attempts to confront the speaker, the latter will deny the message or intent imputed, leaving him in the embarrassing predicament of appearing contentious.

Picture, if you will, the secretary who has become uneasy about the tendency of her knee to come into contact with the hand of her middle-aged boss. She finally decides to confront him and indignantly informs him that she is not that kind of a girl. He responds by feigning hurt innocence: "How could you accuse me of such a thing?" If his innocence is genuine, her misconstrual of the significance of these occasions of body contact possibly comments on her character more than his. She has no way of being certain, and she feels foolish. Now a secretary skilled in the art of signifying could have avoided the possibility of "having the tables turned" by saying "Oh, excuse me Mr. Smith, I didn't mean to get my knee in your way." He would have surely understood her message if he were guilty, and a confrontation would have been avoided. If he were innocent, the remark would have probably been of no consequence.

When there is some ambiguity with reference to the addressee, as in the first example, the hearer must expose himself as the target before the confrontation can take place. The speaker still has the option of retreating and the opportunity, while feigning innocence, to jibe, "Well, if the shoe fits, wear it." The individual who has a well-known reputation for this kind of signifying is felt to be sly and, sometimes, not man or woman enough to come out and say what he means.

Signifying does not, however, always have negative valuations attached to it; it is clearly thought of as a kind of art—a clever way of conveying messages. In fact, it does not lose its artistic merit even when it is malicious. It takes some skill to construct messages with multilevel meanings, and it sometimes takes equal expertise in unraveling the puzzle presented in all of its many implications. Just as in certain circles the clever punster derives satisfaction and is rewarded by his hearers for constructing a multisided pun, the signifier is also rewarded for his cleverness.

4. The following interchange took place in a public park. Three young men in their early twenties sat down with the researcher, one of whom initiated a conversation in this way:

I: Mama, you sho is fine.
R: That ain' no way to talk to your mother.
[Laughter]
I: You married?
R: Um hm.
I: Is your husband married?
[Laughter]
R: Very.
[The conversation continues with the same young man doing most of the talking. He questions me about what I am doing and I tell him about my research project. After a couple of minutes of discussing "rapping," he returns to his original style.]
I: Baby, you a real scholar. I can tell you want to learn. Now if you'll just cooperate a li'l bit, I'll show you what a good teacher I am. But first we got to get into my area of expertise.
R: I may be wrong but seems to me we already in your area of expertise.
[Laughter]
I: You ain' so bad yourself, girl. I ain't heard you stutter yet. You a li'l fixated on your subject though. I want to help a sweet thang like you all I can. I figure all that book learning you got must mean you been neglecting other areas of your education.
II: Talk that talk! [Gloss: *Olé*]
R: Why don't you let me point out where I can best use your help.
I: Are you sure you in the best position to know?
[Laughter]
I: I'mo leave you alone, girl. Ask me what you want to know. Tempus fugit, baby.
[Laughter]

The folk label for the kind of talking engaged in by I is *rapping*, defined by Kochman as "a fluent and lively way of talking characterized by a high degree of personal style," which may be used when its function is referential or directive—to get something from someone or get someone to do something. The interchange is laced with innuendo —signifying because it alludes to and implies things which are never made explicit.

The utterance which initiated the conversation was intended from all indications as a compliment and was accepted as such. The manner in which it was framed is rather stylized and jocularly effusive, and as such makes the speaker's remarks less bold and presumptuous and is permissive of a response which can acknowledge the compliment in a similar and joking impersonal fashion. The most salient purpose of the compliment was to initiate a conversation with a strange woman. The response served to indicate to the speaker that he was free to continue; probably any response (or none at all) would not have terminated his attempt to engage the hearer, but the present one signaled to the speaker that it was appropriate to continue in his original style. The factor of the audience is crucial because it obliges the speaker to continue attempting to engage the addressee once he has begun. The speaker at all points has a surface addressee, but the linguistic and nonlinguistic responses of the other two young men indicate that they are very aware of being integral participants in this interchange. The question "Is your husband married?" is meant to suggest to the hearer, who seeks to turn down the speaker's advances by pleading marital ties, that such bonds should not be treated as inhibitory except when one's husband has by his behavior shown similar inhibition.

The speaker adjusts his rap to appeal to the scholarly learnings of his addressee, who responds by suggesting that he is presently engaging in his area of virtuosity. I responds to this left-handed compliment by pointing out that the researcher is engaging in the same kind of speech behavior and is apparently an experienced player of the game—"I ain't heard you stutter yet"—which is evidenced by her unfaltering responses. At the same time he notes the narrowness of the speaker's interests, and states the evidence leading him to the conclusion that there must be gaps in her knowledge. He benevolently offers his aid. His maneuvers are offensive and calculated to produce defensive responses. His repeated offers of aid are intended ironically. A member of the audience interjects. "Talk that talk!" This phrase is frequently used to signal approval of some speaker's virtuosity in using language skillfully and colorfully and, moreover, in using language which is appropriate and effective to the social context.

The content of the message is highly directive. Those unfamiliar with black cultural forms might in fact interpret the message as threatening. But there are many linguistic cues that suggest that the surface meaning is not to be taken seriously. Note particularly the use of such expressions as "scholar," "cooperate," "area of expertise," fixated on your subject," and "neglecting other areas of your education." All these relatively formal or literary expressions occur in sentences spoken with typically black phonology and black grammar (e.g., "I ain't heard . . ." and "Are you sure you in the best position to know?"). By his code selection and by paralinguistic cues such as a highly stylized leer, the speaker indicates that he is parodying a tête-à-tête and not attempting to engage the reseacher in anything other than conversation. He is merely demonstrating his ability to use persuasive language, "playing a game," as it were. The researcher signals acknowledgement by her use of black forms such as "That ain't no way no way . . .", and ". . . we already in . . .". The speaker indicates that the game is over by saying "I'mo leave you alone," and redirects the conversation. The juxtaposition of the lexical items "tempus fugit" and "baby," which typically are not paired, is meant to evoke more humor by accentuating the stylistic dissonance of the speech sequence.

SIGNIFYING AS A FORM OF VERBAL ART

All other conditions permitting, a style

which has artistic merit is more likely to be selected than one which does not because of positive cultural values assigned to the skillful use of speech. Having discussed some of the characteristics of signifying, I would now like to examine briefly the artistic characteristics of signifying.

No attempt will be made here to formulate an all-encompassing definition of art. That individuals may differ in their conceptions of art is made patently clear, e.g., by Abrahams' summarizing statement that signifying is "many facets of the smart-alecky attitude." That my appreciation differs has, more than likely, been communicated in these pages. For present purposes, what is art is simply what native speakers judge witty, skillful, and worthy of praise. This is a working definition at best. It nevertheless serves to limit our field of discourse and, more importantly, to base our judgments on the native speaker's own point of view.

It is true that poor attempts at signifying exist. That these attempts are poor art rather than non-art is clear from comments with which some of them are met. Needless and extreme circumlocution is considered poor art. In this connection, Labov has made similar comments about sounding. He cites peer group members as reacting to some sounds with such metalinguistic responses as "That's phony" and "That's lame." Signifying may be met with similar critical remarks. Such failures, incidentally, are as interesting as the successes, for they provide clues as to the rules by violating one or more of them while, at the same time, meeting other criteria.

One of the defining characteristics of signifying is its indirect intent or metaphorical reference. This indirection appears to be almost purely stylistic. It may sometimes have the function of being euphemistic or diplomatic, but its art characteristics remain in the forefront even in such cases. Without the element of indirection, a speech act could not be considered signifying. Indirection means here that the correct semantic (referential interpretation) or signification of the utterance cannot be arrived at by a consideration of the dictionary meaning of the lexical items involved and the syntactic rules for their combination alone. The apparent significance of the message differs from its real significance.

Meaning conveyed is not apparent meaning. Apparent meaning serves as a key which directs hearers to some shared knowledge, attitudes, and values or signals that reference must be processed metaphorically. The words spoken may actually refer to this shared knowledge by contradicting it or by giving what is known to be an impossible explanation of some obvious fact. The indirection, then, depends for its decoding upon shared knowledge of the participants, and this shared knowledge operates on two levels.

It must be employed, first of all, by the participants in a speech act in the recognition that signifying is occurring and that the dictionary-syntactical meaning of the utterance is to be ignored. Second, this shared knowledge must be employed in the reinterpretation of the utterance. It is the cleverness used in directing the attention of the hearer and audience to this shared knowledge upon which a speaker's artistic talent is judged.

Topic may have something to do with the artistic merit of an act of signifying. Although practically any topic may be signified about, some topics are more likely to make the overall act of signifying more appreciated. Sex is one such topic. For example, an individual offering an explanation for a friend's recent grade slump quipped, "He can't forget what happened to him underneath the apple tree," implying that the young man was preoccupied with sex at this point in his life and that the preoccupation stemmed from the relative novelty of the experience. A topic which is suggested by ongoing conversation is appreciated more than one which is peripheral. Finally, an act of *signifying* which tops a preceding one, in a verbal dueling sense, is especially appreciated.

Kochman cites such an example in the context of a discussion of *rapping*:

A man coming from the bathroom forgot to zip his pants. An unescorted party of women kept watching him and laughing among themselves. The man's friends hip (inform) him to what's going on. He approaches one woman—"Hey, baby, did you see that big Cadillac with the full tires, ready to roll in action just for you?" She answers, "No, mother-fucker, but I saw a little gray Volkswagen with two flat tires."

As mentioned earlier, signifying may be a tactic used in rapping, defined by Kochman as "a fluent and lively way of talking, always characterized by a high degree of personal style."

Verbal dueling is clearly occurring; the first act of signifying is an indirect and humorous way of referring to shared knowledge —the women have been laughing at the man's predicament. It is indirect in that it doesn't mention what is obviously being referred to. The speaker has cleverly capitalized on a potentially embarrassing situation by taking the offensive and at the same time, displaying his verbal skill. He emphasizes the sexual aspect of the situation with a metaphor that implies power and class. However, he is, as Kochman says, "capped." The woman wins the verbal duel by replying with an act of signifying which builds on the previous one. The reply is indirect, sexual, and appropriate to the situation. In addition, it employs the same kind of metaphor and is, therefore, very effective.

Motherfucker is a rather common term of address in such acts of verbal dueling. The term *nigger* also is common in such contexts, e.g., "Nigger, it was a monkey one time wasn't satisfied till his ass was grass" and "Nigger, I'm gon be like white on rice on you ass."

These two examples are illustrative of a number of points of good signifying. Both depend on a good deal of shared cultural knowledge for their correct semantic interpretation. It is the intricacy of the allusion to shared knowledge that makes for the success of these speech acts. The first refers to the toast "The Signifying Monkey." The monkey signified at the lion until he got himself in trouble. A knowledge of this toast is necessary for an interpretation of the message. "Until his ass was grass"—meaning "until he was beaten up"—can only be understood in the light of its common use in the speech of members of the culture and occurs in such forms as "His ass was grass and I was the lawnmower." What this example means is something like: You have been signifying at me and, like the monkey, you are treading on dangerously thin ice. If you don't stop, I am likely to become angry and beat you!

"Nigger, I'm gon be like white on rice on your ass" is doubly clever. A common way of threatening to beat someone is to say, "I'm gonna be all over your ass." And how is white on rice?—all over it. Metaphors such as these may lose their effectiveness over time due to overuse. They lose value as clever wit.

The use of the term *nigger* in these examples is of considerable linguistic interest. It is often coupled with code features which are far removed from standard English. That is, the code utilizes many linguistic markers which differentiate black speech from standard English or white speech. Frequently, more such markers than might ordinarily appear in the language of the speaker are used. Thus participants in these speech acts must show at least some degree of bidilectalism in black and standard English. They must be able to shift from one code to another for stylistic effect. Note, e.g., that the use of the term *nigger* with other black English markers has the effect of "smiling when you say that." The use of standard English with *nigger,* in the words of an informant, represents "the wrong tone of voice" and may be taken as abusive.

Code selection and terminological choice thus have the same function. They highlight the fact that black English is being used and that what is being engaged in is a black speech act. More is conveyed here than simple emphasis on group solidarity. The hearer is told that this is an instance of black verbal art and should be interpreted in terms of the subcultural rules for interpreting such speech acts.

Code and content serve to define the style being used, to indicate its tone, and to describe the setting and participants as being appropriate to the use of such an artistic style. Further, such features indicate that it should be recognized that a verbal duel is occurring and that what is said is meant in a joking, perhaps also threatening, manner. A slight switch in code may carry implications for other components in the speech act. Because verbal dueling treads a fine line between play and real aggression, it is a kind of linguistic activity which requires strict adherence to sociolinguistic rules. To correctly decode the message, a hearer must be finely tuned to values which he observes in relation to all other components of the speech act. He must rely on his conscious or unconscious knowledge of the sociolinguistic rules governing this usage.

MARKING

A common black narrative tactic in the folk tale genre and in accounts of actual events is the individuation of characters through the use of direct quotation. When in addition, in reproducing the words of individual actors, a narrator affects the voice and mannerisms of the speakers, he is using the style referred to as *marking* (clearly related to standard English "mocking"). Marking is essentially a mode of characterization. The marker attempts to report not only what was said but the way it was said, in order to offer implicit comment on the speaker's background, personality, or intent. Rather than introducing personality or character traits in some summary form, such information is conveyed by reproducing or sometimes inserting aspects of speech ranging from phonological features to particular content which carry expressive value. The meaning in the message of the marker is signaled and revealed by his reproduction of such things as phonological or grammatical peculiarities, his preservation of mispronounced words or provincial idioms, dialectal pronunciation, and, most particularly, paralinguistic mimicry.

The marker's choice to reproduce such features may reflect only his desire to characterize the speaker. It frequently signifies, however, that the characterization itself is relevant for further processing the meaning of the speaker's words. If, e.g., some expressive feature has been taken as a symbol of the speaker's membership in a particular group, his credibility may come into question on these grounds alone.

The marker attempts to replay a scene for his hearers. He may seek to give the implications of the speaker's remarks, to indicate whether the emotions and affect displayed by the speaker were genuine or feigned, in short to give his audience the full benefit of all the information he was able to process by virtue of expressive or context cues imparted by the speaker. His performance may be more in the nature of parody and caricature than true imitation. But the features selected to overplay are those which are associated with membership in some class. His ability to get his message across, in fact, relies on folk notions of the covariance of linguistic and nonlinguistic categories, combined, of course, with whatever special skill he possesses for creating imagery.

The kind of context most likely to elicit marking is one in which the marker assumes his hearers are sufficiently like himself to be able to interpret this metaphoric communication. Since there is, more likely than not, something unflattering about the characterization, and the element of ridicule is so salient, the relationship between a marker and his audience is likely to be one of familiarity and intimacy and mutual positive affect.

An informant quoted a neighbor to give me an appreciation of her dislike for the woman. She quoted the following comment from Pearl in a style carefully articulated to depict her as "putting on the dog," parodying gestures which gave the impression that Pearl is preposterously affected: "You know my family owns their own home and I'm just living here temporarily because it is more beneficial to collect the rent from my own home and rent a less expensive apartment." "That's the kind of person she is," my informant added, feeling no need for further explanation. This is, incidentally, a caricature of a social type which is frequently the object of scorn and derision. The quote was delivered at a pitch considerably higher than was usual for the informant, and the words were enunciated carefully so as to avoid loss of sounds and elision characteristic of fluid speech. What was implied was not that the phonological patterns mimicked are to be associated with affectation in a one-one relationship but that they symbolize affectation here. The marker was essentially giving implicit recognition to the fact that major disturbances in fluency are indexes of "monitored" speech. The presence of the features are grounds for the inference that the speaker is engaged in impression management which is contextually inappropriate. Individuals who are characterized as "trying to talk proper" are frequently marked in a tone of voice which is rather falsetto.

A marker wishing to convey a particular impression of a speaker may choose to deliver a quotation in a style which is felt to best suit what he feels lies underneath impression management or what is obscured by the speaker's effective manipulation of language. In the following example, the marker departs radically from the style of the speaker for purposes of disambiguation. The individuals here, with the exception of S_1, had recently attended the convention of a large corporation and had been part of a group which had been meeting prior to the convention to develop some strategy for putting pressure on the corporation to hire more blacks in executive positions. They had planned to bring the matter up at a general meeting of delegates, but before they had an opportunity to do so, a black company man spoke before the entire body. S_2 said, "After he spoke our whole strategy was undermined, there was no way to get around his impact on the whites."

S_1: What did he say?

S_2: [Drawling] He said, "Ah'm so-o-o happy to be here today. First of all, ah want to thank all you good white folks for creatin so many opportunities for us niggers and ya'll can be sho that as soon as we can git ourselves qualified we gon be filin our applications. Ya'll done done what we been waiting for a long time. Ya'll done give a colored man a good job with the company."

S_1: Did he really say that?

S_2: Um hm, yes he said it. Girl, where have you been. [Put down by intimating S_1 was being literal]

S_1: Yeah, I understand, but what did he really say?

S_4: He said, "This is a moment of great personal pride for me. My very presence here is a tribute to the civil rights movement. We now have ample evidence of the good faith of the company and we must now begin to prepare ourselves to handle more responsible positions. This is a major step forward on the part of the company. The next step is up to us." In other words, he said just what [S_2] said he said. He sold us out by accepting that kind of tokenism.

S_2 attempted to characterize the speaker as an Uncle Tom by using exaggerated stereotyped southern speech coupled with content that was compromising and denigrating. It would certainly be an overstatement to conclude that southern regional speech is taken by anyone as a sign of being an "Uncle Tom," but there is an historical association with the model of this stereotype being southern.

The characterization of individuals according to the way they speak is, of course, not peculiar to black people, although the implicit association of particular ways of speaking with specific social types may be more elaborated than elsewhere.

The parodying of southern regional black speech may sometimes serve as a device for characterizing a speaker as uneducated or unintelligent, and sometimes it is used to underscore the guilelessness of the speaker. The marker encodes his subjective reactions to the speaker and is concerned with the expressive function of speech more than its referential function.

Because marking relies on linguistic expression for the communication of messages, it is revealing of attitudes and values relating to language. It frequently conveys many subtleties and can be a significant source of information about conscious and unconscious attitudes toward language. An individual, on occasion, may mark a nonblack using exaggerated black English, with the emphasis clearly being on communicating that the subject was uneducated and used nonstandard usages. Perhaps more than anything, marking exhibits a finely tuned linguistic awareness in some areas and a good deal of verbal virtuosity in being able to reproduce aspects of speech which are useful in this kind of metaphorical communication.

CONCLUSION

Signifying and marking exemplify the close relationship of message form to content and function which characterizes black verbal behavior. Meaning, often assumed by linguists to be signaled entirely through code features, is actually dependent upon a consideration of other components of a speech act. A remark taken in the spirit of verbal dueling may, e.g., be interpreted as an insult by virtue of what on the surface seems to be merely a minor change in personnel, or a minor change in code or topic. Crucially, paralinguistic features must be made to conform to the rules. Change in posture, speech rate, tone of voice, facial expression, etc., may signal a change in meaning. The audience must also be sensitive to these cues. A change in meaning may signal that members of the audience must shift their responses, and that metalinguistic comments may no longer be appropriate.

It is this focus in black culture—the necessity of applying sociolinguistic rules, in addition to the frequent appeal to shared background knowledge for correct semantic interpretation—that accounts for some of the unique character and flavor of black speech. Pure syntactic and lexical elaboration is supplemented by an elaboration of the ability to carefully and skillfully manipulate other components of the speech act in order to create new meanings.

VI. BEGINNINGS IN CULTURAL ADAPTATION: ARCHEOLOGICAL EXPLORATIONS

WE BEGAN THIS BOOK with an exploration of the gene as the basic mechanism of adaptation; we conclude with man's embarkation on the voyage to civilization. The purpose of this section is not to present a survey of the established archeological records from many parts of the world; it is, rather, to highlight one of the principal themes of this book: The evolution of *Homo sapiens* seems to show that man has striven to free himself from the limitations of his genetic constitution and natural habitats.

This section also represents a shift in our subject matter. Until now we have discussed man's adaptations for coping with different and changing conditions in his habitats, and especially the evolutionary changes in particular features of his architecture. Now we turn directly to the most important mechanism in human adaptation: man's culture. Because of this shift in our focus it is necessary to discuss different processes and introduce new concepts.

At the beginning of this book we stressed that adaptation refers to a population's relationship to its habitat; however, the concept of "habitat" must be qualified, depending on whether I am speaking of physiology, language, or organizations of social relations. Because physiology and culture refer to qualitatively different processes, it is also necessary to speak in different ways of the contexts in which adaptations take place. As I have noted, by "culture" we mean the energy systems, objective and concrete artifacts, organizations of social relations, modes of thought, ideologies, and the total range of customary behavior that are transmitted from one generation to another by a social group and enable it to maintain life in a particular habitat.

When we speak of changing habitats in connection with biological adaptation, we can refer to the migrations of populations from one altitude to another, from forests to plains, from dry to wet areas, or to shifts in mean annual rainfall, changes in climate, and the disappearance of flora and fauna. In other words, biophysical adaptations often can be viewed in terms that suggest that changes in the habitat are "due to natural causes." But other changes are at least of equal significance in connection with man's adaptations. These are the changes that are wrought in his habitats by man himself, which are visible in the feedback between his technology and the natural milieu.

Man does more than effectively exploit the physical habitat. His adaptations are different from those of all other living forms because he makes effective use of energy potentials of the habitat (energy being that which makes work). When the habitat changes—whether due to "natural causes" or to the changes that man himself has wrought in it—the group must change its means of making effective use of energy potentials. Before the emergence of man on the evolutionary scene, this was accomplished almost exclusively by genetic change. Man, however, does this by cultural means: by distinctive processes of extracting energy from the habitat.

Whenever man introduces a new energy system into the habitat he seeks to exploit, he must also change the institutions—that is, the organizations of social relations—in society so that they will be appropriate to the efficient use of the sources of energy on which he relies for maintaining mastery over the habitat and freeing himself from its restricting limitations. Every culture represents a unique strategy for extracting energy from the habitat and for maintaining the social groups with which people equip themselves for making effective use of the energy systems they have developed. Every energy system requires organizations of social relations that will enable people to use it; no energy system can be effective in human society without groups that are designed for using it.

When man began to cultivate land by means of the hoe or digging stick, he had to create new organizations of social relations and ideologies to legitimate these activities, which were different from those of people who gained a livelihood by means of the bow and arrow or spear. As a result, man began to live in an entirely different environment—a total system of components that interact with each other and characterize the group—even though the locale (the habitat) may have been the same. Foragers are not responsible for the presence of the food on which they subsist; they collect and hunt food that is made available by natural forces. Horticulturists, however, create a new environment by their digging and planting, and the environment they produce by means of the hoe or digging stick requires a particular type of care and protection that can be provided only by appropriate organizations of social relations and that is inappropriate in the environment om nomadic foragers.

Pastoralism is another technology that creates an environment in which unique institutions are required, because the animals on which pastoralists depend for their livelihood demand a type of care and protection—herding, pasturage, defense against various predators—that is not required by a horticultural environment. The introduction of terracing or large-scale irrigation networks requires institutional organizations of personnel who will care for these structures, and it produces new environments to which institutional adaptations must be made throughout the social system. Similarly, industrialization must be understood as a technology that creates a new environment to which man must adapt, and this environment also has unique institutional configurations.

Why is it more advantageous to conceptualize the processes of human adaptation in terms of the different environments produced by man than to say simply that each technology requires institutional adaptations throughout the social system? Such a formulation suggests practical considerations, in addition to its theoretical advantages. If modern man is to act on the basis of his self-consciousness with respect to his society (or societies) and try to change institutions that are deleterious to human welfare, he must have a clear notion about priorities and consequences in social change. He must, for example, recognize that any change in technology or social institutions—such as the institutions of the culture of poverty, war, and the curtailment of individual freedom and dignity—will in turn completely change the environment in which he lives. The study of human adaptation teaches us, among other things, that we cannot change institutions like spark plugs in an automobile. The relationship between adaptive changes and habitat is reciprocal, and the feedback between the two is unceasing.

This is not to suggest that man should not change his social institutions merely because he would thereby produce entirely new environments; he should try to produce such changes when they will benefit people. But we must also bear in mind that there appears to be a realistic basis for the uneasy feeling held by many people that the world (environment) "will fall apart" if changes are instituted in the family, in the institutions that perpetuate poverty and indignity, and in political and other centers of power. Correlatively, as long as modern man in industrial society seeks to maintain the environments in which he lives, he willy-nilly preserves the institutions he claims he wants to change, because they are aspects of the environment created by his adaptations. This, in turn, raises the central question about the world in which industrial man lives: What is the nature of his environment? This question has yet to be answered by students of modern society. If we can accurately portray the environments in which premodern man has lived, we will have a model for depicting our own environments, and for this reason studies of prehistoric societies are of great practical importance.

Another advantage to this formulation of the concept of environment is that it enables us to understand the order of priorities in social change at different levels of technological development; each level represents a successive cultural adaptation. The ordering of priorities in social change is different at different stages of cultural evolution; there are stages in which a new environment is created through the harnessing of new sources of energy, while political institutions seem to have priority at other stages. Thus if man is to have control over the evolution of his adaptations, he must adopt policies that are appropriate to specific levels of adaptation: What is true of horticultural society is not necessarily applicable to modern industrial organization. For example, the Great Plains of North America have been exploited by pastoralists, agriculturists, and industrial man. Each of these groups of people lived in an entirely different environment, albeit within the same habitat. Industrial man's adaptations to this habitat have to be understood in terms of the properties that inhere in an industrial environment, not in terms of the physical environment of the Great Plains as such.

It is difficult to capture the full profundity of *Homo's* uniqueness in nature, but the essential difference between man and all other forms of life is that man's adaptations are governed by his cultures. Before the emergence of man, the preservation of life was determined almost exclusively by the interaction of challenges in the habitat with gene mutations, by natural selection. In man, the perpetuation of life is determined by a new kind of natural selection: man's ability to meet challenges in the habitat artificially—culturally—by altering his relationship to his milieu and freeing himself from its limitations, even though he remains biologically constant. To repeat an earlier caveat, we do not say that man's physical evolution has ceased. But his ability to live in a variety of habitats and environments has completely outstripped his rate of genetic change.

Has man realized his evolutionary potential? Clearly and decidedly, the answer to this query must be no, whether we are thinking of physical or cultural adaptation or of both in juxtaposition. Can man realize his evolutionary potential? This is something we are very far from knowing, if only because we do not know what this potential is.

There is no potential—in an individual, a population, or the species as a whole —without accompanying limitations. Man's potential for culture and for making adjustments among the elements of each adaptation is not unlimited. His biological makeup—and this includes his genetic programming—restricts his ability and presents him with a number of imperatives, as in the care of his young. Reciprocally, the adaptations achieved in his cultures exert a direct influence on his biological adaptations, as in stature, respiratory mechanisms, genetic strains, and the like. The disentanglement of this skein is one of the foremost challenges in the study of man, which anthropologists have only recently begun to meet.

The feedback between biology and culture notwithstanding, the development of human culture represents a revolution in nature. As a result of his cultures, man has completely reversed his specific relationship to nature. This can be illustrated in a variety of ways and on many levels, all of which point to the fact that, in man, fitness for survival and the perpetuation of human life is cultural more than biological. Man puts on warm clothes to make the Arctic wastes habitable, even though nature has not seen fit to grow a coat of fur on him. Man is biologically unsuited for living in the ocean depths, but he can do this creating artificial habitats. Man can enter into treaties and other agreements to govern territorial relationships. His languages change instead of remaining genetically fixed signal systems. He can rebel, protest social inequities, and be horrified at his inquisitions, witch hunts, and mass slaughters rather than participate exclusively in biologically ordained pecking and dominance orders.

But let us not forget that we have not created the best of all possible worlds. In nomadic hunting and gathering groups, man's achievements in freeing himself from the limitations of the habitat are minimal, though far from insignificant. Among such groups the organization of social relations is governed largely by seasonal cycles and random fluctuations in available food and water. When food and water are abundant, groups increase in size; during shortages, groups almost invariably divide; and social relations vary accordingly. At such levels of evolution the patterns of human life and behavior are directly tied to cycles in nature. At the other extreme of cultural evolution, one of the outstanding characteristics of an industrial society's adaptation is that the supply of food and the content of the diet remain constant the year round. Correlatively, the institutional structure of the society remains unaffected by seasonal cycles. Cutting across this, however, is the fact that although the standard of living of most people in the United States, for example, has improved at a relatively steady pace, American blacks and other disadvantaged groups have been able to make their greatest economic gains almost exclusively during periods of wartime. These cycles have recently become very short. Thus there seems to be an important element of cyclicality that affects a significant portion of the population, albeit not a seasonal intermittency.

The adaptations that have been achieved in the cultures of industrialized societies are the products of accretions in knowledge over many hundreds of centuries. Of course, the present can be taken for granted, and many people do so, but we cannot understand the present without understanding our past. Nor can we exercise control over our cultural future (if control is to be a part of the civilization of the future) without a grasp of the principles and processes that are locked into the record of the past.

What can we learn about the processes of adaptation from a study of the past? It is often said that present cultural adaptations have grown out of those of previous stages of development, but this is true only in a very limited sense. Cultural adaptations—conceptualized in terms of energy systems, appropriate organizations of social relations, techniques, and value systems—arise from the pressures engendered by the society's relationship with the habitat. Patterns of kinship relations or rules of property, to say nothing of political and legal organization, are influenced by a society's past, but they are most profitably understood as adaptations and adjustments to the relationship currently maintained with the productive milieu. For example, the bilateral kinship systems of modern industrial nations often show subtle traces of patrilineal backgrounds,[1] but they cannot be said to have grown out of those pasts. Instead, how one reckons his kinship affiliations is intimately tied to many other factors in the system.

For example, a bilateral system such as ours is closely related to people's freedom to choose their own spouses rather than have their marriages arranged; they establish their own households regardless of where their parents live, and men and women inherit wealth equally from both parents. However, the choice of spouse, rules governing where people will reside after they have married, and modes of inheritance are very different when people reckon their kinship affiliations patrilineally. Furthermore, kinship patterns are closely tied to the society's

1. In bilateral systems, people reckon their descent (kinship affiliations) equally through the paternal and maternal lines; in patrilineal systems, they reckon their affiliations through the paternal line.

strategy of adaptation. For example, our bilateral emphasis is inseparable from the need to amass large concentrations of capital, which is facilitated by keeping wealth within the small nuclear family of parents and children instead of allowing a large group of kinsmen to have access to each other's wealth, as is often the case in societies with patrilineal systems. Bilateral kinship is often associated with the independence of the nuclear family from wider kinship ties, and this is also an accompaniment of widespread physical and social mobility and of recruitment of individuals for jobs in terms of their competence, rather than in terms of nepotism.

Are we, then, to agree with Henry Ford, who, in an attempt to deify the automobile (or perhaps himself), said "History is bunk"? I hope not, for several reasons. First, if man himself is interesting, his past is of equal intellectual moment with his present. Second, and perhaps of greatest significance from the point of view of a science of man, it is only the study of the past that can tell us what kinds of shifts men have had to make in their thinking to move from one stage of sociocultural development to another. Another way of phrasing this is by posing questions such as the following: How rapidly can people be pushed in the course of social change? Can they be impelled to move more rapidly in one direction than in another? What alternatives are available to man in trying to reach a particular stage of development, or are there some sociocultural integrations that can only be reached along a single path? In view of the fact that the rate of change in culture increases as evolution advances, can we make any educated guesses about the future rate of change from the study of prehistoric adaptations? For example, why is it (apparently) easier for a population to shift from a sedentary to a nomadic way of life than from nomadism to sedentism? Is it more disruptive for a society to shift from a horticultural to an industrial level of integration than from horticulture to agriculture? Is there a necessary relationship between the development of urbanization and statehood and between large-scale irrigation networks and political centralization?

It is in light of such questions, which are among the most important questions confronting contemporary anthropology, that many anthropologists (archeologists among them) maintain that the study of prehistoric cultures is an adjunct of social and cultural anthropology. Archeology provides data for the dimension of time to a degree that is almost entirely lacking in the study of living cultures. More precisely, archeology cannot provide answers to such questions directly, but it can provide data in depth that, when combined with data from historic cultures, can give us greater insight into the processes involved.

It is important to remember, in connection with archeological explorations, that archeologists face many of the same problems as paleontologists in trying to uncover the nature of human adaptation during the prehistoric past. Often, in their interpretations of prehistoric cultural remains, archeologists have to make educated guesses about the adaptive significance of the tools and other goods they uncover. One of the ways in which they make these inferences is by relying heavily on what is known about existing and currently functioning societies. If, for example, an archeologist has uncovered the remains of a prehistoric society with a complex irrigation system, one of the things he can do is compare what is known about that society's climatic conditions with data of contemporary societies that have similar irrigation systems. He can also compare the two societies with respect to settlement patterns, evidences of social stratification, political organization, and religious integration. If an archeologist has uncovered the remains of a prehistoric hunting and gathering society, he can apply what is known about living hunting and gathering groups in his attempt to interpret the prehistoric remains.

Why, then, try to reconstruct prehistoric cultures if the same conclusions can be drawn from living cultures? It must be remembered that these interpretations are hypotheses, and one of the questions an archeologist tries to answer is whether the prehistoric data seem to bear out his hypotheses. For example, is the adaptation of a prehistoric nomadic foraging group basically similar to that of a modern nomadic group? But an even more important question underlies archeological research. How long does it take for a society to adapt, to move from one level of adaptation to another, and what are the conditions under which a culture returns to a less advanced mode or level of adaptation? Similarly, what holds a society to a particular level of adaptation? What is the relationship between its culture and the habitat in which it lives over long periods of time?

Thus archeological explorations are indispensable for understanding the processes involved in man's successive achievements in his attempts to free himself from the restrictions of the physical habitat. As self-evident as this statement seems, it represents a relatively new approach and orientation in archeology. Until quite recently, archeology was almost entirely confined to the excavation of material artifacts: arrowheads, palaces, temples. Intellectual competition among archeologists centered upon uncovering the most exotic projectile point at the deepest stratum of a cave or the largest and most labyrinthine palace with the most undecipherable stone tablets or insignia. Only rarely did archeologists seem to wonder what kinds of people had used these artifacts, what had been required in the organization of social relations to construct these temples and pyramids, and what the modes of life had been among the different groups whose artifactual remains had been dug up.

During the last decade or so, however, there has been an evolution in the concerns of many archeologists. Increasingly often, they are trying to establish relationships between the material artifacts they excavate and the cultures of which these artifacts were parts. Needless to say, the cultures are hypothetical, because they cannot be observed directly. And it is for this reason that archeologists rely more and more on the data gathered by ethnographers, the students of living cultures, to evaluate their hypotheses. Reciprocally, ethnologists are turning to the data of archeologists for the dimension of time that no living group can provide.

Although archeologists do not have living people whom they can question and from whom they can get a valid picture of daily life under different conditions, many of them nevertheless are trying to write the ethnographies of prehistoric cultures. They are doing this by focusing on the formal institutions that are suggested by the material remains that have been uncovered rather than on the daily activities of individuals. A temple that is associated with a palace is no longer merely a feat of architecture; it denotes a particular type of political organization. A temporary hearth in a cave is no longer merely a patterning of stones but, together with other remains that may have been unearthed several miles away, indicative of the organization of social relations of hunters and gatherers or herdsmen. A small figurine with bloated abdomen and pendulous breasts is no longer merely a curio for a museum display or a private collection but evidence of a type of religious organization, a level of adaptation resting on cultivation, and a mode of thought about man's relationship to nature.

In writing the ethnographies of prehistoric cultures and in moving away from an almost obsessive concern with classifications of projectile points and potsherds (although these are not without intrinsic value), archeologists concern themselves with institutions such as kingship rather than with the activities associated with a particular king. They deal with types of settlements (such as urbanization) rather than with the relations among neighbors. They concern themselves with the types of habitats that are exploited rather than with the myths and attitudes associated with life in different areas. An ethnographer may devote considerable attention to the fact that the members of a horticultural society justify the almost perennial leisure of the men with the proverb, "It is sinful for a man to scratch the belly of his mother [the earth]." But an archeologist—not having such data—will focus on the relationship between the level of technological development and the formal aspects of divisions of labor in the society as these are suggested by material remains, such as those that are found in women's graves but not in men's.

Although it may have been startling a decade or so ago, it is no longer surprising to read such hypotheses as this from an archeologist (Grahame Clark):

> That self-awareness was . . . slow in developing is shown by the fact that it was not until the Late Pleistocene that we find explicit recognition of death in the form of the careful burials made by Neanderthal man. Indeed it was not until the appearance of Advanced Paleolithic culture that we find much evidence of ceremony in the disposal of the dead. It is in the context of such ceremonial interments that we find the first evidence for an even more definite manifestation of self-awareness in the form of personal adornment by necklaces, headdresses, bracelets, and girdles made from such things as perforated shells, animal teeth, carved ivory, antler, bone, amber, and lignite.

Thus even phenomena such as increasing self-awareness no longer are the concern only of students of psychological process in culture; they are now of equal interest to the archeologist and to the student of culture, both of whom may be concerned, for example, with the heightening of narcissism in terms of what its indicators can tell him about the culture as a whole.

Continuing with Clark's hypothesis, I will provide one example from contemporary society that can illustrate the importance of such ideas for understanding living cultures. Shortly after 1960 there was a tremendous upsurge of interest among American men in a variety of body adornments and toiletries, although several manufacturers had unsuccessfully attempted to arouse male interest in these products several years earlier. How can we account for the apparently sudden change in American values about masculinity? Surely not in psychological terms exclusively, because the personality structures of American men could not have changed so drastically in such a short span of time, and not in terms of more effective advertising, because almost the very same symbols were used by advertisers during both periods. Nor can the change be explained in terms of greater affluence because the amount of wealth available was not appreciably greater during the post-1960 period than previously.

One of the best clues for understanding this change is provided by Clark's hypothesis that the degree of self-consciousness among the members of a population, as evidenced by their preoccupation with body adornment, is intimately tied to their level of technological development. If we follow Clark's lead, we must assume that changes in technology preceded changes in behavior and values with respect to body adornment. And this is what we do find. Shortly after World War II, electronic control of productive activities (popularly referred to as automation) began to infiltrate American industries. Reliance on electronic control for productive purposes represents a new level of adaptation because it has the potential for displacing a large proportion of human labor and because it requires new organizations of social relations in the organization of labor forces. At first most Americans tended to regard the new electronic controls as little more than ingenious gadgetry, but serious preoccupation with automation began about 1955, approximately a decade before the surge of male interest in body adornments and toiletries.

One hypothesis about the relationship between this new level of technological development and the interest in body adornment is that it took approximately ten to twelve years for people—in this case men (with women's acquiescence)—to begin to sense, though not necessarily consciously, that their society was moving into a new level of adaptation. In addition to new modes of production, this new level is providing greater freedom from the restrictions of the habitat than previous levels, lending further support to the hypothesis that every adaptation is accompanied by greater self-consciousness. If this is true, the exploration of the intervals between technological advance and preoccupation with the self at every level of cultural development from the Pleistocene onward will help us learn whether people always respond to technological advance at the same rate or whether this response varies under different conditions. It also gives us further insight into the adjustments that people make in their personal lives to changing environments.

Similar hypotheses also can be developed about periodic outcroppings of beards, the recent concern in American (and other Western) ideologies for the "worth" and "dignity" of the individual, and the great preoccupation of American women with body adornment. Women who perform manual labor on farms, in factories, or domestically cannot be expected to constitute profitable markets for the materials of personal self-consciousness that are advertised in our daily newspapers and in magazines, some of which are devoted exclusively to feminine body adornment and beautification. It is significant that women were freed from agrarian labor long before the men in our society and that preoccupation with feminine body adornment also preceded the men's. Thus, hypotheses about cultural development, including the effects on the individual of different levels of adaptation, demand a feedback between the data of archeologists and ethnologists. Both groups of anthropologists are becoming increasingly aware of this.

Man's relationship to the natural milieu has not always been of the same kind, and the evolution of his adaptations has not followed a single course. Instead, man has maintained qualitatively different relationships to his habitats at the same as well as at different times. Harnessed energy systems and their appropriate organizations of social relations are at the core of cultural adaptation, but there is no single determining factor in the organization of society. It cannot be said, for example, that technology is always the determinant of social organization; instead, there seem to be different determinants at different stages of cultural development. This phenomenon is one of the principal themes of the accompanying volumes, which deal with the cultural present.

Space limitations preclude a full listing of the record of archeological findings, from the beginnings of man's cultural adaptations through his development of civilization. The hypothesis by Grahame Clark cited above is from his paper "Prehistory and Human Behavior" (in "Archeology: Horizons New and Old," *Proceedings of the American Philosophical Society,* 110 [1966]: 91-99). I consider *The Evolution of Urban Society: Early Mesopotamia and Phehispanic Mexico,* by Robert McC. Adams (Chicago: Aldine, 1966) to contain the finest available examples of the ethnography of prehistoric society based on archeological research. A provocative nonanthropological discussion of the relevance of historical studies to the study of contemporary social systems is Herbert J. Muller's *The Uses of the Past: Profiles of Former Societies* (New York: Oxford University Press, 1952). An excellent introductory survey of this record can be found in *Courses toward Urban Life: Archeological Considerations of Some Cultural Alternates* (edited by Robert J. Braidwood and Gordon R. Willey [Chicago: Aldine, 1962]); this book is a collection of seventeen papers, a few of which are quite technical. The reader who has not yet been initiated into archeological explorations might more profitably begin with William Howell's *Back of History: The Story of Our Own Origins* (Garden City, N.Y.: Doubleday, 1954); this highly readable book begins with the earliest of man's known cultures and ends with his first civilizations in the Americas, Mesopotamia, Egypt, Crete, and Europe. The concept of levels of sociocultural integration was first introduced in the sense in which I have used it here by Julian Steward; his *Theory of Culture Change: The Methodology of Multilinear Evolution* (Urbana: University of Illinois Press, 1955) is one of the cornerstones of modern studies of cultural evolution.

Closely related to Steward's work, although diverging in several important respects is *Evolution and Culture* (edited and in large part written by Marshall D. Sahlins and Elman R. Service [Ann Arbor: University of Michigan Press, 1960]). This book, essentially theoretical and programmatic, has had a considerable influence on contemporary anthropology. Although influenced strongly by Steward, it is more directly in the tradition of evolutionary inquiry that was established by Leslie A. White in *The Evolution of Culture: The Development of Civilization to the Fall of Rome* (New York: McGraw-Hill, 1959), *The Science of Culture* (New York: Farrar Strauss, 1949), and his other writings.

The history, methods, and theories of archeological research are simply explained in *An Introduction to Prehistoric Archeology,* by Frank Hole and Robert F. Heizer (New York: Holt, Rinehart and Winston, 1969). One point of view about the concerns of modern archeology is set forth by Kwang-Chih Chang in *Rethinking Archaeology* (New York: Random House, 1967) and in one of his papers, "Major Aspects of the Interrelationship of Archaeology and Ethnology" (*Current Anthropology,* 8 [1967]: 227-43). This paper also contains a lively exchange between Chang and several other archeologists who comment on his thesis. Examination of this debate will give the reader a clear picture of the several directions modern archeology is taking.

Archeology has undergone a major transformation during the last few years; this development is often referred to by anthropologists as the "new archeology." Though usually associated with younger archeologists, the latter have drawn many of their ideas from archeologists like Lewis Binford and Sally Binford who have been working actively to change the discipline's orientation for more than a decade; many mature scholars are associated with this point of view, though they have not always identified themselves as "new archeologists." Essentially, their approach is to regard archeology as anthropology by trying to reconstruct social systems from the material remains they uncover. Their concern is to formulate their hypotheses in such a way that they are replicable and they seek to explain their findings instead of just describing them.

An excellent introduction to this innovative approach in archeology is to be found in *Contemporary Archeology: A Guide to Theory and Contributions,* edited by Mark P. Leone (Carbondale: Southern Illinois University Press, 1972); many of the papers in this volume have not been previously published. The relationship between material culture and social behavior is provocatively explored in *Analytical Archaeology,* by David Clarke (London: Methuen, 1968). The logical premises of the "new archeology" are explored in *Explanation in Archeology: An Explicitly Scientific Approach,* by Peggy Jo Watson, Steven A. LeBlanc, and Charles L. Redman (N. Y.: Columbia University Press, 1971); this book provides an interesting balance to Clarke's. A wider representation of the "new archeologists" is provided in a collecton of twenty-six essays in *Models in Archaeology,* edited by David Clark (London: Methuen, 1973; U. S. distributor, N. Y.: Barnes and Noble).

35. FOOD PRODUCTION IN PREHISTORIC EUROPE

H. T. WATERBOLK

Reprinted from Science, *Vol. 162 (December, 1968), pp. 1093-1102. Copyright 1968 by the American Association for the Advancement of Science. Dr. Waterbolk was born in 1924. His main field of interest is environmental prehistory and he is presently professor of Prehistory and Director of the Biologisch-Archaeologisch Instituut of the State University of Groningen, the Netherlands.*

■ Archeology demonstrates that the distinction between history and pre-history is false. History, as it is popularly conceived, is based on the deliberate recording of events and ideas. But much of human prehistory is also recorded, though it may be etched in and on the earth without the benefit of writing. The archeologist's task is to decipher the markings left by people who lived many thousands of years before writing—as we know it—was invented. The prehistorical record may not always be able to reveal the roles of especially significant individuals or the precise sequence of events, but we can get a bird's eye view of the major outlines of social life. This record was inadvertently laid out by people in the course of their daily lives—as they disposed of their garbage, arranged their dwellings, discarded or lost their tools and weapons and clothing, buried their dead, and left the remains of hunted animals. In this selection Waterbolk gives us a good example of the reconstruction of human history through the unraveling of these clues, and we are thus able to learn about the cultures of Europe as far back as 10,000 years ago.

As we have seen, the evolution of food production is crucial in social and technological development, and it is the development and spread of different techniques of food production—and their social correlates—with which Waterbolk is concerned. When we seek to understand how the members of a particular group or the inhabitants of a region make a significant change in their means of food production, it is often necessary to focus on the group's relations with others who inhabit other regions, from whom they learn new ways of doing things and on whose techniques they elaborate. Thus Waterbolk not only stresses the consequences of different modes of food production for settlement pattern and population size; he constantly draws our attention to the geographical points at which different traditions met in prehistoric Europe and influenced each other. These contacts provided many of the stimuli for technological advances that made it possible for local groups to alter their relationships with the habitat, to free themselves from the limitations of the local milieu, and to develop more complex social organizations.

I alluded above to the "etchings" that prehistoric people left from which we may decipher many aspects of their social life. This reference was not entirely metaphorical. In "Upper Paleolithic Notation and Symbol" (*Science,* 178 [24 November 1972]: 817-28) and *The Roots of Civilization* (New York: McGraw-Hill, 1972), Alexander Marshack provides the results of his microscopic analyses of Magdalenian art, which offer remarkable insights into European life more than 10,000 years ago. Also relevant is "A Preliminary Analysis of Functional Variability in the Mousterian of Levallois Facies," by Lewis R. Binford and Sally R. Binford (*American Anthropologist,* 68 [1966]: 238-95). For a contrasting picture with prehistoric Europe, see "Human Ecology During the Pleistocene and Later Times in Africa South of the Sahara," by J. Desmond Clark (*Current Anthropology,* 1 [1960]: 307-24). ■

BY ABOUT 300 B.C. all the plains south of the Scandinavian mountains were inhabited by people who lived together in villages of a more or less permanent character. These settlers cut the deciduous forest with stone axes, cultivated a variety of crops, and raised cattle, sheep, goats, and pigs. Hunting was of little importance.

The art of pottery was known everywhere. From highly varied shape and ornamentation, archeologists have been able to distinguish a number of cultures of limited geographical and chronological occurrence and various degrees of relation.

How different is the picture if we go back in time another 5000 years, to about 8000 B.C. The last cold spell of the Ice Age was then almost over. Hunting and gathering were the major means of subsistence. Animal domestication and plant cultivation were unknown. Camp sites were relatively impermanent, shifting from one place to another.

NUCLEAR AND NONNUCLEAR CULTURES

From 8000 to 3000 B.C. in southwestern Asia there was a progression from food collecting to urban civilization, through the levels of incipient cultivation and domestication, of primary effective village farming, and of developed village farming and town life. For the purposes of this discussion, the nuclear development must be accepted as a given fact, though in its later stages some parts of southeastern Europe also participated.

In other parts of the Old World, cultures had, in principle, been developing along lines independent of those of nuclear development. According to Caldwell, who treated eastern North America in its relation to nuclear Meso-America, a nonnuclear type of culture may attain a "primary forest efficiency"—the successful exploitation of the natural food resources that brought about residential stability, great material wealth, development of craftsmanship and art, disposal of goods with the dead, and the building of large earthworks. Forest efficiency could even lead to resistance to the introduction of food production. We shall see that this concept of forest efficiency can be usefully applied to Europe.

The terms Paleolithic, Mesolithic, and Neolithic have little meaning for the study of the introduction of food production. There are very different types of "Mesolithic" economies, and "Neolithic" elements like polished axes, pottery, and domesticated sheep are introduced quite independently; not even the additional presence of cattle and pig and of cultivated wheat and barley need result in full dependence on food production.

CLIMATE AND RELIEF

Any cultural development in Europe has to be seen against the background of two major natural factors, namely climate and relief. The most significant climatic event is the Ice Age.

The last cold stage had a much more intense effect in northern Europe than it did in southern latitudes. Correspondingly, the environmental fluctuations in the early postglacial were of much greater importance in the north than in the south, where the major climatic improvement had started long before and had not been interrupted by pauses and minor readvances.

As to the bearing of relief, mountain areas are less suitable for agriculture than are the plains, where soil conditions generally are good, and internal communications are easy. Not only the lowland plains, but also the higher plains, such as the south German plain, should be considered in this connection. An ordinary contour map gives a false picture of the size of the habitable areas.

The irregularities of coastlines and mountain chains differentiate coastal plains from the continental plains. Continental plains generally communicate with each other along the coast and overseas; the continental plains communicate with the coastal plains and with each other through river valleys or low passes. . . .

Low ridges or even watersheds can be important boundaries between cultural provinces by a combination of factors including soil, climate, and vegetation, as well as by the absence of waterways. For instance, in the

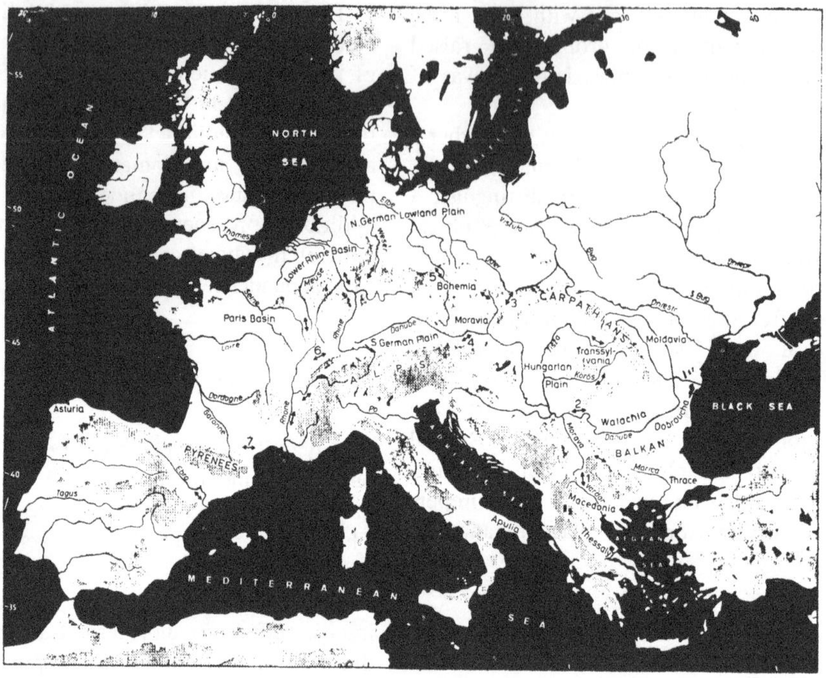

FIGURE 35.1
Major plains of Europe are indicated, irrespective of elevation. High mountain areas (above 1000 meters) are shaded. The interconnections of the plains are shown by a double-headed arrow. 1, Vardar-Morava Gate; 2, Iron Gate; 3, Moravian Gate; 4, Linz Gate; 5, Elbe Gate; 6, Burgundian Gate; 7, Aquitanian Gate.

hills between Bohemia and Moravia there are no geographically determined channels of communication.

FOOD COLLECTORS BEFORE ABOUT 6000 B.C.

Here I emphasize southeastern, central, and northwestern Europe; Iberia, North Africa, Russia, and Scandinavia get less attention.

For a rough chronological arrangement of the evidence, three periods can be distinguished. The appearance of the first village farmers in Greece (about 6000 B.C.) and the expansion of Bandkeramik farmers toward northwestern Europe (about 4500 B.C.) serve as boundaries to separate the three.

In the period in which the first evidence of incipient food production in southwestern Asia is recorded, reindeer hunters still roamed the plains of northern Europe from Belgium to Poland. In this area the Upper Dryas climatic deterioration had again opened up the forest cover which had earlier expanded northward during the preceding Allerod-interstadial (10,000 to 9000 B.C.).

These reindeer hunters belong to the "Tanged Point" tradition, a late phase of which is known as the Ahrensburg culture. In the deep tunnel-valley sediments adjacent to the Stellmoor settlement, remains have been found of at least 650 reindeer, whereas only a few specimens of elk, bison, wild horse, lynx, beaver, fox, and wolf, and of such fowl as ducks and geese have been uncovered. The flint industry comprised tanged points, "Zonhoven" points, gravers, and scrapers. Harpoon points have also been found. Ornamentation is rare: simple geometrical designs are engraved on some reindeer ribs. In the bog, a wooden post with a reindeer skull on its top was found.

Cave sites from a late variety of the Ahrensburg culture with a fauna comprising both tundra and forest animals are known. These sites testify to the adaptation of the reindeer hunters to the forest environment.

Another tradition in the lowlands of western Europe is the Federmesser tradition, which included various local groups such as the Azilian, the Tjongerian, the Creswellian, and others, and probably resulted from an adaptation of Magdalenian reindeer hunters to the forest. In the Allerod period, a northward expansion of these Magdalenian survivals took place. Their sites occur along river valleys or lakes, but there is no indication of any means of subsistence other than the hunting of big game, such as red deer, elk, and aurochs.

Another development within the same tradition can be observed in the western Mediterranean where a number of coastal cave sites are known, the Romanellian (or Grimaldian), which have yielded remains not only of big game, but also of small animals such as hares and birds, and considerable quantities of land snails and of shells collected at the seashore. This is the earliest evidence in Europe for the exploitation of this important natural resource.

Soon after 8000 B.C., the final postglacial climatic improvement started. It caused changes in the environment that were particularly drastic in the northern parts of Europe. In the "Preboreal" the birch-pine woods regained the areas from which they had been pushed during the Upper Dryas period. Deciduous trees, beginning with hazel, moved northward. In Europe nearly everywhere outside the Mediterranean zone a pine-hazel period can be distinguished in the early postglacial ("Boreal"). Into these woods moved elm, oak, lime, and somewhat later ash and alder; after these deciduous trees had gained dominance ("Atlantic") beech, white pine, and hornbeam moved in ("Subboreal"). On the French Mediterranean coast the land snail fauna indicates a change toward a drier climate.

It apparently took some time before man could efficiently adapt to the new environment. In many well-investigated areas, the frequency of sites that can be dated between 8000 and 7000 B.C. (Preboreal) seems to be significantly lower than those from the preceding and following millennia. It is therefore difficult to find a continuous series of intermediate assemblages between the Federmesser and Tanged Point traditions and their successors, and to localize the areas where the transition and readaptation took place.

Perhaps this is a general phenomenon—we may expect a rapid alteration in the environment to result in a change in the quantity of food that can be obtained with traditional methods, and in many cases to bring about diminution of population, or emigration. Only after adjusting the hunting and collecting methods to the new fauna and flora can the population expand again. A similar situation seems to be present in the early Atlantic period in the western Baltic.

In the early postglacial period at least three traditions can be distinguished in the human habitation: Maglemosian, Sauveterrian-Tardenoisian, and Montadian. The center of the Maglemose tradition was in the western Baltic area. It also included parts of Britain and the present North Sea. We know of a seasonal settlement (Star Carr in England), as early as in the Preboreal period, in a river valley bog, where man lived as close to the water as possible during winter and early spring. People hunted red deer (dominant), roe deer, elk, aurochs, wild boar, and other furbearing animals such as fox, marten, and beaver. Birds were not numerous, and fish remains were conspicuously lacking. The preference for settling at the lakeside thus does not seem to have been determined by fishing or hunting of waterfowl. A wooden paddle is evidence of water transport.

Sites dating from the Boreal period are much more frequent. Seasonal culture layers, mainly known from the bogs of Denmark, Schleswig-Holstein, and Scania, now also contain large quantities of hazelnut shells and fish remains.

In all stages of the Maglemosian tradition, macrolithic tools occur with microlithic flint types. Core and flake axes and adzes were

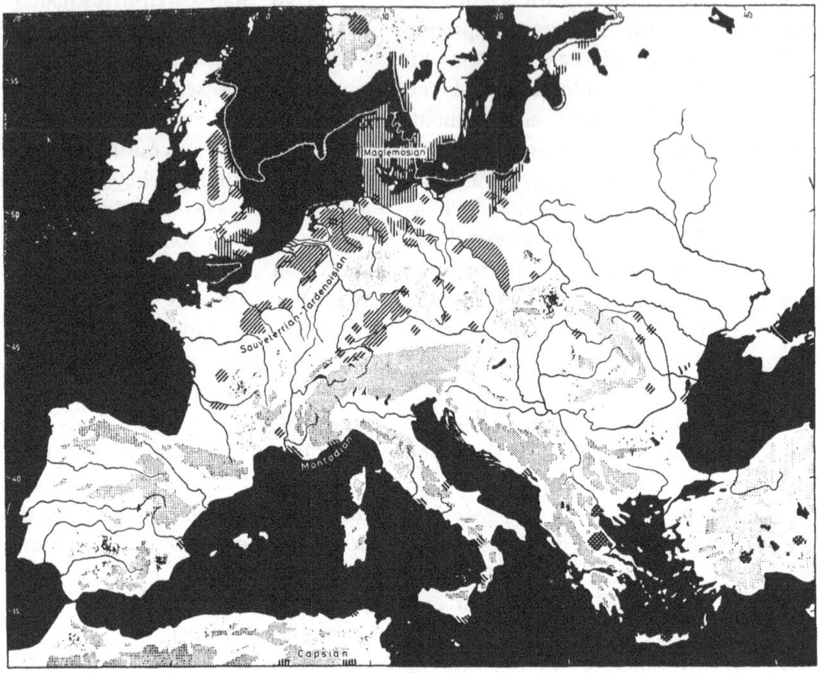

FIGURE 35.2
Distribution of farming cultures (cross-hatched) and hunting-collecting cultures about 6000 B.C.

made, which though still unpolished, could efficiently cut small trees to be used, for example, as substructures for huts. Other stone was also used for making tools, such as mace heads, hammers, and axes, some of them perforated and made by pecking. There was also some art: amber model animals and geometrical ornaments made by incision and pricking on bone and antler tools.

Much less specialized were the groups belonging to the Sauveterrian-Tardenoisian tradition of western Europe, from parts of Britain, the Low Countries, and France to inland Germany. The flint industry was predominantly microlithic.

People lived mainly on game hunting and their settlements were frequently shifted. They would certainly have taken advantage of the hazel, but it is doubtful that fishing was of any importance. In hilly countries cave sites occur rather frequently, and heavy soils are avoided.

Microlithic flint industries of Sauveterrian-Tardenoisian affinity, and therefore probably of Boreal age, occur as far eastward as lower Austria and Slovakia, where they seem, however, to be rare. Southeastern Europe has so far yielded very little that can be attributed with any certainty to this period. Some Rumanian and Moldavian sites may belong here.

Along the Mediterranean coast, habitation continued. In the Rhone delta, a continuous sequence of industries has been established from the Romanellian to the food-producing Chasseen. It is probable that the first stage after the Romanellian, the Montadian, is to be dated from the Preboreal and Boreal periods. The settlements occur in rock shelters. The culture layers are very rich in mollusks, both terrestrial and marine. Further, there are aurochs, red deer, wild boar, and large quantities of hare or rabbit. The flint industry has a degenerated character.

Flint industries of this type also occur elsewhere in the Mediterranean area, as for example, in Liguria, near Salerno, and also along the Adriatic coast of Jugoslavia, in the cave Crvena Stijena.

VILLAGE FARMERS APPEAR IN GREECE

Sometime around 9000 B.C., incipient food production had started locally within what Braidwood calls the natural habitat zone in southwestern Asia among cultures that have a microlithic component strongly reminiscent of that in the later cultures of western and northern Europe. Around 7000 B.C., the first village-farming settlements were present in that area. Pottery was still, however, lacking, and it did not become common for perhaps another 1000 years.

In the prepottery stage, the European continent was hardly influenced by the nuclear development. Only in Thessaly have preceramic levels with remains of a farming culture been reported. Here, people cultivated wheat, barley, millet, lentils, and other legumes. They raised mainly sheep, but also had goats, swine, and cattle. Hunting and collecting were of minor importance.

Such sites have also been found on Crete and Cyprus. The sea apparently posed no more problems: it unites the sites—all in coastal plains—more than it separates them. There are various archeological connections with Anatolia, and it is probable that this area or its lowland flanks in Syro-Cilicia and in the prehistorically little-known western littoral of Turkey was the place of origin of the first colonists on European territory. Admittedly, there is evidence for the presence of hunting and collecting communities in Greece in late-glacial or early postglacial times, but their persistence up to the time immediately preceding the first evidence of farming is still to be proven.

DRAMATIC ENVIRONMENT CHANGE

With the Boreal-Atlantic transition, about 5500 B.C., there was much environmental change north of the Alpine belt. Pine and hazel lost their prominent place to a mixed deciduous forest. Marine transgressions enlarged the North Sea and brought salt water into the Baltic.

Increase in precipitation promoted the vegetation on lakeshores and the filling up of basins. Raised bogs appeared in former lakes or marshes, which was disastrous to fish and waterfowl.

The mixed oak forest was more shady than the fairly open pine woodland, and there was less grazing for large animals. Big game such as aurochs and red deer decreased in number. All these changes required readaptation by man, who now faced a rapid decrease of the natural resources he had been relying on for two millennia.

Danish investigators have noticed an increase in the importance of waterfowl hunting, fishing, and collecting of wild fruits and berries in the Maglemose area, toward the end of the Boreal period, as evidence of elk and aurochs become rare. Only very few habitation sites from the early Atlantic period are known. In contrast to those from the preceding period, these sites all seem to lie on the coast. At the site of Vedbaek, north of Copenhagen, remains were found not only of red deer, roe deer, and wild boar, but also of sea mammals such as grey seal, ring seal, porpoise, and fish such as cod and haddock. Sea hunting and fishing appear here as a new means of subsistence and compensate for the decrease in big game and inland fish. Shellfish, however, were not collected. Pottery was unknown, and there is no evidence for food production.

As was the case in the Preboreal period, there seems to be a decrease in population during the adaptation to the new environment. In the Low Countries, one early Atlantic site is known, de Leijen. It is situated at the shore of a depression which was still open water at the time of habitation. People collected hazelnuts and also water nuts, and must have lived on fish as well (of the possible organic materials, only carbonized fruit remains were preserved at this site). Generally, however, higher grounds had become unsuit-

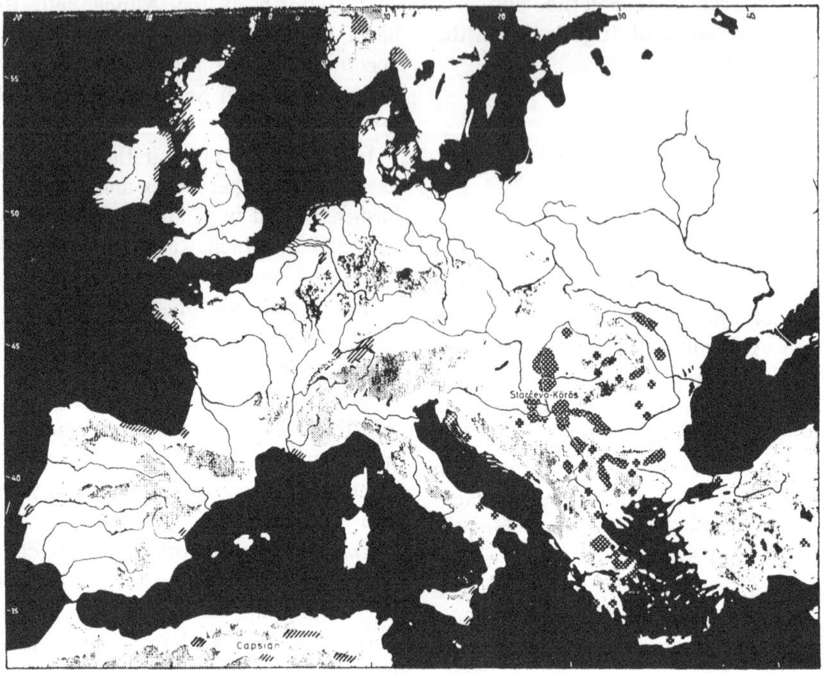

FIGURE 35.3
Distribution of farming cultures (cross-hatched) and hunting-collecting cultures about 5000 B.C.

able for human habitation. But there is reason to believe that parts of the Rhine delta in the North Sea that are now deeply sub-submerged were inhabited during the Atlantic period.

Further south, in Belgium and France, there is as yet no proof of inland habitation during the Atlantic period. But a few shell middens of this age are known from the coast of Brittany, and the rise of the sea level since about 3000 B.C. in that area makes it probable that their number had been much larger. At Téviec and Hoedic, interesting cemeteries have been excavated with graves containing one to three individuals and grave goods—antlers, bone tools, and shell necklaces. Other shell middens of Atlantic age are known from Britain, Portugal, and northwest Spain.

As the Boreal-Atlantic transition set in, people everywhere in western, northwestern, and northern Europe moved toward the coast and added coastal hunting, fishing, and collecting as means of subsistence to the traditional hunting of big game and inland collecting and fishing. The Atlantic forest as such was an unfavorable environment for man, and European man could best survive by adapting himself to the coast. In making this adaptation, however, there was an important consequence. The coastal resources allowed a considerably higher degree of sedentary occupation, freeing man of the necessity of continuous wandering. One precondition for the acceptance of farming—residential stability—was thus automatically fulfilled.

In the narrow zone between Alps and Jura mountains there are many lakes and streams. Here we find a very varied environment, especially if the high mountains are included as suitable area for seasonal hunting. It is not surprising that in this area there is proof of human habitation during the Atlantic period, in caves and open air sites, such as the Lautereck, a rock shelter on the upper course of the Danube where people specialized in carp fishing. Elsewhere in central Europe there is no proof of human habitation between 5500 and 4500 B.C.

Along the western Mediterranean coast, adaptation had, as seen in the preceding section, taken place long before. Possibly it was the Allerod climatic improvement which had here brought about the same necessity of movement toward the coast as the Boreal-Atlantic change was to bring about in the north.

SOUTHEASTERN EUROPE (6000 TO 4500 B.C.)

At about 5500 B.C., in the coastal plains of the Aegean area ceramic farming villages were present everywhere. Large rectangular buildings have been excavated at Neo Nikomedeia. Early sites also occur in the inland valleys of Macedonia and Bulgaria.

At about 5000 B.C. the plains of southeastern Europe were inhabited by agricultural groups of the Starcevo-Koros tradition with their main distribution area in the southern part of the Hungarian plain. Here natural steppe conditions may have prevailed, but sites of this tradition also occur in areas such as Bosnia, which most probably had a forested character. The use of heavy wood in the house construction in Macedonia may be seen as a first adaptation to European forest conditions. This adaptation probably continued in the Hungarian plain, which certainly had gallery forests along the rivers, and was bordered everywhere by wooded hills.

Most of these settlements seem to be completely dependent on food production. There is, however, some evidence for the persistence of a food-collecting economy. On an island in the flood plain of the river Theiss a site has been found with remains of forest animals, many fish, many birds, and layers of shells up to 30 centimeters thick. There was evidence for the local domestication of cattle. At some sites in the Bug-Dnestr area, a "Mesolithic" level is reported below a series of "Neolithic" levels. The lowest Neolithic level contains many bones of wild animals alongside domesticated cattle, and many fish remains. There are sites on river banks. The flint industry is microlithic.

In both areas we must reckon with the possible persistence of communities which did not produce food and instead specialized in fishing and thereby attained sufficient residential stability to be receptive to new ideas on food production. In this connection, mention should also be made of the recently excavated site of Lepenski Vir situated on a bank of the river Danube in the Iron Gate. There were three main horizons. The lower horizon comprises five layers, with a total of 59 trapezoidal house foundations, each with an elaborate hearth. The most spectacular finds are large sandstone pebbles that are worked into human faces with a fishlike mouth. Agricultural tools are lacking. Bones of deer, boar, and various fish species have been found. Fishing and hunting thus seem to have been the major means of subsistence. Pottery appears only in the two uppermost layers of the lower horizon. Horizon III, separated by a sand deposit from the foregoing horizon, belongs to the Starcevo culture.

In large parts of the Hungarian plain, Bulgaria, and Rumania, which were inhabited by the Starcevo-Koros farmers, evidence for the presence of such food-collecting communities is lacking. We therefore assume that there was an actual colonization by people from the narrow coastal plains of the Aegean coast, through the Vardar-Morava gate and parallel valleys, and up the Marcia valley, and from there either westward toward Serbia or northward along the Black Sea coast into the Dobroucha and Walachia. There is also the possibility of migration over sea through the Dardanelles and the Bosporus. Unfortunately, there is little evidence from the coastal plains of western Anatolia and Thrace.

In the Thessalian standard sequence of cultures, Milojcic is found below the Sesklo level, which should be parallel to Starcevo, a level characterized by a pottery that is ornamented by impressions of shells (Cardium). Related pottery occurs in many places along the Mediterranean coasts. It probably originated in the Levant (for example, Mersin) and spread westward. It is found along the Dalmatian coast, in Apulia, Sicily, Liguria, south France, Spain, and even Portugal, always in areas (often at the same sites) that were already inhabited by coastal hunting-collecting communities. From the few radiocarbon dates available it is probable that this pottery had reached the western Mediterranean by 4500 B.C. It may have been preceded by the domesticated sheep and goat, or both. In the section of Chateauneuf-lez-Martiques In the Rhone valley, bones of these animals occur well below the first pottery (of Cardium type). They are also reported in late Tardenoisian layers of Rouffignac and in aceramic shell middens of Portugal, Asturia, Brittany, and Ireland.

THE FINAL PHASE (4500 TO 300 B.C.)

By 4500 B.C., then, the foundations were laid for the final spread of food production throughout Europe. Evidence for this general moment of time suggests (i) that the Hungarian plain was a dominant center which had incorporated all the major achievements of the nuclear area in the Near East, and which had added to these an adaptation to the European deciduous forest; (ii) along the Mediterranean coast, a diffusion sphere connected with the Levant and the Aegean; (iii) along the Atlantic and Baltic coasts, populations of hunters, fishermen, and collectors, who by a many-faceted exploitation of natural resources had attained a fair degree of residential stability; and (iv) finally, locally in inland Europe, small groups of people who had been increasing their economic potential by specializing in river and lake fishing and collecting.

One of the most remarkable events in European prehistory is the explosive spread of the Bandkeramik culture. From about 4400 B.C., large loess areas north and west of the Hungarian plain were colonized in a short time.

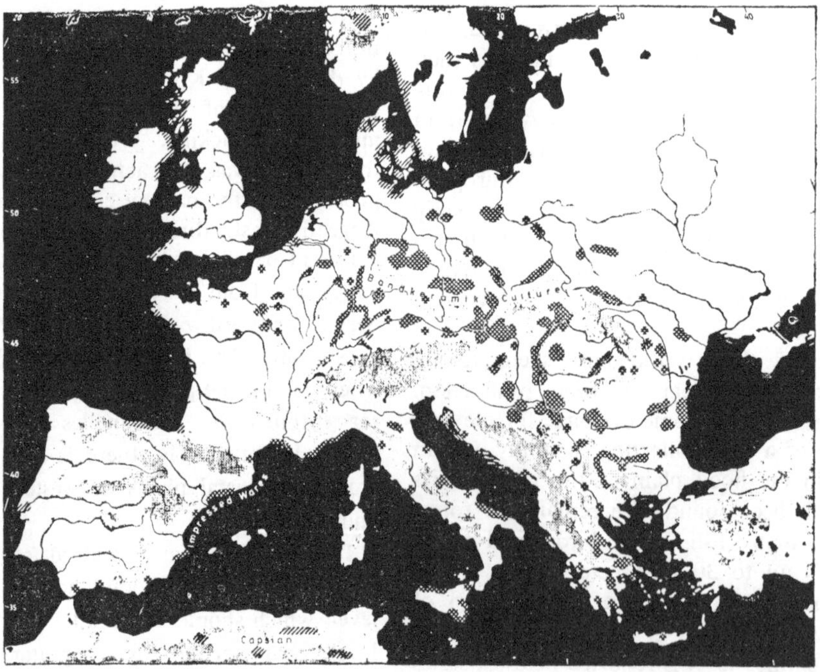

FIGURE 35.4
Distribution of farming cultures (cross-hatched) and hunting-collecting cultures about 4200 B.C.

Permanent settlements were founded everywhere. They consisted of some 10 to 20 wooden buildings up to 40 meters long and 5 to 6 meters wide, the roof of which was supported by three rows of posts inside the houses. In each village, one building is larger and of heavier construction than the others. The people lived by agriculture and husbandry. Hunting was of minor importance; the percentage of wild animals never exceeds 10 percent of the total bone remains.

The pottery is very homogeneous. In a period of several centuries a parallel development of form and ornament took place over the whole area of distribution. Toward the end, regional differentiation increased.

A possible center of this large-scale colonization is the northwestern part of the Hungarian plain. An early pottery stage with Starcevo influences has recently been distinguished in this area. It also occurs in Bohemia, South Germany, and Saxony-Thuringia.

Although the river valleys clearly directed the distribution of the Bandkeramik, it is by no means a riverine culture. Sites occur mainly on well drained loess plateaus, with no direct relation to rivers or streams. In this large area there is no conclusive evidence that earlier populations took over agriculture and husbandry and were assimilated into the Bandkeramik culture.

We mentioned the site of Lautereck. Here, Bandkeramik pottery appeared in the otherwise normal collecting inventory of a carp fishing settlement on the Upper Danube, outside the loess area. In a still later stage, Aichbühl pottery (about 3800 B.C.) also occurred at the same site in an inventory that was otherwise still unchanged. This is one of the few sure contacts between the Bandkeramik culture and an indigenous inland culture, and it certainly does not suggest a fast rate of assimilation.

Contacts of the same kind can be expected at all points where the Bandkeramik culture occurs far down the river valleys so that it meets the coastal communities. The results of such contacts are evident much later, when the Bandkeramik culture had been succeeded by regional cultures such as Rossen in the west, Stichbandkeramik in central regions, and Precucuteni in the east. For this level, dates of 3800 to 3500 B.C. are probable.

For northwestern Europe, the Rossen culture appears to be of great importance. Rossen settlements with long houses are known from the same loess soils and the same areas as were occupied by the Bandkeramik farmers. But isolated finds of Rossen pottery occur at sites along lakes and rivers where the traditional means of subsistence are maintained with small-scale agriculture and husbandry.

In the western Mediterranean area, the communities of coastal collectors which, by way of diffusion, had learned to make pottery and to herd sheep and goat, or both, now also took over agriculture. They no longer lived exclusively on a narrow strip along the coast, but moved onto the coastal plains behind. Minor geographical barriers were crossed. Cardium pottery penetrated from Liguria to the Po plain, from the Provence into the Garonne plain through the Aquitanian gate, from the Catalonian coast into the Ebro valley. The new economy seems to have resulted in an increase of the population, and, with the coastal plains as the starting point, all available open spaces were occupied.

At the same time, the traditional communication lines along the Mediterranean and Atlantic coasts were maintained. A network of connections thus extended all over western Europe, including the British Isles. The coastal communities served as connecting points in this network and as nuclei of cultural innovation. They became starting points for the final spread of farming over the plains of western Europe. In this area a certain cultural uniformity prevailed—the "Western Neolithic" tradition. This tradition is characterized, for example, by certain pottery types, hilltop settlements ("causewayed camps"), and megalithic monuments (menhirs, alignments, collective tombs).

WESTERN AND DANUBIAN TRADITIONS MEET

So far in this account of the spread of food

production into Europe, the new economy has moved into and developed in a succession of regions where food production was unknown. We now have the problem of what happened when the Western Neolithic tradition and the Danubian tradition (as evidenced by the regional successors of the Bandkeramik) met. Keeping in mind the maritime emphasis of the Western tradition and the predilection for loess soils shown by the Danubian tradition, the areas where both traditions can be expected to meet are: the western part of the Alpine foreland, Burgundy, the upper Rhine lowland, the Seine basin, and the coastal plains along the North Sea.

In the Alpine foreland, the Lautereck site shows that the first influence of local groups in this area could have taken place even before 4000 B.C., but probably it was half a millennium later, in Rossen times, that the reorientation became manifest. Here again, the actual transition from food collecting to food production is not observed; but this is a general phenomenon. We see only the result of the adaptation, not the adapting process itself, because it concerned only a small number of people, and the chance of finding their settlements is small. Only where man used caves and rock shelters as settlements—as in Chateauneuf-lez-Martiques—may we expect to observe gradual transitions in the occupation layers. For them the choice of settlement sites was much more restricted; and furthermore, caves and rock shelters are obvious places for archeological exploration.

The first agricultural groups in the Alpine foreland preferred lakeshores or bogs as settlement sites. There are few early sites. The earliest may be those of the Aichbühl culture. Others show Rossen influences (for example, the Egolzwil culture). This level can be dated at about the middle of the 4th millennium B.C.

About 3000 B.C. the population had apparently increased considerably. There were many farming villages, all situated on the marshy shores of the lakes. The old familiar

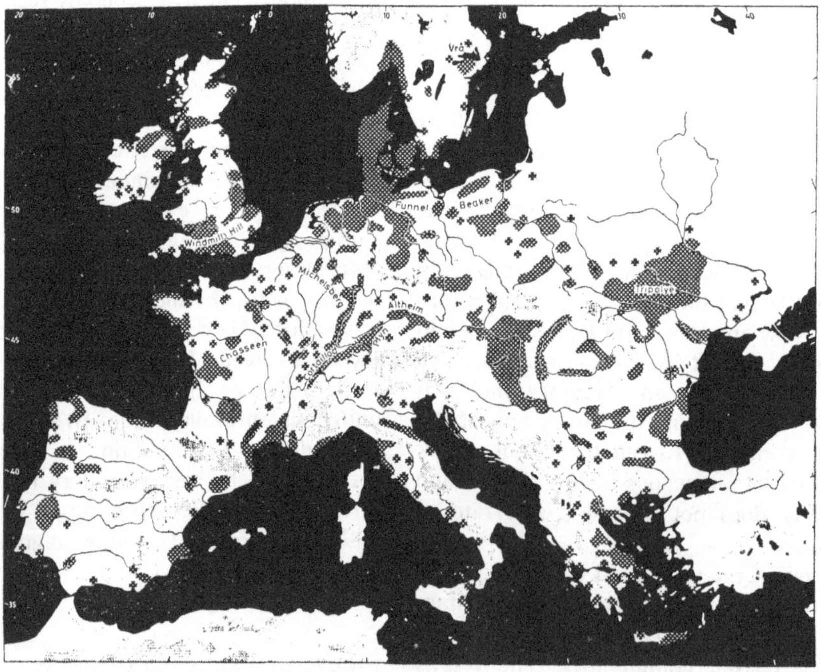

FIGURE 35.5
Distribution of farming cultures (cross-hatched) and hunting-collecting cultures about 3000 B.C.

"Swiss lake dwelling" sites are now known to have been built on the shores of lakes or bogs, but not built out over the water. Cereal grains are common in most of the settlements, but collecting of wild fruits and berries was practiced as well. Locally, fishing remained important. Hunting was a major occupation; in some sites, bones of wild animals dominated over those of domesticated animals. The pottery shows a great diversity in form and ornament. West of Lake Zürich the Younger Cortaillod culture which belonged to the Western tradition developed; east of this lake the Pfyn culture of clearly Danubian affinity developed. This diversity can best be explained by assuming that the hunting and collecting communities in this area had undergone influences from both the west and the east. Two other groups further eastward have the same preference for lakeshores: Schussenried (South Germany) and Mondsee (Austria).

In the North German-Danish area the Ertebolle culture originated at about 3800 B.C. as a successor of both the early Atlantic Vedbaek culture (on the Danish islands) and the Oldesloe culture (a lakeshore, inland hunting and collecting culture in Schleswig-Holstein of Atlantic age). The Ertebolle culture not only combines coastal hunting with inland hunting, fishing, and collecting, it adds new elements: coastal collecting of oysters and other shellfish, and, on a restricted scale, grain growing and cattle raising. Settlements with the same type of flint inventory occur along the coast, on inland high ground, and in freshwater bogs. This situation was maintained for half a millennium; it is not until about 3000 B.C. or even later that food production gained dominance.

The origin of the well-known Ertebolle pottery has been much debated. The Rossen culture certainly influenced the Ertebolle culture, but the pottery could have come from western Europe. In view of the long range of the connections along the Mediterranean and Atlantic coasts it is not surprising that influences should have reached as far as Denmark. The spread of the megalithic "religion" over this same area, soon after the establishment of village farming is another illustration of the strength of this Western diffusion sphere.

The fully food-producing culture of southern Scandinavia is known as the TRB culture. It is probable that the population stock of this culture is of Ertebolle origin, that elements of both the Western and Danubian tradition have been incorporated and transformed, and that an expansion of population had taken place after the establishment of food production. Expansion went northward to south Norway, westward to northwest Germany and Holland, and eastward to Poland. Just as the megaliths illustrate the persistence of the Western tradition in the TRB culture, copper imports from the Hungarian plain, knob-ended battle-axes (*Knaufhammeraxte*), and long houses illustrate the long-distance contacts with the Danubian tradition.

In France the main food-producing culture is the Chasseen, of which Bailloud distinguishes four regional varieties. The southern variety could have developed out of the Cardial. In Brittany the Chasseen is characterized by megalithic grave monuments of clearly Iberian affinity, and by polished axes made of locally quarried greenstone, which were exported over wide areas in France. One megalithic tomb was found on top of a shell midden with a carbon-14 date of 4000 B.C. (La Torche). The presence of goat in the shell middens of Téviec and Hoedic is another indication that the coastal culture of Brittany could have played a part in the development of the western variety of the Chasseen. The eastern variety occurs in Savoy and Burgundy; it is hardly different from the younger Cortaillod culture of Switzerland, and has the same preference for lakeshores.

In the Paris basin, the Chasseen is preceded by the *Groupement de Cerny* which Bailloud thinks followed the late variety of Bandkeramik in this area, and thus belonged to the Danubian tradition. It occurs far to the west and seems to reach the coast of the North Sea. It should be a contemporary of the Rossen culture. In this area the western influences seem to have become more important than the Danubian. The northern variety of the Chasseen is the only one in which the polished flint axe, which it has in common with the Funnel Beaker culture, is known.

For their axes and adzes the Bandkeramik and Rossen cultures depended largely on an amphibolite which occurs naturally in Silesia, Rossen-type adzes are lightly distributed over large parts of the plains of northwestern Europe. It is clear, however, that for a large-scale expansion of farming, a local source of material for axes and adzes would be essential. In the western Baltic area, suitable unweathered flint occurs abundantly both as boulders in cliffs and in the chalk. Here large flint tools had been in use for many millennia and the step toward polishing was a small one.

In north France and Belgium crude macrolithic flint assemblages are known as the "Campignian." Most investigators now agree that it is the flint exploitation "facies" of food-producing cultures (Chasseen, Michelsberg). Both at Spiennes (Belgium) and at Rijckholt (Netherlands), flint mines were in operation before 3000 B.C.

In the British Isles early carbon-14 dates for food-producing cultures have been obtained along the Irish Sea. In this area coastal settlements of Atlantic age are present (Larnian, Obanian) which have yielded evidence for the gradual introduction of elements of food production (polished axes, sheep, or goat). Some of the major stone axe factories are situated in the same area (Tievebulliagh in Ulster, Langdale in the Lake District, Graig Lwyd in North Wales, and Cornwall), and these sent their products as far as Wessex, although flint was available there.

It is therefore possible that the Early Neolithic culture of southern Britain was not entirely due to an invasion from France, but had its origin in the expansion of an Irish Sea population which had gradually acquired the elements of an agricultural economy through its coastal contacts.

In the transitional zone between the Western and Danubian spheres, there is another group whose origin has been much debated —the Michelsberg culture, with its main distribution along the upper Rhine with some outliers as far as Bohemia and Belgium; its pottery occurs in the Spiennes flint-mining area. Radiocarbon dates from Spiennes, Ehrenstein, and Thayngen suggest an age between about 3400 and 2900 B.C. The Michelsberg culture is probably a local successor to the Rossen culture, and its population stock is therefore of ultimate Danubian origin, but it has incorporated various western elements. The relation of Michelsberg with TRB, Altheim, Baalberg, and other contemporary groups north of the Alps need not be explained by one common origin of these groups, but could be due to a phenomenon of acculturation of groups of heterogeneous origin.

By about 3000 B.C. the map of Europe had been filled with a complicated pattern of food-producing cultures of the village-farming type, each with a certain individuality. This individuality seems to have been the result of both environmental and historical factors. The environments called for adaptation to local circumstances of soil, relief, climate. But historical factors, such as migration, diffusion, acculturation, and local tradition were important in establishing boundaries and relations. It will take a long time to unravel the complicated processes which have led to the formation of these cultures.

FINAL CONSIDERATIONS

The major factors in the spread of food production over Europe seem to have been the following. (i) The presence of the nuclear area in southwestern Asia, from which the first fully food-producing groups migrated into Greece, and from which ideas were diffused along the Mediterranean and Atlantic coasts. (ii) The coastal adaptation of the descendants of the upper Paleolithic hunters and collectors. This adaptation was gradual and progressive, and the western Baltic area demonstrates that at least in that area it was provoked by successive environmental changes. Along the Mediterranean coasts such factors had probably been operating too, but at an earlier date. Coastal adaptation meant a greater variety in means of subsistence and, by its very nature, a greater residential stability —both preconditions for the acceptance of the farming way of life. (iii) The successive building up in southeastern Europe of potentials for the expansion of population over large areas. In this process adaptation to a fully forested environment was an essential element. The first wave of expansion comprised large parts of the Hungarian plain (Starcevo-Koros), the second, the loess areas northwest, north, and northeast of the Alpine and Carpathian mountain chains (Bandkeramik). (iv) The slowness in turning to food production exhibited by the sessile coastal and inland groups, doubtless because of the effectiveness of their mode of life. Mixed economies resulted in which agriculture and husbandry played a subordinate role (for example, Ertebolle). (v) The rapid colonization of the coastal plains after food production had gained dominance and a potential for population expansion had been built up. (vi) The interaction of the continental (Danubian) and coastal (Western) movements in a broad zone of contact between the Alps and the North Sea, where cultures with a mixed Danubian and Western character originated.

There is no proof in Europe for the presence of a level of incipient food production in Braidwood's sense. Any domestication of indigenous cattle and pigs ensued after the introduction of domesticated animals in the area. It has been alleged that in the south of France wild sheep were domesticated in the Atlantic period, but this is contradicted by the natural distribution of the species in Asia and by its absence among the animals that were hunted in earlier periods. Only the apparent strong specialization of the Ahrensburg people on the reindeer might have had aspects of domestication, in the same way as with some recent arctic peoples. But with the disappearance of tundra conditions this could not have had any lasting effect, and the successors of the Ahrensburgians were hunters.

There is also no botanical proof for food production before the introduction of wheat and barley from southwestern Asia. Still, there is the hazel, the fruits of which are extremely common in some culture layers, and the postglacial spread of which was so rapid that Firbas has suggested that man played a part in this spread. Of course, it would be easy to promote hazel growth by cutting down other shrubs. But man would only do so if he would be sure of coming back regularly. This might be the case in the Maglemosian area. Here seasonal lakeside settlements are known, where man specialized in fishing. In this rich environment, wanderings were certainly less necessary than in other areas. One might even see the core and flake axes as useful tools for such a purpose. It is doubtful, however, whether we shall ever be able to prove anything like the cultivation of hazel, and it must be said that such a contrivance had no decisive effect, for the addition of coastal collecting was necessary to overcome the effect of the Boreal-Atlantic environmental change.

The development of prehistoric culture in Europe does indeed agree in broad outline and many details with the nonnuclear development in eastern North America. The increasing efficiency in the exploitation of natural resources, the trend toward residential stability, the importance of the shellfish economies, the development of art, the disposal of goods to the dead in the later stages, are all features of the nonnuclear cultures of the eastern United States and southeastern Canada which we also find in northwestern Europe. Admittedly, we do not know a European counterpart for ceremonial centers and earthworks, as there were with the Hopewellian and Adena peoples before food production gained dominance. But the slowness to add animal husbandry and grain growing to an already effective food economy is again a feature which both areas have in common. This cultural slowness—Caldwell calls it resistance—is the main reason why it took as much as 3000 years for village farming to travel by way of diffusion from the Aegean to the North Sea along the coastal route.

The shorter continental route was not much faster, for the obvious reasons that it was barred by mountain chains and depended only on expansion of population and migration. But even so, by 4400 B.C. fully food-producing communities had reached the edge of the north German lowland plain at a distance of 300 kilometers from the heart of the Ertebolle area. A distance of 2000 kilometers—as the crow flies—had already been covered in the same time which thereafter was necessary to cover these last 300 lowland kilometers and to establish fully effective village farming in the Ertebolle area and is surroundings.

SUMMARY

Against the background of the postglacial climatic development and the major physiographic features of Europe, a description is given of the spread of food production over the continent. After the immigration of the first farmers into Greece from Anatolia, adaptation to European forest conditions took place and potentials for successive population expansions were built up. Large parts of continental Europe appear to have been uninhabited at the time immediately preceding the immigration of farmers. The reason was probably that the forest had become unsuitable for human occupation after the deciduous species had gained dominance. Along the Mediterranean and Atlantic coasts, hunting and collecting communities had been able to maintain themselves by adding sea hunting and fishing and shore collecting of shells to the traditional inland hunting, fishing, and collecting.

These communities were sufficiently sessile to be receptive to new ideas on food production that arrived successively by way of diffusion along the coasts. They became centers of population expansion and cultural innovation after food production in the course of time had gained dominance. In a broad zone between the Alps and the North Sea the coastal (Western) and continental (Danubian) traditions interacted.

36. ECOLOGICAL ZONES AND ECONOMIC STAGES

GRAHAME CLARK

Reprinted (without figures and tables) from Prehistoric Europe: The Economic Basis *(London: Methuen & Co., 1952; Stanford, Calif.: Stanford University Press, 1952). Grahame Clark is Disney Professor of Archaeology at Cambridge University. He has worked on the interrelations between the economy and the ecosystem in prehistoric Europe with special reference to the Late-glacial/Post-glacial transition and to the spread of agriculture. He is the author of* Archaeology and Society, Excavations at Star Carr, World Prehistory, *and other books. He has been the editor of the* Proceedings of the Prehistoric Society *(Cambridge) since 1935.*

■ We continue our exploration of the relationship between habitat and culture in prehistoric Europe. If there is an intimate relationship between habitat and culture, it should be possible to find variations in culture that correspond to habitational variations in the same general area of the world. For example, if the European continent during prehistoric times exhibited a series of relatively distinct habitational zones, we should expect to find variations in culture that correspond to those zones.

In this selection Clark describes the culture and the habitat in prehistoric Europe in each of four major zones. It has long been known that there were distinct cultural areas in prehistoric Europe, each based on its own modes of food production, and Clark takes this knowledge a step further and correlates these cultural differences with clear-cut habitational differences. His analysis lays bare the adaptive significance of these cultural patterns.

In the course of his analysis Clark introduces us to several important concepts for the study of cultural adaptation. The first concept is that in the reciprocal relationships between man and nature there is a constant strain toward—an attempt to achieve—an equilibrium between man and nature. Habitats change, as in the climatic variations that were produced by the glacial retreats and melting of the European ice-sheets in prehistoric times. These changes produce a disequilibrium in man's relationship to his habitat, and this leads to a transitional period in an attempt to attain a new equilibrium.

Another important concept in the selection is that, in his attempt to achieve and maintain an equilibrium with his habitat and to free himself from its limitations, man must not only adapt to changes wrought by the milieu itself (such as climatic shifts), but also must cope with alterations in the habitat that he himself has produced. Man not only exploits habitats, he sometimes destroys important parts of them in the course of his adaptation, thereby changing them drastically. For example, when people become cultivators they may deforest a region. This constitutes a fundamental change in the nature of the habitat, and man must adapt to these alterations as well.

At the conclusion of his paper Clark turns to an important point that was raised by Waterbolk: the importance of habitational conditions in the diffusion or transmission of cultural elements from one society to another. If a civilization is to be built on an economic base, which in turn depends on the cultivation of specific crops, these foods must have the proper habitat in which to thrive. They cannot spread to other climatic zones under normal circumstances. But what are "normal" circumstances where man is involved? In the provocative last sentence of his essay Clark subtly and effectively captures the force by which imperialistic expansion—a purely cultural phenomenon—can supersede the limitations of habitational boundaries.

The history of adaptation in Europe is more fully discussed in *Ancient Europe: From the Beginnings of Agriculture to Classical Antiquity,* by Stuart Piggott (Chicago: Aldine, 1964). Still popular among many students of prehistoric cultures, although resting largely on outmoded theories and concepts, are the writings of V. Gordon Childe (without whose work, it should be noted, modern hypotheses might not have developed as rapidly as they did): *Man Makes Himself* (New York: New American Library, 1951), *What Happened in History* (Harmondsworth: Penguin Books, 1946), and *Social Evolution* (New York: H. Schuman, 1951). *Social Evolution* deals at length with European cultural developments. ■

ONE OF THE PRINCIPLE attractions of prehistory is the opportunity it offers for studying the interplay of social aspirations and environing nature over long periods of time. The economy of any community may be considered as an adjustment to specific physical and biological conditions of certain needs, capacities, aspirations and values. There are thus two sides to the equation—on the one hand the character of the habitat, itself to a greater or less degree influenced or even conditioned by culture, and on the other the kind of life regarded as appropriate by the community and the resources, in the form of knowledge, technical equipment and social organization, available for its realization. The relationship between man and external nature is thus a dynamic one and the development of culture viewed in its economic aspect is indeed one of man's growing knowledge of and control over forces external to himself. The history of man differs from that of any other species precisely in that it has been one of progressive emancipation from the thraldom of instinctive conformity with a pattern imposed by external forces: by every advance in his culture man has enlarged the sphere for the exercise of choice for good or for evil.

Yet it remains true that the economy of any community at any moment of time is necessarily the product of an adjustment between culture and environing nature. Were it possible to visit prehistoric Europe the chances are that one would find stable forms of economy prevailing over wide territories, the result of a more or less perfect adjustment between social appetites, technical capacity and organization and the resources of the several regions. Such a state of equilibrium and stability has at any rate often been observed among modern primitive peoples by anthropologists working in the field. For instance, Evans-Pritchard wrote of the Nuer, "Oecological relations appear to be in a state of equilibrium. As long as present relations exist cattle husbandry, horticulture, and fishing can be pursued but cannot be improved. Man holds his own in the struggle but does not advance." As the same authority has shown, the Nuer and their culture could well be considered as part of a mature ecosystem, itself the product of an interaction between biome (the whole complex of living organisms—plants, animals and men) and habitat (the soil and climate). The adjustment between the economic system of the Nuer and their external environment was so perfect that there was no room for any substantial improvement, so long as both these factors remained constant.

Such an equilibrium, with the economic and cultural stability which it implies, was no doubt the normal condition of primitive society. Yet it is equally certain that during prehistoric times there must have been phases of disequilibrium when the pattern of life changed, at times drastically and often quite rapidly. Such periods of change and their underlying causes are obviously a prime concern of economic history, though one should not forget that the tendency must always have been to achieve a new harmony between society and its external environment, so that each phase of disequilibrium invited a fresh adjustment.

The main factors involved in the economic changes whereby the European peoples were ultimately able to reach the stage of recording their own history may be grouped under three main heads. First one should reckon with alterations in the natural environment brought about by processes external to man, changes which were inherent in the natural order of events and over which mankind had no sort of influence, but which enlarged or contracted the range of opportunity open to human societies. The most widespread and comprehensive of these were the climatic changes which marked the transition from late glacial to recent times and are most dramatically exemplified by the retreat of the pleistocene glaciers, by the alterations in land and sea levels due to the melting of the ice-sheets, and by the profound ecological transformations involved in the colonization of open landscape by forest trees and in the successive alterations in the composition of forests brought about by the optimum post-glacial climate and the ensuing deterioration.

Next, and more positively, may be cited changes in the needs and requirements of human societies brought about by a wide range of factors. Some of these, for example the pressure of population on available food supply, were of a purely economic order. Others were primarily cultural, among the most important being the effects of contact between the barbarians of prehistoric Europe and the civilized peoples of the east Mediterranean area. It is indeed axiomatic that the evolution of economic life in prehistoric Europe can only be understood if full account is taken of the influence of the higher forms of economy associated with urban civilization, at first in western Asia and the Nile Valley and later in the Mediterranean. In general it may be urged that once development in any region had gone far enough to give rise to the zoning of societies at varying levels of economy, the mounting needs of the more highly integrated ones affected to an ever-growing degree the history of all the simpler ones within their spheres of influence. So soon as a zoning of societies had been brought about through the attainment of forms of economy more advanced than those based on hunting, fishing and collecting, there was set in motion a process of expansion, whereby the more highly developed spread over the territories of the weaker, less developed communities: farming spread more and more widely over territories formerly given over to food-gathering until it approximated to the limits set by geographical and specifically ecological conditions, and each expansion of the agricultural platform gave more room for enlargements of the urban superstructure. The influence of higher on lower economies was by no means limited to the territories actually incorporated, but through the mechanism of trade spread far beyond these, disturbing the economic—and often the ecological—balance even of distant communities.

Lastly, there are the changes in external environment brought about directly or indirectly by human activities. As a dominant species man must always have affected the balance of nature to some degree, but his importance as an ecological factor grew as he increased in numbers and extended his control over external forces, until he became a principal agent of botanical and zoological change. Each advance in culture increased the ecological dominance of man: the more effectively he was able to intervene in natural processes, the more frequently he must have disturbed by his own activities the adjustment with external nature of his own culture, and the more often in consequence he had to modify his economy to meet the new conditions created by his own exertions. Perhaps the best example of this is deforestation, which is capable not only of altering the character of vegetation and animal life over extensive areas, but also of causing soil-erosion and the formation of mosquito-breeding deltas and even of affecting local climate. Whether as hunter or farmer, the effect of his activities was inevitably to disturb existing relationships between various forms of wild life and between human society and external nature.

The relationships between human society and environing nature were thus reciprocal. The idea that human progress is no more than a product of the inherent dynamism of external nature has only to be formulated for its absurdity to be evident to any whose business it is to observe and classify the manifold diversities of human society. Yet it would be equally wrong, especially when considering societies with a comparatively weak cultural endowment, to minimize the importance of the ecological framework within which all forms of life subsist. As Godwin has recently remarked:

> There is no break in kind between the relationships of the great eco-systems of plant and animal communities, and those which include fewer or more human beings. . . . Since . . . the human agents remain animal in nature, sustain themselves within communities, and subsist on biological materials, they are no less part of the eco-system than were their ancestors.

There is no more faithful indication of

regional differences of soil and climate—and few which define more accurately the limits of economic zones at different stages of prehistory—than vegetation. Vegetation is inevitably the main element in the biome of any terrestrial eco-system, since plants "alone have the power of making organic substance from inorganic, of building up living substances from materials like carbon dioxide, water and mineral salts." All other forms of life, including men, even of the most advanced culture, must needs depend for subsistence, directly or indirectly, on vegetation, whether they eat plants or the animals that feed upon them. Since the quest for food is the central fact of economic life, it follows that plant ecology and its development must be of the utmost relevance to economic history.

Geographers have distinguished four main plant-formations which extend at the present day in broad belts across the European continent. From the north to south these comprise:

1. A *treeless zone,* with an arctic or sub-arctic climate too severe to allow the growth of more than tundra or alpine vegetation and comprising: (a) a belt of circumpolar tundra extending across the northernmost territories of Scandinavia and Russia, (b) a contiguous tract of *fjäll* in the Scandinavian mountains, and (c) isolated alpine pockets above the limit of tree-growth in the Alps and Pyrenees and on a smaller scale on the mountain peaks of the British Isles.

2. *Northern coniferous forest,* comprising pine, spruce and silver fir, associated with such trees as birch and willow, capable of flourishing in a moist, cold climate. Although strongly influenced at the present day by scientific forestry, the northern coniferous forest has not suffered deforestation in connection with agriculture or stock-raising to anything like the same extent as the deciduous forest. It occupies a broad belt, including much of the Scandinavian peninsula, almost the whole of Finland, and most of European Russia between latitude 60°N. and the circumpolar tundra.

3. *Deciduous summer forest,* adapted to a temperate, sub-oceanic climate, and comprising broad-leaved trees, such as oak, elm, lime, alder and beech, which have their period of growth in summer and shed their leaves in autumn. Immense inroads have been made on the deciduous forest, which once occupied a broad zone, defined on the north by the northern coniferous forest and on the south by the Mediterranean evergreens. Today only small patches of deciduous forest survive in what has been transformed into a zone of farmlands interspersed by industrial and urban development.

4. *Mediterranean evergreen forest,* the outcome of a climate of summer drought and winter rains, retains foliage through the winter in order to make growth and is adapted by deep roots and leaves shaped to reduce evaporation to survive the period of enforced rest during the summer heat. At its climax the formation, which is found on the coastal plains and lower mountain slopes of the Mediterranean, was a real forest, though never so dense as those of the temperate and north temperate zones, and comprised such trees as ilex, cypress and olive. Today it mainly survives in forms to which it has been degraded by human activity, namely maquis (scrub of myrtle, juniper, broom, arbutus, oleander, vine, etc.) and garrigue (including aromatic plants like lavender, thyme and sage, and various bulbs).

Since the final phase of the Pleistocene ice age, the limits of these major ecological zones have shifted within wide limits. Exact and comprehensive information on this point can only be available when the method of pollen-analysis has been applied to the total range of pollen-liberating vegetation over the whole continent and in particular over the pivotal area of France, much of which as regards the late glacial and early post-glacial periods remains virtually unexplored from a palaeo-botanical point of view. Yet, provided the limitations are frankly recognized, there is everything to be said for making the attempt to visualize in its broad outlines the course of ecological change. The main sequence of events was determined by climate in which the principal factor was temperature: this rose from the low levels prevailing during most of late glacial times to a peak, constituting the so-called climatic optimum of post-glacial times, from which it fell away during the climatic deterioration marking the closing stages of the prehistoric period. In terms of vegetation this sequence implied, so far as the existing deciduous zone is concerned, a phase of tundra, followed by increasing forest dominance and ultimately by forest revertence and decline. In so far as human activities affected the main course of ecological development during prehistoric times—and they were not sufficient to do so until the third and last phase—they served mainly to speed up and amplify the process of deforestation and to enlarge the areas of open country, whether heath, meadow or cultivated land.

In reviewing the main course of events one should naturally begin with the late glacial period. At the time of the Fenno-Scandian moraines, the retreat from which marks the beginning of the post-glacial period, the ice-sheets, although much reduced in size, still covered a substantial part of the continent including the whole of the zones occupied at the present day by the tundra and northern coniferous formations and even in places the margins of the existing deciduous zone.

Knowledge of the development of vegetation in the ice-free areas of Europe rests to a large extent on material from the beds of late glacial lakes. Wherever these have been investigated over the extensive region from Ireland to East Prussia and from Scandinavia to the Alps the same general sequence has been found to apply, though this naturally varies in detail according to the climate and topographical features of different territories. The story is best illustrated by reference to the well-explored region of Denmark and Schleswig-Holstein. The late glacial period was interrupted in this area by two relatively warm oscillations, the second of which, first noted by Harz and Milthers in 1901 at the Allerod brickworks in north Zealand, has left well-defined traces over the whole of northwestern Europe.

During the cold phases, marked by tundra vegetation with *Dryas octopetala,* arctic birch and dwarf willow, there were no forest trees. Recent work has shown, though, that the vegetation of the Older Dryas periods was more alpine than arctic in character; in particular ericaceous plants were less common and grasses grew more luxuriantly than in the arctic tundra, to which moreover the plant community typified by *Helianthemum, Hippophae* and *Artemisia,* commonly noted in the Older Dryas deposits, is quite foreign. For certain types of herbivorous animals, and therefore for the human societies which preyed upon them, the late glacial vegetation of the zone beyond the ice-sheets provided ideal conditions of life. During the warmer Allerod stage, forest trees began to encroach on the open tundra: even in marginal territories like Ireland, southwest Norway and Denmark birch forests began to establish themselves; further south pine trees spread into the birch woods, even penetrating as far north as Holstein. The Younger Dryas period witnessed a substantial return of open tundra conditions, though birch copses may have survived even in Ireland in sheltered localities. Much of northwestern Europe seems to have presented at the close of the ice age the appearance of "park tundra," like the zone of transition between tundra and forest which exists today in east Russia and Siberia to a depth of several hundred kilometres. The boundary between the forest and park tundra zones would be difficult enough to draw even supposing that the necessary palaeobotanical research had been carried out in France, and it may be assumed to have fluctuated considerably in response to climatic change even during the late glacial period. For much of this time it is considered that continental forest extended over southwestern France, including the richest areas of upper palaeolithic culture. On the other hand there are hints that during the cold spell at the end of late glacial times the Pyrenees may have marked the boundary; at least it is significant that the numerous antlers of reindeer from Late Magdalenian levels as far south as Isturitz have been referred explicitly to the "barren-ground" rather than to the woodland type.

It would appear that towards the end of the

glacial period the areas at present occupied by tundra and northern coniferous forest were still covered by ice-sheets and that the existing zone of deciduous forest then consisted of tundra grading into park tundra, in which the forest element temporarily rose to importance during the warm Allerod oscillation. Beyond the Alps and the Pyrenees forest conditions prevailed, and it may be surmised, thought it cannot yet be fully demonstrated, unless from the evidence of fauna, that coniferous and deciduous forests encroached on territories at present occupied by the Mediterranean evergreen formations.

Now just as the temporary increase in warmth during late glacial times, known as the Allerod oscillation, led to an abortive spread of forest over much of the ice-free zones of northwestern and central Europe, so the much more marked and enduring rise of temperature during the postglacial period had commensurately greater ecological effects. Ice-sheets retreated to their mountain refuges, leaving behind them extensive regions for colonization, first by tundra and alpine vegetation and ultimately, apart from the most northerly latitudes and from mountain heights, by birch and coniferous forest. The former area of tundra and park tundra, as far south as the Alps and the Pyrenees, passed through a series of ecological transformations. At first, during the period of Preboreal climate, while the rise of temperature was still only slight, the forests which began to consolidate themselves were restricted to trees capable of tolerating cold conditions, including willow, birch and pine; indeed, in marginal territories like Ireland and northern Britain, only the first two species were represented. During the Early Boreal climatic period, the forests responded to increasing warmth; pine increased at the expense of birch and even spread to Ireland, though still only in small proportion, and the first deciduous trees, including oak, elm and lime began to infiltrate. In Later Boreal times the deciduous trees, increased in importance, though still subservient to pine, but the most striking feature was the abundance of hazel over the whole territory from Ireland, where its pollen reached proportions five, twelve or even seventeen times that of all the forest trees added together to Esthonia, Transylvania and the Pyrenees. The climax came during the warm, moist Atlantic stage, the period of the post-glacial climatic optimum, when mean annual temperatures in this area reached between 2°-2½ °C. above those now prevailing. The deciduous forests, in which alder was often prominent at this period and into which the beech was beginning to spread from southern France round the western foothills of the Alps, attained their maximum expanison and actually spread into territories such as parts of central Sweden, from which they have since retreated. While the coniferous and deciduous formations were attaining and surpassing their existing areas of distribution the Mediterranean evergreen forest was presumably also advancing to and possibly beyond its present frontiers, though there is little exact information from this zone.

To sum up the situation at the peak of the warm period, it can be said that the major plant formations had occupied their present territories and that the coniferous, deciduous and evergreen forests had even advanced beyond their modern limits. Over the whole territory up to the margins of the tundra, excepting only the windswept islands like the Orkneys, the forests, encouraged by a warm, moist climate and as yet hardly touched by man, spread like a continuous blanket, parted only by rivers and lakes, marshes, sand-dunes, salt-springs and the very sea coasts.

From this point the forest—or at least those of the deciduous and evergreen formations—entered on a period of decline, which led during the historical period to their virtual extermination. This ecological transformation, of such vast significance for the human population of Europe, was the product of climatic and economic forces working together. The spread of farming economy, which first affected Europe at a time when the forests were at their peak, involved inroads on the woodlands which grew as settlement became more dense: not only were trees felled to provide timber for houses, boats, defensive works, fuel and other purposes, but ever-growing tracts of forest were cleared for cultivation (at first temporarily, but later permanently) and domestic livestock grazing on the seedlings and young shoots effectively prevented regeneration. The effects of human interference were particularly catastrophic on the Mediterranean evergreen formation, which at its climax of high forest was notoriously fragile and had but small powers of recuperation; although once "densely forested from snow-line almost to sea-level" the spread of farming soon had the effect of degrading evergreen forest into maquis and garrigue, leaving as the only substantial reserves of timber the deciduous and coniferous forests of the upper mountain slopes. Although the temperate summer forest of northwestern and central Europe had greater powers of resistance, the introduction and above all the intensification of farming exerted a profound influence on the biome.

The most significant feature of the vegetation of the Sub-boreal period was the gradual, if at first very slight, replacement of forest by grasslands and heaths. This development, which was particularly noticeable on the poor sandy soils of northern and northwestern Germany, has commonly been attributed to the onset of a drier, more continental phase of climate, relatively unfavourable to forest growth. Although it may be questioned whether the climate of Sub-boreal times was drier than that of today, the consensus of opinion holds that it was more substantially so than that of Atlantic or Sub-atlantic times. On the other hand, the influence of climatic change on vegetation was certainly enhanced by the effects of farming, for which direct botanical evidence is forthcoming in the pollen of cereals and of the weeds associated with their cultivation.

The influence of human interference on the natural vegetation of the deciduous zone became more and more marked as the area of primary settlement filled up and agriculture entered upon its settled stage. It was precisely at this time that northwestern and central Europe were overtaken by a sustained deterioration of climate. Already during Sub-boreal times there had been recurrent though quite temporary phases of heavy rainfall, but it was not until the Sub-atlantic period that there set in a prolonged phase of heavy rainfall and lowered temperatures. The combined effect of climatic deterioration and of the ever-increasing inroads by man revealed itself clearly in the vegetation: nor only did birch trees and conifers—the latter now including spruce and silver fir, which already during Atlantic times had begun to spread from their respective refuges in southeastern and southwestern Europe—regain some importance in the still extensive forests, but the forest area showed further contraction and blanket-bogs spread over the substantial territories subject to excessive rainfall, like the western parts of Ireland and of the Scottish highlands.

Although the major transformations in plant ecology since late glacial times have continually to be borne in mind, it is equally true that the evolution of European economy during prehistoric times can only be understood in relation to that of the ancient world as a whole. Much remains to be learnt about the origins of farming and the rise of urban civilization in the Near East, but the general position has been established firmly enough and is sufficiently familiar. Only two salient facts need be recalled, namely that in parts of western Asia and the Nile Valley farming was already practised by the fifth and possibly even by the sixth millennium B.C. and that, in both, urban societies, developed in the course of long ages from peasant communities, had already begun to record their own history before even a grain of wheat had been sown in European soil: Europe was so marginal to the main orginative centres of economic change that, Crete excepted, no part of it came within the sphere of neolithic farming economy until literate civilization had already been established in parts of western Asia and the Nile Valley.

From an economic point of view the early history of Europe fell into three main stages: from an age-long and worldwide stage of savagery, under which subsistence was won exclusively by such activities as plant-gathering, hunting, fishing and fowling, it passed by a process of incorporation into the sphere of successively more highly integrated economies, through a stage of barbarism, in which the activities of farming had been added to those of the quest for wild products, to one of urban civilization. Naturally one can trace finer divisions than these: in the field of food-production it is possible to distinguish clearly, at least in the deciduous zone, between the stage of shifting agriculture and that of settled agriculture with fixed fields; or, again, in that of technology, one can point to the transformations wrought by the diffusion from civilized to barbarous societies of copper and bronze metallurgy or iron-working respectively. Yet, the spread of farming, and of the urban civilization based upon it, remain the two chief turning-points in the story and the general pattern of social life in the different parts of Europe during later prehistoric times has been determined more by their relationship to the expansion of these economic zones than by any other factor.

The progress of Europe was therefore in broad essentials a consequence of the expansion of economic zones in the ancient world as a whole. The main driving forces were human needs and aspirations, which in so far as they were economic were inherent in existing forms. The spread of farming was accomplished partly through the gradual appropriation of land by marginal peasants who practised in association with stock-raising an extensive form of agriculture necessitating the continual opening up of new tracts; and partly through the upgrading of indigenous hunter-fishers by a process of acculturation, a process in the course of which higher groups obtained certain economic advantages without the necessity of force and simpler ones assuaged their feelings of inferiority by emulating their more advanced neighbours and adopting in a devolved form their mode of life. The expansion of urban economy was much more deliberate and was normally accompanied by the extension of political authority; yet at bottom the driving impulses were similar and the actual foundation of cities was preceded by a prolonged period of cultural and economic permeation, so that from a comparatively early stage in their history the barbarian communities of Europe were stimulated by the demands of urban societies operating through the medium of trade.

Yet the mere concept of geographical expansion should remind us that the spread of more advanced forms of economy was limited to some degree by the character of the climate, soil and vegetation of the several zones into which Europe has been divided by nature. Thus, the two most northerly zones, those of tundra and northern coniferous forest, were neither of them capable under primitive conditions of supporting societies at any other than a stage of savagery. The frozen sub-soil and arctic-alpine climate of the tundra-*fjall* regions were—and still are—impassable bars to the practice of agriculture, though not of course to reindeer nomadism. The northern coniferous zone was hardly less favourable: the climate, which was too cold and moist for trees of the oak mixed forest, was no more suited at an early stage in the evolution of agricultural practice to the growth and ripening of cereal crops; equally unpropitious were the prevailing podsols, product of a cool, wet climate, under which the downward movement of water exceeded the upward, leading to extreme leaching and eluviation and to the formation of a hard moorpan and an impoverished topsoil; nor was the coniferous forest itself, with its lack of herbaceous undergrowth and its unpalatable foliage, of much use for the grazing of domestic livestock. At best, therefore, the coniferous forest was of marginal value from the point of view of agriculturalists or stock-raisers. By contrast, conditions prevailing in the deciduous and evergreen zones, although differing in some notable respects and subject in the former case to important qualifications as regards the heavier clay soils, were well suited to the spread of primitive farming economy.

A study of prehistoric settlement shows that in point of fact the sphere of farming economy never expanded northwards of the range of the deciduous forest. In this connection it is worth noting that it was only comparatively late in post-glacial times that deciduous replaced coniferous forest over much of northwestern and central Europe and it was not until an advanced stage of

Atlantic times that neolithic farming economy spread over these territories formerly unoccupied or sparsely populated by mesolithic hunter-fishers. Although on the conventional chronology this may seem an academic point, it is worth noting that the ecological circumstances requisite of farming existed much earlier in parts of the Near East and that beyond the northern limits of the deciduous forest they ceased to obtain at all. By the time farming began to spread into Europe the main ecological zones had attained and slightly exceeded their existing northern limits. The new economy spread close up to these within the course of only a few centuries at the turn of the third and second millennia B.C.

Striking confirmation of the role of the northern margin of the deciduous forest as an economic and cultural divide comes from Scandinavia. Here barbarian and savage communities co-existed during the last two thousand years or so B.C. The existence of two distinct provinces has been recognized by Scandinavian archaelogists for over a hundred years, ever since Sven Nilsson published his famous book on the primitive inhabitants of the area. The admirable maps prepared by Oluf Rygh and Oscar Montelius for the Stockholm Congress of 1874 demonstrated that the distributions of the slate artifacts of what was already termed the Arctic culture and of megalithic tombs and stone cists were complementary, and gained acceptance for the view that the later stone age cultures of northern and southern Scandinavia were at least in some measure contemporary. Investigation of the dwelling-places and their associated fauna and of the rock-engravings emphasized the contrast and showed that the differences were economic and conceptual as well as merely formal. As Gutorm Gjessing has brought out so clearly, the Arctic culture of northern and central Scandinavia is an integral part of an immense circumpolar spread, embracing hunter-fisher communities right across northern Eurasia to the northern territories of the New World; whereas the farmer cultures of southern Scandinavia shared in the heritage of central and southern Europe. What no one has yet pointed out, though, is that the distinction between the two provinces is ecological as well as economic and cultural and that the boundary between them coincides with that which demarcates the deciduous from the coniferous forest. That this was so can be seen by comparing the distribution of features characteristic of each of the two zones.

It may first be shown that the early farmers occupied territories within the deciduous forest. Thus, if one plots the distribution of the battle-axe culture in Sweden, it will be found that this broadly coincides with the zone over which the mixed oak forest has been shown by palaeo-botanists to have extended during the Sub-boreal period, while that of megalithic tombs falls well within this limit. Again, in Norway the original colonization by farmers, defined by thin-butted axes of flint and stone, was confined to the deciduous forest zone of the Oslo fjord region, the great valleys into the interior, the coastal strip of southern Norway and the Jaeren region of the southwest; and the later expansion of farming, which was especially marked at the time of the stone cists, was confined to low-lying patches on the islands and along the western coast as far as the Trondheim region and even beyond, which thanks to the influence of the Gulf Stream once supported small areas of deciduous forest. In Finland, also, one finds that apart from a few strays, due no doubt to trade with the trappers of the interior and of the northern territories, finds of objects attributable to prehistoric farmers, whether implements of south Scandinavian flint, battle-axes or bronzes, were virtually confined to the narrow zone of deciduous forest in the south and southwest; indeed, down to the end of the prehistoric Iron Age, around 1300 A.D., the territories occupied by farmer cultures had hardly extended beyond this zone unless in certain coastal areas.

If on the other hand the distribution of characteristic Arctic traits, such as animal-headed slate knives or wooden skis and sledge-runners, is examined, it will be found that these coincide substantially with that of the coniferous birch-forest and the open tundra.

Now, making all allowances for the fact that there was in reality no rigid divide, either between the ecological areas or between the spheres of the two economies, and that the clarity of the picture is impaired in the case of archaeological distributions by the effect of trade, the coincidence of economic and ecological zones is sufficiently marked to justify the hypothesis that the northern margin of the deciduous forest in fact determined the limits of the early spread of farming in the countries of northern Europe.

In conclusion, it is fitting to recall that, as Sir John Myres has taught us, the urban civilization of the Mediterranean represents a remarkably complete and perfect exploitation of the special features of Mediterranean vegetation. Although the primary, barbarian economy, on which several distinctive civilizations were reared, was based on the cultivation of cereals, helped out by fishing and the keeping of livestock, the limited area of cultivable land and the poor conditions for grazing livestock at low altitudes, once the forest had been degraded, set close limits to the potentialities of ordinary mixed farming. The possibility of concentrating substantial numbers of people in cities and of supporting the fabric of civilized existence depended on the cultivation of olive, fig and vine, which in itself involved fixity of settlement. Although all three in their wild forms had been features of the Mediterranean flora since Tertiary times, it would of course be quite wrong to imagine that their domestication took place as it were spontaneously or sporadically at different points within the zone; on the contrary, the domesticated varieties accompanied, as they helped to support, the colonization of the middle and western Mediterranean by civilized peoples. What is true is that this process was immensely facilitated by the perfect adaptation of the three plants to the climate which determined the composition of the middle and western Mediterranean zone. If ecological conditions thus favored the spread of civilization from the eastern Mediterranean to Italy, the south coast of France and eastern Spain, they also served to limit its diffusion. It needed all the power of Rome to break through the boundaries established by ecology and to incorporate within the sphere of Empire a substantial area of deciduous forest, a feat which marks for us the conventional end to prehistoric times.

37. HABITAT, CULTURE, AND ARCHEOLOGY

GEORGE I. QUIMBY

Reprinted from Gertrude E. Dole and Robert L. Carneiro (Eds.), Essays in the Science of Culture in Honor of Leslie A. White *(New York: Thomas Y. Crowell Company, 1960). George I. Quimby is Professor of Anthropology at the University of Washington and Curator of Ethnology at the Thomas Burke Memorial Washington State Museum. He also is Research Associate in North American Archeology and Ethnology at the Field Museum, Chicago, where he was curator for more than twenty years. His special research interests are material culture, ethnohistory, archeology and ethnology of North America, human ecology, underwater anthropology, and historic archeology. He is the author of* Indian Life in the Upper Great Lakes Region *and* Indian Culture and European Trade Goods; *he is co-author of* Indians before Columbus *and* The Tchefuncte Culture: An Early Occupation of the Lower Mississippi Valley.

■ Archeological and ethnological hypotheses must go hand in hand. In this selection Quimby illustrates the procedure of deriving hypotheses about adaptation from living nomadic hunting and gathering societies that can be used to interpret the behavior of prehistoric foraging groups. Although he shows some of the ways in which habitat and culture are related to each other, the intent of his paper is methodical as well as substantive. In discussing the relationship between habitat and foragers' cultures, Quimby focuses on the following hypothesis. Given a society's technology at a particular time in history, the complexity of its social organization is determined by the habitat in which its people live. He bases his hypothesis on what is known about the Naskapi Indians of the Labrador Peninsula and the Kwakiuti of Vancouver Island (on the northwest coast of North America). The Naskapi are true nomadic hunters and gatherers; the Kwakiuti are sedentary fishermen and foragers.

Quimby begins his analysis of adaptation among the Naskapi with their material culture and then seeks to apply the same reasoning to some aspects of their social organization. His discussion of marriage and family organization recalls the hypothesis of Kummer and Kurt (Selection 12) that family organization is an adaptive response to food shortages.

Several very good survey works can be consulted by the reader who wishes to know more about the cultures of prehistoric foragers: *The Prehistory of Southern Africa,* by J. Desmond Clark (Harmondsworth: Penguin Books, 1959); *The Prehistory of East Africa,* by Sonia Cole (New York: Macmillan, 1963); *Social Life of Early Man* (edited by Sherwood L. Washburn [Chicago: Aldine, 1961]); *Prehistory and the Beginnings of Civilization,* by Jacquetta Hawkes and Sir Leonard Wooley (Volume 1 of the History of Mankind series sponsored by UNESCO [New York: Harper & Row, 1962]); *An Introduction to American Archaeology,* by Gordon R. Willey (Volume 1 of this work is entitled *North and Middle America* and Volume 2 is *South America* [Englewood Cliffs, N.J.: Prentice-Hall, 1966]); and—for an overview of world archeology—*Dawn of Civilization,* by Stuart Piggott (New York: McGraw-Hill, 1961). ■

IN THIS PAPER IT is my intention to present an hypothesis concerning the interaction of habitat and culture under certain specified conditions. My ultimate aim is to be able to predict culture from habitat within certain limits to be stated subsequently. Such prediction, although necessarily general, would be of considerable use in the reconstruction of the prehistory of the Upper Great Lakes region in pre-agricultural times, a long period extending back from about 500 B.C. or later to probably 10,000 B.C., for which the archaeological record is sparse.

By habitat I mean the physical environment of a region. Habitat includes the wild flora and fauna, climate and weather, soils, land forms, geological formations and the like. As used in this paper environment and habitat are the same.

Habitat may be relatively stable for long periods or it may change relatively rapidly. In the Upper Great Lakes region the environment has been more or less stable for the last 2,000 years, but during the preceding 10,000 years there were a number of radical environmental changes.

By culture I mean the same thing that Professor [Leslie] White does when he states:

> Culture consists of material objects—tools, utensils, ornaments, amulets, etc.—acts, beliefs, and attitudes that function in contexts characterized by symbolling (making and using symbols). It is an elaborate mechanism, an organization of exosomatic ways and means employed by a particular animal species, man, in the struggle for existence and survival.

Culture as a system may be divided into three aspects: technological, sociological, and ideological. "Social systems are functions of technologies; and philosophies express technological forces and reflect social systems." Thus all of culture is basically geared to the technological system.

The technological system, according to White, "is composed of the material, mechanical, physical, and chemical instruments, together with the techniques of their use, by means of which man, as an animal species, is articulated with his natural habitat." White also states that:

> Man as an animal species, and consequently culture as a whole, is dependent upon material, mechanical means of adjustment to the natural environment. Man must have food. He must be protected from the elements. He must defend himself from his enemies. These three things he must do if he is to continue to live, and these objectives are attained only by technological means. The technological system is therefore both primary and basic in importance; all human life and culture rest and depend upon it.

Culture may be relatively stable for long periods or it may change relatively rapidly. In past times there have been periods of thousands of years duration in which there was very little culture change whereas in more recent times the rate of culture change or development has accelerated tremendously. With regard to the development of culture, White says:

> Assuming the factor of habitat to be a constant, the degree of cultural development, measured in terms of amount of human need-serving goods and services produced per capita, is determined by the amount of energy harnessed per capita and by the efficiency of the technological means with which it is put to work.

For the purposes of this paper I would like to hold energy constant, and consider habitat a variable. I hold energy constant because in the pre-agricultural groups in which I am interested, only human energy, fire for heat and light, and possibly the flow of water to move boats are involved. There is no agriculture, animal husbandry, or harnessing of more elaborate kinds of energy. Under these conditions it is possible to state a corollary of White's formulation which is as follows: Assuming the factor of energy to be a constant, the degree of cultural development is determined by the habitat and the efficiency of the available technological means.

Since for purposes of this paper I am interested only in cultures based upon a hunting and fishing subsistence, cultures with essentially the same kind of technology, I wish also to hold technology constant. With the factors of energy and technology held constant, culture will vary with the habitat. A good example of this corollary lies in the comparison of Kwakiutl culture and Naskapi culture. Both have a subsistence based upon hunting and fishing as a primary part of their technology. Both utilized only human energy. Yet Kwakiutl culture is much more elaborate than that of Naskapi. The reason for this lies in the difference in habitats.

The habitat of the Kwakiutl is much superior to that of the Naskapi. The abundance and behavior of the animal food resources enables a sedentary mode of life with relatively large villages, a large population, the development of specialization, the segmenting of society, the acquisition of wealth, and the development of lineages and classes. Fish can be literally harvested, and the Kwakiutl thereby possess a culture at a level of development usually achieved only by farming peoples.

In contrast, the Naskapi, filled with fears of starvation, freezing, or drowning, live in an unfavorable habitat and eke a living from fishing and the pursuit of caribou. The Naskapi of necessity are nomadic, few in numbers, unspecialized, unsegmented, poor, and classless.

The habitat has placed a definite limit upon the degree of development of Naskapi culture. But a close look at Naskapi culture suggests that the habitat has not only limited the culture: it has been a factor in determining the form of the culture.

My hypothesis is as follows: Under certain conditions a hunting culture will develop to the limits of its habitat and in so doing the culture will acquire forms determined by the habitat. This is possible because, as White says, Man, as an animal species, is articulated with his natural habitat, and culture as a whole is dependent upon material, mechanical means of adjustment to the natural environment. The technological system, which is both primary and basic to human life and culture, is articulated with environment. Furthermore, the three basic functions of the technology—food, protection from the elements, and defense from enemies—are most closely articulated with the environment among hunting peoples.

A hunting weapon, for instance, must be made of materials provided by the habitat. The energy for production and use of the weapon is limited to human energy. The efficiency of the weapon can be increased only by development of skill in its use and by perfection of its form in relation to its function. The perfection of its form depends for the most part upon the inherent qualities of the materials used. It is one of my assumptions (based on long familiarity with material culture) that in a relatively stable environment tools and weapons are developed to a point of perfection or near perfection. It is a further assumption that the whole technology is similarly perfected among hunting peoples, given sufficient time in a relatively stable environment. The habitat may determine the need as well as the form, and the technology must respond or the culture will perish. Life in northern Labrador is absolutely impossible for people of a hunting culture unless they have some efficient means of traveling upon rivers and walking over deep snow in pursuit of animal life for food and clothing.

When the technology of a hunting culture is perfected in relation to its habitat it can be said to be in equilibrium with its environment. With a technology in such equilibrium, it is to be expected that the social system as a function of the technology and the ideological system as a reflection of the social system will also tend to be in equilibrium with the environment.

For such equilibrium to be achieved habitat must remain stable for a sufficient length of time. In an unchanging habitat a nomadic hunting culture achieves at least near equilibrium with the habitat and in so doing acquires or is forced to produce the forms of culture best suited for survival in the habitat. All of this suggests that, under the conditions stated, there may be only one ideal form for any given aspect of a hunting culture in a given habitat. For instance, given the winter habitat of the Central Eskimo in the Hudson's Bay region, there is only one material, snow, readily available for house building, and only one most efficient form such a house could take, namely the dome shape. It seems impossible to conceive of any other kind of house that fulfills the requirements of abundant building material everywhere present in the habitat and a form of shelter that provides the greatest amount of enclosed volume for the least amount of surface.

The Naskapi who live in the northern half of the Labrador Peninsula or New Quebec between the Atlantic Ocean and Hudson's Bay offer excellent illustrations of the articulation of habitat and culture in equilibrium. Here bands of hunters wander after bands of caribou upon which they depend for food, clothing, and shelter. The technology is tied to the habitat to such a degree that one could argue that all aspects of Naskapi technology are perfect or near perfect.

Take, for instance, the Naskapi house. This conical wigwam of poles covered with skin or birch-bark is made of readily available materials. Since over much of the habitat there is little or no soil, the butts of the poles rest against the ground and are braced by a ring of the boulders that abound in this glaciated area. Any house structure that involves posts or poles sunk into the ground is impossible over most of the northern Naskapi area in summer, and over all of it in winter; therefore the dome-shaped houses of the Montagnais and Cree are unsuitable. Given the condition that poles cannot be sunk into the ground and that the poles must be of dwarfed conifers, spruce, or fir, the conical wigwam provides the greatest volume for the least amount of surface cover. Other possible forms of shelter that could be readily constructed by wandering hunters in the Naskapi habitat such as the double lean-to are in one way or another not as efficient as the conical wigwam.

The same sort of situation obtains for snowshoes, toboggans, and canoes. All of these things are absolutely necessary for uncivilized human existence in the Naskapi habitat. And all seem to have reached a point of perfection beyond which they could not have been improved under aboriginal conditions.

In the case of the birch-bark canoe, it seems impossible that any other form made of birch-bark and a wooden frame could fulfill the cultural needs of the Naskapi as dictated by their habitat.

There are, of course, forms of bark canoes other than those of the Naskapi but they would not meet Naskapi requirements. The Kootenay style of canoe with its shovel shaped bow would plunge and sink in the heavy rapids of Labrador rivers and the clumsy canoe of the Iroquois, if it didn't sink, would be hard to maneuver, and if not dashed to pieces on the rocks would be hard to carry over portages. It is true that skin covered canoes of suitable shape are possible and the Naskapi sometimes use these when birch-bark is not available, but they are poor substitutes for bark canoes as they become water logged and are heavy. It seems obvious to me that the Naskapi birch-bark canoe is an ideal tool produced by Naskapi technology in relation to the inherent qualities of birch-bark, spruce gum, spruce root, and specific woods to fulfill a cultural need within a given habitat.

There are also illustrations of the articulation of habitat and culture in the Naskapi social system.

As Speck and Eiseley pointed out, the northern Naskapi who live by following the wandering bands of caribou are socially organized into wandering bands whereas the more southerly dwelling Montagnais, Cree, and Chippewa who hunted the family-like groups of moose and beaver are organized into family groups with concepts of ownership of hunting territories.

Do forms of marriage have any relation to habitat in the context of a culture based upon a hunting subsistence? White made some statements that seem useful here. He wrote:

> Marriage and the family are society's first and fundamental way of making provision for the economic needs of the individual. . . . One set of circumstances will require one definition of incest and one form of marriage; another set will require different customs. The habitat and the technological adjustment to it, the mode of subsistence, circumstances of defense and offense, division of labor between the sexes, and degree of cultural development, are factors which condition the definition of incest and the formulation of rules to prohibit it.

Since, in the above statement, all of the conditioning factors that are cultural are in themselves the product of articulation of habitat and culture (in the limited context of a nonagricultural, hunting society) it seems as if rules of incest and marriage would be (in the same context) the product of interaction between culture and habitat.

The Naskapi bands that come down the rivers leading to Richmond Gulf and Great Whale River, which I observed in the summer of 1939, claimed to marry their cross-cousins, if they marry a known blood relative. If the person to be married is not known to be a blood relative, he or she is called "cross-cousin" anyway. In short, the Naskapi practice cross-cousin marriage. The Naskapi, insofar as I could observe, may marry into their own band, into a band where they have kindred, or into a band where they don't have kindred. Naskapi bands are variable in size and about as unstable as the supply of food.

Since the ideological aspect of culture expresses the technology and also reflects the social system, I would expect to find much of Naskapi religion articulated with the habitat through the interaction of technology and environment. And it seems to me that this is so. Certainly the concepts of animals in special relation to man as well as devination, magical practices, medicinal practices, and charms for hunting are obviously linked directly with survival in the habitat. In fact, one could argue that Naskapi religion is part of Naskapi technology, so closely is it tied to the production of food, production against the elements, and defense against enemies; the three things required for survival. Each Naskapi hunter is his own shaman, and hunting is considered a religious activity.

An environmental explanation of the role of the bear in Naskapi religion is that in the northern habitat no other animal can walk like and look like a man. Eating the bear is close to cannibalism, and cannibalism is a reality in the Naskapi environment.

In the summer of 1939, I frequently heard a whistling which to my untrained ear produced a tune not unlike that of some Naskapi singing. This whistling was the sound of the wind blowing across the rocks and through the shrub-like trees of the high plateaus of the west coast and interior of the Labrador Peninsula. As the wind changed in intensity or direction and as the vegetation moved in the wind, the tones changed, thus producing the whistling sound. The Naskapi account for this whistling and for the unexplained loss of children by attributing both to the Katcimedgizu spirits. Greatly feared by the Naskapi, "They are said to come into the far interior in magical, high-bowed canoes, where they steal the Naskapi children. While their whistling may be heard they are invisible. . . ."

It seems to me that much of Naskapi culture and particularly its technological aspect articulates with habitat in a definite and relatively fixed fashion. Except for language I cannot think of any aspect of aboriginal Naskapi culture that is not to a considerable degree the product of the interaction of habitat and culture.

Even diffusion before European contact would be subject to the dynamics of the interaction of culture and habitat since it occured in a framework of culture and habitat. Wissler's concept of "natural diffusion" seems to fit the picture perfectly. He wrote, "We see that the basis for diffusion of trait-complexes is environmental and, to a large degree, also economic, since . . . man preys upon the organic resources of his habitat. So the immediate factors in the determination of diffusion boundaries are the fauna and flora." Since the upper level of a hunting culture is limited by the habitat, it would be impossible to add by diffusion any trait that would not be useful in the culture that has already achieved equilibrium in the interaction between culture and habitat.

Organized diffusion, such as that produced by outposts of civilization, for example, gold miners, could penetrate the barriers of habitat and culture, but such outposts could not exist without the support of a civilization that controls habitat elsewhere.

So, on a pre-agricultural level of culture, diffusion—itself a cultural product—is controlled by habitat, because the only context in which it can occur is the context of the interaction of culture and habitat.

From a consideration of White's concept of culture, some knowledge of Naskapi and Central Eskimo culture, and Wissler's concept of natural diffusion, it seems to me that under certain conditions a hunting culture will develop over a period of time to the limits of its habitat and in doing so will acquire cultural forms determined by the habitat.

In an unchanging habitat a hunting culture achieves at least near equilibrium with the habitat and in so doing acquires or is forced to produce the forms of culture most suited for survival in the habitat. A primitive hunting culture in a changing habitat tends to achieve a new equilibrium with its new habitat. This new equilibrium is achieved by invention through the interaction of technology and habitat or by diffusion from a culture in a similar habitat. These rules are of limited utility for hunting cultures in an environment where food resources are abundant and are of no use for agricultural societies or cultures with a technology based on the use of elaborate kinds of energy. Such cultures have become varyingly independent of habitat by controlling their environment.

If, at the pre-agricultural level of culture, habitats and cultures may be bound together in rather fixed relationships, then a number of inferences can be made that would be exceedingly useful in the reconstruction of cultural histories and limited cultural developments.

Since the sociological systems of culture depend upon the technological systems, and since the technological system is dependent upon the fauna, and since the fauna is dependent upon the flora, all of these factors could possibly be correlated.

For instance, in a habitat unsuited for agriculture, one might correlate reindeer moss with caribou, caribou with nomadic bands of hunters, whose population and range depended upon caribou population and range, and this factor would ideally correlate with one form of social organization (or at most with one of three possible forms).

Similarly one might be able to correlate water lilies with moose, moose with small groups of nomadic hunters, small groups with patrilineal families, and so on. Or one might link aspen trees to beaver, and beaver to small groups of hunters, and so forth.

The seal seems to correlate well with Central Eskimo size of population, habitation, technology, and social organization. I would even go so far as to suggest that subsistence based upon the seal is most favorable to the Eskimo type of social orgainzation, whereas subsistence based on caribou favors cross-cousin marriage and the composite hunting band—the Yuman type of social organization.

I envision these hypotheses and ideas as archeological aids possibly useful in the reconstruction of culture history in the Upper Great Lakes region where there was a period of about 10,000 years during which culture was based on a hunting economy. The archeological record of this period is sparse. There are, however, adequate reconstructions of habitats based upon data from geology, especially glacial geology; paleontology; pollen analysis; and other fields of natural science. In fact, much more is known about the changing habitats of this period than is known about the cultures. By combining the cultural and environmental data and analyzing and interpreting them in terms of the hypotheses and ideas presented in this paper, it is possible better to reconstruct the prehistoric hunting cultures of the Upper Great Lakes region and to learn more about the kinds of culture changes that take place in hunting cultures.

38. SOME ASPECTS OF PREHISTORIC SOCIETY IN EAST-CENTRAL ARIZONA

WILLIAM A. LONGACRE

Reprinted from New Perspectives in Archeology *edited by Sally R. Binford and Lewis R. Binford (© 1968 by Wenner-Gren Foundation for Anthropological Research, Inc.). William A. Longacre is Associate Professor of Anthropology at the University of Arizona. He received his Ph.D. from the University of Chicago and has also taught at the University of Illinois and at Yale University. A fellow at the Center for Advanced Study in the Behavioral Sciences in 1972-73, he has conducted archeological research in the Southwest for the past 15 years and is Director of the University of Arizona Archeological Field School. He has also done anthropological field work among the Kalinga in northern Luzon, the Philippines.*

■ Among the best known native Americans are those of the Southwest—the Hopi, Zuni, Acoma, and others known by the general term of Pueblo. In this selection Longacre describes adaptive cultural changes that occurred over a period of about 400 to 500 years in a Pueblo group that no longer exists. He concentrates on a single site to assess adaptive change in response to altered habitational conditions.

For the archeologist, some of the most significant clues to the social arrangements of the societies he is studying are provided by the remains of settlement patterns, the spatial distribution of house sites. The number of houses in a settlement, their density, and their apparent functions reflect daily social relationships that are directly recorded in material remains. When combined with information about methods of food production—from pollen remains, remnants of irrigation systems, tools, and so forth—and other kinds of manufacture (such as pottery), we can make educated guesses about the ways in which the members of the group were organized for adaptive purposes. Serving as a background to these inferences is the archeologist's knowledge about the relationship between technology and social organization among living groups.

We also know that every adaptive strategy—hunting-gathering, fishing, and different modes of cultivation—requires a particular organization of personnel who are mobilized in appropriate task units. Cultivators, for example, cannot be organized in the same way as hunter-gatherers or herders. Cultivators must remain where their crops are—to tend, protect, and harvest them; they must have appropriate modes of cooperation, storage methods, lines along which food and other goods are distributed, and the like. Groups of hunter-gatherers and herders, however, often have to split up even though they may ultimately return to their home bases.

Thus as Longacre shows, changes in social organization among the people who inhabited the Carter Ranch Pueblo represented adaptive responses to changing conditions in their milieu. But what must be remembered is that these social transformations were not adaptations to changes in the habitat alone; they occurred within the context of specific methods of cultivation, not in a vacuum. Had these people been nomadic hunter-gatherers or herders, for example, their institutional responses to the altered conditions would have been very different.

The contrasting approaches represented by the selections by Waterbolk and Longacre raise the question of methods in the study of cultural adaptation. One of the best discussions of this —though within an ethnological, rather than an archeological, framework—is found in *Amazonia: Man and Culture in a Counterfeit Paradise,* by Betty J. Meggers (Chicago: Aldine, 1971). One of the most important studies showing how the analysis of pottery provides excellent clues to the investigation of changing social organization is *The Dynamics of Stylistic Change in Arikara Ceramics,* by James D. F. Deetz (Urbana: Illinois Studies in Anthropology, number 4, University of Illinois Press, 1965). Three excellent papers on archeological methods—by John Fritz, Lewis Binford, and Sonia Ragir—are to be found in Part 4 of *Contemporary Archaeology: A Guide to Theory and Contributions,* edited by Mark P. Leone (Carbondale and Edwardsville: Southern Illinois

University Press, 1972), pp. 135-91. A well-known article that attempts to reconstruct prehistoric Pueblo Indian social organization from the archeological analysis of house sites is "A Prehistoric Community in Eastern Arizona," by James N. Hill *(Southwestern Journal of Anthropology,* 22 [1966]: 9-30). ■

THIS PAPER DISCUSSES a series of archeological investigations conducted in east-central Arizona (Figure 38.1) directed toward elucidating the nature of cultural stability and change in the extinct societies of the region. The primary focus is on the Carter Ranch Pueblo (*ca.* A.D. 1050-1200), where we attempted to isolate and explain certain organizational features of the sociocultural system as an initial step toward gaining a better understanding of adaptive changes made by the society to environmental stress. This report must be viewed as a case study and the research as a somewhat crude and initial effort. It is our hope that as the field advances our findings will be rendered obsolete by refinement and growth of methods of data collection and analysis.

In the undertaking of this research we were guided by several assumptions of a basic theoretical nature. First, we adopted a perspective which views culture as a systemic whole composed of interrelated subsystems, such as the social system, the technological system, the religious system, etc. This view of culture has been discussed and described by L. R. Binford. Such a perspective compels the paleoanthropologist to focus on the nature and interrelations of the component parts of the cultural system under study, and to work within an ecological frame of reference. The aim is to isolate and define cultural processes, the means by which cultures remain stable or change.

The second assumption underlying this work is that the patterning of material remains in an archeological site is the result of the patterned behavior of the members of an extinct society and that this patterning is potentially informative as to the way the society was organized. Our first task, then, is to define the structure of the archeological remains at a site and to offer hypotheses as to the organization of the society and associated patterns of individual behavior. The patterned relationships among classes of artifacts should document the context in which they were made, used, and lost or abandoned. It is essential to measure the mutual covariation among all classes and types of archeological data; the structure of this covariation, once delimited, should reflect the organizational and behavioral aspects of the society that produced it.

We will begin by discussing the environmental and cultural setting of the prehistoric pueblo under consideration, and this will be followed by a description of the archeological structure delimited at the site. We then offer some propositions as to the nature of the society represented and the cultural processes operative in the region.

SOCIOCULTURAL BACKGROUND AND THE ENVIRONMENTAL SETTING

This section consists of a brief synthesis of environmental and cultural changes which occurred in east-central Arizona before A.D. 1200 in order to provide a background for the period under study. Data for this synthesis come from excavation and survey.

The area today is environmentally transitional between the White Mountains and the Colorado Plateau. It is a semi-arid, topographically rugged region in the upper drainage of the Little Colorado River. The landscape is dotted with basalt-capped mesas, cinder cones, deep river valleys, and numerous arroyos. This area was occupied from at least 1500 B.C. until approximately A.D. 1350, at which time permanent occupations became restricted to the deepest stream valleys such as Silver Creek and the Little Colorado itself.

At the present time the region is marginal for most forms of agriculture; the most important economic activity is stock-raising. The only economically significant agricultural lands are the irrigated lowlands in the deep stream valleys such as those at Taylor and Shumway on Silver Creek where corn and other cultigens grow readily. The lack of

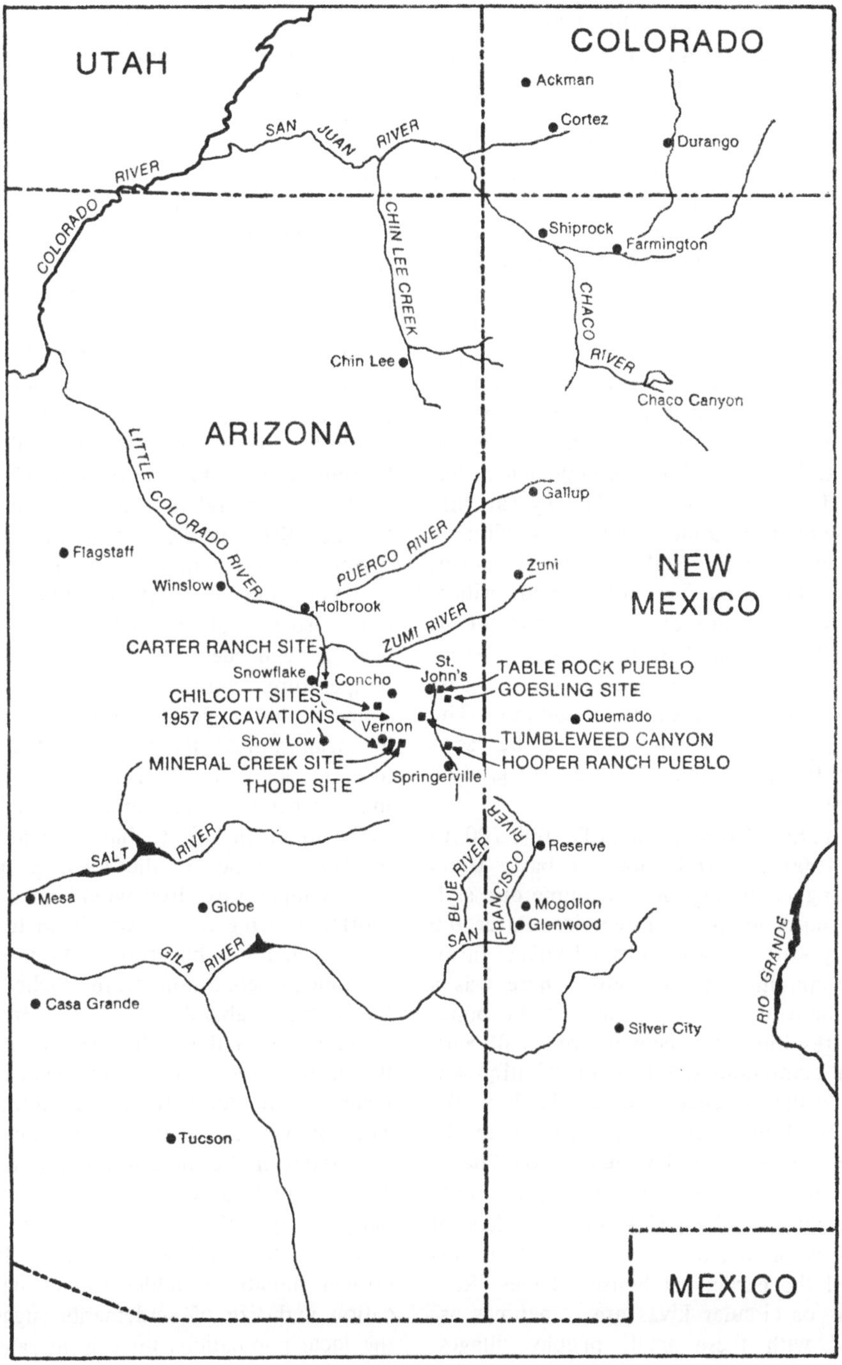

FIGURE 38.1

Area of investigations.

upland agriculture cannot be explained simply by a lack of water; the critical factor seems to be the absence of a regular growing season of at least 120 frost-free days. In the upland town of Vernon, individuals who maintain irrigated garden plots tell us they are lucky to get a crop of corn to mature once every three or four years. It is significant that these upland regions are dotted with prehistoric sites dating from about A.D. 700 to about 1000 and we know that occupants of the sites were dependent upon corn agriculture. Obviously there must have been some important changes in the local environment.

The study of fossil pollens has been a significant part of the investigation of the degree and kind of climatic change which has occurred. Schoenwetter and Hevley can find no palynological evidence of major climatic change in the area. Both workers, however, indicate that there are subtle and critical shifts in the nature of the environment. It is argued here that these minor shifts necessitated major adaptive changes on the part of the area's prehistoric inhabitants. The nature of these changes must be seen against the preceding centuries of gradual development.

The period from about A.D. 600-700 to about 1000 or 1100 shows a basic trend suggesting continuity despite numerous stylistic changes in material remains. After the initial appearance of established village farming communities in the region, there was a regular and impressive increase in the population, resulting in a network of small agricultural communities. Village distribution was linked to the presence of arable lands in the alluvium of the many small streams in the area, streams which today are arroyos. Pueblo architecture appears later, but community size remains unchanged—small hamlets of from three or four to fifteen or twenty rooms replaced the earlier pit house villages. Rectangular or circular kivas are sometimes associated with these small pueblo villages. When the population of a village reached a critical size, budding-off seems to have occurred, and this process resulted in the establishment of increasing numbers of small agricultural communities in previously unoccupied habitats. The newly founded villages probably maintained kinship and ritual ties with the mother community for several generations, but there are indications that through time these ties became attenuated.

Cohesion within the villages was probably maintained along kin lines and might have been reinforced by the emergence of non-kin-based membership groups, a ceremonial sponsorship system, or both. There is no evidence suggesting strong multi-community integration during this period.

At about 1000 or 1100, or perhaps earlier in the eastern portion of the region, there is evidence of major structural change in these cultural systems. The palynological data reflecting minor but critical environmental shifts are correlated with these cultural changes. Schoenwetter argues that the pollen spectra suggest a minor shift in rainfall pattern beginning about A.D. 1000. His report hints also that about 1350 there might have been a slight decrease in the mean annual temperature. Subsequent work by Hevley supports this suggestion and points to a drop in mean annual temperature of about 2°-3°F. at about 1300. This order of change may appear to be insignificant, but it was of sufficient magnitude to have jeopardized the subsistence base of the local people. The small temperature drop would have meant a shorter growing season which, in turn, made corn agriculture hazardous. As modern environmental conditions were reached, the uplands were abandoned and there was a convergence of the prehistoric populations in the deeper valleys, where the lower elevations meant mean temperatures 2°-3°F. higher. The importance of these conditions can be seen today if the agricultural productivity at Taylor and Shumway is compared with the marginal agricultural activity in the uplands.

The coincidental occurrence of changes in critical climatic variables and changes in location and size of settlements suggests that the local populations underwent rather striking adaptive responses. Some of the changes are readily apparent in the data; others are more difficult to delimit.

Two main trends can be detected. First, the beginning of population convergence, with small, single-residence-unit villages coalescing to form larger communities of more than one residence grouping; second, the appearance of Great Kivas at a few sites at about A.D. 1000-1100 documents an attempt to integrate a number of villages via a religious mechanism such as a ceremonial sodality. This pattern of convergence continued and culminated in the presence of a few very large towns composed of units which had previously formed single villages.

This aggregation forming larger population units posed certain integrative problems. A community composed of residence units with strongly traditional kinds of internal cohesion had built-in sharp lines of cleavage. Some of the means available for a new level of integration were religious; some might have involved sodalities such as curing societies which cross-cut residence units; mutual interdependence among social groups could have been created through reciprocal exchange of goods and services.

Carter Ranch Pueblo was investigated in order to determine more precisely the nature of the environmental and cultural changes of this region between A.D. 1050 and 1200.

DESCRIPTION OF THE CARTER RANCH SITE

The Carter Ranch Site consists of a U-shaped block of rooms which face a courtyard with various activity areas such as cooking and storage facilities and religious structures (Figure 38.2). The ceremonial structures include a large D-shaped kiva and a small rectangular kiva adjacent to a detached unit of rooms in the northeast section of the site. A large jug-shaped granary pit lies just east of the large kiva and contained several bushels of charred corn. A detached circular Great Kiva is located about 10 meters northwest of the north wing. This kiva is approximately 18 meters in diameter and has a long lateral entry ramp to the east.

A mound of trash of considerable size occurred just east of the site; this midden yielded a number of burials which clustered in three distinct areas. The burials in the northern cluster were oriented east-west; those in the southern area were oriented north-south; burials in the central area were mixed with respect to orientation.

The site as a whole is oriented toward the east. The interior features of the kivas and most of the rooms are lined up from west to east, and the Great Kiva is also oriented in the same direction, 7°30′ south of true east. This angle corresponds with the angle of sunrise in this location during the first week of March, two weeks before the vernal equinox. Although we have been unable to correlate this orientation with any specific celestial phenomena, such as solstice, eclipses, or constellations, it remains clear that the measuring devices of the site's inhabitants allowed a high order of accuracy in that the Great Kiva and the community itself are aligned exactly.

SEQUENCE OF CONSTRUCTION

The first rooms to be built at the site were in the north end of the central section: rooms 19, 3, 5, 12, 15, 18, 16, and 10. There were four later periods of construction, and the pueblo expanded mainly to the east and the south. The last addition was probably the detached unit in the northeastern part of the site.

THE ASSEMBLAGE

Thousands of potsherds, whole vessels, whole and partial tools of stone and bone as well as ornaments and other kinds of cultural items were recovered. They are described in detail elsewhere.

SUMMARY

The arrangement of the site is that of a "front-facing" plan. The nature of the cultural items from the site suggests that it is stylistically affiliated with what Rinaldo has called "Late Mogollon" culture. The easterly orientation of the site is support for this interpretation.

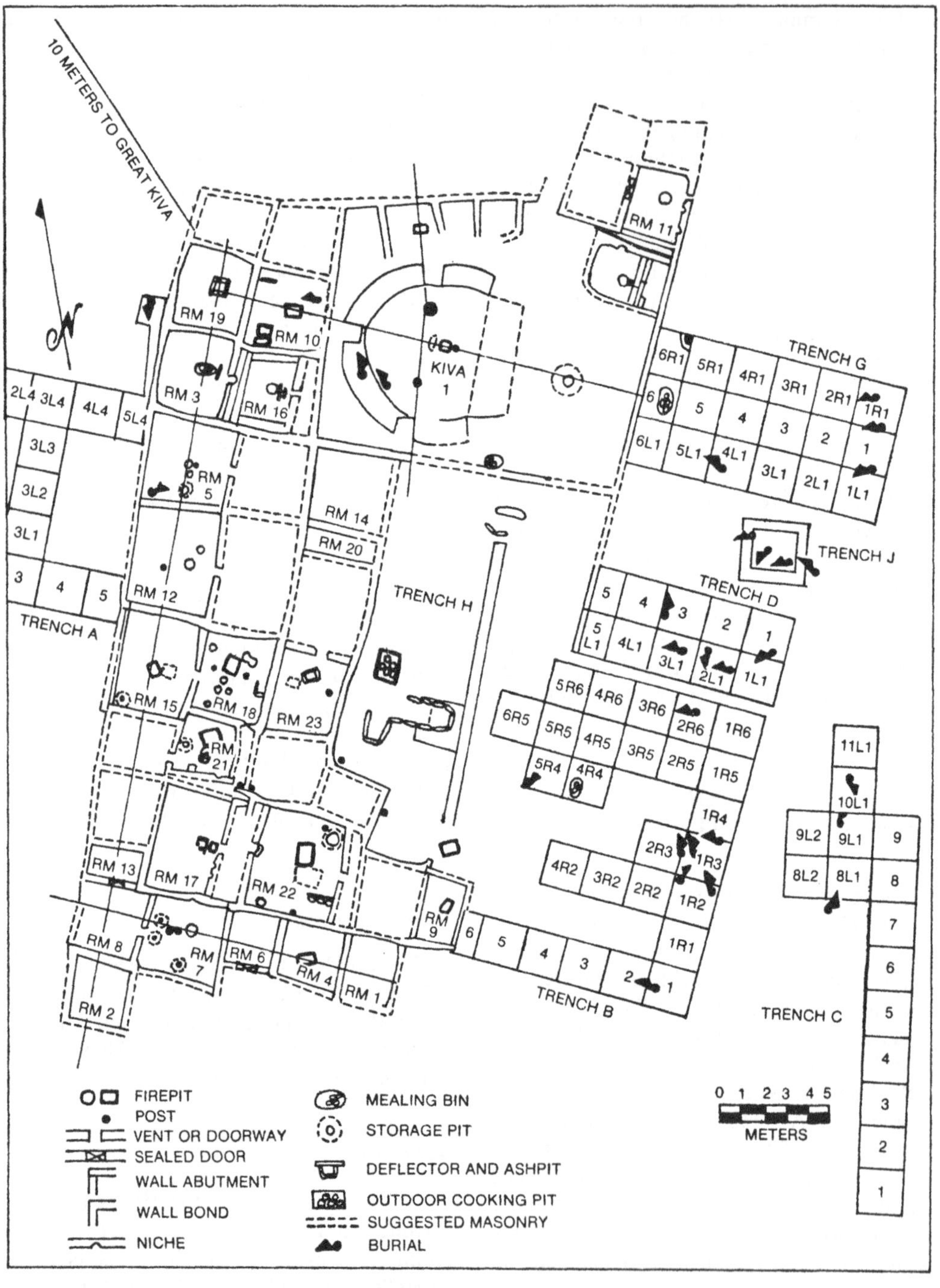

FIGURE 38.2

Carter ranch site.

FORMAL AND SPATIAL CORRELATIONAL ANALYSIS

Forming the initial archeological work at the Carter Ranch Site in 1961, a statistical analysis of a portion of the data was undertaken by L. G. Freeman and James Brown. Frequencies of ceramic types and their distribution were subjected to bivariate regression analysis using a Univac computer at the University of Chicago.

Frequencies of fourteen pottery types in eighteen floor and subfloor samples from twelve features were used in the analysis; frequencies of each of these pottery types were analyzed against each of the other thirteen types. The results showed a strong mutual correlation among certain of the pottery types, a non-random distribution of these covarying ceramic classes among distinct classes of rooms. There were four such clusters of pottery types which varied independently of one another. Brown and Freeman were able to discard temporal variation in pottery styles as a possible explanation for the observed independence and for the related differential distribution of the ceramic classes among different room types. An alternative explanation was advanced—that of functional independence for the ceramic classes. This functional hypothesis was supported by the strong association between the ceramic classes and room types; the defining attributes for the room types were the presence or absence of features such as hearths, mealing bins, etc. The findings indicate that the correlated differences in features and pottery reflect different kinds of activities localized in different kinds of rooms.

Excavations were continued at the Carter Ranch Site during 1962 with finer control of provenience employed. The newly gathered data on ceramic types and their distributions were subjected to multiple regression analysis using the IBM 7094 computer at the University of Chicago. In addition, non-ceramic artifact frequencies and distributions from both the 1961 and 1962 seasons were analyzed in the same way. Multiple regression analysis (IBM Bimed 34) proved to be a more powerful analytical technique and resulted in some refinement of the findings.

The four classes of pottery types originally suggested were expanded to five; these five classes are presented in Table 38.1. One major point of interest concerns class number three which contains a number of smudge types. This corresponds to a class proposed by Freeman and Brown which was noted to occur in high frequency in a ceremonial context (that is, kiva floors and burials). This same correlation was demonstrated in the second multiple regression analysis (Great Kiva and burials) and confirmed the association of this ceramic class with ceremonial activities.

Artifact frequencies by location were also subjected to multiple regression analysis. A number of clusterings were defined, but there was a great deal of overlapping between them. Generalized tools would probably have been used in a number of tasks, and this might account for the less tight artifact clusters when compared to ceramics.

TABLE 38.1

POTTERY CONSTELLATIONS FROM FLOORS OF ROOMS 18-23 AND GREAT KIVA, CARTER RANCH SITE.

I. Snowflake Black-on-White, Snowflake variety.
   Alma Plain.
   McDonald Corrugated, indented.

II. Snowflake Black-on-White, Carterville variety.
   St. Johns Polychrome.
   McDonald Corrugated, plain.

III. St. Johns Black-on-Red.
   Plain Brown Corrugated.
   Brown Indented Corrugated, smudged interior.
   Pattern Corrugated.

IV. Snowflake Black-on-White, Hay Hollow variety.
   Show Low Black-on-Red.
   Brown Indented Corrugated.

V. Plain Brown Corrugated.
   Show Low Black-on-Red.
   St. Johns Polychrome.

There were, however, two strong associations, each between two classes of artifacts. First, the class called "ceremonial" and the weaving implements were highly correlated. Since weaving today is a male activity associated with the kiva among the people of the Western Pueblos, this correlation is consistent with the enthnographic data from the area.

Second, there was an unusually tight association between the chopping artifacts and those used in the manufacture of arrow-shafts. Since both tool kits suggest male activities, this association strengthens the interpretation of a fairly strong division of labor and functionally specific male and female activity areas at the site.

THE DISTRIBUTION OF DESIGN ELEMENTS AND ELEMENT GROUPS

Our pre-excavation research design involved the formulation of a series of testable hypotheses. One of these concerned post-marital residence patterns. Briefly, the argument might be stated as follows: if there were a residence rule which led to related females living in the same locale through several generations, then ceramic manufacture and decoration would be learned and passed down within the context of this residence unit (assuming female potters). Nonrandom preference for certain designs might reflect this social pattern.

One hundred seventy-five design elements and element groups were defined, using more than 6,000 sherds and a number of whole vessels. The first analytical step involved the plotting of distribution of relative frequencies for each design on a site plan. This suggested that there was a nonrandom distribution, but the technique was crude at best. It was decided, therefore, to refine the analysis by using multiple regression analysis. In this operation, actual counts were used rather than relative frequencies.

The counts from each floor were run against every other floor sample. Floors were correlated on the basis of the frequencies of occurrence of the 175 design elements and element groups. The same was done for some of the fill samples. Fourteen floor proveniences were used, and the following clusters emerged: (1) Rooms 2, 4 (both floors), 7, and 8 (both floors) form a tight group; (2) Rooms 3, 5, 10 (both floors), 12, 15, and Kiva I form an equally tight group. Room 11 consistently shows associations with the first group.

When the location of these rooms is noted on the site plan (Figure 38.2), the significance of these clusters becomes clear. The first group consists of a block of adjacent rooms at the south end of the pueblo. The second group is a block of rooms with an adjacent kiva at the north end of the pueblo. Room 11, which associates with the first group, is in the northeastern portion of the site and is part of a group of unexcavated rooms.

The analysis of room fill was less successful, but there were similarities among the fills of Rooms 3, 5, 12, 15, and Kiva I. This suggests continuity in the pattern of design similarities in the floor distributions and argues for the consistent production of similar designs in the northern part of the village, probably over several generations of potters. Therefore, our hypothesis on the spatial distribution of design elements correlating with residence units received support.

THE BURIAL ANALYSIS

Thirty-four burials were excavated at the Carter Ranch Site; most were in the trash east of the pueblo. Grave goods were associated with male and infant burials, while interments of females yielded few or no associated goods.

The burials occurred in three clusters: the north and south clusters each had eight burials, and there were nine in the central area. Seven of the eight burials in the northern clusters were oriented east-west; all eight in the southern group were oriented north-south. Orientation of burials in the central area was mixed.

The design elements and element groups

of the mortuary ceramics were analyzed to see if the burial groups could be related to the architectural units of the site. Our results showed that the northern burial area had pottery with designs that occurred in the northern cluster of rooms at the site, and the southern burial group was similarly linked to the southern group of rooms. The central area was mixed with respect to design distribution, although there are more affinities to the northern block of rooms.

There were many respects in which the central burial cluster differed from the other two. This group was mixed in orientation and in the distribution of design elements and element groups. Further, almost all of the ceremonial items included as grave offerings came from this cluster of interments. There is also a nonrandom distribution of ceramic vessels in the center cluster:

	North	*Center*	*South*
Vessels	19	32	20
Burials	8	9	8

The mixed orientations of the central burials, together with the large number of ceremonial items and vessels, suggest that the central cluster represents high-status individuals from the residence units. It would appear that each of the residence units maintained its own burial area but that a separate portion of the cemetary was reserved for high-status individuals.

SUMMARY AND CONCLUSIONS

At the Carter Ranch, 175 design elements and element groups were analyzed in terms of their distribution in rooms, kivas, burials, and trash. The designs clustered in association with two major architectural units at the site. Kivas, discrete burial areas, and trash deposits were also associated with the architectural units. On the assumption that the females were the potters, this patterned distribution argues for post-marital residence in the vicinity of the wife's female relatives, with ceramic decoration learned and passed down within the residence unit. Time depth is demonstrated by the association of designs on pottery in the architectural units and associated trash in deposits of over one meter deep.

The localization of females in architectural units at the site over a period of several generations suggests in turn that non-portable objects, such as rooms and access to a specific mortuary area, were inherited within the residence units and that this inheritance was probably in the female line. The corporate nature of the residence unit is also suggested by the maintenance of a kiva and by associated mortuary practices. A large bell-shaped storage pit was excavated in association with one of the kivas; this large storage facility contained the charred remains of several bushels of corn. This adds support to the inference regarding the corporate nature of the residence units themselves, as it appears to have been a jointly used storage area which must have been maintained by a social unit larger than a single family.

Although it has not been possible to demonstrate the actual lineality of the residence units, it seems likely that they were localized matrilineal descent groups. The inferred corporate nature of the units and the suggested pattern of matrilineal inheritance support this suggestion. In addition, the maintenance of a discrete cemetery with design elements and element groups clustering in terms of residence units argues for a lineage organization and one which reckoned descent in the female line.

Social groups smaller than the residence unit have not been archeologically determined. Arguing from the ethnographic present, one might expect a household to be present, essentially a familial social unit. Among the Hopi and Zuni, households are localized in adjoining rooms and form the basic local unit among the Western Pueblos. There is today a great deal of variation in the number of individuals who compose a household and variability in the number of rooms constituting the spatial unit. The household today is a basic economic unit and is the one which is the land-holding unit.

The households at Carter Ranch Site

probably consisted of groups of adjoining rooms which formed residence areas. The rooms were apparently multi-functional, but the strong correlations among the ceramic classes with room types suggest that each room had a primary or more usual function. Isolation of specific functions is admittedly difficult, but the ceramic classes themselves offer some clues. The pottery types associated with ritual units such as the kivas are found in relatively low frequencies on habitation room floors. This indicates that some ritual was conducted in some houses, although on a small scale. The same pattern exists among the present day Western Pueblo peoples.

Specific groupings of pottery types may indicate certain generalized room functions. Painted pottery never shows signs of having been used for cooking (absence of sooted exteriors), whereas the smaller brown, textured jars often show such evidence. Brown and Freeman have noted a high frequency of such brown pottery at the site in rooms with circular floor pits, suggesting that these rooms might have had cooking as their primary function. They were probably also used for storage (many had storage pits), general work (such as manufacture of household articles), and other tasks as well.

Painted ceramics occur as bowls, pitchers, and jars. Since they apparently did not serve as cooking vessels, they were probably used in the preparation of food for cooking and in the serving of food. As Brown and Freeman point out, a grouping of painted pottery types consisting principally of pitchers and bowls appears to be associated with rooms with square fire pits and mealing bins. This suggests a function of food preparation for rooms with these features. The smaller, featureless rooms were probably used for storage.

The pattern, then, appears to be one of multi-functional uses for each room but with a more common set of activities for each room type. This is precisely the pattern of the modern Western Pueblos. The multi-functional nature of the rooms makes the delimitation of households at Carter Ranch very difficult. Chang has suggested that the fireplace (kitchen) is the most obvious index of a household. His suggestion is, however, based on ethnographic data from tropical or semi-tropical environments. In the more temperate climates, heating fires would also be necessary. It is, therefore, not surprising to find fire pits in so many of the rooms at the Carter Ranch Site. Some of these fires might well have served also for lighting and food preservation.

The Carter Ranch Site stands as a turning point in the prehistory of this particular region of the Southwest. The occupation of the site by two residence units ushers in an era of population aggregation in the area. Prior to this—over a period of more than 500 years—villages consisted of small groups of people approximating the size of one of the residence units of the Carter Ranch Site. The striking shift in the structure of population culminates in a pattern of fewer but considerably larger towns during the fourteenth century.

The process of aggregation coincides with the onset of a period of environmental stress and would seem to be an adaptive response on the part of these extinct societies. The initial set of adaptive changes documented at the Carter Ranch Site are a prelude to the even greater changes that appear in the region by 1300. Many of the adaptive shifts that have been noted appear to be in the direction of a form of cultural system exemplified by the modern Western Pueblos.

The inferences drawn in this report were largely justified by reference to a set of propositions drawn within a particular theoretical context (general systems theory) which served to structure the field research. The implications of our conclusions have not been stressed here, since they are made explicit elsewhere. The conclusions offered here are tentative and only further testing and research can strengthen, modify, or replace them.

39. CULTURAL ECOLOGY OF NUCLEAR MESOAMERICA

WILLIAM T. SANDERS

Reprinted from American Anthropologist, *64 (1962): 34-44. William T. Sanders is Professor of Anthropology at Pennsylvania State University. His principal research area is Mesoamerica, and he has focused on the ecological factors involved in the evolution of early civilizations. He has conducted a five-year research project in the Teotihuacan Valley, and he is the author of several monographs and papers on research in various areas in Mexico, including* Mesoamerica: The Evolution of a Civilization *(with Barbara Price).*

■ All people participate in culture but only a few societies—at least until relatively recently—have been parts of civilization, and fewer still have created civilizations of their own. In this selection, which deals with Mesoamerica, Sanders distinguishes between culture and civilization. He spells out the characteristic steps that lead to the development of civilization (the effective utilization of natural resources, which produces a relatively dense population; the integration of resources into a large society; and a system of social stratification) and the particular geographical conditions that seem to be necessary for all of this. Urbanization is another important accompaniment of civilization.

In a model of clarity, Sanders notes the principal habitational potentials for the development of civilization: low rainfall, easily controlled natural vegetation, fertile soils, water resources (especially for transportation), and a general deficiency of natural resources except for fertile soils. He also points to the importance of cultural and economic diversity in the growth of civilizations. Sanders suggests that civilization begins in the semi-arid environments that can support urbanization, although it can later spread into humid areas in which urbanization cannot be maintained. Thus he rejects the possibility that civilization can start in a humid area.

The careful reader will note that Sanders adroitly avoids a pitfall that has trapped many other anthropologists who have concerned themselves with the relationship between levels of cultural development and habitational factors. Sanders' inclusion of the variable of social stratification in the development of civilization, in addition to physical factors, expands the concept of environment beyond the natural elements. Man responds to more than the productivity of the soil, rainfall, and temperature in his attempts to free himself from the limitations of the natural milieu.

The reader who wishes to apply Sanders' formulations to other cultures in the New World should look into the four-volume work edited by Julian H. Steward, *The Handbook of South American Indians* (originally published as Bulletin 143 of the Bureau of American Ethnology [New York: Cooper Square, 1963]), as well as Steward's *Theory of Culture Change* (Urbana: University of Illinois Press, 1955). I have explored the concept of civilization in "Schools and Civilizational States" *The Social Sciences and the Comparative Study of Educational Systems,* edited by Joseph Fischer (Scranton, Pennsylvania: International Textbook Company, 1970). ■

THE FOLLOWING PAPER is an attempt to define some of the interrelationships between culture and environment and to demonstrate the value of the concepts of cultural ecology in understanding the process of development of pre-Iron Age civilizations. By cultural ecology I mean simply the study of the interaction of cultural processes with the physical environment. My theoretical position may be stated in the following principles:

1) Each environment offers to human occupation a different set of challenges, and therefore a different set of alternate cultural responses may be expected. There is, of course, some overlapping of both challenges and cultural solutions from environment to environment. One can also say that certain alternative responses are more likely to occur

than others. Some of these responses may be technological, others social, and some even religious.

2) In responding to such challenges, cultural response tends to take the path of greater efficiency in the utilization of the environment.

3) In development of any conceptual scheme in culturology, the environment should be considered as an active, integrated part of the cultural system not as a passive extra-cultural factor.

To most archeologists the term "civilization" has a relatively restricted meaning. Kroeber uses the term more broadly, and even synonymously with "culture," defining it in terms of his pattern concept, and calling it the "total cultural pattern." In this paper I will use the term in the more restricted sense of the archeologist.

A civilization is a particular kind of cultural pattern; as contrasted to cultures as a whole its pattern is broader, less narrowly restricted, and therefore capable of greater elaboration. When one views the growth configuration of a civilization one is impressed by its dynamic quality. All cultures change, but in civilizations change is more rapid, more easily measured, and more apparent to the observer. Certain conditions preclude the development of this type of cultural growth; some of them link specifically with the utilization of the environment. These may be enumerated as follows: (1) an effective utilization of natural resources permitting a relatively dense population; (2) the integration of such resources, natural and demographic, into a relatively large society; (3) a system of social stratification, at least on two levels, in which the surplus production of a large majority is systematically accumulated, controlled, and diverted into culturally specified channels by a small dominant minority.

This third condition seems to be an essential one as we have numerous examples of areas of the world with dense populations, relatively large social groupings, but no systematic manipulation of surplus labor and goods. In the earliest civilizations of the New and Old Worlds, this surplus was apparently controlled first by a priestly bureaucracy and directed towards the construction and glorification of buildings dedicated to the gods. Later periods saw the control exercised by, or at least shared by, a secular ruling class. Archeologists recognize civilizations primarily by the tangible results of the "direction of surplus energy," in the form of permanent architecture and an exceptionally high development of skill in other areas of technology. They use technology as a guide for obvious reasons. There is often a striking difference in the quality of the technology of the folk society that produces the surplus and that of the dominant minority that controls it. Furthermore, although few studies have actually been made, we usually find much greater stability and less dynamism on the folk level.

Before the development of the proletarian metal-iron, and the beginnings of the use of metal tools by both levels in the society, this kind of culture had an extremely limited distribution in the Old World. Its primary area was a large, nearly continuous region embracing the Near East, northwest India, northeast Africa, and southeastern Europe. A secondary, historically derived and later development occurred in North China.

Furthermore, the maximal development of the pre-Iron Age civilizations was in four small areas: the Nile valley in Egypt, the Tigris-Euphrates valleys in Iraq, the Indus valley in India, and the Hwango Ho valley in China, in all cases associated with a major river valley. Each of these centers was a relatively small, compact, densely populated area; the entire Old Kingdom of Egypt, for example, embraced only about 10,000 square miles of territory.

If we look at the ecological settings of the early civilizations, certain fundamental patterns emerge. The environments can be classed into two major types: (1) nearly rainless deserts with exotic major rivers, or (2) semi-arid country with low annual rainfall concentrated in a single season. In the Near East the latter climate falls generally into the type called "Mediterranean" with winter rains; in North China the rainy season falls in the summer. Average annual rainfall in the areas of dense population varies from 200-1,000 mm. In the Near Eastern center another significant characteristic is that there is a great deal of ecological diversity based on altitude and microvariations in climate.

The geographical conditions that seem to be crucial in these centers of the civilizations were: (1) presence of a fertile soil capable of being intensively cultivated and sufficient water for irrigation, (2) scanty plant cover which could be easily controlled (conditions (1) and (2) permitted an effective and intensive use of the land by a peasant population with an essentially Neolithic technology), (3) presence of a major river providing a natural transportation artery, and (4) a general deficiency of natural resources other than good agricultural land (acting as a stimulus to trade).

The situation in an area such as Mesopotamia offers a unique set of problems to a farmer equipped with a Neolithic technology. It is an ecological region of enormous potential, even with a relatively feeble technology. What it does require for successful utilization is a highly organized and cooperative society capable of mass effort in converting swamps and deserts into irrigated cropland. The most effective way to exploit such an environment is the development of a social system of the type we have defined as civilization.

Following V. Gordon Childe's analysis of the spread of civilization from the river valleys into the semi-arid highlands and small coastal plains of the rest of the Near East, it seems to have occurred primarily as a response to ecological condition number (4). The lack of resources in the major river valleys, he argues, led to the establishment of major trade routes, towns grew up at their termini which in turn were the foci of the extended Near Eastern civilization. Furthermore, the expansion occurred into an area where the environmental conditions permitted intensive and/or specialized (olives, dates, grapes, etc.) agriculture, on a small scale.

In the Old World another process developed hand in hand with civilization—urbanization. We have purposefully kept the two processes separate for reasons to be made explicit later.

I define urbanization as a process of evolution of rural communities into urban communities and further define an urban community as possessing the following attributes:

(1) Nucleation—in my analysis of modern urban communities in Mexico, all have population densities exceeding 2,000 persons per square kilometer.

(2) Relatively large size—in another study, based on modern Mexico, I use a specific figure in defining towns and cities, but precise population sizes would not apply to all areas of the world. I would generally state that urban communities would have at least 2,000 to 3,000 inhabitants, and reserve the term city for those with populations exceeding 10,000.

(3) Most of the population are nonfood producers, or at least only part-time food producers, and the majority of the population is composed of part- and full-time specialists in the production and distribution of technology, regulation of social interaction, or administration of services to the supernatural.

(4) A great deal of social differentiation based on occupation, status, control of power, and in some cases, ethnic diversity. In the early development of urban centers in the Near East, the larger communities were political and religious centers, as well as commercial and industrial communities, and the growth of cities was a process directly linked to the growth of the state.

Having set the stage, in terms of basic concepts, definitions, and events in the Old World, I will now attempt to apply these concepts to the New World. As far as present research indicates, the civilizations of the New World developed independently from those of the Old World, from the same kind of folk technological base, thus providing us with a good laboratory test of the above concepts.

Cultures possessing the attributes of civilization occurred in the New World in two regions, the Mesoamerican and the Andean.

The fundamental ecological patterns in the Andean region, in terms of the essential factors that permit the development of a civilization comparable to that of the Near East, correspond to a striking degree. In Peru, the heart of the Andean region, there are three primary, parallel, narrow, north-south, ecological strips. To the east is a humid, slope and foothill zone with exuberant tropical forest cover. In the center is a high mountainous region with numerous small and large valleys and basins varying in altitude from 5,000 to 13,000 feet above sea level. All over this strip the climatic type is similar to our type (2) in the Near East, light to moderate rainfall concentrated in the summer season. Average rainfall over the area varies from 500-750 mm. a year. To the west is a nearly rainless desert crossed east to west by some 25 streams which have their sources in the mountains and flow to the sea, each providing water for a small, compact irrigated plain and isolated from other systems by intervening deserts. The Andean civilization centered in the mountain and desert strip and apparently never penetrated the eastern forest with any degree of success. The ecological principles of the Old World apparently may be applied very successfully to the Andean area, and along the coastal desert, at least, urbanization correlated in its development with civilization.

I will now turn to my primary area of interest and research—Mesoamerica. Here the interrelationships between environment and culture are much more complex. The area is ecologically much more diverse than any other region where pre-Iron Age civilizations have developed. Rainfall varies from 300 mm. in southeastern Puebla to over 5,000 mm. in the northeastern escarpment of Chiapas. Altitude varies from sea level to 2,800 m. (in the area of dense human population) with a corresponding great range of temperature. Vegetation varies from near desert conditions to lush tropical rain forest, soils from siernozems to laterites, hydrography from no surface drainage to small flood season streams, to great rivers with huge tributary basins and extensive flood plains. Within this overall diversity, in terms of problems faced by neolithic farmers, two well defined ecological patterns emerge: (1) a Lowland pattern with heavy rainfall, exuberant vegetation, lower density and more scattered population, with slash and burn cultivation of basic foodstuffs and orchard cultivation of commercial crops; and (2) a Highland pattern with low rainfall, scanty vegetation, dense population living in larger nucleated communities and practicing intensive agriculture.

In actual fact, parts of the Lowlands have subhumid climates, and parts of the Highlands, humid climates, and any detailed analysis of the area should consider these exceptions.

In 1519, the Mesoamerican variant of civilization occurred over the entire region in all of the various ecological zones, including both Highlands and Lowlands. Urbanism, however, has been demonstrated as a corollary trait only in the Highland province, and in reality only definitely in the Central Plateau or Mesa Central of Mexico. In a more expanded paper, to be published in the projected Handbook about Mesoamerican Indians, I have analyzed in detail the role of this area in the development of urbanism and civilization in Mesoamerica. We will present here some of the results of this analysis and apply our basic concepts from Old World archeology.

Within the plateau is a small, compact, centrally located zone we will call the Nuclear Area because of its cultural dominance in the history of Mesoamerica. It includes the Valley of Mexico, Valley of Morelos, and the upper Atoyac-Nejapa drainage basin in Tlaxcala—Western Puebla, an area of approximately 20,000 km. In 1519, 20 per cent of the population of Mesoamerica resided in this demographic heartland, approximately 2½ million people. Both Pan-Mesoamerican empires, Aztec and Toltec, had their capitals in this area, and earlier Teotihuacan seems to have been the center of a third, similar state. It is also one of the main contenders for the scene of the origins of American Indian agriculture and Mesoamerican civilization.

The annual rainfall in the area of heavy human occupation is everywhere below 1,000 mm., generally ranging from 500-800 mm., most of which falls in the summer months. Vegetation cover is sparse and presents no serious challenge to primitive farmers. Within the fundamental unity is a great deal of diversity based upon altitudes ranging from 800 m. to 2,800 m. above sea level, in some parts of the area over a distance of only 50-60 km. Every agricultural plant in the Mesoamerican complex may be grown in some part of the area, and agricultural specialization has probably always been a distinctive feature. Soils generally are classed by Mexican agronomists as Chestnuts or Chernozems and with relatively simple techniques of soil restoration have sustained nearly continuous cropping for at least 3,500 years. The ecology generally is similar to type (2) in the Near East. There is no single great river system (although most of the area is drained by tributaries of the Balsas river) that could have served a single integrated irrigation system or provided a single transportation artery. In the Valley of Mexico, however, where two-thirds of the population resided in 1519, there was a chain of lakes that played the same role as the rivers in Old World centers of civilization.

Having described the general environmental factors of the area, let us now examine them as operational factors in the evolution of urban civilization. In the discussion we will refer back to our previous definitions of urbanization and civilization, and here, as in the Near East, the two processes were correlative and simultaneous.

(1) The ecological conditions noted above are optimal for the development of an intensive system of agriculture. Studies by Palerm, Wolf, Millon, Armillas, and myself have demonstrated conclusively that agriculture in 1519 was as intensive as in the great centers of the Old World civilizations. The combination of low rainfall, easily controlled natural vegetation, fertile soils, water resources (lakes, springs, flood water, melt water from glaciers), and generally limited flat terrain made almost imperative the development of such techniques of soil and water conservation as permanent irrigation, chinampas, flood water irrigation, cajete planting, stone and maguey terracing, and fertilization. In the strip above 2,000 m. the possibilities of early frost and retarded rainy season made irrigation necessary for really effective cultivation, even though 500-800 mm. of summer rains would ordinarily be ample. The application of this body of practices resulted in an extremely dense population as contrasted to other areas of Mesoamerica. I have postulated that civilizations with a Neolithic technological base can only develop with a relatively dense population. I furthermore insist that urbanization can develop only under even more demanding demographic conditions, especially in Mesoamerica, with hand tillage. I doubt that the Neolithic farmer with hand tools can produce more than a 20 per cent surplus. A city of 100,000 population would require a rural population of 500,000 to support it. In the history of the Nuclear Area, two cities, Tenochtitlan and Teotihuacan reached that size.

(2) One of the attributes of an urban community is nucleation. In Gordon Childe's analysis of Mesopotamia, he discusses the process as one of increasing size and social complexity from the Neolithic village to the Copper Age town to the Bronze Age city. In this concept the process begins with a small nucleated community. In terms of the *origin* of urban societies the presence of a rural population living in nucleated as opposed to dispersed communities seems to me a necessary requirement for such a development. Furthermore, in the transition period of growth from village to city, agriculture continued as the primary base, so that intensive agriculture would seem to be a necessary condition for such demographic growth, since the agricultural land still remains relatively accessible even when the population runs into the thousands. My analysis of rural settlement patterns in modern Mesoamerica, and documentary and archeological studies of the pre-Hispanic periods, suggests that the rural

settlement pattern in our Nuclear Area from the early pre-Classic to the modern period was basically one of nucleated villages. In the Nuclear Area today the degree of nucleation of the rural community correlates directly with the intensity of agriculture.

(3) One of the characteristic features of our Nuclear Area, as was stated previously, is ecological diversity. This is true of most of the Mesoamerican Highlands. At the time of the conquest and in some of the Highlands today we find an intensive development of regional trade and specialization on a rural community level. As Sol Tax has pointed out, in the Highlands of southwestern Guatemala, rural communities depend as much on trade for their livelihood as do cities. Although geographical diversity is of course not the only factor involved in this development, certainly it has acted as a powerful stimulus. If we argue that urbanization developed basically out of the agricultural folk society that supported it, and economic specialization is one of the most characteristic traits of urbanization, then the process by which urbanism evolved in Central Mexico seems apparent. Some urban communities grew up at crucial places, in terms of transportation, such as lake shore termini (Chalco) and altitude strips on the border of the Tierra Templada—Tierra Fria. Furthermore, in terms of the origins of cities and towns as centers of trade and craft specialization, the closeness of ecological zoning in Mesoamerica was an important precondition because of the primitiveness of land transportation.

(4) One of the major problems in Mesoamerica as a whole in the support of urban communities was the feeble development of transportation technology. By land, all cargo was hauled by human carriers so that only the closeness of ecological zoning permitted a relatively close spacing of markets and a heavy volume of trade. The chain of lakes in the Valley of Mexico provided a powerful stimulus to trade and undoubtedly was one of the crucial factors in the development of urbanism.

(5) Palerm, Wolf, Millon, and Armillas in various papers have attempted to apply the concept of the Irrigation State to the Nuclear Area, but thus far, with inconclusive results. One of the major problems in setting up a theoretical construct is purely archeological, the lack of data on the age of irrigation in the area. We know, on the basis of documentary and archeological evidence, that it was of great significance during the Aztec period; but it has not been specifically established for the Teotihuacan period where we have evidence of the first large state and urban community in the area. One important difference between our area and the center of Old World civilization is in the hydrographic pattern. In the Nuclear Area there are a great number of separate systems, each of which during the Aztec period provided water for a small, distinct, integrated irrigated system. Of interest with respect to this is the fact that the city-state, made up of a small urban town and its dependent villages, was the largest *stable* political grouping. This is in contrast to Egypt where the normal pattern was political unification of the entire river valley. Supra-city-state aggregations did occur several times in the Nuclear Area, but the constituent city states maintained their separate political and social structure and were never integrated into a state of the Egyptian type.

I am not arguing here, of course, that political states of the size of ancient Egypt, or even larger, could not exist in the environment of Central Mexico (the case of the Inca of Peru obviously makes this position untenable). What I am saying is that, in this type of ecology, city-state political integration is the largest level that one can link directly with ecological factors. Palerm and Wolf, in a brilliant essay on the rise of the state of Acolhuacan on the east shore of the lake in the Valley of Mexico, have pointed out the subtle interrelationships between environment, agricultural technology, and socio-political systems. In this case they postulated the integrative effects of irrigation on the state, but went further and demonstrated, in this area, that the state, once created, regardless of the factors that produced it, acted as a sponsor of extensive irrigation works and as a mechanism of producing a surplus to strengthen its power and further integrate the socio-economic system [Selection 34].

(6) The Nuclear Area is chronically one of overpopulation, a characteristic feature of mountain countries. Population clusters are isolated by high ranges and barren hills into separate compartments. Good flat land is premium land and never abundant so that the topography has always presented serious obstacles to an expanding population. This problem was met by improvement of agricultural technology, but, I suspect, that by Teotihuacan times the basic inventions had all been completed, except probably chinampa agriculture. It could also be met by social techniques, such as a thorough integration of the population by a system of centralized control for the construction of hydraulic works, terraces, and expansion into marginal lands.

Trade itself is one way of meeting the problem, since maguey, a staple food, can grow on even the most barren hillsides, and specialization of crop production would further increase the efficiency of land use. Spanish descriptions of the control of agriculture by the state suggest that, in fact, by 1519 such social techniques were practiced. Another response is, of course, war, conflict, and conquest between the city-states and the expropriation of lands or systematic taxation of conquered groups. This in turn would of course strengthen the power of the state and was one of the factors responsible for the growth of towns into cities. Cook, in an article on Mesoamerican demography, considered war as a demographic safety valve along with human sacrifice.

It has only been in relatively recent times that the concepts of Old World culture history have been applied to Mesoamerica, under the leadership of Julian Steward in the United States and Pedro Armillas in Mexico. H. J. Spinden anticipated this recent development in the 1920's, but he remains a lone pioneer. The lack of such attempts between the years 1930-1950 demands some explanation, since this period was an exceedingly productive one in basic research in Mesoamerican archeology. The primary reason was undoubtedly the development of research in the Maya Lowlands, where archeological exploration revealed the presence of an extraordinarily rich regional variant of Mesoamerican civilization in a humid forested lowland plain. This discouraged attempts to relate the growth of the civilization to environmental factors. Such research developments furthermore tended to obscure the fact that most of Mesoamerica is, in fact, a sub-humid area. Excluding most of the Yucatan Peninsula, parts of the Gulf Coast, narrow strips of escarpment, a few small segments of the Pacific coast, and parts of the Guatemala Highlands, rainfall over most of this huge area is either less than 1,000 mm. a year, or is so irregular from year to year that some techniques of humidity conservation are necessary for an effective enough system of agriculture to provide the demographic basis of a civilization.

The primary agricultural system in most of Lowland Mesoamerica is one called variously in the literature slash and burn, swidden, shifting cultivation, and, in Mexico, roza. It is a system practiced all over the world in tropical areas and, even where iron tools are used, it tends to be correlated with a low population density and a simple folk rural society residing in small, socially autonomous communities. This generalization is even more valid if we apply it to cultures with neolithic technologies. A variant of it was apparently practiced by neolithic societies in humid northern environments as well (northern and central Europe, eastern U.S.).

In the New World, outside of Mesoamerica, this system of farming was practiced all over the lowlands of eastern South America and around the Caribbean; and nowhere in this huge area was it the base of a culture of the type we are calling civilization.

The occurrence of Mesoamerican civilization in the tropical lowlands of Mesoamerica, based on slash and burn agriculture, is a unique one and demands further explanation. On the basis of published settlement pattern studies in the Peten and my own studies in northern Yucatan, Tabasco, central Vera Cruz, and the Huasteca, one can say that urbanism was not a correlative trait with civilization in those areas. The density of house-mounds at sites such as Tikal, Uaxactun, Chichen Itza, and the Puuc sites is well within the range of a rural, or at least suburban, population. Bullard, in his survey of a large area of the northern Peten, found a surprising lack of correspondence of house clusters to major ceremonial complexes. Apparently the settlement pattern was composed of two basic social levels: (1) a ceremonial center with a small resident priest-craftsman population as the dominant level, and (2) hundreds of small dependent rural hamlets occurring in a nearly continuous distribution between the ceremonial centers. This demographic and social pattern clearly relates to slash and burn agriculture and the primitiveness of Mesoamerican transportation. Apparently the system will permit a dense enough population to support a civilization of the Mesoamerican type but not urbanism, and the demands of this system for space tend to produce a dispersed agricultural population.

In recent years there has been a gradual crystallization of two opposed theoretical positions in the interpretation of the culture history of Mesoamerica: (1) in one position the area of the birth and early development is thought to have occurred in the humid lowlands based on slash and burn agriculture; (2) in the other, the development of civilization is seen as a corollary process with urbanization and occurring first in some part of the sub-humid highlands with intensive agriculture. It is postulated that it then spread into the humid lowlands where, because of the ecological conditions, the development of urbanism was aborted.

Archeological data *per se* cannot, at the present state of knowledge, resolve the conflict one way or the other. I have, of course, as the previous discussion indicates, accepted this latter approach. Specifically my position may be elaborated in the following points:

(1) Mesoamerican civilization developed first in Central Mexico as a corollary process with urbanization.

(2) This kind of culture which we are calling civilization must have its roots in the folk society that was its basal level and required the special kind of folk society that intensive agriculture produces. I see very little in the character or personality of a slash and burn folk society or economy that would lead to the development of civilization.

(3) The spread of Mesoamerican civilization into the lowlands was a process directly related to the regional pattern of specialization and symbiosis we discussed previously. As civilizations expanded in the highlands and immediate lowland strips, then the trade orbits were extended all the way to the coast and all of the lowland province shared in the general civilization. This did not occur in the Andean region, primarily because there was a lowland strip along the Pacific coast which had the proper ecological conditions for intensive agriculture. In Mesoamerica, sub-humid lowland areas were more restricted in extent and so the humid areas were incorporated.

(4) The spread of Mesoamerican civilization into the humid lowland was only a partial success, as the spectacular collapse of the Peten Maya civilization demonstrates. The fall of Maya civilization was qualitatively different from the fall of the contemporary Classic civilization of Teotihuacan. In the case of the latter, newer and equally vigorous civilizations replaced it, and the collapse occurred only in the upper, urban level. All of the evidence, archeological and documentary, in the Peten demonstrates that there not only the upper level but the folk agricultural society collapsed as well. There is very little evidence of a post-Classic population in the area, and Spanish records give one the impression that the population in the 16th century was nearly as sparse as the modern.

40. FARMING SYSTEMS AND POLITICAL GROWTH IN ANCIENT OAXACA

KENT V. FLANNERY, ANNE V. T. KIRKBY, MICHAEL J. KIRKBY, AND AUBREY W. WILLIAMS, JR.

Reprinted from Science, *Vol. 158 (October, 1967), pp. 445-54. Copyright 1967 by the American Association for the Advancement of Science. Kent Flannery is a member of the Department of Anthropology at the University of Michigan. Michael Kirkby is presently Professor of Physical Geography at Leeds University, England and Anne Kirkby is Research Associate in Social Psychology at the University of Leeds, England. Aubrey Williams is an ethnologist in the Department of Sociology and Anthropology, University of Maryland.*

■ A major theme running through this volume is the relationship between human populations and the habitat. As we have seen in many of the previous selections, man is part of the environment on which he depends and to which he must respond. As the capacity to harness sources of energy has increased, groups have freed themselves from the constraints of the habitat, expanded their populations, increased their dietary range, and—in the process—accelerated the evolution of culture. Thus a key issue in examining the emergence of social forms in human society is the relationship between habitat and the techniques employed in gaining a livelihood—that is, in mastering the habitat for adaptive ends. This is the theme of the present selection.

Combining physiographic and archeological evidence to reconstruct the development of agriculture and political systems in the Oaxaca Valley of Mexico, the authors show that each of their two dependent variables—agricultural and political development—is related to other factors and to each other. Agricultural potential depends on the valley's structure, the level of the water table, frost conditions, and other physiographic conditions. Political development, while dependent on agricultural potential, seems closely related to the Oaxacans' ability to assimilate new agricultural techniques.

This introduces us to a new dimension in our understanding of cultural evolution. Until now, we have treated adaptive units as more or less self-contained, on the implicit premise that a group's adaptations may be understood entirely in terms of its relationship with the habitat in which it is situated. In recent years, however, evidence has been accumulating rapidly from archeological and ethnohistorical research that points to the importance of trade and other intersocietal relations as influencing a group's technological and social development. The ability to assimilate new agricultural techniques is an aspect of such intergroup relations. As the authors note in their conclusion to this paper, trade seems to have been intimately related to the elaboration of social stratification in Oaxacan communities, and social stratification is always correlated with political complexity.

The most significant agricultural advance in this region was the harnessing of water resources, for which several methods were employed. The so-called "hydraulic theory" of state formation, which was advanced by Karl Wittfogel in his well-known book *Oriental Despotism* (New Haven: Yale University Press, 1957), is not borne out by the data from the Oaxaca Valley; a strong and centralized political system does not seem to be a necessary precondition for elaborate forms of water control. Canal irrigation did have important social consequences, however—for example, leading to inequities in the distribution of valuable land resources and, as a result, to an accentuation of social stratification and political power.

An excellent rebuttal to the "hydraulic theory" of state formation and an exposition of a major source of political centralization is presented in "A Theory of the Origin of the State," by Robert L. Carneiro (*Science,* 169

[1970]: 733-38, reprinted in *Man in Adaptation: The Cultural Present* [2nd edition]). A landmark archeological analysis, describing the dependence of political centralization on intersocietal trade, is "Mesoamerican Trade and its Role in the Emergence of Civilization," by Lee A. Parsons and Barbara J. Price (*Contributions of the University of California Archaeological Research Facility*, Number 11, [1971]: 169-95). ■

DURING THE LAST 15 years an increasing number of anthropologists and geographers have turned their attention to the pre-Hispanic civilizations of Mexico and Guatemala. The evolution of these ancient complex societies is of general theoretical interest because it seems to have taken place independently of the early Old World civilizations. Given the limitations of the archeological data, there has been considerable latitude for varied and competing theories about the origins of the early New World states.

Some authors have theorized that Mesoamerican civilization arose in the arid highlands, because of the need for a strong centralized government to control large-scale irrigation projects. Others have argued that civilization began first in the humid tropical lowlands, where irrigation is not necessary. Still others have sought a middle ground between these positions, maintaining that Mesoamerican civilization began through the "intertwining of many regional strands," both highland and lowland.

One of the most intriguing hypotheses of the evolution of early Mesoamerican civilization was that of Palerm and Wolf, who over a 10-year period in the 1950's sought to find correlations between social systems, agricultural systems, and their environmental settings. A major process they observed at work in ancient Mesoamerica was the formation and dissolution of "key areas" or "regional nuclei." These they defined as "areas of massed power in both economic and demographic terms," which at various points in Mesoamerican prehistory had acted as "nodal points of growth" or as nuclei for "symbiotic areas"; these nuclei were instrumental in stimulating cultural evolution over wide geographic regions. Each time Mesoamerica moved up to a higher level of social and political complexity, this move seems to have been accompanied by a shift of power and influence from one area to another. Some regions were nuclear only in the early (Formative) and middle (Classic) periods; other areas were nuclear only in the late (Post-Classic) periods. Only five regions in Mesoamerica were listed by Palerm and Wolf as having "maintained their key importance from Archaic times right up to the time of the Spanish Conquest." These areas are the Valley of Mexico, the region of Cholula-Puebla, the Mixteca Alta, the Valley of Oaxaca, and the region of Guatemala City (see Figure 40.1).

Why had these five regions attained early nuclearity and retained it throughout the sequence? Palerm and Wolf presented a corollary hypothesis which they felt should be checked by future investigators. They pointed out that all the areas which were nuclear only in the *early* part of the sequence had predominantly slash-and-burn agriculture of lowland (*roza*) type. All areas which rose to prominence only *late* in the sequence were arid regions demanding very efficient irrigation systems. The five *perennially nuclear* regions are ones in which, today at least, virtually every farming technique known in Mesoamerica is applicable. They hypothesized that farming had begun in Mesoamerica as slash-and-burn, and that through time a series of new techniques had been worked out: irrigation, flood-water farming, chinampas, and so on. As such technological innovations appeared, they were "applicable to an ever-decreasing number of areas." The areas in which the greatest variety of techniques could be assimilated remained nuclear; those in which only the older techniques could be applied gradually lost their influence and assumed a marginal role.

None of Palerm and Wolf's "perennially nuclear" areas has ever been investigated with their hypothesis in mind, though a number of related theories have now been tested in the Valley of Mexico. However, excavations

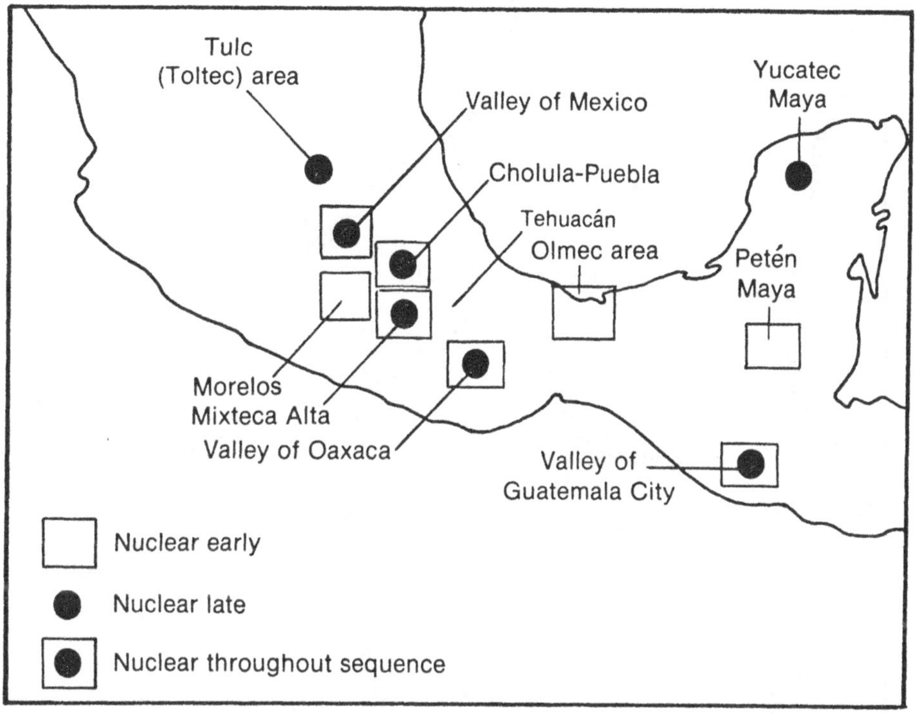

FIGURE 40.1
Outline map of Mexico and Guatemala, showing regions mentioned in the text. "Nuclear" areas listed by Palerm and Wolf are indicated by squares or large black circles; Tehuacán, a "marginal" area, is indicated by the small black dot. For the sake of brevity we have adhered to Palerm and Wolf's original classification, although recent archeological data indicate that it needs to be revised and updated.

in a few of Mesoamerica's "fringe" or "marginal" areas—ones which *never* became nuclear—have been carried back to the very beginnings of agriculture by MacNeish. One of these, in the Valley of Tehuacán, Mexico, has now yielded the longest single stratified sequence in all of Mesoamerica. Publication of the Tehuacán sequence places Mesoamerica in a better position than ever before for the testing of theories about the processes involved in the establishment of village life, and about the evolution of chiefdoms and early states.

In 1966 we selected the Valley of Oaxaca (Figure 40.2) as a natural laboratory in which to investigate a number of these hypotheses. Oaxaca was chosen partly because the outlines of its later prehistoric sequence (see Figure 40.3) had been worked out by Caso and Bernal, and partly because it lay close enough to Tehuacán to be related to it during all or part of its prehistory. This eliminated many of the preliminary steps that would have been necessary were the area totally unknown, and permitted us to concentrate from the very beginning on problems of cultural and ecological process. It also allowed us to compare and contrast the agricultural potential of a nuclear valley (Oaxaca) and a marginal valley (Tehuacán).

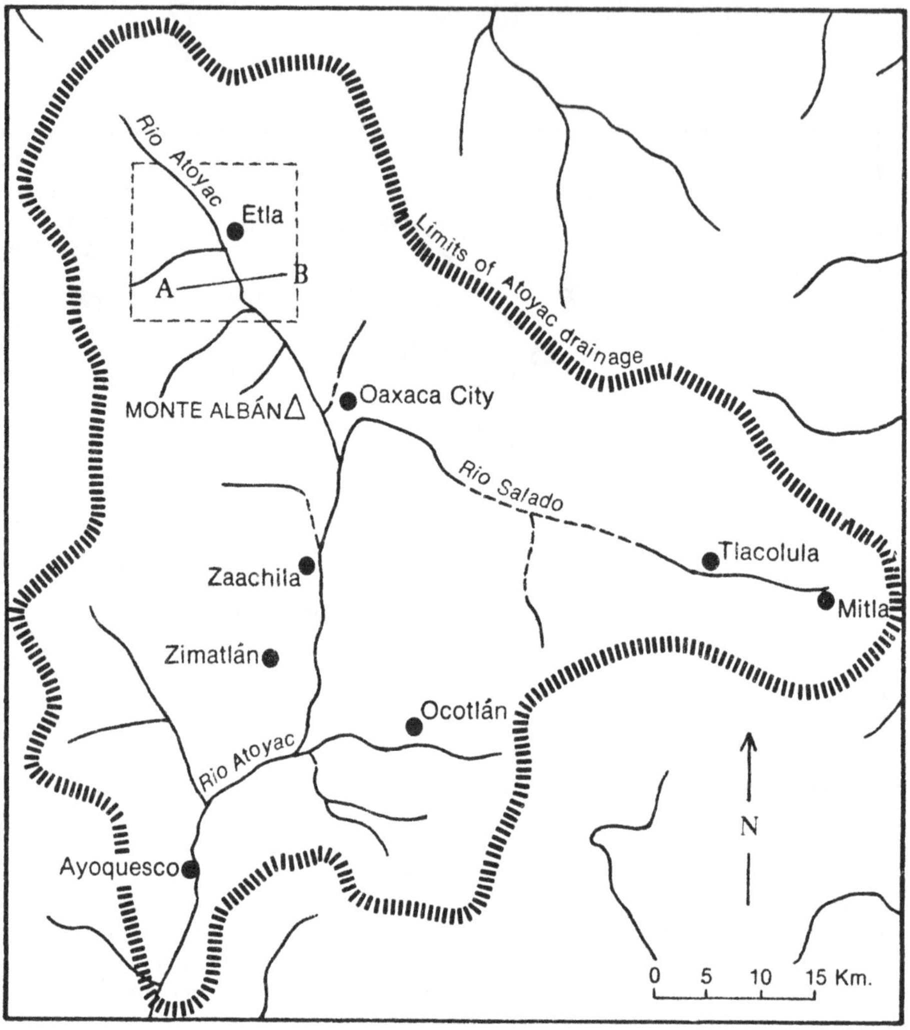

FIGURE 40.2
Map of the upper Atoyac River drainage basin, indicating major towns and archeological landmarks in the Valley of Oaxaca. The rectangle formed by dashed lines delimits the survey area shown in Figure 40.5. Line *A-B* locates the cross section shown in Figure 40.4.

LOCATION OF THE VALLEY OF OAXACA

The Valley of Oaxaca lies in the southern highlands of Mexico, between 16°40′-17°20′N and 96°15′-96°55′W. It is drained by two rivers: the upper Río Atoyac, which flows from north to south, and its tributary, the Río Salado or Tlacolula, which flows westward to join the Atoyac near the present city of Oaxaca. The valley is shaped like a Y or three-pointed star, whose center is Oaxaca City and whose southern limit is defined by the Ayoquesco gorge, where the Atoyac River leaves the valley on its way to the Pacific Ocean. The climate is semiarid, with 500 to 700 millimeters of annual rainfall, confined largely to the summer months. The valley-floor elevation averages 1550 meters.

Situated in the mountainous central part of

DATES	TEHUACAN VALLEY	VALLEY OF OAXACA	CENTRAL CHIAPAS	OLMEC REGION
A.D. 1500		Monte Albán V		
1400				
1300	Venta Salada			
1200				
1100		Monte Albán IV		
1000				
900				
800				
700				
600		Monte Albán III		
500	Palo Blanco		Laguna	
400				Upper Tres Zapotes
300			Jiquipilas	
200				
100		Monte Albán II	Istmo	
0			Horcones	Middle Tres Zapotes
B.C. 100				
200				
300		Monte Albán I	Guanacaste	Lower Tres Zapotes
400	Santa María		Francesa	
500				
600			Escalera	
700		Guadalupe	Dili	La Venta
800				
900				
1000		San José	Cotorra	San Lorenzo
1100	Ajalpan			
1200				
1300				
1400				
1500		?	?	
2000	Purrón			
3000	Abejas			
4000				
5000	Coxcatlán	Coxcatlán		
6000				
7000	El Riego	Guilá Naquitz	Santa Marta	
8000				

FIGURE 40.3

Pre-Hispanic cultural periods in the Valley of Oaxaca, compared with those of adjacent regions, as determined by radiocarbon dating and interregional similarities in artifacts.

the state of Oaxaca, the region is surrounded by valleys with steep sides, narrow floors, and perennially flowing streams. In constrast, the Valley of Oaxaca is a wide, open plain with abundant flat land and streams which are dry most of the year. Yet it was this valley, where moisture is scarce and man must devise means to control it, which became the most powerful nuclear area in the southern highlands. It is generally believed that this development was the work of the Zapotec Indians, who now inhabit the valley and whose history can be traced back many thousands of years in that region.

PHYSIOGRAPHY AND VEGETATION

A typical cross section of the Valley (Figure 40.4) shows four distinct physiographic

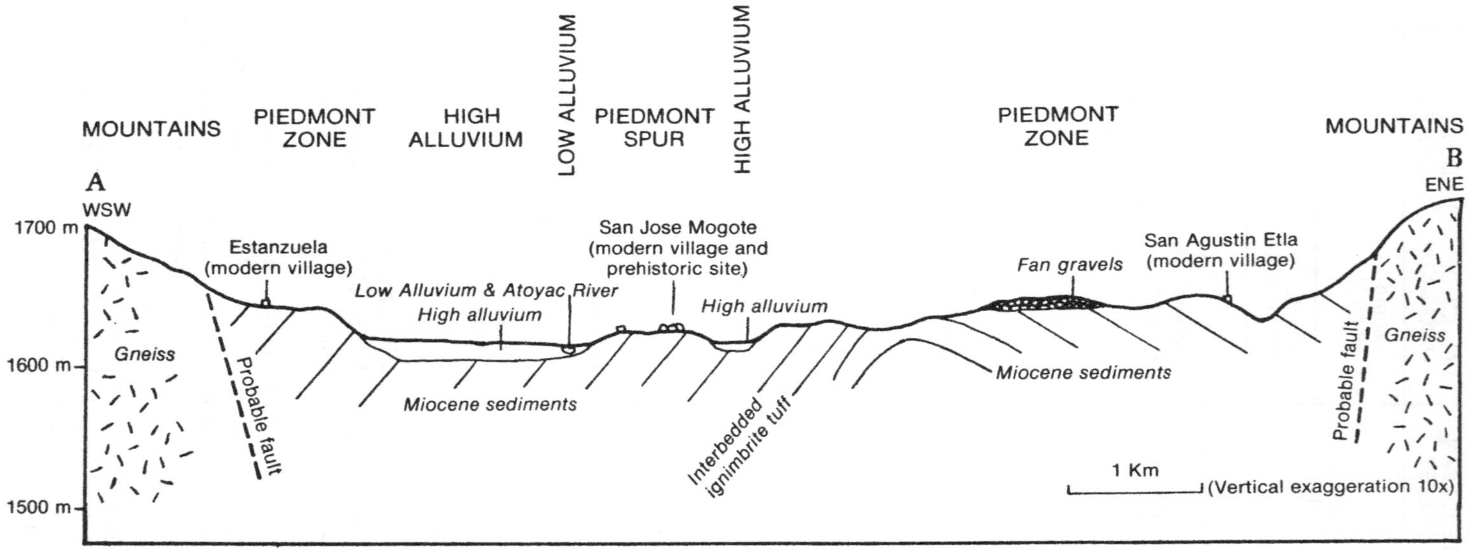

FIGURE 40.4

Cross section of the northwestern Valley of Oaxaca near Etla, showing major physiographic areas discussed in text.

zones: (i) the "low alluvium," or present river flood plain; (ii) a zone of "high alluvium," which is mainly an abandoned flood plain of Pleistocene-to-Recent age, formed by the Atoyac River and its tributaries when they flowed at a higher elevation; (iii) a piedmont zone flanking the high alluvium; and (iv) the surrounding mountains.

The river channel is incised no more than 1 to 2 meters into its present floodplain, which is only locally present and nowhere more than 600 meters wide. The main part of the flat valley floor, which varies from 1 to 15 kilometers in width, is formed by the high alluvium; this zone is separated from the low alluvium by a 1- to 3-meter rise in elevation. Between the high alluvium and the mountains lies the piedmont zones, where the land has a slope of 1 to 2 degrees and has been dissected by tributary streams to form low rounded spurs and isolated hills with up to 30 meters of relief. The piedmont was originally formed as a series of coalescing alluvial fans, and remnants of this deposited material remain as fan gravels of probably Pleistocene age. Later stream dissection has exposed underlying rocks which are pre-Jurassic to Miocene. The piedmont zone grades eventually into the true mountain zone, where valleys have up to 1000 meters of relief and slopes are steep. The mountains are formed mainly of pre-Jurassic metamorphic rocks, Cretaceous limestones, and Miocene ignimbrite tuffs. The tuffs are most extensive in the extreme eastern end of the valley, between Mitla and Tlacolula, where they abound in small caves and rockshelters which were occupied during the earliest periods of Oaxaca prehistory. In the extreme western part of the valley, where rocks of the basal metamorphic complex are most widely exposed, there are deposits of magnetite and mica which were used and traded by the later occupants of the valley as exotic raw materials.

Originally, each of these physiographic provinces would have had its own distinct vegetational cover. Today, after thousands of years of intensive cultivation, so little remains of the original valley-floor vegetation that it can only be hypothetically reconstructed from pollen grains and carbonized seeds in archeological sites of that zone. The present Atoyac floodplain may have had phreatophytic species like bald cypress (*Taxodium*), willow (*Salix*), and wild fig (*Ficus*), while the high alluvium was probably characterized by a more open cover of grasses and woody legumes like mesquite (*Prosopis*). The piedmont is still one of the most complex vegetation zones, with varying communities of tree legumes, prickly pear (*Opuntia*), organ cactus (*Lemaireocereus*), maguey (*Agave*), *Dodonaea*, and—at elevations of 1800 meters and above—scattered oaks (*Quercus* spp.). The high mountains have forests of oak, pine, and manzanita (*Arctostaphylos*). It is to be hoped that future work in Oaxaca will greatly enrich this tentative and oversimplified reconstruction.

AGRICULTURAL POTENTIAL

We feel there are several environmental aspects of the Valley of Oaxaca which make it a better place for agriculture—and especially for primitive, pre-Hispanic types of agriculture—than many adjacent parts of highland Mexico, including some of the other areas described by Palerm and Wolf as nuclear. Second, some of the agricultural techniques worked out by the Zapotec Indian inhabitants of the valley gave them an early advantage over their neighbors. Finally, there are various additional factors, only indirectly related to agriculture, which contributed to the rise of stratified societies in Oaxaca.

In the sections which follow we outline some of the environmental features of the Valley of Oaxaca which may be considered advantageous. We then attempt to show their relevance to the periods of food-collecting and incipient cultivation (8000 to 1500 B.C.), early village farming (1500 to 600 B.C.), and the rise of towns and ceremonial centers (600 to 200 B.C.).

The Valley of Oaxaca has 700 square kilometers of relatively flat land, the largest such expanse in the Mexican highlands south of Cholula-Puebla. Until recently, it was generally believed that the valley was the bed of a former lake, which had dried up prior to 600 B.C. This lake was mentioned in Zapotec legends, and casual inspection of the valley revealed supposed "shorelines" or old "lake terraces," as well as seasonally inundated areas which were reputed to be remnants of the lake. In 1960, Lorenzo presented geological evidence to the contrary. Similarly, our study reveals no evidence of a lake.

The supposed lake "shoreline" is not horizontal but varies in elevation by over 200 meters within a distance of some 40 kilometers. What it really consists of is the break in slope formed where the steeper fan gravels meet the valley alluvium. The fan gravels are clearly fluvial in origin, and no lacustrine deposits or fossils have been found, so the hypothesis of a permanent lake in the valley during the last 10,000 years must be rejected. The seasonally flooded localities reputed to be "lake remnants" are in reality low-lying areas where the water table is close to the ground surface. Many of these occur in clay areas of low permeability, which further tend to maintain standing bodies of water for long periods.

With the rejection of the lake hypothesis, other explanations should be suggested for the Valley of Oaxaca's unusually wide and flat floor. It is known that relatively arid climates favor alluvial-fan deposition, and this tendency toward alluviation in combination with the inability of the upper Río Atoyac to downcut (because of the high resistance of rocks in the Ayoquesco gorge) could explain the great width of the valley. Furthermore, this extensive deposition could have been initiated by downfaulting, for local deformation of Miocene (but not later) sediments shows that some late-Miocene/Pliocene dislocations have occurred in the valley. Thus three factors—aridity, downfaulting, and a low rate of stream degradation—all may have contributed to the alluvial expanses which make the Valley of Oaxaca unique among its neighbors.

LESS SEVERE SOIL EROSION

In the mid-1940's Cook studied a number of the higher valleys just to the north of the Valley of Oaxaca—Tamazulapan, Yanhuitlán, and Nochistlán, in the Mixteca Alta. All these valleys have suffered extremely destructive soil erosion, which Cook traced to agricultural activities extending far back into the pre-Hispanic era.

By comparison, soil erosion is not a severe problem in the Valley of Oaxaca, although there is some local gullying of hillslopes. One reason is that, relatively speaking, little land has so far been cleared in the higher mountains. In the Valley of Oaxaca, accelerated erosion due to clearing of the natural vegetation for agriculture is most severe not only on the steepest slopes but also where the vegetation is densest. This is strikingly shown by our erosion measurements on steep slopes near Mitla, made in 1966 over a period of 2 months during which 237 millimeters of rain fell. Table 40.1 gives the ratios of the rate of erosion on cleared land

TABLE 40.1

COMPARISON OF EROSION RATIOS IN THREE DIFFERENT VEGETATION ZONES NEAR MITLA, IN THE VALLEY OF OAXACA.

Erosion sites	*Elevation (meters)*	*Natural vegetation*	*Ratio for rate of erosion: cleared land/land with natural vegetation cover*
No. 1	1750	Cactus-scrub	1.0
No. 2	2000	Oak-*Dodonaea*	3.8
No. 3	2300	Pine-oak	45.4

Note: [Erosion ratio = C/N, where C = erosion rate on cleared land and N = erosion rate on land with undisturbed natural vegetation. Both C and N were calculated, for slopes of comparable steepness during the same rainy season (1966), by measuring the percentage of surface material moved downslope from a previously established 50-meter line.]

and the rate of erosion on land with natural vegetation.

It would appear that agricultural land clearance in the oak-pine forest zone results in an accelerated erosion of 45 times the natural rate, whereas in the cactus-scrub zone, where slopes are already in equilibrium with a sparse vegetation cover, clearing of land does not appreciably increase erosion. The gentle slopes of the valley floor further discourage erosion, and the presence of an extensive flat area lessens the incentive to clear land higher in the mountains. All these factors combine to keep soil erosion to a minimum in the Valley of Oaxaca. It is perhaps worth noting that the most badly eroded valleys of the Mixteca, like Yanhuitlán, occur at elevations of 2000 meters in the pine-oak zone, where the ratio of acceleration on cleared land is highest.

SOILS AND WATER TABLE

The flat valley floor and the thick alluvial deposits offer clear advantages as a site for early agriculturalists, but these factors are partially offset by the relative aridity, which limits both available water and soil fertility. Soil profiles are poorly developed, and the alluvial structure is retained almost unaltered below the A horizon. Prismatic structure and some salt accumulation is found in the B horizon. Most valley-floor soils belong to the Brown Soils group, but they tend toward Gray Desert soils in the most arid areas, such as the Mitla end of the valley.

Humus and nutrient concentrations in the soils are so low that it is the difference in water availability which constitute the most important determinant of the usefulness of the soils for man. The finest grained soils with the best water-holding characteristics occur on the high alluvium, in a band running parallel to the river, but at distances of 500 to 1500 meters away from it. Except where the high alluvium is more than 2 kilometers wide, this band of fine-grained soils extends to the outer edge of the alluvium, where it meets the piedmont zone.

Soil-grain size influences water retention and is thus important both for (i) dry farming and (ii) commercial crops with high irrigation requirements. In between these two extremes, in cases where more limited types of irrigation are practiced, the depth to water table and the yields from wells are more important than soil texture in determining the value of the soil for farming. This brings to mind recent comments by Stevens on Indian farming in general: it is not necessarily the best soils which are the most intensively used, because factors of technology and water table may be the primary ones.

Within the present flood plain of the Atoyac River, well water is within 3 meters of the surface; within the zone of high alluvium, it lies between 2 and 10 meters down. In both these zones, water yields are usually adequate for small-scale irrigation of the specialized types described below. In the fan gravels of the piedmont zone, water is generally more than 10 meters below the surface, and well yields are only sufficient for immediate household needs.

A RELATIVELY FROST-FREE CLIMATE

At elevations of 2000 to 2800 meters, in areas like the Valley of Mexico or Cholula-Puebla, winter frosts may have been a real deterrent to year-round cultivation of maize until frost-resistant strains were developed, some time after the Middle Formative period. For example, Sanders' figures for the Valley of Mexico indicate that between October and February the area may have temperatures which are detrimental to maize. In contrast, temperatures on the floor of the Valley of Oaxaca are well suited to year-round growing of maize, even the primitive strains of the Early Formative.

At the level of the valley floor (1420 to 1740 meters, with an average of about 1550), the mean annual temperature is 20°C, with an annual range of 6°C and a daily range of 15°C. *Extreme* minimum temperatures over a recent 12-year period are close to 0°C. In any one year there is only slight probability of frost, and this largely in the higher parts of the valley. In the main Atoyac River floodplain south of Oaxaca City, all of which lies below 1550 meters, frosts are virtually non-existent, and the present-day Zapotec of this area cultivates sugarcane, which requires an 18-month frost-free period.

These favorable conditions deteriorate rapidly as one ascends the hills to either side of the Valley of Oaxaca. Above altitudes of about 2300 meters, summer temperatures are low enough to inhibit cultivation of maize, and wheat is at present a more reliable crop. At elevations of 3000 meters, mean daily minima in January are about 0°C, and the dominant cultivar is the potato.

The two ancient indigenous races of maize known so far for the Early Formative period in Mesoamerica—Nal-Tel and Chapalote—do poorly in cold conditions and are sensitive to highland rusts. The essentially frost-free nature of the southern Valley of Oaxaca probably gave it considerable advantage over the higher valleys of the nearby Mixteca at this early period (1500 to 900 B.C.), when only those primitive races of maize were known. In later periods, with frost-resistant strains of maize, this difference was probably less crucial, as high population densities in the Mixteca and the Valley of Mexico suggest.

PRECIPITATION AND HYDROLOGY

Mean annual rainfall on the floor of the Valley of Oaxaca varies from 490 millimeters at Tlacolula to 740 millimeters at Ocotlán. There is a general rise in precipitation with increasing altitude, so that the surrounding mountains at elevations of 3000 meters may receive almost 1000 millimeters annually. Open-water evaporation depends principally on temperature, and decreases with elevation. On the valley floor it averages 2000 millimeters annually (three to five times the precipitation), while at 3000-meter elevations it is only 340 millimeters (one-third the precipitation). Hence the growth of permanent pine forest on the high mountains, and the sparser cactus and mesquite-grassland cover of the lower slopes.

For the growth of annual crops without irrigation, the ratio of rainfall to open-water evaporation must remain close to 1.0 throughout the summer months, with June to August the most critical period. On the floor of the Valley of Oaxaca this ratio ranges from 0.50 near Tlacolula to 0.93 near Ocotlán. This range may be contrasted with similar figures obtained for the floor of the Valley of Tehuacán, which vary between 0.45 and 0.65. Thus conditions in the western part of the Valley of Oaxaca are somewhat more favorable for dry farming than are those in Tehuacán, a fact which was probably important during the early stages of agriculture.

Even more striking contrasts between Oaxaca and Tehuacán may be seen, however, when one examines their irrigation potential. The Tehuacán Valley has an extremely low water table (about 20 meters), but the valley lies just to the south and east of a block of limestone-travertine mountains which constitute a major aquifer. Very large quantities of subsurface water emerge from springs at the base of this range, near the western outskirts of the city of Tehuacán. Thus, shallow-well irrigation is impossible; much more feasible is a large-scale canal-irrigation system to carry water from the prolific springs out to the central and southern parts of the valley. Such a canal system was indeed developed at Tehuacán during later periods of its prehistory, when the population of the valley was already high.

Canal irrigation on a large scale is nowhere practical in the Valley of Oaxaca, where springs are small and surface flows are insufficient for irrigating more than a small area. However, because of the unusually high water table, shallow-well irrigation is widely practiced, and this technique, which requires relatively little effort and can be performed on an individual family basis, can be traced back to at least 700 B.C. and probably earlier.

FOOD-COLLECTING—INCIPIENT-CULTIVATION PERIODS: 8000 TO 1500 B.C.

The oldest archeological materials recovered in 1966 came from a series of caves and rock shelters near Mitla. These shelters occur in volcanic-tuff cliff faces at elevations of 1900 meters, near the transition from the piedmont to the higher mountains.

For years it had been known that this elevated region, 200 meters or more above the valley floor, was richer in surface finds of the food-collecting era than any other; when a recent lake on the valley floor was still considered a possibility, the hypothetical lake was often used to explain the restriction of these early cultures to the upper piedmont. The real reason why this zone was so consistently used in early times is that it has the richest and most varied assemblage of edible wild plants of the entire region. For the most part, shelters immediately overlooking the valley floor were used infrequently or not at all before 1500 B.C.; it was full-time agriculture, with its need for flat land and fine-grained soil, which eventually diverted attention from the upper piedmont and allowed the high alluvium to emerge as the zone of major utilization.

Between 7840 and 6910 B.C. (as estimated on the basis of radiocarbon determinations), the Indians who camped seasonally in Guilá Naquitz Cave collected acorns, pinyon nuts, mesquite beans, prickly pear and organ-cactus fruits, wild onion bulbs, hackberry, maguey *(Agave* sp.), *nanche (Malpighia* sp.), *susi (Jatropha* sp.), and a dozen other species, all of which were preserved within the cave by desiccation. Toward the end of this period, small black beans (*Phaseolus* sp.) and squash seeds *(Cucurbita* sp.) appear in the refuse; thus Guilá Naquitz is added to the list of sites known to belong to the "incipient cultivation" period in ancient Mexico.

A nearby cave, Cueva Blanca, dated at about 3295 B.C. by the radiocarbon technique, yielded a later food-collecting, incipient-cultivation horizon which is in most respects identical to the Coxcatlán phase (5000 to 3000 B.C.) defined by MacNeish at Tehuacán. It would appear that at this period the whole of the southern Mexican highlands was occupied by a series of related, seminomadic bands who moved seasonally from resource area to resource area and engaged in increasingly effective experiments with the growing of maize, beans, and squash.

Not only is it impossible to speak of "key" or "nuclear" regions at this time, it is also virtually impossible to define individual "culture areas" within the southern highlands. This suggests that, while cultures were still primarily food-collecting, the individual peculiarities of the various valleys were not especially significant. It was full-time agriculture which brought about specialized adaptations to local peculiarities of soil, rainfall, and water table and gave each valley its regional character. At this point, even slight differences in agricultural potential may have started certain valleys, like Oaxaca, on the path to nuclearity.

EARLY VILLAGE FARMING PERIOD: 1500 TO 600 B.C.

Several parts of the valley were selected as "pilot areas" in which to survey for early village farming communities, with subsequent test excavations. Chief among these was a 10-kilometer strip in the extreme northwest corner of the valley, near Etla. We concentrated on the Early Formative San José phase (1200 to 900 B.C.) and the Middle Formative Guadalupe (900 to 600 B.C.) and Monte Albán I (600 to 200 B.C.) phases (Figure 40.3).

In the Etla region, the most favorable agricultural land is that part of the high alluvium where the water table is within 3 meters of the surface. As shown in Figure 40.5, most Formative sites thus far located (including all the Early Formative sites) are concentrated in or adjacent to this zone. In the narrow parts of the valley, prime localities were the tips of piedmont spurs, which raised the villages just high enough above the alluvium so they would not flood in the rainy season. In wider parts of the valley, where the piedmont spurs are too far from the 3-meter water-table belt, villages were built on the high alluvium in areas where sandy soils provided them with the best-drained locations available. Our preliminary surveys in other parts of the valley suggest that the pattern observed at Etla is probably typical of the earlier part of the Formative. In areas of low water table, such as north and east of Tlacolula, evidence of Early Formative occupation is correspondingly sparser.

In this belt of high-water-table alluvium, which narrows to 500 meters near Etla and expands to 2 kilometers in the broad plain just south of Oaxaca City, the Zapotec practice a kind of rudimentary water control known as *riego a brazo* or "pot-irrigation." This technique was described in 1960 by Lorenzo, and we have since studied it in detail in the *municipio* of Zaachila.

"Pot-irrigation" involves the digging of a series of shallow wells right in the cornfield, tapping the stratum of water which lies between 1.5 and 3.0 meters from the surface. An acre of land may have ten of these small wells, which are filled in during the plowing season and then reopened when water is needed. Water is drawn up from each well in a 3-gallon pot and poured gently around the individual corn plants. By means of this system, farmers within the 3-meter water-table zone often achieve three harvests a year. At any time of the year, dry season or wet, this belt of pot-irrigated alluvium resembles a huge patchwork of small but highly productive gardens. *Riego a brazo* requires no large labor force or centralized control; it is carried out on an individual-household basis. However, the zone where this technique can be used constitutes a very small percentage of the valley-floor area in Oaxaca, and, as mentioned above, it cannot be used at all in low-water-table areas like the Valley of Tehuacán.

The association of San José and Guadalupe phase villages with this zone of pot-irrigation was very suggestive, but until recently no actual well to demonstrate the existence of the technique in the earlier part of the Formative had been found. In August of 1966, Richard Orlandini and James Schoenwetter of the Oaxaca Project discovered a Formative well which had been exposed by adobe-brickmakers in a bank some 50 meters back from the river at Mitla. Associated pottery dated the well to the Guadalupe phase, considerably strengthening our evidence for water control in the early village farming period.

Agriculture within this high-water-table zone supported villages of large size and material wealth. The best-known site of the San José and Guadalupe phases is San José Mogote, which we tested in 1966 (see Figure 40.5). Here Early Formative artifacts can be picked up over 40 acres of a piedmont spur surrounded on three sides by alluvium. Rows of post molds and burned wall fragments suggest that houses were large and rectangular, with partial stone foundations and wattle-and-daub walls which were plastered with mud and whitewashed. Besides the usual internal features, like hearths and bell-shaped sub-floor cooking pits, one San José phase house had a recessed circular area a meter and a half in diameter, which had been plastered and painted red. Around this circle were scattered fragments of figurines, exotically decorated pottery, fragments of black and white mica, raw chunks and small polished mirrors of magnetite, and ornaments and discarded fragments of imported marine shell. In levels belonging to the early Guadalupe phase, such scatters gave way to an artificial platform of earthen fill with stone retaining walls oriented almost due north-south, and presumably having had a ceremonial function. Such orientations characterize later ceremonial structures in the valley as well.

The evidence of long-distance trade in the San José phase (which is lacking in earlier periods) reflects two things: an increasing interest in status differentiation (with artifacts of imported materials serving as insignia of status) and formalized contacts with other Indian groups in differing environmental zones of Mesoamerica. Marine pearl oyster and *Spondylus* shell were imported from the Pacific, while *Neritina* and pearly freshwater mussels came from the Gulf Coast. *Anomalocardia surugosa,* a mollusk eaten by Formative villagers in the estuary zone of the distant Chiapas-Guatemala coast, were also imported.

Most important are the chunks and mirrors of magnetite, a raw material native to the Valley of Oaxaca, for nodules of this metal are known to have been polished into concave mirrors and buried in ceremonial caches by the "Olmec" peoples of the southern Gulf Coast. At present, the Valley of Oaxaca must be considered a possible source for the Olmec magnetite.

RISE OF TOWNS AND CEREMONIAL CENTERS: 600 TO 200 B.C.

During the later stages of the Middle Formative period, villages within the 3-meter water-table zone increased in size and number. Coupled with this population increase, which we attribute to the success of dry farming and pot-irrigation in that part of the high alluvium, came the first sizable spread of settlement up the more permanent tributaries

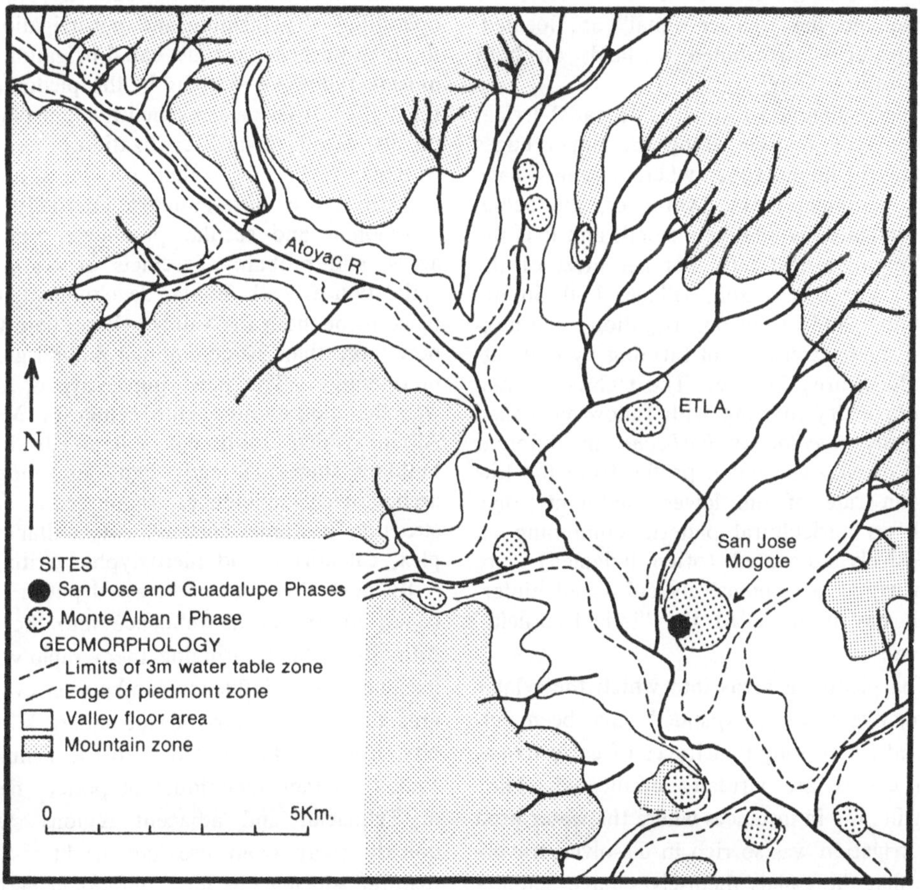

FIGURE 40.5
"Pilot" survey area in the northwestern part of the Oaxaca Valley, showing the distribution of Formative archeological sites with regard to physiographic areas and water resources (see text).

of the Atoyac into the piedmont (Figure 40.5) These latter sites are of two types: "habitation" sites on the first terrace of the stream or a low ridge near it, and "ceremonial" centers on hilltops nearby.

We doubt that this pattern of settlement was random. Most sites outside the high alluvium at this period are on perennial streams, and, like the present villages of the piedmont zone, they are located not downstream, at the point where most water is available, but upstream, where the water can be most effectively diverted for irrigation. Today, these villages divert the water into canals which follow the natural contour of one of the piedmont spurs downstream until they come to the crest of the spur. Here the village and the "master canal" are located, and water is distributed to fields on both sides of the spur.

This technique of small-scale canal irrigation is only feasible along the upper edges of the piedmont zone, where streams have good perennial flows. Moreover, the actual area irrigated is relatively small, and it is the communities upstream that get most of the water. For this reason, villages both in the piedmont and in the pot-irrigation zone augment their water-control farming by cultivating the nearby hillsides. The technique used is simple dry farming with fallowing, called variously *tlacolol* or *barbecho* in different parts of Mexico, and it profits from the low erosion rate of the lower piedmont zone. Such an agricultural pattern, combining an intensively cultivated (often irrigated) core area with a less intensively cultivated hinterland, has been called the "infield-outfield" system.

The piedmont areas into which these later Formative farmers expanded have been cultivated for so long that traces of early irrigation canals are virtually eradicated. They remain only in instances where the water used for irrigation was so rich in dissolved travertine that the canals themselves have actually been "fossilized" through deposition of this calcareous material. In 1966, James Neely of the Oaxaca Project investigated one such area in the mountain zone near Mitla.

This site, called Hierve el Agua, is a complex of "fossilized" ancient irrigation canals covering a square kilometer of hillside below a spring particularly rich in travertine. A series of dry-laid stone terraces had been irrigated by means of small canals which carried the water down to the fields and along the tops of the terrace walls. Neely's 40 test pits dug into these terraces reveal an occupation beginning before 300 B.C. and expanding through all subsequent periods of Oaxaca pre-history. It is probably no accident that this evidence of small-scale canal irrigation begins during the first sizable expansion out of the 3-meter water-table zone and up the perennial tributaries.

With at least four agricultural systems operating—dry farming and pot-irrigation in the high-water-table zone, canal irrigation and hillside fallowing systems in the piedmont—the Valley of Oaxaca reached another plateau on its climb toward civilization. San José Mogote grew to more than 100 acres, and in the process it differentiated internally into ceremonial and secular precincts, cemetery areas, and probably precincts of craft specialization as well. Similar developments took place throughout the valley, where there were now more than 30 ceremonial mound groups in operation. The most impressive of these was the mountaintop elite center of Monte Albán, which (although still in its initial building stages) already had, according to archeological evidence, monumental construction, bas-relief carving, a stela-altar complex, calendrics, and hieroglyphic writing.

By now, this area of "massed power" had begun to extend its influence into the surrounding valleys, bringing them rapidly into its sphere. This influence can be seen over an area of tens of thousands of square kilometers, from the Pacific Coast to the Tehuacán Valley. In fact, techniques of pottery design in Tehuacán and adjacent regions swung quickly away from previous traditions and featured, during this and subsequent periods, provincial imitations of the Valley of Oaxaca styles.

While much of the surrounding area may willingly have entered into a symbiotic relationship with the Valley of Oaxaca for the economic advantages it offered—such as a ready market for their surplus and their locally specialized products—there are hints that not all the marginal valleys joined peacefully. Caso believes that at least one set of glyphs carved in stone at Monte Albán represent conquered neighboring towns, and Coe has recently suggested that an even earlier series of bas-reliefs, the so-called *danzantes* of Monte Albán I, depict slain and mutilated captives. While these war-like interpretations remain to be proved, they are in no way inconsistent with what is known enthnographically of groups of a chiefdom stage of organization.

EVALUATION OF HYPOTHESES

With these data in mind, we can now tentatively evaluate the relevance of various developmental hypotheses to what happened in the Valley of Oaxaca. First of all, it no longer seems likely that slash-and-burn was the sole early-farming technique in Mesoamerica; in fact, given the locations of the cave areas where farming began, such as the moist barrancas of the arid Tehuacán Valley, it is even possible that some kinds of water control (like the terracing of wet arroyos) are as old as agriculture itself. However, our work in Oaxaca strongly supports Palerm and Wolf's view that the ability to assimilate new agricultural techniques through time is a key factor in retaining nuclearity.

Furthermore, the four systems we postulate for the Monte Albán I period did not represent the final stage of Zapotec agriculture. In later periods (A.D. 100 to 900), large mound groups and habitation sites cover areas of the high alluvium where only dry farming or flood-water farming is possible; by A.D. 1300 the area farmed included not only the entire valley floor but also the lower slopes of the mountains, which were frequently terraced. The terraces apparently depended on rainfall and on the lower evaporation rate of the north- and east-facing slopes. These data suggest that the assimilation of new farming techniques was also a way of bringing into cultivation more and more of the previously unproductive physiographic units of the valley: the greater the number of systems used, the greater the acreage producing at top capacity.

Our evidence for early irrigation in the Valley of Oaxaca will undoubtedly please advocates of the "hydraulic theory" of state formation. However, we see no evidence in Oaxaca for irrigation systems so large that they would necessitate a strong centralized authority. In fact, we fear that the hydraulic theory—at least, in its purest form—may at times obscure the real effects of irrigation in ancient society, which are as varied as the techniques themselves. In Oaxaca, canal irrigation of the Hierve el Agua type is applicable to a very tiny portion of the valley, and it is no more productive (in terms of labor input relative to crop yield) than pot-irrigation. Moreover, many canal-irrigating villages are so high in the piedmont that only summer crops can be grown. What canal irrigation *did* do was open up an additional niche within the valley which had not previously been agriculturally productive.

Perhaps most significantly, in canal-irrigation communities there tends to be less equitable distribution of land and property rights than in pot-irrigating communities. As past writers have observed, irrigated land is "improved" land; it represents an investment which makes it a more scarce and competed-for resource than it previously was, and leads to problems of inheritance of property and differential access to good land. If a developing society has tendencies toward status differentiation already, which the Formative peoples of Oaxaca did have, these tendencies can be aggravated by control and inheritance of irrigation systems. It may be that early irrigation in Oaxaca can be most profitably viewed in these terms.

SUMMARY AND CONCLUSIONS

The Valley of Oaxaca's large flat floor, high water table, low erosion rate, and frost-free floodplain give it a higher agricultural potential than that of most surrounding areas.

The development of the pot-irrigation system early in the Formative period gave it a head start over other valleys, where the low water table did not permit such farming; Oaxaca maintained its advantage by assimilating canal irrigation, *barbecho,* infield-outfield systems, flood-water farming, and hillside terracing as these methods arose. With the expansion of population in the high-water-table zone of the high alluvium, competition for highly productive land and manipulation of surpluses may have led to initial disparities in wealth and status; competition probably increased when canal-irrigation systems were added during the Middle Formative, improving some localities to the point where one residental group owned land more valuable than that of its neighbors.

Trade in exotic raw materials, which appear to have served as the insignia for status over much of Formative Mesoamerica, increased the wealth of the Oaxaca communities, and their elite made contact with the elite of other cultures, such as the Olmec. Such contact probably stimulated exchanges of the "lore" known only to the elite—calendrics, hieroglyphic systems, and symbolic art—thus widening the gap between farmer and chief. Through cooperation or coercion, the Oaxaca Valley chiefdom drew together a symbiotic area of 50,000 square kilometers, commemorating its accessions with stone monuments carved in bas-relief. By the start of the Christian Era it was the dominant political entity in the southern highlands of Mexico, and had become a true state.

The Oaxaca Project has only begun, and our reconstruction must remain a tentative one. We are especially aware that we cannot as yet integrate into this scheme the botanical and palynological materials from Oaxaca which are currently undergoing study. They will make the story still more complex, but civilization is a complex process; single-cause theories, no matter how attractive, are inadequate to explain it.

41. IRRIGATION IN THE OLD ACOLHUA DOMAIN, MEXICO

ERIC R. WOLF and ANGEL PALERM

Reprinted from Southwestern Journal of Anthropology *(1955) 11 (3): 265-81.* *Eric R. Wolf is Distinguished Professor of Anthropology, Lehman College, City University of New York. His major interests are the study of peasant societies in general and the investigation of peasants in Latin America and Romance-speaking Europe in particular. He has done field work in Puerto Rico, Mexico, and the Italian Alps, and is the author of* Sons of the Shaking Earth, Anthropology, Peasants, Peasant Wars of the Twentieth Century, *and (with John Cole)* The Hidden Frontier: Ecology and Ethnicity in an Alpine Valley. *Angel Palerm is Professor of Ethnology at the Escuela Nacional de Antropología e Historia de Mexico, and Director of the Escuela de Ciencias Políticas y Sociales at the Universidad Iberoamerican, Mexico City. One of his principal research interests has been the ethnography and ethnohistory of Mesoamerica. He has worked intensively on the origin and development of urban civilization in the New World. Among his books are* The Tajin Totonac, Irrigation Civilizations, *and* La Agricultura y los Origenes de la Civilización en Mesoamerica.

■ Until recently, it was widely believed by students of cultural history that extensive irrigation networks are one of the most important factors in the development of states. The reasoning behind this belief often went as follows. Large-scale irrigation networks introduce such a profound change in man's relationship to the habitat that almost all the institutions of society become dependent in one way or another on this new technology. Irrigation networks often serve many communities and areas; hence there must be some political mechanism to regulate community and regional relationships to assure equitable and regular access to water. Furthermore, because irrigation networks are vulnerable to aggression by external enemies, they must be centrally controlled and protected. These considerations often were thought to be forces responsible for the emergence of states.

In this selection, whose significance was discussed briefly in Sanders' article, Wolf and Palerm work backward in history from present-day to pre-Hispanic Mexico, carefully distinguishing the elements in the equilibrium between man's culture and his habitat. Their historical reconstruction of the history of the Old Acolhua domain, which also can serve as an admirable methodological model, underlies their argument that the construction of the irrigation networks was indeed a profoundly important adaptation but that it was part of a state's attempt to meet the demands resulting from the pressure of population on the land. Equally important, these irrigation networks were an instrument of a state that sought to impose its control over the population by forcing the society to intensify its agricultural production. Irrigation, therefore, is a multifaceted adaptation; it is an aspect of a society's relationship to the food-yielding habitat and an element in the equilibrium achieved by a state in its relationship to the polity.

In the concluding section of their paper Wolf and Palerm introduce the concept of "cultural devolution" referring to the process by which a region changes from a "key" to a "marginal" area. The emergence of statehood and irrigation networks in the Old Acolhua domain can be seen as an example of the steady, if not inexorable, evolution of social organization and technological adaptation in a propitious habitat. If it had been given more time, what more would have evolved?

The inhabitants of Mesoamerica were not given more time, and history is such that hardly any cultures are given "more time." When new historical conditions are created for a society—when, for example, it is overrun by another or loses its dominant position in a civilization—it must adapt to its new environment. We have seen that one such necessary adaptation is in language, although it is a very difficult adaptation to make. In the Old Acolhua domain, a reduction in cultural complexity and in the level of technology in response to the new overall environment created by the Spaniards was an adaptive necessity: a shift from intensive agriculture to sheep raising, wool production, and wheat farming.

The most succinct expression of the debate among anthropologists over the relationship of irrigation to the social and natural environment can be found in comparing Robert McC. Adams' *The Evolution of Urban Society* (Chicago: Aldine, 1966) with Karl Wittfogel's *Oriental Despotism* (New Haven: Yale University Press, 1957). Adams, in his comparative analysis of Mesopotamian and Mesoamerican civilizations, takes the view of Wolf and Palerm. Wittfogel's book has had great influence among many anthropologists and is primarily responsible for their familiarity with the concept of the "hydraulic society." Although the Wolf-Palerm-Adams point of view seems to fit best with the available data, Wittfogel's analysis is an important source for much information that is not readily available elsewhere. The reader may also wish to consult *Irrigation Civilizations: A Comparative Study, A Symposium on Method and Result in Cross-Cultural Regularities* (Washington, D.C.: Pan American Union, 1955). This compact volume is made up of papers by Julian H. Steward, Robert McC. Adams, Donald Collier, Angel Palerm, Karl Wittfogel, and Ralph L. Beals; it was published as Social Science Monographs 1 of the Pan American Union. Volume III in this series also is of relevance: *Studies in Human Ecology*, which is made up of papers by Angel Palerm and Eric Wolf, Waldo R. Wedel, Betty J. Meggers, Jacques M. May, and Lawrence Krader (Washington, D.C.: Anthropological Society of Washington, 1959). ■

INTRODUCTION

IN 1861 SIR EDWARD TYLOR reported on his visit to the Hill of Tetzcutzingo—situated east of the town of Tecoco in the valley of Mexico—in the following words:

> We did not go first to Tetzcotzinco itself, but to another hill which is connected with it by an aqueduct of immense size, along which we walked. The mountains in this part are of porphyry and the channel of the aqueduct was made principally of blocks of the same material, on which the smooth stucco that once had covered the whole inside and out, still remained very perfect. The channel was carried, not on arches, but on a solid embankment a hundred and fifty to two hundred feet high, and wide enough for a carriage road.

He concluded from this visit that:

> the ancient Mexicans were not, it is true, to be compared with the Spanish Arabs or the Peruvians in their knowledge of agriculture and the art of irrigation, but both history and the remains still to be found in the country prove that in the more densely populated parts of the plains they had made considerable progress.

In the summer of 1954, the present writers began a study of the ecology of the area which Tylor described above. We soon discovered that the works of the Tetzcutzingo formed only a small part of a much larger pre-Hispanic system of irrigation. Parts of this system are used by the inhabitants of the area to this day. In this paper we hope to present the essential data collected during our brief survey, and to discuss the theoretical issues raised by them.

The Old Acolhua domain—within which the Tetzcutzingo is situated—extends from the eastern shoreline of Lake Texcoco into the sierra which divides the Valley of Mexico from the valleys to the east. Work in this region offers a great challenge to the modern anthropologist. The analysis of rich archaeological remains tests his archaeological skills; an abundance of pre-Hispanic and post-Conquest documents require historical imagination and linguistic ability for proper interpretation; and his ethnological techniques are required to study the present-day and fully functioning system of irrigation with its roots in the pre-Hispanic past. In an era of increasing specialization, the anthropologist can here employ the repertory of his specialized techniques in a study of culture as a whole.

MODERN SETTLEMENT AREAS

We may divide the area under discussion roughly into four major ecological belts.

THE SIERRA

The highest of these is the sierra, rising to the east above the 2,750 m (9,075 ft) contour towards its highest point in the immediate area, Mt. Tlaloc (13,270 ft). This sierra forms part of the continental divide, and separates the Valley of Mexico from the Valleys of Puebla and Tlaxcala. At the foot of the sierra, just below the 2,750 m contour lies a fringe of villages. These are, from north to south: Santo Tomas Apipilhuasco, San Juan Totolapan, San Gregorio Amanalco, Santa Maria Tecuanulco, Santa Catarina del Monte, and San Pablo Ixayoc. Between San Juan Totolapan and San Gregorio Amanalco are located the springs of San Francisco which feed the larger part of the irrigation system today and probably did so in the past. The main spring supplying the southern canals, called Texapo, issues on the slopes of Mt. Quetzaltepec above San Pablo Ixayoc.

Each of these sierra communities has a church and a small plaza, but dwelling units are widely scattered, each house with its garden, vasiform granary, and sweatbath standing apart from the others. Crops are grown on levelled plots of uneven dimension (*bancos*), hedged about by maguey plants. The main food crops are wheat and corn, both of which are grown primarily for subsistence. Barley and rye are also grown. Wheat is preferred since it resists frosts and is thus more secure than corn. A few flowers are grown for sale in valley towns, but *pulque*—produced from the maguey—and wood burned to charcoal are the main products sold for cash in the lower piedmont and flatland communities. Some hunting is carried on for both food and pleasure.

The system of landed property in the sierra communities must receive further study. Most land seems to be held as *ejidos* by the villages enumerated. Santa Catarina del Monte owns its forest land as part of its *ejido*. Santa Maria Tecuanulco, on the other hand, does not possess *ejido* land.

ARID ZONE

Below the sierra villages lies a belt of arid and eroded country, extending roughly from the 2,750 m (9,075 ft) contour to the lower 2,500 m (8,250 ft) contour. Bare of dwelling units and settlements today, it was heavily populated before the Spanish conquest, and the more impressive earth works built in connection with irrigation are found here. These will receive further discussion below.

PIEDMONT

Where the hills descend to the flat land of the lake shore at the 2,500 m (8,250 ft) contour, we find another cluster of village communities. These piedmont communities, running from north to south, are San Juan Tezontla, Santa Ines, San Joaquin Ixtilxochitl, Purificacion (formerly Tepetetitla), San Miguel, Tlaixpan, San Nicolas Tlaminca, Santa Maria Nativitas, San Diego Nativitas, and Tequesquinahuac. These piedmont villages grow corn for subsistence and flowers for sale in Mexico City flower markets. Fruit trees are cultivated, and fruit is sometimes sold, sometimes bartered for such lake products as moth eggs (*ahuauhitli*) collected by the people on the lake shore. San Joaquin Ixtilxochitl resembles the flat land communities in possessing an artisan specialty, the production of powder for fireworks.

Cultivation seems to be much more intensive in the piedmont than in the sierra, and agricultural methods show greater care. Especially at San Miguel Tlaixpan and San Nicolas Tlaminca, agriculture is carried on irrigated terraces (sing. *kalálli;* pl. *kaláltin*). carefully built of adobe and stone. Plants are allowed to mature in seedbeds before transplanting. Flowers raised for the market are covered to protect them against frost. Animal fertilizer is collected and in general use. Water distribution is handled by a communal body of water commissioners (*regidores de agua*), and each landowner receives water once every Mesoamerican month of twenty days. The amount of water distributed to each person depends on the size of the terraces owned by him. Small storage tanks, situated on the terraces, are used to conserve water for household use. It is interesting to note that absence from the weekly communal work (*faena*) is punishable by withdrawal of water rights for the following period of distribution.

In comparison with the sierra villages, dwelling units in the piedmont communities tend to cluster around the church and plaza, but the settlement pattern in general still shows more dispersion than in the flat land below. At San Miguel Tlaixpan and San Nicolas Tlaminca, houses are usually located on or near agricultural terraces.

Land ownership in these communities represents a complex picture, and many of its aspects must await further research. At San Miguel Tlaixpan and San Nicolas Tlaminca, agricultural terraces are privately owned. One informant reported that a house and the terraces owned by the house owner form an indissoluble complex, and that both carry the same name. *Canoatitla*, for instance, is both the name of a house and of the terraces associated with the house. According to the same source, one cannot be sold without the other. A similar linkage of house and fields has been reported by Redfield for the Nahua of Tepoztlan, although this is denied by Lewis. In both cases we may be dealing either with a reflection of an earlier custom when membership in the community may have been based on rights to land within the community, or simply with an ideal norm serving to restrict transactions of landed property. Whatever the case, today terraces are certainly subject to sale and purchase, and members of neighboring communities can acquire land in San Miguel Tlaixpan or San Nicolas. While terraces and irrigated land are privately owned, the fields watered by rainfall only, which dip down from the village to the flat land below, are owned by the community as *ejido* lands.

THE FLAT SHORE LAND

The flat shore land is characterized by three kinds of building clusters. Close to the piedmont villages are a number of haciendas and fulling-mills constructed during the colonial period. They were apparently built on lands which belonged to the piedmont communities and appropriated some of their water supply. These are La Blanca, El Batan, and Molino de Flores.

The Indian communities of the flat belt are tightly nucleated settlements, in strong contrast to the piedmont and sierra villages. Various kinds of craft production supplement or overshadow the production of maize. The best-known of these artisan communities is probably San Miguel Chiconcuac, famous for its sarapes. These Indian communities are entering into competition for land, labor, and water supply with a growing number of dairy farms that are making inroads into the area. The flat belt is well on the way to becoming part of the dairy belt of Mexico City.

MODERN IRRIGATION

The whole area is tied together by a network of irrigation canals, and by common dependence on the springs of the sierra where the water originates. The major springs of San Francisco at the northern extreme of the area are carefully guarded by the villagers of San Gregorio Amanalco against outside intruders, and in popular belief also receive the protection of supernatural guardians of the water (sing. *anáki;* pl. *anáke*) who punish disturbances of the springs with madness and disease.

Shortly below the springs, the original canal divides in two. A northwestern branch flows through three piedmont communities—San Juan Tezontla, Santa Ines and San Joaquin Ixtlilxochitl—down into the flat land to water Blanco, San Toribio Papalotla, San Miguel Chiconcuac, and others. The piedmont communities can divert or cut off the flow of water during periods of serious water shortage. The southwestern branch of the same canal flows through the three sierra communities of San Gregorio Amanalco, Santa Maria Tecuanulco and Santa Catarina del Monte. At Tecuanulco a separate spring feeds into the system. When the canal reaches the piedmont, it divides in turn. One branch of this second canal feeds Purificacion. The other branch is dammed up and stored temporarily in two small reservoirs, called locally Ocotochco and Apancingo. The water in both reservoirs is now impounded by concrete dams of recent local construction. Further inquiry might determine the manner of their construction before they were rebuilt. It is possible that before reconstruction they were controlled by barriers of vertical posts, branches, and earth fill similar to the one located by Pedro Armillas and ourselves in the Valley of Teotihuacan.

San Pablo Ixayoc and Tequesquinahuac, as well as the villages on the southern side of the Tetzcutzingo, receive their water from the well Texapo, located on Mt. Quetzaltepec. A branch from this southern system also supplies Huexotla, on the southern frontier of the Old Acolhua domain.

PRE-HISPANIC SITES AND WATER WORKS

The strategic archaeological sites lie at the juncture of the piedmont with the flat land below. They are associated in the main with two mountains which stand out prominently from the chain of foothills of which they form a part. The first of these—Mt. San Joaquin—is situated in the north of the area. The second—the Hill of the Tetzcutzingo already mentioned—is situated in the south.

THE MT. SAN JOAQUIN COMPLEX

Mt. San Joaquin is formed by a smaller knoll extending to the northwest and a large mountain mass, running from northwest to southeast. On the smaller knoll, we located an artificial mound topped by a small round platform. It can be approached over a series of terraces which give way to a large round platform and a series of partly destroyed steps.

The saddle which links this knoll with the main mass of the mountain is carefully terraced. The remains of a canal roughly 2,000 yards long follow the western slope of the main mountain mass and linked these terraces with an aqueduct located at the southeastern extremity of the mountain. The canal was shored up by the three retaining walls, which are still visible in parts. The aqueduct is 1,090 yards long, 13 yards wide at its widest point, and 13 yards high at its highest. Its northern terminus is marked by a small artificial mound, locally called Texalco. Both aqueduct and large canal have been abandoned, but the pre-Hispanic canals approaching the aqueduct are still in use to carry water to Purificacion. The aqueduct itself may be the one mentioned in the *Titles of Tetzcotzinco* in the following words: "Third, Mt. Quauhyacac. All of it [water] is borne around the back of the hills in an aqueduct."

The artificial knoll and the associated platforms and terraces on Mt. San Joaquin suggest a ceremonial center or an upper class residence surrounded with a ceremonial aura. The round platforms would seem to link the site with the worship of the god Quetzalcoatl. Religious significance may also have been ascribed to the numerous caves which dot the mountainside. The association of constructions and caves makes it not impossible that Mt. San Joaquin was the *Quauhyacac* of the codices, in whose caves the kings of the Chichimec dynasty were buried.

The water works and the agricultural terraces with which they are connected are located so far up the mountain, however, that they seem to have been built to supply food and water to the occupants of the mountain when these could not be supplied from the flat land below. This suggests that we are dealing not only with a ceremonial center, but also with a military strong point. This suggestion receives support from the geographic location of the mountain. Mt. San Joaquin and the western hill at Tezoyuca—situated directly northwest of it—form ideal defensive positions against possible inroads from the Valley of Teotihuacan to the north, from Tlaxcala to the east, and even from the southern borders of the Old Acolhua domain. We should not forget that the political dominance of Texcoco was challenged by its southern neighbors at Coatlichan and Chalco during the formative phases of the Texcocan state. The hill at Tezoyuca seems to have been surrounded by a defensive system of ditches and escarpments. Its occupants were supplied by a small aqueduct, today badly destroyed, which ran from west to east toward the saddle between the twin hills. This is probably the aqueduct mentioned in the

Titles as "the aqueduct which skirts the hill" at "Tezoyuca, where Cuauhyacac forms a neck."

THE TETZCUTZINGO COMPLEX

South of Mt. San Joaquin and directly across from it lies the Hill of the Tetzcutzingo, once the favorite retreat of the kings of Texcoco, and especially of King Netzahualcoyotl whose name is associated with the so-called Baths of the King on the western side of the hill.

According to Ixtlilxochitl, the Tetzcutzingo was

> The most agreeable of the gardens [of King Netzahualcoyotl] and the one most filled with sights . . . because it not only had a large winding path which permitted people to climb to the top and to walk all over it, but was also terraced, with some terraces built of mortar and others hewn into the very rock; and to lead the water from its spring to the wells, fountains, baths and canals which distributed it to irrigate the flowers and tree clumps of this wood, a strong and very high wall of mortar had to be built from one mountain to the next, at incredible height, to support a canal which led to the very top of the woods.

Padre Davila gives further details:

> Due to the industry and greatness of the Kings of Tezcuco water was brought in canals over a distance of two leagues to irrigate the hill; mountains were levelled and valleys filled to allow the water to flow under its own power until it reached the very top of that hill, where it was allowed to spiral down to irrigate all the trees and plants as it has done until today.

The canals for this "royal town of Tezcotzinco . . . all full of fine obsidian, of jade bracelets . . . of gold" were fed through two aqueducts. The first—rediscovered by Pedro Armillas—is about 327 yards long, 22 yards high at its highest point, and 44 yards wide at its widest; and located at a place called Amanalco north of San Pablo Ixayoc. The contingent canals are nearly all destroyed, but enough traces remain to reconstruct their course. They apparently ran along the southern contour of the hills which extend out to the Tetzcutzingo, "whereby it came alongside Chiconquiauitl, thence straight alongside Tetecatl (Metecatl) [still the name of the hill immediately behind the Tetzcutzingo] thence by Tetzcotzinco." The aqueduct which connected Tetzcutzingo and Metecatl is the one described by Tylor in the paragraphs cited at the beginning of this paper. Then, as now, Huexotla does not seem to have formed an integral part of the system. Netzahualcoyotl left the decision to supply Huexotla to the discretion of his lords Tlazolyahuitl and Maxiuitl. "And the Quetzaltepec water," he is quoted as saying, "perhaps Huexotla will ask . . . for a little, and when you want it you may have it, because it is your water, your property." Huexotla was in fact located in the border area between the Old Northern Acolhua domain and the southern Acolhuacan, and in disputes between the two areas before the final consolidation of the Texcocan state seems to have formed unstable and changing alliances with one or the other.

TERRACING

In a number of places, the irrigation canals and aqueducts were associated with terraces which are today abandoned. Remains of such terraces are best preserved on the reverse eastern slopes of the piedmont, though they may have been more widespread in the past and their traces wiped out by erosion. Such terraces are found on the northeastern slope of Mt. Cocotl, the third cone in the line of hills ending in the Tetzcutzingo; on the reverse slope of Mt. San Miguel; and on the eastern side of the aqueduct leading to Mt. San Joaquin (called Hueytlalli, "big land," by local informants). The terraces on the reverse slope of Mt. San Miguel are located above the present-day canals, thus indicating that the water level was higher at the time of their construction.

PRE-HISPANIC PATTERNS OF SETTLEMENT

We can predict on purely theoretical grounds that the introduction of irrigation will tend to increase agricultural productivity which will lead in turn to an increase in population. In fact, the systems described here must have supported a pre-Hispanic population of great density. The eroded arid belt between the sierra and the piedmont contains enormous quantities of sherds of domestic pottery and numerous traces of domestic construction. The accumulation of such a large number of sherds during what seems to have been a relatively short period of time thus points to a large population where there is not even an isolated settlement today.

The main settlement areas associated with irrigation seem to have been three in number. The first extended from Mt. Cocotl behind the Tetzcutzingo to Amanalco where the aqueduct was located. The second ran from the aqueduct near Mt. San Joaquin to within a mile of Santa Maria Tecuanulco. The third was located between Santa Ines and San Gregorio Amanalco. House sites seem to have been located along the crests of ridges, with fields falling off gradually along the slopes up to the point where these fell off steeply into the deep ravines between the crests.

AGE OF THE PRE-HISPANIC SYSTEM

Two kinds of evidence, pottery and historical documents, may be used to arrive at a rough date for the construction of the system. All shreds collected in the area indicate pottery of Aztec III and IV types. These belong to the century preceding the Spanish Conquest. No pottery belonging to an earlier period has so far come to light in the vicinity of the water works. This late date is supported by the *Titles of Tetzcotzinco*. In this document King Netzahualcoyotl addresses his lords in the following words:

> Let the Chichimeca assemble and go and take the springing water. It rises in Yelloxochitlán, in truth, it is there; fetch it hither. And by virtue of your blood you will obtain it so as to bring it here so as to guide it here, by your toil you will fetch it in order to work in my town, Tetzcotzinco.

This seems to indicate that the system was constructed in association with the royal residence at Tetzcotzinco. The year Tetzcotzinco was built is given in the *Códice en Cruz* as 1 Rabbit, or 1454.

MOTIVES FOR THE CONSTRUCTION OF THE SYSTEM

The entire history of the Acolhua domain is characterized by a strong tendency to intensify food production. We have shown elsewhere that the central political problem of the Texcocan state consisted in establishing the dominance of "Toltec" economic, social, and political patterns over the purely predatory patterns of the "Chichimecas." The kings of the Chichimec dynasty made every effort to force the Chichimeca to adopt the ways of the resident Toltec population by forcing them to carry on agriculture, to change their diet to include more agricultural products, and to settle in nucleated villages. On the other hand, they protected the Toltec agriculturists, married into their noble families, and added to the strength of the Toltec sector by furthering continuous settlement of Toltec groups in their domain. The construction of the Texcocan systems of irrigation thus seems to represent merely the culmination of a long-established trend.

If our estimate of when the system was constructed proves correct, then its construction coincides with the great period of famine in the Valley of Mexico in the middle of the 15th century, when starvation was so severe that men were willing to sell themselves as slaves to the people of the east coast to obtain sufficient food. Indeed, Plate II of the *Códice en Cruz* combines under the date sign of 1 Rabbit (1454) the signs which stand for the construction of Tetzcutzingo with the sign of a naked child vomiting liquid, the symbol of a dry spell and starvation.

We should like to suggest that wide-spread starvation and the building of the irrigation system described bear some sort of relationship to each other. We must keep in mind that the lagoon of Texcoco was considerably larger and the flat belt much smaller in both pre-Hispanic and colonial times than it is today. The expansion of the flat belt is due to the gradual drying up and shrinkage of Lake

Texcoco. In Netzahualcoyotl's time, Tezoyuca, Atenco, Texcoco, and Huexotla were all located on the shores of the lake. Within this narrow corridor, the natural increase of the population alone would inevitably exert pressure on available resources. This pressure would be increased by the repeated waves of Toltec immigrants. Given a tradition of agricultural improvement, it seems likely that the Texcocan systems of irrigation were constructed to meet the problems of increased population pressure. We are probably dealing with an area which was colonized planfully from an already established political center, spurred and organized by the state under the aegis of the "Toltecized" Chichimec kings.

IMPLICATIONS

THE STATE AS ENTREPRENEUR OF IRRIGATION WORKS

If our dating of the Texcocan system is correct, we must recognize that we are dealing with a case in which the state did not grow out of irrigation, but preceded it. There has been a tendency in recent anthropological writing to see in irrigation the primary cause of state formation, rather than one of a number of possible causes. Emphasis on the functional importance of irrigation and a strong interest in cultural causation should not, however, lead us to over-generalize our data. The Old Acolhua domain achieved political consolidation under the fourth king of the Chichimec dynasty, Quinatzin, in the beginning of the 14th century. This unity was reestablished before construction began on the Tetzcutzingo when King Netzahualcoyotl returned from exile in 1431. We have indicated above that the construction of the Texcocan system would seem to represent the culmination of the efforts of the Chichimec dynasty in the economic field. Once established, of course, irrigation probably operated in turn to centralize and intensify political controls.

This raises the question whether we are dealing with an isolated case, or with a more general Mesoamerican pattern. Post-Formative cultural development in Mesoamerica falls into two large over-all stages. The first was characterized by the rise of great ceremonial centers like Teotihuacan, some of which seem to have become "true" cities. The second stage has been labelled "militaristic," and was characterized by the development of a large number of warring city-states. These two stages seem to have occurred also in the growth of hydraulic civilizations in the Old World. But there, it should be noted, both the earlier theocratic stage and the later militaristic stage were characterized by irrigation agriculture. So far no evidence has come to light that the theocratic ceremonial centers of Mesoamerica were in fact based on irrigation. This has been asserted, but never substantiated. Until further research uncovers evidence of irrigation in the vicinity of Teotihuacan and Monte Alban and allows us to judge the scale of such watering systems, we must admit the possibility that the ceremonial centers of Mesoamerica became "true" cities and gave rise to states without a technological basis in irrigation. New and highly productive soils which could support a high density of population, food resources making for an adequate dietary balance, the development of transportation by water and the organization of human carriers, a high development of craft production and trade, a multiplication of the organizational functions exercised by the priesthood may all have been interrelated factors in the development of the large Mesoamerican ceremonial centers. Should irrigation be shown to have been absent or of relatively minor importance during the theocratic stage, Mesoamerican development and the growth of hydraulic civilizations of the Old World would show convergence, rather than parallelism.

THE SCALE OF PRE-HISPANIC IRRIGATION

We believe that our data on irrigation in the Old Acolhua domain can be taken as further evidence that irrigation was of importance in pre-Hispanic times. Even if it should be demonstrated that it had little influence on the development of a theocratic order in Mesoamerica, its established functional importance in other parts of the world requires that attention be paid to it in any analysis of Mesoamerican social and cultural systems. Recent investigations have demonstrated that Kroeber's earlier dictum to the effect that "there is little evidence that irrigation was of basic importance anywhere in Mexico, in pre-Spanish times" is no longer tenable.

At the same time, it would serve little purpose to exaggerate the scale of irrigation in Mesoamerica. The Texcocan system certainly represents one of the largest irrigation systems known to date in Middle America. Yet its construction cannot compare with the massive systems of dykes and dams characteristic of the hydraulic societies of the Old World and Peru. Rather modest in scale, it served to integrate a number of communities or settlement areas into what might be called an "irrigation cluster." Such irrigation clusters appear to have had a wider distribution in Mesoamerica. Armillas has described one such cluster in the headwaters of the Atlixco River, and Palerm makes brief mention of another near Tecomatepec in the State of Mexico. A third formation of this type may have existed in the area of Tacubaya-Coyoacan. Such clusters seem to fit the category of "district irrigation," established by Robert F. Murphy. They differ from "local irrigation" in that the individual community lacks direct access to water, and the problem of allocating water can no longer be met within one community alone. At the same time, such district irrigation lacks the massive character of larger systems comprising several districts. The irrigation cluster discussed here lent a cohesion to the Old Acolhua domain which it might not have acquired otherwise. It is important to note that the limits of the domain coincided closely with the limits of irrigation in the area. Huexotla, marginal to the system of water distribution, was also marginal politically. The domain retained its unity even when the political structure of Texcoco reached out to include a wider area and a larger and more diverse population. Certainly, once the state expanded, the very marginal location of the domain in the narrow corridor leading from north to south along the eastern shore of the lake would have tended to reduce the area to secondary importance, had irrigation not provided a permanent backbone of political cohesion.

THEORETICAL IMPLICATIONS

The modern picture of the area might seem to lend itself to historical reconstructions of the kind used by Redfield in his study of culture change in Yucatan. Today, the communities of the flat belt are clearly heavily involved in an outside market, and through the market with the culture of Mexico City. The sierra communities are much less dependent on outside markets, and more isolated. Other cultural differences seem to go hand in hand with this apparent continuum. More people speak Nahuatl in the sierra; fewer do so in town. "Indian" dress is more common in the mountains; it is rarely met with in the vicinity of Texcoco.

Historical reconstructions based on a mechanical interpretation of the hypothesis that culture change tends to operate from the pole of a "folk society" with a "folk culture" to the opposite pole of an "urban society" with an "urban culture" would, however, lead to serious misinterpretations of the past when applied to the history of the area under consideration. One must not forget that the concepts of the "folk society" and "urban society" represent ideal types, mental constructs to which no known societies correspond precisely. The heuristic value of these concepts is most apparent when social and cultural processes are discussed at the highest level of abstraction, where the universe of discourse is the global history of Man. The conversion of such abstract theoretical models into the supposed termini of real culture change tends to substitute pseudo-historical extrapolations for a scientific inquiry into the various aspects of the problem. In the case of the Old Acolhua domain, both archaeological remains and historical sources indicate that the society of the hill and sierra country was more complex in the past than it is at present.

The area seems to have undergone a process of cultural devolution, a process of reduction in complexity, since the period of the Spanish Conquest. Evidence shows that agriculture was more intensive, covered more ground, and fed more people in pre-Hispanic times than it does today. Organizationally, we also witness a change in complexity, from centralized water controls resting in the hands of the state to decentralization into the hands of local bodies of water commissioners.

Such a process of devolution within the Indian sector of society seems to have occurred many times in Mesoamerica; and if we are to understand the history of Mesoamerican civilization, we must attempt to understand the dynamics of this devolution. It might indeed be said that the Old Acholhua domain experienced a reduction in cultural complexity, precisely because Mesoamerica as a whole achieved a new level of sociocultural integration in the formation of New Spain. Under the Spaniards, Texcoco became a center of wool production, and sheep raising and wheat farming took the place of irrigated corn agriculture. Both sheep raising and deforestation to supply wood to Spanish enterprises and cities further increased erosion, and made the trend away from intensive agriculture irreversible. Thus intensive agriculture and terracing survived in a few villages of the piedmont only.

This type of culture change has not received much attention from the theoretical point of view. Julian Steward's concept of levels of sociocultural integration enables us to analyze major changes in the complexity of sociocultural systems. Yet the problem at hand concerns not so much major transitions from one level of integration to another, but the internal organization of the systems themselves. It seems likely that the different parts of larger emergent systems produced by transitions in level of integration need not undergo integration with the same degree of stringency. Some of the constituent parts may indeed have to undergo wholesale reorganization to make them congruent with the demands set by the new level. Other parts may merely undergo minor adaptations. Still other components cannot be assigned a positive role in the new order, and undergo a reduction in functions and complexity. In other words, some parts of a larger system may become "marginal," and non-functional with regards to the operation of the larger system as a whole, while others acquire new functions which are strategic in its maintenance.

Other disciplines have long made use of the concepts of "key" and "marginal" areas, especially in discussing differences in functional importance of different geographic areas. While these concepts have often been implicit in anthropological discussions, few anthropologists have made them explicit. In discussion of Mesoamerica, Armillas and Sanders have demonstrated the utility of viewing the Valley of Mexico as the key area of Mesoamerica. In the analysis of a particular region within Mesoamerica, we possess two outstanding contributions: Stanislawski's paper on the political geography of the Tarascan area, and Aguirre Beltran's brief discussion of the cultural relationship between the sierra and the lake district throughout Tarascan history. Unfortunately, these contributions have not received the theoretical attention which they would seem to deserve.

Anthropologists are of course familiar with the term "marginal" from diffusion studies. In the present context, however, we are not referring to the distribution of culture traits in space, but to types of sociocultural systems subsumed as parts of larger systems. The emphasis in our use of the term is not on *where* these component systems occur, but on *how* they function, and on how they function with relation to each other. Such systems may in fact occur side by side in the same area, as in many parts of the Valley of Mexico today. It would therefore seem more appropriate to speak of "key" adaptations—strategic relationships between certain aspects of the environment and the culturally available technology—and "marginal" adaptations, and of their relationships to the strategic or key sector of the society under analysis, or to the marginal sector of that society. Such an explicit distinction of the functionally strategic and the functionally secondary components as applied to New Spain would allow us to explain the phenomenon which is the subject of the present discussion. After the Spanish Conquest, the Old Acolhua domain was transformed from a key area into a marginal area. Its social structure was relegated to the marginal sector of the new society. It underwent a cultural decline, because that society was based on new strategic functions which the Old Acholhua domain could not supply. It follows from this conclusion that the condition of the Indian villages of the area today offers no safe guide to their condition in pre-Hispanic times.

42. INVESTIGATING THE ORIGINS OF MESOPOTAMIAN CIVILIZATION

FRANK HOLE

Reprinted from Science *(August 5, 1966), 153 (3736): 605-11. Frank Hole is Associate Professor of Anthropology at Rice University. His principal research interests are the origins of agriculture and the development of complex societies in Southwest Asia. He is the co-author, with Robert E. Heizer, of* Introduction to Prehistoric Archeology.

■ Even several thousand years ago, great oceans and continents separated Mesoamerican and Mesopotamian civilizations. The Mesoamericans and their counterparts in South America could not have known of the Mesopotamians, and they could not have influenced each other. We usually speak of Mesopotamian civilization as the "first" civilization because it appeared first chronologically, but Mesoamerican civilization also was "first" because it developed independently of the Mesopotamians. Hence, at least in the evolution of culture, it is possible to have several "firsts." We now turn to the other "first" civilization, the one that grew and flourished in the Near East.

This selection, by Hole, describes the basic adaptive processes in the development of Mesopotamian civilization; it is an excellent illustration of the reversal in man's relationship to the habitat that ensues as culture develops into civilization. Hole sketches the cultural history of Mesopotamia from nomadic hunting days, when small bands wandered from place to place making use of whatever the habitat provided. Little by little, each major technological advance allowed the people to become increasingly sedentary and to rely on domesticated foods. The turning point (as Hole calls it) in man's relationship with his habitat and his fellows came with the development of centralized political institutions, which, through the sponsorship and control of trade and the redistribution of surpluses, made it possible for people to live in areas (such as southern Mesopotamia) that had few resources. In other words, newly emerging political and urban institutions made it possible for the Mesopotamians to exploit a greater variety of habitats than had been possible before the development of these institutions.

Obviously, such political institutions could not have arisen independently of economic factors; and the most notable of these was the production of large surpluses, which in turn led to the creation of redistributional agencies. The important point here, however, is that large economic surpluses, of themselves, are not the foundation of a civilization. Instead, the key factor is to be sought in the development of political institutions that, at least initially, were adaptations designed to cope with the pressures these economic surpluses presented in society. Many societies developed large economic surpluses, but not all became nation-states or served as the building blocks of civilizations.

Again, Robert McC. Adams' *Evolution of Urban Society: Early Mesopotamia and Prehispanic Mexico* (Chicago: Aldine, 1966) cannot be recommended too highly for those who wish to read more about the role of political and stratification patterns in the development of civilization. An intriguing account of a major kingdom in this civilization is "Ancient Ararat," by Tashin Ozguc (*Scientific American* [March, 1967], 216[3]: 38-46). "Ararat" is a report of the excavation of the mountain on which Noah's ark is believed to have gone aground. Several other excellent works also can serve as an introduction to Mesopotamian civilization: *History Begins at Sumer,* by Samuel N. Kramer (2d ed., rev. and enl.; London: Thames and Hudson, 1961); *The Intellectual Adventure of Ancient Man,* by H. and H. A. Frankfort, John A. Wilson, and Thorkild Jacobsen (Chicago: University of Chicago Press, 1946; reprinted by Penguin Books as *Before Philosophy*); and *The Birth of Civilization in the Near East,* by Henri Frankfort (Bloomington: Indiana University Press, 1951). Each of these books contains excellent bibliographies that direct the reader to other sources. ■

IN SOUTHWEST ASIA, between 8000 and 3000 B.C., human society developed from self-sufficient bands of nomadic hunters to economically and politically integrated city dwellers who specialized in a variety of occupations. A central archeological problem is to try to discover the factors that triggered these fundamental changes in man's way of life. For want of evidence and for want of a satisfactory model of the conditions existing during the period in question, searching for origins and attempting to discover the course of events that led to civilization is difficult. Prehistorians deal with nameless cultures, trusting to reconstructions from physical remains for their picture of life in ancient times. They must work directly with geographic, technological, and demographic factors and only indirectly infer ideologies and philosophical concepts. Archeologists are thus limited in what they can hope to learn by the nature of their data and the tools they have for interpreting them. Within these limits, however, it is possible to construct some plausible theories about the origins of civilization and to test them through controlled programs of excavation and analysis. In this article I define the problem under consideration in ecological terms, review the current evidence, and suggest topics for further study.

Mesopotamian (Sumerian) civilization began a few centuries before 3000 B.C. and was characterized by temples, urban centers, writing, trade, militarism, craft specialization, markets, and art. Inferred characteristics are a class-stratified society and well-defined mechanisms for regulation of production and distribution of resources. To be sure, Sumerian civilization must have had many other important but intangible characteristics, but most of these cannot be inferred from archeological data.

The early Mesopotamian civilizations were restricted to southern Mesopotamia, the alluvial plain that stretches south from Baghdad to the Persian Gulf. Remains of immediately antecedent cultures have been excavated in the same area, and still older cultures have been excavated in the surrounding Zagros mountain valleys of Iraq and Iran and on the steppes at the verge of plain and mountain in Khuzistan, southwest Iran.

Intensive agriculture is a precondition for civilization. The Sumerian societies for which we have some historical records were sustained by cultivation of irrigated barley and wheat, supplemented by crops of dates, and the production of sheep, goats, cattle, pigs, and fish. In 8000 B.C. people were just beginning to plant cereals, raise animals, and live in permanent villages; their societies were small, self-sufficient, egalitarian groups with little differentiation of occupation or status. These people had fewer of the artifacts and qualities of civilization than the Sumerian city dwellers had 5000 years later. In this article I use 8000 B.C. as a convenient base line and attempt to assess some 5000 years of culture history.

THEORIES OF DEVELOPMENT

Recognizing the obvious changes in society that occurred during 5000 years, archeologists and others have proposed causal factors such as characteristics of geography to account for them. The most detailed examination of the relationship between geographic features and social forms has been made by Huntington, but other scholars working with data from Southwest Asia have had more influence on archeologists. For example, in attempting to explain the origins of agriculture, Childe proposed climatic change, specifically desiccation, as the initiating event and set off a chain of thought that is still favored by some authors. Childe argued that "incipient desiccation . . . would provide a stimulus towards the adoption of a food-producing economy. . . ." Animals and men would gather in oases that were becoming isolated in the midst of deserts. Such circumstances might promote the sort of symbiosis between man and beast implied in the word *domestication*. Although Childe's theory is attractive, there is no conclusive evidence that the climate in Southwest Asia changed enough during the period in question to have affected the beginnings of agriculture and animal husbandry.

It was once fashionable to think of culture

as inevitably rising or progressing, and this trend was thought to be analogous to biological evolution. Except in a most general way, however, modern prehistorians do not think of universal stages of cultural development. Rather than focusing on evolutionary stages, many scholars have examined the role of particular social and economic activities in triggering the emergence of complex forms of society. For instance, Marxists have explained the form of society (government, broadly speaking) on the basis of modes of production. Marxist evolutionists even today explain the development of social classes and political states in similar terms. They argue that, as people gained control over the production of food, the concept of private property crept in, and later the mass of people were exploited by the propertied few. "The creation of a state was necessary simply to prevent society from dissolving into anarchy due to the antagonisms that had arisen." Information on the emergence of Sumerian civilization that might support this idea, however, is lacking.

Another attempt to correlate technological systems and social advances was made by Karl Wittfogel in *Oriental Despotism.* He contended that, where people had to depend on irrigation, they inevitably led themselves into an escalating dependence on an organizational hierarchy which coordinated and directed the irrigation activities. "The effective management of these works involves an organizational web which covers either the whole, or at least the dynamic core, of the country's population. In consequence, those who control this network are uniquely prepared to wield supreme political power." Although Wittfogel's analysis seems valid in many instances, archeological investigation in both Mesopotamia and the Western Hemisphere leads to the conclusion that there was no large-scale irrigation at the time of the emergence of the first urban civilization.

AN ECOLOGICAL APPROACH

Single factors such as technology are unquestionably important, but they can be understood only within the cultural, social, and geographic context. A more comprehensive view that takes into account the interrelation of many factors is called human ecology. In a consideration of cultural development, the relevant concept in human ecology is adaptation; hence the approach is to try to discover how particular factors influence the overall adaptation of a society. By means of the general approach, human ecology attempts to understand what happened in the histories of particular cultures. It does not address itself to making general statements about cultural progress or evolution.

In an ecological approach, a human society is treated as one element in a complex system of geography, climate, and living organisms peculiar to an area. To ensure survival, various aspects of a human society must be complementary and the society itself must be successfully integrated with the remainder of the cultural and physical ecosystem of which it is a part. From the ecological view, such factors as technology, religion, or climate cannot be considered apart from the total system. Nevertheless, some parts of the system may be considered more fundamental in the sense that they strongly influence the form of the other parts. Anthropologists, through their study of modern societies, and archeologists, through inference, find that such factors as geographical features, the distribution of natural resources, climate, the kinds of crops and animals raised, and the relations with neighboring peoples strongly influence the forms that a society may take. These factors comprise the major elements of the ecosystem, and societies must adapt themselves to them.

ARCHEOLOGICAL EVIDENCE

For the period 8000 to 3000 B.C., archeological data are scattered and skimpy. This naturally limits the generality of any interpretations that can be made and restricts the degree to which we can test various theories. Ideally we would wish to work with hundreds of instances representing the range of en-

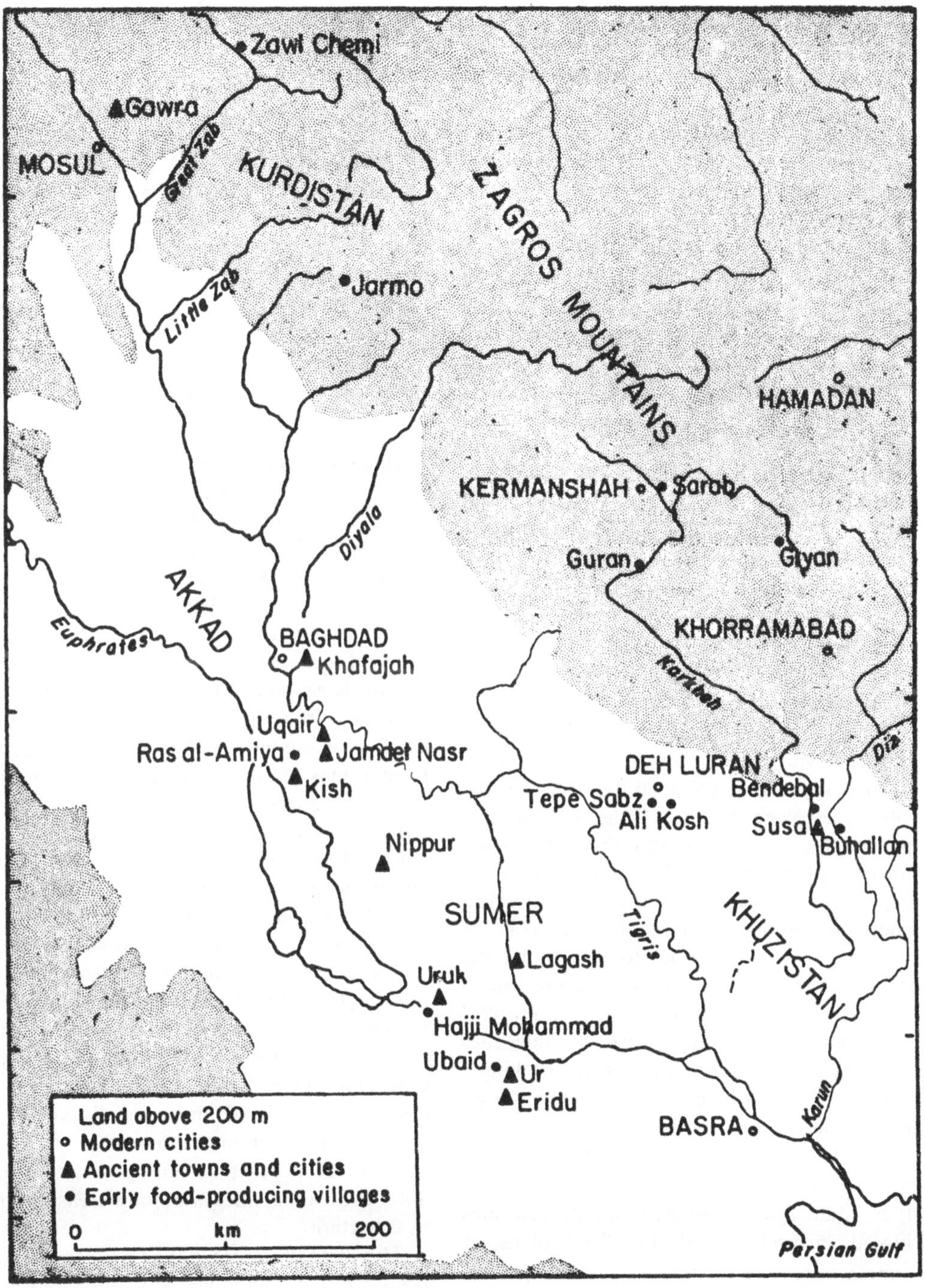

FIGURE 42.1

Archeological sites in the alluvial basin of southern Mesopotamia and in the valleys of the Zagros mountains.

vironmental and cultural variation; instead, for the whole of Southwest Asia we can count fewer than 100 excavated and reported sites for the entire range of time with which we are dealing. Of course the number of unexcavated or unreported sites about which we know something is far greater, but we cannot but be aware of how little we know and how much there is to find out.

In all of Southwest Asia only about 15 villages that date to 8000 B.C. have been excavated, and only two of these, Zawi Chemi and the Bus Mordeh levels at Ali Kosh, give good evidence of the use of domesticated plants or animals. In short, data for the time of our base line are woefully inadequate. We have much fuller information about the villages of 5000 B.C., but, unfortunately, for periods subsequent to 5000 B.C. the *kind* of data we have changes drastically. Thus, although there is historical continuity in the series of known sites, there is discontinuity in some of the data themselves because few archeologists have worked sites spanning the whole period from 8000 to 3000 B.C. Most of the sites dating to about 3000 B.C. were excavated by "historic" archeologists who struck levels that old only incidentally as they plumbed the depths of the cities they were digging. These scholars depended far less on artifacts than on history for their interpretations. The earliest sites were dug by prehistorians who based their inferences on results generated by an array of scientific experts. In order to understand the origins of civilizations, we thus need to bridge two quite different "archeological cultures." Archeologists and their various colleagues working in the early villages painstakingly teased out grains of charred seeds, measured metapodials and teeth of early races of sheep or cattle, and analyzed the chemical and mineral constituents of obsidian and copper; their counterparts working in the historic sites busied themselves with the floor plans of temples, the funerary pottery in the graves, the esthetics of an art style, and the translation of cuneiform impressions in clay.

Bearing in mind the reservations I have already expressed, we can begin to try to pick a coherent path through 5000 years of history. In dealing with Mesopotamia, it is usual to regard the presence of towns, temples, and cities as indicative of civilization. If we do so, we can divide our history into two parts, beginning with small food-producing villages and following with more complex societies that include towns and cities. In the ensuing discussion I assess the available evidence and, for both forms of community, outline the characteristics and indicate how the community developed.

FOOD-PRODUCING VILLAGES

Small food-producing villages have had a long history, but here we are chiefly interested in those that existed between 8000 and 5000 B.C. None of these communities is known thoroughly, and the following descriptions are based on data from several excavated sites and from surface surveys. The fullest data come from the phases represented in Ali Kosh and Tepe Sabz, in southwest Iran, and from Jarmo, Sarab, and Guran in the Zagros mountains. Additional data derive from extensive surveys in Khuzistan and the valleys of Zagros.

During this period villages are small and scattered, typically less than 1 hectare in size and housing perhaps 100 to 300 people. They are situated on the best agricultural land in regions where farming is possible without irrigation. From a handful of sites known to be about 10,000 years old, the number of settlements had increased by 5000 B.C., when many villages were within sight of one another and almost every village was within an easy day's walk of the next. There is no evidence of great migrations or any serious pressure of population during this time. By 4000 B.C. some villages occupy areas as large as 2 hectares.

The increase in population appears to have been a direct consequence of improved agricultural techniques. In 8000 B.C., only primitive, low-yield races of emmer wheat and two-row barley were grown; sheep and goats were both in the early stages of domestication. By

5000 B.C. a modern complex of hybrid cereals and domesticated sheep, goats, cattle, and pigs were being exploited, and irrigation was practiced in marginal agricultural areas such as Deh Luran. The effects of developed agriculture are soon apparent, for, by 4000 B.C., settlement of new areas by prehistoric pioneers can be shown clearly in such places as the Diyala region to the east of Baghdad. The age of the earliest settlements in southern Mesopotamia proper is unknown, but it would be surprising if groups of hunters and fishers had not lived along the rivers or swamps prior to the introduction of agriculture. The oldest settlements, Eridu, has been dated to about 5300 B.C., but there are no contemporary sites. In fact, there are few villages known in southern Mesopotamia that antedate 4000 B.C.

TOWNS AND CITIES

The millennium between 4000 and 3000 B.C. saw the rapid growth of towns and cities. Villages were also abundant, but some evidence suggests that they were less numerous than in earlier periods. "In part at least, the newly emerging pattern must have consisted of the drawing together of the population into larger, more defensible political units." The trends I describe here pertain almost exclusively to southern Mesopotamia; in the north and in the valleys of the Zagros, the pattern remained one of small villages and —emerging later than their counterparts in the south—townships.

From southern Mesopotamia, archeological data for the period before 3000 B.C. are skimpy. Deep soundings at the bases of such sites as Eridu, Ur, Uqair, Tello, Uruk, and Susa, and test excavations at Ubaid, Ras al-Amiya, and Hajji Mohammad are about all we have. Only at Ras al-Amiya is there direct evidence of agriculture, although at Eridu a layer of fish bones on the altar of temple VII suggests the importance of the sea and of fishing. Archeological evidence from several of the remaining sites consists either of temple architecture or pottery, the latter serving more to indicate the age of a site than the social or cultural patterns of its inhabitants. Some temple plans are known, but published data on domestic architecture are few, and the sizes of the communities can be inferred only roughly.

There are extensive enough excavations at sites like Uruk, Khafajah, Kish, Ur, and Nippur to indicate the scale of urbanism and many of its more spectacular architectural and artistic features for the period after 3000 B.C. The largest Early Dynastic site was evidently Uruk, where 445 hectares are enclosed by the city wall; contemporary Khafajah and Ur comprise 40 and 60 hectares, respectively. By contrast, the Ubaid portion of Uqair had about 7 hectares.

HISTORICAL RECONSTRUCTIONS

Pictographic writing began by about 3400 B.C., but is is difficult to interpret, and in any case early writing tells little about society; it is confined to bookkeeping. Nevertheless, by depending on myths, epics, and tales written some 1000 years later, scholars have attempted historical reconstructions of the emerging urban societies.

The oldest texts that characterize the Sumerian community are no earlier than 2500 B.C. and were written at a time when the "Temple-city" had already become the characteristic feature of the Mesopotamian landscape. In the view of many authors, the city was an estate belonging to gods of nature and maintained on their behalf by completely dependent and relatively impotent mortals. Controversy centers around the degree to which the temple controlled the economy. The extreme view is that it controlled everything while the more popular moderate view is that it controlled only part of the economy. In the Early Dynastic period it seems clear, some, if not all, people were responsible to a temple which in turn directed most of the production and redistribution of goods and services. For practical purposes there was no distinction between the economic and the religious roles of the temples, but their administrators may not have had much political influence. Some temples listed large staffs of attendants, craftsmen, laborers, and food producers, but the precise relationship of these people to the temple is by no means clear. Moreover, such staffs would have been associated with the largest temples and not with the host of lesser temples and shrines that seem to have been present in the larger cities. Political control was vested variously in the *en* (lord), *lugal* (great man, or king), or *ensi* (governor-priest), depending on the historical period, the city referred to, and the translator of the text. In early times religious and secular titles seem not to have been held by the same person. Jacobsen describes, for pre-Early Dynastic times, a "primitive democracy" with the leader appointed by and responsible to an assembly of citizens. The arguments about the nature of Sumerian cities are summarized by Gadd: "The issues barely stated here have been discussed with much elaboration and ingenuity, but only a notable increase of contemporary evidence could raise the conclusions to a possibility of much affecting our conception of Sumerian government."

ENVIRONMENT AND SUBSISTENCE

By combining the geographic, economic, and historical data, we can construct some plausible theories about the course of development and the situations that triggered it. The remarkable thing, from an ecological view, is the change in relations between men and products, and then between men and their fellows during the 5000 years. If we return for a moment to the preagricultural ways of life, we find small bands of hunters exploiting the seasonally available resources of a large territory by wandering from one place to another. Each community was self-sufficient, and each man had approximately the same access to the resources as his fellows. The earliest villagers seem to have maintained this pattern, although, as agriculture and stock breeding became more developed and important economically, the villagers tended more and more to stay put. People settled down where they could raise large amounts of grain, store it for the future, and exchange it for products they did not produce. In return for dependability of food supply, people gave up some of their dietary variety and most of their mobility. From a pattern of exploiting a broad spectrum of the environment, there developed a pattern of exploiting a relatively narrow spectrum.

As long as people stayed where they could find sufficiently varied resources through hunting and gathering, they could be self-sufficient. When people settled in villages away from the mountains, out of the zone of rainfall agriculture, they were no longer independent in the sense that they personally had access to the varied resources they desired or needed. Psychologically and sociologically this marked a turning point in man's relations with his environment and his fellows. Southern Mesopotamia is a land with few resources, yet in many ways this was an advantage for the development of society. In a land without timber, stone, or metals, trade was necessary, but the role of trade in the emergence of civilization should not be overemphasized. Date palms and bundles of reeds served adequately instead of timber for most construction, and baked clay tools took the place of their stone or metal counterparts in other areas. On the other hand, travel by boat is ancient, and extensive land and sea trade is attested in early documents. It was easy to move goods in Mesopotamia.

In order to live as well as the farmers in Deh Luran did, the Sumerians had to cooperate through trade, barter, or other means with their fellow settlers. We should remember that the barren vista of modern Mesopotamia on a dusty day does not reveal the full range of geographic variation or agricultural potential of the area. Swamps and rivers provided fish and fowl and, together with canals, water for irrigation and navigation. With sufficient water, dates and other fruits and vegetables could be grown. The unequal distribution of subsistence resources encouraged the beginnings of occupational specialization among the various kinds of food producers, and this trend was further emphasized after craftsmen started to follow their trades on a full-time basis.

ECONOMICS AND MANAGEMENT

Because of the geographic distribution of resources and the sedentary and occupationally specialized population, a social organization that could control production and redistribution was needed. Clearly, any reconstruction of the mechanics of redistribution in emerging Mesopotamian civilization is subject to the severe limitations of the evidence. If we recognize this, however, we may then seek in contemporary societies analogs that may help us imagine appropriate redistributional structures. In modern economies, money markets act as the agency of redistribution, but in virtually all "primitive" societies where surpluses or tradeable goods are produced, a center of redistribution of another kind grows. The "center" can be a person (for example, the chief); an institution, like a temple and the religious context it symbolizes; or a place, like a city with some form of free markets. Jacobsen suggests that in Sumeria temples served as warehouses, where food was stored until times of famine.

Sahlin's studies in modern Polynesia are also relevant to this point. He found that there is a close relation between surplus production and the degree of social stratification in Polynesia—that in a redistributional economy, the greater the surplus is, the greater is the degree of stratification. Of course we can only speculate about Mesopotamia, but, granting this and following Sahlin's findings, we may say that the chief of the Mesopotamian town would have acted as the center of redistribution. In Mesopotamia, most of the surplus labor or food went directly or indirectly into building and maintaining temples. One would also have expected the chief to use a good bit of the surplus to support himself and his family, to pay the wages of craftsmen, and to buy the raw materials that were turned into artifacts, such as jewelry and clothing, that served to distinguish his rank. Others in the lord's biological or official family would also have profited from his control of the resources and ultimately have become recognized as a social class entitled to special prerogatives. This social stratification would have been associated with a similarly burgeoning system of occupational differentiation.

In an emerging system where both technology and governmental forms are relatively simple but susceptible of improvement, there is a maximum opportunity for feedback. That is, if a certain level of production will support a certain degree of social stratification, efficient management by the social elite may result in more productivity. It is interesting to speculate on how much the construction of enormous irrigation systems during later Mesopotamian history may have depended on the rising aspirations of the ruling elite.

Although the need for management of production might in itself have been sufficient cause for a developing social stratification, other factors were probably contributory. Turning now to law and politics, I should point out that, with the establishment of irrigation and the concentration of population in urban centers, man's basic attitudes toward the land must have changed. The construction of irrigation systems, even if primitive, makes the land more valuable to the builders, and this, if it did nothing else, would lead to some notions of property rights and inheritance that had not been necessary when abundant land was available for the taking. An irrigation system also implies that some men may have more direct control over the supply of water than others. This could have led to an increase in the power of individuals who controlled the supply of water, and it certainly must have led to disputes over the allocation of water. It seems inevitable that a working system of adjudicating claims over land would then have been necessary, and the task may have fallen to the chiefs (lords).

The presence of "neighbors" also has ecological implications; it is worth recalling that property invites thievery. Adams argues that the "growth of the Mesopotamian city was closely related to the rising tempo of warfare," and Service points that the integration of societies under war leaders is common, and clearly an adaptation to social-environmental conditions. Several Early Dynastic II cities had defensive walls, attesting to conflict between cities and perhaps between settled farmers and nomadic herders, but the historical evidence for warfare begins only about 2500 B.C.

If we consider both the agricultural system and the wealth, we see conditions that enhanced opportunities for leadership and, ultimately, for direction and control. With these situations, the emerging systems of rank and status are understandable without our resorting to notions of "genius," "challenge and response," or immigration by more advanced peoples.

RELIGION

The role of religion in integrating emerging Mesopotamian society is frequently mentioned. By 3000 B.C. texts and temples themselves attest to the central place of religion in Sumerian life; theoretically, at least, cities were simply estates of the gods, worked on their behalf by mortals. How closely theory corresponds to fact is a question that cannot be answered. Although we cannot date their beginnings precisely, we know that temple centers were well established by 5000 B.C., and that towns and temples frequently go together. Whether towns developed where people congregated because of religious activities or whether temples grew in the market centers where the people were cannot be decided without more data. Both interpretations may be correct. Historic evidence suggests that economic activities were controlled by the temples, but this evidence says nothing about the original relationships between the two. Furthermore, the interpretation of the historical documents is open to questions. As Gadd points out, the picture of Sumerian economy that the various authors use is based on the "detailed records of one temple (Lagash) over a rather short period."

In regard to this limited view of the role of religion it is well to recall that major settlements had several temples. At Khafajah, for example, perhaps as early as 4000 B.C. there were three temples, and a fourth was added later. Our image of the Sumerian temple is nevertheless likely to be that of the large temple oval at Khafajah or Ubaid rather than that of the smaller temples that were contemporary and perhaps just as characteristic. The temple oval appears to have housed a society within a city, but many temples had no auxiliary buildings. More impressive even than the temple ovals were the great ziggurats erected on artificial mounds—at Uruk 13 meters high and visible for many kilometers. Again this was only one of several temples at the same site. In Ubaid, Eridu, and Uqair, for example, where temples were originally associated with residential settlements, the towns were later abandoned and only the temples with cemeteries were maintained.

SUMMARY

It seems unlikely that Mesopotamian society took a single path as it approached the rigidly organized, hierarchal civilization of Early Dynastic times. Rather, we imagine that there was considerable experimentation and variety in the organization of society as people adapted to their physical environment and to the presence of other expanding communities.

Some towns and cities probably arose as the demographic solution to the problem of procuring and distributing resources. It would have made sense to have central "clearing houses." Similarly, it would have made sense to have the craftsmen who turned the raw materials into finished products live close to their supply (probably the temple stores). Temple centers are natural focal points of settlements. Cities and towns, however, are not the only demographic solutions to the problem of farming and maintaining irrigation canals. Both of these tasks could have been carried out by people living in more dispersed settlements. City life in Mesopotamia probably also presented other benefits.

For example, as warfare came to be a recurrent threat, the psychological and physical security of a city must have been a comfort for many. Finally, to judge from some historical evidence, Mesopotamian cities were places of diversity and opportunity, no doubt desiderata for many people as long as they could also gain a suitable livelihood.

In considering the development of civilization, an ecological approach forces us to consider multiple factors. Seeking isolated causes among the many factors possibly involved ignores the central concept of adaptation, with its ramifications of interaction and feedback. Still, we are a long way from fully understanding the emergence of Mesopotamian civilization. In particular, we need a great deal more archeological data that relate to the 2000 years preceding 3000 B.C. in southern Mesopotamia. Specifically, there are three projects which ought to have high priority in the planning of future archeological work in this area. First, we need thorough surveys in order to determine the early history of settlement in Mesopotamia. By means of these surveys in and around the early cities, we would try to determine the duration of occupation, and the variety and location of additional sites. Second, we need extensive excavation of selected smaller sites and portions of larger ones in order to determine the characteristics of different settlements. We would like to know in what way the cities, towns, temple centers, and villages were integrated to form a socioeconomic network. A third question, which gets at the crux of the matter, is What structural form did the emerging Sumerian society take? Answers to this question must depend in large part on the results of future surveys and excavations of the kind suggested above. Then, selective excavations focusing on successive periods should yield data on the relative roles of economic and religious activities and on social differentiation and stratification. These data, after they are eventually pieced together, will comprise the story of the emergence of the world's first civilization.

43. THE STUDY OF ANCIENT MESOPOTAMIAN SETTLEMENT PATTERNS AND THE PROBLEM OF URBAN ORIGINS

ROBERT McC. ADAMS

Reprinted from SUMER, *Vol. 25: 111-124. Robert McC. Adams is Professor of Anthropology and Dean of Social Sciences at the University of Chicago. He is Chairman of the Assembly of Behavioral and Social Sciences of the National Research Council. One of his principal research interests has been the comparative study of early civilizations in the Near East and Middle America. He is the author of* Land behind Baghdad, The Evolution of Urban Society, *and (with H. Nissen),* The Uruk Countryside.

■ In recent years, an influential group of archeologists has been developing techniques and concepts for using the past as a laboratory for the analysis of social and cultural processes; their goal is to contribute to a better understanding of human behavior and social change. They view archeology as part of anthropology rather than as a branch of the humanities. Popularly known as the "new archeologists," this group of scientists are concerned with the processes of adaptation in past societies, with their organization, technology, behavior, and stylistic features. One of the major innovations of this group of archeologists has been to present reports that involve evidence from several sites rather than the traditional report based on a single site. One result is a growing awareness that villages and other sites did not exist as isolated entities but were in constant contact with others and survived as parts of larger social, political, and economic systems. Robert McC. Adams has been in the forefront of these developments.

We have seen in many of the foregoing selections that settlement patterns are among the principal adaptive responses to changing conditions. Cities, too, must be regarded as adaptations to particular pressures. However, many of the early archeological investigations of ancient cities tended to treat each as if it existed in isolation (as do many contemporary studies of modern cities by behavioral scientists). As Adams observes in this selection, ancient Mesopotamian settlement patterns represented systems in which cities, towns, and villages interacted. Moreover, cities need to be understood in terms of their relations with the hinterland; they arise in particular political contexts, and their relationships to the society's central political authority must be examined. But cities must also be seen as arising in particular ecological contexts that, when combined with economic and political factors, give shape to the society in which urban centers are important hubs.

Not only do cities serve as adaptive mechanisms in response to pressures in the milieu, as in ancient Mesopotamia; they are also and always important links in a heterogeneous society in which there are several adaptive strategies. This entails a network of occupational specializations in which urban residents are dependent on farmers and artisans in the hinterland and the latter rely on urbanites for goods and services. An important characteristic of many early civilizations, such as the Mesopotamian, was the presence of pastoralists in the economy and social organization of the countryside. Modern industrial societies may be without pastoralists but their counterparts in the modern hinterland exist and play an important role in the internal organization of modern cities. The material presented by Adams thus not only sheds light on the ancient past, it may serve as a model for those who are concerned with the nature of cities in our own societies.

The relations between different sectors of complex society are further explored in *The Uruk Countryside,* by Robert McC. Adams and Hans Nissen (Chicago: University of Chicago Press, 1972). A very good architectural survey, with important implications for social organization, is provided in *Cities and Planning in the Ancient Near East*, by Paul Lampl (New York: Braziller, 1968). "State Settlements in Tawantinsuyu: A Strategy of Compulsory Urbanism," by Craig Morris (in *Contemporary Archaeology: A Guide to Theory and Contributions*, edited by Mark P. Leone [Carbondale and Edwardsville: Southern Illinois University Press, 1972], pp. 393-401) provides a case from ancient Peru that affords great insight into the exercise of political power in ancient civilizations. A survey of research findings is presented in "Settlement Patterns in Archeology," by K. C. Chang (*Addison-Wesley Module in Anthropology*, number 24. Reading, Mass.: Addison-Wesley, 1972). ■

THERE HAS NEVER been any question that the main feature of the subsistence base on which early Mesopotamian civilization developed during the late fourth and early third millennium B.C. was irrigation agriculture. However, the extent and character of the early irrigation systems—and hence their effect on social and economic institutions—cannot be understood merely through the deductive application of generalizations derived from other parts of the world or from later periods when conditions were demonstrably different. Reliance on such generalizations has, in fact, led to the wide acceptance of a number of stereotypes which distort and obscure both the subsistence economy and the crucial determinants of change in Sumero-Akkadian society at large.

The evidence for early subsistence practices on the Mesopotamian alluvium is still very limited and largely indirect. The earliest pictographic writing provides illustrations of domesticated plants and animals, of main features of the agricultural technology such as the plow, and of other natural food resources such as fish, but the meaning and context of use of these symbols is seldom unambiguous. By the middle of the third millennium there are scattered references to the construction of canals and to the sale of agricultural lands by corporate kin groups and private individuals as well as much more numerous administrative records of temples and other manorial units engaged in agricultural production and redistribution. However, not until toward the end of the third millennium do the greatly expanded numbers and functions of texts written during the Third Dynasty of Ur permit an even partial description and localization of irrigation works constituting particular local systems. Conditions by then were already far removed, particularly in social stability and complexity, from those which accompanied the initial growth of Mesopotamian civilization a thousand years earlier. Moreover, even for that period the available data—much of it still unstudied and unpublished—heavily emphasizes conditions around certain urban centers where important archives have been found, and hence fails to provide a balanced regional picture.

Data on subsistence from archeological excavations is even less relevant and more fragmentary. The prevailing concern in late prehistoric and historic Mesopotamia has been with the recovery of texts, works of art, and monumental architecture. For this reason, and because of the relatively low ratio of scientific staff to the excavation labor force, little evidence has been recovered which bears on the problem. Analysis of soil samples for pollen, from which the changing spectrum of natural and domestic plant communities might be reconstructed, has barely begun. Quantitative studies of faunal remains also are still essentially absent, and for later periods even simple lists of identified species are extremely rare. Limited work has been done on the identification of plant impressions in chaff-tempered bricks and pottery. This has led to the suggestion that the substitution of six-row for two-row barley reflected the onset of irrigation in the alluvium, and contributed to the hypothesis that historic shifts in the wheat/barley ratio reflect soil salinity induced by over-irrigation. But findings of this nature clearly fall far short of providing a basis for understanding the nature and role of irrigation within the larger subsistence system of which it was a part.

In these circumstances, considerable importance attaches to the changing patterns of ancient settlement that can be reconstructed from archeological surface reconnaissance.

The land surface available for agricultural settlement at the beginning of the third millennium B.C. cannot yet be accurately delimited because of uncertainties as to geomorphic processes affecting the position of the head of the Arabian Gulf. It appears however, that the surviving ruins in perhaps two-thirds of it have been more or less systematically mapped and studied during recent years. Not only the historic core-areas of political dominance are represented in these surveys, but also some of the geographically more peripheral areas of the alluvium. Collections of ceramic "index fossils" in most cases permit determination of the sequence of occupation at individual sites, together with a fairly accurate assessment of maximum and terminal settlement size for each site and more speculative estimates of areas of occupation during earlier periods.

Surviving features of former irrigation systems also are a focus of study during these surveys. Only in rare cases can networks of canals that have been out of use for more than the last two thousand years or so be directly followed from air photographs and surface observations. River meanders of almost three times that age have been identified, however, and the major early watercourses can be approximately located from the position of adjoining sites and from the levees of sediment they gradually deposited. Moreover, discontinuities in the distribution of settlements at a given period roughly outline the limits of sedentary occupation. Analysis of the hierarchies of site sizes at different periods provides clues to changing population density, political organization, and underlying ecological instabilities.

As the foregoing implies, there has been substantial progress toward the objective of an essentially complete coverage of the pertinent areas. It must be stressed, however, that there have been substantial differences in the character of the surveys themselves, so that all are not equally applicable to the problem at hand. In particular, there has been a rapid evolution in reconnaissance technique. More intensive studies, involving statistical manipulation of large, genuinely randomized sherd collections, now promise to open a new realm of information on what might be termed "functional differentiation" within and between sites. On the basis of experience both in Iraq and elsewhere, there is reason to believe that specialized craft quarters, centers or wards of administrative and upper-class residence, and regional or local differences in subsistence specialties all will be identified from surface remains in future work. In many cases, this will require renewed surveys and re-samplings of sites previously dated and mapped. Ideally, such re-surveys should be closely integrated with programs of selective soundings and large-scale clearance directed at the solution of specific research problems.

As originally pioneered by Thorkild Jacobsen, the application of systematic archeological surface reconnaissance to Mesopotamian conditions involved a primary concern for the major ancient watercourses. With limited time and resources available, complete coverage of large regions was not originally possible. Greatest attention was given instead to large sites of historical importance, with selective identification of smaller sites where they seemed to confirm the approximate courses followed by the large riverine arteries of antiquity. Only in this way was it possible quickly —if not always irrefutably—to establish the broad outlines of the historical geography of ancient Sumer.

Shortly afterwards, the author and Vaughn Crawford undertook a survey of the northern part of the alluvium, a region roughly coterminous with the land of ancient Akkad. Some advance in technique was made possible by the availability of large-scale (1:50,000, Arabic Series) maps. Moreover, an attempt was made to visit and identify all early sites within the area surveyed. This made possible a clearer understanding of ancient settlement patterns as interacting *systems* of cities, towns and villages. It also brought to light some of the dynamics of change within the systems, arising both from natural fluctuations in the balance between different Euphrates channels and from human action.

In retrospect, there were a number of limitations to the approach then followed.

First, the available maps proved to be of variable quality (particularly for unoccupied regions), and in any case tended to disclose only the larger and more prominent landmarks. Fairly detailed ground reconnaissance partly filled this gap, but could not provide the full information on ancient canal systems (and hence on suitable locations for ancient sites) that would have been (and now is) available only from air photographs. Then too, there were numerous minor—but cumulatively important—respects in which the dating criteria known and used at the time were too limited and imprecise for accurate assessment of some periods of occupation. Finally, the limitation of the Akkad Survey's coverage primarily to the Cassite and earlier periods represented a basic defect in approach. Not only are the settlement patterns of later periods an important research objective in themselves, but in addition the similarities and contrasts between widely different periods subsequently have proved to be a most significant source of insight. . . .

The next stage in the evolution of surface reconnaissance techniques was represented by the author's survey of the Diyala plains east of Baghdad, as part of the Diyala Basin Archaeological Project under the direction of Thorkild Jacobsen. Its results are now available in published form. A general discussion of the methods and limitations of earlier surveys also is available, although already rendered somewhat out-of-date by the rapid pace of recent refinements. Hence it may be sufficient here to note that an opportunity was found to correct the most salient defects of the work in Akkad that were referred to above. On the Diyala plains, and in all subsequent surveys, attention was devoted systematically to all periods of occupation; moreover, since then fundamental reliance has always been placed on aerial photographs. The improvement of chronological controls, of course, goes forward continuously. So also do the improvements in sampling methods and statistical interpretations that were mentioned earlier.

It should be stressed that there is no single, optimal or uniform, method of reconnaissance toward which all of these refinements are tending. Statistical manipulations, in particular, require greatly increased inputs of specialized time for sample collections and processing, thus reducing the area that can be surveyed with given resources in a given period. For certain problems, and as the total surveyed area begins to approach the total area available for settlement, this reduction in geographical scope may be of minor consequence or even irrelevant. For other purposes, particularly where one is concerned with establishing an interpretive framework that embraces an actual historic region rather than an arbitrarily delimited area, it may be more useful to sacrifice the precision and replicability of large, randomly sampled sherd collections for an admittedly more impressionistic coverage of the largest possible terrain. Given that these two choices are always at least partly in contradiction, the author generally has sought to expand the area of coverage. Others have not, and future trends almost certainly will lead increasingly in the direction of intensive studies of small areas. All that is essential is that the methods actually followed be consistent with the problem under investigation, and that they be explicitly described so that their significance for other problems can be independently evaluated.

Results of two survey campaigns by other scholars are highly pertinent to our theme but still await publication. One, conducted by Dr. Henry T. Wright in the environs of Ur, focussed especially on the relationship between Ur and its outlying dependencies at the beginning of the Early Dynastic period. The other, by Dr. McGuire Gibson, advanced well beyond the older Akkad Survey data in an intensive study of the hinterlands of ancient Kish. Both were accompanied by programs of small-scale soundings.

Two other recent surveys will serve to introduce the problems surrounding the initial formation of cities on the Mesopotamian plain. One was carried out in 1967 by the author and Dr. Hans J. Nissen in the region, centering on ancient Uruk, Larsa, Umma and Shuruppak. . . . The second, the Nuffar Survey, involved a relatively small region north and east of ancient Nippur and has just been completed at this writing. The Warka findings, on which only brief preliminary reports are available in publication provides a particularly useful, paradigmatic sequence of change in settlement patterns accompanying the beginnings of urban civilization.

More than a hundred small settlements lay on the plains around what must already have been a substantial shrine or temple center at Uruk by the middle of the fourth millennium B.C. They clustered in irregular groups in which lineal patterning is rarely evident and the empty areas outside of these clusters presumably represent large tracts of unsettled desert or swamp. The internal arrangement of sites within the clusters indicates that most communities were placed along braiding or anastomosing channels that are the natural regime of streams in floodplains, rather than suggesting axially branching, large scale canal systems. There are hints that individual groups of villages may represent some form of territorial unit, for often a cluster appears to be dominated by a single, somewhat larger site. Sites of the latter kind recall the little mud-walled sheikhs' qal'as of the Ottoman period, whose crumbling remains at once-strategic canal-offtakes or weirs still punctuate the desert horizon. But in any case the pattern of irrigation in the main was clearly a shifting, nonintensive one, relying on temporary check-dams, uncontrolled flooding, and very limited use of human- and animal-powered lifting machinery, rather than on heavily capitalized, expensively maintained, bureaucratically administered hydraulic works.

In the closing centuries of the fourth millennium the pattern of settlement was gradually transformed into a quite different one. Small villages remained the most numerous category, but their numbers and aggregate area declined as a new, intermediate class of towns appeared. Detached, relatively formless clusters began to give way to more elongated enclaves centering on linear series of these towns. This suggests a shift away from reliance on essentially natural watercourses, and in the direction of artificial canals that were diked, straightened, and maintained by man. One of these linear enclaves, appearing in a formerly unsettled area south of the later city of Umma in Jemdet Nasr times, may imply the construction of a canal about fifteen kilometers in length for which there was no natural precursor. But on the example of canals dug and maintained by corporate kin groups on their own initiative in the nineteenth and twentieth centuries, growing state intervention and bureaucratic management of hydraulic works were neither prerequisites for this process nor an immediate, necessary outcome of it.

At around 3000 B.C. Uruk underwent a phase of rapid growth, increasing several fold in population and reaching its maximal areal extent soon thereafter. It had become a true city by any reckoning, and now was girt for the first time with a massive enclosing wall. In its hinterlands the decline in smaller settlements quickened, leaving little doubt that the bulk of the urban population formerly had been rural cultivators for whom (or for whose ancestors) the town initially was less of a political capital than a focus of pilgrimages or regional religious observances.

Although attention is directed here primarily toward economic and demographic changes, it would be misleading to isolate such changes from other social trends. This massive shift in settlement patterns coincided with the emergence of contending local dynasties based in individual city-states—politically organized societies in contrast with the greater theocratic emphasis of earlier times. To some degree, urbanization may have been a conscious artifact of royal policy as incipient dynasts strove to consolidate their powers. After all, the formation of large, walled population aggregates would have increased both the king's internal authority and the military effectiveness of the polity he led. On the other hand, the formation of walled centers in an unsettled period would have attracted immigrants from the countryside even without a coherent plan of persuasion or compulsion. And of the increasing dangers that would have been felt in the countryside there can be little doubt. Both historic and archeological records clearly attest to the increasing emphasis on militarism that became characteristic of the area during the early centuries of the third millennium B.C.

Under these circumstances the pattern of settlement throughout southern Mesopotamia had become an essentially urban one by the mid-third millennium. Political relations above the city-state level remained highly unstable, with ephemeral conquests, submissions and alliances, but at any rate the great bulk of the population had taken up permanent residence within walled centers. Of course, these were cities of a kind now rarely encountered —sustained by the agricultural production in which by far the larger part of their own populations was primarily engaged. Fifteen kilometers appears to have been roughly the distance within which cultivation from an urban base was practical. Of the two score villages and towns which at an earlier period had been situated within fifteen kilometers of Uruk, for example, only two remained in the late Early Dynastic period. Beyond that limit regular cultivation presumably came to an end, areas that had been densely inhabited now forming a largely unoccupied buffer zone between hostile city-states. For reasons stemming not from natural conditions but from changes in human society, a new ecological niche had been created which invited only small groups of wary pastoralists.

There must have been corresponding changes within the cultivated radius around each city-state like Uruk. As distance to the fields and the threat of organized, large-scale hostilities became factors, there would have been an unprecedented emphasis on intensifying agricultural production in fields closer to (or even within) the walls. Multiple cropping, labor-intensive summer irrigation, specialized, high-yield gardens and orchards, and permanent hydraulic works all must have been more characteristic of the "green belts" around cities like Uruk than of any pre-urban phase of settlement. But it can also be established from contemporary administrative records that primary reliance continued to be placed —as in fact it still is today—on extensive cultivation under a rotational system involving alternate years in leguminous weed fallow.

As implied above, the concentration of the population in a relatively small number of cities must have occasioned some changes also in the canal system. Larger water storage facilities and more comprehensive irrigation both were necessary to sustain the cities. Both were facilitated by the increasing array of coercive means to draft corvee laborers that were at the disposal of the new political and military authorities who dominated the towns. The construction of city walls was one expression of these new powers to deploy labor, and royal claims to have dredged or opened new canals may be another. It is significant, however, that reference to canals in early royal inscriptions uniformly ignore their possible irrigation functions. Royal claims to have provided new arteries for ship-borne commerce hardly argue very strongly for the exercise of important irrigation responsibilities by political authorities, and in any case more mundane administrative records make clear that state institutions played little part in the actual execution of economic and construction plans.

This is a necessarily much abbreviated account of the sequence of change around some of the most important city-states in southern Mesopotamia. Surveys around others disclose differences in detail; for example, the extent of urbanization, the timing of successive phases of urban growth, and the relative roles of theocratic and dynastic influence all are locally variable. North and east of Nippur, for example, recently collected (and still incompletely analyzed) field data suggest a widespread abandonment of the countryside in or soon after the late Uruk period. Although other interpretations are possible, this may imply that the most rapid phase of growth of urban centers in the vicinity of Nippur was significantly earlier than in the case of Warka. But in most basic respects the paradigm can be said to hold all the way from Sippar, virtually at the northwestern limit of alluvial settlement in early times, to Ur, not far above the swamps and brackish lagoons leading out into the Arabian Gulf.

One noteworthy discontinuity is that, as

we turn northward from Sumer into Akkad, the early cities were generally smaller and more widely separated. A larger proportion of the population continued to reside in villages and other outlying settlements until long after the period we are concerned with here. Small communities not only survived but even were newly founded as the main towns grew and consolidated their hold upon the countryside.

This difference is even more marked if we consider the Diyala plains, on the northeastern periphery of the Mesopotamian alluvium. In that area, the Early Dynastic period saw a vigorous expansion in the frontiers of settlement that had no parallel in Sumer. There was a correspondingly large increase in the number of small communities, without any of the more important centers increasing so much in size as to become more than a moderately large town. Here, then, the territorial unit at the time that the politically organized state made its first appearance seems to have been not a city surrounded by fields cultivated by its own populace, but instead a less centralized system of towns and outlying villages or other dependencies. Undoubtedly the smaller components were subservient to the larger to varying degrees, but at least the forces of attraction or compulsion were insufficient to draw most of the population within the walls of the central towns.

Attention should be drawn to how little we know of the numerous small settlements that remained characteristic of the northern part of the Mesopotamian plain until the late first millennium B.C. Usually they survive as low, inconspicuous, unnamed mounds, and it is to be regretted that in most cases they have attracted no greater interest among archeologists than among the passing Bedouin. Possibly it is enough to describe such settlements as villages, implying only that they form a class of relatively small population aggregates in comparison with towns and cities. On the other hand, there are hints that at least some of them were differentiated as to composition and function. The classical Sumerian Tempelwirtschaft would have been ill-adapted to a region in which most of the population was widely dispersed, and it is likely that the many smaller settlements were organized instead as corporate peasant (or even tribal) communities or as the landed estates of local notables. I. M. Diakonoff has repeatedly pointed out that the predominance of the Tempelwirtschaft in Sumer itself usually has been far too uncritically assumed; probably corporate communities and private estates coexisted alongside the Tempelwirtschaft in all parts of the alluvium as alternative variants in a complex continuum. Nevertheless, it seems likely that the role of the temple as a major land-owning and organizing institution in the agricultural economy was more attenuated in Akkad than in Sumer.

In any case, the suggestions of a partial contrast with Sumer in economic organization are not paralleled by basic differences in the irrigation regime. The lines of Early Dynastic sites on the Diyala plains trace out discontinuous, anastomosing networks of essentially natural watercourses like those in the south, again suggesting primary reliance on temporary weirs and small-scale canals that could be dug and maintained entirely on local initiative. What may have been absent in the Diyala area and Akkad was the somewhat more intensive development of lands immediately outside the larger towns. We may also speculate that the increased proportion of smaller settlements in the northern part of the alluvium reflects an adaptation based on a more nearly balanced mixture of farming and herding on the part of each localized, corporate group or administrative unit. In short, both agriculture and urbanism had become important, stable components of life for the bulk of the Sumerian population by the mid-third millennium. But in spite of a considerable (if fluctuating) degree of political, religious and economic integration with the south, the proportion of the population in the north that had accepted the same course was very much smaller.

The easternmost extension of the Mesopotamian alluvium, the land of Elam around the ancient city of Susa, was still a third regional variant. As a geographical transition zone, intersected by low, barren ridges and then broken off abruptly by the steeply rising flanks of the Zagros Mountains, one might expect this area to have more in common with the Diyala borderlands than with the heartland of southern Sumer. However, the presence on at least the upper Elamite plains of adequate rainfall to permit dry farming, as well as of numerous, relatively small and manageable watercourses to encourage experimentation with small-scale irrigation techniques, led to a widespread and fairly dense population (even by modern standards) already by the end of the fifth millennium B.C. The pattern of settlement at that time consisted largely of clusters of very small villages, possibly with a temple center at Susa corresponding to the center at Uruk and beginning to exercise some form of theocratic hegemony over the surrounding countryside. Subsequently, the same trend we have described at Uruk, the virtual disappearance of rural settlement and the corresponding growth of a major urban center, took place here also. Moreover, this development was not in any sense in retard of the comparable events around Uruk. In spite of its geographically marginal position, the urbanization of Elam may even have slightly preceeded the same trend in southern Sumer.

In Elam, unlike Sumer and Akkad, the historical records do not disclose a pattern of contentious rivalries between neighboring city-states. Susa seems to have maintained its local pre-eminence consistently, at least within the time periods to which written documents refer, although in struggles with more distant Sumerian countryparts the forces of the city met with only varying success. As at Uruk, the growth of Susa as an urban center involved the abandonment of large areas of fertile, well-watered land that lay beyond the reach of cultivators who took up residence within its walls. In this case, however, the explanation that lands were abandoned in order to provide a buffer is made untenable by the absence of hostile neighbors. Presumably, then, urbanization in this case involved the concomitant appearance of new symbiotic relationships with nomads or seminomadic groups. Like the Bakhtiari tribe in the same area during recent times, pastoralists would have used the plains primarily for winter grazing and then moved into higher mountain valleys with the onset of spring. From the very extensive areas apparently devoid of other forms of land use, we might conclude that Elam preserved an unusual balance between sedentary urbanites and transhumant pastoralists as major, interacting components in its political structure.

This brief review of conditions in different parts of the alluvium underlines the diversity of historical and ecological processes that were responsible for the earliest development of urban society. To say that irrigation agriculture was the most essential feature of the subsistence base throughout early lowland Mesopotamia does not provide a paradigm of related economic, demographic and environmental features that applies uniformly to all regions. The one common aspect of these regional variants is, in fact, in a sense a negative one: the scale of the irrigation works was small, the investment of labor and capital in them was low, and their requirements for bureaucratic, despotic management were essentially nonexistent. An explanation of the genesis of state society which focuses on large scale, "hydraulic" characteristics, like Wittfogel's account of what he calls "oriental despotism," accordingly offers neither an accurate characterization of at least early Mesopotamia nor an explanation of the relationship existing there between developing social institutions and ecological processes. Of course this does not deny, and in fact rather emphasizes, other contributions of irrigation to the growth of the Mesopotamian city-state. Especially included among these, as I have argued elsewhere, is the encouragement irrigation offers to processes of internal differentiation and social stratification, as well as to the growth of redistributive institutions.

An equally misleading stereotype involves the visualization of differing subsistence adaptations as polar contrasts. The hostile aspects of relations between nomads and settlers usually are stressed, for example, rather than the continuum of intergrading forms between these two ideal-typical constructs. This probably stems in part from the vivid biblical imagery connected with mutual antagonism of the desert and the sown. A more general, systematic statement of Bedouin life patterns as materially and ideologically "the negation and antithesis of civilization" entered historical scholarship with the great work of Ibn Khaldun.

With the view of nomads and peasants as opposite poles of a single axis of change goes a tendency to overstress the ethnic disjunction between the two groups. Pastoralists are regarded as external predators who generally become significant only in rare cyclical upheavals or in times of weakness, rather than as participants in a continuing process of interaction that shapes the societies of herdsman and farmer alike. This remains characteristic today of much that is written of Sumerians and Akkadians, in spite of the undoubted sedentism of much of the Akkadian population and in spite of Jacobsen's demonstration of the interpretive pitfalls in such an approach. The literate urbanite, whether an ancient chronicler or a contemporary historian or ethnographer, may derive some excuse for doing so from the need to impose whatever limits are most widely acknowledged upon his field of observation and analysis. It cannot be denied that the nomad-peasant contrast still provides the simplest and most general cognitive map held by peoples in the region, and of course this deep ideological distinction profoundly influences many realms of behavior.

On the other hand, the actual distribution of patterns of subsistence and settlement is better characterized as consisting of a variety of fluid, semisedentary adjustments. Lacking uniform, fixed qualities that would facilitate description, the importance of these semi-sedentary patterns has gone largely unrecognized. It lies precisely in their permitting wide, rapid, and remarkably subtle choices among alternatives in order for human groups to survive in an environment that is diverse, unpredictable, and often harshly demanding.

The same tendency to limit and distort our perceptions of the patterns of interaction that are characteristic of the Mesopotamian landscape extends into another ecological niche, the marshes. Marsh Arab or Ma'dan villages frequently are cited not merely as examples of a technologically simple adaptation to conditions of riverine life but as fossilized societies that have preserved its original forms. One implicit, and highly questionable, assumption here is that the marshlands in question still essentially maintain their "pristine" condition as a habitat. Equally questionable is the assumption that a specialized, riverine-oriented pattern of land use has been the central, most durable feature of the Mesopotamian subsistence base, and that it can be regarded as an isolate.

Lower Mesopotamia today—and as far into the past as detailed, trustworthy itineraries, economic or administrative records, and similar documents take us—is not characterized by patterns of settlement in which the archipelagos of the Marsh Arabs play a dominant or even autonomous part. Instead, the characteristic that emerges most strikingly is the diversity of coexisting adaptations to the wide spectrum of conditions ranging between swamp and desert. Of course, special conditions have arisen in recent years as a result of state policies aimed at land registration, extension and improvement of the irrigation system, and an increasing emphasis on mechanization and market-oriented crops. But prior to the implementation of these policies, the sources are uniform in describing a mosaic of local groups that specialized in differing combinations of agricultural and pastoral pursuits, that exhibited great flux in degree of sedentism and place of residence, and that rarely responded consistently or for long periods to a single line of urban or tribal authorities.

There is some evidence that descriptions of this traditional mosaic by literate urbanites, and particularly by European travellers, often have seriously misrepresented its qualities as an adaptive system. The danger and discomfort that unquestionably were encountered by many visiting outsiders led to a gross overemphasis on the supposedly anarchic inconsistency and mutual hostility of intergroup relations. For similar reasons, there was a prevailing skepticism over the capacities of the indigenous system to provide rationally for its own long-term security, or to respond to opportunities for economic growth. Since for the most part it has been based on impressions stemming from short periods of observation, this approach does not take sufficiently into account the wide fluctuations in productivity that until recently have been characteristic of the area. Flood, drouth, destructive silting up and oscillation in river and canal courses, salinization of prime agricultural land, crop blight and insect infestation—to say nothing of agencies more directly linked with human actions, all have been natural forces with which the agricultural regime has had to grapple continuously.

Even a highly successful adaptation under these circumstances will involve an apparent underutilization of resources in favourable years, and varying degrees of hardship for much of the population at other times. Moreover, fluidity and an apparent lack of structure in intergroup relations offer adaptive advantages that should not be ignored, principally, flexibility in pursuing alternative subsistence modes under adverse conditions, or in securing contingent support from a variety of related groups in different circumstances. Even the practice of retaining wealth in herds rather than accepting a permanent commitment to agriculture, often castigated as a rigidly traditional custom that is inconsistent with progress, is better understood in a more positive light. The maintenance and enlargement of herds surely was the investment alternative that was least subject to catastrophic, total loss under the conditions that obtained in the Mesopotamian alluvium until very recent times.

Briefly to summarize, this has been above all a plea for an empirically oriented program of further research on both ancient and modern irrigation, settlement, and subsistence practices in Mesopotamia. Enough is known to rebut some of the grandiose generalizations that too often are accepted as substitutes for a genuine understanding of the detailed, cause-and-effect interplay of ecological and social variables. Enough is known to indicate a very considerable degree of local variability in subsistence patterns, rather than a preoccupation with irrigation agriculture alone. Enough is known to indicate that responsibility for irrigation canal construction, maintenance and administration was retained primarily at the local level, so that the relationship between this activity and the growth of the state was loose and indirect at best. But these largely negative findings merely clear the way for new formulations that must be refined and tested by further research.

It may be assumed that the program of reconnaissance described above will go forward. Some lacunae in our present coverage are particularly in need of early and detailed coverage. Among these, the area comprising the ancient realm of Lagash probably is most important of all. Indeed, it may seem to some that a "complete" survey of the Mesopotamian plain in accordance with reasonably modern standards is entirely sufficient as an objective. However, as one who has been engaged heavily in the task, I can only offer the contrary opinion that the attainment of this objective would produce no increment to knowledge that is commensurate with the very considerable further effort that is needed. Primary attention ought to be turned instead in two quite different directions. One is the refinement and extension of survey techniques that has already been referred to, seeking entirely unprecedented sources of information on the relationships between ancient villagers and city-dwellers from the physical remains that survive on the alluvial surface. The other involves attempting to overcome the inadequacies of the wider intellectual context within which alone the findings of such surveys become relevant. What we are concerned with, after all, is not the abstract geometry of lines and circles on a map, as they shift over time, but with the interplay of natural and social forces that has shaped the course of historical development. To pursue this theme further, neither "complete" coverage nor endless refinements of survey techniques, nor enlarged excavation programs along traditional lines, nor a continuing preoccupation with little more than the political dimensions of history will be enough.

Let us consider the needs that can only be met by further field research. Crucial problems on which new data are necessary include year to year variations in agricultural output and their social consequences; factors contributing to local differences in productivity that might be conducive to the growth of social classes; and population density and distribution in relation to available agricultural resources. Looking beyond agriculture, more information is needed on the long-term dynamics of its interrelationships with pastoralism; on the varying fortunes and places of settlement of social groups which can be traced through time; and on the origins and stability of urban populations in the region. In many cases, generalizations with regard to these questions will be almost equally useful whether they are drawn from the present scene or the ancient past. Such themes are, of course, elementary where social and economic history have become well-developed disciplines. Fully conceding the limited, refractory nature of the evidence, it is time to develop methods and seek out data that will permit us to pursue them in ancient Oriental studies also.

44. EARLY CIVILIZATIONS, SUBSISTENCE, AND ENVIRONMENT

ROBERT McC. ADAMS

Reprinted from Carl H. Kraeling and Robert McC. Adams (Eds.), City Invincible: A Symposium on Urbanization and Cultural Development in the Ancient Near East *(Chicago: University of Chicago Press, 1960). For biographical information on Robert McC. Adams, please see Selection 43.*

■ Several of the preceding selections contain clear-cut hypotheses about the development of civilization in relation to habitational factors. Sanders hypothesizes that civilization begins in areas that have the potential for supporting urbanization, although this does not preclude the spread of civilization to humid areas that cannot support urbanization. State organization is another important accompaniment of civilization; and Wolf and Palerm note that although irrigation can be an important instrument by which a state establishes its authority, large-scale irrigation networks are not necessary preconditions either for urbanization or for statehood. Hole emphazies the political sector of the environment in the emergence of a civilizational social system.

To what extent are these hypotheses applicable to other civilizations, beyond Mesopotamia and Mesoamerica? In this selection Adams examines and compares four civilizations that grew independently of each other: Mesopotamia, Egypt, Peru, and Mesoamerica. As he notes, they represent one type of social system. The problem he sets for himself is the examination of the relationship of this type of social system to the habitat. He finds several differences among the four civilizations, but cutting across the differences are some striking similarities. In all of them he finds evidence that large-scale canal irrigation was a consequence rather than an antecedent condition of intensive cultivation. All of these civilizations had elaborate systems of social stratification and large surpluses of economic products. Such surpluses are products of sedentary life and a high degree of diversification in agriculture.

The similarities also cut across important differences in climate and other habitational features. Juxtaposition of these differences and similarities lead Adams to an important conclusion: There is no one-to-one causal relationship between habitat and civilization. Instead, every civilization is an outgrowth of successive patterns of adaptation to particular habitats over the course of thousands of years. Habitat is not a determining factor; rather, it is a set of potentials and limitations to which men respond dynamically, which they exploit as well as come to terms with—trying, in other words, to make the most of what they have. This is the essence of life.

Another principle that we can draw as we conclude this volume is that an important accompaniment of advances in human adaptation is the progressive freedom of social relations from the restrictions of the natural habitat. We must note, however, that this freedom, even at the most advanced levels of adaptation, is one of degree; probably it can never be complete. Moreover, the increasing freedom from the limitations of their habitats that people have gained in the course of evolution has to be seen as both cause and effect. Each major advance in adaptation represents a greater freedom; it also contains the seeds, though not the assurance, of future adaptations. Civilizations were able to arise in a variety of habitats, even though in a limited variety, because they represented relatively effective adaptations; they were culminations of progressively successful adaptations in their respective pasts, each adding to man's attempts to gain independence from his habitats. Thus another characteristic of civilizations is that they can develop in a variety of habitats because they have exploited the capacities to free themselves from habitational limitations that had been developed during previous eras.

At the same time, however, let us recall something we learned early in this volume, from Medawar in Selection 4: Every adaptation also has its disadvantages. As man has sought to free himself from the restricting confines of the habitat, he has also changed it. In the final section of his paper, Adams picks up a theme that was sounded by Clark in Selection 36: In his exploitation of habitational potentials, man sometimes destroys important features in them, as in deforestation and destruction of soil nutrients by slash-and-burn cultivation. Not only was this a problem for prehistoric man, it is becoming an increasingly important problem in modern man's industrial adaptations.

In addition to Adams' *The Evolution of Urban Society* (Chicago: Aldine, 1966), several other relevant works can be consulted in connection with this paper: "Inca Culture at the Time of the Spanish Conquest," by John H. Rowe (in Volume 2 of *Handbook of South American Indians*, edited by Julian H. Steward [New York: Cooper Square, 1963]); *Civilizations of the Indus Valley and Beyond*, by Sir Mortimer Wheeler (London: Thames and Hudson, 1966); and *The Wonder that was India*, by A. L. Basham (New York: Hawthorne Books, 1963). There is a voluminous literature on other civilizations, such as ancient Egypt, China, Japan, and Roman and post-Roman Europe. An excellent survey, with bibliography and suggestions for further reading, is William H. McNeill's *The Rise of the West: A History of the Human Community* (Chicago: University of Chicago Press, 1963). ■

THIS SYMPOSIUM HAS accepted as its central problem the cumulative, if hardly constant, tendency of human society to grow in size and complexity. Its major substantive foci, of course, are the roots of our own Western tradition in the early civilizations of Egypt and western Asia. At the same time, it is clear that processes and institutions appearing first in the ancient Orient subsequently have recurred, with varying degrees of similarity, in widely separated regions and at different times. A better understanding of some of these recurrent features may help to clarify not only the picture of developing Egyptian and Sumero-Babylonian societies but also the cumulative development of society at large.

My task is to describe briefly some of the major ecological relationships which sustained the growth of civilizations in a number of "nuclear" areas. In addition to Mesopotamia and Egypt, the choice of pre-Spanish Mesoamerica and Peru seems most appropriate. It is supported not only by the volume and historical-archeological depth of relevant data that are available from the latter two areas but also by the likelihood that extreme geographic separation reduced their dependence on Old World precursors to a minimum. In spite of this separation there is a striking similarity, in scope and form, of nuclear American sociopolitical attainments to those of the Fertile Crescent area at a much earlier time.

J. H. Steward has argued convincingly that even the demonstrated fact of diffusion between two cultural traditions is insufficient to "explain" their likenesses. "One may fairly ask," he maintains, "whether each time a society accepts diffused culture, it is not an independent recurrence of cause and effect." From this point of view, it is possible to regard all four areas as historically distinct examples regardless of the ultimate "origins" of particular traits. This is especially true for our purposes, since cultural-environmental relationships within an area are preeminently a matter of independent adjustment to local conditions and resources.

Moreover, the substantive evidence in these cases for the presence of diffusion from some outside source as a determinative factor is either lacking or at best equivocal. Each of the four areas stood out over its surroundings as a highly creative rather than a passively receptive center. While the complete absence of trans-Pacific stimuli for New World high cultural development cannot be assured, the conclusion of most Americanists today is that the latter "stands clearly apart and essentially independent from the comparable culture core of the Old World." There is certainly no suggestion of any New World—Old World contact as important as the relatively brief but catalytic influence of Mesopotamia on Egypt at about 3000 B.C., yet in the latter case Frankfort took pains to point out the selective, qualified, and generally transient character of the borrowing. With respect to interrelations between Peru and Mesoamerica, it is sufficient to state that not a single object or record of influence or contact between these areas has been accepted as authentic from the long time span between the Formative (or Early Village) period and the coming of the Spaniards, although the overall tempo of development in each is remarkably similar. In short, it is both reasonable in *a priori* theoretical grounds and justified by present evidence to use Mesopotamia, Egypt, Mesoamerica, and Peru as essentially independent examples for a discussion of their internal ecological relationships.

Within the limits of this discussion it is neither possible nor necessary to explore fully the similarities in cultural development among these four areas. All clearly became civilizations, in the sense in which that term is defined here as a functionally interrelated set of social institutions: class stratification, marked by highly different degrees of ownership or control of the main productive resources; political and religious hierarchies complementing each other in the administration of territorially organized states; a complex division of labor, with full-time craftsmen, servants, and officials alongside the great mass of primary peasant producers. Each was a complex, deeply rooted cultural tradition displaying most of all of V. G. Childe's more inclusive civilizational criteria as well: monumental public works, the imposition of tribute or taxation, "urban" settlements, naturalistic art, the beginnings of exact and predictive sciences, a system of writing suitable at least for rudimentary records and accounts. The attainment of civilization, from a diachronic point of view, was expressed in each of the four areas by a series of parallel trends or processes: urbanization, militarization, stratification, bureaucratization, and the like. Of course, these processes were truncated in the New World by the Spanish Conquest—as a plausible approximation, after a level of development had been reached which was functionally equivalent to Old Kingdom Egypt or southern Mesopotamia under the Dynasty of Agade. However, this does not affect our comparisons here, which will be limited to earlier periods in the Near East for which New World equivalents are available.

It thus seems possible to group the four civilizations as representatives of a single type or class of social system. (Other members of the class would include the unknown Indus Valley polity of Harappa and Mohenjo Daro, Shang China, and perhaps certain West African city-states.) To be sure, this stress on structural and functional similarities needs supplementing by the traditional humanistic emphasis on the unique and relatively timeless qualities of each civilization for a properly balanced view. One example of the latter emphasis is the invocation of particular environmental features of different civilizations to account in part for their differing views of the natural world as reconstructed from work of ancient literature or art, for the distinctive structuring of their formal cosmologies, and perhaps even for dominant psychological attitudes. A typological approach necessarily neglects, although certainly cannot deny the unique total patterning of every culture irrespective of what proportion of its constituent elements may have close parallels elsewhere. Probably this patterning is expressed most systematically, concisely, and impersonally in stylistic or configurational terms. But in any case these widely ramifying, largely ideational, aspects of the interrelations between man and the natural world are beyond the scope of this paper. Here we are concerned only with the generalized social order common to a group of autochthonous civilizations and with its relations to the environment.

CLIMATE, PHYSIOGRAPHY RESOURCES, AND POPULATION

Beyond the limitation of each of the nuclear areas to subtropical latitudes, the combined gross catalogue of environmental features is characterized mainly by its diversity. If Egyptian and Sumero-Babylonian civilizations are restricted to great arid or semi-arid river valleys, no such uniform description holds for the zones occupied by either Mesoamerican or Peruvian civilization. Both of the latter range from sea level to high mountain slopes, with tropical, temperate, or even cold-temperate climates corresponding to their altitudes.

If coastal Peru and much of highland Mesoamerica are sufficiently dry to be closely comparable with the Old World centers, this is progressively less true in the Peruvian sierra with increasing altitude and distance from the Pacific coast and not true at all in the Gulf Coastal lowlands of Middle America.

Both of the New World areas lack great inclusive river systems comparable to Egypt and the Nile or Mesopotamia and the Tigris-Euphrates. Instead, short, steeply descending watercourses that drain relatively small watersheds are common, and many of the largest of these are reduced in their pre-Hispanic importance by geographic factors. The main valley of the Rio Balsas and the intermontane basins of the Bajio on the Rio Lerma in Mexico, for example, were lightly occupied before the Spanish introduction of draft animals and the iron-tipped plow made it possible for agriculturalists to deal with heavy soils and sod. The Amazon headwaters in the eastern sierra and Montana of Peru may be found to provide a more significant exception when they have been explored more adequately, but at least the lowland rain forest of the Amazon basin proper acted as a major ecological barrier to the expansion of Peruvian civilization. Since the potentialities of the Old World rivers for disastrous floods, for large-scale irrigation, and as arteries of commerce are often thought to have promoted political unification and the growth of trade in the ancient Orient, it is worth noting that the same cultural phenomena appeared independently in regions where these potentialities were absent or at least far less important.

With respect to natural resources, it is sufficient to recall the absence of even stone in the alluvial soil of southern Mesopotamia, as well as the extremely poor quality for building of the soft and quick-growing woods that alone were available locally. In contrast, parts at least of the New World nuclear regions were well favored, although with great altitudinal variation, local self-sufficiency was often replaced by patterns of regional specialization and exchange. As with climate and terrain, then, we cannot identify a fixed constellation of raw materials which acted as a necessary precondition (much less as a "cause") for the emergence of civilization in every area.

While relatively continuous settlement in linear pattern coinciding with the positions of the watercourses was possible in southern Mesopotamia and Egypt, enclaves of dense occupation separated by stretches of relatively inhospitable terrain were more characteristic of Mesoamerica and Peru. The best known and largest of the Mesoamerican enclaves is the interior drainage basin called the Valley of Mexico, which has provided the bulk of population and subsistence resources successively for the religious center of Teotihuacan, the Toltec realm with Tula as its capital, the widespread conquests and incipient empire formation of the Aztecs, and present-day Mexico City. Yet in spite of the unparalleled importance of this region its area does not exceed 8,000 sq. km. In Peru the areas of intensive settlement and cultivation were all still smaller. Perhaps the largest of the mountain basins able to support a concentrated population is that of Huancayo, in the central highlands, with an area of only 1,200 sq. km. The arable area of the Chicama Valley, the largest in the North Coastal lowlands, is approximately the same.

In all of nuclear America, only along the Gulf Coast and on the low-lying Yucatan Peninsula were the conditions suitable for relatively uniform and continuous settlement. There, too, the rivers most nearly resemble the Nile or the Euphrates in regularity of flow and ease of control. But the lateritic soils and heavy rain-forest vegetation impose a very long recovery period after brief use for slash-and-burn agriculture, which materially reduces population density and perhaps helped to postpone for a considerable time the onset of urbanization processes which had been initiated in adjacent Mesoamerican highlands. A sharper contrast would be hard to imagine than that between Sumerians clustering in cities and Classic Mayans living in dispersed, essentially rural, hamlets while only a small elite permanently inhabited the elaborate religious centers. Yet both were civilized. In short, the distribution of population and settlements within the nuclear areas appears to have been as variable as the general environmental conditions within which they occurred, although average density in each case was surely much higher than in surrounding areas.

VARIATIONS IN AGRICULTURAL SUBSISTENCE PATTERNS

While the essential basis for subsistence in every civilization is obviously to be found in sedentary agriculture, this rubric covers impressive technical, botanical, and zoological differences when it is applied to the high cultures of both the New and the Old World. Largely following C. O. Sauer, we may summarize these differences briefly.

New World agriculture, in the first place, essentially did not involve stockbreeding or the utilization of such animal products as dung fertilizer or milk. Domesticated Andean camelids such as the llama were used mainly for transport and were largely confined to the higher slopes; hence they cannot be regarded as important exceptions. Also missing in nuclear America, therefore, is the unique and powerful ambivalence of relations between herdsman and farmer, involving both symbiosis and hostility, which has shaped the social life, tinctured the history, and enriched the literature of the civilizations of the Fertile Crescent.

Second, nuclear American agriculture involves an entirely different range of cultivated plants, which nonetheless seem to have provided as balanced and adequate a diet as the cereal-date-vegetable-livestock complexes of the ancient Orient.

Third, basically different methods of cultivation were employed in the New World. In the absence of draft animals, the major implements were the digging stick and the hoe instead of the plow. Instead of a definite brief harvest season, crop-gathering was prolonged by the use of the major food crops also as green vegetables during earlier stages of their growth and by the widespread practice of interspersing different crops within a single field.

Finally, corresponding to the greater variations in climate because of altitude, New World agriculture was far more variable. There is little difference in at least the potential yields of the Assyrian uplands and the Mesopotamian alluvial plain other than that due to the inability of the date palm to flourish beyond the northern limit of the alluvium and to the greater (but not exclusive) reliance in barley rather than wheat south of that limit. By contrast, coastal Peruvian agriculture essentially revolved around a maize-beans-squash-cotton-fruits complex, while in the sierra subsistence depended on an entirely different complex composed of root crops like potatoes, oca, and quinoa. Similarly, maize, beans, and squash were the staple foods in both highland and lowland Mesoamerica, but they had been differentiated very early into altitudinally specialized varieties. Moreover, the cultivation of cotton, cacao, and many fruits was restricted to the lowlands.

SIMILARITIES IN SUBSISTENCE PATTERNS

In spite of these profound differences common features are not lacking. Perhaps something can be learned of the general place of subsistence in the growth of civilizations by outlining three common elements which seem to be of greatest importance.

One such significant common feature is that "farmers were persuaded or compelled to wring from the soil a surplus above their own domestic requirements and [that] this surplus was made available to support new economic classes not directly engaged in producing their own food." It must be understood that the notion of a surplus is related to fixed biological needs and the level of productive efficiency only in very general terms and that both the kinds and the quantities of available surpluses were determined to a considerable degree by the broad social contexts—"non-economic" as well as "economic"—within which they occurred. Yet the institutional forms for the concentration and redistribution of surpluses show a high degree of uniformity among the early civilizations and serve to distinguish the latter sharply from societies in which no full-time activity other than primary food production finds sanction. Although it is impossible to quantify, it is only reasonable to assume that the proliferation of nonagricultural specialists common to all the early civilizations was correlated with a general increase in agricultural efficiency. It is, of course, quite another matter to assume that improved efficiency was independent of and prior to the whole ramifying network of concurrent social changes. Even purely technological advances, which in most instances these increased surpluses probably do not reflect, are usually linked with the social and cultural milieu, as Kroeber's study of independent and relatively simultaneous inventions was first to show.

A second common feature of some importance may be the complexity of the subsistence base on which each of the civilizations seems to have rested. We are dealing in no case with a single-crop economy or with one in which the bulk of the population normally could supply the entire range of agricultural produce for themselves. Perhaps the diversity of resources is partly to be understood as the protection against natural calamity necessary for long-term cultural growth. But also in part it must have been responsible for the development of trade, exchange, and redistributive institutions which in turn enhanced the growth of some form of centralized authority.

Mesopotamia is perhaps the best-documented example. The complementarity of dates and grain finds symbolic expression in the alabaster "Uruk vase" of late Protoliterate date, where alternate palm and cereal shoots in the bottom register figuratively support the abundant ceremonial life illustrated above. Fishing was another essential subsistence pursuit; of the 1,200 or so members of the Baba temple community in Girsu in the mid-third millennium B.C., more than 100 were fishermen. The precise role of fishing in earlier times is difficult to ascertain, but quantities of fish offerings found in a late Ubaid temple at Eridu may indicate that it had already attained considerable importance by that remote period. Slightly less numerous than the Baba temple fishermen were its shepherds and herdsmen, but their numbers in that specific case do not adequately reflect the crucial position of sheep, donkeys, and oxen in the mixed economy of ancient Mesopotamia for plowing, transport, wool, and fertilizer as well as meat. Surely the prominence of the shepherd-and-byre motif in protoliterate glyptic art reflects a high antiquity for husbandry, an essential part of the configuration of subsistence activities. In all of these cases it is interesting to note that the temple and state institutions played a vital part in the collection and redistribution of the agricultural produce.

To the far more limited degree to which there are pertinent data in diversification and specialization of subsistence in Old Kingdom Egypt, the picture is at least not inconsistent with what has been described for Mesopotamia. The idealized representations in the tombs of life on the estates of court officials record a great variety of craft activities and subsistence pursuits; since an organization of the work under foremen is sometimes illustrated, there must have been at least a partial specialization of function in the real world as well. While the great bulk of the peasant's caloric intake may always have been derived from grain, the cultivation of vegetables and fruits and fowling, fishing, and animal husbandry also play a substantial part in the tomb scenes of Old Kingdom officials. The importance of herding, in particular, may have been obscured by its limited modern role, under very different conditions of land use. For obvious reasons the main center of husbandry was in the Nile Delta, and the close concern of the state for husbandry is clearly to be seen in the emphasis on livestock in lists of claimed tribute and loot, in periodic censuses of the herds, and in the appointment of numerous officials charged with responsibility of one kind or another for domestic animals.

In the New World the differentiation of

subsistence pursuits seems to have been mainly on a regional basis, perhaps as a consequence of the greater environmental diversity that has previously been alluded to. But the necessity for a wide interchange of agricultural products remained the same, and the organization of this interchange similarly must have helped to expand and consolidate the position of centralized social authority. In North Coastal Peru, for example, llamas from the sierra were already being ceremonially buried in a community shrine or public building in Late Formative times (*ca.* 800 B.C.) In another case, the only llama bones from a contemporary site of the same period were found in association with the burial of an individual whose relatively elaborate *Beigaben* suggest a priestly status. By the succeeding Florescent era, the relative abundance of llama bones, wool, and droppings indicates that trading contacts with the highland centers of domestication for these animals had been regularized and enlarged. Presumably cotton, maritime products, peppers, fruits, and coca were among the commodities moving in the reverse direction, as they were at the time of the Conquest. To some degree, regional specialization with regard to subsistence extended into craft production as well, as is implied by the importation of a colony of Chimu craftsmen to work for the Inca government in Cuzco. It is interesting to note that a high degree of specialization still characterizes the Quechua community.

Similar patterns of differentiation in specialized production can be identified in Mesoamerica. Cotton from the lower-lying valleys of Puebla and Morelos was already being interchanged with the Valley of Mexico in Early Formative times, and the securest archeological dating horizons of later periods are provided by distinctive pottery wares that were traded widely from their different centers of manufacture. For the Conquest period these traces of evidence can be greatly amplified with eyewitness accounts of, for example, the great and diversified market at Tlatelolco with its separate vendors for many varieties of fruit, meat, maize, vegetables, and fish and with a reputed daily attendance of 60,000 persons. From a different point of view, the heterogeneity of native resources is also underlined by the *matricula de tributos*. Although it accounts for tribute levied by the Aztecs rather than for trade, the general concentration of assignments for particular kinds of produce (other than the ubiquitous mantles) to a very few provinces surely reflects earlier patterns for the interchange of normal regional surpluses. And by Aztec times, if not earlier, the integration of interregional trading with the needs and policies of the expanding state is well known.

A third significant feature common to the agricultural pursuits of the early civilizations was the development of some degree of intensive land use. Whether or not this was accompanied by a general increase in agricultural efficiency (output/labor input), certainly it must have increased at least the total agricultural output. However, the point of current interest is not so much the effect of intensive methods of cultivation on the volume of available surplus as their effect directly on social organization. The argument, following Ralph Linton's lucid portrayal of the introduction of wet rice cultivation in Madagascar, is that under conditions of intensive cultivation plots of land acquire different values based, for example, on cumulative improvements and the availability of water. Since water, or good bottom land, or some other similar resource was almost always relatively scarce, well-favored and improved plots came to be regarded as capital investments. While unimproved land was allotted equitably among all members of the village or extended kin group, under conditions of intensive cultivation the cohesiveness of the older social units broke down and tended to be replaced by a small number of individual families as the hereditary land-holding units. The emergence of an authoritarian "king," of rudimentary social classes including nobles, commoners, and war-captive slaves, and increasing expenditures on warfare are some of the further consequences which Linton traces to the basic shift in cultivation practices. Under at least some circumstances, in other words, the social processes we have identified with the beginnings of civilization are closely interconnected with the beginning of intensive agriculture. No necessary distinction into "cause" and "effect" is implied, be it understood, between subsistence change and institutional change. The investment of labor in land improvement and the adoption of intensive cultivation techniques were as much influenced by contemporary social forms as they influenced the latter.

Intensive agriculture, in the case of the earlier civilizations, usually is taken to be roughly synonymous with irrigation. Indeed without some kind of irrigation agriculture is and probably always was impossible in southern Mesopotamia, Egypt, and coastal Peru. But we shall attempt to show that in most cases irrigation was part of a broader range of intensive techniques and that some of the assumed implications of irrigation as a single, gross category are misleading when applied to the four nuclear areas where the civilizations with which this paper is concerned had their beginnings. Here, then, irrigation is subsumed under the general rubric of intensive cultivation rather than equated with it.

It is important to distinguish between the functional significance of different kinds of irrigation if we are to understand better the relations between ecology and cultural growth. Small-scale irrigation, including floodwater techniques and the construction of short lengths of canal serving small landholdings, does not seem essentially different in its social effects from those observed by Linton in Madagascar. It may make available for agricultural purposes only a fraction of the potentially irrigable land surface, since it will seldom extend very far from the streams and since short canals will not be sufficient everywhere to bring the water to fields at a high enough level. Alluvial situations, in which rivers tend to rise their beds above the level of the surrounding land, are particularly favorable for small-scale irrigation. For the same reason, they invite destruction of existing canals by silting and flooding, although this is not critical where canals do not represent a heavy investment in labor and can be quickly replaced. The construction and maintenance of this kind of irrigation, we submit, requires no elaborate social organization and does not depend on labor resources larger than those at the disposal of the individual community, kin group, or even family—or, at most, those easily available locally through patterns of reciprocity. To the extent that this kind of irrigation is important, its chief influence on social development would seem to arise from its encouragement of stratification based on differentiation of landholdings. Perhaps also it encouraged the growth of militarism associated with increasing competition for developed canal networks and the most fertile and easily irrigated lands.

Large-scale irrigation, on the other hand, imposes technical and social demands of a different order. Masses of labor must be mobilized from many scattered communities, and their activities need close coordination. The problem of maintenance and supervision is a continuous one and again demands a superordinate authority. Some kind of equitable distribution of the available irrigation water must be imposed on many competing communities, and disputes must be adjudicated. Since downstream users are inherently at the mercy of those higher up, large-scale irrigation networks are only durable where the entire area they serve is a politically integrated unit. As has often been observed, large-scale canal networks can only be associated with formal state superstructures in which the ultimate authority rests wtih an administrative elite.

The problem for us is an absolutely basic one, however sparse, refractory, and ambiguous most of the present evidence may be. To the extent that large-scale irrigation is found to have begun very early, its social requirements may be adduced as a convincing explanation for the origin of primitive states in the ancient civilizations. Processes of class stratification associated with intensive agriculture then might be a secondary and derivative phenomenon on this reconstruction; because of its monopoly over hydraulic facilities, the state bureaucracy is identified as the strongest social force. Largely following Karl Wittfogel, Julian Steward took this position in a recent symposium with respect to Mesopotamia and Peru although not to Mesoamerica. Our view is firmly to the contrary. It is beyond the scope of a paper dealing with cultural ecology to argue that the primitive state is mainly linked instead with the emergence of a stratified society, but at least it will be suggested here that the introduction of great irrigation networks was more a "consequence" then a "cause" of the appearance of dynastic state organizations—however much the requirements of large-scale irrigation subsequently may have influenced the development of bureaucratic elites charged with administering them. The admittedly still inadequate evidence for this proposition now needs to be briefly summarized.

Our present understanding of the antiquity of irrigation in Mesopotamia is derived mainly from surface reconnaissance in Akkad and the Diyala basin and is obscured by the heavy and continuous alluviation with which the northern part of the alluvial plain has been particularly affected over the milleniums intervening since Sumerian times. At least in this region, however, there appears to have been little change in settlement pattern between the beginning of widespread agricultural occupation in the Ubaid period and the end of the third millennium B.C. or even later. There is historical documentation for the construction of occasional large canals and irrigation works as early as the Protoimperial period, but on the whole the settlements followed closely the shifting, braided channels of the major rivers.

In other words, for a long time irrigation seems to have been conducted principally on an *ad hoc* and small-scale basis, which would have involved periodic cleaning and perhaps straightening of clogged natural channels, adjusting the location of fields and settlements in the closest possible conformity with the existing hydraulic regime, and for the most part constructing and maintaining only relatively small-scale field and feeder canals that were wholly artificial. Where the king explicitly claims credit for initiating dredging operations on either a canal or a natural watercourse (as in modern Iraq, the same word is used for both!), it is noteworthy that the aspect of canals as providers of irrigation water is entirely unmentioned. Moreover, whatever the rhetoric of the king's claimed responsibilities, the necessary labor forces for the maintenance work were apparently organized and directed by the individual temples. No Early Dynastic or Protoimperial record has survived of the mode of allocation of irrigation water, but at least in Ur III times this was separately handled in each temple constituency by a special official in charge of sluice gates. In short, there is nothing to suggest that the rise of dynastic authority in southern Mesopotamia was linked to the administrative requirements of a major canal system.

There are very few data yet available on the character or extent of Egyptian irrigation during the period for which it might be compared with New World equivalents, that is, up to the beginning of the Middle Kingdom. Prior to the opening of the Fayyum depression to irrigation in the Twelfth Dynasty, there is nothing less ambiguous to demonstrate state responsibility for irrigation than the statement of a Sixth-Dynasty royal architect that he had dug two canals for the king. Unfortunately, the inscription fails to make clear whether the canals were intended for irrigation or only for the movement of royal supplies like building stone, as was the case with five contemporary canals dug to bypass the First Cataract of the Nile. Still another possible explanation of the significance of the passage is that it refers to land reclamation by swamp drainage, much as a very late (and therefore doubtful) tradition credits Menes with having drained the territory around Memphis. Yet swamp drainage began long before any pharaoh appeared on the scene—if the obvious meaning is attached to the claim of a Third-Dynasty official that he "founded" twelve estates in nomes of Lower Egypt—and continued afterward without the necessity of royal initiative. In considering alternatives other than irrigation we are also confronted with a protodynastic scorpion macehead ostensibly showing the king breaking ground for a waterway of some kind. Again, an immunity charter of Pepi I protects the priesthood of the two pyramids of Snefru against any obligation for labor service on what may be a canal; here it is neither clear that the putative canal was for irrigation nor that the pharaoh was responsible for its construction. Interestingly enough, the same charter continues with an injunction against enumerating canals, lakes, wells, hides, and trees belonging to the priesthood for tax purposes and thus suggests that all of those categories were under purely local jurisdiction.

In short, considering the number of known records of royal building activity in the Old Kingdom, it seems only fair to regard their silence on the construction of irrigation works as strange if the demands of large-scale irrigation had indeed been responsible for the initial emergence of a pharaoh at the head of a unified state. On the assumption of a centrally administered irrigation system, the failure of officials with long and varied careers of public service to refer to administrative posts connected with canal maintenance or water distribution is equally puzzling. To the degree that an *argumentum es silentio* ever carries conviction, the Egyptian case parallels that of Mesopotamia.

Although there is serious danger of overgeneralizing from it, the data on Peruvian irrigation are reasonably consistent with what has been adduced from Mesopotamia and Egypt. Drawing principally from Gordon Willey's pioneer study of settlement patterns in a typical small valley transecting the arid North Coastal strip, we cannot presently trace large-scale irrigation earlier than the Florescent era (beginning probably at about the time of Christ). The distribution of Late Formative sites suggests, however, that small-scale experimentation with canal-building had begun in a few advantageous locales several centuries prior to this time, and some success with at least flood-water irrigation on the river flats is implied by the slow expansion inward from the valley mouth which began a millennium earlier. The Early Florescent (Gallinazo) canals, it is interesting to note, were built as integral parts of an elaborate and impressive complex of monumental construction which included fortifications and ceremonial pyramids as well; on present evidence, both of the latter types of monumental construction antedated the large canals. By mid-Florescent times at least, valley wide systems of irrigation were in use on the North Coast (although our particular example comprises only 98 sq. km. of arable land!), and some individual canals are large by any standards: the canal of La Cumbre in the Chicama Valley, for example, is 113 km. long. A subsequent development, probably dating only from the Militaristic era (beginning after A.D. 700), was the still more extensive reshaping of natural drainage patterns through the introduction of intervalley irrigation systems in which urban zones occupied by a governing elite were set off from areas for agricultural exploitation.

Irrigation apparently developed more slowly in highland Peru than on the North Coast, although the sharpness of the contrast may be a reflection in part of the lesser amount of archeological attention that the sierra has received. Terraces for soil conservation have been reported first for the Tiahuanaco horizon, at the outset of the Militaristic era. In the characteristically steep and narrow Andean valleys rapid runoff was perhaps a more serious problem than paucity of rainfall, but in general the later terraces seem to have been associated with irrigation channels as well. The elaborate, well-cut, and extensive terrace-irrigation systems for which Peru is famous all were products of the labor-service obligation imposed by the Inca state as a tax in the final century or so of its successful expansion before the coming of the Spaniards. Even the Early Inca terraces, probably postdating the onset of the Tiahuanaco horizon by four or more centuries, have been described as "small and irregular, and probably the work of individual family groups." As in North Coastal Peru, Egypt, and southern Mesopotamia, we seem to have evidence here of a very gradual evolution of irrigation practices beginning with local and small-scale terracing which emphatically did not require political organization embracing a large group of communities. Large-scale, integrated programs of canalization and terracing apparently were attempted only after the perfection of the Inca state as a political apparatus controlling the allocation of mass-labor resources. They are consequences, perhaps, of the attainment of a certain level of social development; we repeat that they cannot be invoked to explain the process by which that level was attained.

For Mesoamerica the situation is more complex and not a little contradictory. The traditional view is that "there is little evidence that irrigation was of basic importance anywhere in Mexico, in pre-Spanish times, and that it is erroneous to speak of maize culture as having flourished most in arid or subarid regions of that country." Recently this conclusion has been controverted effectively by a number of investigators, although the full significance of their empirical findings is still open to dispute. On the whole though, the situation seems to be quite similar to that described for the other nuclear areas; in fact, it was primarily the recent findings in Mesoamerica which stimulated the reconsideration of irrigation that this paper represents.

The question of the role of irrigation in the formation of Mesoamerican civilization takes us back at least to the beginning of the Classic era (*ca.* A.D. 100?), if not earlier, and revolves particularly around the population and ceremonial center of Teotihuacan in the Valley of Mexico. The Pyramid of the Sun there, one of the largest pre-Hispanic structures in Mesoamerica, apparently antedates that era. It has been estimated that before its abandonment in Late Classic times (*ca.* A.D. 700) the site occupied 750 hectares or more of religious and civic buildings, residential "palaces," workshops, and clusters of ordinary rooms and patios housing "at least" 50,000 inhabitants. True, the observed limits of surface debris may reflect only the aggregate area of the center over a period of several centuries and not its maximum size at any one period. Moreover, the proportion of residential units within the built-up area of the site is still not at all clear. But even if the estimate is scaled down considerably, it certainly reflects an urban civilization in being. To what extent, if at all, did it depend on irrigation agriculture? No direct evidence for canal irrigation has yet been reported. Instead, we have the observations that irrigation is necessary today for cultivation of even a single yearly crop in the subregion of which Teotihuacan is a part, that according to paleoclimatic studies based on pollen analysis and fluctuating lake levels it was even more necessary during the time of emergence of Teotihuacan as a great center, and hence that the use of irrigation must be assumed. The difficulty is that a center of the enormous size of Teotihuacan must have developed on a sustaining area far larger than its immediate subregion and that a major contribution from its immediate surroundings cannot be assumed to have been indispensable for the growth of the site. Monte Alban, Xochicalco, and other examples can be found which approach Teotihuacan in size but which lie at some distance from their main agricultural hinterland. A second argument is still less conclusive. It consists of the suggestion that irrigation is implied by representation of cacao and fruit trees along the banks of streams or canals in a mural from a Teotihuacan "palace." Even if the identification of cacao is accepted as correct, the location of the scene is unknown and the crucial question of whether the waterways are natural or artificial is unanswered. There remains only a distributional argument, based on the wide extent of Mesoamerican irrigation practices at the time of the Conquest. Like all distributional arguments, it is loaded with presuppositions and provides no real clue to the antiquity of the trait in question. And so for Formative and Classic times the existence of canal irrigation still remains to be demonstrated.

For the final, or Historic, era (beginning *ca.* A.D. 900 with the founding of Tula), on the other hand, the evidence for large-scale irrigation agriculture and other hydraulic works is incontrovertible. Perhaps such works are already implied by the legendary account of the formation of Tula in the Codex Ramirez which describes the damming-up of a river in order to form an artificial lake stocked with fish and waterfowl. In any case, the Spanish conquerors were full of admiration for the scale and intricacy of the system of dikes and aqueducts that by 1519 was both supplying Tenochtitlan with potable water and controlling fluctuations in the salt- and fresh-water levels of the lakes surrounding the city. The sequence of construction of these works can be traced in some detail in historical sources, and the conclusion seems justified that they should be viewed "not so much as the result of many small-scale initiatives by small groups, but as the result of large-scale enterprise, well-planned, in which an enormous number of people took part, engaged in important and prolonged public works under centralized and authoritative leadership." Elsewhere in the Valley of Mexico, an irrigation complex in the Old Acolhua domain has been described that was roughly contemporary with the Aztec construction and also seems to have been initiated by a dynastic authority and carried out as a planned large-scale enterprise. Finally, an impressive list of places, with a wide distribution throughout Mesoamerica outside the Maya area, can be assembled for which irrigation is definitely identified or can reasonably be inferred in Spanish contact sources. In short, the position that irrigation was not important anywhere or at any period in pre-Spanish Mexico no longer seems tenable.

It needs to be stressed again, however, that distribution is a highly unreliable index to antiquity and that even the examples from the Valley of Mexico appertain only to the final century before the Conquest. Moreover, with the exception of the above-mentioned Aztec system all the known Mesoamerican irrigation networks are quite small in comparison with those of the Old World and Peru. On present evidence, then, Wolf and Palerm rightly tend to regard planned large-scale canal irrigation not as a primary cause of Mesoamerican civilization but merely as its culminating activity in the economic sphere. They recognize, to be sure, that political controls in turn probably were centralized and intensified by the introduction of major irrigation works.

But if larger-scale canalization is late in Mesoamerica, there are indications that other forms of irrigation and intensive cultivation—as in Peru and Mesopotamia also—can be traced to a more remote antiquity. Canal irrigation probably never became as important a technique in the Valley of Mexico as chinampa agriculture, that is, the cultivation of artificial islands made out of plant debris and mud scooped from the lake beds. Modern chinampas are largely devoted to truck gardening, but, since the tasks of construction and maintenance do not require extensive organization and capital, they may have been used aboriginally as highly productive subsistence plots for kin groups or even families. The only example of an apparent chinampa so far subjected to archeological scrutiny contained occupational refuse dating to about the beginning of the Classic period and suggests that the technique is sufficiently old to have been a factor in the subsistence of Teotihuacan. The means were at hand early enough, in other words, for differential return from specialized farming to have provided the material basis for the growth of a stratified society.

Since chinampas were unknown elsewhere in Mesoamerica (or depended on conditions not repeated elsewhere), their high and perennial productivity may not have been a direct factor in the development of civilization throughout the whole area. At the same time, the Valley of Mexico was in many other respects the key area of development for the greater part of Mesoamerica, for a very long time the center of its most advanced political forms, its widest and most closely intercommunicating trade network, its densest population. To a degree, then it may have set the course of development which elsewhere was merely followed with more or less local innovation. To that degree, chinampa agriculture may far exceed in importance its highly circumscribed geographical limits. Unfortunately, having largely set aside simple diffusion studies, anthropologists are only beginning to develop more functional approaches to the analysis of interregional relations, through which the supposed primacy of the Valley of Mexico might be understood and evaluated.

Another, and broader, aspect of intensive cultivation in Mesoamerica is perhaps to be seen in the maintenance of dooryard garden plots in close symbiosis with individual houses, which augment the production of foodstuffs through the use of leavings as fertilizer and encourage stability of residence. Although not subject to archeological confirmation at present, this practice was apparently well established at the time of the Conquest and is possibly very old. Again, crudely made terraces for erosion-control purposes have been observed at many places in highland Mesoamerica and in at least one instance in the lowland rain forest of the Yucatan Peninsula. Certainly in many cases of considerable pre-Spanish antiquity, they suggest agricultural regimes of greater intensity than the milpa system as it is practiced today. Although at present impossible to document for pre-Conquest times, a more intensive application of labor in the form of hand-weeding would have prolonged cultivation and increased output, particularly in the tropical lowlands. This might make less inexplicable or even "explain" the extraordinary cultural achievements of the Classic Maya in the lowlands.

By assisting in the establishment of residential stability and in the production of surpluses, all the above-mentioned practices would have provided at least a receptive hinterland within which the new and more complex social forms could expand and consolidate. The origin of innovations such as the primitive state might then be sought in a few small strategic regions such as the Valley of Mexico where the inducements to accumulate surpluses and institutionalize class differences were probably greatest. In a wider sense, it may be granted, the florescence of the state could only take place where conditions in the hinterland were also propitious, so that the pinpointing of precise points of origin is probably misleading.

Briefly to recapitulate, we have attempted to show that developments in modes of subsistence within Mesoamerica were substantially similar to those in Mesopotamia, Egypt, and Peru in that large-scale canal irrigation was a culminating, rather than an early and persistent, form of intensive cultivation. It is conceded that differences in the rate of development existed, probably in large part because of the fewer inducements and opportunities to depend on irrigation that Mesoamerica offered. But these, we suggest, are quantitative and not qualitative differences. In North Coastal Peru the culmination came in the mid-Florescent era—or even later, in the Militaristic era, if the introduction of intervalley irrigation systems is accepted as a significant later innovation. In Mesoamerica it came in late Historic or Militaristic times, as it also seems to have done in *highland* Peru. According to our Mesopotamian data, admittedly inadequate in detail and based on a possibly retarded Akkad instead of Sumer, the onset of large-scale artificial canalization did not occur until after the time of Hammurabi. Even in Sumer itself there is no justification for supposing that this process began any earlier than the late Early Dynastic or the Protoimperial period—a sound equivalent for the New World Historic or Militaristic era. In *no* area, then, at least on present evidence, was large-scale irrigation early enough to "explain" the emergence of the great theocratic centers of the Classic era or the dynastic states which closely followed them. The concern of Wolf and Palerm, and latterly of Steward, over the distinction between "Theocratic Irrigation States" (Protoliterate Mesopotamia and Florescent Peru) and "Ceremonial Trade States" (Classic Mesoamerica) thus seems groundless.

RECIPROCAL EFFECTS OF HUMAN CULTURE ON ENVIRONMENT

This discussion so far has assumed that the natural physiography and resources of the four nuclear areas were relatively stable. The different cultural traditions have been regarded implicitly as evolving successive patterns of ecological adjustment and land use entirely according to some internal dynamic of their own. The effect of environment, in these terms, is merely that of providing a fixed framework of potentialities and limiting conditions which somehow is then exploited selectively by the creative cultural growth within it. Such a view is obviously an oversimplification of the processes of interaction between man and the natural world, even if decisive climatic shifts no longer are regarded as likely to have occurred during the span of time that led to the emergence of any civilization.

Unfortunately the reciprocal effects of changing patterns of human activity on the land and flora cannot be traced continuously for any area. Perhaps the clearest and best-documented example is provided by recent work in central Mexico, where it has been shown that intensive hill-slope cultivation during the last centuries of Aztec dominance had gone far to destroy the capacity of the soil to sustain agriculture even before the arrival of the Spaniards. But the more remote history of occupance in even this relatively well studied region is still insufficiently known for its environmental effects to be understood. The abandonment of the central Peten region by the lowland Classic Maya furnishes an even more dramatic case, with ecological processes such as sheet erosion, the silting-up of fresh-water sources, and the gradual replacement of forest vegetation by uncultivable savanna the course of slash-and-burn agriculture, all having been suggested as contributing factors. But in spite of a generation of speculation and interest these factors still exist only as hypotheses, and in a recent general work on the Maya it is interesting to note that they are largely rejected in favor of an explanation of the collapse of at least the elaborate ceremonial life in purely historical terms.

In the alluvial valleys of the Old World civilizations, processes of erosion are less likely to have affected directly the course of cultural development. It is not impossible, however, that deforestation at the headwaters of the Tigris and Euphrates increased both the silt loads carried by those rivers and their flooding potential. In turn, this would have affected the continuity of occupation in the alluvium and the problems associated with constructing and maintaining irrigation systems. But, although deforestation undoubtedly went on, there are no empirical data at present on its rate nor on its consequences for the alluvial plain as a whole. Even the traditional assumption that the area of the plain has been continuously enlarged by the deposition of silt along the margin of the Persian Gulf has now been challenged by evidence that extensions of the land have been roughly counterbalanced by subsidence.

On the other hand, a group of different and important reciprocal effects is likely to have been initiated directly by the introduction of various techniques of intensive cultivation. Depletion of soil nutrients by inadequate crop rotation or following cycle is one example. Salinization of poorly drained land as a result of continuous irrigation is another. Still a third may be the disturbance of natural patterns of drainage by the slow rise of canal beds and banks as a result of silting. To some degree all of these processes must have gone on, but their importance can only be gauged against the background of a far better understanding of ancient agriculture than we have at present for any area. To begin with, empirical studies are necessary of changes in the intensity of land use and of the exact nature of the full agricultural cycle over a long period in the past. At the time of this writing, a study along these lines has been undertaken for a small section of the Mesopotamian plain but not for any other nuclear area.

For the present, therefore, the distortions of a picture in which cultures are conceived as having evolved within a static environmental framework must remain uncorrected. If several possible types of correction have been mentioned, their effects cannot even be demonstrated satisfactorily with the evidence available from most areas, and in any case they are virtually impossible to quantify. One can only conclude that attempts to invoke changing ecological factors as "causes" of cultural development—however convenient they may appear as heuristic hypotheses—are still no more than *a priori* speculations.

In a broader sense, the lack of data on population density and land use underlines the purely speculative character of all those heuristic hypotheses which regard cultural change as an adaptive response to direct environmental forces. One account of the rise of militarism, for example, sees it as a consequence of the displacement of a population surplus, although there is absolutely no evidence of a concurrent reduction in the sustaining capacity of the environment or of a trend toward overpopulation in any of the nuclear areas. Another recent synthesis, going still farther, attributes not only the rise of large-scale warfare but also the cyclical character of the early empires in large part to population pressure. How population "pressure" can be defined usefully except by reference to real patterns and intensities of land utilization and settlement pressing against clearly defined ecological limits—for which, we must emphasize again, the evidence is still almost entirely lacking—is not apparent.

There is always an attraction for explanations of historical and cultural phenomena that stem from "outside" the immediate field of study. They have the advantage of providing fixed points from which analysis may proceed in a straightforward chain of cause-and-effect processes. But on closer inspection many such fixed points will be found to dissolve into shifting relationships which are not as separate and distinct from cultural influences as they may appear. Premature dependence upon explanations in terms of the external environment only diverts the historian or anthropologist from unraveling the complex stresses within human institutions. In all but the simplest societies, it is forces within the social order rather than direct environmental factors which have provided the major stimulus and guide to further growth.

CONCLUSION

In retrospect, the significant common features of land use among the early civilizations of the Old and the New World are so general that they are almost trite. If we have attempted to define the terms more closely than is usual, there is certainly nothing unusual about finding that all the great civilizational traditions rested on surpluses made available through sedentary, diversified, intensive agriculture. In addition, of course, it is implicit in this discussion that the common social institutions and processes of development identified in each of the four civilizations were bound up together with this general constellation of subsistence practices in a functionally interacting network which characterizes early civilization as a sort of cultural type.

Against this simple and limited finding of regularity, the diversity of other environmental subsistence features and the hugh proliferation of cultural forms stand in sharp contrast. History is not a mathematical exercise in the application of "laws," and the meaning of human experience is not to be found by suppressing its rich variety in the search for common, implicitly deterministic, denominators. From this ponit of view, perhaps the lack of closer specificity in the ecological relationships that are common to the early civilizations is the single most important point to be made. Much of sociocultural development seems to proceed very largely on its own terms, including even some important aspects of ecological adjustment. Societal growth is a continuously creative process, conditioned far more by past history than by directly felt environmental forces. On the whole, then, one may reasonably conclude that for an understanding of the meaning of the early civilizations—both in their own terms and for the modern world—the natural environment serves as no more than a backdrop.

INDEX